Modern Acoustics and Signal Processing

More information about this series at http://www.springer.com/series/3754

The ASA Press

The ASA Press imprint represents a collaboration between the Acoustical Society of America and Springer dedicated to encouraging the publication of important new books in acoustics. Published titles are intended to reflect the full range of research in acoustics. ASA Press books can include all types of books published by Springer and may appear in any appropriate Springer book series.

John L. Butler • Charles H. Sherman

Transducers and Arrays for Underwater Sound

Second Edition

John L. Butler
Chief Scientist
Image Acoustics, Inc.
Cohasset, MA, USA

Charles H. Sherman
Image Acoustics, Inc.
Cohasset, MA, USA

ISSN 2364-4915 ISSN 2364-4923 (electronic)
Modern Acoustics and Signal Processing
ISBN 978-3-319-81802-3 ISBN 978-3-319-39044-4 (eBook)
DOI 10.1007/978-3-319-39044-4

Printed on acid-free paper

This Springer imprint is published by Springer Nature
The registered company is Springer International Publishing AG Switzerland

Acoustical Society of America

The mission of the **Acoustical Society of America** (www.acousticalsociety.org) is to increase and diffuse the knowledge of acoustics and promote its practical applications. The ASA is recognized as the world's premier international scientific society in acoustics, and counts among its more than 7,000 members, professionals in the fields of bioacoustics, engineering, architecture, speech, music, oceanography, signal processing, sound and vibration, and noise control.

Since its first meeting in 1929, The Acoustical Society of America has enjoyed a healthy growth in membership and in stature. The present membership of approximately 7,500 includes leaders in acoustics in the United States of America and other countries. The Society has attracted members from various fields related to sound including engineering, physics, oceanography, life sciences, noise and noise control, architectural acoustics; psychological and physiological acoustics; applied acoustics; music and musical instruments; speech communication; ultrasonics, radiation, and scattering; mechanical vibrations and shock; underwater sound; aeroacoustics; macrosonics; acoustical signal processing; bioacoustics; and many more topics.

To assure adequate attention to these separate fields and to new ones that may develop, the Society establishes technical committees and technical groups charged with keeping abreast of developments and needs of the membership in their specialized fields. This diversity and the opportunity it provides for interchange of knowledge and points of view has become one of the strengths of the Society.

The Society's publishing program has historically included the *Journal of the Acoustical Society of America*, the magazine *Acoustics Today*, a newsletter, and various books authored by its members across the many topical areas of acoustics. In addition, ASA members are involved in the development of acoustical standards concerned with terminology, measurement procedures, and criteria for determining the effects of noise and vibration.

Series Preface for Modern Acoustics and Signal Processing

In the popular mind, the term "acoustics" refers to the properties of a room or other environment—the acoustics of a room are good or the acoustics are bad. But as understood in the professional acoustical societies of the world, such as the highly influential Acoustical Society of America, the concept of acoustics is much broader. Of course, it is concerned with the acoustical properties of concert halls, classrooms, offices, and factories—a topic generally known as architectural acoustics, but it is also concerned with vibrations and waves too high or too low to be audible. Acousticians employ ultrasound in probing the properties of materials, or in medicine for imaging, diagnosis, therapy, and surgery. Acoustics includes infrasound—the wind-driven motions of skyscrapers, the vibrations of the earth, and the macroscopic dynamics of the sun.

Acoustics studies the interaction of waves with structures, from the detection of submarines in the sea to the buffeting of spacecraft. The scope of acoustics ranges from the electronic recording of rock and roll and the control of noise in our environments to the inhomogeneous distribution of matter in the cosmos.

Acoustics extends to the production and reception of speech and to the songs of humans and animals. It is in music, from the generation of sounds by musical instruments to the emotional response of listeners. Along this path, acoustics encounters the complex processing in the auditory nervous system, its anatomy, genetics, and physiology—perception and behavior of living things.

Acoustics is a practical science, andmodern acoustics is so tightly coupled to digital signal processing that the two fields have become inseparable. Signal processing is not only an indispensable tool for synthesis and analysis but it also informs many of our most fundamental models about how acoustical communication systems work.

Given the importance of acoustics to modern science, industry, and human welfare Springer presents this series of scientific literature, entitled Modern Acoustics and Signal Processing. This series of monographs and reference books is intended to cover all areas of today's acoustics as an interdisciplinary field. We expect that scientists, engineers, and graduate students will find the books in this series useful in their research, teaching, and studies.

William M. Hartmann

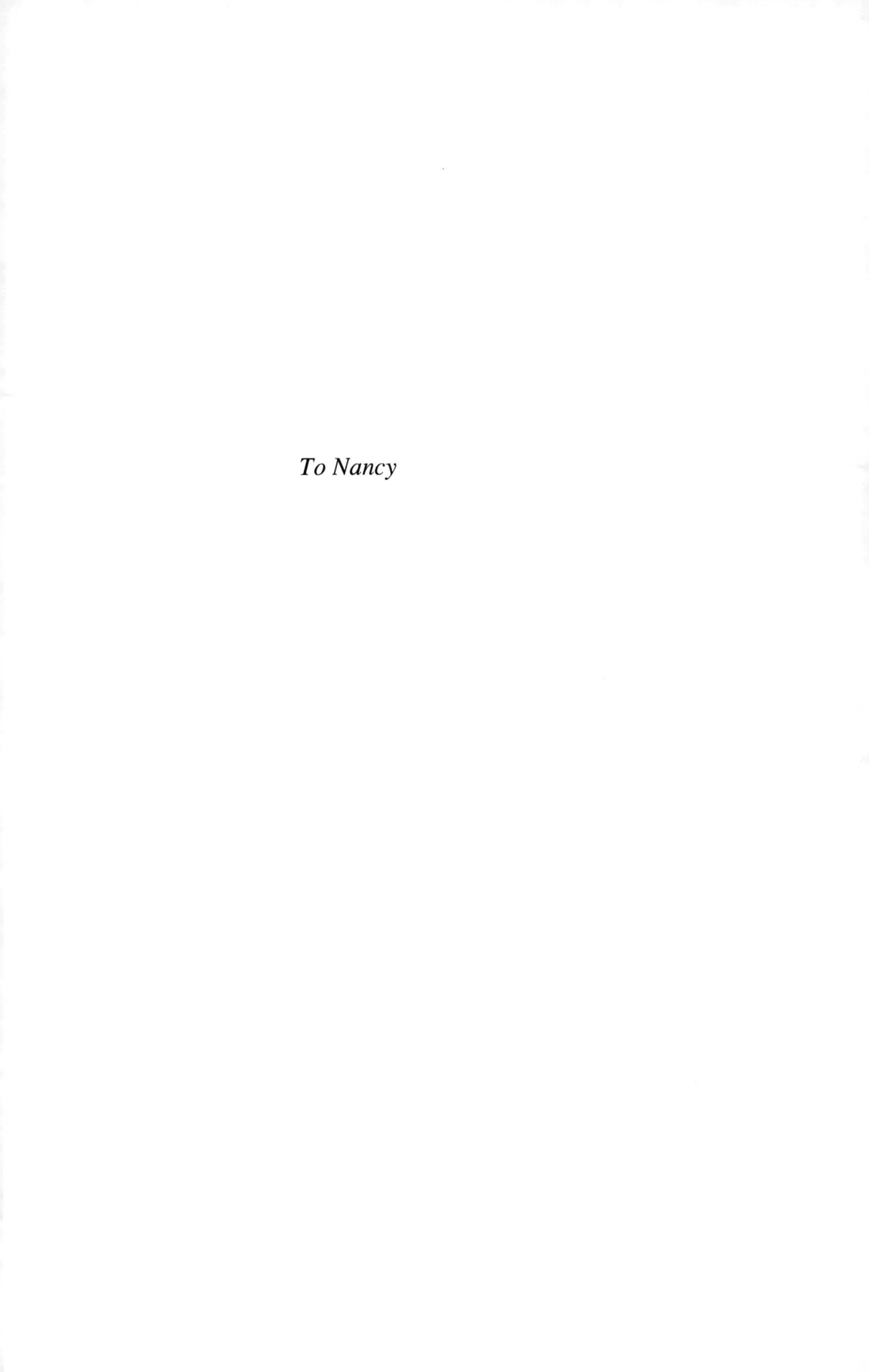

To Nancy

Preface to Second Edition

This second edition presents the theory and practice of underwater sound electroacoustic transducers and arrays as developed during the last half of the twentieth century and into the initial part of the twenty-first century. This second edition has been reorganized into a form suitable for students as well as engineers or scientists who use or design transducers and arrays and includes new design concepts, analysis, and data.

Comprehensive coverage is presented on the subject of transducers and arrays for underwater sound. The most important basic concepts of electroacoustic transduction are introduced in Chap. 1, after a brief historical review and a survey of some of the many applications of transducers and arrays. Chapter 2 describes and compares the six major types of electroacoustic transducers, presents additional transducer concepts and characteristics, and introduces the equivalent circuit method of transducer analysis. Chapter 3 describes the principal methods of transducer modeling, analysis, and design, including an introduction to the finite element method. Chapter 4 gives further discussion of the most important transducer characteristics.

Chapters 5–8 contain the main body of results on modern transducers and arrays. Chapters 5 and 6 cover transducers as projectors, which produce sound, and transducers as hydrophones, which receive sound, including many details of specific transducer designs as they are used in current applications as well as new designs. Chapters 7 and 8 explain the benefits of combining large numbers of transducers in arrays that often contain hundreds of individual transducers. These large arrays are necessary in many sonar applications, but they introduce other problems that are also discussed and analyzed. Chapter 9 is a summary of the major methods of measurement used for the evaluation of transducer and array performance.

Chapter 10 presents the basic acoustics concepts and analysis necessary for determining those acoustical quantities, such as directivity patterns and radiation impedance, which are essential to transducer and array analysis and design. It also includes useful results for such quantities in several typical cases. Chapter 11 extends the discussion of acoustical quantities by introducing more advanced methods of analysis that can be applied to more complicated cases including a brief introduction to numerical methods. Chapter 12 describes the principal nonlinear mechanisms that occur in all the transducer types and presents methods of analyzing important nonlinear effects such as harmonic distortion. The book ends with an extensive Appendix containing several types of specific information that can be used in transducer analysis and design along with a Glossary of Terms and Solutions for the Odd-Numbered Exercises given in the 12 chapters of the book. Instructors who have adopted the text in their courses can contact the publisher for Solutions to the Even-Numbered Exercises.

This second edition has been structured to be more suitable for students and teachers as well as practitioners. Although some parts of this book may be useful to undergraduates, it is written on a graduate level for engineers, scientists, and students in the fields of electrical engineering, mechanical engineering, physics, ocean engineering, and acoustical engineering. The book uses SI (MKS) units in general, but English units are also occasionally used to clarify the relationship to practical devices.

Cohasset, MA
March 2016

John L. Butler

Preface to First Edition

We have written this book as part of the underwater acoustics monograph series initiated by the Office of Naval Research (ONR), Department of the Navy of the United States. The ONR objective for this series is publication of in-depth reviews and analyses of the state of understanding of the physics of sound in selected areas of undersea research. This monograph presents the theory and practice of underwater sound electroacoustic transducers and arrays as developed during the last half of the twentieth century and into the initial part of the twenty-first century.

We have attempted a comprehensive coverage of the subject of transducers and arrays for underwater sound starting with a brief historical review and a survey of some of the many modern applications. Descriptions of the six major types of electroacoustic transducers are presented in a unified way that facilitates their comparison and explains why some types are better suited than others for producing and receiving sound in the water. The characteristics of transducers used as both projectors and hydrophones, and the methods available for predicting and measuring transducer performance, are presented in detail. The reasons for combining large numbers of transducers in arrays are explained, and the special problems that must be considered in such arrays are analyzed. The nonlinear mechanisms that exist in all transducers are described, and analyses of some of their most important effects are given. Many different acoustical quantities play essential roles in the design and performance of electroacoustic transducers and arrays, and the methods for determining these quantities are presented. Analytical modeling and understanding are emphasized throughout the book, but it is also made clear that numerical modeling is now an essential part of transducer and array design. Non-electroacoustic types of transducers that are used in certain underwater applications, such as explosive sources, spark sources, hydroacoustic sources, and optical hydrophones, are not included in this book.

The monograph is organized in a manner that brings the reader quickly to the main body of results on current transducers and arrays in the first six chapters with a minimum of background material. The most important basic concepts of electroacoustic transduction are introduced in Chap. 1, after a brief historical review and a survey of some of the many applications of transducers and arrays. Chapter 2 describes and compares the six major types of electroacoustic transducers, presents additional transducer concepts and characteristics, and introduces the equivalent circuit method of transducer analysis. Chapters 3–6 contain the main body of results on modern transducers and arrays. Chapters 3 and 4 cover transducers as projectors, which produce sound, and as hydrophones, which receive sound, including many details of specific transducer designs as they are used in current applications. Chapters 5 and 6 explain the benefits of combining large numbers of transducers in arrays that often contain more than 1000 individual transducers. These large arrays are necessary in many sonar applications, but they introduce other problems that are also discussed and analyzed.

The remaining six chapters, Chaps. 7–12, support the earlier chapters and carry the discussion of concepts and methods into much more detail for those who seek a deeper understanding of transducer operation. Chapter 7 describes all the principal methods of transducer modeling, analysis, and design, including an introduction to the finite element method. Chapter 8 gives further discussion of the most important transducer characteristics. Chapter 9 describes the principal nonlinear mechanisms that occur in all the transducer types and presents methods of analyzing important nonlinear effects such as harmonic distortion. Chapter 10 presents the basic acoustics necessary for determining those acoustical quantities, such as directivity patterns and radiation impedance, that are essential to transducer and array analysis and design. It also includes useful results for such quantities in several typical cases. Chapter 11 extends the discussion of acoustical quantities by introducing more advanced methods of analysis that can be applied to more complicated cases including a brief introduction to numerical methods. Chapter 12 is a summary of the major methods of measurement used for the evaluation of transducer and array performance. The book ends with an extensive Appendix containing several types of specific information that can be used in transducer analysis and design and with a Glossary of Terms.

We have attempted to make this monograph suitable for beginners to learn from and for practitioners in the transducer field to learn more from. In addition those concerned in any way with undersea research may find useful guidance regarding applications of transducers and arrays. Although some parts of this book may be useful to undergraduates, it is written on a graduate level for engineers and scientists in the fields of electrical engineering, mechanical engineering, physics, ocean engineering, and acoustical engineering. The book uses SI (MKS) units in general, but English units are also occasionally used to clarify the relationship to practical devices.

Cohasset, MA John L. Butler
Cohasset, MA Charles H. Sherman
January 2006

Acknowledgments

The first and second editions are based on the experience of the authors in both government and industrial organizations. We are grateful to our early teachers in the fields of transducers and acoustics: Dr. R. S. Woollett, E. J. Parsinnen, and H. Sussman of the Navy Underwater Sound Laboratory (now Naval Undersea Warfare Center, NUWC); Dr. W. J Remillard, Northeastern University; Dr. R. T. Beyer, Brown University; Dr. T. J. Mapes, NUWC; G. W. Renner, Hazeltine Corporation (now Ultra Ocean Systems, Inc.); B. McTaggart, NUWC; Frank Massa, Massa Products Corporation; and Stan Ehrlich, Raytheon Company. We would also like to thank Dr. W. Thompson, Jr. and W. J. Marshall for their review of the first edition. Discussions with the following colleagues were also very helpful at various points: S. C. Butler, D. T. Porter, and Drs. H. H. Schloemer, S. H. Ko, W.A. Strawderman, R. T. Richards, A. E. Clark, J. E. Boisvert, M. B. Moffett, and R. C. Elswick.

I am grateful to the many other people who have contributed in specific ways to this second edition: to Jan F. Lindberg for originally encouraging us to take on the task of the first edition and to Alexander L. Butler and Victoria Curtis of Image Acoustics, Inc. for their help with the illustrations, graphs, and analysis. I would also like to thank William J. Marshall, Jan F. Lindberg, and Dr. Harold C. Robinson for their comments, suggestions, and review of the draft of this second edition.

I am especially grateful to my wife Nancy Clark Butler, for her encouragement and understanding. This second edition was written in memory of Charlie Sherman, coauthor, mentor, colleague, and friend.

John L. Butler

Contents

About the Authors

John L. Butler is chief scientist at Image Acoustics, Inc. and has had over 40 years of both practical and theoretical experiences in the design and analysis of underwater sound transducers and arrays. He has worked for and consulted to a number of underwater acoustics firms as well as Parke Mathematical Laboratories and the US Navy. He has also taught courses in acoustics at Northeastern University, Naval Air Development Center, Raytheon Company, Harris Transducer Products, Hazeltine Corporation (now Ultra Ocean Systems, Inc.), Massa Products Corporation, Etrema Products, Plessey Australia, and Lund Institute of Technology, Sweden. He holds 27 patents and has presented or published well over 30 papers on electroacoustic transducers. In 1977 he was elected fellow of the Acoustical Society of America and has received their 2015 Silver Medal Award for advancing the field of acoustic transducers and transducer arrays. His education includes Ph.D., Northeastern University, Boston, MA, and Sc.M., Brown University, Providence, RI.

Charles H. Sherman (1928–2009) received a B.S. degree in physics from the Massachusetts Institute of Technology in 1950. After his first job at TracerLab, Inc. in Boston, he became a research physicist at the Navy Underwater Sound Laboratory in New London, CT. He received M.S. and Ph.D. degrees from the University of Connecticut and was elected Fellow of the Acoustical Society of America in 1974. He became a prominent expert in underwater transducers and arrays, presenting and publishing over 30 papers related to underwater acoustics. He also worked at Parke Mathematical Laboratories in Carlisle, MA, and taught advanced acoustics at the University of Connecticut and in the Ocean Engineering Department of the University of Rhode Island. He received the prestigious Decibel Award, which is presented to a scientist or engineer for outstanding contributions to sonar and underwater acoustics. After his retirement from the Navy Underwater Sound Laboratory in 1988, he worked for Image Acoustics, Inc. and, in 2007, coauthored the first edition of *Transducers and Arrays for Underwater Sound*, a technical monograph commissioned by the Office of Naval Research and the most comprehensive treatment to date of underwater transducers and arrays.

Chapter 1
Introduction

The development of underwater electroacoustic transducers expanded rapidly during the twentieth century, and continues to be a growing field of knowledge that combines mechanics, electricity, magnetism, solid state physics, and acoustics with many significant applications. In the most general sense a transducer is a process or a device that converts energy from one form to another. Thus, an electroacoustic transducer converts electrical energy to acoustical energy or vice versa. Such processes and devices are very common. For example, a thunderstorm is a naturally occurring process in which electrical energy, made visible by the lightning flash, is partially converted to the sound of thunder. On the other hand, a familiar man-made transducer is the moving coil loudspeaker used in radio, television, and other sound systems. Loudspeakers are so common that they probably outnumber people in developed parts of the world. The familiar designations loudspeaker and microphone for transducers used as sources and receivers of sound in air become projector and hydrophone for sources and receivers in water. The term SONAR (SOund Navigation And Ranging) is used for the process of detecting and locating objects by receiving the sounds they emit (passive sonar), or by receiving the echoes reflected from them when they are "insonified" in echo ranging (active sonar). Every use of sound in the water requires transducers for its generation and reception, and most are based on electroacoustics. Several non-electroacoustic transducers also find applications in water, e.g., projectors based on explosions, sparks, and hydroacoustics as well as optical hydrophones, but they are not included in this book.

This book presents the theory and practice of underwater sound electroacoustic transducers at the beginning of the twenty-first century. Chapter 1 begins with a brief historical survey of the development of electroacoustics and its many applications to underwater sound. It also introduces the basic concepts of electroacoustic transduction in a general way applicable to all types of electroacoustic transducers. Chapter 2 describes and compares major types of electroacoustic transduction mechanisms and shows why certain piezoelectric materials now dominate the field of underwater transducers. Chapter 3 introduces the transducer models and

J.L. Butler, C.H. Sherman, *Transducers and Arrays for Underwater Sound*,
Modern Acoustics and Signal Processing, DOI 10.1007/978-3-319-39044-4_1

analysis methods and Chap. 4 discusses transducer characteristics. Specific projector and hydrophone designs are presented in Chaps. 5 and 6 and arrays of projectors and hydrophones are discussed in Chaps. 7 and 8. Chapter 9 presents means for transducer evaluation and measurement. Chapter 10 presents a discussion of acoustic radiation from transducers and Chap. 11 discusses and implements advanced mathematical models for acoustic radiation. And finally, Chap. 12 provides an analysis of nonlinear effects in transducers which was introduced in Chap. 2.

1.1 Brief History of Underwater Sound Transducers

Electroacoustics began to develop more than 200 years ago with observations of the mechanical effects associated with electricity and magnetism, and found an important place in underwater sound early in the twentieth century. F. V. Hunt has given the most complete historical survey of the development of electroacoustics including a section entitled Electroacoustics Goes to Sea [1]. R. J. Urick's brief historical introduction concentrates on underwater applications of electroacoustics [2]. R. T. Beyer's history of the past 200 years of acoustics also contains many references to underwater sound transducers [3]. A few historical items, taken from these books, will be briefly described here. Daniel Colladon and Charles Sturm collaborated in 1826 on the first direct measurement of the speed of sound in the fresh water of Lake Geneva in Switzerland [3]. They had no electroacoustic transducer to generate sound in the water; instead their projector was a mechanoacoustic transducer—the striking of a bell under water. At one point on the lake the bell was struck simultaneously with a flash of light, while an observer in a boat 13 km away measured the time interval between the flash and the arrival of the sound. The observer also had no electroacoustic transducer for detecting the arrival of the sound; his hydrophone consisted of his ear placed at one end of a tube with the other end in the water. Their measured value at a water temperature of 8 °C is given by Beyer as 1438 m/s [3], and by Rayleigh as 1435 m/s [4]. A modern value for fresh water at 8 °C is 1439 m/s [3, 5]. This is remarkable accuracy for a first measurement and for a propagation time of less than 10 s.

Interest in telegraphy in the latter part of the eighteenth and the first part of the nineteenth centuries provided the first practical impetus for the development of electrical transducers. Acoustics was not involved at first; a mechanical input causing an electrical signal could be visually observed at the other end of the telegraph wires as another mechanical effect, e.g., the motion of a needle. The devices used at each end of the system were electromechanical or magnetomechanical transducers. Electroacoustic transducers were introduced into telegraphy by Joseph Henry in 1830 using a moving armature transducer (now often called variable reluctance transducer) in which the transmitted signal was observed by the sound of the armature striking its stops. These developments led to the invention of the telephone, primarily by Alexander Graham Bell in 1876, using moving armature electroacoustic transducers on both ends of the line and making possible transmission of the human voice.

James Joule is usually credited with the discovery of magnetostriction based on his quantitative experiments between 1842 and 1847 including measurement of the change in length of an iron bar when it is magnetized, although various manifestations of magnetostriction had been observed earlier by others [1]. In 1880 piezoelectricity was discovered in quartz and other crystals by Jacques and Pierre Curie [1]. The discoveries of magnetostriction and piezoelectricity would eventually have tremendous importance for underwater sound, since materials with such properties are now used in most underwater transducers. Magnetostrictive and piezoelectric materials change dimensions when placed in magnetic or electric fields, respectively, and have other properties that make them very suitable for radiating or receiving sound in water. Interest in the mechanical effects of electric and magnetic fields was also closely associated with the development during the nineteenth century of a theoretical understanding of electricity, magnetism, and electromagnetism.

The first application of underwater sound to navigation was made by the Submarine Signal Company (later a Division of the Raytheon Company) early in the twentieth century. It required the crew of a ship to measure the time interval between hearing the arrival of an underwater sound and an airborne sound. A bell striking underwater was the source of sound in the water while the simultaneous blast of a foghorn at the same location provided the sound in air. Early shipboard acoustic devices included mechanical means for generating sound, as illustrated in Fig. 1.1, and binaural means for determining the direction of sound as shown in Fig. 1.2.

Fig. 1.1 Early simple underwater signaling system using hammer, rod, and piston, courtesy Raytheon Company [6]

Fig. 1.2 Early binaural detection and localization air tube underwater sensor, courtesy Raytheon Company [6]

L. F. Richardson filed patent applications with the British Patent Office for echo ranging with both airborne and underwater sound in 1912, soon after the Titanic collided with an iceberg. He apparently did not implement these ideas, probably because suitable transducers were not available. However, R. A. Fessenden, a Canadian working in the United States, soon filled that need by developing a new type of moving coil transducer which, by 1914, was successfully used for signaling between submarines and for echo ranging. On 27 April, 1914 an iceberg was detected by underwater echo ranging at a distance of nearly 2 miles. These "Fessenden Oscillators" operating at 500 and 1000 Hz were installed on United States submarines during World War I. This was probably the first practical application of underwater electroacoustic transducers [6–8].

Before the start of World War I it was understood that electromagnetic waves were absorbed in a short distance in water, except for extremely low frequencies and also for blue-green light. Thus sound waves were the only means available for practical signaling through the water. For the first time a significant submarine menace existed [9], and many underwater echo ranging experiments were initiated. In France, Paul Langevin and others started work early in 1915 using an electrostatic transducer as a projector and a waterproofed carbon microphone as a hydrophone. Although some success was had in receiving echoes from targets at short range, numerous problems made it clear that improved transducers were necessary. When the French results were communicated to the British, a group under R. W. Boyle (the Allied Submarine Detection Investigation Committee, ASDIC)

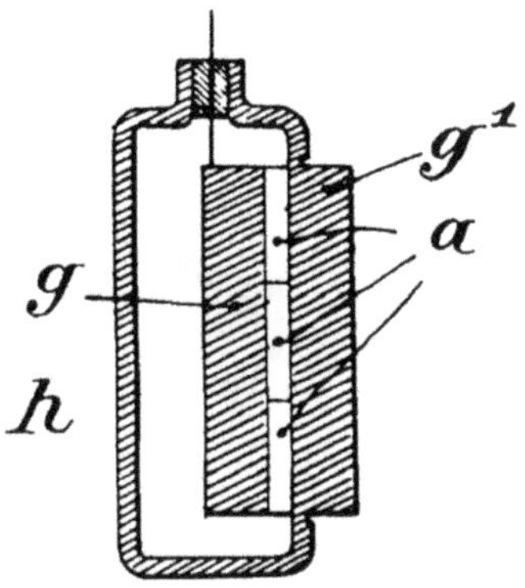

Fig. 1.3 British patent 145,691 July 28, 1921 of P. Langevin invention showing steel (g)-quarts(a)-steel(g′) sandwich transducer

began similar experiments in 1916. While both sides realized that use of the piezoelectric effect in quartz had the potential for improved transducers, it was Langevin who demonstrated the value of piezoelectricity as soon as he found suitable samples of quartz. Improved results were obtained first by replacing the carbon hydrophone with a quartz hydrophone, and again in early 1917 when quartz transducers were used for both projector and hydrophone. After further improvements in the design of the quartz transducers, echoes were heard from a submarine in early 1918. The major design improvement consisted of making a resonator by sandwiching the quartz between steel plates (see Fig. 1.3), an approach still used in modern transducers.

These successes greatly improved the outlook for effective echo ranging on submarines, and the efforts increased in France, Great Britain, the United States and also in Germany. Boyle's group developed equipment, referred to as "ASDIC gear," for installation on some ships of the British fleet. In the United States an echo ranging program was initiated at the Naval Experimental Station in New London, Connecticut with supporting research, especially on piezoelectric materials, from several other laboratories. Although none of this work progressed rapidly enough to have a significant role in WWI, it did provide the basis for continued research in echo ranging that would soon be needed in WWII.

Between the World Wars depth sounding by ships underway was developed commercially, and the search for effective echo ranging on submarines was continued in the United States, primarily at the Naval Research Laboratory under H. C. Hayes. One of the main problems was the lack of transducers powerful enough to achieve the necessary ranges. It was found that magnetostrictive transducers could produce greater acoustic power, while their ruggedness made them very suitable for underwater use. However, both electrical and magnetic losses in magnetostrictive materials resulted in lower efficiency compared to piezoelectric transducers. Other transducer concepts were also explored including one that used the extensional motion of magnetostriction to drive a radiating surface in flexure (called a flextensional transducer; see Fig. 1.4).

After WWI Rochelle salt, which was known to have a stronger piezoelectric effect than quartz, also became available in the form of synthetic crystals to provide

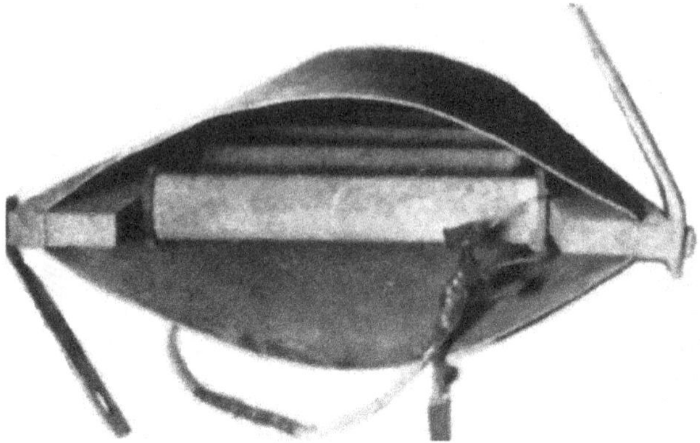

Fig. 1.4 An experimental flextensional transducer built at NRL in May 1929 for in-air operation with magnified shell motion driven by three magnetostrictive tubes [10]

another possibility for improved transducers. Synthetic Rochelle salt was probably the first example of what would become the most important type of innovation in the field of electroacoustic transducers: new man-made materials with improved electromechanical properties.

Early in World War II these accomplishments in transducers, combined with advances in electronics and better understanding of the propagation of sound in the ocean, provided the basis for development of sonar systems with useful but limited capability. The potential for significant improvement was clear, and, with German submarines causing serious damage to shipping off the east coast of the United States, the need was great [9]. The work had already started in 1941 with Columbia University's Division of War Research in New London, Harvard University's Underwater Sound Laboratory (HUSL) in Cambridge, and the University of California's Division of War Research in San Diego. Their work resulted in many American ships being equipped with echo ranging and passive listening systems. Other types of equipment employing transducers and underwater sound were also developed such as acoustic homing torpedoes, acoustic mines, and sonobuoys. A large amount of practical experience was accumulated from the use of all this equipment, and it provided a firm basis for many new developments during and after the war [10].

At the end of WWII the Columbia work at New London continued under the direction of the Naval Research Laboratory [11]. The New London facility was called the Navy Underwater Sound Laboratory with John M. Ide as Technical Director and J. Warren Horton as Chief Consultant. Later in 1945 the sonar projects at the Harvard Underwater Sound Laboratory, and about half the personnel, went to New London to join the Navy Underwater Sound Laboratory, while the Harvard ordnance projects, and the remaining personnel, went to a new Ordnance Research Laboratory at Pennsylvania State University. Major developments in sonar and electromagnetics continued at New London for many years including a wide variety of research and development on transducers and arrays [11]. During the same period similar research on transducers and arrays was conducted at the Naval Research Laboratory in Washington, DC and Orlando, Florida and also at the Navy Electronics Laboratory in San Diego.

WWII, and the Cold War that followed, strongly motivated the search for new man-made transduction materials which led to ammonium dihydrogen phosphate (ADP), lithium sulfate, and other crystals in the early 1940s. Then, in 1944, piezoelectricity was discovered by A. R. von Hippel in permanently polarized barium titanate ceramics [12], and in 1954 even stronger piezoelectricity was found in polarized lead-zirconate-titanate ceramics [13]. The discovery of these materials initiated the modern era of piezoelectric transducers in time to play an important role in, once again, meeting the threat of submarines off the east coast of the United States, this time Soviet submarines armed with long range nuclear missiles.

At the end of the twentieth century lead-zirconate-titanate (PZT) ceramic compounds are still being used in most underwater sound transducers. However, other similar materials such as lead manganese niobate (PMN), textured ceramic and single crystals of related compounds, such as lead indium niobate-lead magnesium niobate-lead titanate (PIN-PMN-PT) and the magnetostrictive materials Terfenol-D and Galfenol, have been developed which have the potential for improvement over PZT in some applications. Piezoelectric ceramics and ceramic-elastomer composites can be made in a great variety of shapes and sizes with many variations of composition that provide specific properties of interest. The characteristics of these materials have led to the development and manufacture of innovative, relatively inexpensive transducer designs that would have been unimaginable in the early days of electroacoustics.

1.2 Underwater Transducer Applications

The useful spectrum of underwater sound extends from about 1 Hz to over 1 MHz with most applications in large (but sometimes shallow) bodies of water. For example, acoustic communication over thousands of kilometers is possible in the oceans, but frequencies below about 100 Hz are required because the absorption of sound increases rapidly as the frequency increases [2]. On the other hand, depth sounding in water as shallow as 1 m is important for small boats, but it requires short pulses of sound at a few hundred kHz to separate the echo from the transmission. High resolution, short-range active sonar has used frequencies up to 1.5 MHz. Applications over this wide frequency range require many different transducer designs.

Naval applications of underwater sound require a large number and variety of transducers. Acoustic communication between two submerged submarines requires a projector to transmit sound and a hydrophone to receive sound on each submarine; echo ranging requires a projector and a hydrophone usually on the same ship;

passive listening requires only a hydrophone. However, hydrophones and projectors are often used in large groups of up to 1000 or more transducers closely packed in planar, cylindrical, or spherical arrays mounted on naval ships.

Other naval applications include acoustic mines activated by the voltage from a hydrophone sensitive to the low frequency sound radiated by a moving ship. Special projectors and hydrophones are required for acoustic communication between submerged submarines or from a surface ship to a submarine. Torpedoes with active acoustic homing systems require high frequency, directional arrays, while those with passive homing require lower frequency capability to detect ship radiated noise. Submarines are usually equipped with other specialized hydrophones to monitor their self-noise or to augment the major sonar systems. Sonobuoys are expendable hydrophone/radio transmitter combinations dropped into the water from an aircraft. The radio floats on the surface, with the tethered hydrophone at a suitable depth for detecting submarines. Some types of sonobuoys listen passively, while others echo range, but both radio information back to the aircraft. Urick [2] discusses most of these naval applications.

As sonar technology matured it began to have significant commercial applications such as depth sounding, a form of active sonar in which echoes are received from the bottom. Accurate knowledge of the water depth under the boat is important not only to the Navy but also to all mariners from those aboard the largest ships to those on small recreational boats. Sonar can do more than find the water depth at the point where the ship is located. It can be extended to provide detailed bottom mapping, and with good bottom maps navigation by depth sounding is feasible. Bottom maps now exist for much, but not all, of Earth's 140 million square miles of ocean. In a similar way sounding on the lower surface of ice is critical for submarines navigating under the Arctic ice cap. Bottom mapping techniques can be readily extended to exploration and search for sunken objects that vary from ship and aircraft wreckage to ancient treasure. Active sonar has commercial importance in the fishing industry where systems have been developed specifically for locating schools of fish. Underwater transducers can even be used to kill mosquito larva by irradiating them with ultrasonic energy [14].

Bottom mapping with sonar is an important part of oceanography, and it can be extended to sub-bottom mapping and determination of bottom characteristics. For example, the bottom of Peconic Bay, Long Island, New York has been studied by sonar in an attempt to determine the reasons for the decrease in the scallop population [15]. Acoustic propagation measurements can be used for modeling ocean basins using echo sounding and tomographic techniques.

Underwater sound is useful in ocean engineering in many ways. The precise location of specific points or objects is often crucial when drilling for oil and gas deep in the ocean or laying underwater cables or pipelines. A combination of underwater and seismic acoustics is needed for finding deposits of oil or gas under the oceans. Networked underwater communication systems involving many acoustic modems, each with a projector and a hydrophone, are important for naval operations and other underwater projects.

Several research projects make use of underwater sound to gather data related to a wide variety of topics. The Acoustic Thermometry of Ocean Climate project (ATOC) measures the acoustic travel time over ocean paths thousands of kilometers long to determine whether the average sound speed is increasing as time passes. Since an increasing sound speed in ocean waters means an increasing average temperature over a large portion of the earth, it may be one of the best measures of global warming. This project requires very low frequency projectors and hydrophones as well as very careful signal processing [16]. The Sound Surveillance System has been used to study the behavior of sperm whales by detecting the "clicks," which they emit, and to detect earthquakes and volcanic eruptions under the sea [17]. In what may become the ultimate version of the sonobuoy concept, plans have been made to land acoustic sensors on Jupiter's moon Europa in about the year 2020. Cracks, which are thought to occur naturally in the ice that covers the surface of Europa, generate sound in the ice and in the ocean that may lie beneath the ice. The sounds received by the acoustic sensors may be interpretable in terms of the ice thickness and the depth and temperature of the underlying ocean. Such information may give clues about the possible existence of extraterrestrial life [18]. Underwater sound may even play a role in the field of particle physics if physicists succeed in showing that hydrophone arrays are capable of detecting the sounds caused by high energy neutrinos passing through the ocean [19].

All these applications of underwater sound require large numbers of transducers, with a great variety of special characteristics for use over a wide range of frequency, power, size, weight, and water depth. The problems raised by the variety of applications and the numerous possibilities for solutions continue to make underwater sound transducer research and development a challenging subject. Figures 1.5, 1.6, 1.7, 1.8, 1.9, 1.10, 1.11, 1.12, 1.13, 1.14, 1.15, 1.16, and 1.17 illustrate several more recent underwater sound transducers, Fig. 1.9 shows an early low frequency array, Figs. 1.10, 1.11, and 1.12 illustrate more current arrays, Figs. 1.13 and 1.14 show flextensional transducers and Figs. 1.15, 1.16, and 1.17 illustrate transducers that use newer materials, Terfenol-D and single crystal PMN-PT.

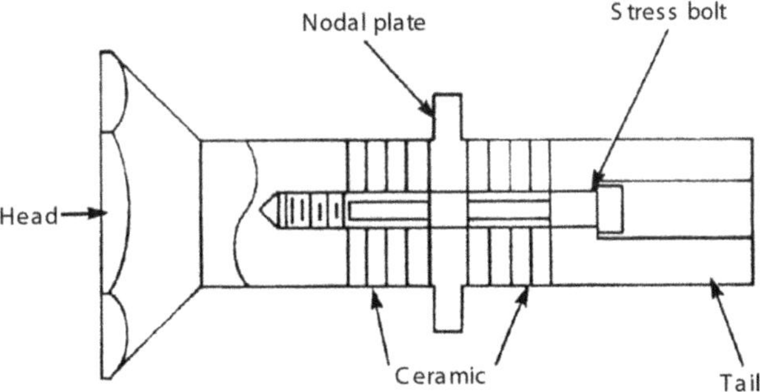

Fig. 1.5 Sketch of a Tonpilz transducer with nodal plate mounting system

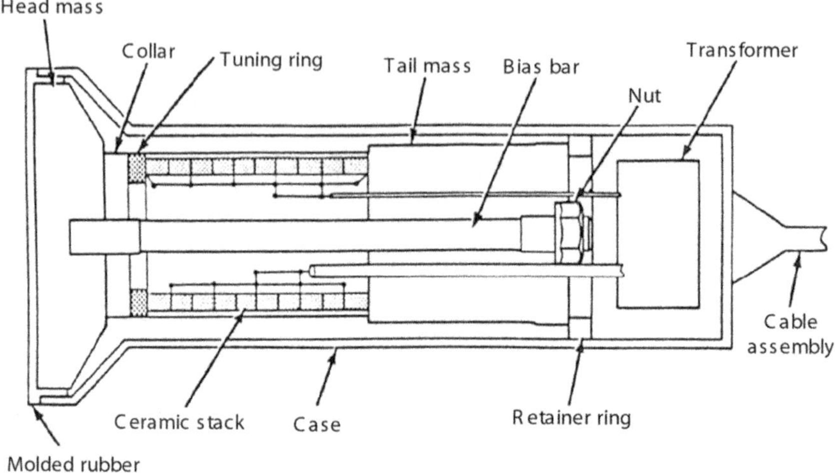

Fig. 1.6 Sketch of high power low frequency Tonpilz transducer with head and tail mounting system and steel housing case

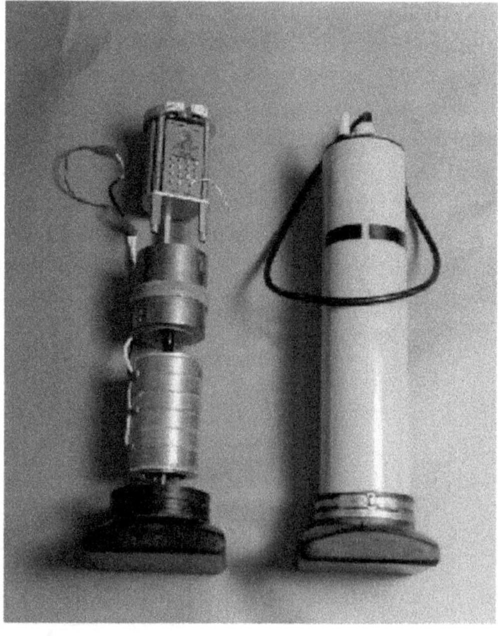

Fig. 1.7 Photograph of a low frequency high power Tonpilz transducer showing rubber molded piston, fiberglass wrapped drive stack of six piezoelectric ceramic rings, tail mass, transformer, and housing

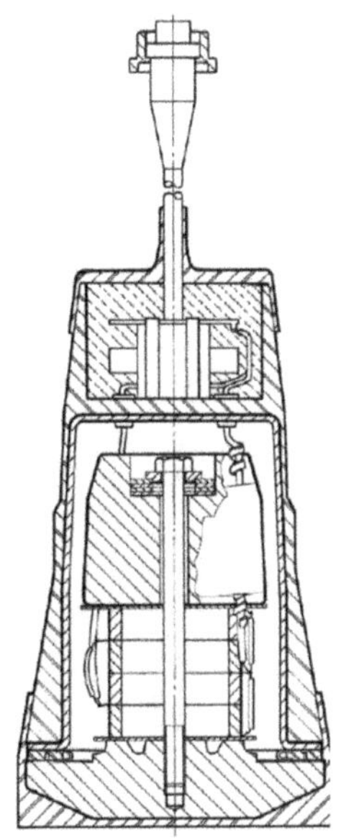

Fig. 1.8 Photograph of cutaway housing and cross-sectional sketch of a mid-frequency Tonpilz transducer showing detail of compression bolt as well as molded rubber encapsulation housing and water tight "pig tail" connector, courtesy of Massa Products [30]

Fig. 1.9 Photograph of large array of very low frequency magnetic variable reluctance dipole "shaker box" transducers ready for deep submergence testing, courtesy of Massa products [30]

Fig. 1.10 Cylindrical scanning array of Tonpilz transducers

Fig. 1.11 Submarine sonar spherical array undergoing tests

Fig. 1.12 A panel of a submarine conformal array during testing

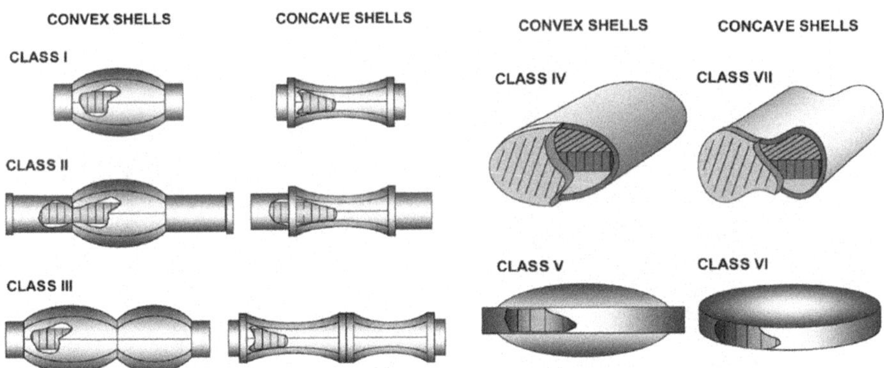

Fig. 1.13 Sketch of various classes of flextensional transducers [31]

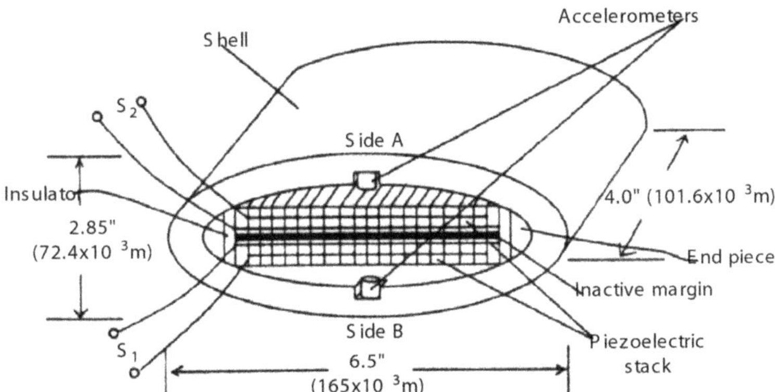

Fig. 1.14 Sketch of a Class IV flextensional transducer with inactive central section for operation in a dipole mode [32]. Shell and interface by John Oswin, British Aerospace England

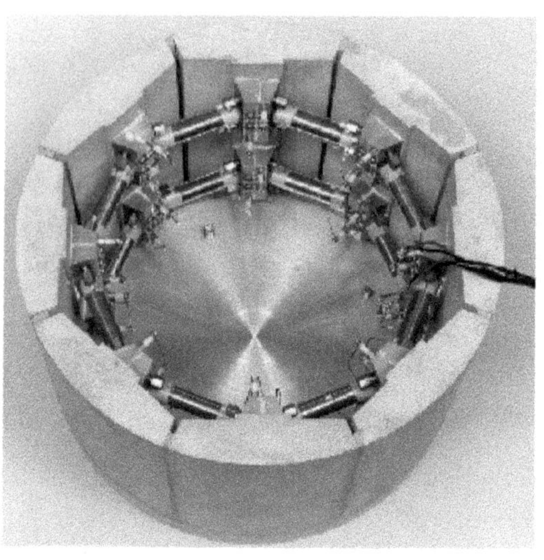

Fig. 1.15 An experimental ring mode magnetostrictive transducer (without top end cap) driven by 16 Terfenol-D magnetostrictive rods [33]

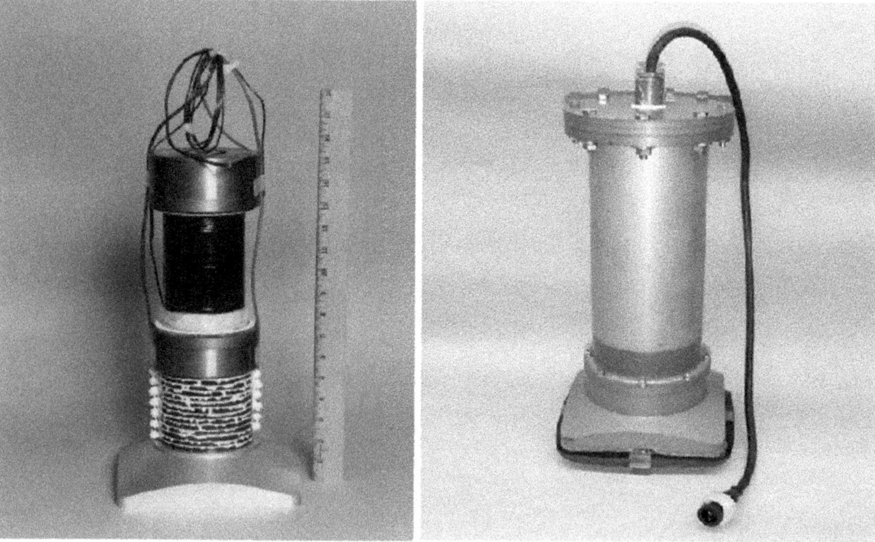

Fig. 1.16 A high power magnetostrictive/piezoelectric hybrid transducer with square piston, piezoelectric ceramic drive, centermass, magnetostrictive drive, and tail mass along with water-tight housing and electrical connector [34]

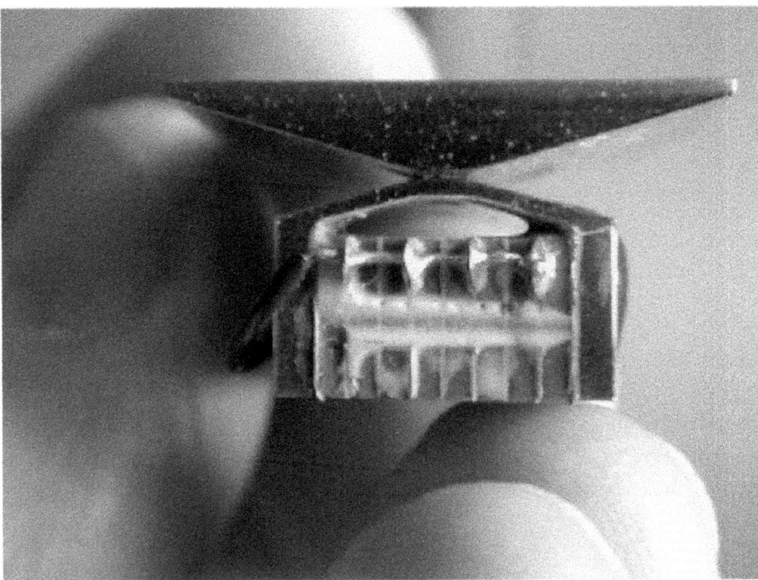

Fig. 1.17 A flextensional amplified single-side one-inch-square piston transducer driven by eight single crystal PMN-PT plates driven into a bending mode. Designed by Image Acoustics and further refined by Northrop Grumman, fabricated by Harris Acoustic Products for Northrop Grumman

1.3 General Description of Linear Electroacoustic Transduction

There are six major types of electroacoustic transduction mechanisms (piezoelectric, electrostrictive, magnetostrictive, electrostatic, variable reluctance, and moving coil), and all have been used as underwater sound transducers. Although the details of these mechanisms differ considerably, the linear operation of all six can be described in a unified way. Three of the six involve electric fields; the other three involve magnetic fields. The piezoelectric, electrostrictive, and magnetostrictive mechanisms are called body force transducers since the electric or magnetic forces originate throughout the active material, while the electrostatic, variable reluctance, and moving coil mechanisms are called surface force transducers since the forces originate at surfaces. The piezoelectric and moving coil transducers have linear mechanisms for small amplitude of vibration, but the other four transducers are inherently nonlinear and must be polarized or biased (see Chap. 2) to achieve linear operation even for small amplitudes. We will concentrate on the body force transducers, because of their more common usage in underwater sound transduction and in particular focus our efforts on the most commonly used piezoelectric transduction materials.

When the nonlinearities are ignored, any electroacoustic transducer can be idealized as a vibrator, with mass, M, stiffness, K_m, and internal resistance, R,

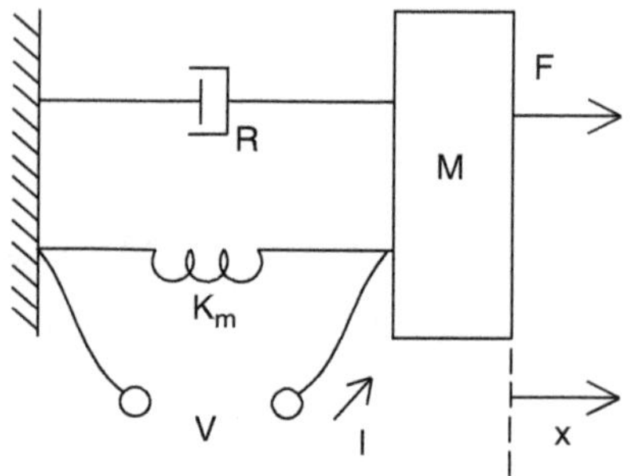

Fig. 1.18 The one-dimensional simple harmonic oscillator

(see Sect. 13.12) subjected to an acoustic force, F, and also connected to a source of electrical energy that provides an electrical force as shown in Fig. 1.18.

In the electric field case the electric force is proportional to voltage, V, and can be represented by $N_{em}V$ where N_{em} is a constant. The motion of the mass under the influence of these forces is given by Newton's Law:

$$M\ddot{x} = -K_m x - R\dot{x} + F + N_{em}V, \tag{1.1a}$$

where x is the displacement of the mass. When the driving forces vary sinusoidally with time as $e^{j\omega t}$, at angular frequency ω, this equation becomes (see Sect. 13.17)

$$(j\omega M - jK_m/\omega + R)u = Z_m u = F + N_{em}V, \tag{1.1b}$$

where $u = j\omega x$ is the velocity of the mass and the mechanical impedance $Z_m = (j\omega M + 1/j\omega C_m + R)$ is the ratio of force to velocity with the mechanical compliance $C_m = 1/K_m$. The solution for the resulting velocity may then be written as

$$u = (F + N_{em}V)/Z_m. \tag{1.1c}$$

And for the case of no external forces, F, we get simply $u = N_{em}V/Z_m$. The maximum output velocity, $u = N_{em}V/R$, is obtained at resonance, $\omega_r = (K_m/M)^{1/2}$, where $\omega_r M = K_m/\omega_r$ and the impedance, $Z_m = R$.

For example: if $K_m = 1 \times 10^6$ N/m and $M = 2.5$ g $= 0.0025$ kg, then $\omega_r = 20{,}000$ and $f_r = 20{,}000/2\pi = 3.18$ kHz. Here, at resonance, for $N_{em} = 2$ N/V, $V = 1$ V, and $R = 2000$ N s/m we get $u = 1 \times 10^{-3}$ m/s and, as discussed below, a power of $W = u^2 R = 2 \times 10^{-3}$ W for 1 V in and 20 W for an input voltage of 100 V. In this example we have assumed the value for the voltage V is RMS and consequently the velocity is RMS.

The essential characteristic of a linear transducer is a relationship, such as Eq. (1.1b), that couples the mechanical and electrical variables F, u and V, with a similar relationship between the current, I, and u and V, which can be written as

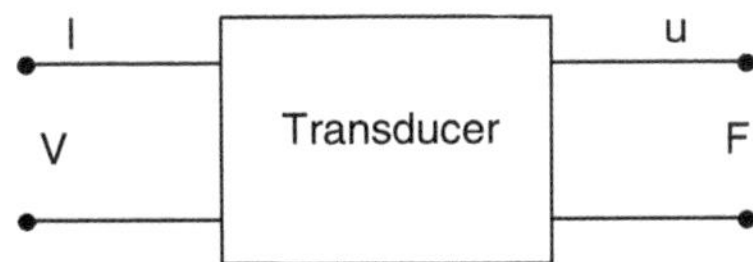

Fig. 1.19 General representation of voltage, current, velocity, and force for any linear transducer

$$I = N_{me}u + YV, \tag{1.2}$$

where Y is the electrical admittance, the ratio of current to voltage, and N_{me} is another constant. The use of these equations, and the cases where $N_{me} = \pm N_{em}$, will be discussed fully after we define the variables more carefully. We note here that N_{em} is the quantity that connects the electrical part to the mechanical part of the transducer and vice versa. And that for $N_{em} = 0$ there is no coupling and the device is not a transducer.

A transducer can also be considered as a mechanism, with unspecified details, that has one electrical port (two wires) and one acoustical port (a surface that can vibrate in an acoustic medium). In Fig. 1.19, the acoustical port is represented in the same way as the electrical port, i.e., as two wires, and F is made analogous to voltage with u analogous to current.

This is called the classical, or impedance, analogy, and is the analogy most often used. The other analogy, called the mobility analogy, is also used with F and u being analogous to I and V, respectively.

The variables V and I are clearly defined as the voltage across the ends of two wires and the current through the wires. However, since F and u are associated with an extended surface in contact with an acoustic medium, their definitions require more consideration. In general $F = F_r + F_b$ where F_r is the reaction force of the medium to the motion of the surface, and F_b is an external force such as a sound wave from another source. The velocity u is well defined when the motion of the surface is uniform, with the same normal velocity, $u = u_0 e^{j\omega t}$, at every point on the surface. In that case F_r is the integral of the acoustic pressure over the surface

$$F_r = -\iint_S p(\vec{r})\, dS, \tag{1.3a}$$

where the negative sign indicates that F_r is the reaction force of the medium on the transducer and $\vec{r}$ is the position vector on the surface. (If the pressure were uniform over the entire area S, then $F_r = -pS$) The radiation impedance is defined as

$$Z_r = -F_r/u_0 = \frac{1}{u_0}\iint_S p(\vec{r})\, dS = R_r + jX_r, \tag{1.4a}$$

where R_r, the radiation resistance, and X_r, the radiation reactance, are critical parameters of any electroacoustic transducer. The time average of $F_r u$ is the radiated acoustic power and is equal to $\frac{1}{2}R_r u_0^2 = R_r u_{rms}^2$ (see Sect. 13.3). Thus R_r

is the resistance associated with transfer of power from the transducer to the external acoustic field, which is the basic function of the projector. The role of radiation resistance in the operation of a transducer is entirely different from that of the internal mechanical resistance, R, which represents loss of power inside the transducer. High values of R_r and low values of R are needed for efficient radiation of sound. The radiation reactance, X_r, represents a mass, $M_r = X_r/\omega$, added to the vibrator, since the vibrating surface accelerates the adjacent medium and transfers kinetic energy to it. The added mass affects the operation of the transducer by changing the resonant frequency, since resonance occurs at the frequency where the mass reactance and the stiffness reactance cancel [see Eq. (1.1b)]. Chapter 10 discusses radiation impedance in detail for various types of radiators.

For a surface with a specified, but nonuniform, normal velocity, $u(\vec{r})$, a particular value of $u(\vec{r})$ at some point on the surface must be chosen to serve as the reference velocity for defining the radiation impedance. The value at the center, or the spatial average value over the surface, is most often used. The chosen value of reference velocity will be denoted by u_0 and used to define F_r and Z_r for the case of nonuniform motion in a manner consistent with Eq. (1.4a):

$$F_r = -Z_r u_0, \tag{1.3b}$$

where

$$Z_r = (1/u_0 u_0^*) \iint_S p(\vec{r}) u^*(\vec{r}) \, dS \tag{1.4b}$$

is the radiation impedance referred to u_0 (* indicates complex conjugate). The basis for this more general definition of radiation impedance is that the integrand in Eq. (1.4b) is related to the time average acoustic intensity (see Chap. 10 and Sect. 13.3), and, therefore, the integral over the surface gives the radiated acoustic power. The time average power is then equal to $\frac{1}{2} R_r u_0^2$ in the same familiar way as for uniform velocity transducers. Equation (1.4b) also reduces to the uniform velocity case when $u(\vec{r})$ is a constant over the surface of the transducer. Foldy and Primikoff [20, 21] give a more general discussion of a transducer vibrating in an acoustic medium.

The above definitions only apply when it is a good approximation to assume that the transducer has a fixed velocity distribution [21]. A fixed velocity distribution means that the velocity amplitude is unrestricted, but its relative spatial distribution does not change because of changes of water depth or acoustic loading or under the influence of acoustic waves from external sources, although it may change with frequency. This concept also shows that the velocity distribution chosen to describe a transducer must be consistent with the medium in which the transducer will be used. For example, a mode of vibration for an object vibrating in a vacuum changes when that object is vibrating in water because of the water loading [22].

Although the fixed velocity distribution is an idealization that is never met exactly, most underwater transducers are mechanically quite stiff and are well approximated by this concept. However, in some cases the velocity distribution does change significantly over the range of operating conditions, and it is then necessary to consider other methods such as modal analysis of the vibrating surface (see Sect. 7.4.1).

Thus, the problem of the reaction of the medium, at least for fixed velocity distribution transducers, comes down to determination of the radiation impedance, which will be treated in Chap. 10. Until then it is only necessary for analysis purposes to consider the mass and internal mechanical resistance of the transducer to be augmented by the radiation mass and radiation resistance. However, it should be noted that the radiation reactance, the radiation resistance, and the radiation mass are all generally frequency dependent, whereas M and R in Eq. (1.1a) are often treated as frequency-independent parameters. At low frequency the radiation mass is constant and the resistance is small, while at high frequency the resistance is constant and the mass is small, and the frequency-independent approximation for the total mass and resistance is often adequate over a small band of frequencies.

For projectors $F_b = 0$, and the only external force is the radiation force, $F = F_r = -Z_r u_0$. For hydrophones the external force, F_b, resulting from the pressure of an incident sound wave, is the driving force, but the reaction of the medium is also present because the wave sets the surface of the hydrophone into vibration. Thus, for hydrophone operation

$$F = -Z_r u_0 + F_b, \tag{1.5}$$

where F_b is the surface integral of the sum of the incident free field pressure, p_i, and the scattered incident pressure, p_s, weighted by the velocity distribution:

$$F_b = \frac{1}{u_0^*} \iint_S [p_i(\vec{r}) + p_s(\vec{r})] u^*(\vec{r}) \mathrm{d}S. \tag{1.6}$$

The scattered pressure consists of reflections from the hydrophone and the structure in which it is mounted when the hydrophone surface is clamped and cannot vibrate. The diffraction constant, D_a, is defined [23] by

$$D_a = F_b / A p_i, \tag{1.7}$$

where A is the area of the moveable surface of the hydrophone. In hydrophone applications at low frequency scattering is usually negligible, making $F_b = A p_i$ and D_a approximately unity. (See Sects. 6.6 and 11.3 for more on the diffraction constant.)

Pairs of linear equations relating the four time harmonic variables V, I, F, and u that we have now defined can be used to describe the performance of any linear transducer. The natural pair for electric field transducers, where the electric force depends on voltage, is Eqs. (1.1b) and (1.2) rewritten as:

$$F_b = Z^E_{mr}u + N_{em}V, \tag{1.8}$$

$$I = N_{me}u + Y_0V, \tag{1.9}$$

where the short circuit ($V = 0$) mechanical impedance, including the radiation impedance, is, with $K^E_m = 1/C^E_m$,

$$Z^E_{mr} = (R + R_r) + j\left[\omega(M + M_r) - K^E_m/\omega\right].$$

The superscript notation here, and in the following cases, refers to the field involved; i.e., $V = 0$ means the electric field, $E = 0$. This will result in superscript consistency between transducer equations and the equations of state of materials to be introduced in Chap. 2. This condition means that the time harmonic part of the electric field is zero, but the total electric field may be nonzero but constant, since time-independent biases also exist in most cases (see Chap. 2). Y_0 is the clamped ($u = 0$) electrical admittance. The transduction coefficients N_{em} and N_{me} (also called electromechanical transfer ratios or turns ratios) are the key quantities that determine the coupling between the mechanical and electrical variables. The natural pair of equations for magnetic field transducers, where the electric force depends on current, gives F_b and V as functions of I and u. An advantage of both these forms of the equations is that two of the coefficients have familiar meanings, one coefficient is a mechanical impedance, one is an electrical admittance (or impedance in the magnetic case) while only the transduction coefficients, N_{em}, have both electrical and mechanical characteristics.

Equations (1.8) and (1.9), or similar equations, are the basis for analysis of transducer performance. They will be derived in the following chapters for different transducers, which will give specific results for all the parameters in the equations in terms of material properties and dimensions. As an example of their use consider projector operation, where $F_b = 0$. Equation (1.8) gives the velocity produced by a voltage, V, as

$$u = -N_{em}V/Z^E_{mr},$$

which, using the radiation resistance, gives the radiated power as $\frac{1}{2}R_r|u|^2$. With R the mechanical loss due to the motion u, the power lost is $\frac{1}{2}R|u|^2$.

For example: The mechanical efficiency, η_{ma}, may be calculated from the ratio of the radiated power out to the total power in and easily shown to be $\eta_{ma} = R_r/(R + R_r)$. If $R = R_r$ the efficiency is 50 % while if $R = R_r/3$ the mechanical efficiency $\eta_{ma} = 3R/(R + 3R) = 75$ % which is not unusual for sonar transducers.

If $N_{em} = N_{me}$, the coupling is called symmetric or reciprocal; if $N_{em} = -N_{me}$, it is called antisymmetric or sometimes anti-reciprocal. However, the term anti-reciprocal is misleading, since essentially the same reciprocal properties exist in both cases [1]. It will be seen in Chap. 2 that the equations for electric field transducers with u and V independent, and for magnetic field transducers with u and I independent, all have transduction coefficients that are equal in magnitude but opposite in sign. The transduction coefficient N_{em} being equal to either plus or minus N_{me} is a property called electromechanical reciprocity, and it holds for all the major transducer types. It is important because it allows an idealized transducer to be described by three parameters instead of four and it means that energy flow through a transducer is independent of direction. The second characteristic is especially important because, combined with acoustic reciprocity, it forms the basis for reciprocity calibration (see Sects. 9.5 and 11.2.2). Foldy and Primikoff [20] show that reciprocity calibration is valid when $|N_{em}| = |N_{me}|$.

Rewriting Eqs. (1.8) and (1.9) to make u and I, rather than u and V, the independent variables, gives, with $N_{me} = N$ and $N_{em} = -N$ for electric field transducers,

$$F_b = Z^D_{mr} u - (N/Y_0) I, \tag{1.10}$$

$$V = -(N/Y_0) u + (1/Y_0) I, \tag{1.11}$$

where $Z^D_{mr} = Z^E_{mr} + N^2/Y_0$ is the open circuit ($I = 0$) mechanical impedance and D is the electric displacement. Note that the transduction coefficient has changed to N/Y_0, showing that its value depends on which variables are dependent. Note also that the antisymmetric equations (1.8) and (1.9) have become symmetric (1.10) and (1.11), although both sets of equations apply to the same transducer. Thus the difference between symmetric and antisymmetric equations has little physical significance, although it does affect the types of electrical circuits that can represent a given transducer. Woollett [24] has discussed the different sets of transducer equations and the electrical circuits that can represent them.

As a second example of use of the transducer equations consider hydrophone operation with $I = 0$ and $F_b = AD_a p_i$ from Eq. (1.7). Combining Eqs. (1.10) and (1.11) then shows that the voltage output per unit of acoustic pressure is $V/p_i = (NAD_a)/(Y_0 Z^D_{mr})$, which is the hydrophone open circuit receiving sensitivity.

Although a reciprocal transducer can be described by three parameters, such as Z^E_{mr}, N, and Y_0, it is convenient to define other parameters, such as Z^D_{mr} in the previous paragraph. Thus rewriting Eqs. (1.8) and (1.9) again to make F_b and V the independent variables gives, with $N_{em} = -N$ and $N_{me} = N$,

$$u = (1/Z^E_{mr}) F_b + (N/Z^E_{mr}) V, \tag{1.12}$$

$$I = (N/Z^E_{mr}) F_b + Y_f V, \tag{1.13}$$

which introduces another parameter, $Y_f = Y_0 + N^2/Z^E_{mr}$, the free ($F_b=0$) electrical admittance. The differences between free and clamped electrical admittance, $(Y_f - Y_0)$, and between open circuit and short circuit mechanical impedance, $(Z^D_{mr} - Z^E_{mr})$, are important indicators of electromechanical coupling. If these differences were zero, the device would not be a transducer.

This formulation of the linear transducer equations applies to all the major types of electroacoustic transducers, but it is restricted to transducers that are passive in the sense that they have no internal energy source. For example, the carbon button microphone, which contains an electrical source and a pressure-dependent resistance, is not a passive transducer.

1.4 Transducer Characteristics

1.4.1 Electromechanical Coupling Coefficient

While the transduction coefficients in Eqs. (1.8) and (1.9) that relate electrical and mechanical variables are measures of electromechanical coupling, a more general measure is needed that facilitates comparison of different types of transducers and different designs within a given type. The electromechanical coupling coefficient (also called electromechanical coupling factor), denoted by k, fills this need. The definition and physical significance of k has been discussed at some length [1, 24–26]. Here we will introduce two consistent definitions with different physical interpretations.

The coupling coefficient is defined for static (not moving) or quasistatic (moving slowly) conditions, although it is often determined from quantities that are measured under dynamic conditions. Quasistatic conditions occur at low frequency where, after neglecting the resistance, the mechanical impedances, Z^D_{mr} and Z^E_{mr}, reduce to the stiffness reactances, $K^D_m/j\omega$ and $K^E_m/j\omega$ (or $1/j\omega C^D_m$ and $1/j\omega C^E_m$) and the electrical admittances, Y_0 and Y_f, reduce to the capacitive susceptances, $j\omega C_0$ and $j\omega C_f$. Thus the relationships $Z^D_{mr} = Z^E_{mr} + N^2/Y_0$ and $Y_f = Y_0 + N^2/Z^E_{mr}$ derived in the previous section for reciprocal transducers reduce to

$$K^D_m - K^E_m = \frac{N^2}{C_0} \tag{1.14}$$

and

$$C_f - C_0 = \frac{N^2}{K^E_m} = N^2 C^E_m. \tag{1.15}$$

The electromechanical coupling causes these changes in stiffness and capacitance when the electrical boundary conditions, or the mechanical boundary conditions, are changed. Eliminating N^2 from the two relations shows that $C_0\left(K_m^D - K_m^E\right) = K_m^E$ $(C_f - C_0)$ and thus that $C_0 K_m^D = C_f K_m^E$. It follows that the relative changes are equal and can be defined as k^2:

$$k^2 \equiv \frac{K_m^D - K_m^E}{K_m^D} = \frac{C_f - C_0}{C_f} = N^2 C_m / C_f. \tag{1.16}$$

This definition follows Hunt's suggestion [1] that k^2 could be defined such that its physical meaning is the change in mechanical impedance caused by the coupling, but this is not the only physical meaning of k^2, as we will see in Eq. (1.19). Equation (1.16) makes k^2 a dimensionless quantity that applies only to linear, reciprocal electric field transducers, but a similar definition will be given below for magnetic field transducers. One of the two equivalent definitions in Eq. (1.16) involves only mechanical parameters, the other only electrical parameters. Alternative expressions for k^2 involving both electrical and mechanical parameters can be derived from the relations above, for example,

$$k^2 = \frac{N^2}{K_m^D C_0} = \frac{N^2}{K_m^E C_f} = \frac{N^2 / K_m^E}{C_0 + N^2 / K_m^E} \tag{1.17a}$$

Since it is common in transducer work to use both mechanical stiffness and compliance, the reciprocal of stiffness, Eq. (1.17a) will also be written in terms of compliance, $C_m = 1/K_m$:

$$k^2 = \frac{N^2 C_m^D}{C_0} = \frac{N^2 C_m^E}{C_f} = \frac{N^2 C_m^E}{C_0 + N^2 C_m^E}. \tag{1.17b}$$

The definition in terms of the stiffness change is especially useful because it relates k^2 to the measurable resonance/antiresonance frequencies of a transducer if we assume that the stiffnesses, K_m^D and K_m^E, do not depend on frequency. The definition based on the capacitance change does not offer a convenient method of measuring k^2, because it is usually difficult to clamp an underwater sound transducer as required to measure C_0. However, it can easily be done in the ideal world of finite element simulations.

If we start from the equations for magnetic field transducers in which F and V are the dependent variables, and the magnetic field variables are H and B, the definition of the coupling coefficient in terms of stiffness changes, inductance changes, and transduction coefficients is, with the open circuit mechanical compliance $C_m^H = 1/K_m^H$

$$k^2 \equiv \frac{K_m^B - K_m^H}{K_m^B} = \frac{L_f - L_0}{L_f} = \frac{N^2}{K_m^B L_0} = \frac{N^2}{K_m^H L_f} = N^2 C_m^H / L_f \tag{1.18}$$

The quantities $(C_f - C_0)$ and $(L_f - L_0)$ are called the motional capacitance and inductance since they are the differences between the values when the transducer is free to move and when it is clamped.

Mason [25] pointed out, in connection with the electrostatic transducer, that "k^2 has the significance that it represents the portion of the total input electrical energy stored in mechanical form for a static or DC voltage." A statement of this kind is often used as a definition of k^2 for all types of electroacoustic transducers, because its simple physical meaning is very appealing. For example, voltage, V, applied to a reciprocal transducer described by Eqs. (1.8) and (1.9) with $F_b = 0$ causes a displacement $x = NV/K_m^E$ at low frequency. The transducer is mechanically free and the input electrical energy is ½ $C_f V^2$, the energy stored in mechanical form is ½$K_m^E x^2$ and Mason's definition gives, with the short circuit mechanical compliance $C_m^E = 1/K_m^E$

$$k^2 = \frac{\text{converted mechanical energy}}{\text{input electrical energy}} = \frac{K_m^E x^2/2}{C_f V^2/2} = \frac{N^2}{K_m^E C_f} = N^2 C_m^E / C_f \qquad (1.19)$$

as in Eq. (1.17a) obtained from the other definition of k^2. We can see from this definition that we would expect transduction materials to satisfy the condition that $0 < k < 1$. Piezoelectric ceramics have a $k \approx 0.7$ while piezoelectric single crystals have a $k \approx 0.9$.

For example: For piezoelectric ceramics $k \approx 0.7$ and $k^2 \approx 0.5$ yielding a conversion of approximately 50 % of the electrical energy into mechanical energy. On the other hand, for single crystals piezoelectric material $k \approx 0.9$ and $k^2 \approx 0.8$ yielding a higher energy conversion of approximately 80 %.

The importance of k^2 is the concept of the ratio of energy converted to energy stored. It is different from another important concept, efficiency, which is the ratio of the power output to the power input. The former depends on energy conversion from one form to another while the latter depends on the power radiated (which depends on the power lost to heat).

1.4.2 Transducer Responses, Directivity Index, and Source Level

The function of an electroacoustic transducer is to radiate sound into a medium such as air or water or to detect sound that was radiated into the medium by some other source. The transducer responses are measures of a transducer's ability to perform these functions. They are defined as the transducer output per unit of input as a function of frequency, for fixed drive conditions. Transducers generally radiate sound in a directional manner which changes with frequency and with distance from the transducer. At a given frequency the far field is the region beyond which the directional characteristics become independent of distance, and the sound pressure becomes inversely proportional to distance. The distance to the far field is an important concept that will be made more quantitative in Chaps. 9 and 10. The variation of acoustic intensity, $I(r, \theta, \phi)$, with the polar and azimuthal angles, θ, ϕ, at a given distance, r, in the far field is called the far-field directivity function. It is essentially an interference pattern consisting of angular regions of high intensity (lobes) separated by angular regions of low intensity (nulls) and characterized quantitatively by the directivity factor and the directivity index. The direction in which the maximum acoustic intensity, $I_0(r)$, occurs is called the acoustic axis or maximum response axis (MRA). The directivity factor is the ratio of the maximum acoustic intensity to the acoustic intensity averaged over all directions, $I_a(r)$, at the same distance, r, in the far field. The average intensity is the total radiated acoustic power, W, divided by the area of a sphere at the distance, r, i.e., $I_a = W/4\pi r^2$. (See Sect. 10.1 for the full definition of acoustic intensity.) Thus the directivity factor is defined as

$$D_f \equiv I_0(r)/I_a(r) = \frac{I_0(r)}{W/4\pi r^2} = \frac{I_0(r)}{(1/4\pi r^2)\int_0^{2\pi}\int_0^{\pi} I(r,\theta,\phi) r^2 \sin\theta \, d\theta \, d\phi}, \tag{1.20}$$

and the directivity index is defined as the directivity factor expressed in dB,

$$\mathrm{DI} = 10 \log D_f. \tag{1.21}$$

When the area, A, of the vibrating surface of a transducer is large compared to the acoustic wavelength, λ, squared and the normal velocity of the surface is uniform, a convenient approximation for the directivity factor is $D_f \approx 4\pi A/\lambda^2$ (see Sect. 10.2.2).

The source level of a transducer is a measure of the far-field pressure it is capable of producing on its maximum response axis. The total radiated power is independent of distance in a lossless medium, but, since the pressure varies inversely with distance, it is necessary to define a reference distance for the source level, conventionally, 1 m from the acoustic center of the transducer. Let $p_{rms}(r)$ be the far-field rms pressure magnitude on the MRA at a distance, r, and $p_{rms}(1) = r\, p_{rms}(r)$ be the far-field pressure extrapolated back to 1 m. The source level is defined as the ratio of $p_{rms}(1)$ to one micropascal expressed in dB:

$$\mathrm{SL} = 20 \log \left[p_{rms}(1) \times 10^6\right]. \tag{1.22}$$

The source level can be written in terms of the total radiated acoustic power and the directivity index by use of the following relations:

$$W = 4\pi I_a(1) = 4\pi I_0(1)/D_f, \tag{1.23}$$

$$I_0(1) = [p_{rms}(1)]^2/\rho c, \tag{1.24}$$

where ρ is the density and c is the sound speed in the medium (see Sect. 10.1). Then the source level in water, where $\rho c = 1.5 \times 10^6\,\text{kg/m}^2\,\text{s}$, can be written as

$$\text{SL} = 10 \log W + \text{DI} + 170.8 \text{ dB referred to } 1\,\mu\text{Pa at } 1\text{ m}, \tag{1.25}$$

where W, the output power in watts, is the input electrical power reduced by the electroacoustic efficiency. The source level corresponding to the maximum acoustic power that can be reliably radiated is usually the most important measure of a projector. The acoustic power is related to the radiation resistance and the magnitude of the reference velocity of the transducer's radiating surface by $W = ½R_r u_0^2$.

For example: Consider a transducer which radiates 1000 W and has a DI = 5 dB then the SL = 205.8 dB and this would be the same source level for a transducer that radiates only 500 W but has a DI = 8 dB. An increase of 3 dB in the DI requires only one-half the power from a transducer to attain the same source level.

The velocity, given by Eq. (1.8) for an electric field transducer, is

$$u_0 = N_{em} V / Z^E_{mr}, \tag{1.26}$$

and then the source level can be expressed as

$$\text{SL} = 10 \log \left[½\, R_r \left| N_{em} V / Z^E_{mr} \right|^2 \right] + \text{DI} + 170.8 \text{ dB re } 1\,\mu\text{Pa at } 1\text{ m}. \tag{1.27}$$

Equation (1.27) expresses one of the most important properties of an electroacoustic projector in a single equation involving the acoustical parameters, R_r and DI, an electrical drive amplitude, V, an electromechanical transduction coefficient, and the mechanical impedance including the radiation impedance. (Note: Often RMS values are used for the voltage V and in this case the factor ½ would not appear in Eq. (1.27).)

Transmitting voltage and current responses are defined as the source level for an input of one rms volt or one rms amp. Other transmitting responses, such as the source level for an input of one watt or one volt-amp, are also used in some cases. The free field voltage receiving response is defined as the open circuit voltage output for a free field pressure input of one micropascal in a plane wave arriving on the MRA. These responses will be discussed more fully in Chaps. 5 and 6, where a variety of projector and hydrophone designs will be described, and in Chap. 9 on measurements.

1.5 Transducer Arrays

Large arrays of transducers are needed to achieve the directivity required for accurate bearing determination and noise rejection and, in active arrays, to achieve sufficient power for range determination of distant targets. Arrays also provide flexibility in shaping and steering both active and passive acoustic beams. Range determination is done in active sonar by timing the return of an echo, but passive determination of range is also possible by triangulation with two arrays or measurement of wavefront curvature with three arrays if there is sufficient distance between the arrays. Figures 1.10, 1.11, and 1.12 show examples of arrays used in sonar: cylindrical and truncated spherical active arrays and a conformal passive array. Schloemer has given a comprehensive review of hull-mounted sonar arrays [27].

The U.S. Navy uses hundreds of thousands of transducers in its principal active and passive sonar systems and many others in smaller systems for specialized purposes. Active arrays for medium range detection usually have a bandwidth of about one octave and operate in the 2–10 kHz region, those for shorter range applications, such as mine or torpedo detection, use frequencies up to 100 kHz, while high resolution applications may go up to 1.5 MHz [28]. Passive naval arrays containing hundreds of hydrophones are designed in many configurations from those that conform to the ship's hull to line arrays that are towed far astern.

Some passive arrays are designed for surveillance at fixed installations in the ocean. This usually has an advantage over arrays mounted on a ship in that the self-noise of the ship is removed, and long range capability is feasible. However, long range active surveillance requires a combination of low frequency and high power that raises two major design problems. The first concerns the individual projectors, because they must operate near resonance to radiate high power, and resonant low frequency transducers are large, heavy, and expensive [29]. The second problem arises when projectors are close together in an array, since then the sound field of each one affects all the others (see Chap. 7). These acoustic interactions, or couplings, complicate array analysis and design, but are important because they can cause severe problems. For example, coupling may reduce the total mechanical impedance of some transducers to the point that their velocity becomes high enough to cause mechanical failure.

While characteristics such as beam width and side lobe level are fixed for an individual transducer, they can be changed for arrays by adjusting the relative amplitudes and phases of the individual transducers. For example, the amplitudes of individual transducers can be adjusted to reduce the side lobes relative to the main lobe (shading) or the phases can be adjusted to steer the beam. Both active and passive arrays also present the problem of grating, or aliasing, lobes near the high end of a frequency band, where the transducer outputs may combine to form lobes as large as the main lobe in undesired directions, especially when steered (see Chaps. 7 and 8). Increasing the bandwidth of active arrays is especially challenging because the high frequency end is limited by grating lobes, while the low frequency end is limited by acoustic interactions between transducers.

In fixed passive arrays, the ability to discriminate against internal noise and ambient sea noise is critical. In hull-mounted passive arrays, flow noise and structural noise excited by flow and machinery are more important than ambient sea noise except at very low speed. Towing passive arrays reduces the ship noise, but towed arrays are limited by their own flow noise and flow excited structural noise. Ambient noise, flow noise, and structural noise are very large subjects that are not treated comprehensively in this book, but some aspects directly related to array design will be discussed in Chap. 8.

1.6 Summary

In this chapter we have presented a brief history of underwater sound transducers, their applications, along with a general description of transduction, transducer characteristics, and transducer arrays. One of the first applications was installation of Fessenden moving coil transducers on World War I submarines. Since then magnetostrictive and mostly piezoelectric projectors and hydrophone have dominated underwater sound transducer systems from 500 Hz to above 1 MHz. The most common types of transducers include rings, cylinders, piston (Tonpilz), and flextensional types. In its most basic form the transducer may be represented by a two port network with a voltage, V, and current, I, on one side and a force, F, and velocity, u, on the other side. Pairs of linear equations relating the four time harmonic functions are defined and can be used to describe the performance of the transducer. A measure of the conversion of electrical to mechanical energy and vice versa is the square of the coupling coefficient, k, which lies within the range $0 < k < 1$ and depends on the active material and specific transducer design.

As a projector, the voltage drives the transducer which draws a current, I, based in the input transducer impedance $Z = V/I$ resulting in an output force which creates a velocity, u, based the radiation impedance Z_r and the efficiency of the transducer. This velocity radiates sound into the water and creates a far-field acoustic pressure, p, in the medium as well as a near-field pressure on the radiating area of the transducer. As a hydrophone, an incoming acoustic wave of pressure p_i creates a force $F = \int p_i dA$ on the transducer of area A resulting in an open circuit voltage. The diffraction constant $D_a = F/p_iA$ is useful in determining the force on the active transducer surface for a given geometry and input wave pressure, p_i. The directivity index $DI = I_0(r)/I_a(r)$ where I_0 is the on-axis intensity and I_a is average intensity on a virtual sphere of far-field radius r. The far-field intensity is simply $I = |p|^2/\rho c$ where ρ is the density and c is the sound speed of water ($\rho c = 1.5 \times 10^6$ kg/m^{2}s).

An important equation for the source level, SL, referenced to 1 m and a pressure of 1 μPa is $SL = 20 \log [p_{rms}/10^{-6}] = 10 \log W + DI + 170.8$ dB where p_{rms} is the rms acoustic pressure referenced to 1 m and W is the output acoustical power of the transducer or array of transducers and equal to $\eta_{ea}W_i$ where η_{ea} is the electroacoustic efficiency and W_i is the input electrical power. Large arrays of transducers allow a means of increasing the source level by increasing the DI as result of the narrower beam generated by the array of transducers arranged to add the pressure in a desired direction while mostly cancelling the pressure in other directions. This increase in DI also helps hydrophone arrays by adding the desired signal impinging on each transducer of the array and cancelling signals that do not come from the desired radiating object.

Exercises (Degree of Difficulty: *Lowest, **Moderate, ***Highest)

1.1** The free simple harmonic oscillator is basic to most transducers. Its equation of motion is given by Eq. (1.1a) with the external mechanical and electrical forces removed by setting $F = 0$ and $V = 0$. Show that the solution for the displacement of the mass represents oscillations at the angular frequency $\left[(K_m/M) - (R/2M)^2\right]^{1/2}$, and that these oscillations diminish in amplitude exponentially with time with a decay factor $R/2M$.

1.2** Calculate the directivity factor of a transducer that has a far-field intensity directivity function of $(A + B\cos\theta)^2$. What is the DI if $A = B = 1$? What value of the ratio B/A gives the maximum DI? Sketch the directivity pattern for some specific values of A and B. See Sect. 6.5.6 for other examples of similar directivity patterns.

1.3* Calculate the time average of the product of the two harmonically time varying quantities $x_1 = X_1 e^{j(\omega t+\phi_1)}$ and $x_2 = X_2 e^{j(\omega t+\phi_2)}$ (see Sect. 13.3), and show that it is equal to $\frac{1}{2}\mathrm{Re}\left(x_1 x_2^*\right)$.

1.4** A ship approaching a harbor in poor visibility, before the invention of radio direction finders, Loran or GPS, knows its position only within a circle of 2 mile diameter. At the harbor entrance are a bell buoy and a nearby foghorn and underwater sound source located close together that blast simultaneously at 1 min intervals. The navigator has his eyes, ears, and an omnidirectional underwater listening device. Derive a formula for the distance of the ship from the buoy, d, in terms of the speeds of sound in water, c_w, and air, c_a, and the time interval between hearing the two sounds.

1.5*** For magnetic field transducers the equations

$$F_b = Z_{mr}^H u + N_{em} I$$

$$V = N_{me} u + Z_0 I$$

correspond to Eqs. (1.8) and (1.9) for electric field transducers, where H is the magnetic field (proportional to I), Z_0 is the clamped electrical impedance, and Z_{mr}^H is the open circuit mechanical impedance. Go through the steps following Eqs. (1.8) and (1.9) to derive the short circuit mechanical impedance, $Z_{mr}^B = Z_{mr}^H + N^2/Z_0$, and the free electrical impedance, $Z_f = Z_0 + N^2/Z_{mr}^H$. Then follow the steps in Sect. 1.4.1 to derive the expressions for the coupling coefficient squared, k^2, for magnetic field transducers in Eq. (1.18).

1.6** Use the Table in Sect. 13.2 and list the ratio of the ρc values to that of water for tungsten, steel, PZT-4, aluminum, and magnesium. Consider a sandwich transducer with PZT-4 in the center as described in Sect. 5.4.1. Of the listed materials which would you put on the water side and which on the opposite side for the best match to water and, separately, for a less expensive "cost-effective" match to water?

1.7* Show that Eq. (1.25) may be written as the expression given in Sect. 13.13, Eq. (13.61). Calculate the source level for a transducer with an input power of 1000 W and efficiency of 50 % operating as an omnidirectional radiator with DI = 0 dB and also as a directional radiator with DI = 6 dB. What is the reduced value of power needed in the directional case to achieve the source level of the omnidirectional case?

1.8* Show that we may write the far-field intensity as $I_0 = D_f W/4\pi r^2$ and also as $I_0 = D_f u_r^2 R_r/4\pi r^2$, where u_r is the rms velocity.

1.9** Use the pressure expression for a piston in a rigid baffle, Sect. 13.13, Eq. (13.64), to obtain the on-axis ($\theta = 0$) intensity expression. For the case of a piezoelectric ceramic transducer, eliminate the velocity and write the expression in terms of the voltage, mechanical impedance, and the electromechanical turns ratio.

1.10** Of the two intensity expressions, given in Exercises 1.8 and 1.9, one is directly dependent on the directivity factor while the other is not, but depends on the velocity or voltage. Does this mean that if the intensity is calculated using the velocity or voltage that the D_f should not be additionally used? Show that both expressions are equivalent by use of Sect. 13.13, Eq. (13.50a) for a piston in a rigid baffle.

References

1. F.V. Hunt, *Electroacoustics* (Wiley, New York, 1954)
2. R.J. Urick, *Principles of Underwater Sound*, 3rd edn. (Peninsula, Los Altos Hills, CA, 1983)
3. R.T. Beyer, *Sounds of Our Times* (Springer/AIP Press, New York, 1999)
4. J.W.S. Rayleigh, *The Theory of Sound*, vol. 1 (Dover, New York, 1945), p. 3
5. L.E. Kinsler, A.R. Frey, A.B. Coppens, J.V. Sanders, *Fundamentals of Acoustics*, 4th edn. (Wiley, New York, 2000), p. 121
6. H.J.W. Fay, *Sub Sig Log—A History of Raytheon's Submarine Signal Division 1901 to Present.* (Raytheon Company, 1963)
7. G.W. Stewart, R.B. Lindsay, *Acoustics* (D. Van Nostrand, New York, 1930), pp. 249–250
8. I. Groves (ed.), *Acoustic Transducers, Benchmark Papers in Acoustics*, vol 14 (Hutchinson Ross Publishing, Stroudsburg, PA, 1981)
9. T. Parrish, *The Submarine: A History* (Viking, New York, 2004)
10. National Defense Research Committee, Div. 6, Summary Technical Reports (1946) vol 12, *Design and Construction of Crystal Transducers*, vol 13, *Design and Construction of Magnetostrictive Transducers*

11. J. Merrill, L.D. Wyld, *Meeting the Submarine Challenge—A Short History of the Naval Underwater Systems Center* (U.S. Government Printing Office, Washington, DC, 1997)
12. M.S. Dresselhaus, Obituary of A. R. von Hippel. Phys. Today, p. 76, September (2004); see also R.B. Gray, US Patent 2,486,560, Nov 1, 1949, filed Sept 20 (1946)
13. B. Jaffe, R.S. Roth, S. Marzullo, J. Appl. Phys. **25**, 809–810 (1954); J. Res. Natl. Bur. Standards **55**, 239 (1955)
14. New Mountain Innovations, 6 Hawthorne Rd. Old Lyme, Connecticut
15. R. Ebersole, Sonar takes bay research to new depths. Nat. Conserv. Mag. **52**, 14 (2002)
16. P.F. Worcester, B.D. Cornuelle, M.A. Dziecinch, W.H. Munk, B.M. Howe, J.A. Mercer, R.C. Spindel, J.A. Colosi, K. Metzger, T.G. Birdsall, A.B. Baggeroer, A test of basin-scale acoustic thermometry using a large-aperture vertical array at 3250 km range in the eastern North Pacific Ocean. J. Acoust. Soc. Am. **105**, 3185–3201 (1999)
17. Acoustics in the news. Echoes Acoust. Soc. Am. **13**(1) (2003)
18. N. Makris, Probing for an ocean on Jupiter's moon Europa with natural sound sources. Echoes Acoust. Soc. Am. **11**(3) (2001)
19. T.D. Rossing, Echos, scanning the journals. Acoust. Soc. Am. **12**(4) (2002)
20. L.L. Foldy, H. Primikoff, General theory of passive linear electroacoustic transducers and the electroacoustic reciprocity theorem. J. Acoust. Soc. Am., Part I, **17**, 109 (1945); Part II, **19**, 50 (1947)
21. L.L. Foldy, Theory of passive linear electroacoustic transducers with fixed velocity distribution. J. Acoust. Soc. Am. **21**, 595 (1949)
22. M. Lax, Vibrations of a circular diaphragm. J. Acoust. Soc. Am. **16**, 5 (1944)
23. R.J. Bobber, Diffraction constants of transducers. J. Acoust. Soc. Am. **37**, 591 (1965)
24. R.S. Woollett, *Sonar Transducer Fundamentals.* (Naval Undersea Warfare Center, Newport, Rhode Island, Undated)
25. W.P. Mason, *Electromechanical Transducers and Wave Filters*, 2nd edn. (D. Van Nostrand, New York, 1948), p. 390
26. J.F. Hersh, *Coupling Coefficients.* Harvard University Acoustics Research Laboratory Technical Memorandum No. 40, Nov 15 (1957)
27. H.H. Schloemer, Technology development of submarine sonar hull arrays. *Technical Digest.* Naval Undersea Warfare Center-Division Newport, Sept 1999, see also Presentation at the Undersea Defense Technology Conference and Exhibition, Sydney, Australia, Feb 7–9 (2000)
28. C.M. McKinney, The early history of high frequency, short range, high resolution, active sonar. Echos Acoust. Soc. Am. **12**, 4 (2002)
29. R.S. Woollett, Power limitations of sonic transducers. IEEE Trans. Sonics Ultrason. **SU-15**, 218 (1968)
30. Massa Products Corporation, 280 Lincoln Street, Hingham, MA
31. D.F. Jones et al., Performance analysis of a low-frequency barrel-stave flextensional projector. ONR Transducer Materials and Transducers Workshop, March, 1996, Penn Stater Conference Center, State College, PA, Artwork by Defence Research Establishment Atlantic, DREA, Dartmouth, Nova Scotia, CANADA B2Y 3ZY
32. S.C. Butler, A.L. Butler, J.L. Butler, Directional flextensional transducer. J. Acoust. Soc. Am. **92**, 2977–2979 (1992)
33. J.L. Butler, S.J. Ciosek, Rare earth iron octagonal transducer. J. Acoust. Soc. Am. **67**, 1809–1811 (1980)
34. S.C. Butler, F.A. Tito, A broadband hybrid magnetostrictive/piezoelectric transducer array. Oceans 2000 MTS/IEEE Conference Proceedings, Providence, RI, vol 3, September (2000), pp. 1469–1475

Chapter 2
Electroacoustic Transduction

This chapter will describe the six major electroacoustic transduction mechanisms in a unified way using one-dimensional models to derive pairs of linear equations specific to each mechanism as discussed in general in Sect. 1.3. Important characteristics of the transducer types will be summarized and compared to show why piezoelectric and magnetostrictive transducers are best suited for most applications in water.

In piezoelectric, electrostrictive, and magnetostrictive materials applied electric or magnetic fields exert forces on charges or magnetic moments contained within their crystalline structure. In these body force transducers the electric or magnetic energy is distributed, with the elastic energy and some of the kinetic energy, throughout the active material. Thus the stiffness, mass, and drive components are not completely separated as implied by Fig. 1.18; instead the stiffness, drive, and some of the mass are included in the active material as suggested by Fig. 2.5. Figure 1.18, where the symbol M is only mass and the symbol K_m is only stiffness, represents the so-called lumped-parameter approximation. This very frequently used approximation holds for body force transducers only when the dimensions of the active material are small compared to the wavelength of stress waves in the material and the mass and stiffness are independent of frequency. Lumped-parameter models will be used in this chapter because, although relatively simple, they still include the basic characteristics of each transduction mechanism. A lumped-parameter equivalent circuit transducer model will also be used for evaluating the heating of transducers.

The transducer models in this chapter do not include important dynamic effects that will be discussed in Chaps. 3 and 4, nor do they include nonlinear effects. The latter are also important in some cases, since most transduction mechanisms are inherently nonlinear, and those that are not become nonlinear for high amplitudes. In this chapter some of the nonlinearities will be included initially in the equations for each mechanism, but the equations will then be reduced to the linear form used in most transducer work. In Chap. 12 we will return to nonlinear mechanisms and methods for calculating their effects.

J.L. Butler, C.H. Sherman, *Transducers and Arrays for Underwater Sound*,
Modern Acoustics and Signal Processing, DOI 10.1007/978-3-319-39044-4_2

This chapter will also introduce additional transducer characteristics such as resonance, quality factors, characteristic impedance, efficiency, and power limits as well as a brief summary of transducer modeling by means of equivalent circuits. This will prepare the reader for Chaps. 5 and 6 on specific underwater electro-acoustic projector and hydrophone designs and Chaps. 7 and 8 on projector and hydrophone arrays. Chapters 3 and 4 will present the details of all the major methods of modeling transducers and further discussion of transducer characteristics. We begin this chapter with a discussion of piezoelectricity in biased electrostrictive materials, such as piezoelectric ceramics, the most commonly used underwater sound transduction material.

2.1 Piezoelectric Transducers

2.1.1 General

Although this section is concerned with piezoelectric transducers it will be helpful first to clarify the distinction between electrostriction and piezoelectricity. Cady [1] states it succinctly: "It is this reversal of sign of strain with sign of field that distinguishes piezoelectricity from electrostriction." In other words, piezoelectricity displays a linear relationship between mechanical strain and electric field while electrostriction displays a nonlinear relationship between the same variables as shown in Fig. 2.1.

Although the natural piezoelectric materials have a linear response to small electric fields they still display nonlinearity when the field is large enough. The origin of such nonlinearity is, at least in part, the weak electrostriction which occurs in all materials and becomes more important relative to piezoelectricity at higher electric fields.

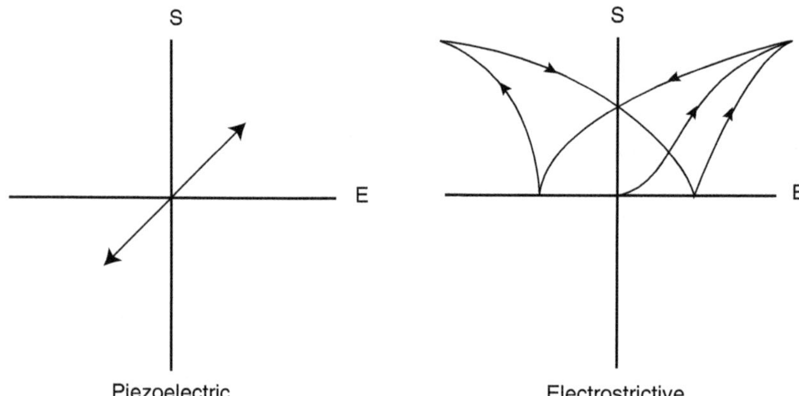

Fig. 2.1 Strain vs. electric field for piezoelectric and electrostrictive materials

Piezoelectricity occurs only in crystal structures that lack a center of symmetry [1, 2]. If an electric field is applied parallel to the length of a bar of piezoelectric crystal in one direction the bar will get longer; if the field is applied in the opposite direction the bar will get shorter. The change of length in an electric field is called the converse (sometimes reciprocal or inverse) piezoelectric effect, while the appearance of electric charge caused by mechanical stress is called the direct piezoelectric effect, because it was observed first by the Curie brothers. A polycrystalline piezoelectric material, composed of randomly oriented piezoelectric crystallites, displays no macroscopic piezoelectric effect because the effects in the individual crystallites cancel, and, therefore, it would not be useful for transducers. Quartz was the first piezoelectric material discovered, followed by others such as Rochelle salt, ammonium dihydrogen phosphate (ADP), and lithium sulfate.

Electrostriction occurs in all dielectric materials including solids, liquids, and gases, but the effect is only large enough for practical use in the ferroelectric materials that contain domains of oriented electric dipoles [1]. In these materials an applied electric field aligns the domains and causes significant dimensional changes as shown in Fig. 2.2. A bar of electrostrictive material gets longer (in most cases [1]) when an electric field is applied parallel to its length, regardless of the direction of the field. Thus the mechanical response is nonlinear, since it is not directly proportional to the electric field. The response depends on the square and higher even powers of the field, and there is no reciprocal electrical response to a mechanical stress. To achieve a linear response to an applied alternating drive field a much larger, steady polarizing, or bias, field must first be applied. The bias field establishes a polar axis of symmetry and causes a fixed displacement along that axis. Then a superimposed alternating drive field causes a variation of the total field

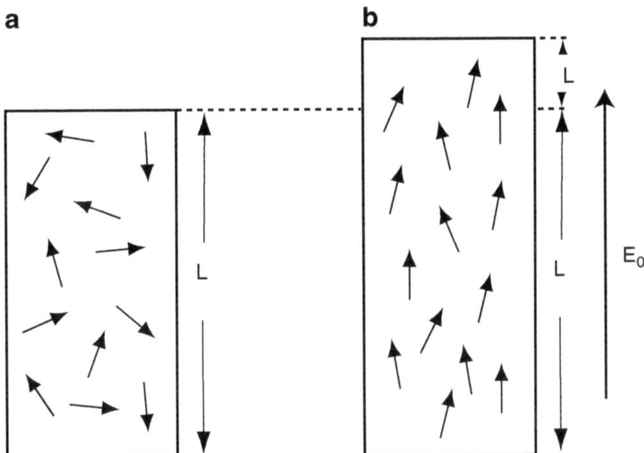

Fig. 2.2 Polarization of a ferroelectric, electrostrictive material occurs when randomly oriented dipole moments (**a**) are approximately aligned (**b**) by a strong, steady electric field, E_0. The material also increases in length in the direction of E_0

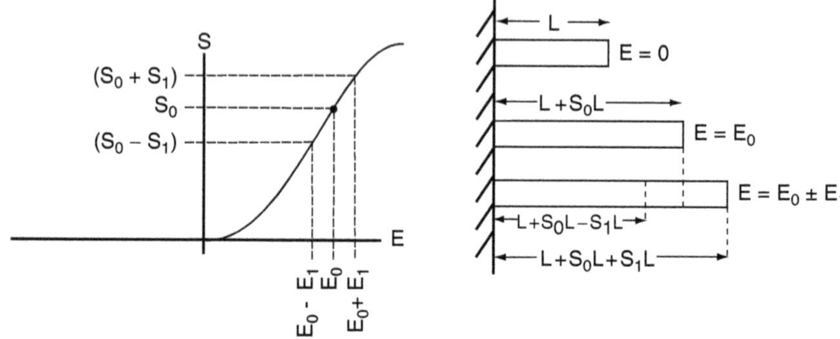

Fig. 2.3 Electrostrictive material with maintained static bias field E_0 and static strain S_0. An alternating field E_1 follows a minor hysteresis loop and causes an alternating strain S_1. If $E_1 \ll E_0$ the variations are approximately linear

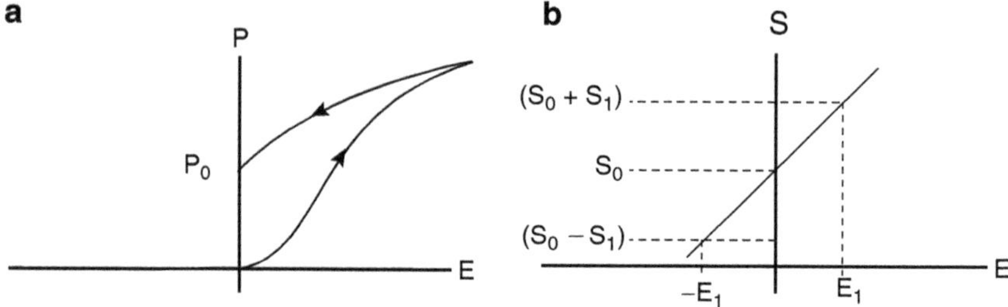

Fig. 2.4 Polarization and strain vs. electric field showing remanent polarization, P_0 (**a**), and remanent strain, S_0 (**b**), when the field is removed. Application of the alternating field E_1 then causes alternating strain S_1 about the remanent strain S_0

accompanied by an alternating displacement about the fixed displacement. The result is an approximately linear, reciprocal mechanical response to the drive field as shown in Fig. 2.3. The bias converts an isotropic polycrystalline electrostrictive material to a material with planar isotropy in planes perpendicular to the polar axis. This type of symmetry has the same elasto-piezo-electric properties as crystals of the class C_{6V} [1,2], and the biased electrostrictive material is, for transducer purposes, equivalent to a piezoelectric material. A general description of the properties of materials has been given by Newnham [3].

The electrostrictive materials can be divided into two groups: those with high coercive force that retain a significant remanent polarization when the bias E_0 in Fig. 2.2b is removed, and those with low coercive force that require the bias to be maintained in order to maintain the polarization. Barium titanate ($BaTiO_3$) and lead zirconate titanate (PZT) are ferroelectric electrostrictive materials with a high coercive force in which the domains remain partially aligned after temporary application of a high polarizing field at temperatures near the Curie temperature. The resulting remanent polarization gives an approximately linear response to an alternating electric field as shown in Fig. 2.4. In ceramic $BaTiO_3$ and PZT the remanent polarization is very stable and large enough to give a strong piezoelectric effect. However, depolarization can result from operating temperatures too near the Curie temperature, from high static pressure cycling in deep water applications [4], from high alternating electric fields and to a slight extent from the passage of time (see Sect. 13.14). Thus, while the properties of true piezoelectrics are determined by their internal crystal structure and cannot be changed, the piezoelectric properties of polarized electrostrictive materials depend on the level of remanence achieved in the polarization process, and may also be changed by operating conditions. Despite these limitations these materials are now used more than any other material for underwater transducers, but the limitations must be considered during the design process. Except for these depolarization problems "permanently" polarized $BaTiO_3$ and PZT can be considered to be piezoelectric with the symmetry associated with crystal class C_{6V}. Since they can also be conveniently made in the form of ceramics they are known as piezoelectric ceramics (sometimes shortened to piezoceramics). The properties of the most commonly used piezoelectric ceramics are given in Sect. 13.5.

Some ferroelectric materials have strong electrostrictive properties but do not have high coercive force. The remanent polarization in these materials may be sufficient for low field applications, such as hydrophones, but not sufficient for high field projector applications. In the latter cases a steady electric bias must be maintained to achieve linear operation. Lead magnesium niobate (PMN), lead zirconium niobate (PZN), and mixtures with lead titanate, (PZN-PT) or (PMN-PT), are promising materials of this type. With indium, PIN-PMN-PT has a high coercive force and does require an external steady electric bias.

It will be seen in Sects. 2.4–2.6 that analysis of the surface force transduction mechanisms is based directly on fundamental physical laws applied to macroscopic objects. However, the body force transducers present a different situation since the electromechanical effects result from interactions on the atomic level. Macroscopic descriptions of these effects take a phenomenological form similar to Hooke's law of elasticity. Since this section is restricted to linear effects, the description is a set of linear equations that relate stress, T, strain, S, electric field, E, and electric displacement, D, all of which are functions of position and time. For most transducer work adiabatic conditions can be assumed, and temperature and entropy variables can be omitted from the equations [2], but it must be understood that the coefficients in the equations are generally temperature dependent in a way that varies from one material to another.

Since the second rank tensors T and S are symmetric it is simpler to consider only their six independent components and write the phenomenological equations of state as two matrix equations since the coefficients then have two subscripts rather than four:

$$S = s^E T + d^t E, \tag{2.1a}$$

$$D = dT + \varepsilon^T E. \tag{2.1b}$$

In these equations S and T are 1×6 column matrices, E and D are 1×3 column matrices, s^E is a 6×6 matrix of elastic compliance coefficients, d is a 3×6 matrix of piezoelectric coefficients (d^t is the transpose of d), and ε^T is a 3×3 matrix of permittivity coefficients [2]. Each of these coefficients is proportional to a partial derivative where the superscript gives the variable that is held constant. For example, s^E is the partial derivative of S with respect to T with E held constant, and s^E can be measured from the slope of a curve of strain versus stress while holding electric field constant. The superscript is omitted from d, because $d = (\partial D/\partial T)_E = (\partial S/\partial E)_T$ as can be derived from thermodynamic potentials [2]. This is the origin of the electromechanical reciprocity discussed briefly in Sect. 1.3.

When the coefficient matrices in Eqs. (2.1a) and (2.1b) are combined they form a symmetric 9×9 matrix with 45 unique coefficients in general. However, for piezoelectric crystals of class C_{6V}, and for permanently polarized electrostrictive materials, many of the coefficients are zero and others are related, leaving only ten independent coefficients. Thus, for this symmetry Eqs. (2.1a) and (2.1b) expands to:

$$
\begin{aligned}
S_1 &= s_{11}^E T_1 + s_{12}^E T_2 + s_{13}^E T_3 + d_{31} E_3, \\
S_2 &= s_{12}^E T_1 + s_{11}^E T_2 + s_{13}^E T_3 + d_{31} E_3, \\
S_3 &= s_{13}^E T_1 + s_{13}^E T_2 + s_{33}^E T_3 + d_{33} E_3, \\
S_4 &= s_{44}^E T_4 + d_{15} E_2, \\
S_5 &= s_{44}^E T_5 + d_{15} E_1, \\
S_6 &= s_{66}^E T_6, \\
D_1 &= d_{15} T_5 + \varepsilon_{11}^T E_1, \\
D_2 &= d_{15} T_4 + \varepsilon_{11}^T E_2, \\
D_3 &= d_{31} T_1 + d_{31} T_2 + d_{33} T_3 + \varepsilon_{33}^T E_3,
\end{aligned}
\tag{2.2}
$$

where $s_{66}^E = 2(s_{11}^E - s_{12}^E)$ and the subscripts 4, 5, 6 refer to shear stresses and strains.

For example: Consider in Eq. (2.2) the special case where there are no loads or forces in the 1 and 2 direction and we wish to attain the strain in the 3 direction. Here

$$S_3 = s_{33}^E T_3 + d_{33} E_3.$$

For a short circuit across the electrodes $E_3 = 0$ we are left with the mechanical condition $S_3 = s_{33}^E T_3$. From Sect. 13.5 the elastic modulus $s_{33}^E = 15.5 \times 10^{-12}$ m^2/N and for a stress $T_3 = 0.2 \times 10^6$ N/m^2 (29 psi), we get a mechanically derived train of $S_3 = 3.10 \times 10^{-6}$. For a length, L, of 1 m a displacement of

(continued)

$X_3 = S_3 L = 3.10 \times 10^{-6}$ m. If now we apply an electric field of 10 kV/m, without a load so that $T_3 = 0$, the electrically derived strain, S_3, is nearly the same at 2.89×10^{-6} based on $d_{33} = 289 \times 10^{-12}$ C/N given for PZT-4 in Sect. 13.5.

Three other equation pairs relating S, T, E, and D are also used, depending on which variables are more convenient to make independent in a given application:

$$\begin{aligned} T &= c^E S - e^t E, \\ D &= eS + \varepsilon^S E, \end{aligned} \tag{2.3}$$

$$\begin{aligned} S &= s^D T + g^t D, \\ E &= -gT + \beta^T D, \end{aligned} \tag{2.4}$$

and

$$\begin{aligned} T &= c^D S - h^t D, \\ E &= -hS + \beta^S D, \end{aligned} \tag{2.5}$$

where c^E, s^D, and c^D are 6×6 matrices of stiffness and compliance coefficients, h, g, and e are 3×6 matrices of piezoelectric coefficients and ε^S, β^T, and β^S are 3×3 matrices of permittivity and impermittivity coefficients. General relationships exist among these coefficients [2] which can be used to convert results obtained with one pair of equations to notation associated with another pair (see Sect. 13.4). There are ten coefficients in each pair, Eqs. (2.2)–(2.5), for a total of 40 different (but not independent) coefficients; 36 of them are given in Sect. 13.4 for several piezoelectric ceramics and one single crystal. The remaining four coefficients are the impermittivities ($\beta' s$), which are the reciprocals of the permittivities ($\varepsilon' s$) as indicated in Sect. 13.4. Five different coupling coefficients, the dielectric loss factor, $\tan \delta$ (defined in Sect. 2.8.5), and the density, ρ, for each material are also given in Sect. 13.5.

2.1.2 *The 33 Mode Longitudinal Vibrator*

Now an idealized one-dimensional longitudinal vibrator transducer shown in Fig. 2.5 will be analyzed.

A piezoelectric ceramic bar of length L is fixed at one end and attached to a mass M at the other end. The other side of the mass, of area A, is in contact with an acoustic medium. The length of the bar is assumed to be less than a quarter

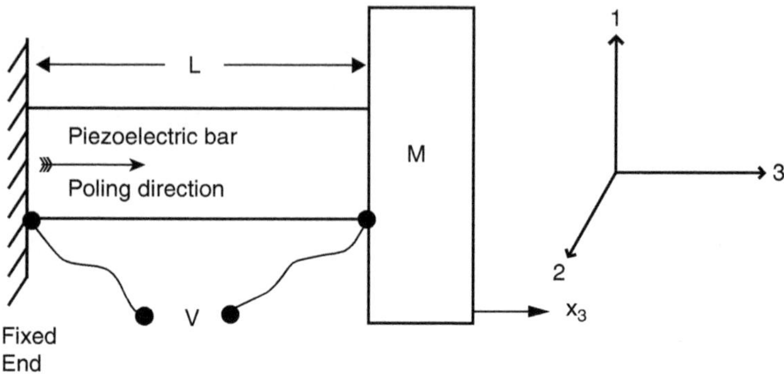

Fig. 2.5 The 33 mode piezoelectric longitudinal vibrator with working strain parallel to poling direction and electrodes on the ends of the bar

wavelength of longitudinal waves in the ceramic to validate the lumped-parameter approximation (see Chap. 4). Assume that the bar has been polarized by using electrodes on the ends to establish the polar axis (by convention called the x_3 axis, but also sometimes the z-axis) parallel to the length of the bar. An alternating voltage, V, is applied between the same electrodes creating an alternating electric field, E_3, parallel to the polarization. The electric fields $E_1 = E_2 = 0$ on the electrodes which are equipotential surfaces, and, if fringing of the field is neglected, these components can be assumed to be zero throughout the bar. It is also assumed that this purely longitudinal electric field does not excite shear stresses, i.e., $T_4 = T_5 = T_6 = 0$. Then the equations of state in Eq. (2.2) reduce to

$$\begin{aligned} S_1 &= s_{11}^E T_1 + s_{12}^E T_2 + s_{13}^E T_3 + d_{31} E_3, \\ S_2 &= s_{12}^E T_1 + s_{11}^E T_2 + s_{13}^E T_3 + d_{31} E_3, \\ S_3 &= s_{13}^E T_1 + s_{13}^E T_2 + s_{33}^E T_3 + d_{33} E_3, \\ D_3 &= d_{31} T_1 + d_{31} T_2 + d_{33} T_3 + \varepsilon_{33}^T E_3. \end{aligned} \tag{2.6a}$$

Let the cross-sectional area of the bar be A_0 and assume that the sides of the bar are free to move. Then the stresses T_1 and T_2 are both zero on the surfaces of the sides, and, if the lateral dimensions are small, T_1 and T_2 are zero throughout the bar, and the state equations reduce further to:

$$S_1 = s_{13}^E T_3 + d_{31} E_3, \tag{2.6b}$$

$$S_2 = s_{13}^E T_3 + d_{31} E_3, \tag{2.6c}$$

$$S_3 = s_{33}^E T_3 + d_{33} E_3, \tag{2.6d}$$

$$D_3 = d_{33} T_3 + \varepsilon_{33}^T E_3. \tag{2.6e}$$

The first two equations show that $S_1 = S_2$. These lateral strains are caused by a Poisson ratio effect modified by the piezoelectric strain, and they play no role in the operation of the transducer, because the sides are not usually in contact with the acoustic medium in a real transducer. This case illustrates the importance of choosing a convenient pair of equations. Note that, in this case, it is convenient to make stress an independent variable, because the unimportant strains, S_1 and S_2, are then analytically separated from the important strain, S_3.

Equations (2.6d) and (2.6e) provide the basis for the two transducer equations. If the bar is short enough, the displacement varies linearly along its length from zero at the fixed end to a maximum at the end attached to the mass. The stress and strain, T_3 and S_3, are then constant along the length, and the force exerted on the mass by the bar is A_0T_3. The equation for S_3 can be solved for T_3 and inserted directly into the equation of motion of the mass giving

$$M_t\ddot{x}_3 + R_t\dot{x}_3 + A_0T_3 = M_t\ddot{x}_3 + R_t\dot{x}_3 + \left(A_0/s_{33}^E\right)[S_3 - d_{33}E_3] = F_b \qquad (2.7)$$

where x_3 is the displacement of the mass, $M_t = M + M_r$ and $R_t = R + R_r$ where R_r and M_r are the radiation resistance and mass and F_b is an external force. The strain in the bar is $S_3 = x_3/L$ and the electric field in the bar is $E_3 = V/(L + x_3)$ resulting in

$$M_t\ddot{x}_3 + R_t\dot{x}_3 + \left(A_0/s_{33}^E L\right)x_3 = \left(A_0d_{33}/s_{33}^E\right)V/(L + x_3) + F_b, \qquad (2.8)$$

where $A_0/s_{33}^E L$ is equal to the short circuit stiffness, K_m^E. This equation shows that the piezoelectric ceramic bar provides the spring force (proportional to x_3) and the electric drive force (proportional to V). The drive force is nonlinear because x_3 appears in the denominator, but $x_3 \ll L$, and the equation will be linearized by neglecting x_3 compared to L. For sinusoidal drive, and omitting the factor $e^{j\omega t}$ appearing in all the variables, Eq. (2.8) becomes

$$\begin{aligned} F_b &= \left[-\omega^2 M_t + j\omega R_t + \left(A_0/s_{33}^E L\right)\right]x_3 - \left(A_0d_{33}/s_{33}^E L\right)V \\ &= Z_{mr}^E u_3 - \left(A_0d_{33}/s_{33}^E L\right)V, \end{aligned} \qquad (2.9a)$$

where $u_3 = j\omega x_3$ and

$$Z_{mr}^E = (R + R_r) + j\omega(M + M_r) + \left(A_0/Ls_{33}^E\right)/j\omega, \qquad (2.9b)$$

is the total mechanical impedance including the radiation impedance.

The lumped-parameter spring constant of the short bar is $K_m^E = A_0/Ls_{33}^E$, where $1/s_{33}^E$ is Young's modulus for constant voltage. Although the bar is the spring, it also has mass; one end is moving with the same velocity as the radiating mass, while the other end is not moving causing some fraction of the bar's mass to be involved in the total kinetic energy. This dynamic mass will be discussed in Chaps. 4 and 5, where it will be shown that the effective mass of a short bar is 1/3 the static mass of the bar. It is more important to consider the dynamic mass of a short bar than the dynamic stiffness. For a very short bar the dynamic stiffness approaches the static stiffness, for a 1/8 wavelength bar it is only 1 % greater and for a 1/4 wavelength bar 23 % greater (see Chap. 4). Thus the lumped-parameter approximation of the short piezoelectric ceramic bar consists of a spring, a portion of the mass of the bar and a force.

The other member of the pair of transducer equations comes from Eq. (2.6e). Since electric displacement is charge per unit area, the current is

$$I = A_0 \frac{\mathrm{d}D_3}{\mathrm{d}t} = A_0 \frac{\mathrm{d}}{\mathrm{d}t}\left[d_{33}T_3 + \varepsilon_{33}^T V/L\right]. \tag{2.10}$$

Later another term will be included in this equation to represent electrical losses in the piezoelectric material. To put Eq. (2.10) in terms of u_3 it is necessary to express T_3 in terms of S_3 using Eq. (2.6d) which gives, for sinusoidal drive,

$$I = \left(A_0 d_{33}/s_{33}^E L\right) u_3 + Y_0 V, \tag{2.11}$$

where

$$Y_0 = j\omega\left(\varepsilon_{33}^T A_0/L\right)\left[1 - \left(d_{33}^2/\varepsilon_{33}^T s_{33}^E\right)\right] = j\omega C_0 \tag{2.12}$$

is the clamped $(x_3 = 0)$ electrical admittance, the clamped capacitance is

$$C_0 = \left(\varepsilon_{33}^T A_0/L\right)\left[1 - \left(d_{33}^2/\varepsilon_{33}^T s_{33}^E\right)\right], \tag{2.13}$$

and the first factor in Eq. (2.13) is the free $(F_b = 0)$ capacitance,

$$C_f = \varepsilon_{33}^T A_0/L \tag{2.14}$$

Using Eq. (1.16), in the form $C_0 = C_f\left(1 - k_{33}^2\right)$, identifies the coupling coefficient as

$$k_{33}^2 = d_{33}^2/\varepsilon_{33}^T s_{33}^E \tag{2.15}$$

It also follows from Eq. (1.16) that

$$K_m^D = K_m^E/\left(1 - k_{33}^2\right) = K_m^E + N_{33}^2/C_0. \tag{2.16}$$

The transduction coefficient, appearing in Eqs. (2.9a) and (2.11) with opposite signs, is $N_{33} = A_0 d_{33}/s_{33}^E L$; note that N_{33} is positive since d_{33} is positive. The subscripts on k_{33} indicate that this is the value that applies when the working strain, S_3, and the electric field, E_3, are both parallel to the polar axis, the situation that usually gives the highest electromechanical coupling. Such a transducer is called a 33 mode longitudinal vibrator.

For example: From Sect. 13.5 the free ($T = 0$) dielectric constant for PZT-4 is $\varepsilon_{33}^T = \varepsilon_0 K_{33}^T = 8.842 \times 10^{-12} \times 1300 = 11.49 \times 10^{-9}\,\mathrm{C/mV}$ and along with the d_{33} and s_{33}^E values we get from Eq. (2.15) $k_{33} = 0.70$, which, incidentally, is also listed in Sect. 13.5. If the piezoelectric piece has an area $A_0 = 400 \times 10^{-6}\,\mathrm{m}^2$ and a length $L = 0.01\,\mathrm{m}$, the ratio $A_0/L = 40 \times 10^{-3}$. From Eq. (2.14), $C_f = 11.49 \times 10^{-9} \times 40 \times 10^{-3} = 460\,\mathrm{pF}$ and $C_0 = 234\,\mathrm{pF}$ and from the discussion above, an electromechanical transducer turns ratio value of $N_{33} = 40 \times 10^{-3}(289 \times 10^{-12}/15.5 \times 10^{-12}) = 0.746$.

Equations (2.9a) and (2.11) lead to the source level and transmitting response by solving for u_3 with $F_b = 0$ and using Eq. (1.27). The open circuit receiving response is given for $F_b = D_a A p_i$ by setting $I = 0$ and solving for V. However, such calculations cannot be completed until the acoustical parameters, radiation impedance, directivity index, and diffraction constant are determined. Equations (2.9a) and (2.11) will also be used to obtain other transducer parameters in Sect. 2.8.

2.1.3 *The 31 Mode Longitudinal Vibrator*

Another example of a piezoelectric longitudinal vibrator transducer will be given, a variation on the case just discussed that uses the same ceramic bar in a different way called the 31 mode. The 31 mode has lower coupling but has the advantage of being less susceptible to depoling by static pressure cycling because the polarization is perpendicular to the static stress [4, 5]. (In single crystals, certain poling orientations have a high 32-mode coupling and the treatment deriving the equivalent circuit is the same except the field is now E_2 and d_{31} is not equal to d_{32}.) Consider the 31 mode bar to have lateral dimensions h and w, with $A_0 = hw$, and to have been polarized using electrodes on the sides of area hL as shown in Fig. 2.6.

The polar axis is now perpendicular to the length and parallel to the side of dimension w and is still called the x_3 axis. One end of the bar is fixed with the other end attached to the mass as before. The only non-zero stress component is still parallel to the length of the bar, but it will now be called T_1, with $T_2 = T_3 = 0$. The driving voltage is applied between the electrodes used for polarizing, and E_3 is the only electric field component. In this case Eq. (2.2) becomes

$$\begin{aligned} S_1 &= s_{11}^E T_1 + d_{31} E_3, \\ S_2 &= s_{12}^E T_1 + d_{31} E_3, \\ S_3 &= s_{13}^E T_1 + d_{33} E_3, \\ D_3 &= d_{31} T_1 + \varepsilon_{33}^T E_3. \end{aligned} \tag{2.17}$$

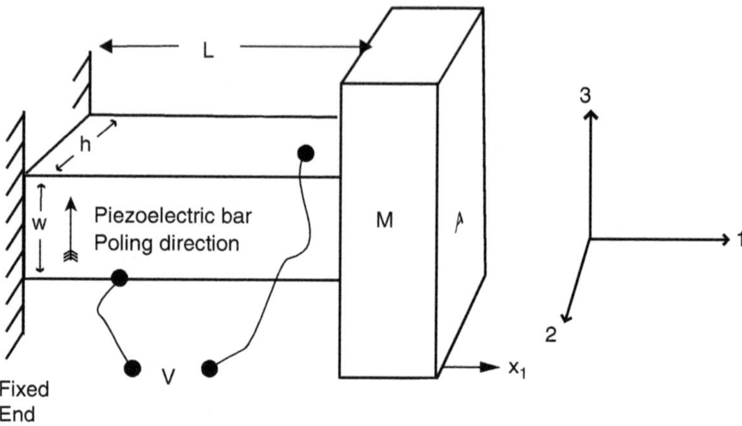

Fig. 2.6 The 31 mode piezoelectric longitudinal vibrator with working strain perpendicular to poling direction and electrodes on the side of the bar

The lateral strains, S_2 and S_3, again play no role, but they are not equal, as S_1 and S_2 were in the 33 mode case, because one of them is parallel, and the other perpendicular, to the polar axis. The equation for S_1 can be solved for T_1 and inserted in the equation of motion of the mass:

$$M_t\ddot{x}_1 + R_t\dot{x}_1 = -A_0T_1 + F_b = -\left(A_0/s_{11}^E\right)[S_1 - d_{31}E_3] + F_b. \tag{2.18}$$

The strain is $S_1 = x_1/L$ and the electric field is $E_3 = V/w$, for $x_3 \ll w$, giving

$$M_t\ddot{x}_1 + R_t\dot{x}_1 + \left(A_0/Ls_{11}^E\right)x_1 = \left(A_0 d_{31}V/ws_{11}^E\right) + F_b \tag{2.19}$$

or, for sinusoidal drive,

$$F_b = Z_{mr}^E u_1 - \left(A_0 d_{31}/ws_{11}^E\right)V. \tag{2.20}$$

Note that Z_{mr}^E is the same as before except for the spring constant, which is now $K_m^E = A_0/Ls_{11}^E$, depends on s_{11}^E rather than s_{33}^E. The transduction coefficient is also different, depending on d_{31} and s_{11}^E rather than d_{33} and s_{33}^E,

$$N_{31} = A_0 d_{31}/ws_{11}^E = hd_{31}/s_{11}^E. \tag{2.21}$$

Since d_{33} is considered to be positive and the lateral dimensions decrease when the length increases, d_{31} and N_{31} are considered to be negative.

The other transducer equation comes from the equation for D_3:

$$I = hL\frac{dD_3}{dt} = hL\frac{d}{dt}\left[d_{31}T_1 + \varepsilon_{33}^T V/w\right]. \tag{2.22}$$

After expressing T_1 in terms of S_1 this becomes

$$I = \left(A_0 d_{31}/ws_{11}^E\right)u_1 + Y_0 V, \tag{2.23}$$

$$\text{where} \quad Y_0 = j\omega\left(\varepsilon_{33}^T hL/w\right)\left[1 - d_{31}^2/\varepsilon_{33}^T s_{11}^E\right] = j\omega C_0 \tag{2.24}$$

is the clamped admittance and C_0 is the clamped capacitance. The quantity $C_f = \varepsilon_{33}^T hL/w$ is the free capacitance and comparison with Eq. (1.16) shows that the coupling coefficient is

$$k^2 = 1 - C_0/C_f = d_{31}^2/\varepsilon_{33}^T s_{11}^E = k_{31}^2. \tag{2.25}$$

The piezoelectric coefficient d_{31}, with working strain perpendicular to the electric field, is usually about half of d_{33}, the controlling coefficient for the 33 mode where the working strain is parallel to the electric field. Since s_{11}^E and s_{33}^E have similar magnitude (see Sect. 13.5) k_{31} is significantly less than k_{33}, and the 33 mode transducer outperforms the 31 mode transducer in most respects.

It is important to note that k_{33} and k_{31} depend only on properties of the piezoelectric material. They are material coupling coefficients that characterize active materials, but do not necessarily characterize complete transducers. The idealized analysis that led to k_{33} and k_{31} assumed that the bar was short compared to the wavelength. This corresponds to optimum use of the piezoelectric ceramic in that the strain and electric field are the same throughout the active bar, and the coupling coefficients have their maximum values. Chapters 3–5 will discuss transducers where wave motion makes the strain and electric field vary within the active material or where a portion of the transducer structure, other than the active material, stores either electric or elastic energy. In such cases the transducer can be characterized by an effective coupling coefficient that is less than the material coupling coefficient [6]. It is usually a valid design goal to make the effective coupling coefficient as close as possible to the material coupling coefficient. Piezoelectric ceramic is also the current favored material for many other transducer configurations, in addition to longitudinal resonators, as will be seen in Chap. 5.

2.2 Electrostrictive Transducers

The permanently polarized piezoelectric ceramics can be analyzed and used, for most purposes, as though they were piezoelectric as discussed in the previous section. Now we will continue the discussion of electrostriction in order to include the electrostrictive materials with low coercive force that must be used with a maintained bias for projector applications. In this category PMN ceramics and PMN-PT single crystals [7] are promising new materials for underwater transducers. Properties of PMN-PT (as well as high coercive force polarized

PIN-PMN-PT crystals and textured PMN-PT) are given in Sect. 13.5. The electric field dependence and the temperature dependence of some of these properties, as shown for PZT in Sect. 13.5, probably need further investigation. In this section we will derive the effective piezoelectric constants of electrostrictive materials from nonlinear electrostrictive equations of state to show how they depend on the bias.

Figures 2.3 and 2.4 compare strain vs. electric field when operating with a maintained bias and when operating at remanence. The need to provide a biasing circuit and electric power is an added burden that reduces overall efficiency. The feasibility and value of maintained bias has been clearly demonstrated by a study that showed the benefits of adding a maintained bias to already permanently polarized PZT [8]. However, the promising electromechanical properties of the PMN based materials indicate that their use may be advantageous in spite of the need for bias in some applications. PMN ceramic requires maintained bias for almost all applications, but PMN-PT has some remanence, depending on the degree of alignment of the crystallites achieved in the processing [9], while PIN-PMN-PT single crystals have considerable remanence and may be used with considerably higher drive voltages.

To show the dependence of the effective piezoelectric constants on the bias it is necessary to start with approximate nonlinear equations of state, which can be derived from thermodynamic potentials [10] as extensions of linear equations such as Eq. (2.2). For simplicity the one-dimensional case with parallel stress and electric field will be considered, as in the 33 mode transducer in Sect. 2.1.2. When only linear and quadratic terms are kept the nonlinear equations can be written [10]:

$$S_3 = s_{33}^E T_3 + d_{33} E_3 + s_2 T_3^2 + 2 s_a T_3 E_3 + d_2 E_3^2, \tag{2.26a}$$

$$D_3 = d_{33} T_3 + \varepsilon_{33}^T E_3 + s_a T_3^2 + 2 d_2 T_3 E_3 + \varepsilon_2 E_3^2. \tag{2.26b}$$

Some of the coefficients are the same in these two equations as a result of their derivation from a thermodynamic potential that is an exact differential [10]. For the present purpose these equations will be specialized to electrostrictive materials in which the strain is an even function of the electric field by setting d_{33} and s_a equal to zero. Furthermore, since elastic and dielectric nonlinearities are not of interest for present purposes, s_2 and ε_2 will also be set equal to zero, leaving

$$S_3 = s_{33}^E T_3 + d_2 E_3^2, \tag{2.27a}$$

$$D_3 = 2 d_2 T_3 E_3 + \varepsilon_{33}^T E_3. \tag{2.27b}$$

The unfamiliar coefficient in these equations, d_2, is a property of the material that can be determined by measuring the strain as the field is varied with stress held constant. These are the simplest possible nonlinear electrostrictive equations. Note that they are non-reciprocal, since application of E_3 with $T_3 = 0$ gives S_3, but application of T_3 with $E_3 = 0$ does not give D_3 as it would for a piezoelectric material. Note also that the strain does not change sign when E_3 changes sign.

Now consider a bias electric field, E_0, and a static prestress, T_0, both applied parallel to the length of the bar. A prestress is usually necessary in high power applications of ceramics to prevent the dynamic stress from exceeding the tensile strength. When only E_0 and T_0 are applied the static strain and static electric displacement that result are given by Eqs. (2.27a) and (2.27b) as:

$$S_0 = s_{33}^E T_0 + d_2 E_0^2, \tag{2.28a}$$

$$D_0 = 2d_2 T_0 E_0 + \varepsilon_{33}^T E_0. \tag{2.28b}$$

When a small alternating field, E_a, is applied, in addition to the bias field, and hysteresis is neglected, the equations that determine the alternating components, S_a, T_a, and D_a, are

$$S_0 + S_a = s_{33}^E (T_0 + T_a) + d_2 (E_0 + E_a)^2, \tag{2.29a}$$

$$D_0 + D_a = 2d_2 (T_0 + T_a)(E_0 + E_a) + \varepsilon_{33}^T (E_0 + E_a). \tag{2.29b}$$

Cancelling the static components by using Eqs. (2.28a) and (2.28b), and neglecting the small nonlinear terms, gives for the alternating components

$$S_a = s_{33}^E T_a + (2d_2 E_0) E_a, \tag{2.30a}$$

$$D_a = (2d_2 E_0) T_a + \left(\varepsilon_{33}^T + 2d_2 T_0\right) E_a. \tag{2.30b}$$

These are the linearized electrostrictive equations; if the bias field was zero there would be no electromechanical effect. These equations are reciprocal and in exactly the same form as the piezoelectric equations, Eqs. (2.6d) and (2.6e), with the effective $d_{33} = 2d_2 E_0$ and the permittivity augmented by $2d_2 T_0$. Thus, if a bias field is maintained the electrostrictive material can be used as though it was piezoelectric, and the analysis of the 33 mode transducer in Sect. 2.1 applies with the understanding that the effective d constant depends on the bias field, E_0, through $d_{33} = 2d_2 E_0$, neglecting the effects of saturation.

Starting from equations with T and D as independent variables and assuming that S is an even function of D gives a different set of nonlinear electrostrictive equations:

$$S_3 = s_{33}^D T_3 + Q_{33} D_3^2, \tag{2.31a}$$

$$E_3 = -2Q_{33} T_3 D_3 + \beta_{33}^T D_3, \tag{2.31b}$$

where Q_{33} is a material property. After applying a bias electric field and a prestress, cancelling the static components and linearizing as before, these equations become:

$$S_{\mathrm{a}} = s_{33}^{D} T_{\mathrm{a}} + 2Q_{33} D_0 D_{\mathrm{a}}, \tag{2.32a}$$

$$E_{\mathrm{a}} = -2Q_{33} D_0 T_{\mathrm{a}} + \left(\beta_{33}^{T} - 2Q_{33} T_0\right) D_{\mathrm{a}}, \tag{2.32b}$$

which are similar to those used by Mason [11]. The resulting effective piezoelectric constant [see Eq. (2.4)] is $g_{33} = 2Q_{33} D_0$ where D_0 is the static electric displacement caused by the bias field. This result for g_{33} is also the same as that obtained by Berlincourt [2] that was said to apply to electrostrictive material with high coercive force operating at remanence with D_0 the remanent electric displacement.

Piquette and Forsythe have developed a more complete phenomenological model of electrostrictive ceramics that explicitly includes saturation of the polarization [12–14] as well as remanent polarization. Their one-dimensional equations are

$$S_3 = s_{33}^{D} T_3 + Q_{33} D_3^2, \tag{2.33a}$$

$$E_3 = (D_3 - P_0)\left[\left(\varepsilon_{33}^{T}\right)^2 - a(D_3 - P_0)^2\right]^{-1/2} - 2Q_{33} T_3 D_3, \tag{2.33b}$$

Using the notation of [14], except that permittivity, rather than relative permittivity, is used here. The quantity P_0 is the remanent polarization and a is called the saturation parameter. A small value of a means that saturation is not significant until the electric field is high. For $P_0 = 0$ and $a = 0$ these equations reduce to Eqs. (2.31a) and (2.31b). Other phenomenological models of electrostriction in PMN and similar materials have been proposed [15–17], including comparisons with experimental data. A similar model of electrostriction in polyurethane with measurement of some of the coefficients is also available [18].

The Piquette–Forsythe equations include the nonlinearity associated with both saturation and electrostriction. In this section saturation will be neglected by setting $a = 0$, and the equations will be applied to low level operation of piezoelectric ceramics at remanence by applying an alternating electric field E_{a}. The equations then become $D_3 = P_0 + D_{\mathrm{a}}$, where D_{a} is the alternating electric displacement associated with E_{a},

$$S_3 = s_{33}^{D} T_3 + Q_{33}(P_0 + D_{\mathrm{a}})^2, \tag{2.34a}$$

$$E_{\mathrm{a}} = \beta_{33}^{T} D_{\mathrm{a}} - 2Q_{33} T_3 (P_0 + D_{\mathrm{a}}). \tag{2.34b}$$

Assuming no prestress ($T_3 = T_{\mathrm{a}}$) and linearizing by neglecting products of variable quantities gives for the alternating components

$$S_a = s_{33}^D T_a + 2Q_{33}P_0 D_a, \tag{2.35a}$$

$$E_a = -2Q_{33}P_0 T_a + \beta_{33}^T D_a. \tag{2.35b}$$

These equations are essentially the same as Eqs. (2.32a) and (2.32b) with P_0 in place of D_0. At remanence $g_{33} = 2Q_{33}P_0$, whereas with maintained bias $g_{33} = 2Q_{33}D_0$. The impermittivity β_{33}^T is not modified here as it was in Eq. (2.32b), but it would be modified if prestress had been applied.

These results show how the piezoelectric constants of linearized electrostrictive materials depend on the remanent polarization or on the maintained bias. Since the bias is controlled by the transducer designer it can be optimized as discussed by Piquette and Forsythe [13, 14].

2.3 Magnetostrictive Transducers

Magnetostriction is the change in dimensions that accompanies a change in magnetization of solid materials. In many respects it is the magnetic analog of electrostriction with the largest effects occurring in ferromagnetic materials. Both positive and negative magnetostriction occur in nature, e.g., an iron bar gets longer when magnetized, while a nickel bar gets shorter. The mechanical response of magnetostrictive materials to an applied magnetic field is nonlinear and depends on even powers of the field. Thus for small fields it is essentially a square law, and a magnetic bias is required to obtain a linear response. The bias can be obtained by direct current windings on the magnetostrictive material or by an auxiliary permanent magnet forming part of the magnetic circuit. The remanent magnetization of materials with high magnetostriction is usually not sufficient for operation at remanence.

In the development of high power transducers for active sonar before and during WWII [19] magnetostrictive nickel was the most useful transducer material available and even after the advent of piezoelectric ceramics it still had the advantage of high tensile strength and low input electrical impedance. Nickel scroll-wound ring transducers (see Sect. 5.2.4) were built and tested after WWII for low frequency, high power applications, including rings up to 13 ft in diameter, probably the largest individual transducers ever built [20]. However, PZT has much lower electrical losses and higher coupling coefficient, and its effective tensile strength can be increased by prestressing.

Since WWII several other magnetostrictive materials have been investigated. The nonmetallic ferrites [21] were of interest because their low electrical conductivity reduced eddy current losses and Metglas [22] was promising because of its high coupling coefficient. But none were competitive with the piezoelectric ceramics until, in the 1970s, rare earth-iron compounds were discovered with magnetomechanical properties that surpassed the piezoelectric ceramics in some respects [23]. These new materials, especially Terfenol-D (Terbium, Tb, Dysprosium, Dy, Iron, Fe) rekindled interest in magnetostriction, and have made possible new transducer designs such as the hybrid piezoelectric-magnetostrictive transducer (see Sect. 5.3.2 and Fig. 1.16). The most recent rare earth-iron magnetostrictive material is the higher strength Galfenol [24]. The properties of magnetostrictive materials of current interest are given in Sect. 13.7.

Although there is no magnetic analog of the piezoelectric effect, magnetostrictive materials are sometimes called piezomagnetic after being biased. They then have the same symmetry as the piezoelectric ceramics, and can be described to a good approximation by linear equations of state analogous to the piezoelectric equations of state [2]. The major difference is that the magnetic variables B and H replace D and E, but otherwise it is convenient to keep the notation analogous to that for piezoelectric materials [25]. Thus one of the sets of piezomagnetic matrix equations is

$$S = s^H T + d^t H, \tag{2.36a}$$

$$B = dT + \mu^T H, \tag{2.36b}$$

where s^H is a 6×6 matrix of elastic compliance coefficients, d is a 3×6 matrix of piezomagnetic coefficients, and μ^T is a 3×3 matrix of permeability coefficients. These matrices are analogous to the piezoelectric matrices, the complete set of equations has the same form as Eq. (2.2), and other sets of equations corresponding to Eqs. (2.3)–(2.5) are also used. The piezomagnetic coefficients are related to the bias in a way analogous to the discussion in Sect. 2.2.

An idealized longitudinal vibrator will be used to illustrate magnetostrictive transducers as was done for the piezoelectric transducers. Since magnetic fields occur in closed loops the structure must differ from the piezoelectric case by providing a closed magnetic circuit, e.g., two thin bars of magnetostrictive material, as shown in Fig. 2.7, with the magnetic circuit (see Sect. 13.9) completed at both ends by high permeability magnetic material.

The length of each bar is L, the total cross-sectional area of both bars is A_0 and the coil surrounding both bars has n turns per bar. It will be assumed that the magnetostrictive material is operated with a bias current that determines the value of the effective d_{33} constant [26].

In the configuration shown in Fig. 2.7 the stress T_3 is parallel to the bars, the stresses T_1 and T_2 are zero and the only magnetic field components are H_3 and B_3 parallel to the bars. Under these conditions the equations of state reduce to

$$S_1 = S_2 = s_{13}^H T_3 + d_{31} H_3, \tag{2.37a}$$

$$S_3 = s_{33}^H T_3 + d_{33} H_3, \tag{2.37b}$$

$$B_3 = d_{33} T_3 + \mu_{33}^T H_3. \tag{2.37c}$$

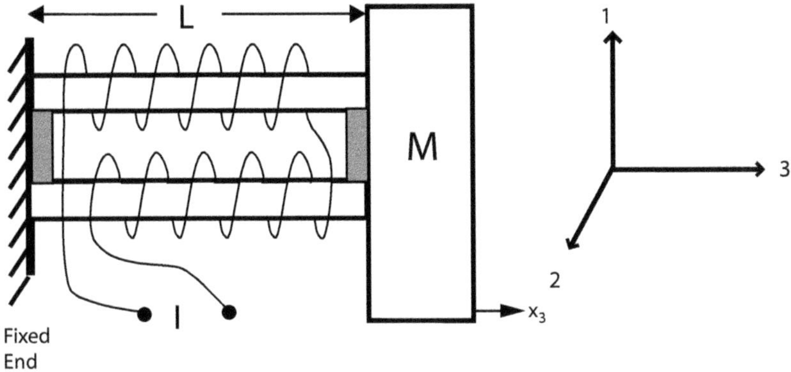

Fig. 2.7 The 33 mode magnetostrictive longitudinal vibrator

Determination of the transducer equations is analogous to that for the 33 mode piezoelectric case. The equation for S_3 is solved for T_3, $S_3 = x_3/L$ and $H_3 = nI/L$, where I is the current through the coil (using Ampere's Circuital Law with $2n$ turns in the path length of $2L$ and neglecting the path length in the high permeability end pieces). Then the equation of motion is

$$M_t\ddot{x}_3 + R_t\dot{x}_3 = -A_0T_3 + F_b = -\left(A_0/s_{33}^H L\right)[x_3 - d_{33}nI] + F_b, \tag{2.38}$$

which becomes for sinusoidal drive

$$F_b = Z_{mr}^H u_3 - \left(nA_0d_{33}/s_{33}^H L\right)I. \tag{2.39}$$

The Faraday induction law and the equation for B_3 give the electrical equation

$$V = (2n)(A_0/2)\frac{dB}{dt} = (j\omega nA_0)\left[d_{33}x_3/s_{33}^H L + \left(\mu_{33}^T - d_{33}^2/s_{33}^H\right)nI/L\right], \tag{2.40}$$

or

$$V = \left(n\,A_0\,d_{33}/s_{33}^H\,L\right)u_3 + Z_0I. \tag{2.41}$$

The impedance Z_{mr}^H is the open circuit mechanical impedance with $K_m^H = A_0/s_{33}^H L$ the open circuit stiffness, including both magnetostrictive bars. The clamped electrical impedance is $Z_0 = j\omega L_0$ where $L_0 = \left(\mu_{33}^T A_0 n^2/L\right)\left(1 - d_{33}^2/\mu_{33}^T s_{33}^H\right)$ is the clamped inductance and $L_f = \mu_{33}^T A_0 n^2/L$ is the free inductance. It is evident from the definition of the coupling coefficient in Eq. (1.18) that

$$k^2 = 1 - L_0/L_f = d_{33}^2/\mu_{33}^T s_{33}^H = k_{33}^2 , \quad (2.42)$$

The transduction coefficient is $N_{m33} = nA_0 d_{33}/s_{33}^H L$, which appears with opposite signs in Eqs. (2.39) and (2.41). Equation (1.18) also shows that

$$K_m^B = K_m^H + N_{m33}^2/L_0. \quad (2.43)$$

These results are completely analogous to the piezoelectric case with k_{33} the material coupling coefficient of the magnetostrictive material. The analogy follows from assuming that the bias linearizes the magnetostrictive mechanism and from neglect of electric and magnetic dissipation mechanisms. Electrical losses in the windings, as well as magnetic hysteresis and eddy current losses, have also been neglected, although they are usually more important than dielectric losses in piezoelectric materials. These losses will be included in the more complete models to be developed in Chaps. 3 and 5.

2.4 Electrostatic Transducers

The electrically generated force in the electrostatic transducer acts at the surfaces of condenser plates; it has the simplest force law of all the transducers, the attraction between opposite electric charges on the two condenser plates. A detailed description of electrostatic transducers, often called capacitive transducers, is given by Hunt [27]. Electrostatic transducers are now very important in micro-electromechanical systems (MEMS) [28], but they have found little use in underwater sound, although they do have a place in its history. Early in World War I Langevin did his first echo-ranging experiments in water with electrostatic transducers, but he soon replaced them with quartz piezoelectric transducers [27].

In the idealized model of this transducer one condenser plate is considered fixed while the other is the vibrating mass in contact with an acoustic medium, and the plates are held apart by a spring of spring constant K_m as shown in Fig. 2.8.

Consider the plates to have area A_0 and the separation between them to be L before voltage is applied. A voltage V gives a charge $\pm Q$ on the plates, which causes one plate to move toward the other a distance x, considered a negative value when the plates move closer together. Neglecting fringing, the electric field between the plates is assumed to be uniform and given by

$$E = \frac{V}{L + x}, \quad (2.44)$$

and the electric displacement is $D = \varepsilon E$ where ε is the permittivity of the material between the plates (usually air). The energy stored in the field is

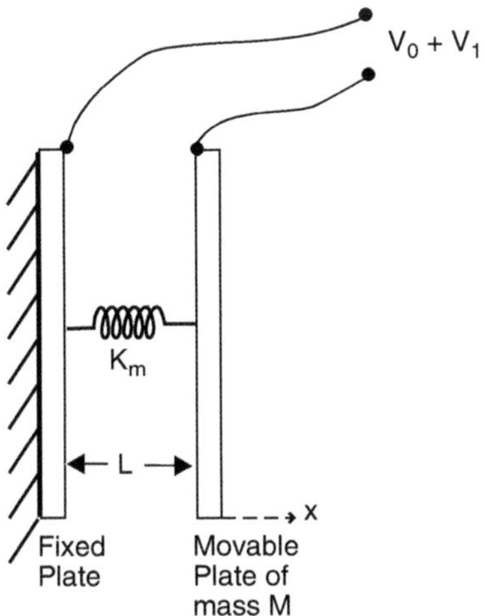

Fig. 2.8 Basic elements of the electrostatic transducer

$$U = \tfrac{1}{2} QV = \tfrac{1}{2} EDA_0(L+x) = \frac{\varepsilon A_0 V^2}{2(L+x)}. \tag{2.45}$$

The force attracting the plates is given by the rate of change of energy with position:

$$F = \frac{\mathrm{d}U}{\mathrm{d}x} = -\frac{\varepsilon A_0 V^2}{2(L+x)^2}, \tag{2.46}$$

and the equation of motion of the moveable plate with mass M is

$$M_t \ddot{x} + R_t \dot{x} + K_m x = -\varepsilon A_0 V^2 / 2(L+x)^2 + F_b. \tag{2.47}$$

The dependence of the electric force on V^2 shows that a bias voltage, V_0, is required to achieve a linear mechanical output. When the bias voltage is applied alone with no external force ($F_b = 0$) Eq. (2.47) reduces to the static case in which the plates reach an equilibrium separation, x_0, where the electric and spring forces are equal:

$$K_m x_0 = -\frac{\varepsilon A_0 V_0^2}{2(L+x_0)^2}. \tag{2.48}$$

The solutions of this equation give the equilibrium positions about which vibrations can occur. But when the bias voltage is increased beyond a certain value it is found that no solutions exist, indicating that the electric force overcomes the spring force, and the gap between the plates closes as a result of nonlinearity (see Sect. 12.2.3).

When the drive voltage V_1 is superimposed on the bias voltage the moveable plate vibrates with amplitude x_1 about the value x_0. The total displacement, x, can be written as $x_0 + x_1$ and Eq. (2.47) becomes

$$M_t\ddot{x}_1 + R_t\dot{x}_1 + K_m(x_1 + x_0) = -\frac{\varepsilon A_0(V_0 + V_1)^2}{2(L_0 + x_1)^2}, \tag{2.49}$$

where $L_0 = L + x_0$. The nonlinearity appears in Eq. (2.49) in two different ways: the electric force is a nonlinear function of x_1, and it depends on the square of the total voltage, but for now it will be linearized by expanding $(V_0 + V_1)^2$ and $(L_0 + x_1)^{-2}$ and dropping the nonlinear terms. The result, using Eq. (2.48) to cancel the static terms, is

$$M_t\ddot{x}_1 + R_t\dot{x}_1 + \left(K_m - \varepsilon A_0 V_0^2/L_0^3\right)x_1 = -\left(\varepsilon A_0 V_0/L_0^2\right)V_1 + F_b, \tag{2.50}$$

which, because of the linearization, is a good approximation only for $x_1 \ll L_0$. Since the electric force depends on both V_1 and x_1, linearizing it still leaves two linear forces, an electric force proportional to V_1 and a mechanical force proportional to x_1. The latter combines with the usual spring force to give an effective spring constant of $K_m^V = \left(K_m - \varepsilon A_0 V_0^2/L_0^3\right)$ for the biased transducer; $(\varepsilon A_0 V_0^2/L_0^3)$ is called the negative stiffness because it represents a force that opposes the spring force. Equation (2.50) can be written for sinusoidal drive as

$$F_b = Z_{mr}^V u_1 + \left(\varepsilon A_0 V_0/L_0^2\right)V_1, \tag{2.51}$$

where $u_1 = j\omega x_1$ and $Z_{mr}^V = R_t + j\omega M_t + (1/j\omega)K_m^V$ is the total short circuit mechanical impedance. Expressing the force in terms of voltage clearly shows the physical origin of two important features of the electrostatic mechanism, the negative stiffness and the instability.

The other transducer equation for the electrostatic transducer gives the current, I_1, caused by the drive voltage, V_1:

$$I_1 = A_0\frac{dD}{dt} = A_0\,\varepsilon\frac{dE}{dt} = A_0\,\varepsilon\frac{d}{dt}\left[(V_0 + V_1)/(L_0 + x_1)\right]. \tag{2.52}$$

After linearizing, this equation becomes

$$I_1 = -\left(\varepsilon A_0 V_0/L_0^2\right)u_1 + (\varepsilon A_0/L_0)dV_1/dt, \tag{2.53}$$

or, for sinusoidal drive,

$$I_1 = -\left(\varepsilon A_0 V_0/L_0^2\right)u_1 + Y_0 V_1, \tag{2.54}$$

where $Y_0 = j\omega C_0$ is the clamped electrical admittance, and $C_0 = \varepsilon A_0/L_0$ is the clamped capacitance. The transduction coefficient in Eq. (2.54), $N_{ES} = \varepsilon A V_0/L_0^2$, is the same as that in Eq. (2.51), with a negative sign.

When Eqs. (2.51) and (2.54) are solved to make u_1 and I_1 the independent variables it is found that $K_m^I = K_m$. Using the above expression for K_m^V in Eq. (1.16) shows that the coupling coefficient is

$$k^2 = 1 - K_m^V/K_m^I = \varepsilon A_0 V_0^2/K_m L_0^3. \tag{2.55}$$

It can be seen that k^2 increases as V_0 increases. It will be shown in Chap. 9 that $k = 1$ when V_0 has reached the value at which the plates collapse together. Thus $k < 1$ is a condition for physical realizability of the electrostatic transducer (see Chap. 4).

As with electrostrictive and magnetostrictive transducers the transduction coefficient and coupling coefficient depend on the bias, raising the question of the optimum bias. For example, Eq. (2.55) shows that the coupling coefficient increases with V_0, but increasing V_0 leads to instability. Thus the optimum value of V_0 cannot be found without considering the nonlinear mechanisms that determine static and dynamic stability and harmonic distortion (see Chap. 12).

2.5 Variable Reluctance Transducers

The variable reluctance transducer (also called electromagnetic or moving armature transducer) is a magnetic field transducer in which the magnetically generated force acts at the surfaces of gaps in a magnetic circuit [26, 27]. Some of what has been said about the electrostatic transducer can also be said about the variable reluctance transducer, since, when both are sufficiently idealized, the variable reluctance is the magnetic analog of the electrostatic.

Figure 2.9 shows the essential components of a variable reluctance transducer in which an electromagnet is separated into two parts by two narrow air gaps, each of length L and area $A_0/2$ held apart by a spring of spring constant K_m. When current flows through the windings of $2n$ turns it creates a magnetic field H and a magnetic flux $BA_0/2$ in the gaps which causes an attractive force between the two poles of the magnet. One part of the magnet is attached to a moveable plate in contact with an acoustic medium, the other part is fixed. Let x represent displacement of the moveable part with negative values corresponding to closing the gaps. The magnetic reluctance, the ratio of magnetomotive force to magnetic flux (see Sect. 13.9) of each gap is $2(L + x)/\mu A_0$ which varies as the moveable part vibrates, giving this transducer its name. The quantity μ is the permeability of the air in the gaps, and it will be assumed that the permeability of the magnetic material is much greater than that of air, making the reluctance of the rest of the magnetic circuit negligible.

The force between the two poles can be found from the magnetic energy. Neglecting fringing and the energy in the magnetic material, the energy in the gaps is

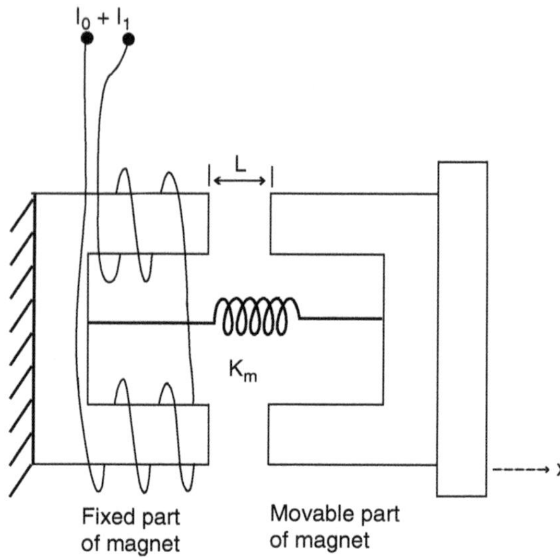

Fig. 2.9 Basic elements of the variable reluctance transducer

$$U = ½(BH)2(A_0/2)(L + x) = \mu A_0(nI)^2/2(L + x), \tag{2.56}$$

and the force between the two poles is

$$\mathrm{d}U/\mathrm{d}x = -\mu A_0(nI)^2/2(L + x)^2. \tag{2.57}$$

With Eq. (2.57) as the force the equation of motion of the moving part of the transducer is

$$M_t\ddot{x} + R_t\dot{x} + K_m x = -\mu A_0(nI)^2/2(L + x)^2 + F_b. \tag{2.58}$$

In this case M_t is the sum of the radiation mass, the mass of the moveable part of the magnetic material, and the mass of the plate in contact with the medium. This equation for current drive of the variable reluctance transducer has the same form as Eq. (2.47) for voltage drive of the electrostatic transducer with permeability in place of permittivity and nI in place of V, including nonlinearities of the same kind. Thus a bias current is needed in addition to a drive current to achieve a linear output.

A fundamental difference between the variable reluctance and the electrostatic transducers exists since lines of magnetic flux form closed loops while lines of electric flux end on the plates of the condenser. Thus a closed magnetic circuit is needed in the variable reluctance transducer, a significant disadvantage because of hysteresis and eddy current losses in the magnetic material and the copper losses in the windings.

The instability found in the electrostatic transducer also occurs in the variable reluctance transducer, and it has been put to practical use; the magnetic relay is a variable reluctance transducer operating in the unstable region. When only a steady bias current is applied ($I = I_0$), and there is no external force, Eq. (2.58) shows that the stable equilibrium positions are given by solutions of

$$K_m x_0 = -\mu A_0 (nI_0)^2 / 2(L + x_0)^2. \tag{2.59}$$

Equation (2.59) has the same form as Eq. (2.48) for the electrostatic transducer, and the analysis of stability and other nonlinear effects applies to both transducers.

When a drive current, I_1, is added to the bias current I_0, and x is written as $x_0 + x_1$, Eq. (2.58) can be linearized in the same way as Eq. (2.49). The result can be written for sinusoidal drive with $L_x = L + x_0$ as,

$$F_b = Z^I_{mr} u_1 + (\mu A_0 n^2 I_0 / L_x^2) I_1, \tag{2.60}$$

where the open circuit mechanical impedance is $Z^I_{mr} = R_t + j\omega M_t + K^I_m / j\omega$, and $K^I_m = K_m - \mu A_0 n^2 I_0^2 / L_x^3$, showing a negative stiffness for current drive analogous to the negative stiffness in the electrostatic case for voltage drive.

The electrical equation for the variable reluctance transducer is the Faraday induction law. Using the steady bias current I_0 and driving current I_1, and neglecting losses, the electrical equation becomes

$$V_1 = 2n(A_0/2) \frac{dB}{dt} = nA_0 \frac{d}{dt} [\mu n (I_0 + I_1)/(L_x + X_1)]. \tag{2.61}$$

When this equation is linearized and the static terms are cancelled the voltage is

$$V_1 = -(\mu A_0 n^2 I_0 / L_x^2) u_1 + Z_0 I_1. \tag{2.62}$$

$Z_0 = j\omega L_0$ is the clamped electrical impedance, $L_0 = \mu A_0 n^2 / L_x$ is the clamped inductance, and the transduction coefficient is $N_{VR} = \mu A_0 n^2 I_0 / L_x^2$ with the same magnitude but opposite signs in Eqs. (2.60) and (2.62). The latter equations also show that $K^V_m = K_m$. Using the relation obtained from Eq. (2.60), that $K^I_m = K^V_m - \mu A_0 n^2 I_0^2 / L_x^3$, the stiffness definition of k^2 in Eq. (1.18) gives

$$k^2 \equiv (K^V_m - K^I_m)/K^V_m = \mu A_0 n^2 I_0^2 / K_m L_x^3. \tag{2.63}$$

This result for k^2 is analogous to that for the electrostatic transducer in Eq. (2.56), including $k = 1$ when nI_0 equals the value that causes the gaps to collapse.

2.6 Moving Coil Transducers

The moving coil transducer (also called the electrodynamic transducer) is probably more familiar than any other transducer because it is used as the loudspeaker in most music and speech reproduction systems [27]. But moving coil transducers have also found an important place in underwater acoustic calibration where low

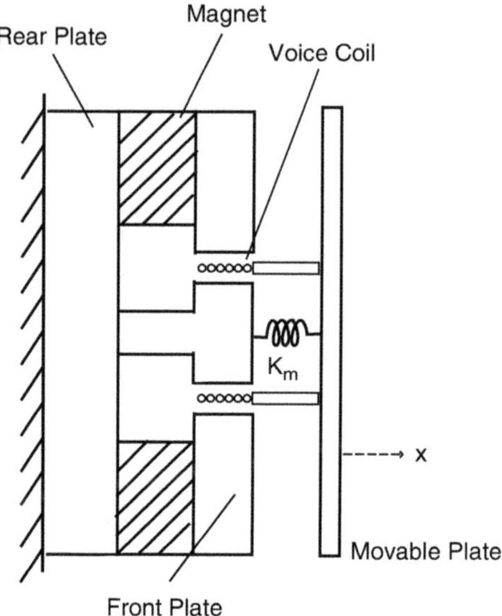

Fig. 2.10 Basic elements of the moving coil transducer

frequency, broadband sound sources of moderate power are needed [29]. This is based on the feasibility of making their resonance frequency very low, thus providing a broad, flat transmitting response above resonance, if other resonances can be avoided (see the J9 transducer, Sect. 9.8).

The transducer consists of a circular coil of wire suspended in an annular gap in a permanent magnet where the radial magnetic field exerts a force on the coil when a current passes through it as shown in Fig. 2.10.

This is the Lorentz force on moving electric charges in a magnetic field, and since it is perpendicular to the magnetic field and to the current it causes motion in the x direction in Fig. 2.10. If the static magnetic field produced by the magnet is B_0, the total length of wire in the coil is l_c, and the current through the coil is I the magnitude of the force is

$$F = B_0 l_c I. \tag{2.64}$$

The basic configuration of the transducer is completed in most air loudspeakers by attaching the coil to a paper or plastic cone that serves as the radiating surface. The stiffness of the cone edge suspension (the surround) combined with the stiffness of the centering device, or spider, serves as the spring; it is shown in Fig. 2.10 as K_m. For use in water, with characteristic mechanical impedance much greater than air, a rigid piston radiator is used in place of the paper cone, as suggested by Fig. 2.10.

The moving coil mechanism differs significantly from the electrostatic or variable reluctance mechanism in that the basic force law is a linear function of the current. This basic linearity is one of the reasons why the moving coil transducer is so widely used, but it must also be noted that there have probably been more studies of nonlinear effects in this transducer than in any of the other transducer types.

This interest in small nonlinear effects shows that the moving coil transducer has great potential for linearity and is very suitable for high fidelity music systems and underwater transducer calibration where linearity is especially important. But the linearity of the moving coil transducer is limited, as the vibration amplitude increases. Here Eq. (2.64) will be used as the only drive force in the equation of motion:

$$M_t \ddot{x} + R_t \dot{x} + K_m x = B_0\, l_c\, I + F_b, \tag{2.65}$$

where M_t is the radiation mass plus the mass of the coil plus the effective mass of the paper cone or piston. For sinusoidal drive Eq. (2.65) becomes

$$F_b = Z^I_{mr} u - B_0\, l_c I, \tag{2.66}$$

where $Z^I_{mr} = Z_{mr}$, $K^I_m = K_m$ and, when $F_b = 0$, the velocity is $u = B_0 l_c I / Z_{mr}$.

It may appear that the moving coil transducer differs from the electrostatic and variable reluctance transducers by achieving linear operation without bias. However, the magnetic field, B_0, plays the role of a bias, although it provides no static force as the bias does in the variable reluctance transducer. The transduction coefficient, $B_0 l_c$, is proportional to B_0, and there is no linear transduction if $B_0 = 0$. The field, B_0, is provided either by a permanent magnet or by direct current in a magnetizing coil.

The electric equation for the moving coil transducer comes from the Faraday induction law. Since the current is changing in the coil and the coil is moving at right angles through the radial magnetic field there are two sources of voltage. With L_b the clamped inductance of the coil, and again neglecting losses, the voltage equation is

$$V = L_b(\mathrm{d}I/\mathrm{d}t) + B_0 l_c(\mathrm{d}x/\mathrm{d}t), \tag{2.67}$$

or,

$$V = B_0 l_c u + Z_b I, \tag{2.68}$$

where $Z_b = j\omega L_b$ is the clamped electrical impedance.

The moving coil transducer differs significantly from the variable reluctance transducer in that it has no negative stiffness because the internal electrical force is independent of displacement for small displacements as can be seen from Eq. (2.64). Thus using $K^I_m = K_m$ and rearranging the equations to make V a dependent variable shows that

$$Z^V_{mr} = Z^I_{mr} + (B_0 l_c)^2 / Z_b, \tag{2.69}$$

$$\text{and} \quad K^V_m = K_m + (B_0 l_c)^2 / L_b. \tag{2.70}$$

Thus the short circuit stiffness for the moving coil transducer increases as B_0 is increased, since L_b is usually approximately proportional to l_c^2.

Using the values of K_m^I and K_m^V above in the stiffness definition of k^2 in Eq. (1.18) gives

$$k^2 = \left(K_m^V - K_m^I\right)/K_m^V = (B_0 l_c)^2 / \left[K_m L_b + (B_0 l_c)^2\right]$$
$$= 1/\left[1 + K_m L_b/(B_0 l_c)^2\right]. \tag{2.71}$$

The second form of k^2 was given by Hersh [30] and the third by Woollett [26]. Equation (2.71) shows that k^2 can be made to approach unity by decreasing the stiffness, K_m, for a given B_0. Woollett [26] has pointed out that this condition corresponds to a poorly defined equilibrium position for the coil, i.e., a drift type of instability. But for the moving coil transducer k^2 can be made close to unity (about 0.96) [30] without becoming unstable, by making $(B_0 l_c)^2 \gg K_m L_b$.

Note that for the variable reluctance transducer $K_m^V = K_m$, while for the moving coil transducer $K_m^I = K_m$. The difference arises from the different physical nature of the magnetic forces in the two cases. In the variable reluctance the magnetic force is a function of x and I which, after linearizing, has a component proportional to x (the negative stiffness force) in addition to the component proportional to I. In the moving coil for small amplitude the magnetic force does not depend on x and consists only of the component proportional to I. In the variable reluctance the bias reduces the stiffness K_m^I while K_m^V remains at K_m, but in the moving coil K_m^I remains at K_m while B_0 (the bias) increases the stiffness K_m^V.

2.7 Comparison of Transduction Mechanisms

The fundamental features of the major transduction mechanisms will now be summarized and compared to identify those most suitable for underwater sound applications. The main results of this chapter are the equation pairs for each transducer type, which have been derived with force and current (or voltage) as dependent variables. They can be summarized as:

Electric Field Transducers:

$$F_b = Z_{mr}^V u - NV, \tag{2.72a}$$

$$I = Nu + Y_0 V. \tag{2.72b}$$

Magnetic Field Transducers:

$$F_b = Z_{mr}^I u - NI, \tag{2.73a}$$

$$V = Nu + Z_0 I. \tag{2.73b}$$

Table 2.1 Comparison of transduction mechanisms

Transducer type	Transduction coefficient, N	Coupling coefficient, k^2	$(N/k)^2$
Electrostatic[a]	$-\varepsilon A_0 V_0/L_0^2$	$\dfrac{\varepsilon A_0 V_0^2}{K_{\mathrm{m}}^I L_0^3}$	$K_{\mathrm{m}}^V C_{\mathrm{f}}$
Piezoelectric/electrostrictive	$d_{33}A_0/s_{33}^E L$	$d_{33}^2/\varepsilon_{33}^T s_{33}^E$	$K_{\mathrm{m}}^E C_{\mathrm{f}}$
Magnetostrictive	$d_{33}nA_0/s_{33}^H L$	$d_{33}^2/\mu_{33}^T s_{33}^H$	$K_{\mathrm{m}}^H L_{\mathrm{f}}$
Variable reluctance[a]	$-\mu A_0 n^2 I_0/L_x^2$	$\dfrac{\mu A_0 n^2 I_0^2}{K_{\mathrm{m}}^V L_x^3}$	$K_{\mathrm{m}}^I L_{\mathrm{f}}$
Moving coil	$B_0 l_{\mathrm{c}}$	$\dfrac{(B_0 l_{\mathrm{c}})^2}{K_{\mathrm{m}}^V L_{\mathrm{b}}}$	$K_{\mathrm{m}}^I L_{\mathrm{f}}$

[a]N must be considered negative for the electrostatic and variable reluctance mechanisms to be consistent with the transducer equations above. In these cases an increase in V or I, with $F_{\mathrm{b}} = 0$, gives a negative displacement; in the other cases an increase in V or I gives a positive displacement

Table 2.1 lists the transduction coefficients, N, and the coupling coefficients squared, k^2, for each transducer type. These quantities are expressed in terms of material constants, dimensions, and bias to show the dependence on the parameters that are critical in transducer design. Note that the expressions for N give the turns ratios of the ideal transformers in the electrical equivalent circuits that will be discussed in Sect. 2.8 and Chap. 3.

Table 2.1 shows that bias is the common factor that determines the values of both transduction coefficients and coupling coefficients. It is most obvious in the surface force transducers where the bias fields appear explicitly as V_0/L_0, nI_0/L_0, and B_0. In electrostrictive and magnetostrictive transducers the d_{33} constants depend on bias as shown in Sects. 2.2 and 2.3. In the true piezoelectric materials the measured d_{33} is a naturally occurring property related to the crystalline structure in a way that could be interpreted in terms of an internal bias. The last column in Table 2.1 emphasizes the similarity among the linearized transducer types.

The term NV or NI in the transducer equations determines the ability of a transducer to produce force, and the most critical parts of N are the material properties and the bias fields. For example, d_{33}/s_{33}^E and $\varepsilon_0 V_0/L_0$ are critical for piezoelectric and electrostatic transducers, where $d_{33}/s_{33}^E \sim 20$ for PZT-4, while $\varepsilon_0 V_0/L_0 \sim 10^{-5}$ with a bias of 10^6 V/m. Therefore, a piezoelectric transducer has about a million times more capability for producing force than an electrostatic transducer of the same size. Similar considerations show that the order of increasing force capability is: electrostatic, moving coil, variable reluctance and then magnetostrictive and piezoelectric, which are about equal [31]. This conclusion is relevant for high power projectors but does not apply to underwater applications in general; e.g., at very low frequency, where the acoustic loading is low, large displacement is more important than large force.

The other critical factor that determines the potential of a transducer to produce the force required as a projector is the limiting drive voltage or current that can be applied without damage (see Sects. 2.8.5 and 13.14). This is a difficult concept to

Table 2.2 Comparison of force per unit area capability

NV_1/A_0 or NI_1/A_0 in Pa		
Electrostatic	3	$[E_0 = 10^6\,\text{V/m}, E_1 = 3 \times 10^5\,\text{V/m}]$
Moving coil	8000	[200 turns, radius = 0.1 m, $B_0 = 1\text{T}, I_1 = 2\,\text{A}$]
Variable reluctance	10^5	$[H_0 = 5 \times 10^5\,\text{A/m}, H_1 = 1.7 \times 10^5\,\text{A/m}]$
Piezoelectric (PZT)	8×10^6	$[E_1 = 4 \times 10^5\,\text{V/m}]$
Magnetostrictive (Terfenol)	8×10^6	$[H_1 = 2 \times 10^4\,\text{A/m}, H_0 = 4 \times 10^4\,\text{A/m}]$

quantify because limiting drives are determined by phenomena such as electrical breakdown, mechanical failure, and overheating which depend on material properties and the details of design and construction. Therefore the estimates about to be made are intended to show the relative, but only approximately the ultimate, capabilities of the transducer types. The transducer equations show that with maximum drive NV_1/A_0 or NI_1/A_0 is the maximum dynamic force per unit area that can be produced; e.g., for the piezoelectric case $NV_1/A_0 = V_1 d_{33}/s_{33}^E L$. These forces per unit area are shown in Table 2.2 for specified high drive conditions. For comparison some underwater applications require acoustic pressures at the radiating face of the transducer that approach the static pressure (e.g., 1–10 atm or 10^5–10^6 Pa). Higher pressures would be limited by cavitation (see Chap. 10). For optimum acoustic loading at resonance the pressures produced in the water by the forces per unit area in Table 2.2 would be reduced by the ratio of radiating area (A) to drive area (A_0), a value in the range 1–5.

Table 2.2 shows that the electrostatic transducer is not suitable for high power applications in water, the moving coil transducer is marginal, the variable reluctance transducer is much better, and the piezoelectric and magnetostrictive transducers are fully capable of producing the required forces. The characteristic impedance of piezoelectric, electrostrictive, and magnetostrictive transducers also favors their use in water ($\rho c = 22 \times 10^6\,\text{kg/m}^2\text{s}$ for PZT), and, combined with their high force capability, makes them superior for most underwater sound applications.

2.8 Equivalent Circuits

2.8.1 Equivalent Circuit Basics

Electrical equivalent circuits are an alternative representation of transducers that can be combined with power amplifier circuits, and other electrical circuits, for a more complete systems representation. In addition, they provide a visual alternative to an analytical representation of the individual parts and interconnections of the transducer. We present here a brief introduction to equivalent circuits which will be used extensively in Chaps. 5 and 6 to represent specific projector and hydrophone designs. Equivalent circuits will be developed more fully in Chap. 3 as one of the major methods of modeling transducers. The simplest equivalent circuit uses lumped electrical elements such as inductors, resistors, and capacitors to represent mass, resistance, and compliance (reciprocal of spring constant), respectively, and voltage, V, and current, I, to represent force, F, and velocity, u. The analogy is based on the following similarities between the laws of electricity and magnetism and the laws of mechanics:

- For an electrical resistance R_e, the voltage $V = R_e I$.
  For a mechanical resistance R, the force $F = Ru$.
- For a coil of inductance L, the voltage $V = L\,dI/dt = j\omega LI$.
  For an ideal mass M, the force $F = M\,du/dt = j\omega Mu$.
- For a capacitor C the voltage $V = (1/C)\int I\,dt = I/j\omega C$.

  For a compliance C_m the force $F = (1/C_m)\int u\,dt = u/j\omega C_m$.
- For an electrical transformer of turns ratio N the output voltage is NV.
  For an electromechanical transformer the force $F = NV$.
- Electrical power is $W = VI = |V|^2/2R_e = |I|^2 R_e/2$.
  Mechanical power is $W = Fu = |F|^2/2R = |u|^2 R/2$.

  Angular resonance frequency, ω_r, and quality factor, Q, analogies are:

- For inductance L, capacitor C, resonance is $\omega_r = (1/LC)^{1/2}$
  For mass M, compliance C_m, resonance is $\omega_r = (1/MC_m)^{1/2}$.
- For inductance L, resistance R_e, the Q is $Q = \omega_r L/R_e$
  For mass M, mechanical resistance R, Q is $Q = \omega_r M/R$

where terms containing $j\omega$ apply to sinusoidal conditions. The mechanical compliance $C_m = 1/K_m$ acts like a capacitor and is used in equivalent circuits rather than the stiffness K_m.

Thus, an electrical circuit can represent a mechanical vibrating system by replacing the voltage V, with a force, F, and the current, I, with a velocity u. Since electroacoustic transducers involve both electrical and mechanical parts, one circuit can be used to represent the entire transducer with an ideal electromechanical transformer connecting the electrical and mechanical parts as shown in Fig. 2.11.

The electromechanical turns ratio of the ideal transformer, $N = F/V$, is the quantity called the transduction coefficient earlier in this chapter. Figure 2.11 is an example of the impedance analogy in which the force acts like a voltage and the velocity acts like a current; it is directly applicable to piezoelectric ceramic and other electric field transducers. Because of the widespread use of piezoelectric ceramic in underwater sound transducers, only the impedance analogy will be considered in this introductory section. However, as will be shown in Chap. 3, magnetic field transducers are more readily represented by a mobility analog circuit in which the velocity acts like a voltage and the force acts like a current, since the magnetic forces are derived from a current in the associated coil.

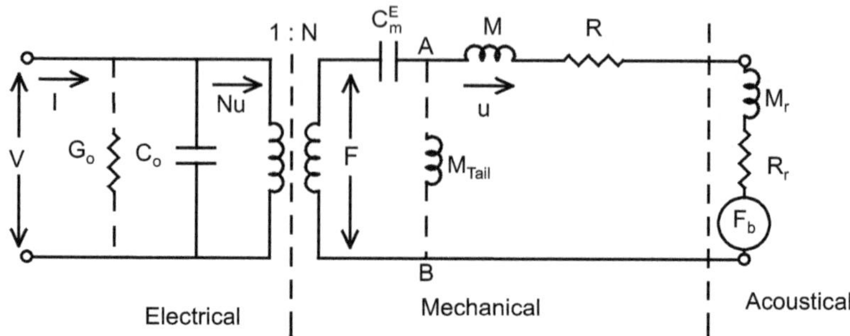

Fig. 2.11 Lumped equivalent circuit. Tail mass inductor between AB is for the case of a finite reaction mass

The circuit of Fig. 2.11 is an equivalent way of representing Eqs. (2.9a) and (2.11) with K_{m} replaced by the compliance $C_{\mathrm{m}}^{E} = 1/K_{\mathrm{m}}^{E}$. Accordingly, Eq. (2.9a) may be written as

$$j\omega(M + M_{\mathrm{r}})u + (R + R_{\mathrm{r}})u + u/j\omega C_{\mathrm{m}}^{E} = uZ_{\mathrm{mr}}^{E} = NV + F_{\mathrm{b}}, \tag{2.74}$$

which gives the sum of the forces around the mechanical and acoustical part of the circuit of Fig. 2.11. It has the same form as an electrical series "RLC" circuit driven by a voltage of $NV + F_{\mathrm{b}}$, where $F = NV$, justifying the representation of the mechanical and acoustical parts of the equivalent circuit of Fig. 2.11.

The electrical part of Fig. 2.11 can be characterized through the input current in Eq. (2.11) written as

$$I = j\omega C_0 V + Nu. \tag{2.75}$$

Equation (2.75) gives the sum of two currents and satisfies the electrical part of the equivalent circuit where $j\omega C_0 V$ is the current through the capacitor, C_0, and Nu is the current entering the ideal transformer of turns ratio N. The capacitor, C_0, would be the only reactive component if the acoustical part were open circuited with the velocity $u = 0$. Without this capacitor the transduction would be perfect, all the electrical energy would be converted to mechanical form, and the coupling coefficient would be unity.

The ideal transformer turns ratio N is proportional to the coupling coefficient and connects the electrical and mechanical parts of the circuit by the relations

$$F = NV \quad \text{and} \quad u = I/N. \tag{2.76}$$

The power passing through this transformer is identical on both sides, since $VI = Fu$. The circuit represents an acoustic projector for $F_{\mathrm{b}} = 0$, where the power out is $W = |u|^2 R_{\mathrm{r}}/2$. The output velocity

$$u = NV/\{(R + R_r) + j[\omega(M + M_r) - 1/\omega C^E_m]\} = NV/Z^E_{mr}, \tag{2.77}$$

can be obtained directly from the equivalent circuit parameters using common circuit theorems and is consistent with Eq. (2.74). The source level may then be obtained from Eq. (1.25) or Eq. (1.27) if the *DI* is known. The source level may also be obtained without use of the *DI* from Eq. (2.77) and analytical expressions for the far-field acoustic pressure as a function of the transducer normal velocity (see Chap. 10).

2.8.2 Circuit Resonance

At resonance the transducer output velocity is magnified by the Q of the transducer for a given electrical input. Accordingly, projectors are usually operated in a frequency band around the fundamental resonance and in this region important characteristics can be obtained from the electrical admittance of electric field transducers. Resonance may be defined as the frequency at which the kinetic and potential energies are equal. (See Chap. 4 for a more extensive discussion of resonance.) In the case of a circuit this is equivalent to the frequency at which the reactance vanishes. Because of the shunt capacitance, C_0, in Fig. 2.11 there are two resonance frequencies, one at which the susceptance, B, vanishes and another at which the reactance, X, vanishes. These two frequencies play an important role in characterizing a transducer.

The input electrical admittance, $Y = G + jB$, may be obtained from the circuit of Fig. 2.11, with the external force $F_b = 0$, and written as

$$\begin{aligned} Y = I/V &= j\omega C_0 + N^2/\{(R + R_r) + j[\omega(M + M_r) - 1/\omega C^E_m]\} \\ &= j\omega C_0 + N^2/Z^E_{mr}. \end{aligned} \tag{2.78}$$

The mechanical resonance frequency occurs under short circuit conditions (which shorts out the capacitor C_0) where the reactance cancels, $\omega(M + M_r) - 1/\omega C^E_m = 0$, and gives

$$\omega_r = 1/[(M + M_r)C^E_m]^{1/2}. \tag{2.79}$$

This is the frequency of maximum response for projectors driven with a constant voltage and also the frequency of maximum conductance. Under open circuit conditions the capacitor C_0/N^2 is in series with C^E_m reducing the compliance to

$$C^D_m = (C^E_m C_0/N^2)/(C^E_m + C_0/N^2) = C^E_m/(1 + N^2 C^E_m/C_0),$$

and yielding a higher open circuit antiresonance frequency given by

$$\omega_a = 1/\left[(M + M_r)C_m^D\right]^{1/2}. \tag{2.80}$$

This is the frequency of maximum response for hydrophones and also approximately the frequency of maximum impedance. Use of the stiffness definition of the coupling coefficient k in Eq. (1.16) with $K_m^E = 1/C_m^E$ and $K_m^D = 1/C_m^D$ leads to

$$k^2 = \left[1 - (\omega_r/\omega_a)^2\right] = \left[1 - (f_r/f_a)^2\right] \tag{2.81}$$

which is often used to determine the effective coupling coefficient of a transducer by calculation or measurement of the resonant and antiresonant frequencies. This important quantity, k, is a measure of transduction and ranges between zero and unity (see Chaps. 1 and 4). If k was near unity, C_0 would be near zero and would no longer effectively shunt the circuit and we would have a near perfect transducer. In practice C_0 is the given and $C_f = C_0/\left(1 - k^2\right)$ so that for a very high coupling coefficient we get $C_f \gg C_0$.

2.8.3 *Circuit* Q *and Bandwidth*

The mechanical quality factor, Q_m, (See Chap. 4 for a more extensive discussion of Q_m.) is a measure of the sharpness of a resonant response curve and may be determined from

$$Q_m = f_r/(f_2 - f_1) = \omega_r M^*/R^*. \tag{2.82}$$

In the first expression f_r is the mechanical resonance frequency and f_2 and f_1 are the frequencies at half-power relative to the power at resonance. Traditionally, the bandwidth is considered to be $\Delta f = f_2 - f_1$, which covers the response region where the output power is within 3 dB of the value at resonance and may be extended to cover the case of multiply resonant transducers. Occasionally the bandwidth is determined by the input power or intensity response (which differs from the output power by the efficiency and *DI*, respectively). At the band edge frequencies, f_1 and f_2, the phase angle is $\pm 45°$ in the combined mechanical and acoustical (motional) part of the circuit of Fig. 2.11.

Both expressions in Eq. (2.82) are equivalent for the simple circuit of Fig. 2.11 where there is only a single mechanical resonance frequency ω_r. The first is most useful for measurement while the second is more useful for analysis. In the second expression M^* is the effective mass and R^* is the effective mechanical resistance. The latter form may be obtained from the energy based definition:

$$Q_m = 2\pi(\text{Total Energy})/(\text{Energy dissipated per cycle at resonance}).$$

In the electric field transducer circuit of Fig. 2.11 the quality factors at mechanical resonance, Q_m, and at antiresonance, Q_a, are

$$Q_m = \omega_r(M + M_r)/(R + R_r), \quad Q_a = \omega_a(M + M_r)/(R + R_r). \tag{2.83}$$

With the expressions for ω_r, ω_a, Q_m, and Q_a, the admittance, given by Eq. (2.78), may be recast in the form

$$Y = G_0 + j\omega C_f\left[1 - (\omega/\omega_a)^2 + j\omega/\omega_a Q_a\right]/\left[1 - (\omega/\omega_r)^2 + j\omega/\omega_r Q_m\right], \tag{2.84}$$

by using $N^2 = k^2 C_f K_m^E$. Equation (2.84) shows that the admittance approaches the electrical loss conductance, G_0, plus $j\omega$ times the free capacity, $C_f = C_0/\left(1 - k^2\right)$, at frequencies well below mechanical resonance, ω_r. Equation (2.84) also shows that for low losses, with G_0 small and, more importantly, Q_m and Q_a large, the magnitude of Y becomes a maximum at mechanical resonance, $\omega = \omega_r$, and a minimum at antiresonance, $\omega = \omega_a$. As discussed in Chap. 9, these conditions may be used to determine resonance and antiresonance of piezoelectric ceramic transducers and are usually accurate under air-loaded conditions if the mounting does not introduce significant stiffness or damping. Under water loading conditions where Q_m and Q_a are not necessarily high, the frequency of maximum conductance may be used to obtain mechanical resonance, ω_r, and the frequency of maximum resistance may be used to obtain antiresonance, ω_a, for the common case of small electrical loss conductance, G_0.

A low Q_m is usually desirable, but, for efficient radiation, it must be based on a large radiation resistance, R_r, rather than a large mechanical loss, R. The Q_m can be related to the coupling coefficient, k, by starting from the electrical admittance at mechanical resonance, ω_r, where Eq. (2.78) becomes

$$Y = G_0 + j\omega_r C_0 + N^2/(R + R_r). \tag{2.85}$$

Using $k^2 = N^2/K_m^E C_f$, from Eq. (1.17a), the motional conductance, $G_m = N^2/(R + R_r)$, becomes

$$G_m = k^2 \omega_r C_f Q_m. \tag{2.86}$$

If we ignore the usually small electrical loss conductance, G_0, in the expression for Y at resonance, the electrical quality factor, Q_e, evaluated at mechanical resonance, is defined as

$$Q_e = \omega_r C_0/G_m. \tag{2.87}$$

The quantity Q_e is a measure of the ratio of the susceptance to the conductance and also a measure of the power factor which is an important consideration when the transducer is connected to a power amplifier. Broadband power operation is difficult if Q_e is high, since then a large volt-ampere capacity from the power amplifier is needed for a given power requirement as a result of the shunted current through C_0. Substituting G_m from Eq. (2.86) into Eq. (2.87) gives $Q_e = C_0/(k^2 C_f Q_m)$ which, with $C_0 = C_f(1 - k^2)$, leads to the important general transduction expression,

$$Q_m Q_e = (1 - k^2)/k^2, \tag{2.88}$$

which fixes the relationship between Q_m and Q_e for a given coupling coefficient, k. This formula may also be used to determine the effective coupling coefficient, k_{eff}, (see Chap. 4) from measurements of Q_m and Q_e as discussed in Chap. 9.

For example: Equation (2.88) may be rewritten as $k^2 = 1/(1 + Q_m Q_e)$. If the transducer $Q_m = 9$ and $Q_e = 1$ or vice versa, then $k = 0.32$. In both cases the bandwidth is narrow either from a mechanical or electrical point of view. However, if $Q_m = Q_e = 3$ we still get a coupling of 0.32 along with a more desirable bandwidth.

While the quantity Q_m is a measure of the sharpness of the constant voltage drive response curve, the Q_e is a measure of the reactive electrical susceptance. Since low values of both are usually desirable for broadband response, it is useful to define a total quality factor,

$$Q_t = Q_m + Q_e = Q_m + (1 - k^2)/(k^2 Q_m), \tag{2.89}$$

where Eq. (2.88) was used. The minimum Q_t occurs for $\mathrm{d}Q_t/\mathrm{d}Q_m = 0$, which yields

$$Q_m = (1 - k^2)^{1/2}/k, \tag{2.90}$$

as the optimum Q_m for broadband response. Using Eq. (2.88) again shows that the optimum $Q_m = Q_e$. Thus a low optimum Q_m, which corresponds to broadband performance, requires a high coupling coefficient (see, Mason [11] for further discussion of this concept). Table 2.3 lists the values for optimum Q_m and values of 1.25 Q_m (see Stansfield [32] for the factor 1.25) with broader response but with slight ripple.

Low Q_m is commonly achieved by matching the impedance of the transducer to the medium. The characteristic impedance is an important concept in acoustics and also in electroacoustics. The characteristic specific acoustic impedance (i.e., the mechanical impedance per unit area) of a fluid medium such as air or water is ρc where ρ is the density and c is the sound speed in the medium. The value for water is $1.5 \times 10^6\,\mathrm{kg/m^2s}$ or 1.5×10^6 rayls. At boundaries between two different media the relative values of ρc (see Sect. 13.2) determine how a sound wave is divided into reflected and transmitted waves. A similar situation occurs at the interface between the vibrating surface of a transducer and the medium in which it is immersed.

Table 2.3 Values of optimum Q_m

k	Q_m	$1.25Q_m$
0.1	9.9	12.4
0.2	4.9	6.1
0.3	3.2	4.0
0.4	2.3	2.9
0.5	1.7	2.1
0.6	1.3	1.6
0.7	1.0	1.2
0.8	0.8	0.9
0.9	0.5	0.6

The characteristic mechanical impedance of a transducer is a measure of the approximate average mechanical impedance in a frequency band near resonance (see Sect. 4.3). For good performance over a broad frequency band the characteristic mechanical impedance of the transducer must be similar to the characteristic mechanical impedance of the medium.

2.8.4 *Power Factor and Tuning*

The ultimate power capabilities of a transducer may not be achieved if the electrical power source is inadequate because of excessively high transducer voltage and current requirements. The power factor is a way of assessing this capability, and its relationship to transducers may be understood by referring to the equivalent circuit of Fig. 2.11. As discussed in Sect. 2.8.3, the clamped capacitance, C_0, stores energy that is not transformed into mechanical motion and shunts reactive current through it. Without this capacitance, the transducer would achieve a coupling coefficient of 100 %. This capacitive reactive power does not contribute to the power radiated and corresponds to a power factor less than unity even at resonance. Improvement may be obtained by tuning out the capacitor, C_0, with a series or shunt inductor (see Sect. 9.6). Operation at frequencies off resonance additionally reduces the power factor as a result of non-cancelled mechanical reactive power in the spring of compliance C_m^E and the mass, M.

The power factor, P_f, is given by the cosine of the phase angle between the voltage and the current. With phase angle $\varphi = \tan^{-1}(B/G)$, the power factor $\cos\varphi = W/VI$, where W is the power absorbed and VI is the product of the input voltage and current magnitude. Accordingly, the electrical power into a transducer is $W = VI\cos\varphi$ and a transducer with a high power factor is desirable. Since the input power is $W = V^2G$ and $VI = V^2|Y|$,

$$P_f = G/|Y| = G/|G + jB| = 1/\left[1 + (B/G)^2\right]^{1/2}, \tag{2.91}$$

allowing evaluation of P_f from the input electrical admittance, Y, such as given by Eq. (2.78) for the equivalent circuit of Fig. 2.11. Equation (2.91) demonstrates the need for a small B/G ratio for large P_f, since the power factor attains a maximum value of unity for $B/G = 0$.

Without electrical tuning Eq. (2.78) may be rewritten as

$$Y = j\omega C_0 + G_m/[1 + jQ_m(\omega/\omega_r - \omega_r/\omega)]. \tag{2.92}$$

On rationalizing the denominator in Eq. (2.92) and expressing the result as $Y = G + jB$ we get

$$B/G = \left[Q_e + Q_e Q_m^2(\omega/\omega_r - \omega_r/\omega)^2\right]\omega/\omega_r - Q_m(\omega/\omega_r - \omega_r/\omega). \tag{2.93}$$

At resonance the ratio $B/G = Q_e$ and $P_f = 1/\left(1 + Q_e^2\right)^{1/2}$ showing the desirability of a low Q_e (and, consequently, low value of C_0). Note that a power factor of 0.707 is obtained for the case of $Q_e = 1$. Equation (2.93) also shows the desirability of a low Q_m and high k for off resonance operation where

$$\begin{aligned} B/G &\approx Q_m(\omega_r/\omega)/k^2, \quad \omega \ll \omega_r \quad \text{and} \\ B/G &\approx Q_m(\omega/\omega_r)^3\left(1 - k^2\right)/k^2, \quad \omega \gg \omega_r. \end{aligned} \tag{2.94}$$

The power factor can be improved with electrical tuning (for more on tuning see Sect. 9.6.1) of the clamped capacitance, C_0, by a parallel inductor, L_p, across the electrical terminals of Fig. 2.11 which gives the admittance

$$Y = j\omega C_0 + 1/j\omega L_p + G_m/[1 + jQ_m(\omega/\omega_r - \omega_r/\omega)]. \tag{2.95}$$

With L_p chosen to tune out C_0 at resonance $L_p = 1/\omega_r^2 C_0$ and Eq. (2.95) gives

$$B/G = \left[Q_e - Q_m + Q_e Q_m^2(\omega/\omega_r - \omega_r/\omega)^2\right](\omega/\omega_r - \omega_r/\omega). \tag{2.96}$$

At frequencies well removed from resonance Eq. (2.96) becomes

$$\begin{aligned} B/G &\approx -Q_m(\omega_r/\omega)^3\left(1 - k^2\right)/k^2, \quad \omega \ll \omega_r \quad \text{and} \\ B/G &\approx Q_m(\omega/\omega_r)^3\left(1 - k^2\right)/k^2, \quad \omega \gg \omega_r. \end{aligned} \tag{2.97}$$

And the inductor lowers the value of B/G for $\omega \ll \omega_r$ more than in the case of no inductor, Eq. (2.94), but has no affect at the high end of the band. At resonance $\omega = \omega_r$ the second factor in Eq. (2.96) vanishes yielding $B/G = 0$, and $P_f = 1$.

There are two other possible frequencies under which $B/G = 0$ and these occur when the first factor (in brackets) vanishes. Under this condition Eq. (2.96) may be written as

$$(\omega/\omega_r)^4 + \left[1/Q_m^2 - k^2/(1-k^2) - 2\right](\omega/\omega_r)^2 + 1 = 0. \quad (2.98)$$

Equation (2.98) is a quadratic equation and the solution may be written as

$$(\omega/\omega_r)^2 = -b/2 \pm \left[(b/2)^2 - 1\right]^{1/2}, \quad (2.99)$$

where $b = \left[1/Q_m^2 - k^2/(1-k^2) - 2\right]$ and $\omega/\omega_r = f/f_r$. For example for $k = 0.5$ and $Q_m = 3$ we get $b = -20/9$ with solutions $f = 0.792f_r$ and $f = 1.263f_r$ in addition to $f = f_r$ as the frequencies for $B/G = 0$ and $P_f = 1$. At these frequencies the admittance locus crosses the G axis three times as illustrated in Fig. 9.19 and produces a P_f of unity, as illustrated in Fig. 2.12a with $Q_m = 3$. Equation (2.90) yields an optimum $Q_m = \sqrt{3} = 1.73$, for $k = 0.5$, as listed in Table 2.3 and illustrated by the power factor response in Fig. 2.12a.

The optimum Q_m value shows a single resonance with a broad response while the case for the lower value, $Q_m = 1$, shows a sharper single resonance. On the other hand, the case with the higher value, $Q_m = 2$, appears to be slightly broader than the optimum case, but with a very small fluctuation. Power factor results are also shown in Fig. 2.12b for $k = 0.7$ and in Fig. 2.12c for $k = 0.9$ with optimum Q_m values from Eq. (2.90) of 1 and approximately 0.5, respectively, as listed in Table 2.3. As seen, the higher coupling coefficient cases yield a broader bandwidth and smother response with lower values of Q_m. It may also be seen that Q_m values slightly higher than optimum values of 1.73, 1, and 0.5 give an even broader bandwidth with a small ripple, as shown in Fig. 2.12a–c for Q_m equal to 2, 1.73, and 0.7 (for $k = 0.5$, 0.7, and 0.9, respectively). This result is consistent with Stansfield [32] that a Q_m value 25 % higher than that given by Eq. (2.90) yields an even broader bandwidth. These higher Q_m values are also listed in Table 2.3.

The curves of Fig. 2.12a–c may also be used to represent the effective coupling coefficient, k_{eff}, which, for practical piston type transducers, can be about 25 % less than the material coupling coefficient (see Chap. 4). The case of $k_{eff} = 0.5$ is typical for the effective coupling coefficient of a practical Tonpilz transducer using PZT piezoelectric ceramic (see Sect. 5.3). The case of $k_{eff} = 0.7$ could represent a Tonpilz transducer which, instead, uses a higher coupling coefficient single crystal PIN-PMN-PT material. However, the case of $k_{eff} = 0.9$ indicates what might be achieved from a Tonpilz transducer using a new transduction material with material coupling well above 0.9 plus design improvements that reduce the effects that lower k_{eff} and also achieves a low Q_m. On the other hand, other types of transducers, such as low frequency flextensional transducers (see Sect. 5.5), typically have an effective coupling coefficient from 0.3 to 0.36 which, in the lower case, is roughly 50 % lower than the material coupling coefficient, leading to an optimum Q_m design goal of about 4.

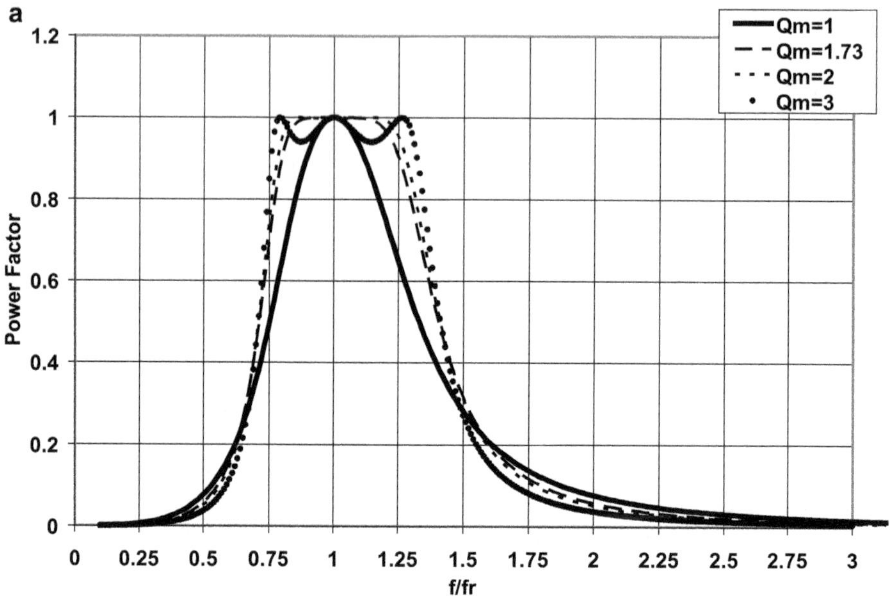

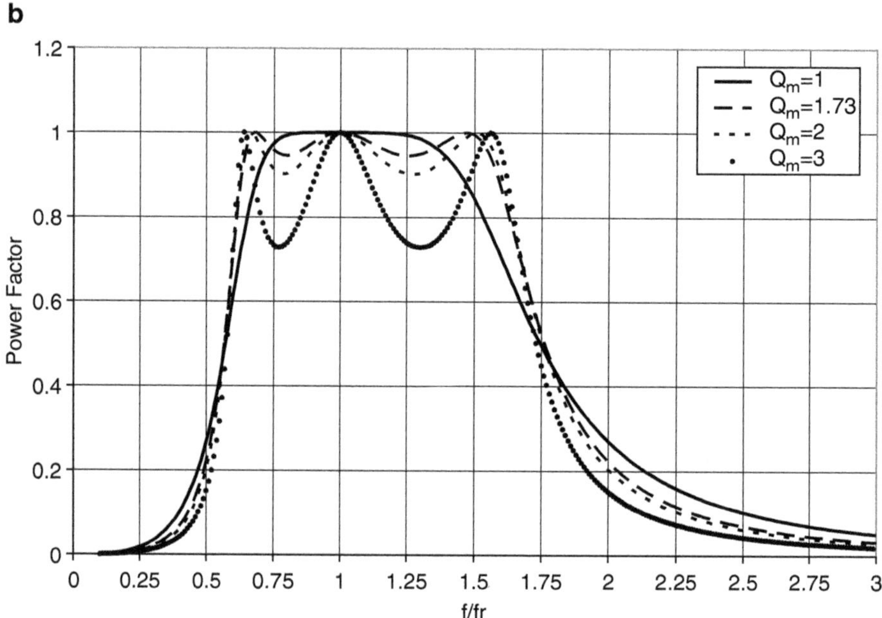

Fig. 2.12 (**a**) Power factor as a function of frequency, related to resonance, for $k = 0.5$ and $Q_m = 1$, 1.73, 2, and 3. (**b**) Power factor as a function of frequency, relative to resonance, for $k = 0.7$ and $Q_m = 1$, 1.73, 2, and 3. (**c**) Power factor as a function of frequency, relative to resonance, for $k = 0.9$ and $Q_m = 0.5$,0.7, 1, and 2

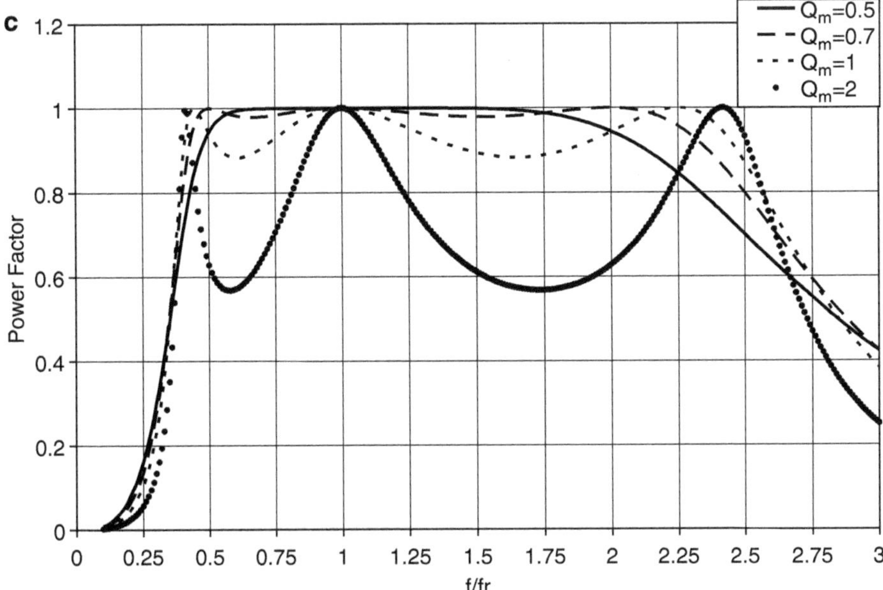

Fig. 2.12 (continued)

2.8.5 *Power Limits*

The value of Q_m for maximum output power [2, 33] in the vicinity of resonance can differ from the Q_m value for optimum bandwidth given by Eq. (2.90). This can be seen from calculating both the electrical and mechanical stress limited power. The total input power, W, for a transducer with negligible electrical losses may be calculated at resonance from Eq. (2.86) as $W = ½V^2G_m = ½V^2\omega_r C_f k^2 Q_m$. Substituting the free capacity $C_f = \varepsilon_{33}^T A_0/L$, yields an expression for the input power density in the piezoelectric material in terms of the electric field $E = V/L$,

$$P = W/(A_0 L) = ½\, k^2 \omega_r \varepsilon_{33}^T E^2 Q_m. \tag{2.100}$$

The maximum value of E that can be used, without excessively high electrical dissipation, is often a limitation on transducer power output unless Q_m is quite high. For example, the maximum rms value of electric field, E_m, for piezoelectric ceramics typically ranges from 2 to 4 kV/cm (approximately 5–10 V/mil). Figures in Sect. 13.14 show the increase of electrical dissipation with increasing electric field for PZT. This is an important property for high power applications that must also be considered for new materials.

An expression for the power density in terms of stress, rather than electric field, may also be obtained, since at resonance the stress in the ceramic is due to the force of the mass, $(M + M_r)$, moving with acceleration $\omega_r u = \omega_r NV/(R + R_r)$. Using the value of N from Eq. (2.9a), this force gives the stress

$$T = Q_m d_{33} E/s_{33}^E. \tag{2.101}$$

Solving for E and substituting into Eq. (2.100), with $k^2 = d_{33}^2/s_{33}^E \varepsilon_{33}^T$, then yields the stress limited input power density

$$P = W/(A_0 L) = ½\,\omega_r s_{33}^E T^2/Q_m. \tag{2.102}$$

The maximum safe tensile stress, T_m, in piezoelectric ceramic transducers with glue bonds is about 2 kpsi $= 1.4 \times 10^7$ Pa (see Sect. 13.14). A compressive prestress (typically by means of a stress rod or fiberglass wrap) is often used to increase this limit by about a factor of 3, allowing a ninefold increase in power output if the transducer is stress limited.

Equations (2.100) and (2.102) show an inverse dependence on Q_m. Transducers with a high Q_m are usually stress limited at resonance while transducers with a low Q_m are usually electric field limited. Maximum input power occurs when the transducer is operated at maximum safe values of both E_m and T_m. The optimum Q_m, in this sense, can be found by equating Eqs. (2.100) and (2.102) which gives

$$Q_m = \left(s_{33}^E/d_{33}\right)(T_m/E_m), \tag{2.103}$$

as the value of Q_m that would achieve the electrical and mechanical limits simultaneously. For PZT this value of Q_m is about 3, a value that corresponds to a reasonable bandwidth for underwater projectors. Substituting this optimum Q_m, back into Eq. (2.100) or (2.103) gives the maximum power density as

$$P = ½\,\omega_r d_{33} E_m T_m. \tag{2.104}$$

The maximum mechanical energy density, U_m, for particular materials may also be calculated from the maximum strain, S, at the maximum accepted drive field and the Young's modulus, Y, as $U_m = YS^2/2$. [Since the strain $S = dE$ and the mechanical energy density $U_m = YS^2/2$, the energy density may be also directly written as $U_m = d^2E^2/2s$, where E is the peak electric field intensity $E = E_{rms}\sqrt{2}$ and the Young's modulus $Y = 1/s$. For piezoelectric materials $s = s^E$ and may be obtained from Sect. 13.5 along with the corresponding value of the piezoelectric d constant.] Energy density values [34] of some common transduction materials are listed below in Table 2.4.

The Young's modulus values in Table 2.4 are under short circuit conditions for electric field materials, PZT, PMN, and PIN-PMN-PT, and open circuit conditions for magnetic field material, Terfenol-D. Although PZT-4 has a higher energy density than PZT-8, the value is limited to lower duty cycle operation. Terfenol-D,

Table 2.4 Energy density values for transducer materials

Material	Young's modulus	Field (E_{rms})	Strain (0-pk)	Energy density
PZT-8	74 (GPa)	10 V/mil	125 (ppm)	578 (J/m^3)
PZT-4	66	10	159	830
PMN	88	15	342	5150
Terfenol-D	29	64 kA/m	582	4910
PIN-PMN-PT	17.5	10 V/mil	735	4720

PMN, and PIN-PMN-PT appear to offer significantly greater energy density, although heating caused by electrical dissipation in PMN and undoped PIN-PMN-PT is an issue for high power applications, especially under high duty cycle operation where heating can be an issue. We note that the greatest strain is obtained from the single crystal material PIN-PMN-PT.

2.8.6 *Efficiency*

The acoustic output power radiated into the far field is equal to the input power reduced by the overall efficiency of the transducer. As discussed in Chap. 1 or as seen from the circuit of Fig. 2.11, the acoustic power is given by $W_a = ½R_r u^2$, while the total power mechanically dissipated is $W_m = ½\,(R + R_r)u^2$. Thus the mechanoacoustical efficiency is defined as

$$\eta_{ma} = W_a/W_m = R_r/(R + R_r), \tag{2.105}$$

showing that the radiation resistance, R_r, must significantly exceed the internal mechanical resistance, R, for high efficiency. Typically η_{ma} increases with frequency since the radiation resistance increases up to a maximum value of about ρcA (see Sect. 10.4). The radiation resistance R_r and radiation mass M_r may often be approximated by an equivalent spherical radiator of the same area, A, as the radiating surface of the transducer. Therefore the radius of the sphere is $a = (A/4\pi)^{1/2}$ and, with the wave number $k = \omega/c$, we have from Sect. 10.4.1

$$R_r = \rho cA(ka)^2/\left[1 + (ka)^2\right] \quad \text{and} \quad M_r = \rho 4\pi a^3/\left[1 + (ka)^2\right]. \tag{2.106}$$

The expression for R_r is a good approximation for transducers which are small compared to the wavelength in water where $ka \ll 1$ and also in the other extreme where $ka \gg 1$. The internal resistance depends on a variety of dissipation mechanisms related to the details of the encapsulated transducer structure and also tends to increase with frequency. Typical values of η_{ma} range from 60 to 90 %.

For example: At very low frequencies $R_r \approx \rho cA(ka)^2$ and $\eta_{ma} \approx \rho cA(ka)^2/R$ when $R_r \ll R$. In this region the efficiency will fall off rapidly as the frequency ($k = \omega/c$) is reduced. The fall off rate will be greater for dipole and quadrupole transducer which have an even higher order dependence on frequency.

The other factor in the overall efficiency of a transducer is the power lost to electrical or magnetic dissipation. With W_e as the input electrical power and W_m as the power delivered to the mechanical part of the transducer the electromechanical efficiency is defined as

$$\eta_{em} = W_m/W_e, \tag{2.107}$$

and the overall electroacoustic efficiency is defined as

$$\eta_{ea} = W_a/W_e = \eta_{em}\eta_{ma}. \tag{2.108}$$

For electric field transducers the electrical input power is $W_e = ½GV^2$, where G is the total input conductance which can be written as

$$G = G_0 + Re\left(N^2/Z^E_{mr}\right) = G_0 + G_e, \tag{2.109}$$

where the electrical loss conductance $G_0 = \omega C_f \tan\delta$ for piezoelectric ceramics, and $\tan\delta$, the electrical dissipation factor, is defined by this relationship. The power delivered to the mechanical part of the transducer is $½G_eV^2$ yielding the electromechanical efficiency

$$\begin{aligned}\eta_{em} &= G_e/(G_0 + G_e)\\ &= 1/\left[1 + \left(\tan\delta/k^2_{eff}\right)\left\{\omega/\omega_r Q_m + (Q_m\omega_r/\omega)\left(1 - \omega^2/\omega_r^2\right)^2\right\}\right],\end{aligned} \tag{2.110}$$

as shown in Sect. 13.15, Eq. (13.87).

The influence of all the transducer parameters on the electroacoustic efficiency can best be seen by evaluating Eq. (2.110) in different frequency regimes, and then multiplying by η_{ma} to get the overall electroacoustic efficiency as given by Eq. (2.108). The results, expressed in terms of the effective coupling coefficient k_{eff} (see Sects. 4.4.2 and 4.4.3) are:

$$\eta_{ea} = \omega R_r k^2_{eff} C^E_m/\tan\delta, \quad \omega \ll \omega_r \tag{2.111}$$

$$\eta_{ea} = \left[k^2_{eff}Q_m/\left(k^2_{eff}Q_m + \tan\delta\right)\right]\left[R_r/(R_r + R)\right], \quad \omega = \omega_r \tag{2.112}$$

$$\eta_{ea} = R_r k^2_{eff}(\omega_r/\omega)^3/\tan\delta\,\omega_r(M + M_r). \quad \omega \gg \omega_r \tag{2.113}$$

The efficiency always improves as the radiation resistance and effective coupling coefficient increase and always decreases as the electrical dissipation increases. Below resonance the efficiency increases as the mechanical compliance increases, while above resonance the efficiency increases as the mass decreases. This indicates a greater efficiency for a low impedance transducer both below and above resonance. We note that near resonance the first factor in brackets is nearly unity since normally $k_{eff}^2 Q_m \gg \tan\delta$, and the efficiency is mainly determined by R_r and R. The efficiency well above resonance falls off as ω^3, while below resonance it appears from Eq. (2.111) to fall off only as ω. However, at low frequency the radiation resistance decreases as ω^2 causing the efficiency to fall off as ω^3, as it does above resonance.

Similar expressions for the efficiency of magnetic field transducers are complicated by eddy currents (see Sect. 3.1.5). However, if eddy current losses are negligible, simple expressions for η_{ea} may be obtained at resonance and well above and below resonance. The results are, on using $N_m^2 = k_{eff}^2 L_f / C_m^H$ and with the coil quality factor defined as $Q_0 = \omega_r L_f / R_0$:

$$\eta_{ea} = (\omega/\omega_r)^2 \omega_r R_r k_{eff}^2 Q_0 C_m^H, \quad \omega \ll \omega_r \tag{2.114}$$

$$\eta_{ea} = \left[k_{eff}^2 Q_m Q_0 / \left(k_{eff}^2 Q_m Q_0 + 1\right)\right] [R_r/(R_r + R)], \quad \omega = \omega_r \tag{2.115}$$

$$\eta_{ea} = R_r k_{eff}^2 Q_0 (\omega_r/\omega)^2 / \omega_r (M + M_r). \quad \omega \gg \omega_r \tag{2.116}$$

It can be seen that the efficiency is proportional to the radiation resistance, R_r, as well as k_{eff}^2 as in the electric field case, while the coil quality factor has the same effect as the reciprocal of $\tan\delta$. It turns out that Eq. (2.114) is valid even with eddy currents if the frequency is low enough, and Eq. (2.115) is approximately true if f_r is low enough. Although the rate of reduction in efficiency is only quadratic with frequency in Eq. (2.116), compared to cubic in Eq. (2.113), one should expect a greater reduction rate due to actual eddy current losses at high frequencies.

In active sonar systems with a high duty cycle, where the transducer is turned on 20 % of the time or more, heating from electrical losses, and in some cases mechanical losses in the transduction material, can often be the limit on the output power [35]. In these cases it is important to choose the transduction mechanism with the lowest dissipation factor under high drive conditions (see Sect. 13.14). In some cases specific measures may be taken to extract heat from the transducer; e.g., by filling the housing with oil, providing external fins or potting in a conductive epoxy [36]. See Sect. 2.9 for more discussion and a transducer thermal model. Pulsed sonar systems with a short pulse length and slow repetition rate can be either stress or electric field limited and the value of Q_m for maximum power, of Sect. 2.8.5, becomes an important design factor.

2.8.7 Hydrophone Circuit and Noise

The equivalent circuit can also be used to represent a hydrophone which converts an input free field pressure p_i to an open circuit output voltage, V, with blocked force

$$F_b = D_a A p_i, \tag{2.117}$$

where A is the sensitive area of the hydrophone and D_a is the diffraction constant. The circuit of Fig. 2.11 may be solved for the open circuit output voltage V as a voltage divider circuit (see Sect. 13.8) with input voltage F_b/N yielding

$$V = (F_b/N)(1/j\omega C_0)/\left[1/j\omega C_0 + Z^E_{mr}/N^2\right]. \tag{2.118}$$

Substitution of Z^E_{mr} from Eq. (2.9b) gives

$$V = (F_b/N)/\left[1 + C_0/C^E_m N^2 - \omega^2 C_0(M + M_r)/N^2 + j\omega C_0(R + R_r)/N^2\right]. \tag{2.119}$$

At frequencies well below antiresonance where the impedance becomes $1/j\omega C_f$, the output voltage becomes a constant with value k^2F_b/N, which is typically independent of frequency for pressure sensitive electric field hydrophones. At antiresonance the output voltage, $V = -jQ_a k^2 F_b/N$, is magnified by the Q at antiresonance, Q_a

The hydrophone sensitivity, $M = V/p_i$, is a common measure of the hydrophone response and may be obtained from the equivalent circuit as

$$M = AD_a V/F_b. \tag{2.120}$$

The open circuit receiving response and constant voltage transmitting response are related to each other through the input impedance and reciprocity, as will be shown in Chap. 9.

Just as mechanical and electrical losses inside the transducer are a limitation on the output of projectors, these losses are also a limitation on the performance of hydrophones due to the internal thermal noise generated by the equivalent series resistance. This condition allows the calculation of the hydrophone self-noise through the thermal noise voltage developed in the series resistance, R_h, of the hydrophone (see Sect. 6.7). This resistance then determines the equivalent total Johnson thermal mean squared noise voltage, $\langle V_n^2\rangle$, through

$$\langle V_n^2\rangle = 4KTR_h\Delta f, \tag{2.121}$$

where K is Boltzman's constant $\left(1.381 \times 10^{-23}\,\text{Joule/Kelvin}\right)$, T is the absolute temperature, and Δf is the bandwidth. At 20 °C Eq. (2.121) leads to

$$10\log\langle V_n^2\rangle = -198\,\text{dB} + 10\log R_h + 10\log \Delta f. \tag{2.122}$$

The value of the noise voltage in Eq. (2.122) becomes more useful when converted to an equivalent mean squared noise pressure, $\langle p_{\mathrm{n}}^2\rangle = \langle V_{\mathrm{n}}^2\rangle / M^2$ by means of the hydrophone sensitivity M. The resulting level of the equivalent noise pressure is then

$$10\log\langle p_{\mathrm{n}}^2\rangle = -198\,\mathrm{dB} + 10\log R_{\mathrm{h}} - 20\log M + 10\log\Delta f, \tag{2.123}$$

which emphasizes the fact that the noise increases with the bandwidth. The signal to self-noise ratio is equal to unity or greater when the signal pressure $p_{\mathrm{s}} \geq \langle p_{\mathrm{n}}^2\rangle^{1/2}$.

The electrical component of the noise is given by the loss conductance $G_0 = \omega C_{\mathrm{f}} \tan\delta$, shunted across electrical input terminals in the equivalent circuit of Fig. 2.11, where $\tan\delta$ is the electrical dissipation factor and C_{f} is the free capacity. It may be shown (see Sect. 6.7.2) that at low frequencies

$$R_{\mathrm{h}} = (\tan\delta/\omega C_{\mathrm{f}})/(1 + \tan^2\delta) \approx \tan\delta/\omega C_{\mathrm{f}}, \tag{2.124}$$

where the approximation $\tan\delta \ll 1$ holds for most electric field transducer. Under this $\tan\delta$ approximation, the input series resistance of the entire circuit of Fig. 2.11 may be shown to be (see Sect. 13.15)

$$R_{\mathrm{h}} = R_{\mathrm{r}} k^2 C^E / C_{\mathrm{f}} \eta_{\mathrm{ea}} \left[(\omega/\omega_{\mathrm{a}} Q_{\mathrm{a}})^2 + \left(1 - \omega^2/\omega_{\mathrm{a}}^2\right)^2 \right]. \tag{2.125}$$

This expression includes the noise contributions from G_0, the mechanical loss resistance, R, and the radiation resistance R_{r} and may be used in Eq. (2.122) and with, M, Eq. (2.123). Equation (2.123) may also be used with measured values of M and R_{h}.

Alternatively, as shown in Sects. 6.7.1 and 13.15, if the transducer efficiency, η_{ea}, and DI are known the equivalent mean square self-noise may be determined from

$$10\log\langle p_{\mathrm{n}}^2\rangle = 20\log f - 74.8 - 10\log\eta_{\mathrm{ea}} - DI, \tag{2.126}$$

at 20 °C and in dB//(μPa)^2 Hz. It should be noted that the DI does not increase the sensitivity of a hydrophone, but it does decrease the isotropic acoustic equivalent of internal noise for hydrophones with directivity, thus increasing the signal-to-noise ratio. The hydrophone self-noise should be less than the many other noise sources in the medium and hydrophone platform, as discussed in Chaps. 6 and 8.

2.9 Thermal Considerations

Excessive transducer heating, often a result of continuous or high duty cycle operation, may be mitigated by improving the efficiency of a transducer and improving the conduction and convection of the heat to the surrounding medium.

In this section we extend the use of a lumped mode equivalent circuit representation (see Fig. 2.11) as a guide in determining heat sources, and develop an equation set that yields a more accurate means for evaluation of piezoelectric material losses. These results are shown to yield the heat powers in the piezoelectric drive section and the transducer mounting, which can then be used in a finite element model or heat flow equivalent circuit model to obtain the temperature rise in the transducer.

There are limitations to this lumped model representation including assumed linearity as well as uniform electric and mechanical field. In addition to this the piezoelectric stack length and lateral dimensions are assumed small, compared to the wavelength of the compression wave in the piezoelectric stack. However, with this model the fraction of power delivered to the piezoelectric stack and the transducer mounting system may be readily determined from common parameters, such as resonance frequency, mechanical Q, effective coupling coefficient, and dissipation factors.

This section also provides an example of the use and manipulation of transducer circuit structure to provide a simple representation that can serve as a model for analysis. Additional circuit representations and discussion for spherical, cylindrical, and Tonpilz transducers are given in Chaps. 3 and 5. The main results from the detailed equivalent circuit development of Sect. 2.9.1 are Eqs. (2.128) through (2.131) which are used in Sect. 2.9.2 to develop the power lost to heat given by Eqs. (2.133) and (2.134).

2.9.1 Transducer Thermal Model

The stiffness and mass of the common Tonpilz transducer, illustrated in Fig. 2.13, may be derived from a T network transmission line distributed model and reduced to a single degree of freedom lumped model with piezoelectric stack velocity $u = u_1 + u_2$. This single degree of freedom lumped circuit is a generalization of a ring equivalent circuit for heat analysis presented by Butler et al. [36]. The

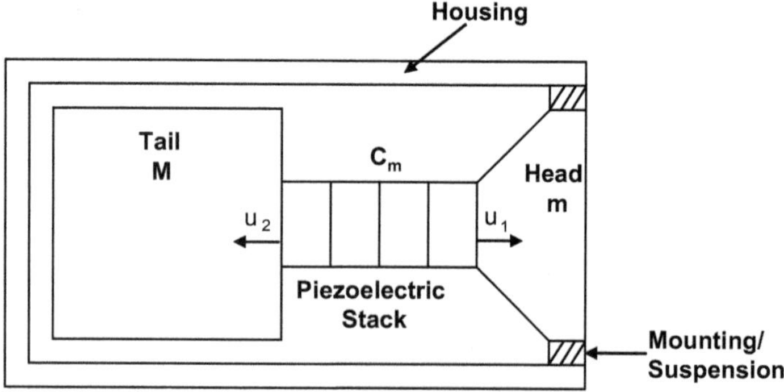

Fig. 2.13 Sketch of a Tonpilz transducer with head and tail velocities, u_1 and u_2, respectively

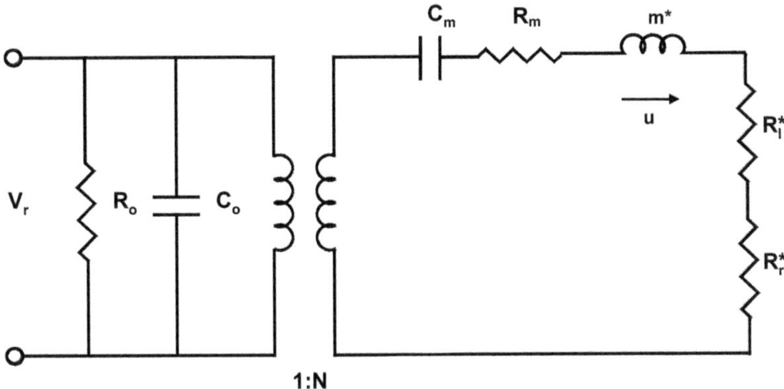

Fig. 2.14 Lumped 1 degree of freedom equivalent circuit $\omega_r M \gg R_l + R_r$ with R_o the clamped shunt resistance, R_m the equivalent-stack mechanical series-loss resistance, R_l^* the equivalent mechanical mounting loss resistance, and R_r^* the equivalent radiation resistance

generalized circuit [37] is illustrated in Fig. 2.14. See Sects. 3.1.3 and 5.3.1 for further discussion on the equivalent circuit of the Tonpilz transducer.

Series equivalent components of Fig. 2.14 may be derived from the equivalent circuit of Fig. 2.11 with replacement of those circuit components C_m^E, M_{Tail}, M, R, and M_r by C_m, M, m, R_l, and m_r, respectively, along with $F_b = 0$. Note also that the conductance $G_0 = 1/R_0$ where R_0 is the shunt electrical loss resistance and $R = R_l$ is the mechanical loss resistance. The circuit of Fig. 2.14 is valid for the typical Tonpilz conditions $\omega_r L/c \ll 1$ and $\omega_r M \gg R_l + R_r$ where ω_r is the angular resonance frequency, c is the speed of sound in the piezoelectric stack of length L, M is the tail mass, R_l is the mounting loss, and R_r is the radiation resistance. Under these conditions the series equivalent head mass, m^*, mechanical loss resistance, R_l^*, and radiation resistance, R_r^*, are, respectively, given by

$$m^* = m(1 + m/M), \quad R_l^* = R_l/(1 + m/M)^2, \quad R_r^* = R_r/(1 + m/M)^2.$$

The mass of the head, m, can include the radiation mass, m_r, and the portion of the stack mass, $m_s/2$. The tail mass, M, can also include $m_s/2$ along with an effective "mass," $Z/j\omega$, should there be a mounting impedance of value Z. And finally the mass—$m_s/6$ can be inserted in series with the stack compliance for a more complete representation of the stack mass. The short circuit mechanical compliance is C_m, the electromechanical turns ratio is N, the clamped capacitance is C_0, and the clamped electrical shunt resistance is $R_0 = 1/\omega C_0 \tan\delta_0$, where $\tan\delta_0$ is the clamped electrical dissipation factor. The free capacity of the piezoelectric material, C_f, the effective coupling coefficient, k_e, and the clamped capacitance, C_0, are, respectively, given by

$$C_f = N^2 C_m + C_0, \quad k_e^2 = N^2 C_m / C_f, \quad C_0 = C_f(1 - k^2).$$

At mechanical resonance the potential and kinetic energies are equal and $j\omega_r m^* + 1/j\omega_r C_m = 0$ leaving only resistive mechanical components. And here the details of the representation of the piezoelectric stack and other components are not as important, as long as the stack heat sources are uniformly distributed and the series representation of the resistive mechanical losses is maintained. The equivalent-stack series-loss resistance R_{ms} of Fig. 2.14 is obtained from the shunt stack mechanical loss resistance $R_m = 1/\omega C_m \tan\delta_m$ where the mechanical dissipation factor $\tan\delta_m = 1/Q_p$ and Q_p is the measured or published mechanical quality factor of the piezoelectric material or assembled stack. The series representation of the parallel shunt compliance, C_m, and loss resistance R_m can be obtained through the series equivalent $C_{ms} = C_m(1 + \tan^2\delta_m) \approx C_m$ and

$$R_{ms} = \tan\delta_m / \left[\omega C_m \left(1 + \tan^2\delta_m\right)\right] \approx \tan\delta_m / \omega C_m = 1/\, Q_p \omega C_m.$$

The approximations indicated are generally good since, typically, $\tan\delta_m \ll 0.1$. The approximation is better for PZT-8 ceramic type materials than for some single crystal materials where the mechanical dissipation factor can be higher. With this series resistance representation all the mechanical dissipations in Fig. 2.14 can be conveniently summed into one quantity

$$R = R_{ms} + R_l^* + R_r^*.$$

Here, the first two mechanical resistances represent mechanical sources of heat, while the last represents the acoustic radiation.

The single degree of freedom transducer equivalent circuit of Fig. 2.14 is similar to the equivalent circuit for a ring, cylinder, or sphere and certain other transducers operating in their fundamental mode. The total electrical input admittance, Y, of the circuit may be written as

$$Y = G + jB = j\omega C_0 + 1/R_0 + \left[N^2/R\right]/[1 + j\,(\omega/\omega_r - \omega_r/\omega)Q_m], \qquad (2.127)$$

where the angular mechanical resonance frequency is given by $\omega_r = \left(1/m^* C_m\right)^{1/2}$ and here the mechanical Q is defined as $Q_m = (1/\omega_r C_m R)$ with R a function of frequency. The input power is given by the product of the square of the rms voltage, V_r, and the electrical conductance, G, which from Eq. (2.127) and the effective coupling coefficient, $k^2 = N^2 C_m / C_f$, may be written as

$$G = 1/R_0 + k_e^2 \omega_r C_f Q_m / \left[1 + (\omega/\omega_r - \omega_r/\omega)^2 Q_m^2)\right], \qquad (2.128)$$

with the electrical input power $W_i = V_r^2 G$.

It follows from Eq. (2.128) that at frequencies well below resonance, $\omega \ll \omega_r$, the conductance is

$$G \approx \omega C_f \left[\left(1 - k^2\right) \tan\delta_0 + k^2 \tan\delta_m\right], \qquad (2.129)$$

where we have used $C_0 = C_f\left(1 - k^2\right)$. Assumptions were made in the development of Eq. (2.129) that in the quantity Q_m of Eq. (2.128) the resistances R_l and R_r are negligible compared to R_{ms} at low frequency. This is so since $R_{ms} \approx \tan\delta_m/\omega C_m$ which increases as the frequency decreases while R_l is a constant and radiation resistance, R_r, decreases. At these low frequencies the electrical input conductance is the sum of an electrical and a mechanical component, which is coupled to the electrical component through the effective coupling coefficient, k_e. We call and write the sum of these contributions, given by the bracketed term of Eq. (2.129), the total free electrical dissipation factor, $\tan\delta$, in the piezoelectric; i.e.,

$$\tan\delta = \left(1 - k^2\right)\tan\delta_0 + k^2\tan\delta_m. \tag{2.130}$$

Equation (2.130) shows how the total dissipation is distributed between electrical and mechanical components depending on the value of the coupling coefficient, k. And we note that for $k \approx 1$, $\tan\delta \approx \tan\delta_m$; however, for $k \approx 0$, $\tan\delta \approx \tan\delta_0$.

Substitution of Eq. (2.130) into Eq. (2.129) yields $G = \omega C_f \tan\delta$, which is the value that is electrically measured at frequencies well below resonance under free conditions. Since $\tan\delta$ and $Q_p = 1/\tan\delta_m$ are often specified for piezoelectric material we may obtain the clamped value, $\tan\delta_0$, through Eq. (2.130) written as

$$\tan\delta_0 = \left(\tan\delta - k^2\tan\delta_m\right)/\left(1 - k^2\right). \tag{2.131}$$

Note that for small values of k and $\tan\delta_m$, the clamped dissipation factor $\tan\delta_0 \approx \tan\delta/\left(1 - k^2\right)$. This has been used in the past [2] and is based on a clamped resistive loss of $R_0 = \left(1 - k^2\right)/\omega C_0 \tan\delta$, rather than the exact representation $R_0 = 1/\omega C_0 \tan\delta_0$ presented here. Although the original approximation appears to be good for piezoelectric ceramics, it may not be as good for single crystal piezoelectric materials with higher k values near 90 % and possible higher values of $\tan\delta_m$

Equation (2.131) allows us to accurately calculate the clamped electrical loss for either PZT or single crystal materials based on published or measured values of $\tan\delta$ and $\tan\delta_m$. And Eq. (2.131) also allows us to develop an equation set for determining the heat power at resonance in terms of available well known dissipation values, $\tan\delta$ and $\tan\delta_m$.

2.9.2 Power and Heating at Resonance

At mechanical resonance the reactive mechanical components $j\omega m^*$ and $1/j\omega C_m$ of Fig. 2.14 cancel, the conductance is a maximum and the transducer input and output powers can be large. Operation at resonance is typical for high drive transducer applications and we have developed an equation set for heat power ratios at this frequency based on the Eq. (2.130) relationship. At mechanical resonance, $\omega = \omega_r$, and from Eqs. (2.128) and (2.130) the conductance is

$$G_r = \omega_r C_f \left[\tan \delta + k^2 (Q_m - 1/Q_p) \right], \tag{2.132}$$

where, as before, $Q_p = 1/\tan \delta_m$ and $Q_m = 1/\omega_r C_{ms} R$ and is evaluated at resonance. Since the above and following equations are based on Eq. (2.130) they are applicable to single crystal piezoelectric material as well as ceramic piezoelectric material.

The above considerations may be used to calculate the power into the heat dissipation mechanisms characterized by $\tan \delta$, $\tan \delta_m$, and R_l. At resonance the total input power is $W_i = V_r^2 G_r$, the power delivered to the clamped electrical part of the piezoelectric driving stack is $W_0 = V_r^2/R_0$, the power delivered to the mechanical part of the piezoelectric stack is $W_m = u^2 R_{ms}$, the power delivered to the head mounting loss is $W_l = u^2 R_l^*$, and finally the acoustic output radiated power delivered to the water is $W_r = u^2 R_r^*$. It is convenient to evaluate the total loss in the piezoelectric stack as $W_p = W_0 + W_m$ in our lumped model. The velocity u in the power calculations for W_m and W_l may be written in terms of the voltage using $u = V_r N/R$. With this, the ratio of the dissipated power to the input power is independent of the voltage, the free capacitive susceptance $\omega_r C_f$ and the turns ratio N on using $N^2 = k^2 C_f / C_m$.

After an identification of common transducer parameters, the fractional heat power delivered to the piezoelectric element at resonance may be written as

$$W_p/W_i = \left[\left(\tan \delta - k^2/Q_p \right) + \left(k^2 Q_m^2/Q_p \right) \right] / \left[\tan \delta + k^2 (Q_m - 1/Q_p) \right]. \tag{2.133}$$

The first term in parenthesis in the numerator of Eq. (2.93) represents the electrical dissipation while the second term in parenthesis represents the mechanical dissipation.

The fractional heat power delivered to the mounting loss system may also be shown to be

$$W_l/W_i = k^2 Q_m \left(1 - \eta_{ma} - Q_m/Q_p \right) / \left[\tan \delta + k^2 (Q_m - 1/Q_p) \right]. \tag{2.134}$$

The mechanoacoustic efficiency $\eta_{ma} = R_r^*/R \approx 1 - Q_m/Q_a$ where Q_a is the mechanical Q under in-air loading conditions. The quantity Q_m/Q_p is usually small and can often be ignored. Finally, the fractional acoustical power delivered to the water may be written as

$$W_r/W_i = \left(k^2 Q_m \eta_{ma} \right) / \left[\tan \delta + k^2 (Q_m - 1/Q_p) \right]. \tag{2.135}$$

The input power given by $W_i = V_r^2 G_r$ may be used in Eqs. (2.133)–(2.135).

Equations (2.133) and (2.134), along with Eq. (2.132) and $W_i = V_r^2 G_r$, are our desired results and characterize the two dissipative sources of heat of the transducer: power loss to the piezoelectric stack, W_p, and power loss to the piston head mounting structure, W_l, which may be used as the dissipative power sources in a thermal model. The power dissipated in the piezoelectric stack is of paramount interest under high duty cycle conditions, as high operating temperatures can affect the performance of the material. An example of using this equation set with a transducer finite element model has been given by Butler et al. [37].

2.10 Extended Equivalent Circuits

The lumped equivalent circuit of Fig. 2.11 is a good representation for a thin-walled piezoelectric ring transducer and may be used as an approximation for other transducers operating in the vicinity of resonance. On the other hand, Tonpilz transducers are structured with a radiating piston of mass M, a piezoelectric drive section with approximate compliance C_m^E, some distributed mass, and an inertial tail mass. As will be shown in Chap. 3, the rigid wall restriction shown in Fig. 2.5 may be removed by including a tail mass, M_{Tail}, between the nodes A, B of Fig. 2.11. In this case, as will also be shown in Chap. 3, the frequencies, ω_r, ω_a, of Eqs. (2.79) and (2.80) are increased by approximately $[1 + (M + M_r)/M_{Tail}]^{1/2}$ while the Q's, Q_m and Q_a of Eq. (2.83) are increased by approximately $[1 + (M + M_r)/M_{Tail}]$. In typical Tonpilz designs the tail mass, M_{Ttail}, is approximately three times the head mass, M, and moves with approximately one third the velocity of the radiating head mass.

The distributed nature of the piezoelectric drive may also be included by replacing the pure spring compliance element, C_m^E, with a distributed compliance element,

$$C_m^E = \left(s_{33}^E L/A\right)(\sin kL)/kL, \tag{2.136}$$

and two distributed mass elements,

$$m = ½\rho A_0 L(\tan kL/2)/(kL/2), \tag{2.137}$$

one in series with the head mass, M, and the other in series with the tail mass, M_{tail}. The quantity ρ is the density of the piezoelectric bar, the wave number $k = \omega/c^E$, and the sound speed c^E is the short circuit sound speed in the bar. This model is typically used where a number of piezoelectric elements are wired in parallel and cemented together to make a longer drive section to obtain greater output, and it is assumed that the distance between the electrodes is small compared to the wavelength. Much more will be said about equivalent circuit representation of transducers in Chap. 3.

The examples of the piezoelectric transducer operating as a projector and as a hydrophone illustrate voltage drive and acoustic drive conditions. Current drive could also be used for electric field transducers, but it is more appropriate for magnetic field transducers. These drives refer to idealizations used for analytical work in which the drive variable is constrained to have a sinusoidal waveform with constant amplitude as frequency is varied. The two electrical drives correspond to the extremes of power amplifier internal impedance with very low impedance giving approximate voltage drive and very high impedance giving approximate current drive. Real amplifiers give intermediate conditions, but usually nearer voltage drive. In terms of electrical boundary conditions voltage drive is approximately short circuit while current drive is approximately open circuit.

2.11 Summary

This chapter began with a discussion of the six major means of electroacoustic transduction: piezoelectric, electrostrictive, magnetostrictive, electrostatic, variable reluctance, and moving coil. Associated electro mechanical equations of motion were given. It was shown why piezoelectric and magnetostrictive are best suited for most underwater transducer applications.

It is important to note that once electrically biased, electrostrictive materials become linear and perform the same as (or even better than) natural piezoelectric crystal and in this state are referred to as "piezoelectric." The 33 and 31 modes of operation, where the first subscript is the direction of the electric field and the second is the direction of the mechanical motion, were presented. There is nearly a factor of two performance increase in the 33 mode of operation.

A basic lumped equivalent circuit for transducers was presented along with a discussion of circuit resonance, f_r, antiresonance, f_a, mechanical Q, Q_m, and electrical Q, Q_e, coupling coefficient, k, power factor, P_f, tuning, bandwidth optimization and electrical and mechanical power limits. Projector mechanoacoustical, η_{ma}, and electromechanical, η_{em}, efficiencies, thermal considerations along with an equivalent circuit model and hydrophone noise were also discussed. The resonant and antiresonant frequencies and Q_m and Q_e were shown to be related to the coupling coefficient through $k^2 = 1 - (f_r/f_a)^2 = 1/(1 + Q_m Q_e)$.

With W the power absorbed by a transducer and VI the product of the input voltage and current the power factor $P_f = W/VI$ with unity being the best possible condition, which usually occurs at resonance and under tuned conditions and over a wide band, if k is high. It was also shown that if the Q_m is high, the transducer may be mechanically stress limited while if it is low, the transducer may be limited by the electric field. The overall transducer electroacoustical efficiency is the power-out/power-in and equal to $\eta_{ea} = \eta_{em}\eta_{ma}$. Transducers can also be limited by the temperature rise due to electrical and mechanical losses. This can become more of an issue as the duty cycle is increased beyond 10 %. A thermal model and means for calculating the power delivered to resistive elements was also presented.

Exercises (Degree of Difficulty: *Lowest, **Moderate, ***Highest)

2.1* Perform a few calculations to become familiar with some of the important parameters of transducers. Consider an ideal transducer operating in the 33 mode of piezoelectric ceramic Type I material (PZT-4, see Sect. 13.5) of bar length 1.27 cm and cross sectional area 1.6 cm^2 with rigid backing on the rear end and an ideal piston of area 8 cm^2 on the front end. Assume the piston face radiates into the water and the remaining part is isolated from the water by a water-tight housing. Calculate the free, $C_f = \varepsilon_{33}^T A_0/L$, and clamped, C_0, capacitance.

2.2* Ignore the mass of the piezoelectric ceramic of Exercise 2.1 and calculate the short circuit mechanical compliance $C^E = s_{33}^E L/A_0$, the open circuit compliance C^D, the short circuit resonance frequency, f_r, and open circuit antiresonance frequency, f_a, for a total piston mass and radiation mass of $M = 0.1$ kg. Also, calculate the effective coupling coefficient from f_r and f_a. Why does $k_{eff} = k_{33}$ in this case?

2.3** Calculate the electromechanical turns ratio, N, and the corresponding mechanical force, F, for 1 V and 5 kV drive for the transducer of Exercise 2.1. Then determine the piston velocity, u, at mechanical resonance for a total mechanical loss resistance $R = 1.2 \times 10^3$ and radiation resistance $R_r = 1.2 \times 10^3$ Ns/m. What is the mechanoacoustic efficiency?

2.4** Calculate the power output, W, and average surface intensity, $I_s = W/A_0$, at resonance for 5 kV for the transducer of Exercise 2.1. Assume a directivity factor of $D_f = 2$ and calculate the source level in dB re 1 μPa and the far-field intensity, I_0, at 1 m and compare with the surface intensity. Explain the difference in the values of the intensities I_s and I_0.

2.5* To gain a little more familiarity with transduction consider Eq. (2.9a) with external force $F_b = 0$ and at low frequency where the mass and resistance terms are negligible compared with the stiffness term in the impedance. Show that under the usual linear case where $x_3 \ll L$ the strain $S_3 = d_{33}E_3$ and the displacement $x_3 = d_{33}V$. Calculate this linear strain for Type I piezoelectric ceramic with typical maximum electric field of $E_3 = 4$ kV/cm (10.2 kV/in.).

2.6* Using the results of Exercise 2.5 calculate the corresponding displacement and voltage for a length, L, of 1.27 cm.

2.7* Consider the piezoelectric and magnetostrictive cases of Table 2.1 and show that the third column can be obtained from the first and second columns.

2.8* Calculate the values of the electromechanical turns ratio N for piezoelectric and magnetostrictive cases of Table 2.1 with $A_0 = 1.6$ cm^2 and $L = 1.27$

cm for Type I piezoelectric material (see Sect. 13.5) and Terfenol-D magnetostrictive material (see Sect. 13.7) with $n = 100$ coil turns.

2.9* Calculate the piezoelectric force for voltage $V = 10$ V and the magnetostrictive force for current $I = 0.10$ A for the N values of Exercise 2.8.

2.10** Derive Eq. (2.82), $Q_m = \omega_r M/R$, from the definition $Q_m = 2\pi$ (Total Energy)/(Energy dissipated per cycle at resonance). See Eq. (4.5) and Sect. 4.2.1 for further discussion of Q_m.

2.11* Show that, on using resonance relations, the expression $Q_m = \omega_r M/R$ may also be written as $Q_m = 1/\omega_r R C_m^E$ or $Q_m = \left(M/C_m^E\right)^{1/2}/R = \left(MK_m^E\right)^{1/2}/R$ where C_m^E is the short circuit mechanical compliance and K_m^E is the short circuit mechanical stiffness.

2.12* Although the mechanical Q may be calculated for a 100 % efficient transducer, the mechanoacoustical efficiency, η_{ma}, usually must be measured or estimated because the internal mechanical resistance is not known. Accordingly, show that the mechanical Q may be written as $Q_m = Q_m' \eta_{ma}$ where $Q_m' = \omega_r(M + M_r)/R_r$ is the mechanical Q for $\eta_{ma} = 100\%$. If the calculated transducer Q_m' is 5 what is Q_m for $\eta_{ma} = 80\%$?

2.13** Show that the mechanical Q may be written as $Q_m = [\omega_r M/R_r + \omega_r M_r/R_r]$ η_{ma} where the first term in the brackets is the Q_m of the transducer without radiation mass loading and the second term is called the radiation Q and is due to the radiation mass and radiation resistance alone. Also show that for the case of a spherical radiator that the second term may be written as $1/k_r a$ where the wave number $k_r = \omega_r/c$ and a is the radius of the sphere. Determine the Q_m for a spherical transducer where $\omega_r M/R_r = 4$ and $\eta_{ma} = 80\%$ for the cases where $k_r a = 0.1$, 1, and 10.

2.14* Calculate the electroacoustic efficiency of an electric field transducer at resonance for $\eta_{ma} = 80\%$, $k_{eff} = 0.5$, for $Q_m = 1$ and 10 for $\tan\delta = 0.01$ and, under high electric drive conditions, for $\tan\delta = 0.10$. Make the same calculation for a magnetostrictive transducer with coil $Q_0 = 100$ and 10 (instead of $\tan\delta = 0.01$ and 0.10, respectively) but with negligible eddy current losses.

2.15* Construct a table or graph of the power factor, P_f, as a function of B/G from 0 to 10. Show that in particular for $B/G = 0$, 1, and 10 the power factor $P_f = 1$, 0.707, and ≈ 0.1, respectively. Show that in the limit $P_f \approx 1$ for $B/G \ll 1$ and $P_f \approx G/B$ for $G/B \ll 1$.

2.16** The effective coupling coefficient may be obtained from measurable quantities by use of the equations $k^2 = 1 - (f_r/f_a)^2$ or $k^2 = 1/(1 + Q_m Q_e)$. Prove the last very important equation.

2.17* Calculate the input power at resonance for a transducer with $k = 0.5$, $C_f = 10$ nF, $Q_m = 5$, and $f_r = 10$ kHz for an RMS input voltage of 1 and 100 V assuming that electrical losses are negligible. What is the efficiency and output power if the internal mechanical resistance is one half the radiation resistance?

References

1. W.G. Cady, *Piezoelectricity*, vol 1. (Dover Publications, New York, 1964), p. 177. See also, *Piezoelectricity,* ed. by C.Z. Rosen, B.V. Hiremath, R. Newnham (American Institute of Physics, New York, 1992)
2. D.A. Berlincourt, D.R. Curran, H. Jaffe, Piezoelectric and Piezomagnetic Materials and Their Function in Transducers, in *Physical Acoustics*, ed. by W.P. Mason, vol. 1 (Academic, New York, 1964)
3. R.E. Newnham, *Properties of Materials* (Oxford University Press, Oxford, 2005)
4. E.J. Parssinen (verbal communication), The possibility of depoling under pressure cycling was the reason for choosing the 31 mode over the 33 mode in the first use of PZT for submarine transducers of NUWC, Newport, RI
5. E.J. Parssinen, S. Baron, J.F. White, *Double Mass Loaded High Power Piezoelectric Underwater Transducer*. Patent 4,219.889, 26 Aug 1980
6. R.S. Woollett, Effective coupling factor of single-degree-of-freedom transducers. J. Acoust. Soc. Am. **40**, 1112–1123 (1966)
7. W.Y. Pan, W.Y. Gu, D.J. Taylor, L.E. Cross, Large piezoelectric effect induced by direct current bias in PMN-PT relaxor ferroelectric ceramics. Jpn. J. Appl. Phys. **28**, 653 (1989)
8. M.B. Moffett, M.D. Jevenager, S.S. Gilardi, J.M. Powers, Biased lead zirconate titanate as a high-power transduction material. J. Acoust. Soc. Am. **105**, 2248–2251 (1999)
9. S. Trolier-McKinstry, L. Eric Cross, Y. Yamashita (eds.), *Piezoelectric Single Crystals and Their Application.* (Pennsylvania State University and Toshiba Corp., Pennsylvania State University Press 2004).
10. V.E. Ljamov, Nonlinear acoustical parameters in piezoelectric crystals. J. Acoust. Soc. Am. **52**, 199–202 (1972)
11. W.P. Mason, *Piezoelectric Crystals and Their Application to Ultrasonics* (Van Nostrand, New York, 1950)
12. J.C. Piquette, S.E. Forsythe, A nonlinear material model of lead magnesium niobate (PMN). J. Acoust. Soc. Am. **101**, 289–296 (1997)
13. J.C. Piquette, S.E. Forsythe, Generalized material model for lead magnesium niobate (PMN) and an associated electromechanical equivalent circuit. J. Acoust. Soc. Am. **104**, 2763–2772 (1998)
14. J.C. Piquette, Quasistatic coupling coefficients for electrostrictive ceramics. J. Acoust. Soc. Am. **110**, 197–207 (2001)
15. C.L. Hom, S.M. Pilgrim, N. Shankar, K. Bridger, M. Massuda, R. Winzer, Calculation of quasi-static electromechanical coupling coefficients for electrostrictive ceramic materials. IEEE Trans. Ultrason. Ferroelectr. Freq. Control **41**, 542 (1994)
16. H.C. Robinson, A comparison of nonlinear models for electrostrictive materials. Presentation to 1999 I.E. Ultrasonics Symposium, Oct 1999, Lake Tahoe, Nevada
17. J.C. Piquette, R.C. Smith, Analysis and comparison of four anhysteretic polarization models for lead magnesium niobate. J. Acoust. Soc. Am. **108**, 1651–1662 (2000)
18. F.M. Guillot, J. Jarzynski, E. Balizer, Measurement of electrostrictive coefficients of polymer films. J. Acoust. Soc. Am. **110**, 2980–2990 (2001)
19. NDRC, *Design and Construction of Magnetostriction Transducers*. Div 6 Summary Technical Reports, vol 13 (1946)
20. R.J. Bulmer, L. Camp, E.J. Parssinen, Low Frequency Cylindrical Magnetostrictive Transducer for Use as a Projector at Deep Submergence. *Proceedings of 22nd Navy Symposium on Underwater Acoustics*, October 1964. See also T.J. Meyersm, E.J. Parssinen, Broadband Free Flooding Magnetostrictive Scroll Transducer, Patent No. 4,223,401, 16 Sept 1980
21. C.M. van der Burgt, Phillips Res. Rep. **8**, 91 (1953)
22. M.A. Mitchell, A.E. Clark, H.T. Savage, R.J. Abbundi, Delta E effect and magnetomechanical coupling factor in $Fe_{80}B_{20}$ and $Fe_{78}Si_{10}B_{12}$ glassy ribbons. IEEE Trans. Mag **14**, 1169–1171 (1978)

23. A.E. Clark, Magnetostrictive rare earth-Fe_2 compounds. Ferromag. Mater. **1**, 531–589 (North Holland Publishing, 1980). See also, A.E. Clark, H.S. Belson, Giant room temperature magnetostriction in $TbFe_2$ and $DyFe_2$. Phys. Rev. B **5**, 3642 (1972)
24. A.E. Clark, J.B. Restorff, M. Wun-Fogle, T.A. Lograsso, D.L. Schlagel, Magnetostrictive properties of b.c.c. Fe-Ga and Fe-Ga-Al alloys. IEEE Trans. Mag. **36,** 3238 (2000). See also, A.E. Clark, K.B. Hathaway, M. Wun-Fogle, J.B. Restorff, V.M. Keppens, G. Petculescu, R.A. Taylor, Extraordinary magnetoelasticity and lattice softening in b.c.c. Fe-Ga alloys. J. Appl. Phys. **93**, 8621 (2003)
25. S.L. Ehrlich, Proposal of piezomagnetic nomenclature for magnetostrictive materials. Proc. Inst. Radio Eng. **40**, 992 (1952)
26. R.S. Woollett, *Sonar Transducer Fundamentals.* (Naval Undersea Warfare Center Report, Newport Rhode Island, Undated)
27. F.V. Hunt, *Electroacoustics* (Wiley, New York, 1954)
28. A. Caronti, R. Carotenuto, M. Pappalardo, Electromechanical coupling factor of capacitive micromachined ultrasonic transducers. J. Acoust. Soc. Am. **113**, 279–288 (2003)
29. R.J. Bobber, *Underwater Electroacoustic Measurements* (US Government Printing Office, Washington, DC, 1970)
30. J.F. Hersh, *Coupling Coefficients.* Harvard University Acoustics Research Laboratory, Technical Memorandum No. 40, 15 Nov 1957
31. C.H. Sherman, Underwater sound transducers—a review. IEEE Trans. Sonics Ultrason. **Su-22**, 281–290 (1975)
32. D. Stansfield, *Underwater Electroacoustic Transducers* (Bath University Press, Bath, UK, 1991)
33. R.S. Woollett, Power limitations of sonic transducers. IEEE Trans. Sonics Ultrason. **SU-15**, 218–229 (1968)
34. J.F. Lindberg, The application of high energy density transducer material to smart systems, Mat. Res. Soc. Symp. Proc. vol 459 (Materials Research Society, 1997). See also D.F. Jones, J.F. Lindberg, Recent transduction developments in Canada and the United States. Proc. Inst. Acous. **17**(Part 3), 15 (1995)
35. J. Hughes, High power, high duty cycle broadband transducers; R. Meyer High power transducer characterization. ONR 321 Maritime Sensing (MS) Program Review, 18 August, 2005, NUWC, Newport, RI
36. S.C. Butler, J.B. Blottman III, R.E. Montgomery, A thermal analysis of high drive ring transducer elements. NUWC-NPT Technical Report 11,467, 15 June 2005, see also R. Montgomery, S.C. Butler, Thermal analysis of high drive transducer elements (A). J. Acoust. Soc. Am. **105**, 1121 (1999)
37. J.L. Butler, A.L. Butler, S.C. Butler, Thermal model for piezoelectric transducers. J. Acoust. Soc. Am. **132**, 2161–2163 (2012)

Chapter 3
Transducer Models

The previous two chapters presented the physics and associated equations for underwater sound transduction. The present chapter begins to fill in the background with detailed discussions of the models and methods used in transducer analysis and design [1–6]. The following chapters will continue with further discussion of important transducer characteristics followed by a description of projectors and hydrophones along with implementation in arrays. This will then be followed by measurement methods for the evaluation of transducers and calculation of acoustic radiation from transducers.

The underwater sound electroacoustic transducer is a vibrating device which, as a projector, is set into motion by electrical means causing it to alternately push and pull on the water and radiate sound. As a hydrophone, sound waves in the water set the transducer into motion generating an electrical signal. The electroacoustic transducer is part acoustical at its moving surface in contact with the acoustic medium, part mechanical as a moving body controlled by forces, and part electrical as a current controlled by voltage. Thus electrical equivalent circuits representing the acoustical and mechanical parts make possible an electrical simulation of the whole transducer. This facilitates analysis and design, and is especially useful since transducers are always connected to electrical components such as a preamplifier for a hydrophone or a power amplifier with tuning and transformer circuits for a projector.

In Sect. 2.8 we indicated how an electroacoustic transducer may be represented by an electrical equivalent circuit. In this chapter we will discuss this representation in much more detail including electrical equivalents of mechanical and electromechanical systems and all the parameters and physical concepts needed for transducer design and analysis. The equivalent circuits will initially be limited to lumped parameters consisting of pure masses and pure springs and then extended to include distributed systems that support acoustic waves. Although we will concentrate on transducers as projectors of sound, the analysis developed is also applicable to hydrophones. Other details specific to hydrophones are covered in Chap. 6. Since piezoelectric materials are most commonly used in underwater sound, this

J.L. Butler, C.H. Sherman, *Transducers and Arrays for Underwater Sound*,
Modern Acoustics and Signal Processing, DOI 10.1007/978-3-319-39044-4_3

mechanism will be our main focus, but models for magnetostrictive transducers will be included and the applicability to other electric field and magnetic field transducers will be indicated.

Some transducers are not fully realizable as an equivalent circuit. For example, it will be shown that magnetostrictive transducers represented by an impedance analogy do not yield the proper phase of the output velocity although they yield the correct amplitude. In this case a mobility analogy or a gyrator must be used. Alternatively the relationship between the input and output terminals of each part of a transducer may be represented by a transfer matrix. The overall matrix representation of the connected parts that form the complete transducer is then obtained by matrix multiplication of the individual matrices. This mathematical approach has no limitations since there are no circuits to realize, and it may also be adapted for array interaction analysis. In this chapter we will also develop the foundations for finite element modeling of transducers along with equivalent circuit and ABCD parameter models based on FEA models.

Although not presented here, an alternative energy approach may be used to develop models for transducers. Aronov has presented this energy method along with examples of its use in modeling a pulsating sphere and a multimodal cylinder [7].

3.1 Lumped-Parameter Models and Equivalent Circuits

3.1.1 *Mechanical Single Degree of Freedom Lumped Equivalent Circuits*

We begin with the simplest possible case of a single degree of freedom mechanical resonator with a spring and resistance (dashpot) attached to a rigid boundary at one end with a mass attached at the other end. This is called a lumped-parameter model since it is assumed that the physical elements are each smaller than about one quarter of a wavelength, that the mass is perfectly rigid and undergoes no compression or bending, and that the spring has stiffness but no mass. The dynamic situation is illustrated in Fig. 3.1 with a force, F, proportional to voltage or current,

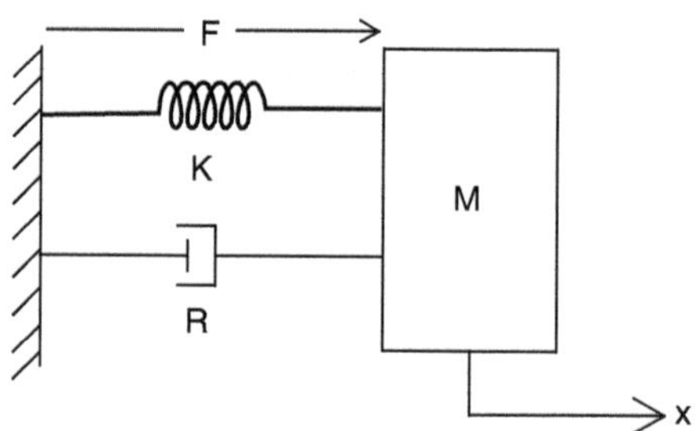

Fig. 3.1 Lumped-parameter vibrator with mass M, spring of stiffness K, mechanical resistance R, and driving force F

a mass, M, a spring of stiffness $K = 1/C$ (where C is the compliance) and a mechanical resistance, R. (See Sect. 13.12 for more information regarding these components.) The displacement of the mass is x, and the velocity is $u = \mathrm{d}x/\mathrm{d}t$.

We can develop an electrical equivalent circuit for the simple mechanical case of Fig. 3.1 from the mechanical equation of motion. In a typical idealized voltage driven transducer, such as a piezoelectric transducer, the force $F = N_\mathrm{v}V$ where V is the voltage, N_v is the transduction coefficient or ideal electromechanicaltransformer ratio, the spring is the piezoelectric material, and the mass is the radiating piston. In a typical idealized current driven transducer, such as a magnetostrictive transducer, the force $F = N_\mathrm{I}I$ where I is the current, N_I is an ideal electromechanical transformer, and the spring is the magnetostrictive material.

The equation of motion for the model of Fig. 3.1 may be written in terms of the velocity as

$$M\mathrm{d}u/\mathrm{d}t + Ru + (1/C)\int u\,\mathrm{d}t = F. \tag{3.1a}$$

Under sinusoidal drive where $F = F_0 e^{j\omega t}$, the velocity has the form $u = u_0 e^{j\omega t}$ where ω is the angular frequency of vibration and Eq. (3.1a) becomes

$$j\omega M u_0 + Ru_0 + u_0/j\omega C = F_0. \tag{3.1b}$$

The solution for the velocity of the mass may then be written as

$$u_0 = F_0/Z, \tag{3.2}$$

where the mechanical impedance is $Z = R + j(\omega M - K/\omega) = R + jX$. (See Sect. 13.17 for more information on complex algebra.) At resonance the reactance, X, vanishes and the angular mechanical resonance frequency is given by $\omega_\mathrm{r} = (K/M)^{1/2} = (1/MC)^{1/2}$ (see Sects. 2.8.2 and 4.1).

The response characteristics of Eq. (3.2) for a given force, F, may be appreciated by referring to the curves of Fig. 3.2.

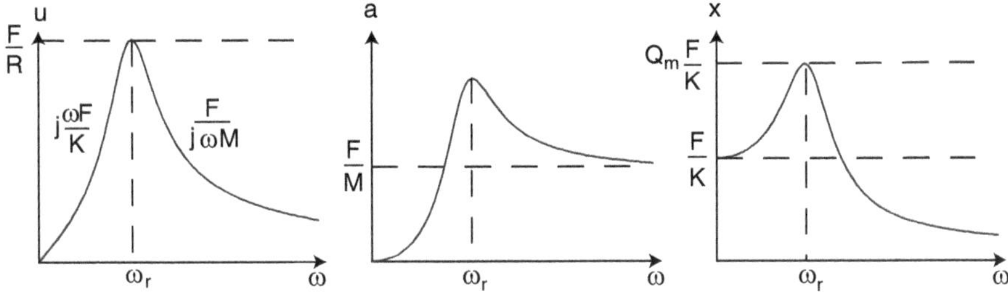

Fig. 3.2 Velocity, u, acceleration, a, and displacement, x, response for the lumped-parameter vibrator of Fig. 3.1

At resonance the velocity, u, is F/R. Well above resonance the velocity is $F/j\omega M$, decreasing with frequency, while the acceleration, a, is the constant, F/M. Well below resonance the velocity is $j\omega F/K$ increasing with frequency while the displacement, x, is the constant, F/K. At resonance the low frequency displacement is magnified by the quantity $Q_m = \omega_r M/R$, which is usually called the mechanical quality factor but also referred to as the mechanical storage factor. The quantity Q_m is an important transducer parameter because it is a measure of both the displacement amplification at resonance and the bandwidth (see Sects. 2.8.3 and 4.2).

The equation of motion, Eq. (3.1a), for the velocity has the same form as a series electrical circuit equation for the current, I, with a resistor, R_e, inductor, L_e, and capacitor, C_e, driven with voltage V as shown in Fig. 3.3. The corresponding circuit equation for the voltage drops around the circuit is

$$L_e \mathrm{d}I/\mathrm{d}t + R_e I + 1/C_e \int I \, \mathrm{d}t = V. \tag{3.3}$$

Equations (3.1a) and (3.3) are analogous if we make the analogies of force to voltage (F:V), velocity to current (u:I), compliance to capacitance (C:C_e), mass to inductance (M:L_e), and resistance to resistor (R:R_e). Thus, we may represent Eq. (3.1a) by the electrical equivalent circuit of Fig. 3.4, and then may use electrical circuit theorems to solve mechanical problems.

We note that the electrical and mechanical power relations are also similar: the input powers are VI and Fu, and the time average output or dissipative powers are $I^2_{rms} R_e$ and $u^2_{rms} R$. The electrical impedance is $Z_e = V/I$ and the analogous mechanical impedance is $Z = F/u$. Further similarities include the angular electrical resonance $\omega_e = 1/(L_e C_e)^{1/2}$, analogous to $\omega_r = 1/(MC)^{1/2}$, as well as the electrical

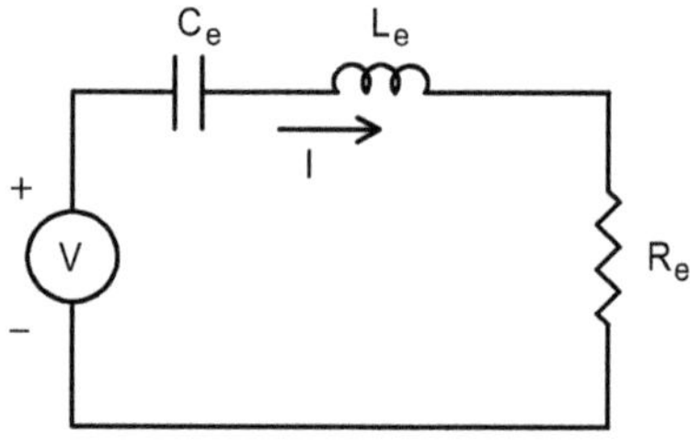

Fig. 3.3 Electrical circuit for series inductor, L_e, capacitor, C_e, and resistor, R_e, and voltage, V

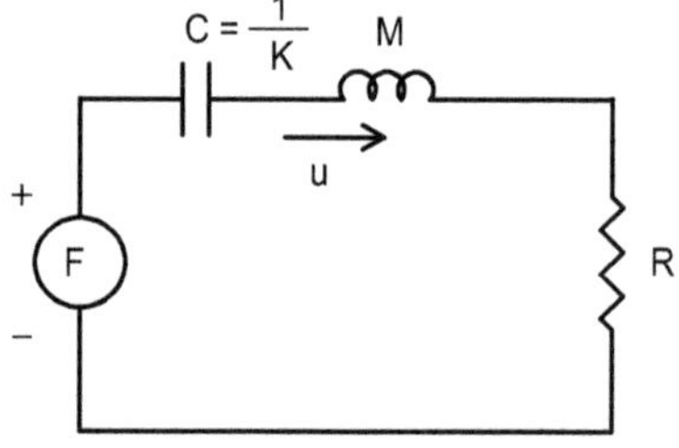

Fig. 3.4 Electrical equivalent circuit of lumped-parameter vibrator shown in Fig. 3.1 with velocity $u = j\omega x$

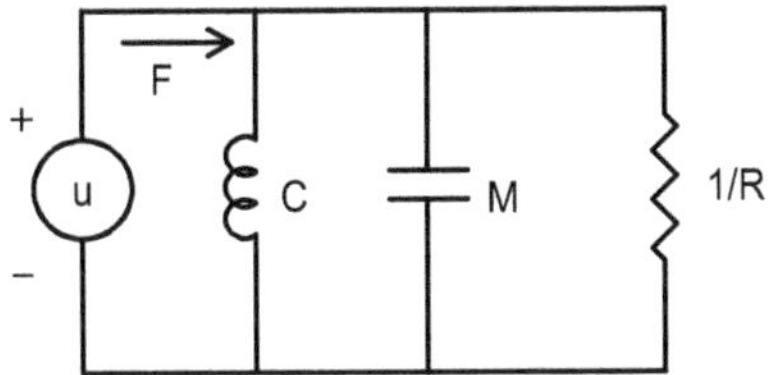

Fig. 3.5 Mobility representation of Fig. 3.1

quality factor $Q_e = \omega_e L_e / R_e$, analogous to $Q_m = \omega_r M/R$. The electrical equivalent representation is particularly useful since a transducer is part mechanical and part electrical, and other electrical systems are connected to the transducer through its electrical port.

The circuit of Fig. 3.4 is a representation of the impedance analogy where F:V and u:I. Further inspection of Eq. (3.1b) reveals another interpretation called the mobility analogy where F:I and u:V. In this case the three terms on the left of Eq. (3.1b) act as currents which when summed equal the total current, represented by F. The summing of three branch currents is indicative of three electrical elements in parallel. The current through a capacitor is $j\omega C_e V$, the current through an inductor is $V/j\omega L$, and the current through a resistor is V/R. Equation (3.1b) shows that the mass, M, acts as a shunt capacitor, the compliance, C, acts as a shunt inductor, and the resistance R acts as a shunt conductance. Thus, we may also represent the mechanical system of Fig. 3.1 by the mobility equivalent circuit of Fig. 3.5.

The circuit in Fig. 3.5 is called the dual of the circuit in Fig. 3.4. It may also be obtained by a topological transformation where series elements are replaced by parallel elements, capacitors by inductors, inductors by capacitors, resistors by conductances, voltage by current, and current by voltage. The mobility representation is natural and convenient for magnetic field transducers where the force is derived from a current instead of a voltage.

The model of Fig. 3.1 and corresponding equivalent circuit of Fig. 3.4 is the simplest form of a vibrating system; it is also directly applicable to vibrating systems with symmetry, such as a ring, or systems with a single location of no motion or node.

3.1.2 Mechanical Lumped Equivalent Circuits for Higher Degrees of Freedom

The model of Fig. 3.1 and the circuit of Fig. 3.4 are not realistic unless the spring, K, is very soft, which allows the assumption of a rigid boundary or "rigid wall," where the displacement is zero, to be realistically implemented. Piezoelectric materials are, on the contrary, quite stiff and a rigid wall is not easily implemented. A more realistic model is shown in Fig. 3.6 where the rigid wall is replaced by a mass M_1 moving with displacement x_1 and a load represented by a resistance R_1.

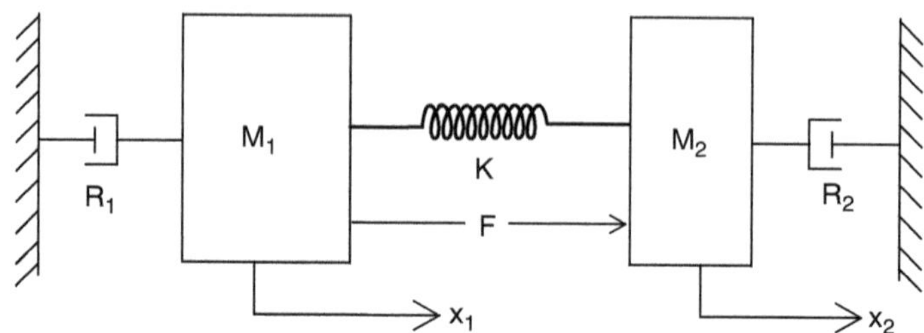

Fig. 3.6 Dual mass two-degree of freedom vibrator

The original mass is now represented by mass M_2, moving with displacement x_2 and loaded by resistance R_2. Both masses are connected by the spring of stiffness K and driven by a force $F = N_v V$ applied between the two masses. If the acoustic radiation is from M_2, this mass will be considered the piston radiator, or "head" mass, and the mass M_1 will be considered the so-called "tail" or inertial reaction mass, which, if massive enough, approaches a rigid wall condition.

As the force pushes the two masses, they are displaced by the amounts x_1 and x_2 and the spring is stretched by the amount x_2–x_1. Thus, the equations of motion of the two masses are

$$M_2 \mathrm{d}^2 x_2/\mathrm{d}t^2 = F - K(x_2 - x_1) - R_2 \mathrm{d}x_2/\mathrm{d}t, \tag{3.4a}$$

$$M_1 \mathrm{d}^2 x_1/\mathrm{d}t^2 = -F - K(x_1 - x_2) - R_1 \mathrm{d}x_1/\mathrm{d}t. \tag{3.4b}$$

Note that the stiffness term, representing the spring connected to both masses, couples the two equations of motion. For a sinusoidal force these equations can be written as

$$R_2 u_2 + j\omega M_2 u_2 = F - (K/j\omega)(u_2 - u_1), \tag{3.5a}$$

$$-R_1 u_1 - j\omega M_1 u_1 = F - (K/j\omega)(u_2 - u_1). \tag{3.5b}$$

We see that the right-hand sides of the equations are the same, and equal to the driving force reduced by the force drop across the spring. Use of Kirchhoff's rules shows that the equivalent circuit in Fig. 3.7 is consistent with Eqs. (3.5a) and (3.5b). The difference, u_2–u_1, is the relative velocity of the two masses.

When the tail mass M_1 is vibrating in air the loss resistance R_1 is negligible, and the equivalent circuit simplifies to the form in Fig. 3.8. If R_2 is also small, M_1 and M_2 are in parallel and the effective series mass is $M^* = M_1 M_2/(M_2 + M_1)$ leading to the resonant frequency

$$\omega_r = \left(K/M^*\right)^{1/2} = (K/M_2)^{1/2}(1 + M_2/M_1)^{1/2}.$$

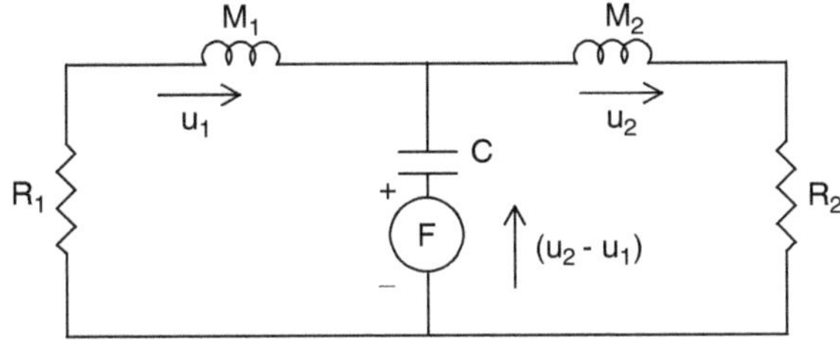

Fig. 3.7 Electrical equivalent circuit of dual mass vibrator shown in Fig. 3.6

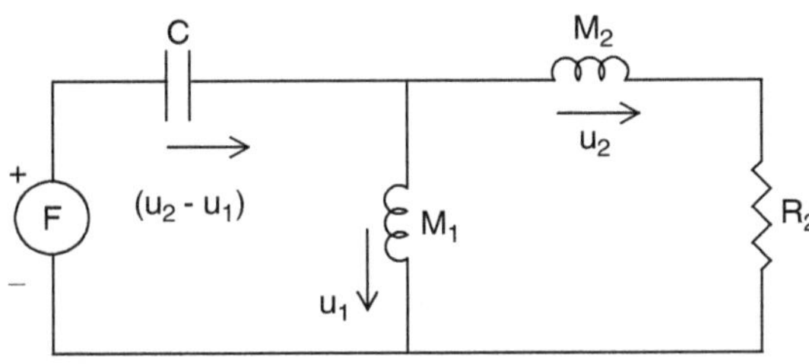

Fig. 3.8 Equivalent circuit of Fig. 3.7 with resistor $R_1 = 0$

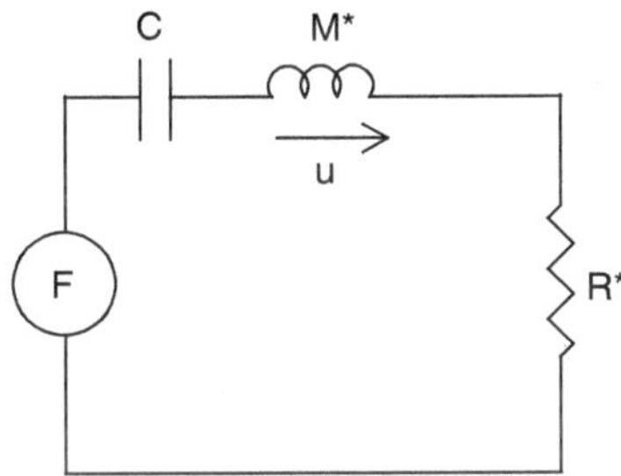

Fig. 3.9 Reduced mass representation of Fig. 3.8

This result can also be obtained by solving Eqs. (3.5a) and (3.5b) with $R_1 = R_2 = 0$. The solution in which the two masses vibrate at the same frequency is a normal mode which can be described by a single degree of freedom with effective mass M^*. It can be seen that if $M_1 \gg M_2$, this resonant frequency approaches the resonant frequency for a single mass spring system mounted on a rigid wall as in Fig. 3.1. As the tail mass M_1 is made smaller, the resonant frequency increases, and when $M_1 = M_2$ the resonance is 41 % higher than the rigid wall case.

The mechanical Q_m of the two mass vibrating system may be found by converting the parallel combination of masses and resistance in Fig. 3.8 to the series form in Fig. 3.9.

For the usual condition of $R_2 \ll \omega(M_1 + M_2)$, the effective series resistance $R^* \approx R_2/(1 + M_2/M_1)^2$. Then, using the effective mass M^* given above,

$$Q_m = \omega_r M^*/R^* = (\omega_r M_2/R_2)(1 + M_2/M_1) = \left[(KM_2)^{1/2}/R_2\right](1 + M_2/M_1)^{3/2}.$$

This result for Q_m applies because the system vibrates in a normal mode with a single degree of freedom. The term in brackets is the mechanical Q_m for the simpler case of a rigid wall replacing the tail mass M_1.

For example: For $M_1 = M_2$, Q_m is 100 % greater than it would be for the rigid wall case where $M_1 \gg M_2$. For $M_1 = 2M_2$ the increase is 50 % while for $M_1 = 3M_2$ the increase is 33 %. This latter case is often used in practical Tonpilz designs.

The mechanical model shown in Fig. 3.6 and its equivalent circuit of Fig. 3.7 may be extended to other one-dimensional vibrators by considering Fig. 3.7 without the loading resistors as shown in Fig. 3.10. We can then represent a series of masses and springs by a cascade of mass and stiffness elements as in Figs. 3.11 and 3.12 which corresponds to a driven resonator composed of elements M_1, K_1 and M_2 driving a second resonator composed of a spring of stiffness K_2 and mass, M_3, into a resistive load, R_3. Such an interface between the transducer and the load is used as a matching device or as a means of amplifying the velocity of the transducer, u_2, at an angular parallel resonant frequency given by $(K_2/M_3)^{1/2}$. A continuous distribution of these elements may also be used to represent a distributed system that supports standing waves.

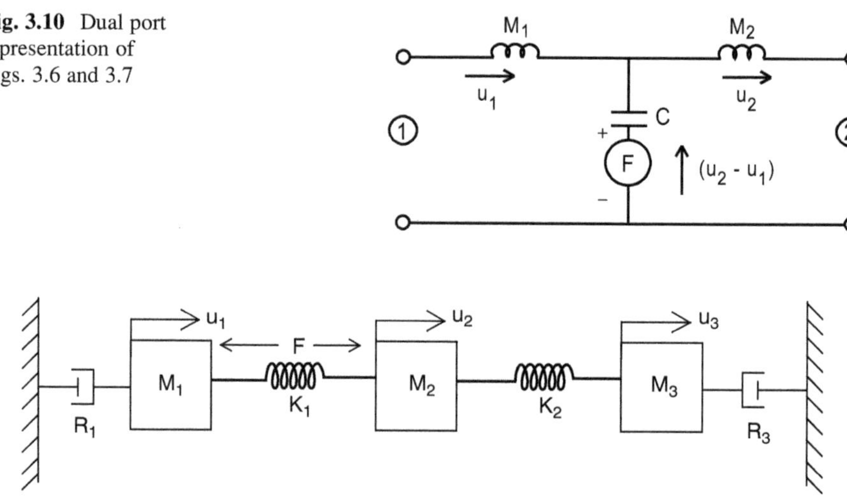

Fig. 3.10 Dual port representation of Figs. 3.6 and 3.7

Fig. 3.11 Three degree of freedom mechanical vibrator

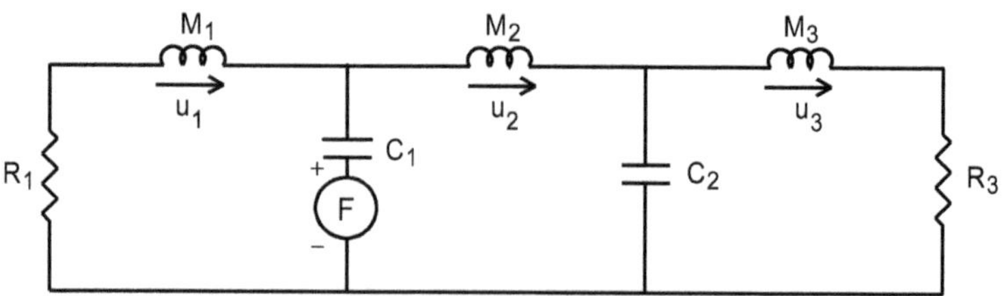

Fig. 3.12 Electrical equivalent circuit of Fig. 3.11

3.1.3 *Piezoelectric Ceramic Lumped-Parameter Equivalent Circuit*

As discussed in Chap. 2 the term piezoelectric ceramic applies to electrostrictive ceramic materials, such as lead zirconate titanate (PZT), that have been biased or permanently polarized for linear operation. The specific case of piezoelectric ceramic transduction will be further developed from Eqs. (2.6d) and (2.6e) which are repeated here:

$$S_3 = s_{33}^E T_3 + d_{33} E_3, \tag{3.6a}$$

$$D_3 = d_{33} T_3 + \varepsilon_{33}^T E_3. \tag{3.6b}$$

These equations are based on an ideal one-dimensional case with both motion and electric field in the direction of polarization for operation in the so-called 33 mode of vibration illustrated in Fig. 3.13. (The equations also apply to high coercive single crystal materials such as PIN-PMN-PT.)

It is also assumed that the lateral dimensions of the ceramic bar in Fig. 3.13 are very small compared to the wavelength of longitudinal waves in the material and that there are no loads on the sides of the bar making the stresses T_1 and T_2 essentially zero. The electric field components E_1 and E_2, which are zero on the electrodes, are also assumed to be zero throughout the bar. If the bar is short compared to a quarter wavelength, motion does not modify the electric field, and E_3 is approximately constant along the length of the bar.

As pointed out in Chap. 2 the physical meaning of the coefficients in Eqs. (3.6a) and (3.6b) can best be seen from their definitions as partial derivatives. For example, the short circuit elastic modulus $s^E = \partial S/\partial T|_E$ and the free dielectric

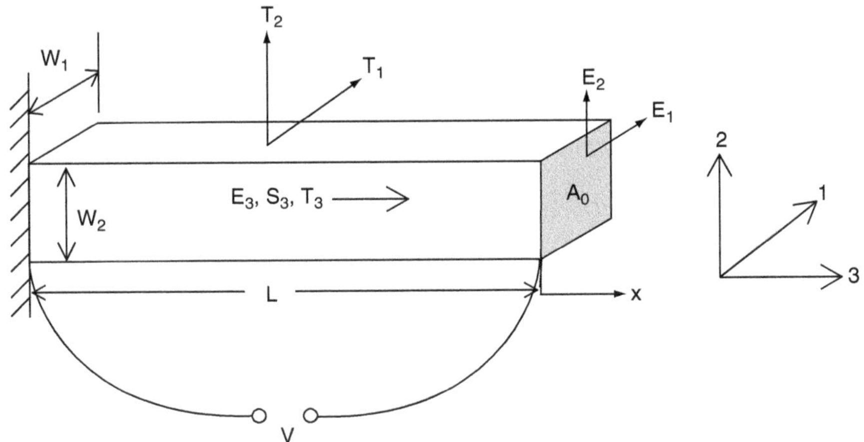

Fig. 3.13 End electroded piezoelectric ceramic bar rigidly mounted on one end

constant $\varepsilon^T = \partial D/\partial E|_T$ are the slopes of S vs. T and D vs. E curves measured with constant field and constant stress, respectively. We also see that the piezoelectric charge coefficient, d, usually called the "d" constant, may be evaluated as $\partial D/\partial T|_E$ or more commonly as $\partial S/\partial E|_T$ from the slope of the S vs. E curve with constant stress.

If Eq. (3.6a) is divided by s_{33}^E, Eq. (3.6b) is divided by ε_{33}^T, and T_3 is eliminated in Eq. (3.6b), Eqs. (3.6a) and (3.6b) may be rewritten as

$$S_3 = s_{33}^D T_3 + g_{33} D_3, \tag{3.7a}$$

$$D_3 = e_{33} S_3 + \varepsilon_{33}^S E_3, \tag{3.7b}$$

revealing two additional piezoelectric constants $g_{33} = d_{33}/\varepsilon_{33}^T$ and $e_{33} = d_{33}/s_{33}^E$ and the open circuit $(D = 0)$ elastic modulus, s_{33}^D and clamped $(S = 0)$ dielectric constant ε_{33}^S given by

$$s_{33}^D = s_{33}^E\left(1 - k_{33}^2\right),$$

and

$$\varepsilon_{33}^S = \varepsilon_{33}^T\left(1 - k_{33}^2\right).$$

As discussed in Chap. 2, $k_{33}^2 = d_{33}^2/s_{33}^E \varepsilon_{33}^T$ is the electromechanical coupling coefficient for this case. It is a measure of the electrical or mechanical energy converted by the transducer relative to the total electrical and mechanical energy stored in the transducer with values ranging from 0 to less than 1. (Other definitions and interpretations of the coupling coefficient are discussed in Sect. 4.4.)

These relations show that the open circuit elastic compliance, s_{33}^D, is less than the short circuit elastic compliance, s_{33}^E, and that the clamped dielectric constant, ε_{33}^S, is less than the free dielectric constant, ε_{33}^T. If k_{33} is large there will be a large change in the elastic modulus as the electrical boundary conditions are changed from short to open. Likewise there will be a large change in the dielectric constants as the mechanical boundary conditions are changed from free to clamped.

Equations (3.6a) and (3.7b) may be rewritten in terms of the variables force, F, displacement, x, charge, Q, and voltage, V, by referring to Fig. 3.13 which shows an idealized piezoelectric ceramic bar of area, A_0, and length, L, with a voltage V across the length. Thus we have $x = SL$, $F = TA_0$, $Q = DA_0$, $V = EL$ and Eqs. (3.6a) and (3.7b) become

$$x = C^E F + C^E N V, \tag{3.8}$$

$$Q = C_0 V + N x, \tag{3.9a}$$

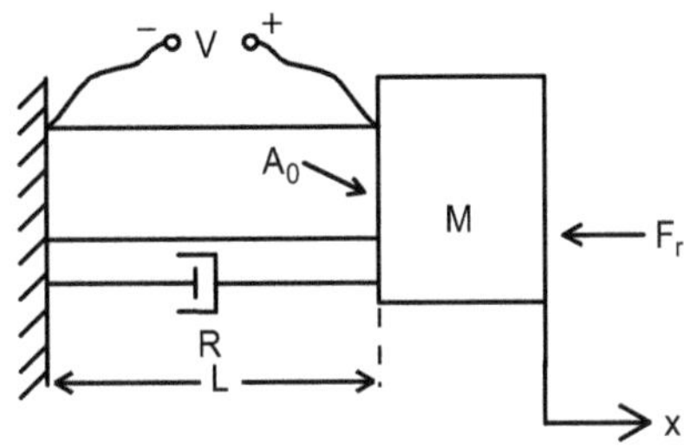

Fig. 3.14 One-degree of freedom mass, M, driven by a piezoelectric ceramic bar of cross section area A_0 and length L with voltage V applied

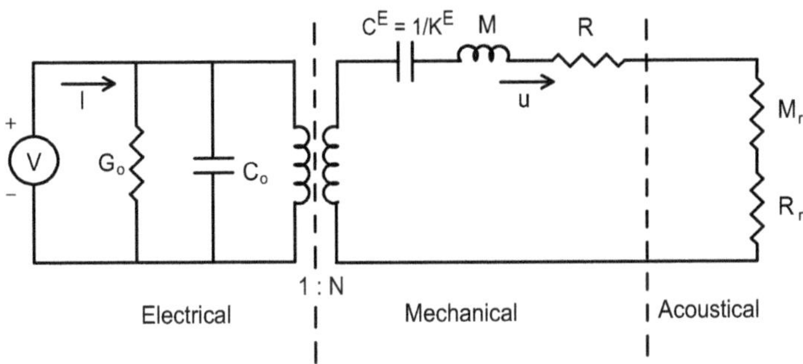

Fig. 3.15 Electrical equivalent circuit of the single degree of freedom piezoelectric vibrator shown in Fig. 3.14

where the short circuit compliance $C^E = s_{33}^E L/A_0$, the clamped (or "blocked") capacitance $C_0 = \varepsilon_{33}^S A_0/L$, and the electromechanical turns ratio $N = d_{33}A_0/s_{33}^E L$.

Consider now Fig. 3.14 where we have added to the bar of Fig. 3.13 a resistance R, a radiating piston of mass M, and a radiation reactive loading force F_r. At very low frequencies where the acceleration and velocity are small, the reactive loads are small and the piezoelectric ceramic bar is free to move. In this case $F = 0$, Eq. (3.8) reduces to $x = C^E NV$ and Eq. (3.9a) becomes

$$Q = \left(C_0 + N^2 C^E\right)V = C_f V. \tag{3.9b}$$

where N^2C^E is the motional capacitance and C_f is the free capacitance, i.e., the capacitance that would be measured at very low frequencies.

If we now let F in Eq. (3.8) include the mechanical and radiation loads, M_r and R_r, on the piston, as discussed in Chaps. 1 and 2 and shown in Fig. 3.15, Eq. (3.8) becomes

$$(M + M_r)\mathrm{d}u/\mathrm{d}t + (R + R_r)u + \left(1/C^E\right)\int u\,\mathrm{d}t = NV. \tag{3.10}$$

Equation (3.10) has the same form as Eq. (3.1a), previously used to develop the equivalent circuit analogy. Now, however, we have a force equal to NV with

$N = d_{33}A_0/s_{33}^E L$ for the specific case of piezoelectric ceramic transduction. If the voltage V is sinusoidal, the solution for the velocity is

$$u = NV/Z. \tag{3.11}$$

where the mechanical impedance is $Z = R + R_r + j[\omega(M + M_r) - 1/\omega C^E]$.

The input electrical admittance, Y, may now be obtained by writing Eq. (3.9a) in terms of the current I,

$$I = \mathrm{d}Q/\mathrm{d}t = C_0\mathrm{d}V/\mathrm{d}t + Nu. \tag{3.12a}$$

Under sinusoidal conditions, and with electrical dissipation, G_0, added, Eq. (3.12a) becomes

$$I = G_0V + j\omega C_0V + Nu, \tag{3.12b}$$

where $G_0 = \omega C_f \tan\delta$ is the electrical loss conductance and $\tan\delta$ is the electrical dissipation factor for the piezoelectric ceramic material (typically 0.004–0.02 under low field, see Sect. 13.5). On using Eq. (3.11) in Eq. (3.12b) the electrical admittance, I/V, can be written as

$$Y = G_0 + j\omega C_0 + N^2/Z, \tag{3.13}$$

where $G_0 + j\omega C_0$ is the clamped admittance and N^2/Z is the motional admittance. Equations (3.11) and (3.13) for the velocity u and the input electrical admittance are the results needed for constructing the equivalent circuit of Fig. 3.15 which is consistent with these equations and includes the essential parts of a piezoelectric ceramic transducer. One could use either the equation pair, Eqs. (3.11) and (3.13), or the equivalent circuit of Fig. 3.15 for transducer analysis.

Although it is somewhat arbitrary to divide an electroacoustic transducer into separate electrical and mechanical parts, Fig. 3.15 illustrates one common division. The radiation into the medium takes place in the acoustical section representing the mechanical radiation impedance. The section to the left of the transformer is electrical; it represents the clamped electrical admittance under conditions where $u = 0$. The section to the right of the transformer is called the motional part of the circuit; it has an associated motional impedance or admittance. Alternatively, the circuit representation may be considered as a "Van Dyke" electrical circuit [8] in which the motional elements are converted to electrical elements (with subscripts e) through the ideal transformer of turns ratio N as shown in Fig. 3.16. We will show in Chap. 9 how the electrical elements may be evaluated.

The rigid wall in Fig. 3.14 may be replaced by a more realistic tail mass as shown in Fig. 3.17, with the equivalent circuit in Fig. 3.18, in the same way as we replaced Figs. 3.1 and 3.4 by Figs. 3.6 and 3.8. The circuit shown in Fig. 3.18 adds the tail mass, M_1, as a shunt element and displays a relative velocity u–u_1 through the compliance, C^E. It can be seen that in this representation the multiple mass/resistive

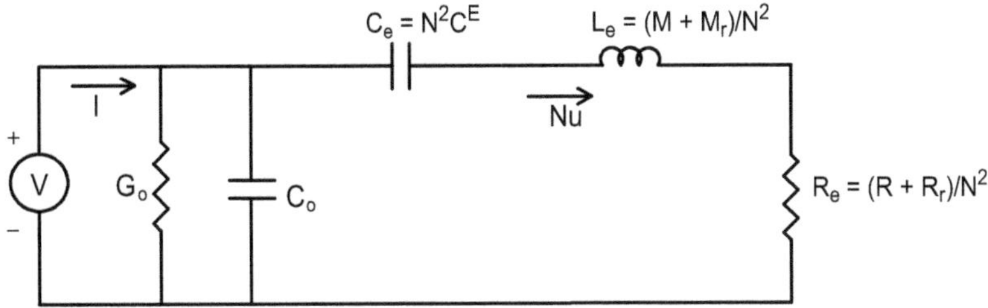

Fig. 3.16 All electric circuit representation of the circuit of Fig. 3.15

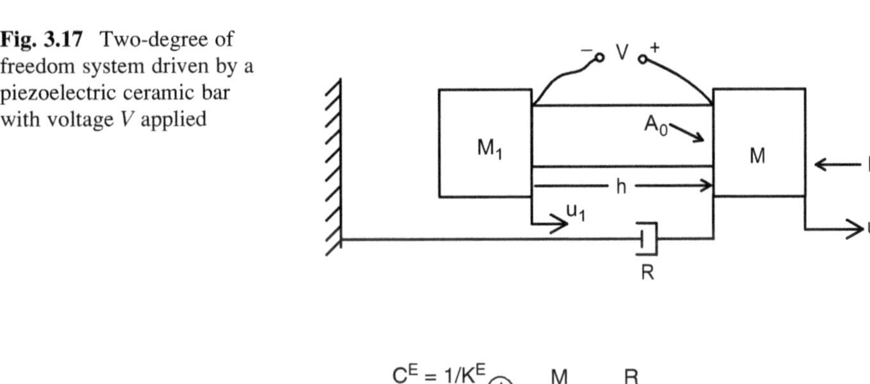

Fig. 3.17 Two-degree of freedom system driven by a piezoelectric ceramic bar with voltage V applied

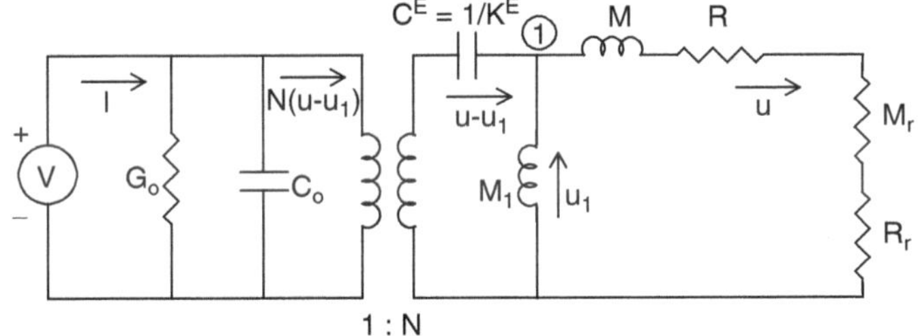

Fig. 3.18 Electrical equivalent circuit of the two-degree of freedom piezoelectric vibrator shown in Fig. 3.17

circuit to the right of terminal 1 is the load on the piezoelectric ceramic which is located to the left of terminal 1. The tail mass M_1 reduces the velocity of the head mass, M, unless M_1 is much larger than M. Typical ratios of M_1 to M are approximately 3. With R, R_r and M_r small, the force is nearly the same on both masses, $j\omega M_1 u_1 \approx j\omega M u$, and the tail velocity relative to the head velocity is M/M_1; e.g., for $M_1 = 3M$ the head velocity is three times the tail velocity.

The above analysis may be applied to other electric field transducers which are operated in a region where equations of the form of Eqs. (3.11) and (3.13) apply. The equations for magnetostrictive transduction are similar in form but different enough in physical basis to warrant a separate development as follows.

3.1.4 Magnetostrictive Lumped-Parameter Equivalent Circuit

The large strain capabilities of the highly active rare earth magnetostrictive transduction material Terfenol-D [9, 10] and the high strength magnetostrictive material Galfenol [11] (see Sect. 13.7) brought about a resurgence of interest in the design and use of magnetostrictive transducers and a need for practical transducer models. The equivalent circuit for a biased magnetostrictive transducer may be developed in a manner similar to that for a piezoelectric transducer except in this case the driving force on the mechanical system is proportional to the current rather than the voltage [12–15]. This difference yields an equivalent circuit that is the dual of the piezoelectric circuit. We start with the magnetostrictive equations introduced in Sect. 2.3 and assume one-dimensional motion in the 3 direction with $T_1 = T_2 = 0$ and $H_1 = H_2 = 0$. Equations (2.37b) and (2.37c) are repeated here:

$$S_3 = s_{33}^H T_3 + d_{33} H_3, \tag{3.14}$$

$$B_3 = d_{33} T_3 + \mu_{33}^T H_3, \tag{3.15}$$

where S_3 is the strain, B_3 is the flux density, T_3 is the mechanical stress, H_3 is the magnetic field intensity, s_{33}^H is the elastic compliance under open circuit conditions ($H_3 = 0$), μ_{33}^T is the permeability under free conditions ($T_3 = 0$), and the magnetostrictive "d_{33}" constant, $d_{33} = \partial S_3/\partial H_3|_T$. As shown in Sect. 2.3, the coupling coefficient squared is $k_{33}^2 = d_{33}^2/s_{33}^H \mu_{33}^T$. Equations (3.14) and (3.15) may be rewritten as

$$S_3 = g_{33} B_3 + s_{33}^B T_3, \tag{3.16}$$

$$B_3 = e_{33} S_3 + \mu_{33}^S H_3, \tag{3.17}$$

where $g_{33} = d_{33}/\mu_{33}^T$ and $e_{33} = d_{33}/s_{33}^H$. The short circuit ($B_3 = 0$) elastic compliance, s_{33}^B, and clamped ($S_3 = 0$) permeability μ_{33}^S are given by

$$s_{33}^B = s_{33}^H (1 - k_{33}^2),$$

$$\mu_{33}^S = \mu_{33}^T (1 - k_{33}^2),$$

showing that in the magnetostrictive case the open circuit elastic modulus s_{33}^H is greater than the short circuit elastic modulus s_{33}^B, making the open circuit resonance frequency lower than the short circuit resonance frequency, in contrast to the piezoelectric case. Also note that the clamped permeability μ_{33}^S is less than the free permeability μ_{33}^T, as is the permittivity in the piezoelectric case.

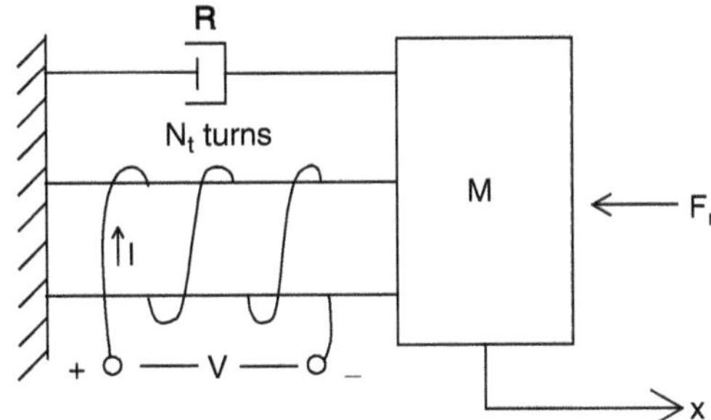

Fig. 3.19 One-degree of freedom lumped magnetostrictive model

The model is illustrated in Fig. 3.19 with a magnetostrictive rod of length, L, and cross-sectional area, A_0, with I the current through the coil of N_t turns. A closed magnetic return path is assumed, but not shown, in Fig. 3.19. Such return paths are an important part of magnetostrictive transducers and are discussed in Sect. 13.9 and Chap. 5. Ampere's circuital law gives $H = IN_t/L$ and Eq. (3.14) may be written as

$$x = C^H F + C^H N_m I, \tag{3.18}$$

where the load $F = A_0 T_3$, $C^H = s_{33}^H L/A_0$ and the magnetostrictive turns ratio $N_m = N_t d_{33} A_0/L s_{33}^H$. Note that the electromechanical ratio, N_m, is analogous to the piezoelectric electromechanical ratio $N = d_{33} A_0/L s_{33}^E$. We will show in Sect. 3.2.3 that if the piezoelectric bar of length L is divided into n equal segments, each of thickness h, then $N = n d_{33} A_0/L s_{33}^E = d_{33} A_0/h s_{33}^E$ and the analogy to the magnetostrictive case is even stronger. Inclusion of the reactive and resistive forces and the radiation load into the load F, as in the piezoelectric case, yields

$$(M + M_r)\mathrm{d}u/\mathrm{d}t + (R + R_r)u + \left(1/C^H\right)\int u\,\mathrm{d}t = N_m I. \tag{3.19}$$

Under sinusoidal conditions the velocity response, for a given current, I, is given by

$$u = N_m I/Z, \tag{3.20}$$

where $Z = R + R_r + j\left[\omega(M + M_r) - 1/\omega C^H\right]$. A comparison with the corresponding piezoelectric response of Eq. (3.11) shows that the voltage V has been replaced by the current I, the electromechanical turns ratio is now, N_m, and the mechanical impedance contains the open circuit compliance C^H instead of the short circuit compliance C^E. Equation (3.17) may be used to obtain the input electrical impedance with the help of Faraday's law where the voltage across the coil of N_t turns is

$$V = N_t A_0 (\mathrm{d}B/\mathrm{d}t) = \left(d_{33} N_t A_0 / L s_{33}^H\right) u + \left(\mu_{33}^S N_t^2 A_0 / L\right) \mathrm{d}I/\mathrm{d}t. \tag{3.21}$$

With $N_m = N_t d_{33} A_0 / L s_{33}^H$ and the clamped inductance $L_0 = \mu_{33}^S N_t^2 A_0 / L$, Eq. (3.21) becomes

$$V = L_0 \mathrm{d}I/\mathrm{d}t + N_m u. \tag{3.22}$$

Under sinusoidal conditions and with u given by Eq. (3.20), Eq. (3.22) yields the electrical impedance

$$Z_e = V/I = R_e + j\omega L_0 + N_m^2/Z, \tag{3.23}$$

where we have added a resistor, R_e, to account for resistive losses in the coil. Equations (3.20) and (3.23) are sufficient to describe the general behavior of a magnetostrictive transducer if we ignore, for the moment, eddy current and hysteresis losses.

Comparison of Eqs. (3.20) and (3.23) with their piezoelectric counterparts, Eqs. (3.11) and (3.13), shows the strong similarities in form if we replace V with I, C_0 with L_0, Y with Z_e, and N with N_m. Inspection of Eqs. (3.20) and (3.23) leads directly to the equivalent circuit with a mechanical mobility representation shown in Fig. 3.20 where χ is the eddy current factor (not included in the equations but defined below). In mobility representations force is analogous to current and velocity is analogous to voltage. The mobility representation is awkward especially if piezoelectric and magnetostrictive models are combined (see Hybrid Transducer, Chap. 5, Sect. 5.3.2) or if array interaction impedance models are used for the radiation load. An alternative representation may be developed through additional inspection of Eqs. (3.20) and (3.23) or through the dual of Fig. 3.20 which leads to the equivalent circuit of Fig. 3.21. This circuit has a mechanical impedance representation consistent with Fig. 3.18, but now a dual representation of the electrical section; that is, voltage, V, is now interpreted as the "flow" or "through" quantity and current, I, as the "potential" or "across "quantity.

An additional alternative representation is possible with the use of a gyrator [16], which converts an input voltage to an output current and an input current to an output voltage. In the electromechanical case it converts voltage to velocity and current to force. Thus replacing the transformer of Fig. 3.20 by a gyrator of transformation factor N_m (often referred to as γ) gives the representation of Fig. 3.22.

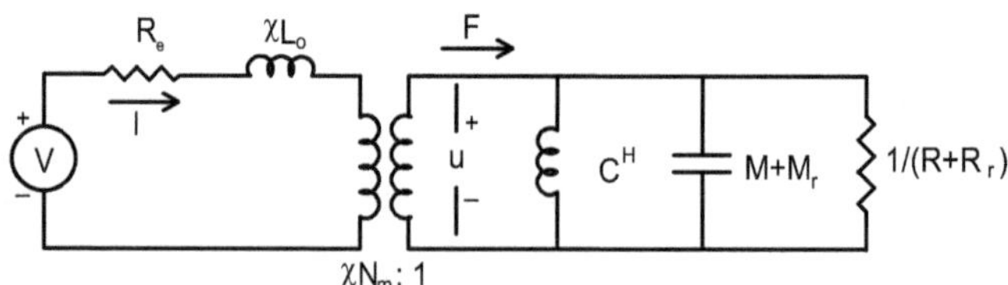

Fig. 3.20 Mobility equivalent circuit of Fig. 3.19

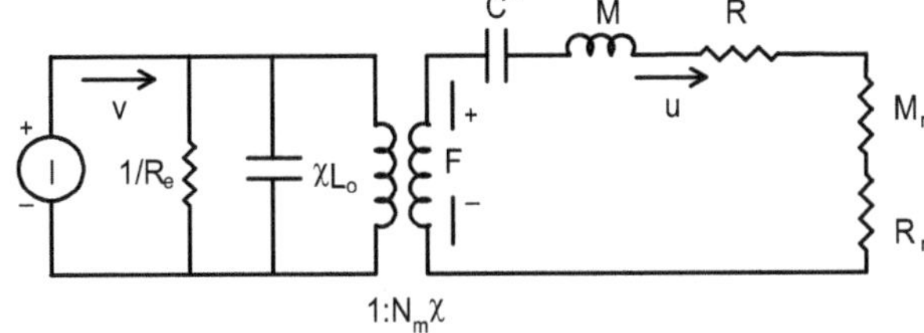

Fig. 3.21 Dual equivalent circuit representation of Fig. 3.20

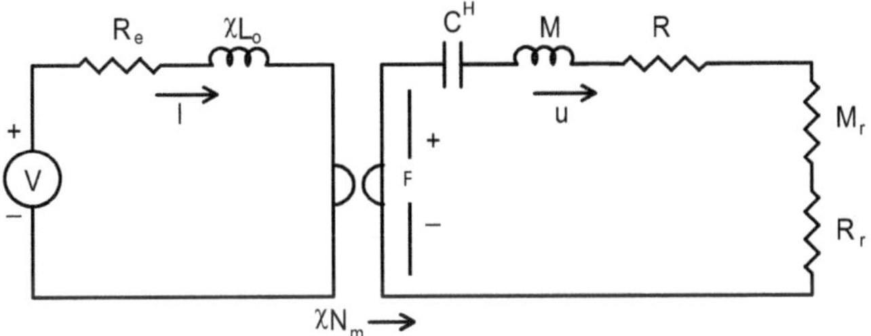

Fig. 3.22 Gyrator circuit representation of Fig. 3.19

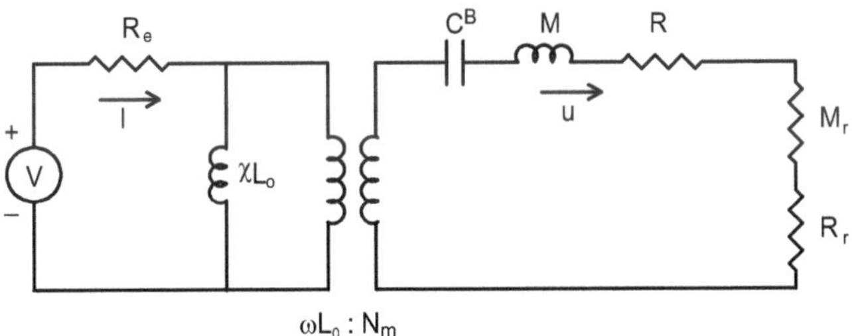

Fig. 3.23 Impedance circuit representation of Fig. 3.19

There is even a further representation, which includes a frequency-dependent electromechanical transformer, which produces the correct velocity magnitude but not the correct phase. This representation is shown in Fig. 3.23 where we note that the compliance is the short circuit value, C^B, rather than the open circuit value, C^H, and that the inductance, χL_0, is now in parallel, rather than in series, with R_e. Although the equivalent circuits of Figs. 3.20 and 3.23 appear dissimilar, they both give the same electrical input impedance. This may be shown from the input electrical impedance of the circuit in Fig. 3.23 written as

$$
\begin{aligned}
Z &= R_e + Z_0 Z_m^E / \left[Z_m^E + Z_0 (N_m/\omega L_0)^2 \right] \\
&= R_e + Z_0 - Z_0^2 (N_m/\omega L_0)^2 / \left[Z_m^E + Z_0 (N_m/\omega L_0)^2 \right]
\end{aligned}
$$

Substitution of $Z_0 = j\omega\chi L_0$ and $Z_m^E = R + j\omega M + 1/j\omega C^B$, and use of $k^2 = N_m^2 C^B/L_0$ leads to

$$Z = R_e + j\omega\chi L_0 + (\chi N_m)^2/\left[R + j\omega M + \left(1 - \chi k^2\right)/j\omega C^H\left(1 - k^2\right)\right].$$

For small eddy current effects, where $\chi = 1$, the impedance Z becomes

$$Z = R_e + j\omega\chi L_0 + (\chi N_m)^2/\left(R + j\omega M + 1/j\omega C^H\right),$$

which equals the input electrical impedance of the circuit in Fig. 3.20. Note that in Fig. 3.20 the clamped inductance, L_0, and turns ratio, N_m, are multiplied by the eddy current factor χ while in the case of Fig. 3.23 only the clamped inductance is multiplied by χ.

The circuits of Figs. 3.21, 3.22, and 3.23 have the advantage of a mechanical impedance branch (rather than the mechanical mobility branch in Fig. 3.20) which allows a direct connection to self and mutual mechanical radiation impedance models (see Chaps. 7 and 11). The circuit implementation disadvantages are that the circuit of Fig. 3.21 requires the current source to act like a zero impedance voltage source and the circuit of Fig. 3.22 requires a gyrator element instead of a transformer. While the circuit of Fig. 3.23 is readily implemented and used, there is a 90° phase error in the output. This is not a problem if only the pressure amplitude or intensity is desired. On the other hand, the circuit of Fig. 3.23 should not be combined with a piezoelectric transducer circuit where the phase relationship between the two is important, as in the hybrid transducer design, discussed in Sect. 5.3.2. The lack of a 90° phase shift associated with the circuit of Fig. 3.23 is related to the current-induced force of magnetic transduction and associated "burdens of antisymmetry" [3], which can limit circuit realization of the analytical representation.

3.1.5 Eddy Currents

Eddy currents circulating in magnetostrictive, and other magnetic, materials reduce the inductance and introduce a loss resistance. In the case of a rod with a magnetic field along its length, caused by a surrounding coil, a circulating current is set up in the rod in a direction opposite to the current in the coil. This induced current causes a power loss due to the electrical resistivity of the magnetostrictive material and also produces a magnetic field that cancels part of the magnetic field produced by the coil. Both these effects are included in the complex eddy current factor defined as

$$\chi = \chi_r - j\chi_i = |\chi|e^{-j\xi} = \left(R' + j\omega L_0'\right)/j\omega L_0,$$

where R' and L_0' are the added resistance and modified coil inductance caused by eddy currents.

For the case of a circular rod the expressions for χ may be written in terms of Kelvin functions [17] or in terms of the rapidly convergent series [18]

$$\chi_\mathrm{i} = (2/p)\left\{\sum \left[(p/4)^{2q}(2q)/(q!)^2(2q)!\right]\right\}/D_\mathrm{r}, \tag{3.24a}$$

$$\chi_\mathrm{r} = \left\{\sum \left[(p/4)^{2q}/(q!)^2(2q+1)\right]\right\}/D_\mathrm{r}, \tag{3.24b}$$

$$D_\mathrm{r} = \sum \left[(p/4)^{2q}/(q!)^2(2q)!\right],$$

where the sum is over q ranging from 0 to typically 15. The quantity $p = f/f_\mathrm{c}$ with f the drive frequency and f_c the characteristic frequency given by

$$f_\mathrm{c} = 2\rho_\mathrm{e}/\pi\mu^S D^2\ , \tag{3.24c}$$

where ρ_e is the resistivity of the magnetostrictive material and D is the diameter of the circular rod.

A high characteristic frequency, f_c, lowers the value of p and reduces the eddy current loss. The case of high resistivity and low permeability yields a high characteristic frequency and extends the frequency range of acceptable performance. Terfenol-D has this combination with a resistivity of $60 \times 10^{-8}\,\Omega$m and a low relative permeability of approximately 5; i.e., $\mu^S = 5\left(4\pi \times 10^{-7}\right)$ h/m. Significant eddy current effects can also lead to a lower value of the effective coupling coefficient for the transducer (see Sect. 4.4.2).

In the case where the eddy current loss is small and $f_\mathrm{c} \gg f$, Eqs. (3.24a) and (3.24b) may be approximated by the simpler expression $\chi = 1 - jf/8f_\mathrm{c}$. In this case the impedance associated with the clamped inductance is approximately [13]

$$j\omega L_0(1 - jf/8f_\mathrm{c}) = j\omega L_0 + \omega^2 L_0/16\pi f_\mathrm{c}.$$

At these low frequencies the eddy current loss may be represented by a single large resistor $R_\mathrm{s} = 16\pi f_\mathrm{c} L_0$ in shunt with the clamped inductor L_0 as illustrated in Fig. 3.24. This may be seen from the impedance of the circuit of Fig. 3.24 given by

$$j\omega L_0 R_\mathrm{s}/(R_\mathrm{s} + j\omega L_0) \approx j\omega L_0 + \omega^2 L_0^2/R_\mathrm{s} = j\omega L_0 + \omega^2 L_0/16\pi f_\mathrm{c}.$$

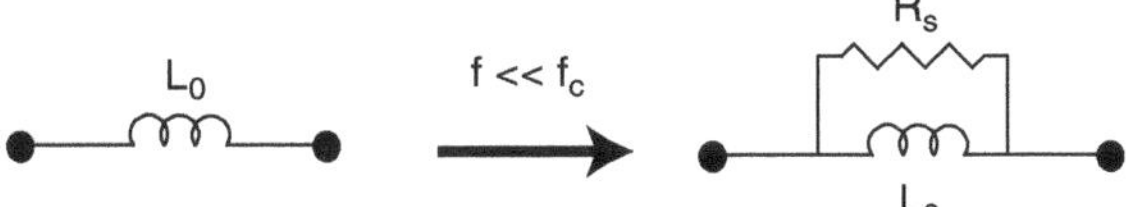

Fig. 3.24 Low frequency shunt resistor representation for eddy current losses

For the case of a rectangular magnetostrictive rod with laminations of thickness, t, the real and imaginary parts of the eddy current factor are [17]:

$$\chi_i = (\sinh\theta - \sin\theta)/D,$$

$$\chi_r = (\sinh\theta + \sin\theta)/D,$$

where $D = \theta(\cosh\theta + \cos\theta)$ and $\theta = t(\pi f \mu^S/\rho_e)^{1/2}$.

Another effect which introduces losses in magnetostrictive transduction is hysteresis which occurs in most ferromagnetic materials. As the magnetic field, H, increases the flux density, B, also increases. However, as H decreases the curve of B versus H differs from the curve for increasing H, resulting in loss of energy absorbed in the magnetic material. Hysteresis loss may be represented by a series resistor with value given by $\omega L_f \tan\delta$ where L_f is the free inductance and $\tan\delta$ is the hysteresis dissipation factor determined from the area of the magnetic hysteresis curve [17, 19]. Hysteresis, eddy current, and wire losses also occur in other magnetic field transducers.

3.2 Distributed Models

The lumped-parameter transducer models developed in Sect. 3.1 are useful for understanding, designing, and evaluating transducers where the identifiable moving parts are small compared to the wavelength in the material and may be considered to be represented by a mass, spring, or dashpot. These simple models may also be used to represent more complicated distributed models by determining equivalent lumped elements from kinetic and potential energy considerations (see Sects. 4.2 and 4.3). The lumped approach is very useful, but it has limited accuracy and does not predict higher order resonant modes. Moreover, with today's computational tools it is no more difficult to numerically evaluate a distributed model than a lumped model. In this section we will develop distributed mechanical and electromechanical models and corresponding equivalent circuits and matrix models. The term distributed system is used to emphasize that the mass and stiffness are distributed continuously throughout the structure, rather than being specified at a small number of points as in a lumped-parameter model. The distributed (sometimes called transmission line) models presented here are limited to one-dimensional longitudinal wave motion. In the case of bars we assume thin bars of cross-sectional area A_0 with, for example, circular, tubular, square, or rectangular cross sections.

The various types of flextensional and bender transducers discussed in Chaps. 5 and 6 are based on flexural wave motion which is more complicated to analyze than longitudinal wave motion. Therefore distributed models of flexural wave transducers are not developed here, but references to flexural wave transducer modeling are given in Sects. 5.5 and 5.6 where the results of such modeling are used.

3.2.1 Distributed Mechanical Model

We begin with a distributed representation of a simple bar of length L, cross section area A_0, and total mass M as illustrated in Fig. 3.25 with port 1 at one end and port 2 at the other end. If the bar is very small compared to the wavelength in the bar, it could be represented as a mass $M = \rho A_0 L$ or as a spring of stiffness $K = YA_0/L$ where ρ is the density and Y is the Young's modulus. If the bar were free to move it would act like a mass; however, if it were clamped on one end and slowly compressed on the other end it would act like a spring. In general the bar has both qualities, as will be shown by the development of a distributed model based on wave motion in the bar.

A lumped spring-mass representation of a vibrating system may be used to develop the theory of longitudinal wave motion in the bar by first dividing the bar into small adjacent lumped elements. Consider the small element of length Δz subjected to a longitudinal force and shown in Fig. 3.25. Lateral forces may be ignored if the lateral dimensions are small compared to the length. If the element Δz is very small compared to the wavelength, it may be represented as a mass and spring. A series of masses and springs for adjacent small elements also of length Δz is shown in Fig. 3.26 where $\Delta z = L/n$ and n is the total number of elements in the bar. The mass of each element is $M = \rho A_0 \Delta z$ and the stiffness is $K = YA_0/\Delta z$. The longitudinal displacement of the ith element is ζ_i as shown in Fig. 3.26. The corresponding equivalent circuit representation of Fig. 3.27 shows the masses, M, and compliances, $1/K$, along with the displacements and relative displacements.

As seen in Fig. 3.26, the acceleration of the center mass is opposed by the two springs on opposite sides with a force proportional to their relative displacements. The equation of motion of the ith mass is

$$M\partial^2\zeta_i/\partial t^2 = -K(\zeta_i - \zeta_{i-1}) - K(\zeta_i - \zeta_{i+1}),$$

which, since $K/M = Y/\rho(\Delta z)^2$, may be rewritten as

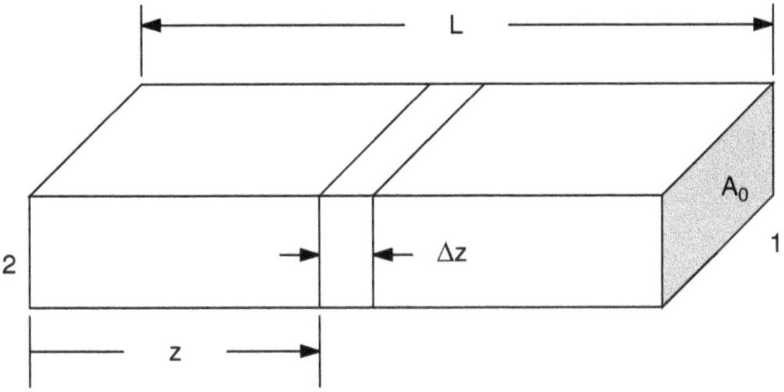

Fig. 3.25 Mechanical bar with small element Δz located at z

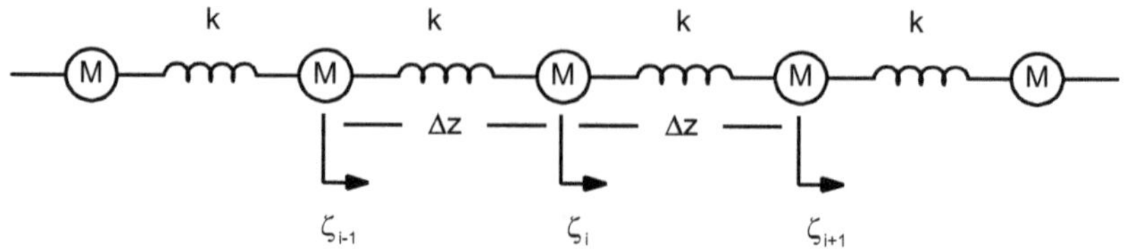

Fig. 3.26 Lumped mode representation of the bar of Fig. 3.25

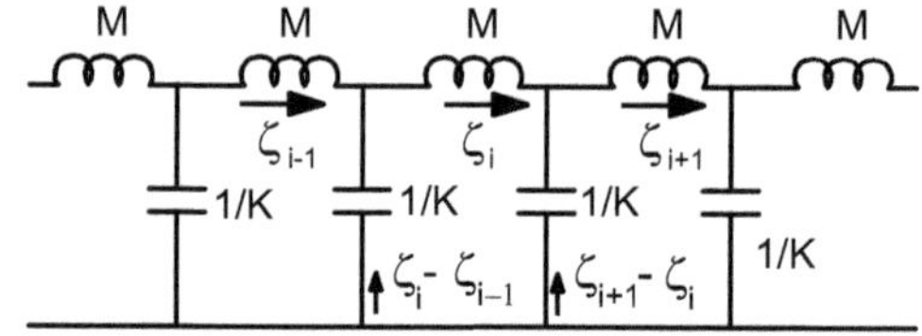

Fig. 3.27 Equivalent circuit representation of the bar of Fig. 3.25

$$\partial^2\zeta_i/\partial t^2 = (Y/\rho)[(\zeta_{i+1} - \zeta_i)/\Delta z - (\zeta_i - \zeta_{i-1})/\Delta z]/\Delta z.$$

Since the bar length is $L = n\Delta z$, a continuous distribution may be obtained by letting Δz shrink in size and approach ∂z as the number of elements, n, increases to maintain L constant. As Δz shrinks, the displacement differences also become small. The difference between the difference in the displacements on each side of the mass then, in the limit, leads to the wave equation

$$\partial^2\zeta/\partial t^2 = c^2\partial^2\zeta/\partial z^2, \tag{3.25}$$

where the constant $c = (Y/\rho)^{1/2}$ will be shown to be the speed of longitudinal waves in the bar. The stress in the bar is proportional to the strain and given by $T = Y\partial\zeta/\partial z$. Substitution into the wave equation yields

$$\partial T/\partial z = \rho\partial^2\zeta/\partial t^2, \tag{3.26}$$

which can be interpreted as a statement of Newton's second law per unit volume of material. The term on the right is the mass per unit volume times the acceleration while the term on the left (the gradient of the stress) is the net force per unit volume.

The general solution of the one-dimensional loss-less wave equation, Eq. (3.25), may be written in terms of two arbitrary function f and g as

$$\zeta(z,t) = f(z - ct) + g(z + ct), \tag{3.27}$$

with the first term representing a longitudinal wave moving to the right (i.e., in the direction of increasing z) with speed c and the second term representing a wave moving to the left with speed c.

For example: The approximate value of c is 5100 m/s for steel or aluminum and 3000 m/s for PZT (see Sect. 13.2). For comparison bulk sound speeds that depend on the bulk modulus, B, are 1500 m/s for water and 343 m/s for air. The relationship between the bulk modulus, B, and Young's modulus, Y, is $Y = 3B(1 - 2\sigma)$ where σ is Poisson's ratio; for fluids $\sigma = ½$ and $Y = 0$. There is no bar wave speed for liquids and gases since they cannot support the free lateral boundary conditions that hold for a thin bar.

In most instances the response as a function of frequency is desired and a Fourier transform solution may be obtained. Accordingly, and more simply, we may assume a complex time dependent sinusoidal solution of the form $\zeta(z,t) = \zeta(z)$ $e^{j\omega t}$ (see Sect. 13.17). Substitution into the wave equation, Eq. (3.25), then yields the one-dimensional Helmholtz differential equation (also known as the reduced wave equation)

$$d^2\zeta/dz^2 + k^2\zeta = 0, \tag{3.28}$$

where the wave number $k = \omega/c$.

The general solution for the displacement, as may be verified by substitution into Eq. (3.28), may be written as

$$\zeta(z) = Be^{-jkz} + De^{jkz}, \tag{3.29}$$

where the coefficients B and D are constants. The time dependent displacement solution is

$$\zeta(z,t) = \zeta(z)e^{j\omega t} = Be^{-j(kz-\omega t)} + De^{j(kz+\omega t)}, \tag{3.30}$$

where the two terms represent longitudinal waves traveling in the $+z$ and $-z$ directions.

The particle velocity, $u = \partial\zeta/\partial t$, is given by

$$u(z) = j\omega\zeta(z) = j\omega\left[Be^{-jkz} + De^{jkz}\right], \tag{3.31}$$

and the force, $F = -A_0T = -A_0Y\partial\zeta/\partial z$ is

$$F(z) = -jkYA_0\left[Be^{-jkz} - De^{jkz}\right]. \tag{3.32}$$

Equations (3.31) and (3.32) are the basis for developing models for the bar. The coefficients B and D may be obtained by imposing boundary conditions on the ends of the bar, e.g., $F = 0$, or $u = 0$ or more generally $F_i/u_i = Z_i$ where Z_i is the impedance at the bar ends ($i = 1, 2$).

The mechanical impedance at any point along the bar may be written as $Z(z) = F(z)/u(z)$:

$$Z(z) = \rho c A_0 \left[B e^{-jkz} - D e^{jkz}\right] / \left[B e^{-jkz} + D e^{jkz}\right]. \tag{3.33}$$

If there is no wave in the $-z$ direction then $D = 0$ and $Z(z) = \rho c A_0$. If the bar has an impedance of Z_0 at $z = 0$ and Z_L at $z = L$ the coefficients B and D may be determined in terms of Z_0 and Z_L yielding the very useful transmission line equation

$$Z_0 = \rho c A_0 [Z_L + j\rho c A_0 \tan kL] / [\rho c A_0 + jZ_L \tan kL]. \tag{3.34}$$

Thus, if the load impedance, Z_L, is zero, this free condition yields $Z_0 = j\rho c A_0 \tan kL$ as the impedance at the other end of the bar. Moreover if the impedance, Z_0, is also zero then $\tan kL = 0$. This occurs for $kL = n\pi$, $(n = 1, 2, 3, \ldots)$, resulting in the free-free bar half wavelength resonance condition $f_n = nc/2L$ and yielding a series of even and odd standing wave modes with harmonic resonance frequencies. On the other hand, if Z_L is very large compared to $\rho c A_0$ then this clamped condition yields $Z_0 = -j\rho c A_0 \cot kL$. If in addition to this the impedance $Z_0 = 0$, then $\cot kL = 0$. This occurs for $kL = \pi(2n - 1)/2$, resulting in the free-clamped bar quarter wavelength resonance condition $f_n = (2n - 1)c/4L$ and yielding a series of odd numbered resonance frequencies and standing waves with a fundamental resonance at one-half the frequency of the free-free bar. The first two displacement modes of vibration of the free-free and clamped-free bars are shown in Fig. 3.28. Although the magnitude of the displacement is depicted by the curves, actual displacement is along the length of the bar in the direction of the longitudinal wave.

It may be shown, after some algebra and use of trigonometric identities, that the equivalent circuit or "T Network" shown in Fig. 3.29 is consistent with Eq. (3.34) with $F_1/u_1 = Z_0$ at port 1 and $F_2/u_2 = Z_L$ at port 2. This circuit then becomes an equivalent circuit or transmission line representation of plane waves traveling along the bar subject to the boundary conditions at ports 1 and 2. There is also an equally good "π Network" representation (see, for example, Woollett [2]) that may be useful in some cases.

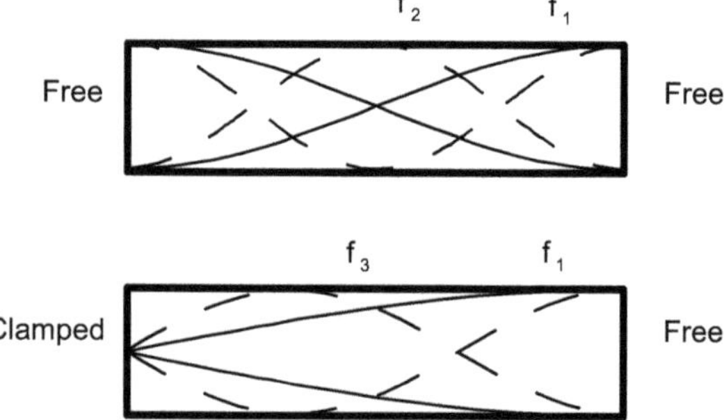

Fig. 3.28 Displacement distribution for the fundamental, f_1, second harmonic, f_2, and third harmonic, f_3, resonant frequencies

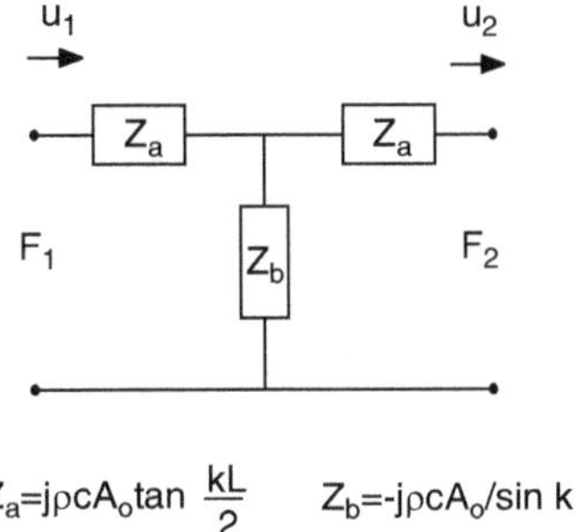

Fig. 3.29 Distributed T network circuit representation of the bar of Fig. 3.25

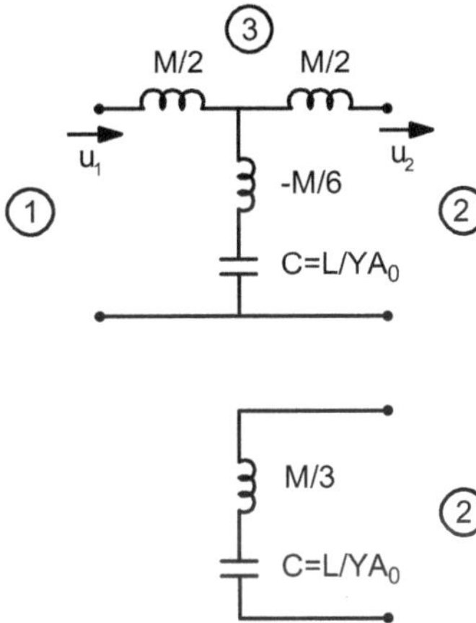

Fig. 3.30 Lumped mode circuit representation of the bar of Fig. 3.25

Fig. 3.31 Representation with one end clamped

If the bar length, L, is small compared to the wavelength, λ, such that $kL \ll 1$ then $j\rho cA_0 \tan(kL/2) \approx j\rho cA_0 kL/2 = j\omega M/2$ and $-j\rho cA_0/\sin(kL) \approx -j\rho cA_0 (1/kL + kL/6) = YA_0/j\omega L - j\omega M/6$ resulting in the lumped equivalent circuit of Fig. 3.30 [20]. The $-M/6$ element is not included in some representations as it is often negligible at low frequencies. However, it can be important, as in the case where the bar terminal 2 is clamped such that u_2 is zero and port 2 is therefore open circuited. In this case the masses $M/2$ and $-M/6$ add resulting in an effective mass of the bar of $M/3$ in the equivalent circuit of Fig. 3.31. (An alternative energy calculation of the effective mass, $M/3$, of a fixed-end bar is given in Sect. 4.2.) Piezoelectric ceramic materials have a high density and their mass usually needs to be included in lumped equivalent circuit representations as in Fig. 3.30.

3.2.2 Matrix Representation

The loop equations for the circuit of Fig. 3.29 may be written with impedance coefficients as

$$F_1 = Z_c u_1 - Z_b u_2, \tag{3.35a}$$

$$F_2 = Z_b u_1 - Z_c u_2, \tag{3.35b}$$

where $Z_c = Z_a + Z_b = -j\rho c A_0 \cot kL$ and $Z_b = -j\rho c A_0 / \sin kL$. It can be seen that with $u_2 = 0$, Z_c is the input impedance at port 1. This pair of equations may be solved for u_1 and u_2 for a given load at one end of the bar. If Z_2 is the load at port 2 then $F_2 = Z_2 u_2$ and Eq. (3.35b) gives $u_2 = u_1 Z_b / (Z_c + Z_2)$. Substitution into Eq. (3.35a) yields, for a given F_1, the solution for the velocities u_1 and u_2 with load Z_2:

$$u_1 = (Z_2 + Z_c)\, F_1 / \left(Z_c^2 + Z_c Z_2 - Z_b^2\right), \tag{3.35c}$$

$$u_2 = Z_b F_1 / \left(Z_c^2 + Z_c Z_2 - Z_b^2\right). \tag{3.35d}$$

Thus, for a force F_1, the input and output velocities, u_1 and u_2, may be determined from the circuit of Fig. 3.29 with load, Z_2.

The above impedance equation set, Eqs. (3.35a) and (3.35b), may also be written in the transfer form:

$$F_2 = (Z_c / Z_b)\, F_1 + \left(Z_b - Z_c^2 / Z_b\right) u_1,$$

$$u_2 = (-1 / Z_b)\, F_1 + (Z_c / Z_b) u_1,$$

and represented as

$$F_2 = a_{11} F_1 + a_{12} u_1, \tag{3.36a}$$

$$u_2 = a_{21} F_1 + a_{22} u_1, \tag{3.36b}$$

where $a_{11} = Z_c / Z_b = \cos kL, a_{12} = \left(Z_b - Z_c^2 / Z_b\right) = -j\rho c A_0 \sin kL, a_{21} = -1 / Z_b = -j[\sin kL] / \rho c A_0$, and $a_{22} = Z_c / Z_b = \cos kL$.

This equation set for the transmission line bar may be written in a matrix form which transforms the input vector $\underline{F}_1, \underline{u}_1$ to the output vector $\underline{F}_2, \underline{u}_2$ through a matrix $\mathbf{A}$ with elements a_{ij}. The matrix equation for Eqs. (3.36a) and (3.36b) is

$$\begin{pmatrix} F_2 \\ u_2 \end{pmatrix} = \begin{pmatrix} a_{11} & a_{12} \\ a_{21} & a_{22} \end{pmatrix} \begin{pmatrix} F_1 \\ u_1 \end{pmatrix}.$$

A series of bars connected together, as in Fig. 3.32, can be modeled as a series of Fig. 3.29 circuits connected together as shown in Fig. 3.33a. The output is then related to the input by multiplication of the matrices $\mathbf{A}$ and $\mathbf{A}'$ for each bar.

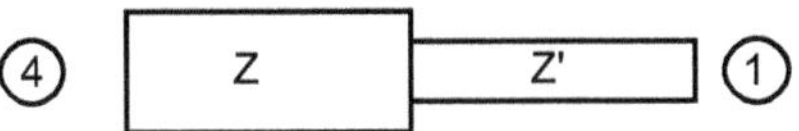

Fig. 3.32 Two bars mechanically connected in series

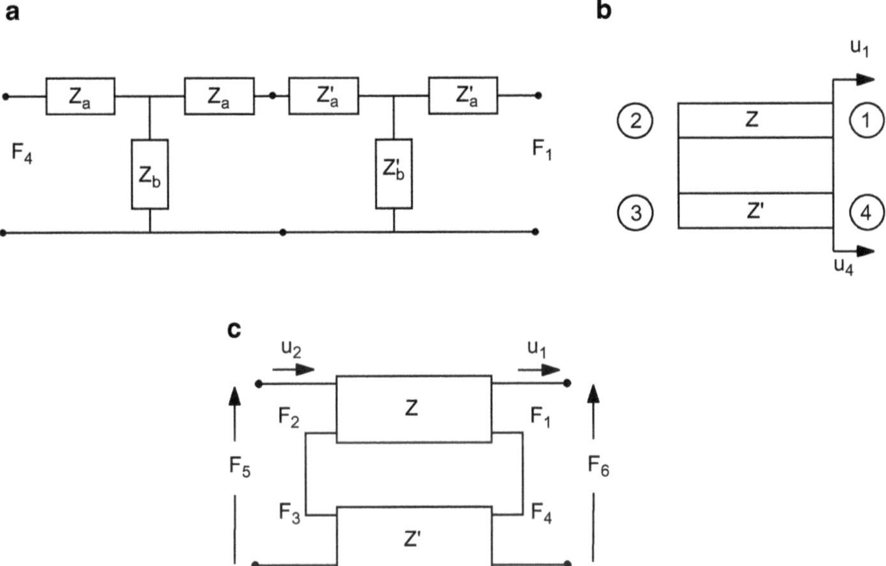

Fig. 3.33 (**a**) Distributed network representation for the two series bars of Fig. 3.32. (**b**) Two bars mechanically connected in parallel. (**c**) Distributed network representation for the two parallel bars of (**b**)

For a series of four bars, for example, the transfer matrix is $\mathbf{A_1A_2A_3A_4}$. The matrix product may then be written as a single matrix, and the output velocity u_4 for an input force F_1 can be obtained for the given load, $Z_4 = F_4/u_4$, on terminal 4 as

$$u_4 = F_1(a_{11}a_{44} - a_{14}a_{41})/(Z_4a_{44} - a_{14}),$$

where the a_{ij} are now the elements of the product matrix $\mathbf{A_1A_2A_3A_4}$.

A parallel arrangement of two ideally connected bars is illustrated in Fig. 3.33b. This situation occurs, for example, when a stress bolt is used to compress piezoelectric ceramics to prevent cracking under tension. Because of the ideal direct connections, the connecting boundary conditions are $u_4 = u_1$ and $u_3 = u_2$. With these parallel connections the total force on each end is the sum of the two forces as illustrated in Fig. 3.33c so that $F_5 = F_2 + F_3$ and $F_6 = F_1 + F_4$. This case is most easily solved if the equations are rewritten in the impedance matrix form

$$F_1 = z_{11}u_1 - z_{12}u_2, \tag{3.37a}$$

$$F_2 = z_{21}u_1 - z_{22}u_2, \tag{3.37b}$$

where $z_{11} = z_{22} = Z_c = Z_a + Z_b = -j\rho c A_0 \cot kL$ and $z_{12} = z_{21} = Z_b = -j\rho c A_0/\sin kL$. With a similar set of z's for the Z' bar of Fig. 3.33b written as

$$F_4 = z'_{11}\, u_1 - z'_{12}\, u_2, \tag{3.37c}$$

$$F_3 = z'_{21} u_1 - z'_{22} u_2, \tag{3.37d}$$

we get, on adding the two forces on each end

$$F_6 = \left(z_{11} + z'_{11}\right) u_1 - \left(z_{12} + z'_{12}\right) u_2, \tag{3.37e}$$

$$F_5 = \left(z_{21} + z'_{21}\right) u_1 - \left(z_{22} + z'_{22}\right) u_2, \tag{3.37f}$$

which can be compacted and converted to a single transfer matrix form for a further cascade process with other elements.

Multiple transmission line equivalent circuits or cascading matrix computer programs are commonly used for solving these one-dimensional multiple-element transducer wave problems [21–25].

3.2.3 Piezoelectric Distributed Parameter Equivalent Circuit

The mechanical model developed in Sect. 3.2.1 will now be extended to include piezoelectric excitation. Use will be made of the lumped electromechanical equivalent circuit and the wave based distributed mechanical model to develop the matrix and equivalent circuit models for a piezoelectric ceramic bar [1, 2]. We will consider four piezoelectric ceramic and one magnetostrictive configuration. Each case is unique because of the different electric and mechanical boundary conditions and requires different transduction constants (see Sects. 13.5 and 13.7 for a complete listing of piezoelectric ceramic and magnetostrictive constants). Accordingly, we present separate development for each model. The casual reader may wish to take an overview approach and concentrate on the results and the differences between the models.

Although all the models use a one-dimensional acoustic plane wave analysis, the electric field can be oriented along the direction of motion or perpendicular to the direction of motion. Moreover, in two cases the electric field is constant (Segmented 33 bar and Unsegmented 31 bar) while in the other two cases (Length Expander Bar and Thickness Mode Plate) the electric filed changes along the length of the bar as a result of the wave motion along the length of the bar. This results in a different equivalent circuit representation for the two cases. In the four cases, represented by bars (including the magnetostric bar), the transduction equations are written in a form where the stress, T, is an independent variable and can be set equal to zero in the lateral direction, perpendicular to the direction of the wave motion. This conforms to the case of a free lateral boundary condition allowing wave propagation at the bar speed determined from the Young's modulus. In the case of the thickness mode plate the transduction equations are written in a form where the strain, S, is an independent variable and set equal to zero in the lateral direction conforming to a rigid boundary, resulting in bulk wave speed propagation.

3.2.3.1 Segmented 33 Bar

Transducers are often fabricated from a number of piezoelectric ceramic segments cemented together and wired in parallel to yield lower electrical impedance than a single bar of the same total length [2]. This arrangement allows a lower drive voltage for the same electric field. It is also necessary in order to avoid extremely high voltage poling fields. For example, consider the case of a piezoelectric bar in the 33 mode as shown in Fig. 3.34 with four segments wired in parallel.

Each segment has thickness h and cross-sectional area $A_0 = w_1 w_2$ with an electric field E_3 in the 3, or z, direction causing expansion and contraction in the 3 direction. Silvered surfaces serving as electrodes at the ends of the segments are cemented together with the polarization direction P_0 as shown for additive motion in the 3 direction. It is assumed that the cement bonds are ideal and negligibly thin so that the velocity, u, and the force, F, are continuous across them. Loading forces F_2 and F_1 and corresponding impedances Z_2 and Z_1 are assumed at the bar terminations at $z = 0$ and $z = L$.

It is also assumed that the thickness, h, between the electrodes is considerably less than one-quarter wavelength although the total length L is not restricted. Additionally it is assumed that $w_1 \ll L$ and $w_2 \ll L$ with no loads in the 1 and

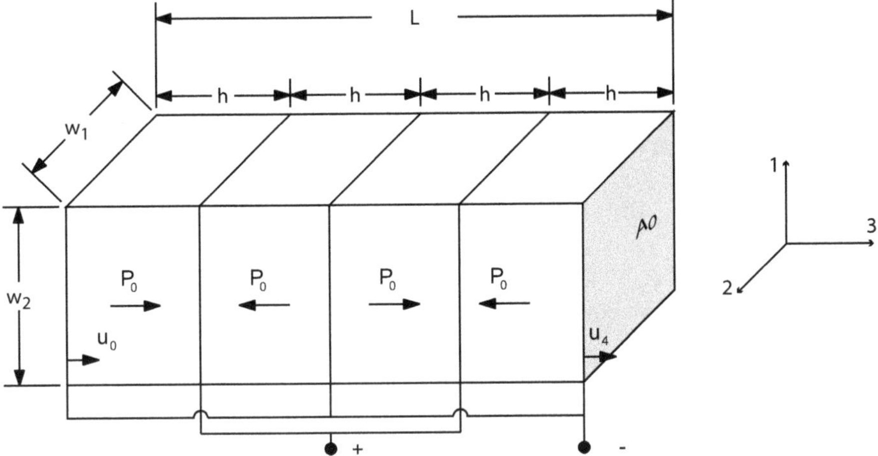

Fig. 3.34 Four piezoelectric bars stacked mechanically in series and wired in parallel

2 directions so that $T_1 = T_2 = 0$ on the sides of the bar and within the material. With no lateral clamping the lateral motion is uninhibited yielding the lowest stiffness, based on the Young's modulus, and greatest coupling in the 3 direction. Also, since the electrode surfaces are equipotentials, there is no electric field in the 1 and 2 directions and since h is small $E_1 = E_2 = 0$ within the piezoelectric ceramic material. Maximum performance is generally achieved with both the stress and applied field in the direction of the polarization vector, i.e., the 3 direction. The 33 mode is the most active mode of vibration with typical coupling coefficients on the order of 0.7 for PZT materials and 0.9 for single crystal PMN-PT and PIN-PMN-PT materials.

For example: Under free end conditions the strain, S, is proportional to the d constant for a given electric field E. The d_{33} for PZT is about 300×10^{-12} m/V while d_{31} (see Sect. 3.2.3.2) is only about -135×10^{-12} m/V (see Sect. 13.5 for piezoelectric constants).

The piezoelectric equation pair of interest under the above assumptions is the one used before, Eqs. (3.6a) and (3.7b), which can be rewritten as

$$S_3 = s_{33}^E T_3 + d_{33} E_3, \tag{3.38a}$$

$$D_3 = \left(d_{33}/s_{33}^E\right) S_3 + \varepsilon_{33}^S E_3, \tag{3.38b}$$

where the clamped dielectric constant $\varepsilon_{33}^S = \varepsilon_{33}^T\left(1 - k_{33}^2\right)$.

If we let ζ be the displacement of a longitudinal wave along the z direction as a result of an electric field E_3, the strain may be written as $S_3 = \partial\zeta/\partial z$. Accordingly, the derivative of the first equation of the pair may be written as

$$\partial^2\zeta/\partial z^2 = s_{33}^E \partial T_3/\partial z + d_{33} \partial E_3/\partial z. \tag{3.39}$$

It was found in the previous section that $\partial T/\partial z = \rho \partial^2\zeta/\partial t^2$. Substitution into Eq. (3.39) then leads to the wave equation in the form

$$\partial^2\zeta/\partial z^2 = s_{33}^E \rho \, \partial^2\zeta/\partial t^2 + d_{33} \partial E_3/\partial z. \tag{3.40}$$

With V the voltage, $E_3 = V/h$ is the same in each segment and $\partial E_3/\partial z = 0$ which leads to the usual wave equation

$$\partial^2\zeta/\partial t^2 = c^2 \partial^2\zeta/\partial z^2, \tag{3.41}$$

with the longitudinal wave speed $c = 1/\left(s_{33}^E \rho\right)^{1/2}$.

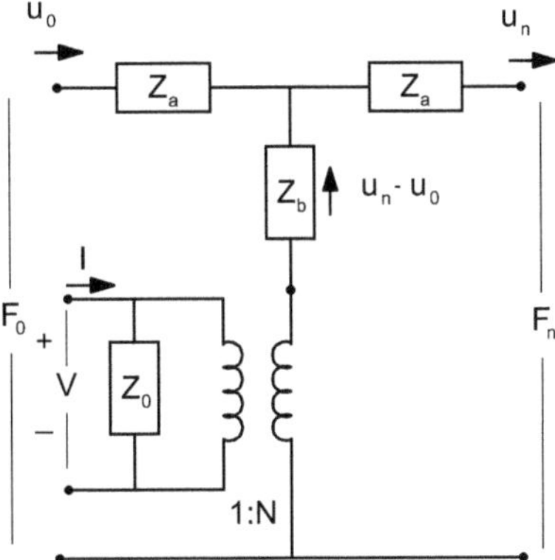

Fig. 3.35 Distributed network representation of the model shown in Fig. 3.34

The analysis in the previous section showed that solutions of this wave equation maybe be represented by an equivalent circuit of the form shown in Fig. 3.29. This equivalent circuit is also applicable to piezoelectric devices, as was shown by Mason [26]. The lumped-parameter equivalent circuit model for piezoelectric ceramics suggests that the full equivalent circuit, with voltage V and current I, can be represented by Fig. 3.35 which, under short circuit conditions, is the same as Fig. 3.29.

Evaluation of the electromechanical turns ratio, N, and the shunt electrical impedance Z_0 may be obtained from Eq. (3.38b), which may be written as

$$D_3 = \left(d_{33}/s_{33}^E\right)\partial\zeta/\partial z + \varepsilon_{33}^S E_3. \tag{3.42}$$

The electric displacement $D_3 = Q/A_0$, where Q is the charge and A_0 is the surface area of the electrodes. Since the time derivative of the charge is the current I, and the time derivative of the displacement, ζ, is the velocity, u, the time derivative of Eq. (3.42) is

$$I(z) = \left(A_0 d_{33}/s_{33}^E\right)\,\partial u/\partial z + j\omega\left(\varepsilon_{33}^S A_0/h\right)V. \tag{3.43}$$

Since the distance between electrodes is small compared to the wavelength, we would not expect the velocity, u, or the current, I, to vary in each segment as illustrated in Fig. 3.36. Thus on replacing $\partial u/\partial z$ with the approximation $(u_i - u_{i-1})/h$ the current through the ith segment is

$$I_i = \left(A_0 d_{33}/hs_{33}^E\right)(u_i - u_{i-1}) + j\omega\left(\varepsilon_{33}^S A_0/h\right)V. \tag{3.44}$$

As seen in Fig. 3.36, since the electrodes are wired in parallel and the voltage V is the same across each electrode pair, the total current is the sum of the currents through each segment. The total current for n segments, for $i = 0$ to n, is

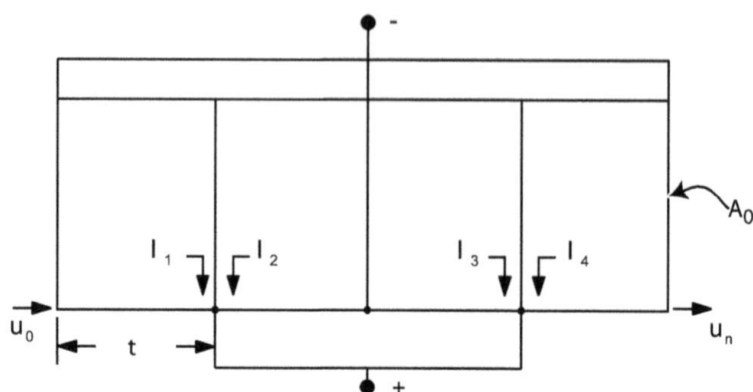

Fig. 3.36 Illustration of current summation

$$I = \sum I_i = \left(A_0 d_{33}/hs_{33}^E\right)(u_n - u_0) + j\omega n\left(\varepsilon_{33}^S A_0/h\right)V. \tag{3.45}$$

Thus the total current is the sum of two currents: one resulting from the relative velocities at the ends of the bar and the other from the voltage V across the electrical impedance, $1/j\omega\ n(\varepsilon_{33}^S A_0/h)$, of the n capacitors wired in parallel. The current from the velocity difference, u_n–u_0, is transformed to the electrical side through the transformer of turns ratio N. A comparison with the equivalent circuit of Fig. 3.35 then yields

$$N = A_0 d_{33}/hs_{33}^E, \quad Z_0 = 1/\ j\omega\ nC_0, \quad C_0 = \varepsilon_{33}^S A_0/h, \quad c = 1/\left(s_{33}^E \rho\right)^{1/2}, \tag{3.46}$$

which is consistent with Eqs. (3.41) and (3.45). The wave speed value of c from Eq. (3.46) is to be used in $k = \omega/c$ and in $Z_a = j\rho c A_0 \tan(kL/2)$ and $Z_b = -j\rho c A_0 / \sin(kL)$ of Fig. 3.35. The quantity C_0 is the clamped capacitance of a single segment of the bar, nC_0 is the total clamped capacitance of the bar and the electromechanical turns ratio, N, for the bar is the same as the turns ratio for a single segment. The coupling coefficient for each segment is given by $k_{33}^2 = d_{33}^2/s_{33}^E \varepsilon_{33}^T$, but the effective coupling coefficient for the whole bar at resonance is reduced by dynamic effects (see Sect. 4.4.3). The 33 mode bar equivalent circuit of Fig. 3.35 has played an important role in the design of high power longitudinal vibrator transducers for sonar arrays.

3.2.3.2 Un-segmented 31 Bar

The equivalent circuit of Fig. 3.35 may also be used to represent a thin bar operated in the 31 mode [1, 26] (or the 32 mode when E_2 is the driving field) as shown in Fig. 3.37.

Under the same assumptions as before, this mode may be represented through the equation pair (see Sect. 2.1.3):

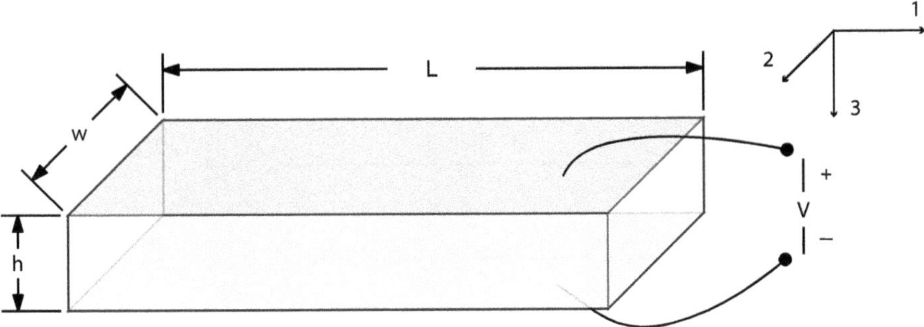

Fig. 3.37 Piezoelectric ceramic side electrode 31 mode bar

$$S_1 = s_{11}^E T_1 + d_{31} E_3,$$

$$D_3 = d_{31} T_1 + \varepsilon_{33}^T E_3.$$

On eliminating T_1 between these equations, the second equation may be rewritten as

$$D_3 = \left(d_{31}/s_{11}^E\right) S_1 + \varepsilon_{33}^S E_3.$$

In this mode of operation the electric field E_3 is in a direction perpendicular to the strain S_1 and stress T_1 and consequently is indirectly coupled to the motion in the 1 direction. It can then be readily shown, following the previous 33 mode development, that

$$N = wd_{31}/s_{11}^E, \quad Z_0 = 1/j\omega C_0, \quad C_0 = \varepsilon_{33}^S wL/h, \quad c = 1/\left(s_{11}^E \rho\right)^{1/2}. \tag{3.47}$$

Note the different wave speed value of c from Eq. (3.47), to be used in $k = \omega/c$ and in $Z_a = j\rho c A_0 \tan\,(kL/2)$ and $Z_b = -j\rho c A_0/\sin\,(kL)$ of Fig. 3.35. The quasistatic coupling coefficient is given by $k_{31}^2 = d_{31}^2/s_{11}^E \varepsilon_{33}^T$, which is reduced by dynamic effects at resonance (see Sect. 4.4.3). The coupling coefficient, k_{31}, with value of about 0.33 and the d_{31} constant with value of about $-135 \times 10^{-12}\,\text{m/V}$ for PZT are approximately half the 33 mode values.

3.2.3.3 Length Expander Bar

Other cases of interest include the 33 mode length expander bar (without segmentation) and the thickness mode plate where the distance between the electrodes is not small compared to the wavelength in the material. In these cases the electric field is not constant but is a function of the propagation distance z and consequently $\partial E(z)/\partial z$ does not vanish [1]. Thus, the most appropriate equation pair is one in

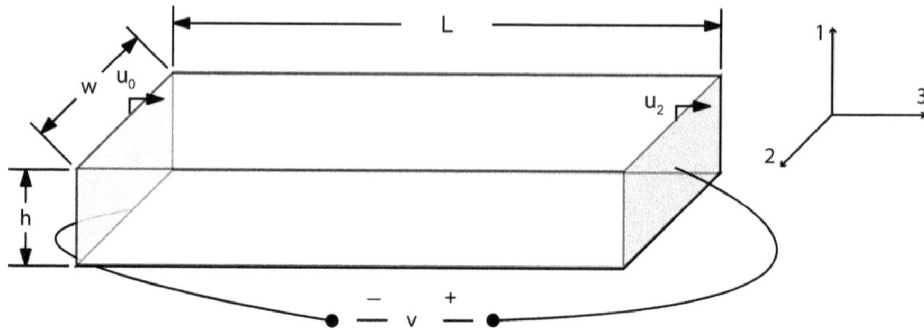

Fig. 3.38 Piezoelectric ceramic end electrode 33 mode bar

which D_3 rather than E_3 is the independent variable. Normally because of the comparatively high dielectric constant, there is no fringing so $D_1 = D_2 = 0$. Also, since $\operatorname{div} D = 0$ for a nonconductive dielectric with no free charge, $\partial D_3(z)/\partial z = 0$. For the case of a long thin bar with electrodes on the ends as illustrated in Fig. 3.38, the equations may be written as

$$S_3 = s_{33}^D T_3 + g_{33} D_3, \tag{3.48a}$$

$$E_3 = -\left(g_{33}/s_{33}^D\right) S_3 + \left(1/\varepsilon_{33}^S\right) D_3, \tag{3.48b}$$

where use has been made of the relation $s_{33}^D = s_{33}^E\left(1 - k_{33}^2\right)$.

The longitudinal displacement and velocity along the 3 direction are ζ and u. The strain $S_3 = \partial\zeta/\partial z$, and the gradient of the stress $\partial T_3/\partial z = \rho\partial^2\zeta/\partial t^2$. Differentiating Eq. (3.48a) yields the wave equation

$$\partial^2\zeta/\partial z^2 = s_{33}^D\rho\, \partial^2\zeta/\partial t^2\,, \tag{3.49}$$

since $\partial D_3(z)/\partial z = 0$.

Thus, in this case the wave speed $c = 1/\left(\rho s_{33}^D\right)^{1/2}$ is higher than the wave speed in the segmented bar because $s_{33}^D < s_{33}^E$. To be consistent with this, the electro-mechanical equivalent circuit for the open circuit case must take the form of Fig. 3.39 rather than Fig. 3.35. In this form it can be seen that under electrical open circuit conditions, Z_0 and $-Z_0$ cancel causing a short circuit in the mechanical branch leaving $c = 1\left(\rho s_{33}^D\right)^{1/2}$ as the wave speed. On the other hand, under electrical short circuit conditions, Z_0 is shorted out and only the negative impedance $-Z_0$ remains in the circuit. This negative impedance when transformed to the mechanical side (see Fig. 3.40) reduces the stiffness of the system and yields a lower resonance frequency, as may be expected, under electrical short circuit conditions.

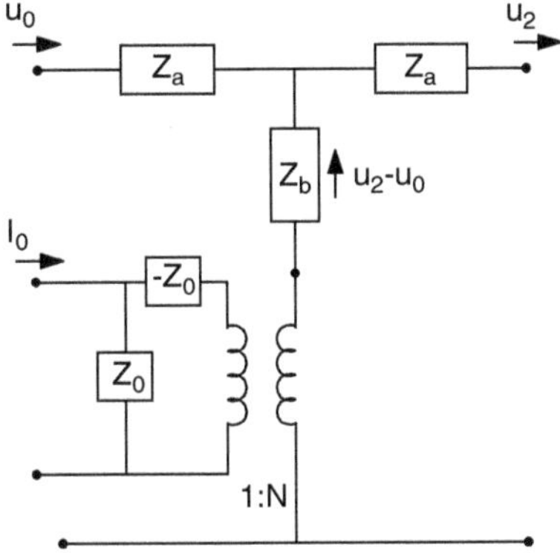

Fig. 3.39 Distributed equivalent circuit of the end electrode bar of Fig. 3.38

The value of Z_0 and the turns ratio N can be obtained from Eq. (3.48b) written as

$$E_3 = -\left(g_{33}/s_{33}^D\right)\partial\zeta/\partial z + \left(1/\varepsilon_{33}^S\right)D_3. \tag{3.50}$$

Since $E_3(z) = \partial V/\partial z$, integration from $z = 0$ to $z = L$ yields an expression for the voltage. Furthermore, since the current $I = A_0\partial D_3/\partial t$, $u = \partial\zeta/\partial t$ and $\partial V/\partial t = j\omega V$ for sinusoidal drive, it follows that

$$I = j\left(\omega A_0\varepsilon_{33}^S/L\right) V + \left(d_{33}A_0/Ls_{33}^E\right)[u(L) - u(0)]. \tag{3.51}$$

This result and the wave equation then show that

$$N = A_0 d_{33}/Ls_{33}^E, \quad Z_0 = 1/j\omega C_0, \quad C_0 = \varepsilon_{33}^S A_0/L, \quad c = 1/\left(s_{33}^D\rho\right)^{1/2}, \tag{3.52}$$

for the equivalent circuit of Fig. 3.39. The wave speed value of c from Eq. (3.52) is to be used in $k = \omega/c$ and in $Z_a = j\rho c A_0 \tan(kL/2)$ and $Z_b = -j\rho c A_0/\sin(kL)$ of Fig. 3.39. The quasistatic coupling coefficient is k_{33} which is reduced by dynamic effects at resonance, and the reduction is greater than it is for the segmented bar (see Sect. 4.4.3).

The alternative form of the equivalent circuit, shown in Fig. 3.40, has the negative impedance transformed to the mechanical branch in series with Z_b. When L is considerably less than the wavelength, $kL \ll \pi$, these two elements combine to give a compliance equal to $L/A_0 s_{33}^E$, and the circuit of Fig. 3.40 reduces to the circuit of Fig. 3.30 where M is the total mass of the bar. Martin [27] has also shown that a segmented series of n expander bars of equal length h, with $kh \ll \pi$ and total length $L = nh$, reduces to the representation shown in Fig. 3.35 with accompanying Eq. (3.46) as originally developed in Sect. 3.2.3.1.

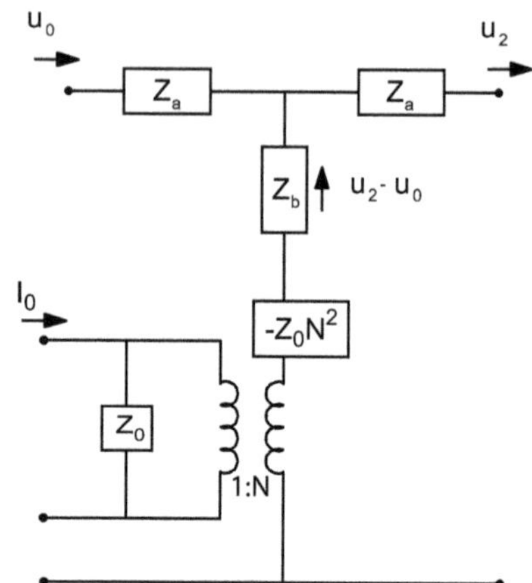

Fig. 3.40 Alternative equivalent circuit representation of the end electrode bar of Fig. 3.38

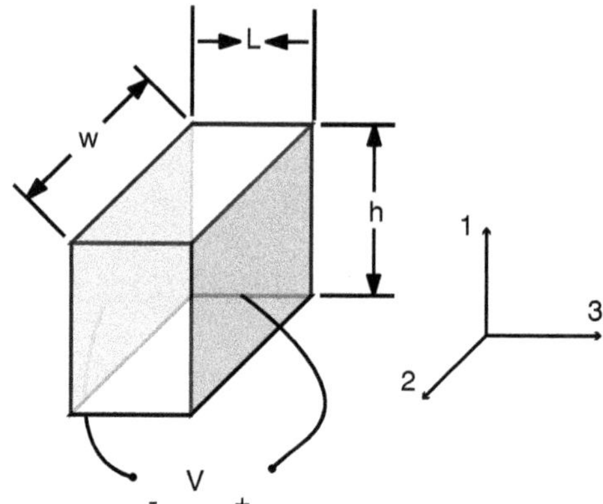

Fig. 3.41 End electrode piezoelectric plate

3.2.3.4 Thickness Mode Plate

The circuit of Fig. 3.40 may also be used to represent a plate, as in Fig. 3.41. However, because of the large lateral dimension the stiffness in the thickness mode is greater than the stiffness of a bar and the coupling coefficient is lower.

The equations of interest for this case use the piezoelectric constant $h_{33} = g_{33}/s_{33}^D$ and the clamped impermittivity $\beta_{33}^S = 1/\varepsilon_{33}^S$,

$$T_3 = c_{33}^D S_3 - h_{33} D_3, \tag{3.53a}$$

$$E_3 = -h_{33} S_3 + \beta_{33}^S D_3, \tag{3.53b}$$

where the coefficients can be identified by setting S_3 and D_3 separately equal to zero. The values for the equivalent circuit of Fig. 3.40 may be obtained by following the above procedure for the length expander bar. The result is

$$N = C_0 h_{33}, \quad Z_0 = 1/j\omega C_0, \quad C_0 = \varepsilon_{33}^S A_0/L, \quad c = \left(c_{33}^D/\rho\right)^{1/2}. \tag{3.54}$$

The value of c from Eq. (3.54) is to be used in $k = \omega/c$ and in $Z_a = j\rho c A_0 \tan(kL/2)$ and $Z_b = -j\rho c A_0/\sin(kL)$ of Fig. 3.39. The quasistatic coupling coefficient is the thickness coupling coefficient, $k_t = h_{33}\left(\varepsilon_{33}^S/c_{33}^D\right)^{1/2}$. Often large plates are cut or "diced" to reduce the lateral stiffness, improve the coupling coefficient, and reduce lateral mode interference.

3.2.3.5 Magnetostrictive Rod

The equivalent circuit of a magnetostrictive transducer discussed in Sect. 3.1.4 and shown in Fig. 3.23 may be extended to include wave effects in a magnetostrictive rod transducer. The usual wave equation can be derived from Eq. (3.16) using the fact that $\operatorname{div} B = 0$, and assuming that the lateral components of B can be neglected. In this case we may use the circuit of Fig. 3.35 with $N_m = \left(N_t d_{33} A_0/L s_{33}^H\right)$ and

$$N = N_m/\omega L_0, \quad Z_0 = j\omega L_0, \quad L_0 = \chi \mu_{33}^S N_t^2 A_0/L, \quad c = 1/\left(s_{33}^B \rho\right)^{1/2}, \tag{3.55}$$

where χ is the eddy current factor and the turns ratio $N = N_m/\omega L_0$ as a result of the circuit configurations shown in Figs. 3.23 and 3.35. The value of c from Eq. (3.55) is to be used in $k = \omega/c$ and in $Z_a = j\rho c A_0 \tan(kL/2)$ and $Z_b = -j\rho c A_0/\sin(kL)$ of Fig. 3.35. This circuit has an output velocity phase limitation but is useful for obtaining the electrical input impedance and the magnitude of the output velocity. The quasistatic coupling coefficient, $k_{33} = \left(d_{33}^2/s_{33}^B \mu_{33}^T\right)^{1/2}$, is reduced by dynamic effects at resonance in the same way as the reduction for the piezoelectric ceramic length expander bar (see Sect. 4.4.3).

The phase limitation for this shunt clamped inductance representation, noted in Sect. 3.1.4, still applies. For magnetic field transducers $F = N_m I$. Since $V/I = Z_0 = j\omega L_0$ under clamped conditions, we could write the clamped force $F = N_m V/Z_0 = (N_m/j\omega L_0)V$ in the representation of Fig. 3.23. However, ideal transformers must be real for proper impedance transformation and the factor j is accordingly dropped in Eq. (3.55), leading to a $-90°$ phase error in the voltage to velocity transfer function for this representation. It will be shown in Sect. 3.3 that the corresponding matrix representation has no such equivalent circuit transfer phase limitation.

3.3 Matrix Models

3.3.1 Three Port Matrix Model

Consider now the matrix representation of the distributed circuit of Fig. 3.35 where the electromechanical turns ratio N transforms the voltage V into a force F as illustrated in Fig. 3.42.

This allows a convenient means for incorporating the drive voltage into the matrix representation. The corresponding impedance equation pair in Eqs. (3.35a) and (3.35b) may now be rewritten as

$$F_1 = Z_c u_1 - Z_b u_2 + NV, \tag{3.56a}$$

$$F_2 = Z_b u_1 - Z_c u_2 + NV, \tag{3.56b}$$

where, as before, $Z_c = Z_a + Z_b = -j\rho c A_0 \cot kL$ and $Z_b = -j\rho c A_0/\sin kL$ with $k = \omega/c$ and c is the short circuit longitudinal wave speed.

The equation pair may also be rewritten in the transfer equation form as

$$F_2 = a_{11}F_1 + a_{12}u_1 + a_1 V, \tag{3.57a}$$

$$u_2 = a_{21}F_1 + a_{22}u_1 + a_2 V, \tag{3.57b}$$

or in transfer matrix form as

$$\begin{pmatrix} F_2 \\ u_2 \end{pmatrix} = \begin{pmatrix} a_{11} & a_{12} \\ a_{21} & a_{22} \end{pmatrix} \begin{pmatrix} F_1 \\ u_1 \end{pmatrix} + \begin{pmatrix} a_1 \\ a_2 \end{pmatrix} V, \tag{3.57c}$$

where

$$a_1 = N(1 - Z_c/Z_b), \quad a_2 = N/Z_b, \tag{3.57d}$$

$$a_{11} = Z_c/Z_b, \quad a_{12} = Z_b - Z_c^2/Z_b, \quad a_{21} = -1/Z_b, \quad a_{22} = Z_c/Z_b. \tag{3.57e}$$

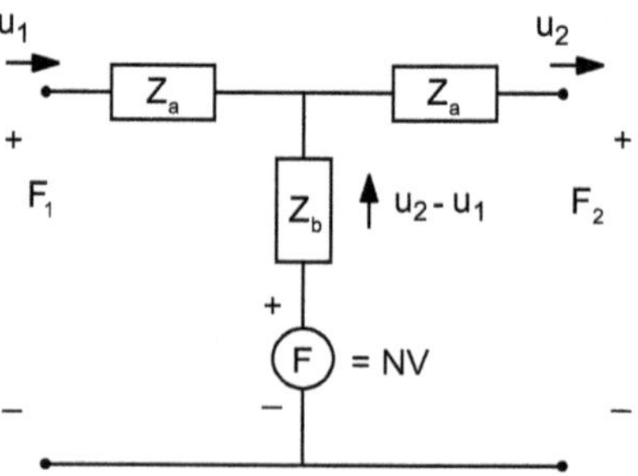

Fig. 3.42 Force driven representation of Fig. 3.35

Substitution for Z_c and Z_b yields

$$a_1 = N(1 - \cos kL), \quad a_2 = jN \sin kL/\rho c A_0, \tag{3.58a}$$

$$\begin{aligned} a_{11} &= \cos kL, \quad a_{12} = -j\rho c A_0 \sin kL, \\ a_{21} &= -j \sin kL/\rho c A_0, \quad a_{22} = \cos kL, \end{aligned} \tag{3.58b}$$

where the electromechanical turns ratio N is specific to the particular transduction device and configuration.

Since the turns ratio, N, transforms a velocity to a current the total input current in Fig. 3.35 is the sum of the velocity (with $u_0 = u_1$ and $u_2 = u_n$), transformed to a current, and the current through Z_0 for the voltage V, which can be written as

$$I = N(u_1 - u_2) + (1/Z_0)V, \tag{3.59}$$

where $1/Z_0 = Y_0$, the clamped electrical admittance. For the n parallel wired segmented piezoelectric ceramic transducer, with a small distance between electrodes,

$$N = nd_{33}A_0/Ls_{33}^E, \quad Z_0 = 1/j\omega nC_0, \quad c = 1/\left(s_{33}^E\rho\right)^{1/2}.$$

For a magnetostrictive transducer of coil turns, N_t, the parameters N, Z_0, and c are

$$N = \left(N_t d_{33}A_0/Ls_{33}^H\right)/j\omega L_0, \quad Z_0 = j\omega L_0, \quad c = 1/\left(s_{33}^B\rho\right)^{1/2}.$$

Since the clamped inductance $L_0 = N_t^2\mu_{33}^S A_0/L$, the magnetostrictive electromechanical turns ratio may also be written as $N = d_{33}/\left(j\omega N_t s_{33}^H\mu_{33}^S\right) = d_{33}/\left(j\omega N_t s_{33}^B\mu_{33}^T\right)$.

Consider the transducer example of Fig. 3.43 where Z_r is the radiation load impedance on the piston head, Z_m is the mounting impedance on the tail mass and the piezoelectric ceramic drive is sandwiched between the head and the tail. An equivalent T-network representation is illustrated in Fig. 3.44a, which allows a solution for the output by use of a circuit analysis program. A matrix representation is shown in Fig. 3.44b where 1 refers to the tail, 2 to the piezoelectric ceramic, and 3 to the head.

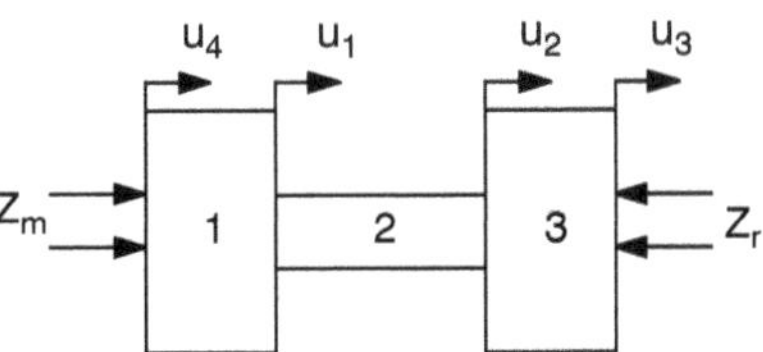

Fig. 3.43 Three part transducer with mounting impedance Z_m and radiation impedance Z_r

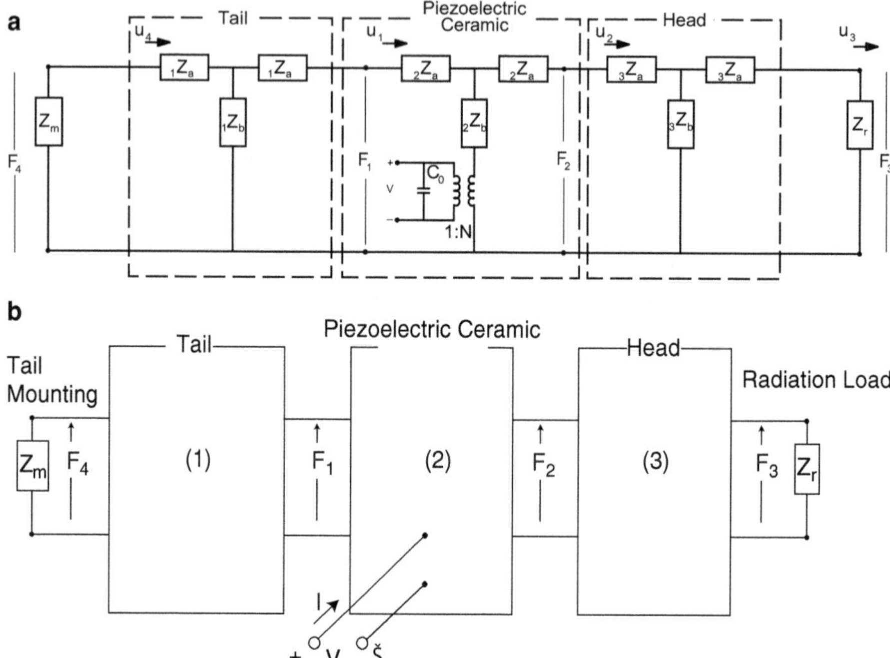

Fig. 3.44 (**a**) Cascade T network representation of Fig. 3.43. (**b**) Cascade block matrix representation of Fig. 3.43

The matrix approach relies on the equation pair, Eqs. (3.57a) and (3.57b), given above for the piezoelectric ceramic and the equation sets below for the head and tail, respectively.

$$F_3 = {}_3a_{11}F_2 + {}_3a_{12}u_2, \quad F_1 = {}_1a_{11}F_4 + {}_1a_{12}u_4, \tag{3.60a}$$

$$u_3 = {}_3a_{21}F_2 + {}_3a_{22}u_2, \quad u_1 = {}_1a_{21}F_4 + {}_1a_{22}u_4, \tag{3.60b}$$

where the pre-subscript, m, on ${}_ma_{ij}$ identifies a particular matrix, $\mathbf{A_m}$. If we let $\underline{\mathrm{Fu}}_i$ be the ith force velocity vector and $\mathbf{A_m}$ the matrix of the a_{ij} elements we can write $\underline{\mathrm{Fu}}_3 = \mathbf{A_3}\underline{\mathrm{Fu}}_2$ for the head, $\underline{\mathrm{Fu}}_1 = \mathbf{A_1}\underline{\mathrm{Fu}}_4$ for the tail, and $\underline{\mathrm{Fu}}_2 = \mathbf{A_2}\underline{\mathrm{Fu}}_1 + \mathbf{a_2}V$ for the piezoelectric ceramic where $\mathbf{A_2}$ and $\mathbf{a_2}$ correspond to Eqs. (3.57a) and (3.57b). Substitution for $\underline{\mathrm{Fu}}_2$ and $\underline{\mathrm{Fu}}_1$ in the equation for $\underline{\mathrm{Fu}}_3$ yields

$$\underline{\mathrm{Fu}}_3 = \mathbf{A_3A_2A_1}\underline{\mathrm{Fu}}_4 + \mathbf{A_3a_2}V. \tag{3.61a}$$

After matrix multiplication the above may be compacted and rewritten as

$$\underline{\mathrm{Fu}}_3 = \mathbf{A_4}\underline{\mathrm{Fu}}_4 + \mathbf{a_4}V, \tag{3.61b}$$

with $\mathbf{A_4} = \mathbf{A_3 A_2 A_1}$ and $\mathbf{a_4} = \mathbf{A_3 a_2}$. The set of equations relating the motion at the back of the tail, the voltage on the piezoelectric ceramic element and the motion at the front of the head may then be written as

$$F_3 = {}_4a_{11}F_4 + {}_4a_{12}u_4 + {}_4a_1 V, \tag{3.62a}$$

$$u_3 = {}_4a_{21}F_4 + {}_4a_{22}u_4 + {}_4a_2 V. \tag{3.62b}$$

A solution may be obtained for a given set of boundary conditions such as the loading impedances shown in Fig. 3.44b with $F_4 = Z_m u_4$ and $F_3 = Z_r u_3$. Substituting F_3 and F_4 into Eqs. (3.62a) and (3.62b) yields

$$Z_r u_3 = ({}_4a_{11}Z_m + {}_4a_{12})u_4 + {}_4a_1 V, \tag{3.63a}$$

$$u_3 = ({}_4a_{21}Z_m + {}_4a_{22})u_4 + {}_4a_2 V. \tag{3.63b}$$

Eliminating u_4 between the two equations leads to the desired solution for the piston velocity, u_3, as a function of the voltage, V. Typically transducers are designed to make the mounting impedance, Z_m, small. The case where $Z_m = 0$ gives

$$u_3 = V({}_4a_{22}a_1 - {}_4a_{12}a_2)/({}_4a_{22}Z_r - {}_4a_{12}). \tag{3.64}$$

In this section we have presented circuit analysis and cascade matrix models for transducers of arbitrary length in the direction of motion. Although the lumped-parameter models provide physical insight, they lack the accuracy of the distributed models unless the components are very short. The one-dimensional models discussed in this section agree well with measured results if all the important pieces (e.g., cement joints and electrodes) are included and the lateral dimensions are small compared to the wavelength in the material. The cases where the lateral dimensions are comparable to the wavelength in the material, e.g., where there is head flexure, are more accurately described using finite element modeling (FEM) techniques (see Sect. 3.4).

3.3.2 *Two Port ABCD Matrix Model*

The circuit representations given above can often be reduced to equivalent two port systems with voltage, V, and current, I, on the electrical side and force, F, and velocity, u, on the side that radiates into the water. The force F is external and depends on the radiation impedance and incident acoustic pressure, if present. We can picture this arrangement as shown in Fig. 3.45 where the A, B, C, D parameters represent the transducer alone, without the radiation impedance, and are given through the equation set

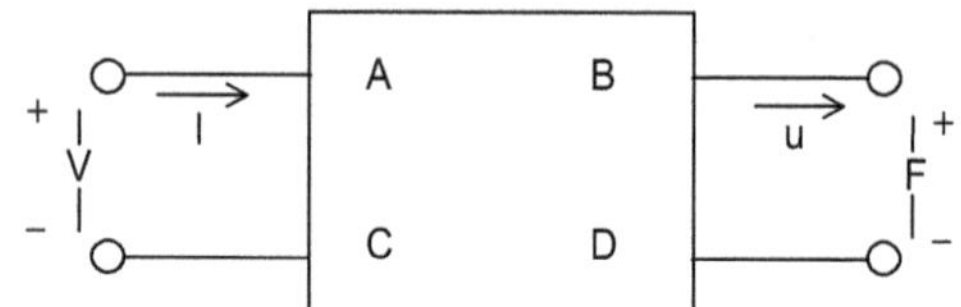

Fig. 3.45 *ABCD* electromechanical representation

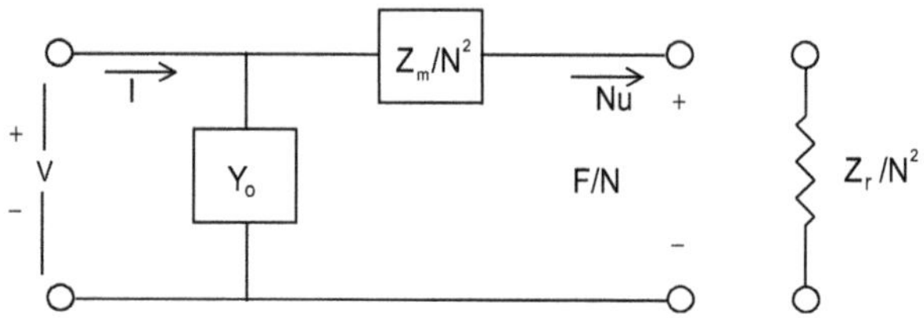

Fig. 3.46 Alternative electrical representation of equivalent circuit of Fig. 3.15

$$V = AF + Bu, \tag{3.65a}$$

$$I = CF + Du. \tag{3.65b}$$

One can see from Eqs. (3.65a) and (3.65b) that the coefficients may be written as

$$A = V/F|_{u=0}, \quad B = V/u|_{F=0}, \quad C = I/F|_{u=0}, \quad \text{and} \quad D = I/u|_{F=0}. \tag{3.66}$$

Equation (3.65a) divided by Eq. (3.65b) gives the electrical impedance, $Z_e = V/I$,

$$Z_e = (AF + Bu)/(CF + Du). \tag{3.67}$$

As a projector F/u is the radiation load, Z_r, on the transducer, and Eq. (3.67) becomes

$$Z_e = (AZ_r + B)/(CZ_r + D). \tag{3.68}$$

The *ABCD* representation of transducers is particularly useful in transducer array analysis because it separates the radiation impedance of each transducer from the other transducer parameters. Inspection of the piezoelectric equivalent circuit of Fig. 3.15 allows us to construct Fig. 3.46. The mechanical side has been transformed to the electrical side through the electromechanical transformer of turns ratio N, the admittance is $Y_0 = G_0 + j\omega C_0$ and the mechanical impedance of the transducer is $Z_m = R + j(\omega M - 1/\omega C^E)$. Thus the *ABCD* parameters represent the transducer alone, and the radiation impedance, Z_r, is a separate component that could be attached to the electrical circuit of Fig. 3.46 as Z_r/N^2. In a typical array all the transducers are identical with the same *ABCD* parameters, but the radiation impedance varies from one transducer to another because of the mutual impedances (see Sect. 7.2). Thus, it is convenient to have the variable part separated from the fixed part.

The *ABCD* parameters may be calculated for any two port transducer using the definitions given in Eq. (3.66). Figure 3.46 is also a convenient representation for calculating the *ABCD* values for cases that can be reduced to a simple lumped circuit model as in Fig. 3.15. This is often valid in the vicinity of the fundamental resonance of the transducer. Under these conditions Eq. (3.66) yields

$$A = 1/N, \quad B = Z_m/N, \quad C = Y_0/N \quad \text{and} \quad D = Y_0 Z_m/N + N. \tag{3.69}$$

None of the *ABCD* parameters are purely electrical or purely mechanical, all have electromechanical characteristics, and except for A, are strongly frequency dependent through Y_0 and Z_m. The electromechanical transformer ratio, N, is independent of frequency in most cases.

3.4 Finite Element Models

Lumped parameter, distributed circuit, and cascade matrix models have been presented for mechanical and electromechanical one-dimensional systems. It was demonstrated how multiple lumped models may be extended to represent distributed transducer systems, which could then be used for further analysis. The finite element method [28–30] takes the opposite approach and reduces distributed systems to a three-dimensional array of a large number of lumped or discrete elements spatially distributed throughout the transducer. Accurate models of complicated transducers can be readily obtained through the reduction of element size with a proportional increase in the number of elements. Readily available high speed computers with large memory and user oriented finite element computer programs that include piezoelectric, magnetostrictive, moving coil, MEMS and acoustic radiation elements, have revolutionized the design of transducers. It is possible to develop and design complicated transducers and expect predicted results to agree very well with measured results, if accurate material properties are known. We briefly describe the finite element method (FEM, also called FEA for finite element analysis) and related transducer analysis techniques in this section.

3.4.1 A Simple FEM Example

An introductory finite element model technique is presented here to illustrate the underlying principles. A more extensive development is beyond the scope of this book but can be found in a number of references [28–30]. We begin by considering the example of a tapered bar [28] of Young's modulus Y, density ρ, total length L, and variable cross-sectional area A_i. The bar is illustrated in Fig. 3.47 divided into a number of discrete elements of length, L_i, connected at nodes, i, moving with displacement, x_i. In contrast to wave notation where nodes are planes of no

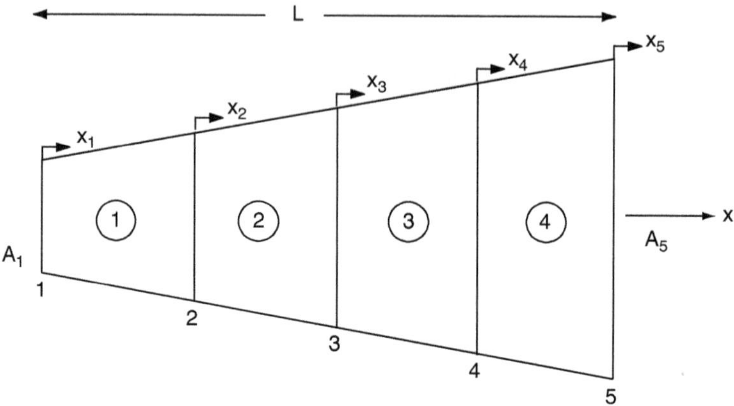

Fig. 3.47 Bar with variable cross-sectional area A_i

vibration, finite element nodes are the points of motion and appropriate forces and masses are associated at these nodal points.

The stiffness occurs between two nodal points and is part of the discrete element. The stiffness of each element is $K_i = YA_i^*/L_i$ where the average area $A_i^* = (A_i + A_{i+1})/2$. The stiffness reaction force of each element is then

$$F_i = K_i(x_{i+1} - x_i). \tag{3.70}$$

The mass of each element is the product of the density, element length, and the average area, i.e.,

$$m_i = \rho L_i A_i^* = \rho L_i A_i/2 + \rho L_i A_{i+1}/2. \tag{3.71a}$$

The scheme for this type of element is to associate a mass, M_i, with each node, x_i, by adding one half the element mass to the left of the node, m_{i-1} and one half the element mass to the right of the node, m_i:

$$M_i = m_{i-1}/2 + m_i/2. \tag{3.71b}$$

At the first and last nodes of Fig. 3.47 we get only $M_1 = \rho A_1 L_1/2$ and $M_5 = \rho A_5 L_4/2$ since there is no structure to the left of node x_1 and none to the right of node x_5. Thus the nodes in the middle of the bar have an associated mass that is the average of the mass of the elements on each side, while the nodes on the ends have an associated mass that is one-half the mass of the end elements.

The mechanical lumped model representation of the tapered bar of Fig. 3.47 is shown in Fig. 3.48 where the stiffness $K_i = YA_i^*/L_i$. A simple representation of one of the elements is shown in Fig. 3.49 where we have added excitation or loading forces F_i and F_{i+1}. The model may be generalized by associating a separate Young's modulus, Y_i, and density ρ_i with each element of length L_i and average cross-sectional area A_i^*.

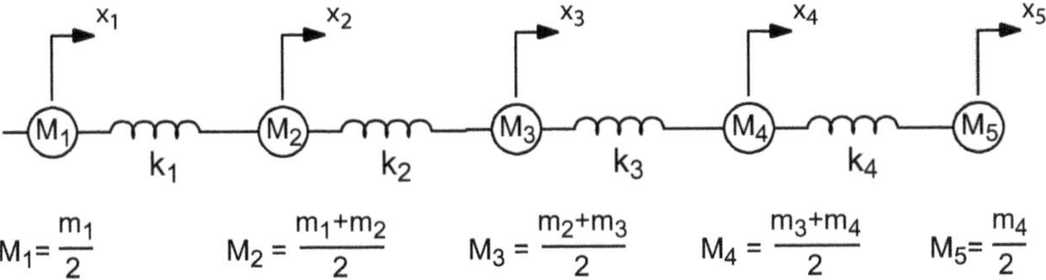

Fig. 3.48 Mechanical lumped mode model of the bar of Fig. 3.47

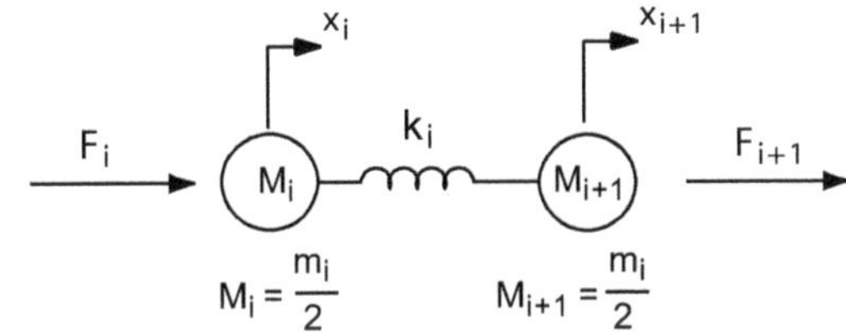

Fig. 3.49 Lumped representation of a single element

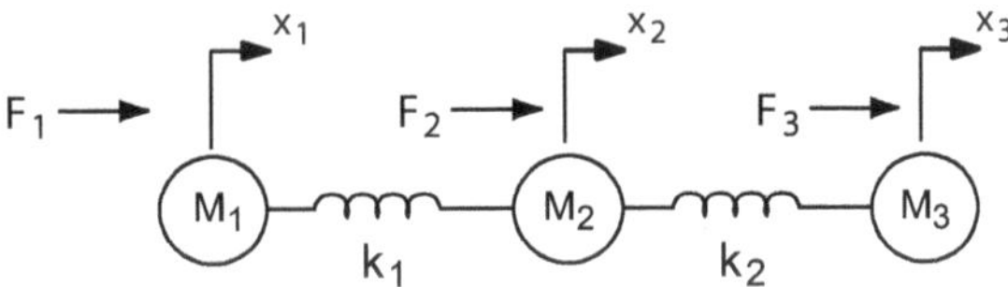

Fig. 3.50 Two spring lumped element representation of a bar

3.4.2 *FEA Matrix Representation*

The determination of the displacements for a given force generally requires solution of a set of simultaneous equations. Many powerful equation solvers have been developed for this most computationally intensive part of the computer program. The equation set is initially cast in a matrix form in terms of "global" coordinates, which are relative to a coordinate reference system. Consider, for simplicity, a two-spring-element representation of a bar as illustrated in Fig. 3.50 with forces F_1, F_2, F_3 at nodes 1, 2, and 3 with displacements x_1, x_2, and x_3.

The equations of motion for the three masses M_1, M_2, and M_3 may be written as

$$M_1 \mathrm{d}^2 x_1/\mathrm{d}t^2 = F_1 - K_1(x_1 - x_2), \tag{3.72a}$$

$$M_2 \mathrm{d}^2 x_2/\mathrm{d}t^2 = F_2 - K_2(x_2 - x_3) - K_1(x_2 - x_1), \tag{3.72b}$$

$$M_3 \mathrm{d}^2 x_3/\mathrm{d}t^2 = F_3 - K_2(x_3 - x_2). \tag{3.72c}$$

The above equation set may be recast into a more systematic form as

$$M_1 \mathrm{d}^2 x_1/\mathrm{d}t^2 + K_1 x_1 - K_1 x_2 + 0x_3 = F_1, \tag{3.73a}$$

$$M_2 \mathrm{d}^2 x_2/\mathrm{d}t^2 - K_1 x_1 + (K_1 + K_2) x_2 - K_2 x_3 = F_2, \tag{3.73b}$$

$$M_3 \mathrm{d}^2 x_3/\mathrm{d}t^2 + 0x_1 - K_2 x_2 + K_2 x_3 = F_3, \tag{3.73c}$$

which is readily identified with the matrix equation

$$\begin{pmatrix} M_1 & 0 & 0 \\ 0 & M_2 & 0 \\ 0 & 0 & M_3 \end{pmatrix} \begin{pmatrix} \mathrm{d}^2 x_1/\mathrm{d}t^2 \\ \mathrm{d}^2 x_2/\mathrm{d}t^2 \\ \mathrm{d}^2 x_3/\mathrm{d}t^2 \end{pmatrix} + \begin{pmatrix} K_1 & -K_1 & 0 \\ -K_1 & K_1 + K_2 & -K_2 \\ 0 & -K_2 & K_2 \end{pmatrix} \begin{pmatrix} x_1 \\ x_2 \\ x_3 \end{pmatrix} = \begin{pmatrix} F_1 \\ F_2 \\ F_3 \end{pmatrix}. \tag{3.74}$$

We see the presence of both a mass and a stiffness matrix with acceleration, displacement, and force vectors which may be written, in bold matrix notation, as

$$\mathbf{M}\mathbf{d}^2\mathbf{x}/\mathbf{d}\mathbf{t}^2 + \mathbf{K}\mathbf{x} = \mathbf{F}, \tag{3.75}$$

and, under sinusoidal conditions, as

$$\left(\mathbf{K} - \omega^2 \mathbf{M}\right)\mathbf{x} = \mathbf{F}. \tag{3.76}$$

Inspection of the elements of the stiffness matrix in Eq. (3.74) shows the inclusion of the separate stiffness sub-matrices for the springs of stiffness K_1 and K_2 in the form

$$\begin{pmatrix} K_1 & -K_1 \\ -K_1 & K_1 \end{pmatrix} \quad \text{and} \quad \begin{pmatrix} K_2 & -K_2 \\ -K_2 & K_2 \end{pmatrix},$$

which are associated with the "local" element stiffness matrices with nodal displacements x_1, x_2 and x_2, x_3. This example illustrates the combination of two "local" spring elements into the "global" matrix of Eq. (3.74). Often hundreds or thousands of elements are combined to form the global matrix from which the solution is obtained.

A solution for the displacement $\mathbf{x}$ for a given force is usually desired and in this case Eq. (3.76) yields the harmonic response. Equation (3.76) takes a homogenous form with $\mathbf{F} = \mathbf{0}$ and the solution of $\left(\mathbf{K} - \omega^2 \mathbf{M}\right) = \mathbf{0}$ yields the modal resonant frequencies or eigenvalues of the transducer under air (or more strictly, vacuum) loading conditions. Equation (3.76) may be extended, through the force, F, to include damping with resistance, R, drive voltage, V, and electromechanical turns ratio, N, written as

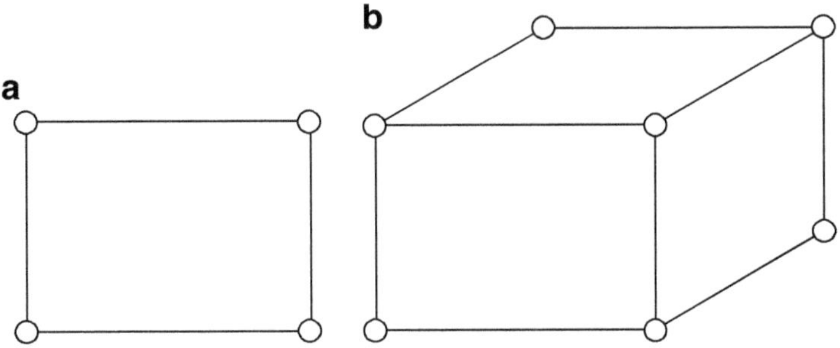

Fig. 3.51 (**a**) Two-dimensional 4-node element. (**b**) Three-dimensional 8-node element

$$\left[\mathbf{K} + \mathbf{j}\omega\mathbf{R} - \omega^2\mathbf{M}\right]\mathbf{x} = \mathbf{N}\mathbf{V}. \tag{3.77}$$

This one-dimensional model is a simplified introduction to the finite element method; however, the matrix formalism can be readily extended to two and three dimensions. The spring and mass elements are only two of the many elements available in the commercial programs, which include beams, plates, shells, 2-dimensional 4 node structural elements, 3-dimensional 8 node structural solids, piezoelectric elements, fluid elements, and acoustic absorbers. Some of these elements include interpolation functions, based on variational procedures, to improve the accuracy of a single element. Two-dimensional four node, with two degrees of freedom per node, and three-dimensional eight node, with three degrees of freedom per node, elements are illustrated in Fig. 3.51a, b. The discussion that follows applies to one-, two-, and three-dimensional systems.

The fluid elements use pressure as the dependent variable rather than displacement and require a fluid structural interface element, FSI, when connected to a structural element. The gradient of the pressure is used to obtain the velocities in given directions. Computational speed is increased by use of the inherent symmetry of the transducer under consideration. A three-dimensional case that can be reduced to an axi-symmetric case will run as fast as two-dimensional cases. Other symmetries may be imposed using rigid boundary conditions at the nodes as well as built in symmetry planes. Automatic meshing and post-processing are useful aids in setting up the model and obtaining the desired output.

3.4.3 Inclusion of a Piezoelectric Finite Element

Some commercial finite element programs contain piezoelectric elements that can be used for transducer modeling [31–33]. The finite element piezoelectric model can be based on the matrix equation set

$$\mathbf{T} = \mathbf{c}^{\mathbf{E}}\mathbf{S} - \mathbf{e}\mathbf{E}, \tag{3.78a}$$

$$\mathbf{D} = \mathbf{e}^{\mathbf{t}}\mathbf{S} + \boldsymbol{\varepsilon}^{\mathbf{S}}\mathbf{E}, \tag{3.78b}$$

where $\mathbf{c}^{\mathbf{E}}$ is the short circuit elastic stiffness matrix, $\boldsymbol{\varepsilon}^{\mathbf{S}}$ is the clamped permittivity matrix, $\mathbf{e}$ is the piezoelectric constant matrix, and $\mathbf{e}^{\mathbf{t}}$ is the transpose of $\mathbf{e}$. Equations (3.78a) and (3.78b) may be rewritten as

$$\mathbf{K}^{\mathbf{E}}\mathbf{x} - \mathbf{N}\mathbf{V} = \mathbf{F}_{\mathbf{t}}, \tag{3.79a}$$

$$\mathbf{N}^{\mathbf{t}}\mathbf{x} + \mathbf{C}^{\mathbf{S}}\mathbf{V} = \mathbf{Q}, \tag{3.79b}$$

where $\mathbf{K}^{\mathbf{E}}$ is the short circuit stiffness matrix, $\mathbf{N}$ is the electromechanical turns ratio matrix, and $\mathbf{C}^{\mathbf{S}}$ is the clamped electrical capacitance matrix. The force, $\mathbf{F}_{\mathbf{t}}$, is the total force vector which can include the radiation loading force and the blocked acoustic force $\mathbf{F}_{\mathbf{b}}$, $\mathbf{Q}$ is the charge vector, $\mathbf{x}$ is the displacement vector, and $\mathbf{V}$ is the voltage vector.

With $\mathbf{M}$ the mass and $\mathbf{R}$ the resistance $\mathbf{F}_{\mathbf{t}} = \mathbf{F}_{\mathbf{b}} - \mathbf{M}\mathbf{d}^2\mathbf{x}/\mathbf{dt}^2 - \mathbf{R}\mathbf{dx}/\mathbf{dt}$, and Eq. (3.79a) may be written as

$$\mathbf{M}\,\mathbf{d}^2\mathbf{x}/\mathbf{dt}^2 + \mathbf{R}\,\mathbf{dx}/\mathbf{dt} + \mathbf{K}^{\mathbf{E}}\mathbf{x} - \mathbf{N}\mathbf{V} = \mathbf{F}_{\mathbf{b}}. \tag{3.80a}$$

We may also rewrite Eq. (3.79b) in a similar form as

$$\mathbf{0}\,\mathbf{d}^2\mathbf{x}/\mathbf{dt}^2 + \mathbf{0}\,\mathbf{dx}/\mathbf{dt} + \mathbf{N}^{\mathbf{t}}\mathbf{x} + \mathbf{C}^{\mathbf{S}}\mathbf{V} = \mathbf{Q}. \tag{3.80b}$$

The above equation set leads to the coupled matrix equation written as

$$\begin{pmatrix} M & 0 \\ 0 & 0 \end{pmatrix}\begin{pmatrix} \ddot{x} \\ \ddot{V} \end{pmatrix} + \begin{pmatrix} R & 0 \\ 0 & 0 \end{pmatrix}\begin{pmatrix} \dot{x} \\ \dot{V} \end{pmatrix} + \begin{pmatrix} K^E & -N \\ N^t & +C^S \end{pmatrix}\begin{pmatrix} x \\ V \end{pmatrix} = \begin{pmatrix} F_{\mathrm{b}} \\ Q \end{pmatrix}. \tag{3.81}$$

In this representation each matrix is composed of four sub-matrices and the vectors are composed of two vectors each. This is the matrix representation for coupling the mechanical and piezoelectric elements in a finite element model. For $N = 0$ the equations uncouple and become separate electrical and mechanical equations.

3.4.4 Application of FEA Without Water Loading

In practice the user need not be concerned with the details of the mathematical process any more than one needs to be concerned with the mathematical details behind a circuit analysis computer program. We consider now the application and utility of FEA (FEM) transducer models under air loading conditions.

The transducer design process usually begins with a goal or specification, which sets the operating frequency band as well as the desired output and approximate size. In order to obtain high acoustic output over a frequency band it is usually necessary to make the resonance occur near the center of the band, although doing so for a given size can be a problem. Although a particular design might achieve the target resonance frequency, it may not meet other requirements unless it has a significant effective coupling coefficient. The modal analysis section of FEA programs allows the evaluation of the modal resonance and antiresonance frequencies from which the effective coupling coefficient is obtained (see Chaps. 2, 4, and 9).

A mechanical based FEA program that lacks a piezoelectric model may still be used to evaluate transducer designs in cases where the piezoelectric drive stack exhibits one-dimensional motion. In the modulus substitution method [34] the model is constructed with the piezoelectric section treated as a mechanical section but with the density and Young's modulus of the piezoelectric material. The FEA program is run twice: the first time with the short circuit value of the Young's modulus, Y^E, yielding a fundamental modal resonance frequency f_r and then run again, with the open circuit value of the Young's modulus, Y^D, yielding the antiresonance frequency f_a. As presented in Chaps. 2 and 9, the effective coupling coefficient may then be calculated from

$$k_{\mathrm{eff}} = \left[1 - (f_{\mathrm{r}}/f_{\mathrm{a}})^2\right]^{1/2} \tag{3.82}$$

If the FEA program has a piezoelectric element, the piezoelectric parameters are entered and the resonant frequencies, f_r and f_a, are evaluated through the modal option under short circuit and open circuit conditions, respectively. This evaluation is more accurate and is not limited to one-dimensional motion of the piezoelectric elements. The modal option often allows a useful dynamic animation of the vibration which gives a detailed picture of the motional physics involved. An example of the static and peak dynamic motion is shown in Fig. 3.52a, b for a Tonpilz transducer with and without head flexure computed for two different wave speeds in the head.

The number of modal frequency results can be overwhelming unless limited to the frequency band of interest. A simple circuit or matrix model computer program is often used initially to obtain estimates of the fundamental resonance and other performance features. Animation of vibration modes may also be obtained under a frequency sweep response option where the actual coupling to the modes may be evaluated.

Although Eq. (3.82) is the most commonly used dynamic method for obtaining the coupling coefficient, alternative static methods are available. Since the effective coupling coefficient is also given by $k_{\mathrm{eff}} = \left[1 - K^E/K^D\right]^{1/2}$ we may use the FEA static option to obtain k_{eff} by evaluating the displacement of the motional surface, x, under short circuit, x^E, and open circuit, x^D, conditions. Consider the displacement, x, of the simple electric field transducer example of Fig. 3.53 with the radiation force, F_r, in the figure replaced by a static force, F_s.

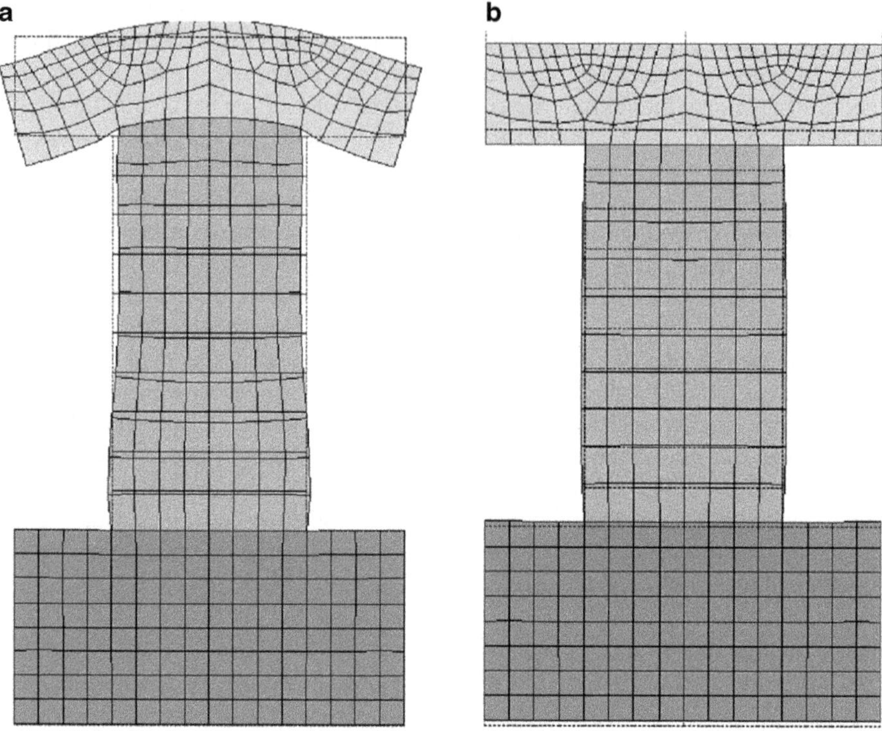

Fig. 3.52 (**a**) Transducer dynamic motion with flexing head. (**b**) Transducer dynamic motion with stiffened head

Fig. 3.53 Lumped model of one-degree of freedom piezoelectric transducer

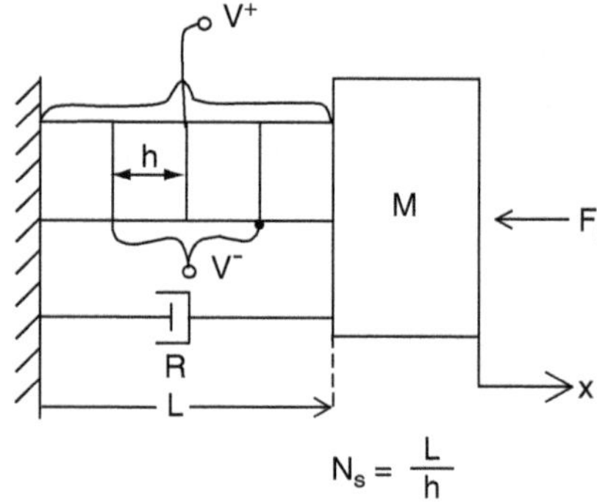

For a given F_s, the stiffness $K^E = F_s/x^E$ and $K^D = F_s/x^D$ and the above formula yields

$$k_{\mathrm{eff}} = \left[1 - x^D/x^E\right]^{1/2}, \tag{3.83}$$

where the displacements under short and open circuit conditions, x^E and x^D, are easily determined from the FEA program.

We may also obtain the coupling coefficient from the clamped (blocked) and free electrical capacitances C_0 and C_f since $k_{eff} = [1 - C_0/C_f]^{1/2}$. Again consider Fig. 3.53, but this time with $F_r = 0$. With the clamped charge, $Q_0 = C_0V$, and the free charge, $Q_f = C_fV$, the formula now takes the form

$$k_{eff} = [1 - Q_0/Q_f]^{1/2}. \tag{3.84}$$

Thus, evaluation of the charge with the motional surface clamped (the mass M of Fig. 3.53 supported with finite element "rollers") and with the surface free to move yields the effective coupling coefficient.

The static option may also be used to obtain an effective d constant. Under static application of voltage V, the displacement, x, of the free motional surface ($F_r = 0$ in Fig. 3.53) is related to the effective d constant by

$$d_{eff} = x/V. \tag{3.85}$$

Evaluation of x with a voltage of $V = 1$ yields d_{eff}, determined directly by FEA. A good design requires the d_{eff} values to be close to the material d_{33} values. (The material $d_{33} = S_3/E_3|_{T=0} = x_3/V$ where x_3 is the displacement of the piezoelectric drive section with voltage V and field E_3). In deep submergence applications the ability of the transducer to withstand high pressure can also be evaluated using the static option.

FEA transducer models include mechanical damping, but mechanical loss parameters are often not well known. However, if in-air measurements of the conductance are available, the value of the mechanical dampers or dissipation factors in the FEA transducer model may be adjusted to attain the same mechanical Q_m as that given by the measured conductance curves (see Chap. 9). Once matched, this allows the FEA prediction of the mechanical efficiency of the transducer under water-loaded conditions.

3.4.5 Application of FEA with Water Loading

The finite element acoustic medium uses fluid elements that describe the pressure field with pressure values at the nodes of the elements, in contrast to the mechanical elements with displacement values at the nodes. Consequently a fluid surface interface, FSI, element is needed to join the mechanical and fluid elements at the surface of the transducer and its housing. At the outer side of the fluid field "ρc" matched absorbers are often used to satisfy the radiation condition of no reflection from the far field. Three-dimensional acoustic fluid elements are available that can satisfy all three conditions of interface, fluid element, and absorber. There are also special spherical elements that apply infinite acoustic continuation of the wave in addition to absorbing the wave. These spherical wave elements require a spherical fluid field with a specific coordinate center.

The fluid-structure interaction is coupled through a pair of mechanical and fluid matrix equations that may be written as [31]

$$[M_s]\{\ddot{x}\} + [K_s]\{x\} = \{F_s\}+[R]\{P\}, \tag{3.86}$$

$$[M_f]\{\ddot{P}\} + [K_f]\{P\} = \{F_f\} + \rho[R]^t\{\ddot{x}\}, \tag{3.87}$$

where [] means matrix and { } means column vector, $[M]$ is the mass matrix, $[K]$ is the stiffness matrix, $\{F\}$ is the force vector, $\{P\}$ is the pressure vector, $\{x\}$ is the displacement vector, ρ is the density of the medium, and $[R]$ represents the coupling matrix and the effective surface area at each node between the fluid equivalent and mechanical structure with subscripts f and s, respectively. These equations imply that the nodes on a fluid structure interface have both displacement and pressure degrees of freedom that allows the connection between the two systems. The effective fluid loads can be generated from

$$F_f = -AL\rho\partial^2 x_n/\partial t^2, \tag{3.88}$$

where A and L are the area and distance associated with the nodes of the fluid element with density ρ while x_n is the outward normal displacement.

Two-dimensional and axi-symmetric FEA transducer models may be constructed initially without a fluid field to simulate operation in air and evaluated to determine the motion at the fundamental mode and other higher modes of interest. This is usually done at resonance, f_r, with the electrodes set at zero voltage and at antiresonance, f_a, with the electrodes removed. Then the dynamic effective coupling coefficient is given by $k_{eff} = \left[1 - (f_r/f_a)^2\right]^{1/2}$. This procedure is followed by the application of 1 unit of static pressure on the radiating surface that would be in contact with the medium to determine the stress on critical parts of the transducer.

Finally, the FEA fluid field is added to the model and terminated with absorbers at an appropriate distance from the transducer. This distance is normally in the far field but can be made closer if near-field to far-field techniques are used, reducing the number of water elements and computational time. An example of an axi-symmetric finite element model of a Tonpliz transducer with fluid field, rigid baffle, and absorbers is shown in Fig. 3.54.

The absorbers have a characteristic impedance equal to that of the fluid. The rigid baffle condition is obtained by no imposed condition on the nodes of the fluid element which is a condition of no displacement for a fluid element. On the other hand, for solid elements, as in the transducer, no imposed condition on the element nodes is a free or zero stress condition. The rigid baffle case is an intermediate step which is normally followed by a water field which surrounds the transducer with an enclosure simulating a possible transducer housing. In the case of three-dimensional FEA models the fluid field is initially built as part of the transducer,

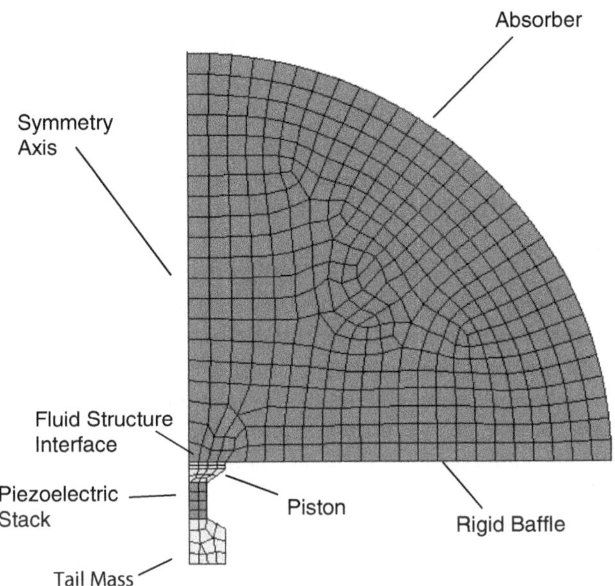

Fig. 3.54 Axi-symmetric finite element model of a Tonpilz transducer set in a rigid baffle with fluid field and absorbers

temporarily removed to obtain the unloaded results and then replaced to obtain the fluid-loaded results. This procedure is necessary because of the complexity of the 3-D structure and its interface with the fluid load.

Near-field FEA results for a small array of 16 close-packed piezoelectric Tonpilz transducers with square radiating faces vibrating in water are shown in Fig. 7.17a, b of Projector Arrays, Chap. 7. The contour plots in Fig. 7.17a show the pressure magnitude, while those in Fig. 7.17b show the normal displacement magnitude, on the surface of the array. Both plots are symmetrical about the two lines that bisect the array and about the two diagonals because all the transducers are driven in phase with the same voltage magnitude. This symmetry was used to reduce the size of the computation to one quadrant of the array with appropriate boundary conditions on the sides of the quadrant that join the rest of the array.

The fluid field must be large enough to include the essential parts of the near field for an accurate determination of the radiation mass loading. For example, an acoustically small circular piston of radius a has a radiation mass of equivalent volume equal to $8a^3/3$ and thus, a height of approximately $0.85a$ (see Sect. 13.13). Thus, in this case the FEA water field must extend at least a distance equal to $0.85a$. If fluid absorber elements are used at the termination of the FEA water field they should be small and on a surface that conforms to the wave-front of the radiated field since the absorption is most effective at normal incidence. Ideally the water field is terminated with absorbers on a spherical surface at a distance greater than the near- to far-field transition, i.e., at the Rayleigh distance of $R \approx 2a^2/\lambda$ (see Chap. 10). This avoids boundary effects in the near field and also allows the acoustic pressure to be evaluated at the nodes of fluid elements on a sphere to obtain a beam pattern, as would be done by measurements in a test facility. The size of the fluid elements should ideally be less than a tenth of a wavelength for accurate results. Unfortunately, the above conditions often require long computational run times, especially if the transducer surface is acoustically large or an array of transducers is being evaluated. In such cases the size of some of the fluid elements can be increased to nearly a quarter wavelength to achieve practical run times, but with reduced accuracy.

Some transducer-specific finite element programs, such as Atila [32], avoid the need for large fluid fields by evaluating the pressure and velocity on a closed surface near the transducer, and then using a Helmholtz Integral approach (see Chap. 11) for calculating the far-field pressure and beam patterns. This is an example of the feasibility of combining analytical methods with FEA to reduce run time or increase the size of the problems that can be handled. A Helmholtz Integral subroutine could be added to other FEA programs after determining the minimum fluid field required.

The near-field measurement techniques, to be discussed in Chap. 9, can also be combined with FEA. For example, if the pressure, $p(0)$, on the center of a transducer that can be approximated as a circular piston of radius $a \ll \lambda$ is determined by FEA, the far-field pressure, p_f, is given by $p(0)a/2$ with a rigid baffle or $p(0)a/3$ without a rigid baffle. The pressure on the center can be found with a smaller FEA fluid field than that required to find the far field. If the piston is acoustically larger and set in a rigid baffle, Rayleigh's integral, Eq. (10.26) with $\theta = 0$ for on-axis field points,

$$p_f(r) = -2\pi\rho f^2 \left(e^{-jkr}/r\right) \iint x_n dA, \tag{3.89}$$

may be used to obtain the on-axis far-field pressure from the normal displacement, x_n, on the surface of the piston. A relatively small FEA fluid field is sufficient to find the surface displacement. The surface integral may be replaced by a summation for the average displacement, $\langle x_n \rangle$. If the total area is A, we then have

$$p_f(r) = -2\pi\rho f^2 \left(e^{-jkr}/r\right) A \langle x_n \rangle, \tag{3.90}$$

and at a reference distance of $r = 1$ meter we get, simply, $|p_f| = 2\pi\rho f^2 A \langle x_n \rangle$. In these approximate methods it is essential to determine the minimum fluid field required for satisfactory accuracy.

3.4.6 *Water Loading of Large Arrays*

The FEA run time of a large array can be very long and therefore the analysis is usually limited to representative sections of a large array. Alternatively, large planar array conditions can be approximately simulated for closely packed arrays by assuming the array to be large enough that, except for the transducers on the edge, each transducer experiences the same loading condition, and only one transducer needs to be analyzed. For example, the case of a large rectangular array of square piston transducers could be analyzed by considering each transducer to be surrounded by a square rigid-walled waveguide with an absorbing element at the open end. If there were gaps between the transducers, the waveguide walls would be positioned in the center of the gaps. The finite element program could then be run at frequencies of interest for the displacement distribution on the surface of the piston and for the pressure in the waveguide water column.

The far-field pressure of the array could be estimated from the displacement or velocity distribution using Eq. (10.25a) of Chap. 10, or from the pressure in the waveguide. The latter is easier because the waveguide constrains the pressure to form an approximate plane wave with a value p_p, determined by the FEA calculation, and with an associated particle velocity of $p_p/\rho c$. Then combining all the waveguides makes the entire array have the same uniform normal velocity and, assuming full ρc loading, a total radiated power of $W = NA\left(p_p^2/2\rho c\right)$ where N is the number of transducers in the array and A is the area of each transducer including gaps. The power is related, through the directivity factor, to the far-field intensity, and consequently the pressure, p, at a distance r in the broadside direction by

$$p^2/2\rho c = D_f W/4\pi r^2. \tag{3.91}$$

Using $D_f = 4\pi NA/\lambda^2$ for the array directivity factor from Sect. 7.1.2 gives

$$p = NAp_p/\lambda r = NAfp_p/cr, \tag{3.92}$$

as a simple relation between the FEA calculated value of p_p and the broadside response of the array, p. This is an approximate solution since the transducers on the edges have less acoustic loading than the central transducers. Thus the validity depends on the relative number of transducers on the edges (e.g., in a square array of 400 transducers, nearly 20 % are on the edges). If these elements receive half the loading, the assumed full loading estimate might be high as 10 %.

3.4.7 Magnetostrictive FEA

There is a limited availability of FEA programs with magnetostrictive elements [32, 33], possibly because magnetostrictive devices are not used as much as piezoelectric devices. On the other hand, since one transducer is the dual of the other, piezoelectric elements [31] may be used to simulate magnetostrictive elements [35]. Consider the simple magnetostrictive model of Fig. 3.19 and the corresponding Eqs. (3.19) and (3.22) repeated here as

$$(M + M_r)\mathrm{d}u/\mathrm{d}t + (R + R_r)u + \left(1/C^H\right)\int u\,\mathrm{d}t = N_m I, \tag{3.93a}$$

$$V = L_0 \mathrm{d}I/\mathrm{d}t + N_m u, \tag{3.93b}$$

with

$$N_m = N_t dA/Ls^H,\ \ L_0 = \mu^S N_t^2 A/L,\ \ C^H = s^H L/A,\ \ \text{and}\ \ N_t = \text{number of turns.} \tag{3.93c}$$

The corresponding piezoelectric model was given as Fig. 3.14 and is repeated here as Fig. 3.53 with the piezoelectric bar now composed of four sections. The original associated set of equations, Eqs. (3.10) and (3.12a), was restricted, for simplicity, to one piezoelectric section. Multiple parallel wired sections, presented in Sect. 3.2.3.1, can readily be included, and Eqs. (3.10) and (3.12a) can then be written in terms of the number of sections, $N_s = L/h$, as

$$(M + M_r)\mathrm{d}u/\mathrm{d}t + (R + R_r)u + \left(1/C^E\right)\int u\,\mathrm{d}t = NV, \tag{3.94a}$$

$$I = C_0 \mathrm{d}V/\mathrm{d}t + Nu, \tag{3.94b}$$

with

$$N = N_s dA/Ls^E,\ \ C_0 = \varepsilon^S N_s^2 A/L,\ \ C^E = s^E L/A,\ \ \text{and}\ \ N_s = \text{number of sections.} \tag{3.94c}$$

A term-by-term comparison of these equations shows that a piezoelectric element may be used to represent a magnetostrictive element if we let voltage represent current, current represent voltage, and admittance represent impedance for a transduction section of the same length and cross-sectional area. Further details include the replacement of the clamped dielectric constant by the clamped permeability ($\varepsilon^S \Rightarrow \mu^S$), the short circuit elastic modulus by the open circuit elastic modulus ($s^E \Rightarrow s^H$), and the piezoelectric d constant by the magnetostrictive d constant $\left(d_{\text{piezo}} \Rightarrow d_{\text{mag}}\right)$. In addition the number of piezoelectric sections must equal the number of coil turns ($N_t = N_s$). But, if the number of coil turns is large, the number of piezoelectric pieces may be made smaller as long as the conditions $N_s d_{\text{piezo}} = N_t d_{\text{mag}}$ and $\varepsilon^S N_s^2 = \mu^S N_t^2$ are maintained. The equivalence may also be seen through a comparison of the equivalent circuits of Figs. 3.15 and 3.21. These equivalent circuits remind us of the limitation of this FEA substitution model in that wire heating losses and eddy current losses are not included as well as flux leakage and demagnetizing effects discussed in Sect. 3.1.4. However, these effects may be separately evaluated through a magnetic finite element model and incorporated into the FEA magnetostrictive parameters.

We have presented an introductory foundation for finite element modeling of piezoelectric transducers and also a piezoelectric based model for magnetostrictive transducers. A more detailed development of finite element modeling and its capabilities can be found in references [28–30]. The finite element programs can provide accurate models of most transducers and can significantly reduce the number of experimental models that need to be fabricated and tested. The best approach is to use FEA together with circuit analysis or cascade matrix models to provide cross checks and reveal possible modeling errors, should they exist.

3.4.8 Equivalent Circuits for FEA Models

FEA models can be accurate and provide the desired performance on transducers and arrays over a wide range of frequencies. However, an equivalent circuit can provide a simplified picture of the transducer operation, a dummy load for amplifier testing, a means for rapid design optimization, and a transducer element model for rapid evaluation of large array interaction. We present here a simple means for determining the lumped parameters of an equivalent circuit for the motion along the major axis of an FEA transducer model.

Examples of a one degree of freedom transducer are: a ring or spherical transducer where the main motion is radial (see Sect. 5.2) or a Tonpilz with a very massive tail mass where the main motion is the lighter head mass (see Sects. 3.1.3 and 5.3.1). With essentially just one mass moving, the electromechanical circuit of Fig. 5.4 of Chap. 5 is applicable. The reactive equivalent circuit parameters are most easily evaluated under low loading impedance conditions, such as in air. At very low frequencies the mechanical impedance of the compliance, C^E, is $1/j\omega C^E$ and becomes very large compared to the other impedances, $j\omega M$, R_{m}, and Z_{r}. In this case the compliance becomes the dominant component and essentially the only component across the secondary terminals of the transformer of turns ratio N. Under these conditions, the total free capacity at the input terminals is $C_{\mathrm{f}} = C_0 + N^2 C^E$.

We wish to evaluate the equivalent circuit components C_0, N, C^E, and M of Fig. 5.4 for a given FEA model. The effective coupling coefficient (see Sects. 1.4.1 and 4.4) may be written as $k^2 = N^2 C^E / C_{\mathrm{f}}$ so that the turns ratio, N, may be obtained from

$$N^2 = k^2 C_{\mathrm{f}} / C^E. \tag{3.95}$$

The free capacitance can be obtained from the FEA model from the low frequency susceptance $B = \omega C_{\mathrm{f}}$, with

$$C_{\mathrm{f}} = B/\omega \tag{3.96}$$

And here the short circuit mechanical compliance, C^E, may be also obtained from the FEA model by applying a force $F = 1$ Newton on the piezoelectric and obtaining the displacement, x, in the direction of the desired motion at the location of the radiating surface or attached piston. Under short circuit conditions the compliance is then given by

$$C^E = x/F, \tag{3.97}$$

For force $F = 1\,\mathrm{N}$ we get $C^E = x$. The effective coupling coefficient, k, is obtained from the FEA fundamental mechanical resonance, f_r, and antiresonance, f_a, through

$$k^2 = 1 - (f_r/f_a)^2. \tag{3.98}$$

With k^2, C_f and C^E we can get N through Eq. (3.95) and the clamped capacity, C_0, from

$$C_0 = C_f(1 - k^2). \tag{3.99}$$

Finally the structural mass, M, of the vibrating surface or piston may be obtained from the mechanical resonance condition $(2\pi f_r)^2 = 1/MC^E$ in the form

$$M = 1/C^E(2\pi f_r)^2. \tag{3.100}$$

The conductance $G_0 = \omega C_f \tan\delta$ may be calculated from the electrical dissipation factor $\tan\delta$ or, with more accuracy, the mechanical dissipation factor using the method described in Sect. 2.9. Mounting losses may be then evaluated from the FEA conductance. If the transducer is a two degree of freedom Tonpilz type with a tail mass of M_1, as in Fig. 3.18, the reduced mass is

$$M^* = MM_1/(M + M_1) = M/(1 + M/M_1) \tag{3.101}$$

and here $M^* = 1/C^E(2\pi f_r)^2$ (see Figs. 3.8 and 3.9). The mass ratio, M/M_1, may be obtained from the respective displacements x and x_1 since the momentum $Mx = M_1 x_1$ yielding

$$M/M_1 = x_1/x, \tag{3.102}$$

and this ratio may be evaluated from the FEA model displacements, x_1/x, of the masses for a given voltage, V, or force, $F = NV$.

Although a lumped equivalent circuit model can be useful, the accuracy is limited to frequencies in the vicinity of resonance and frequencies well below resonance with element sizes small compared to the associated wavelength. At frequencies above resonance the transducer often excites other modes of vibration, such as head flexure, distributed extensional higher order modes or lateral motions, which are not available in the simple lumped mode model. Moreover, transducer mounting-to-housing resonances can occur at frequencies below resonance. These modes of vibration and their effect on performance may be important in wide band systems. Although a more elaborate equivalent circuit model could be used, it may be better to use the two-port *ABCD* parameter system (see Sect. 3.3.2) to characterize the FEA modeled transducer at the frequencies of interest. If there are other modes, the *ABCD* parameters can then change to accommodate these modes as a function of frequency and affect the performance of the transducer element and, consequently, the array under consideration. The A, B, C, D values may be obtained at each frequency from the FEA model by using the Fig. 3.45 two-port block representation along with the boundary condition values of Eq. (3.66), repeated here as Eq. (3.103), with I the current and u the velocity.

$$A = V/F|_{u=0}, \quad B = V/u|_{F=0}, \quad C = I/F|_{u=0}, \quad \text{and} \quad D = I/u|_{F=0}. \tag{3.103}$$

3.5 Summary

In this chapter we discussed the current models used to describe piezoelectric and magnetostrictive underwater sound transducers including: lumped-parameter equivalent circuits, distributed circuits, matrix models, and finite element models. We started with the mechanical resonator and developed one-dimensional equivalent circuits for one, two and three degrees of freedom. A 33 mode piezoelectric bar was then considered as a lumped element part of the equivalent circuit, as used in Chap. 2. A lumped equivalent circuit for a magnetostrictive transducer was also developed.

Distributed circuit and matrix representations were introduced for mechanical systems and developed from a one-dimensional distribution of alternating masses and springs. The distributed circuit was shown to be in the form of a "T" network with a tangent function $[j\rho cA_0 \tan(kL/2)]$ in the "arms" and sine function $[-j\rho cA_0/\sin(kL)]$ in the central "body." And at low frequencies these functions reduce to a mass $M/2 = \rho cA_0L/2$ in the "arms" and a compliance $C = L/YA_0$ along with (an often neglected) negative mass term, $-M/6$, in series in the "body." In the model A_0 is the cross-sectional area, L is the length, and here k is the wave number, $2\pi/\lambda = \omega/c$, ρ is the density and c is the compressional sound speed in the material. Distributed wave models were then developed for a segmented 33 mode piezoelectric bar, an un-segmented 31 mode bar, a length expander 33 mode bar, a thickness mode plate, and a magnetostrictive rod.

A three-port distributed piezoelectric matrix model was developed and it was shown how a mechanical and electromechanical series of these models can be used to represent transducers and that an overall representation is obtained by matrix multiplication. A two-port model was then presented for voltage, V, and current, I, and force, F, and velocity, u, along with A, B, C, D transducer parameters, which can then be written as $V = AF + Bu$ and $I = CF + Du$ and used to represent a transducer with parameters determined by $A = V/F|_{u=0}$, $B = V/u|_{F=0}$, $C = I/F|_{u=0}$, and $D = I/u|_{F=0}$ at each frequency.

Finite element models offer accurate performance predictions of complicated transducer models. In this chapter an introductory finite element model technique was presented with a bar of variable cross section followed by the case of three masses, two springs along with a corresponding set of equations which was cast into a matrix form, allowing a matrix solution. This mechanical exercise was then augmented by the inclusion of piezoelectric finite elements. It was also shown how the FEM (Finite Element Model), also often called FEA (Finite Element Analysis) model, can be used to investigate unwanted flexural vibration of the transducer piston and means for obtaining effective coupling coefficient and "*d*" constant. These in-air means were followed by a discussion of in-water loading. A method for evaluating magnetostrictive transducers using piezoelectric modeling was also presented. Finally a means for determining equivalent circuit parameters from piezoelectric FEA models was also presented.

Exercises (Degree of Difficulty: *Lowest, **Moderate, ***Highest)

3.1.** Consider the dual mass mechanical resonator of Fig. 3.6. Assume loss resistance R_1 is negligible and $R_2 \ll \omega_r M_1$. Calculate the approximate resonance frequency for stiffness $K = 4 \times 10^{10}\,\mathrm{N/m}$ and head mass $M_2 = 1\,\mathrm{kg}$ for tail masses $M_1 = 1$, 2, 4, and 8 kg. Also calculate the mechanical Q_m with $R_2 = 6 \times 10^4\,\mathrm{Ns/m}$ for the above values of M_1. Does there appear to be much benefit gained in doubling the tail mass from 4 to 8 kg, considering the weight increase in the transducer?

3.2.* The equivalent circuits of Figs. 3.15 and 3.16 show a single resonance under short circuit conditions (or constant voltage drive) at the mechanical resonance frequency where $\omega_r M = 1/\omega_r C^E$. Show that antiresonance, $\omega_a = \omega_r/\left(1 - k^2\right)^{1/2}$ is obtained under open circuit conditions where, in this case, C_0/N^2 is now in series with C^E reducing the compliance to C^D raising the stiffness and the resonance frequency to the value ω_a.

3.3.** Expand the function $-j\rho c A_0/\sin(kL)$ and obtain $-j\rho c A_0(1/kL + kL/6)$ for small kL using trigonometric and binomial expansions; and thus, verify the $-M/6$ in Fig. 3.30.

3.4.* Demonstrate that the equivalent circuit of Fig. 3.30 may be used to represent an ideal mass, M, or any mass at low enough frequencies. Show that at very low frequencies the reduced equivalent circuit for a force F at terminals 2 and an air-loaded free condition, i.e., a short circuit, at terminals 1 is simply a mass M, as it should be.

3.5.** Show that Eqs. (3.35a) and (3.35b) may be written in the transfer matrix form of Eqs. (3.36a) and (3.36b).

3.6.* Show that the transmission line equivalent circuit of Fig. 3.35 may be reduced to the simple lumped mode representation of Fig. 3.18 (without G_0) for $kL = \omega L/c \ll 1$ if the $-M/6$ term is ignored, with a free condition at one end ($F_0 = 0$) and a radiation load of $F_n/u_n = R_r + j\omega M_r$ on the other end.

3.7.* Evaluate the A, B, C, D parameters of Eq. (3.69) using the electrical and mechanical parameters of Fig. 3.15.

3.8.* The finite element evaluation of the broad side far-field pressure from a large planar array (with most of the elements under array-loaded conditions) could require very large run times at each frequency. Consider the alternative approach of first evaluating the pressure of a single element by approximating its environment by a rigid fluid-filled waveguide with a ρc absorber at the end (see Sect. 3.4.6). Then calculate the far-field pressure, reduced to 1 m, for an array of 100 elements each of area 6.45 cm^2 (1 in.^{2}) for a single element waveguide pressure of 0.01 Pa at 10 kHz.

References

1. D.A. Berlincourt, D.R. Curran, H. Jaffe, Ch. 3, Piezoelectric and Piezomagnetic Materials, in *Physical Acoustics*, ed. by W.P. Mason, vol. I (Part A) (Academic, New York, 1964)
2. R.S. Woollett, *Sonar Transducer Fundamentals*. (Naval Underwater Systems Center, Newport, RI, Undated)
3. F.V. Hunt, *Electroacoustics* (Harvard University Press, New York, 1954)
4. O.B. Wilson, *Introduction to Theory and Design of Sonar Transducers* (Peninsula, Los Altos Hills, CA, 1988)
5. D. Stansfield, *Underwater Electroacoustic Transducers*. (Bath University Press, Bath, UK, 1990). See also G.W. Benthien, S.L. Hobbs, *Modeling of Sonar Transducers and Arrays*. Tech Doc. 3181, April, 2004, available on CD, Spawar Systems Center, San Diego, CA
6. A. Ballato, Modeling piezoelectric and piezomagnetic devices and structures via equivalent networks. IEEE Trans. Ultrason. Ferroelectr. Freq. Control **48**, 1189–1240 (2001)
7. B. Aronov, The energy method for analyzing the piezoelectric electroacoustic transducers. J. Acoust. Soc. Am. **117**, 210–220 (2005)
8. K.S. Van Dyke, The piezoelectric resonator and its equivalent network. Proc. IRE **16**, 742–764 (1928)
9. A.E. Clark, H.S. Belson, Giant room temperature magnetostriction in $TbFe_2$ and $DyFe_2$. Phys. Rev. **B5**, 3642 (1972)
10. A.E. Clark, Magnetostrictive Rare Earth-Fe_2 Compounds, in *Ferromagnetic Materials*, ed. by E.R. Wohlforth, vol. 1 (North-Holland, Amsterdam, 1980), pp. 531–589
11. A.E. Clark, J.B. Restorff, M. Wun-Fogle, T.A. Lograsso, D.L. Schlagel, Magnetostrictive properties of b.c.c. Fe-Ga and Fe-Ga-Al alloys. IEEE Trans. Mag. **36**, 3238 (2000). See also, A.E. Clark, K.B. Hathaway, M. Wun-Fogle, J.B. Restorff, V.M. Keppens, G. Petculescu, R.A. Taylor, Extraordinary magnetoelasticity and lattice softening in b.c.c. Fe-Ga alloys. J. Appl. Phys. **93**, 8621 (2003)
12. S. Butterworth, F.D. Smith, Equivalent circuit of a magnetostrictive oscillator. Proc. Phys. Soc. **43**, 166–185 (1931)
13. *Design and Construction of Magnetostrictive Transducers*. Summary Technical Report of Division 6, vol 13 (National Defense Research Committee, 1946)
14. L. Camp, *Underwater Acoustics* (Wiley-Interscience, New York, 1970)
15. E.L. Richardson, *Technical Aspects of Sound II, Ch. 2* (Elsevier, Amsterdam, 1957)

16. B.D.H. Tellegen, The gyrator, a new network element. Phillips Res. Rep. **3**, 81–101 (1948)
17. R.M. Bozorth, *Ferromagnetism* (D. Van Norstrand, New York, 1951)
18. J.L. Butler, N.L. Lizza, Eddy current factor series for magnetostrictive rods. J. Acoust. Soc. Am. **82**, 378 (1987)
19. W. Weaver, S.P. Timoshenko, D.H. Young, *Vibration Problems in Engineering* (Wiley, New York, 1990)
20. J.L. Butler, *Underwater Sound Transducers*. (Image Acoustics, Cohasset, MA, Course Notes, 1982), pp. 217 and 231
21. TAC Program User's Manual, General Electric, Syracuse, NY (1972). Developed for NUWC, Newport, RI
22. E. Geddes, *Audio Transducers*. (Geddes Associates LLC, 2002)
23. TRN Computer Program (NUWC, Newport, RI 02841). Developed by M. Simon and K. Farnham with array analysis module by Image Acoustics, Inc., Cohasset, MA. The computer programs TRN and TAC are based on the program SEADUCER, see H. Ding, L. McCleary, J. Ward, *Computerized Sonar Transducer Analysis and Design Based on Multiport Network Incterconnection Techniques*. TP-228. (Transducer and Array Systems Division, Naval Undersea Research and Development Center, San Diego, CA, 1971)
24. R. Krimholtz, D.A. Leedom, G.L. Matthaei (KLM Transducer Model), New equivalent circuit for elementary piezoelectric transducers. Electron. Lett., **6**(13), 398 (1970)
25. TAP Transducer Analysis Program. (Image Acoustics, Cohasset, MA)
26. W.P. Mason, *Electro-Mechanical Transducers and Wave Filters* (D. Van Nostrand, New York, 1942), p. 205
27. G.E. Martin, On the theory of segmented electromechanical systems. J. Acoust. Soc. Am. **36**, 1366–1370 (1964)
28. W.B. Bickford, *A First Course in the Finite Element Method* (Irwin, Boston, MA, 1990)
29. O.C. Zienkiewicz, *The Finite Element Method* (McGraw-Hill Book Company, Maidenhead, 1986)
30. K.J. Bathe, *Finite Element Procedures in Engineering Analysis* (Prentice-Hall, Englewood Cliffs, 1982)
31. ANSYS, Inc., Canonsburg, PA
32. ATILA, MMech, State College, PA (Also models magnetostriction)
33. COMSOL, Burlington, MA (Magnetostrictive can be added)
34. K.D. Rolt, J.L. Butler, *Finite Element Modulus Substitution Method for Sonar Transducer Effective Coupling Coefficient*, ed. by M.D. McCollum, B.F. Harmonic, O.B. Wilson. Transducers for Sonics and Ultrasonics (Technomic Publishing, PA, 1992). See also J.L. Butler, M.B. Moffett, K.D. Rolt, A finite element method for estimating the effective coupling coefficient of magnetostrictive transducers. J. Acoust. Soc. Am. **95**, 2533–2535 (1994)
35. J.L. Butler, A.L. Butler, *Analysis of the MPT/Hybrid Transducer*. March 11, 2002, Image Acoustics, Inc., NUWC Contract N66604-00-M-7216

Chapter 4
Transducer Characteristics

Important transducer characteristics such as resonant frequency, mechanical quality factor, characteristic impedance, electromechanical coupling coefficient, and figure of merit (FOM) will be discussed in more detail in this chapter. One reason for more discussion is that some aspects of transducer theory are not standardized. For example, several definitions of electromechanical coupling coefficient and of mechanical quality factor are in use. In particular, it is important to extend the discussion of the electromechanical coupling coefficient in Sect. 1.4.1 to include other definitions, properties, and interpretations. Another reason for more discussion of transducer characteristics is the need to present certain practical considerations that have not been fully developed in the previous chapters. It was pointed out in Chap. 2 that transducer performance is determined by an effective coupling coefficient that is usually less than the material coupling coefficient of the active material used in the transducer. There are numerous causes of this reduced effective coupling, such as inactive transducer components and dynamic operating conditions, that occur in all transducers, and eddy currents that occur in magnetostrictive transducers. We will present practical methods of determining the effective coupling coefficients that result from many of these causes, with some specific examples. Finally, we will present a parameter based figure of merit (FOM) as a metric for power transducers.

4.1 Resonance Frequency

Projectors are usually operated at or near a resonance frequency in order to obtain the most power output for a given driving force. In general, a resonance frequency is defined as the frequency at which some quantity reaches a maximum when the driving frequency is varied while the driving force amplitude is held constant. At this frequency, the displacement is amplified by the mechanical quality factor, Q_m, of the resonator (see Sect. 4.2). The resonances associated with power or velocity

J.L. Butler, C.H. Sherman, *Transducers and Arrays for Underwater Sound*,
Modern Acoustics and Signal Processing, DOI 10.1007/978-3-319-39044-4_4

are usually of most interest, and occur at the same frequency when the mechanical resistance is independent of frequency. Displacement and acceleration resonances occur at frequencies that differ only slightly from the velocity resonance frequency unless the mechanical resistance is very high.

In the single degree of freedom system of Fig. 3.1, the lumped mechanical impedance is $Z_m = R + j(\omega M - K/\omega)$ where R, M, and K are all independent of frequency (see Sect. 13.17 for a review of complex algebra). The magnitude of the impedance, $|Z_m|$, is minimum and the velocity, $F/|Z_m|$, is maximum, with a value F/R (see Fig. 3.2a), when the driving frequency is such that $\omega M = K/\omega$, making the velocity resonance frequency, $\omega_r = (K/M)^{1/2}$. Since the power is proportional to Ru^2, the power resonance occurs at the same frequency as the velocity resonance in this case. The resonance frequency can also be found from the fundamental relationship between the kinetic energy of the mass, $\frac{1}{2}Mu^2 = \frac{1}{2}M\omega^2x^2$, and the potential energy of the spring, $\frac{1}{2}Kx^2$. At the point in the vibration cycle where $x = 0$, the potential energy is zero, the kinetic energy is maximum, and the mass moves with a maximum velocity. On the other hand, at the point of maximum displacement where $u = 0$, the kinetic energy is zero and the potential energy is maximum. For sinusoidal drive, resonance occurs when there is a complete exchange of energy and the peak kinetic and potential energies are equal, $\frac{1}{2}Kx^2 = \frac{1}{2}M\omega^2x^2$, which gives the resonance frequency above. The natural frequency of free vibration also occurs with equal peak kinetic and potential energies at the same frequency as velocity resonance, except as modified by resistance.

In the case of Fig. 4.1 displacement resonance occurs at $(\omega_r^2 - R^2/2M^2)^{1/2}$ where $\omega|Z_m|$ is minimum, and acceleration resonance occurs at $\omega_r^2/(\omega_r^2 - R^2/2M^2)^{1/2}$ where $|Z_m|/\omega$ is minimum. Figure 4.1 compares the displacement, velocity and acceleration as functions of frequency for a case with rather

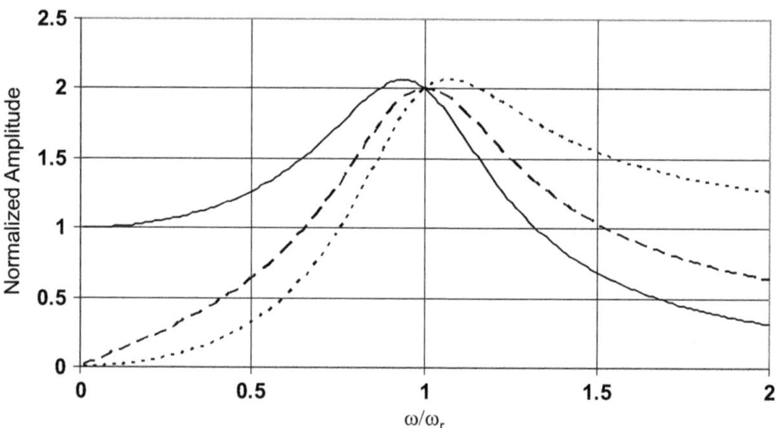

Fig. 4.1 Normalized displacement, x_n, (*solid line*), velocity, u_n, (*dashed line*), and acceleration, a_n, (*dotted line*) vs. normalized frequency for the simple vibrator in Fig. 3.1 for $Q_m = 2$

large resistance ($R = \omega_r M/2$ or $Q_m = 2$) in order to display the difference in the resonance frequencies.

In electroacoustic transducers, where the stiffness depends on electrical boundary conditions, a velocity resonance and a velocity antiresonance occur. For electric field transducers resonance occurs for short circuit conditions, and is designated ω_r (or ω_r^E), while antiresonance occurs for open circuit conditions, and is designated ω_a (or ω_r^D). For magnetic field transducers resonance occurs for open circuit conditions and antiresonance occurs for short circuit conditions. For both transducer types, $\omega_r < \omega_a$. The lumped-parameter circuits of Figs. 3.15 or 3.16 illustrate the two frequencies for electric field transducers: when the electrical terminals are shorted the mechanical resonance occurs at $\omega_r = (1/C_e L_e)^{1/2}$; when they are open it occurs at $\omega_a = [(C_0 + C_e)/C_0 C_e L_e]^{1/2}$.

As an example of resonance and antiresonance frequencies in a distributed transducer model consider a bar of piezoelectric ceramic of length L and cross-sectional area A_0 as shown in Fig. 4.2.

Consider the bar fixed on one end with a load on the other end consisting of a lumped mass M and a resistance R. The equivalent circuit in Fig. 3.29 applies with open circuit mechanical terminals on the left side, since $u_0 = 0$, and mechanical impedance of $R + j\omega M$ connected to the right side. Equation (3.34) with $Z_L = \infty$ shows that the mechanical impedance of the bar is $-j\rho c A_0 \cot kL$, and the impedance of the transducer is

$$Z_m = R + j\omega M - j\rho c A_0 \cot kL. \tag{4.1}$$

For kL small the last term of Eq. (4.1) reduces to the lumped stiffness reactance, $-jK/\omega$, where $K = \rho c^2 A_0/L$. When R is independent of frequency the velocity resonance frequency, ω_r, occurs when $\omega_r M = \rho c A_0 \cot k_r L$, which can be written as

$$k_r L \tan k_r L = M_b/M, \tag{4.2}$$

where $k_r = \omega_r/c$ is the wavenumber at resonance, c is the longitudinal wave speed in the bar, and $M_b = \rho A_0 L$ is the mass of the bar.

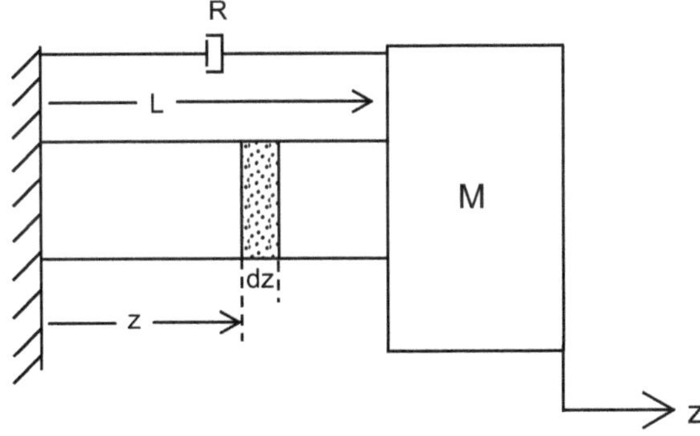

Fig. 4.2 Calculation of kinetic or potential energy in a mass-loaded vibrating bar

Each of the three piezoelectric bar electrode arrangements discussed in Sect. 3.2 involves a different elastic constant (s_{33}^E, s_{11}^E or s_{33}^D) and a different longitudinal wave speed. Equation (4.2) applies to any of these cases, but each solution for $k_r L$ is related to a different value of $\omega_r = (c/L)(k_r L)$ that depends on the wave speed for the specific electrode arrangement. As an example consider the segmented bar in Sect. 3.2.3.1 with both R and M zero, i.e., with one end free. Equation (3.29) with $\zeta(0) = 0$ and $\zeta(L) = \zeta_0$ shows that the displacement is given by

$$\zeta(z) = \zeta_0 \frac{\sin kz}{\sin kL},$$

where $k = \omega/c^E$ and $c^E = \left(\rho s_{33}^E\right)^{-1/2}$. It follows from Eq. (3.38a), by using $S_3 = \partial\zeta/\partial z$, that the displacement of the free end, where the stress is zero, is related to the voltage by

$$k\zeta_0 \cot kL = d_{33}V/h. \tag{4.3a}$$

The short circuit resonance condition occurs for $V=0$, or for $\cot k_r^E L = 0$, where $k_r^E = \omega_r/c^E$, the same condition as Eq. (4.2) for $M=0$. The solutions are $k_r^E L = m\pi/2$, $m = 1, 3, 5\ldots$ and the resonance frequencies are $\omega_{rm} = m\pi c^E/2L$.

The current, I, for this case is given by Eq. (3.45) with $u_0=0$ and $u_n = j\omega\zeta_0$ (see Fig. 3.29). The antiresonances occur when $I=0$, for which Eq. (3.45) can be simplified by use of Eq. (4.3a) and written as

$$k_a^I L \cot k_a^I L = -\frac{k_{33}^2}{1 - k_{33}^2}, \tag{4.3b}$$

where $k_a^I = \omega_a/c^E$. This equation has a set of solutions for $k_a^I L$ that give the antiresonance frequencies as $\omega_{am} = (c^E/L)k_a^I L$. For example, for $k_{33}^2 = 0.5$, the first solution is $k_a^I L \approx 2.03$, and the ratio of resonance to antiresonance frequency for the fundamental mode is $\omega_{r1}/\omega_{a1} = (\pi/2)/2.03 = 0.773$. The frequencies ω_{rm} and ω_{am} differ because of the different electrical boundary conditions, but they depend on the same longitudinal wave speed, $c^E = \left(\rho s_{33}^E\right)^{-1/2}$. Note that for very small k_{33} the two frequencies are approximately equal. See the discussion following Eq. (4.35b) regarding use of the resonance/antiresonance frequencies to determine the effective coupling coefficient.

Transducers are usually capable of vibrating in more than one mode, each with a different resonance frequency, but in most cases they are used in the vicinity of the fundamental mode. In some cases unwanted higher modes interfere with fundamental mode operation, while in other cases the radiation from higher modes is deliberately combined with radiation from the fundamental mode to achieve greater bandwidth or special directional characteristics (see Chaps. 5 and 6). It should be noted that not all the mechanical resonance frequencies are necessarily excited in typical transducers (see Sect. 5.4.2).

4.2 Mechanical Quality Factor

4.2.1 *Definitions*

Resonances are also characterized by the width of the peak in the power versus frequency curve, and the mechanical quality factor, or storage factor, Q_m, is a dimensionless, real, positive number that is the usual measure of this width. For a specific mode of vibration with resonance frequency ω_r, Q_m is often defined by

$$Q_m \equiv \frac{\omega_r}{\omega_2 - \omega_1}, \tag{4.4}$$

where ω_2 and ω_1 are the frequencies above and below ω_r at which the power is half the maximum power as shown in Fig. 4.3 [1, 2].

For example, in the case of frequency-independent resistance, ω_2 and ω_1 occur at the frequencies where $|Z_m|^2 = 2|Z_m(\omega_r)|^2$. The physical meaning of this definition is clear since $(\omega_2 - \omega_1)$ is a direct measure of bandwidth, i.e., high Q_m means narrow band, low Q_m means wide band. When the term quality factor is applied to crystal vibrators used as narrow band filters, high Q_m is associated with high quality, because it means low internal resistance. On the other hand, underwater sound applications usually require a wide band, and in these cases low Q_m is associated with high quality where it means high bandwidth and high radiation resistance. The term quality factor is still appropriate, however, since when underwater sound transducers are measured in air, high Q_m is desirable because it indicates low internal losses and, consequently, high efficiency in water.

Some writers [3] consider that a definition of Q_m in terms of stored and dissipated energies is more basic than the bandwidth definition in Eq. (4.4); the energy definition can be expressed as:

$$Q_m \equiv \omega_r \frac{U_s(\omega_r)}{W_d(\omega_r)} = \frac{2\pi U_s(\omega_r)}{T_r W_d(\omega_r)}, \tag{4.5}$$

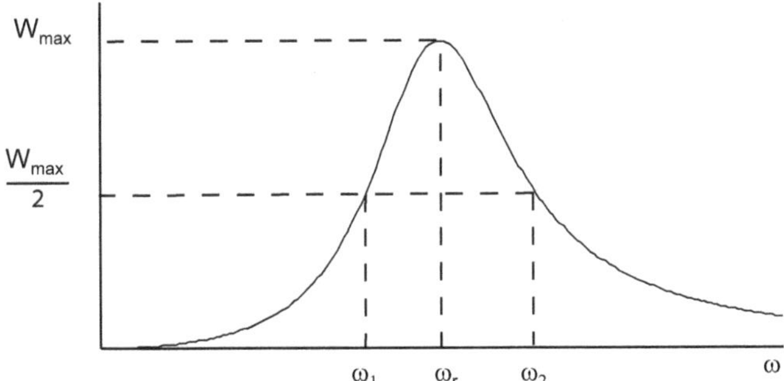

Fig. 4.3 Power vs. frequency showing half power frequencies

where $U_s(\omega_r)$ is the peak kinetic energy stored in the mass at ω_r, $W_d(\omega_r)$ is the time average dissipated power at ω_r, and T_r is the period of oscillation at resonance. This definition also has a clear physical meaning; since $T_r\ W_d$ is the energy dissipated in one period, it must exceed the peak stored energy to make $Q_m < 2\pi$. Low Q_m means that the input energy is radiated, and partially dissipated internally, with relatively little of it being stored.

Another expression for Q_m is

$$Q_m = \frac{\omega_r}{2R(\omega_r)} \left|\frac{dZ_m}{d\omega}\right|_{\omega_r}, \tag{4.6}$$

where Z_m is the mechanical impedance and $R(\omega_r)$ is the total mechanical resistance at resonance. The origin of this expression for Q_m is uncertain; it was introduced to one of the authors (JLB) by Professor W. J. Remillard at Northeastern University about 1965, and probably came from Professor F. V. Hunt at Harvard. This expression has been found to give results in certain cases that are either exactly the same as, or in close agreement with, the other two expressions for Q_m. It can be approximately justified for frequency-independent resistance by writing $Z_m = R + jX$ and using the mean value theorem for the derivative which gives

$$\left|\frac{dZ_m}{d\omega}\right|_{\omega_r} = \left|\frac{dX}{d\omega}\right|_{\omega_r} \approx \frac{|X(\omega_2) - X(\omega_1)|}{\omega_2 - \omega_1}.$$

When R is independent of frequency ω_r occurs when $X = 0$, and ω_2 and ω_1 occur when $R^2 + X^2 = 2R^2$ or $X = \pm R$ and $|X(\omega_2) - X(\omega_1)| = 2R$. Using these results in Eq. (4.6) shows that it is equivalent to Eq. (4.4) under the conditions assumed. Equation (4.6) appears to be a convenient means for calculating Q_m in some cases, rather than a physical definition of Q_m. Equation (4.6) can also be useful in other calculations, e.g., see Sect. 7.1.6. When an expression for Z_m is known Eqs. (4.5) or (4.6) are usually more convenient than Eq. (4.4) for calculating Q_m. But Eq. (4.4) is needed for determining Q_m from power or conductance measurements as a function of frequency (see Chap. 10) and for calculating bandwidth when Q_m is determined in some other way.

In the simple lumped transducer model of Fig. 3.1 where $Z_m = R + j(\omega M - K/\omega)$, with R, M, and K all constant, the three expressions above for Q_m all give the same result: $Q_m = \omega_r M/R$. This result comes most easily from the energy expression in Eq. (4.5) since $U_s = ½Mu_r^2$ and $W_d = ½Ru_r^2$, where u_r is the velocity amplitude at resonance. It also follows easily from the derivative expression in Eq. (4.6) since the derivative is $M + K/\omega^2$ which equals $2M$ at $\omega = \omega_r$. The bandwidth expression in Eq. (4.4) requires calculation of ω_2 and ω_1 from $X = (\omega M - K/\omega) = \pm R$ which gives $(\omega_2 - \omega_1) = R/M$. The expression,

$$Q_m = \omega_r M/R,$$

can be used as the definition of Q_m [4] whenever consistent effective values of the parameters ω_r, M, and R are known (see Sect. 4.2.2).

4.2.2 *Effect of the Mass of the Bar*

We now return to the example of the fixed-end, mass-loaded piezoelectric bar to show how the mass of the bar affects the Q_m. The bar provides the spring and the driving force, but it also contributes mass that may be significant for dense materials such as piezoelectric ceramic, e.g., the density of PZT is about 7500 kg/m^3. The kinetic energy in the bar is distributed along its length with the energy in a thin slice of thickness dz located at z being (see Fig. 4.2)

$$dU_K = \tfrac{1}{2}\rho A_0 dz\, u(z)^2,$$

where $u(z)$ is the longitudinal velocity at z and ρ is the density. Equation (3.31) shows that for $u(0)=0$ and $u(L)=u_0$

$$u(z) = u_0 \sin kz/\sin kL.$$

Combining these expressions and integrating from 0 to L gives the total kinetic energy in the bar:

$$U_K = \frac{1}{2} M_b u_0^2 \left[\frac{1 - \sin 2kL/2kL}{2\sin^2 kL}\right].$$

The effective, lumped dynamic mass of the bar, M_d, referred to the velocity at the free end of the bar, is defined by equating U_K to $\tfrac{1}{2}M_d u_0^2$ which gives:

$$M_d = M_b \left[\frac{1 - \sin 2kL/2kL}{2\sin^2 kL}\right]. \tag{4.7a}$$

The dynamic mass is frequency dependent, and cases of special interest occur when $kL \ll 1$ and $M_d = M_b/3$, when $kL = \pi/4$ and $M_d = M_b(1 - 2/\pi)$, and when $kL = \pi/2$ and $M_d = M_b/2$. Some of these results were also obtained in Chap. 3 from the equivalent circuits in Figs. 3.24 and 3.25. The added dynamic mass of the piezoelectric bar reduces the resonance frequency and raises Q_m; for example, when $kL \ll 1$ the stiffness is $K = YA_0/L$ and

$$\omega_r = \left[\frac{K}{M + M_b/3}\right]^{\frac{1}{2}} \quad \text{and} \quad Q_m = \omega_r \frac{M + M_b/3}{R}.$$

Since there is stored kinetic energy in both the mass, M, and the dynamic mass of the bar, the energy expression for Q_m in Eq. (4.5) becomes, using Eq. (4.7a), and with u_r as the velocity of M at resonance:

$$Q_m = \omega_r \frac{M_d u_r^2/2 + M u_r^2/2}{R u_r^2/2} = \omega_r \frac{M_d + M}{R} = \frac{\omega_r M}{R}\left[1 + \frac{M_b}{M}\left\{\frac{1 - \sin 2k_r L/2k_r L}{2\sin^2 k_r L}\right\}\right].$$

This equation can be simplified by expanding sin $2k_rL$ as 2 sin k_rL cos k_rL and using the resonant condition in Eq. (4.2) with the result:

$$Q_m = \frac{\omega_r M}{R}\left[\frac{1}{2} + \frac{M_b/M}{2\sin^2 k_r L}\right]. \tag{4.7b}$$

The same result can be found, and more easily, from the derivative expression for Q_m in Eq. (4.6), using the mechanical impedance in Eq. (4.1), to show that

$$\left|\frac{dZ_m}{d\omega}\right|_{\omega_r} = M + \rho A_0 L/\sin^2 k_r L.$$

The dynamic mass can be used in the vicinity of the fundamental resonance to give a simple lumped equivalent circuit of the mass-loaded bar and its associated Q_m. The same could be done in the vicinity of one of the higher resonant modes given by Eq. (4.2). However, the distributed model is necessary when an accurate knowledge of the transducer performance is desired at frequencies where more than one mode may be involved, and the usual meaning of Q_m does not apply.

4.2.3 Effect of Frequency-Dependent Resistance

In the above the mechanical resistance was considered to be independent of frequency, but at low frequency the radiation resistance is proportional to frequency squared (see Chaps. 2 and 10) and is also the major part of the resistance in well-designed transducers. Therefore, a reasonable way of investigating the effects of frequency-dependent resistance at low frequency is to consider the mechanical impedance,

$$Z_m = R_0(\omega/\omega_0)^2 + j(\omega M - K/\omega), \tag{4.8}$$

where R_0, M, and K are constants and $\omega_0^2 = K/M$. For example, for a spherical transducer of radius a, $R_0 = 4\pi\omega_0^2\rho a^4/c$. The resonance frequency and Q_m are both affected when the resistance changes with frequency. The velocity resonance occurs when $|Z_m| = \left[R_0^2(\omega/\omega_0)^4 + (\omega M - K/\omega)^2\right]^{1/2}$ is a minimum, which is determined by the cubic equation for $(\omega/\omega_0)^2$:

$$\left(2R_0^2/KM\right)(\omega/\omega_0)^6 + (\omega/\omega_0)^4 - 1 = 0.$$

If $2R_0^2/KM$ is small, we can find an approximate solution by writing $\omega_r \approx \omega_0(1-\alpha)$ with $\alpha \ll 1$ and approximating the cubic equation by

$$\left(2R_0^2/KM\right)(1-6\alpha) + (1-4\alpha) - 1 = 0.$$

This equation gives $\alpha \approx R_0^2/2KM$ and

$$\omega_r \approx \omega_0\left(1 - R_0^2/2KM\right) = \omega_0\left(1 - 1/2Q_{m0}^2\right), \tag{4.9}$$

where $Q_{m0} = \omega_0 M/R_0$ is the Q_m when the resistance is constant with value R_0. Thus frequency-dependent resistance lowers the velocity resonance frequency.

Frequency-dependent resistance affects Q_m both directly and indirectly through the change in resonance frequency. The result given by the energy expression for Q_m is

$$Q_{me} = \omega_r\left(Mu_r^2/2\right)/\left[R_0(\omega_r/\omega_0)^2 u_r^2/2\right] = Q_{m0}/\left(1 - 1/2Q_{m0}^2\right), \tag{4.10}$$

which shows that frequency-dependent resistance raises the Q_m. In this case the derivative expression for Q_m gives essentially the same result, but the two are not exactly the same.

4.3 Characteristic Mechanical Impedance

The mechanical impedance for a simple lumped vibrator can be written as

$$Z_m = R + j(\omega M - K/\omega) = (MK)^{1/2}[1/Q_m + j(\omega/\omega_r - \omega_r/\omega)],$$

where the factor $(MK)^{1/2}$ determines the general level of the impedance magnitude and is called the characteristic mechanical impedance of the transducer [3], $Z_c = (MK)^{1/2}$. The mechanical quality factor is closely related to Z_c since

$$Q_m = \omega_r M/R = (K/M)^{1/2}(M/R) = Z_c/R, \tag{4.11}$$

which shows that Q_m is low when Z_c and R are similar in magnitude.

The characteristic impedance for a distributed system can be determined from effective lumped parameters such as the dynamic mass in Eq. (4.7a) for the mass-loaded piezoelectric bar fixed at one end. In a similar way the lumped dynamic stiffness can be found from the potential energy in the bar. The differential element of potential energy in a thin slice of the bar is (see Fig. 4.2)

$$dU_p = \frac{1}{2}Y\left(\frac{\partial \zeta}{\partial z}\right)^2 A_0 dz,$$

where Y is Young's modulus, and, from Eq. (3.29), the displacement $\zeta(z) = \zeta_0 \sin kz/\sin kL$ with ζ_0 the displacement at the end of the bar. Integrating over the length of the bar gives

$$U_{\mathrm{p}} = \left(\frac{YA_0}{L}\right)\left(\frac{\zeta_0 kL}{2\sin kL}\right)^2\left[1+\frac{\sin 2kL}{2kL}\right].$$

The effective, lumped dynamic stiffness of the bar, K_{d}, referred to the displacement at the free end of the bar, is defined by equating $½K_{\mathrm{d}}\zeta_0^2$ to U_{p} which gives

$$K_{\mathrm{d}} = \frac{1}{2}\left(\frac{YA_0}{L}\right)\left(\frac{kL}{\sin kL}\right)^2\left[1+\frac{\sin 2kL}{2kL}\right], \tag{4.12}$$

where $YA_0/L = K_{\mathrm{b}}$ is the static stiffness of the bar. Some specific cases are: $kL \ll 1$, $K_{\mathrm{d}} \approx K_{\mathrm{b}}$; $kL = \pi/4$, $K_{\mathrm{d}} = (\pi/8)(1+\pi/2)K_{\mathrm{b}} = 1.0095K_{\mathrm{b}}$; $kL = \pi/2$, $K_{\mathrm{d}} = (\pi^2/8)$ $K_{\mathrm{b}} = 1.234\ K_{\mathrm{b}}$.

Equation (4.12) explains why the lumped approximation for the bar is reasonable when the length is less than a quarter wavelength. When $L = \lambda/4$, K_{d} is 23 % greater than K_{b}. But for L less than $\lambda/4$, K_{d} quickly approaches K_{b}, being only 1 % greater than K_{b} when $L = \lambda/8$, and essentially constant at lower frequencies. Thus it is reasonable to ignore the dynamic stiffness for bars less than $\lambda/4$ long, but it is more important to consider the dynamic mass which is never less than one third the static mass.

The characteristic impedance at resonance can now be expressed in terms of the dynamic quantities M_{d} and K_{d} evaluated at ω_{r}, M_{dr}, and K_{dr}. The result is, using $[M_{\mathrm{b}}K_{\mathrm{b}}]^{1/2} = \rho c A_0$,

$$Z_{\mathrm{cr}} = [(M+M_{\mathrm{dr}})K_{\mathrm{dr}}]^{1/2} = \left[\left(\frac{M}{M_{\mathrm{b}}}+\frac{M_{\mathrm{dr}}}{M_{\mathrm{b}}}\right)\frac{K_{\mathrm{dr}}}{K_{\mathrm{b}}}\right]^{1/2}\rho c A_0. \tag{4.13a}$$

When a value of M/M_{b} is specified $k_{\mathrm{r}}L$ and ω_{r} are determined by Eq. (4.2); then $M_{\mathrm{dr}}/M_{\mathrm{b}}$ and $K_{\mathrm{dr}}/K_{\mathrm{b}}$ are obtained from Eqs. (4.7a) and (4.12). The characteristic impedance can also be expressed in another way by combining Eqs. (4.11) and (4.7b) to get

$$Z_{\mathrm{cr}} = \omega_{\mathrm{r}}M\left[\frac{1}{2}+\frac{M_{\mathrm{b}}/M}{2\sin^2 k_{\mathrm{r}}L}\right]. \tag{4.13b}$$

For three values of M/M_{b}, the values of $Z_{\mathrm{cr}}/\rho c A_0$ and Q_{m} are given in Table 4.1.

Table 4.1 Characteristic impedance at resonance and quality factor for the mass-loaded, fixed-end bar

M/M_{b}	Q_{m}	$Z_{\mathrm{cr}}/\rho c A_0$
$\gg 1$	$\frac{\omega_{\mathrm{r}}M_{\mathrm{b}}}{R}\left(\frac{M}{M_{\mathrm{b}}}+\frac{1}{3}\right)$	(M/M_{b})
$4/\pi$	$\frac{\omega_{\mathrm{r}}M_{\mathrm{b}}}{R}[1+2/\pi]$	1.29
$\ll 1$	$\frac{\omega_{\mathrm{r}}M_{\mathrm{b}}}{2R}=\frac{\pi\rho c A_0}{4R}$	$\pi/4$

For example: Table 4.1 shows, adding mass to the end of the bar raises the Q_m and the characteristic impedance, but when $M \leq M_b$, $Z_{cr} \approx \rho c A_0$. For PZT, $\rho c = 22 \times 10^6$ kg/m^{2}s, and for full acoustic loading $R \approx (\rho c)_w A$ where $(\rho c)_w = 1.5 \times 10^6$ kg/m^{2}s for water, and A is the radiating area of the mass. As an example, for $A/A_0 = 5$,

$$Q_m = \frac{Z_{cr}}{R} \approx \frac{\rho c A_0}{(\rho c)_w A} \approx 3,$$

showing that low Q_m is possible with materials such as PZT when full acoustic loading is achieved, as it sometimes is in closely packed arrays (see Chap. 7).

As discussed qualitatively at the end of Sect. 2.8 the effective transmission of energy from the vibrating surface of a transducer into the water requires that the characteristic impedance of the transducer be similar to that of the water. The results in this section and the distributed transducer model developed in Chap. 3 quantifies those ideas and shows that the stress waves generated in a longitudinal vibrator are radiated effectively, with a low Q_m, when Z_{cr} is approximately the same as $(\rho c)_w A$.

4.4 Electromechanical Coupling Coefficient

We will now discuss other properties and definitions of the electromechanical coupling coefficient, a very important real, positive, dimensionless measure of transducer performance. The definitions in terms of the stiffness change or the capacitance change in Eq. (1.16) were applied in Chap. 2, to all the basic linearized transducer types, but other definitions are often used, especially those involving energy. Equation (1.16), or one of its variants, will also be applied to more complicated transducer structures containing components that participate in the vibration but do not convert energy, and to cases where dynamic effects are significant. It is important to consider other definitions of electromechanical coupling, because Eq. (1.16) applies only to linear transduction mechanisms, while definitions based on energy are capable of being generalized to include nonlinearity.

4.4.1 Energy Definitions of Coupling and Other Interpretations

4.4.1.1 Mason's Energy Definition

Hunt [5] discussed the importance of the electromechanical coupling coefficient and suggested a definition of the type in Eq. (1.16) based on the change in mechanical stiffness associated with changing the electrical conditions from open circuit to short circuit. However, he only considered the three surface force transducers and only discussed the coupling coefficient for one of them, the electrostatic transducer. Hersh [6] made a comprehensive technical and historical study of the coupling coefficient in which numerous approaches to defining coupling were reviewed, including coupling in electric circuits as well as coupling in electromechanical systems. Hersh also discussed most of the transducer types and developed a new definition of coupling based on feedback. This feedback definition has not been used much, although Woollett mentions it in one of his studies [3]. Many writers [2, 7, 8] have emphasized the value of the coupling coefficient for comparing different types of transducers and different design concepts.

Woollett [3] considers one of the basic physical meanings of k to be its indication of the limits of physical realizability. The definition of k^2 in terms of the stiffness in Eq. (1.16) has this property. Repeating the definition here:

$$k^2 = 1 - K_{\mathrm{m}}^E / K_{\mathrm{m}}^D, \quad \text{for electric field transducers,} \tag{4.14a}$$

$$k^2 = 1 - K_{\mathrm{m}}^H / K_{\mathrm{m}}^B, \quad \text{for magnetic field transducers,} \tag{4.14b}$$

shows that $k \rightarrow 1$ corresponds to $K_{\mathrm{m}}^E \rightarrow 0$ for electric field transducers, or $K_{\mathrm{m}}^H \rightarrow 0$ for magnetic field transducers. A transducer with vanishing stiffness is unrealizable in some sense. In the electrostatic and variable reluctance transducers the unrealizability takes the form of a dramatic instability when the condenser plates or pole faces crash together as $k \rightarrow 1$. In the moving coil transducer the unrealizability appears as an uncertain equilibrium position as $K_{\mathrm{m}}^I \rightarrow 0$, but values of k very close to unity are practical [6]. In the body force transducers k is apparently prevented from exceeding unity by internal nonlinear mechanisms related to saturation of polarization or magnetization. The highest value of a material k_{33} found so far is about 0.96 in single crystal PMN-33%PT [9]. If compositions such as this are close to a phase boundary their properties may not be stable enough for use in operational transducers where temperature and stress variations are unavoidable.

Thus the realizable range of k is $0 < k < 1$. This well-defined range of values for all transducers is the property that makes k so useful for comparison purposes. However, when comparing different transduction mechanisms other conditions must be considered before a valid comparison of coupling coefficients can be made [7]. For example, moving coil transducers, with limited usefulness as projectors in water, can easily have coupling coefficients close to unity when designed for air.

Mason's energy definition of k [10], introduced in Sect. 1.4.1 and Eq. (1.19), is widely used and will be discussed more fully here. We will calculate the ratio of converted energy to input energy under different conditions and compare it with k^2 as defined by Eq. (4.14a). Equations (2.72a) and (2.72b) for electric field transducers when reduced to quasistatic conditions, as required for determining the quasistatic coupling coefficient, are:

$$F_{\mathrm{b}} = K_{\mathrm{m}}^{E} x - NV, \tag{4.15}$$

$$Q = Nx + C_0 V. \tag{4.16}$$

We will assume that the biased state is the zero of mechanical and electrical energy and first consider projector operation where $F_{\mathrm{b}} = 0$. The electrical input energy, U_{e}, resulting from an increase of voltage, $\mathrm{d}V$, can be calculated by multiplying Eq. (4.16) by $\mathrm{d}V$, using Eq. (4.15) to put $\mathrm{d}V$ in terms of $\mathrm{d}x$ and integrating from 0 to V and 0 to x. The result is

$$U_{\mathrm{e}} = \int Q\,\mathrm{d}V = ½ K_{\mathrm{m}}^{E} x^2 + ½ C_0 V^2 = ½ K_{\mathrm{m}}^{E} x^2 \left(1 + C_0 K_{\mathrm{m}}^{E}/N^2\right), \tag{4.17}$$

which shows that the electrical input energy associated with raising the voltage from 0 to V is divided into two parts: one part is the converted mechanical energy, $½K_{\mathrm{m}}^{E}x^2$, while the other part, $½\, C_0V^2$, remains as electrical energy. Thus the fraction of the electrical input energy converted to mechanical energy can be written, using the relations in Sect. 1.4.1, as

$$\left(½ K_{\mathrm{m}}^{E} x^2\right)/U_{\mathrm{e}} = \left(1 + C_0 K_{\mathrm{m}}^{E}/N^2\right)^{-1} = 1 - C_0/C_{\mathrm{f}} = 1 - K_{\mathrm{m}}^{E}/K_{\mathrm{m}}^{I} = k^2, \tag{4.18}$$

in agreement with Eqs. (4.14a).

When the energy input is acoustic $F_{\mathrm{b}} = p_{\mathrm{i}}A$, where p_{i} is the average incident pressure over the transducer surface A. Then Eqs. (4.15) and (4.16) become, with $Q = 0$ for operation as an open circuit hydrophone,

$$p_{\mathrm{i}}A = K_{\mathrm{m}}^{E} x - NV, \tag{4.19}$$

$$0 = Nx + C_0 V. \tag{4.20}$$

The integral of $p_{\mathrm{i}}A\mathrm{d}x$ from 0 to x gives the input mechanical energy,

$$U_{\mathrm{m}} = ½ K_{\mathrm{m}}^{E} x^2 + ½ C_0 V^2 = ½ C_0 V^2 \left(1 + C_0 K_{\mathrm{m}}^{E}/N^2\right). \tag{4.21}$$

The first form of Eq. (4.21) appears to be the same as Eq. (4.17), but the relationship between x and V is different, being given by Eq. (4.20), rather than Eq. (4.15) with $F_b = 0$. Under these conditions with $Q = 0$ the converted energy is electrical, ½ C_0V^2, and the fraction of the mechanical energy converted is

$$\tfrac{1}{2}C_0V^2/U_m = \left(1 + C_0K_m^E/N^2\right)^{-1} = k^2, \qquad (4.22)$$

as it was when energy was converted from electrical to mechanical. These results, showing that Mason's energy definition of k^2 agrees with the definition in Eq. (4.14a), also apply to the simple models of magnetic field transducers.

Another way of relating k^2 to energy starts with the capacitance change definition:

$$k^2 = \frac{C_f - C_0}{C_f}, \qquad (4.23)$$

which, for an applied voltage, V, can be written as

$$k^2 = \frac{C_fV^2/2 - C_0V^2/2}{C_fV^2/2}. \qquad (4.24)$$

The denominator of Eq. (4.24) is the free energy (or total input energy), while the numerator, which is the difference between the free energy and the clamped energy, is the converted mechanical energy or the motional energy. Thus the energy ratio in Eq. (4.24) has the same physical meaning as the ratio in Eq. (4.18).

Some comments regarding the energy definition of the electromechanical coupling coefficient by Hersh [6] and Woollett [3] may be worth noting. Hersh concludes that basing a definition on the division of energy within a coupled system, as is required to determine the stored energies, "cannot be made without ambiguity and confusion even in simple systems." This comment may have some merit, since a recent paper on the coupling coefficient of electrostatic transducers uses two different expressions for the converted energy and calculates two different values for k^2 [11]. Woollett seems to agree with Hersh's conclusion when he questions the generality of the energy definition and points out that the stored energies are not observables. These comments may be important for more complicated transducer structures with one or more internal degrees of freedom which are not observable at the transducer ports. The Mason energy definition of k^2 will be discussed further in connection with more complicated cases in Sects. 4.4.2 and 4.4.3.

4.4.1.2 Mutual Energy Definition

A quite different energy definition of the coupling coefficient (apparently introduced by Vigoureux in 1950 [6]) is also frequently used [12–17]. According to this definition

$$k^2 = U_{\text{mut}}^2 / U_{\text{mech}} U_{\text{el}}, \tag{4.25}$$

where U_{mech} is the "mechanical energy," U_{el} is the "electrical energy," and U_{mut} is the "mutual energy." The conditions under which these energies are to be determined, or the physical meaning of mutual energy, have not been well defined. The usual approach when applying this definition is to write an expression for the total energy and identify U_{mech} as those terms that involve only mechanical variables, U_{el} as those terms that involve only electrical variables, and U_{mut} as those terms that involve both kinds of variables (see example below). Note that the total input energies calculated in Eq. (4.17) for projector conditions ($F_{\text{b}} = 0$) and in Eq. (4.21) for hydrophone conditions ($Q = 0$) were clearly separated into electrical and mechanical energy with no mutual energy.

The mutual energy concept probably arose by analogy to coupling of two electrical circuits involving, for example, mutual inductance. Berlincourt et al. [12] used Eq. (4.25) and pointed out that it agrees with the other definitions of k^2 only when equations with homogeneous variables are used, while equations with mixed variables give $k^2/(1 - k^2)$. Thus, when the energies in Eq. (4.25) are identified as described above they depend on the way the equations are written, as will be illustrated below. An example may clarify this situation. Equations (4.15) and (4.16) have mixed independent variables, because x is extensive and V is intensive. It follows that the dependent variables are also mixed; F_{b} is intensive, and Q is extensive (see Glossary). But these equations can be rewritten in the following homogeneous form:

$$F_{\text{b}} = K_{\text{m}}^D x - (N/C_0)Q,$$
$$V = -(N/C_0)x + (1/C_0)Q,$$

and then we will see that Eq. (4.25) will give the same k^2 as the other definitions. If an external force, F, and a voltage, V, are applied together the differential change in total energy is

$$\text{d}U_{\text{tot}} = F\,\text{d}x + V\,\text{d}Q = K_{\text{m}}^D x\,\text{d}x - (N/C_0)(Q\,\text{d}x + x\,\text{d}Q) + Q\,\text{d}Q/C_0.$$

which includes two mutual terms involving both x and Q. Integrating this expression gives the total energy:

$$U_{\text{tot}} = ½ K_{\text{m}}^D x^2 - (N/C_0)xQ + ½ Q^2/C_0 = U_{\text{mech}} - 2U_{\text{mut}} + U_{\text{el}}.$$

With the indicated identifications of the mechanical, mutual, and electrical energy Eq. (4.25) gives

$$k^2 = N^2 / K_{\text{m}}^D C_0,$$

which agrees with the definition of k^2 in Sect. 1.4.1. Note that the term identified here as mechanical energy, $½K_{\text{m}}^D x^2$, differs from the mechanical energy in

Eq. (4.17), $\frac{1}{2}K_{\mathrm{m}}^{E}x^2$, because the stiffnesses and the displacements are different. In one case the displacement is caused by applied voltage only, in the other case by voltage and force applied together. It should be noted that the value used for U_{mut}, in order to get the correct value of k^2, was one half the value of the mutual term.

A somewhat different perspective on U_{mut} can be gained by rewriting the expression for U_{tot} above in terms of the other pairs of variables:

$$U_{\mathrm{tot}}(x,Q) = \tfrac{1}{2}K_{\mathrm{m}}^{D}x^2 - (N/C_0)xQ + \tfrac{1}{2}Q^2/C_0,$$
$$U_{\mathrm{tot}}(x,V) = K_{\mathrm{m}}^{E}x^2/2 - NxV + NxV + C_0V^2/2,$$
$$U_{\mathrm{tot}}(F,Q) = F^2/2K_{\mathrm{m}}^{D} + FNQ/C_0K_{\mathrm{m}}^{D} - FNQ/C_0K_{\mathrm{m}}^{D} + Q^2/2C_{\mathrm{f}},$$
$$U_{\mathrm{tot}}(F,V) = F^2/2K_{\mathrm{m}}^{E} + NFV/K_{\mathrm{m}}^{E} + C_{\mathrm{f}}V^2/2.$$

We find that when U_{tot} is expressed in terms of mixed variables (x, V or F, Q) two equal mutual terms appear with opposite signs. These terms cancel out of U_{tot}, leaving no U_{mut} and no way to apply Eq. (4.25). When U_{tot} is expressed in terms of homogeneous variables there is a positive mutual term for F, V, and a negative mutual term for x, Q. In these cases, if the negative sign is ignored and U_{mut} is considered to be half of the mutual term, Eq. (4.25) gives a value for k^2 in agreement with the other definitions. Equation (4.25) has recently been discussed and evaluated by Lamberti et al. [16] who conclude that it does not always have the same physical meaning as the definition based on the converted energy. Woollett also discussed this method of calculating the coupling coefficient and concluded that other methods based on a more definite rationale are preferable [3].

4.4.1.3 Other Features of the Coupling Coefficient

Some writers define the coupling coefficient as the ratio of cross products of the coefficients in the transducer equations [1, 14], an approach which also includes the k^2 vs. $k^2/(1-k^2)$ ambiguity. For example, using the pair of homogeneous equations at the beginning of the previous section this ratio is $N^2/K_{\mathrm{m}}^{D}C_0 = k^2$. Use of the related mixed equations, Eqs. (4.15) and (4.16), gives the ratio $N^2/K_{\mathrm{m}}^{E}C_0 = k^2/(1-k^2)$. Using this approach without first going to the low frequency limit [1, 14] also leads to a coupling coefficient that depends on frequency and resistance. The usual definitions make the coupling coefficient independent of driving frequency for any individual mode of a vibrator, although it does differ for different modes of the same vibrator. The effect of electrical or mechanical dissipation on the coupling is usually not considered, although dissipation clearly reduces the converted stored energy.

The IEEE Standard on Piezoelectricity abandoned the definition of coupling coefficients based on interaction (mutual) energies in 1978, and defined static coupling coefficients by energy ratios related to prescribed cycles of stress-strain or electric field-displacement. This definition appears to be equivalent to the definition based on converted energies. This change in the Standard was continued in the 1987 revision [18].

Although k_{33} and k_{31} are the most commonly used piezoelectric coupling coefficients for underwater sound transducers there are many others corresponding to different stress systems or different combinations of free and loaded surfaces of the active material. Tables of such formulae for specific piezoelectric configurations that might be used in underwater sound transducers are available [3, 12, 18]. Three specific cases that often occur are: the thickness coupling coefficient (e.g., a disc that expands in the thickness direction, see Sects. 3.2.3.4, 5.4.3, and 5.4.4),

$$k_t^2 = h_{33}^2/\beta_{33}^S c_{33}^D;$$

the planar extensional coefficient (e.g., a disc that expands radially, see Sect. 5.4.3),

$$k_p^2 = 2d_{31}^2/\varepsilon_{33}^T\left(s_{11}^E + s_{12}^E\right);$$

and the shear mode coefficient (a mode used in some accelerometer designs),

$$k_{15}^2 = d_{15}^2/s_{44}^E\varepsilon_{11}^T.$$

Values of these coefficients are included in Sect. 13.5 for specific piezoelectric ceramics.

Baerwald [19] has shown that piezoelectric crystals can be characterized by three (for piezoelectric ceramics, two) invariant coupling coefficients, one of which is the maximum possible coupling coefficient. It is fortunate that k_{33} for piezoelectric ceramics is only a few percent less than the maximum possible coupling coefficient [3, 12], since k_{33} can be easily achieved with the simple stress system $T_1 = T_2 = T_4 = T_5 = T_6 = 0$, and only T_3 non-zero.

4.4.2 The Effect of Inactive Components on the Coupling Coefficient

All transducers contain inactive components such as cable capacitance, stray capacitance, leakage inductance, and waterproofing seals that do not convert energy, but do influence energy conversion if they are capable of storing energy. In most cases transducers also necessarily contain inactive components such as stress rods between the head and tail masses, glue bonds between piezoelectric segments, and insulators between the ceramic and the head mass. And in some cases, such as the shell of flextensional transducers, a major part of the vibrating system is inactive. In this section we will calculate the reduction of the coupling coefficient caused by inactive components [3, 20].

It is convenient to start from an expression for the coupling coefficient in terms of the electromechanical turns ratio, N, the short circuit mechanical compliance, C^E, and the clamped capacitance, C_0, that follows from Eq. (1.16) in Sect. 1.4.1:

$$k^2 = N^2C^E/\left(N^2C^E + C_0\right). \tag{4.26}$$

An effective coupling coefficient, k_e, can then be defined by replacing all the parameters in Eq. (4.26) by effective values, N_e, C_e^E, and C_{0e}:

$$k_e^2 = N_e^2C_e^E/\left(N_e^2C_e^E + C_{0e}\right). \tag{4.27}$$

This generalization is valid as long as the transducer, including the inactive components, can be reduced to a system with one degree of mechanical freedom and one degree of electrical freedom with the basic equivalent circuit in Fig. 3.15 containing the effective circuit parameters. Under these conditions the effective values of the circuit parameters replace the basic values, but the relationships between them remain the same. Any of the other expressions for k^2 given in Sect. 1.4.1 involving other parameters could also be generalized in the same way. And, furthermore, since Eq. (4.26) is consistent with Mason's energy definition of k^2, that consistency must also hold for Eq. (4.27) under these conditions, although some of the converted energy may be stored in inactive components. We will now use Eq. (4.27) to derive expressions for k_e that show how specific inactive components modify the original coupling coefficient, k.

As a first example consider cable capacitance and some forms of stray capacitance which can be lumped together as C_c, in parallel with the clamped capacitance, C_0, without making any changes in the motional part of the circuit as shown in Fig. 4.4.

Thus $C_{0e} = C_0 + C_c,\ \ N_e = N,\ \ C_e^E = C^E$ and Eqs. (4.27) and (4.26) give

$$k_e^2 = \frac{N^2C^E}{N^2C^E + C_0 + C_c} = \frac{k^2}{1 + \left(1 - k^2\right)(C_c/C_0)} = \frac{k^2}{1 + C_c/C_f}, \tag{4.28}$$

where k is the coupling coefficient when there are no inactive components.

Cable capacitance reduces the effective coupling coefficient but has no effect on Q_m or the resonance frequency. The energy interpretation holds for the effect of

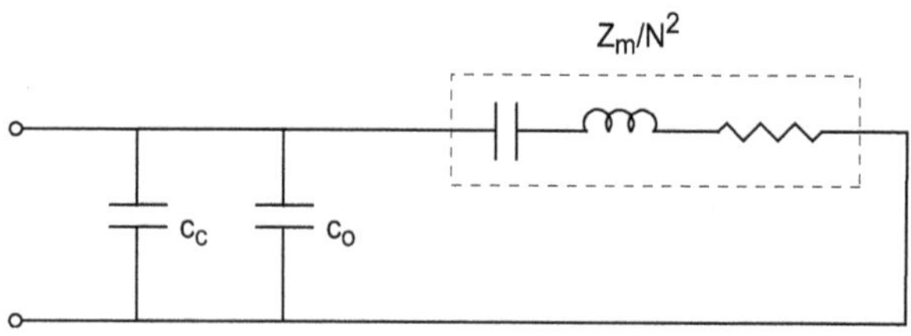

Fig. 4.4 Circuit with cable and/or stray capacitance, c_c

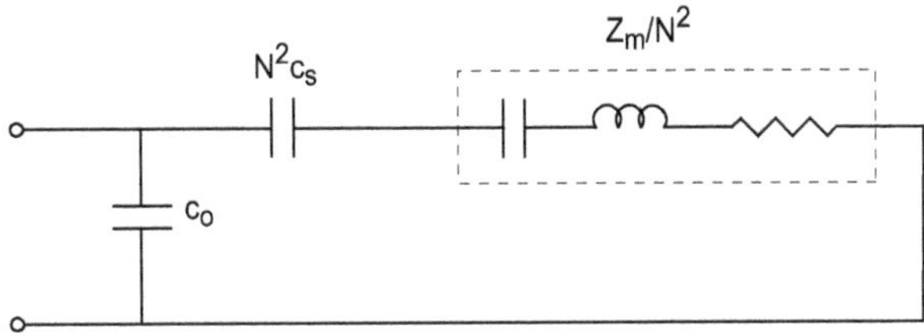

Fig. 4.5 Circuit with stress rod of stiffness $K_s = 1/c_s$

cable capacitance on the coupling coefficient; i.e., for a given applied voltage the input electrical energy is increased from ½ $(C_0+N^2C^E)V^2$ to ½ $(C_0+C_c+N^2C^E)V^2$, while the converted mechanical energy, ½$N^2 C^E V^2$, is not changed. Thus the cable decreases the coupling coefficient by storing some of the input energy.

Another example occurs for piston transducers driven by a piezoelectric ceramic bar under compression by a stress rod of stiffness K_s as shown in the equivalent circuit of Fig. 4.5.

In this case the stiffnesses of the ceramic bar and the stress rod add which makes $1/C_e^E = K_e^E = K^E + K_s$, while $N_e = N$ and $C_{0e} = C_0$. Equation (4.27) then gives

$$k_e^2 = \frac{N^2/(K^E + K_s)}{N^2/(K^E + K_s) + C_0} = \frac{k^2}{1 + (1 - k^2)K_s/K^E} = \frac{k^2}{1 + K_s/K^D}, \tag{4.29}$$

showing that the effective coupling coefficient decreases as the stiffness of the stress rod increases. In this case, for a given voltage, the input electrical energy and the converted mechanical energy are both reduced by the same amount, ½N^2V^2 $[1/K^E - 1/(K^E + K_s)]$, which decreases the coupling coefficient. The stress rod stores some of the input energy and also increases the resonance frequency by the factor $(1 + K_s/K^E)^{1/2}$.

The cable and stress rod examples illustrate the meaning of an inactive component in a transducer; it is a part of the transducer that can store only one kind of energy, electrical energy as in the cable or mechanical energy as in the stress rod, but it cannot convert energy. When some of the input energy is stored in an inactive component the coupling coefficient is reduced, even when that energy is converted energy as it is in the stress rod. Equations (4.28) and (4.29) show that the cable and the stress rod reduce the coupling coefficient in the same way when each is acting alone.

A layer of elastic material between the ceramic and the head mass, used as an electrical insulator, or glue bonds between ceramic segments also introduce inactive components (see Fig. 4.6).

This case is more complicated because the end of the ceramic bar and the head mass can move independently which means another degree of freedom exists. This internal degree of freedom must be eliminated to make Eq. (4.27) applicable. The three equations for the displacement of the mass, the displacement at the end of the ceramic, and the voltage can be combined to eliminate the displacement at the end of the ceramic. This leaves the following pair of static equations for the

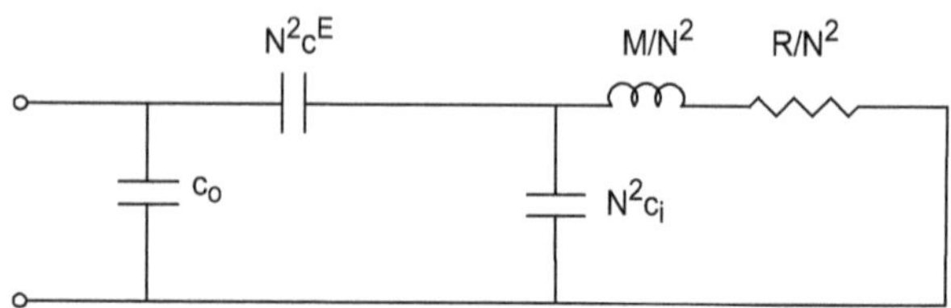

Fig. 4.6 Circuit with glue bond or insulating layer of stiffness $K_i = 1/c_i$

displacement of the mass, x, and the voltage, V, in the usual form which identifies the effective circuit parameters:

$$F_b = \frac{K^E K_i}{K^E + K_i} x - \frac{N K_i}{K^E + K_i} V = K_e^E x - N_e V,$$

$$Q = \frac{N K_i}{K^E + K_i} x + \left[C_0 + \frac{N^2}{K^E + K_i}\right] V = N_e x + C_{0e} V,$$

where K_i is the stiffness of the insulating layer or glue bonds. These equations show that all three of the circuit parameters are changed. The effective stiffness depends on both K_i and K^E. The clamped capacity is changed since clamping the mass still leaves the ceramic free to move. In addition, the force transferred from the ceramic to the mass is reduced, making $N_e < N$. Since the model has been transformed to one degree of freedom, Eq. (4.27) can be used to find the effective coupling coefficient with the result

$$k_e^2 = \frac{k^2}{1 + K^E/K_i} = \frac{k^2}{1 + (1 - k^2)(K^D/K_i)}. \tag{4.30}$$

Since the effective coupling coefficient decreases as the inactive layer becomes more compliant, insulating layers or glue bonds should be made as thin as possible. The resonance frequency is also reduced by the factor $(1 + K^E/K_i)^{-1/2}$.

Calculating the energies in terms of the effective circuit parameters shows that, for a given voltage, a glue bond or an insulating layer does not change the input energy. But the converted mechanical energy is reduced to ½ $[N^2V^2/K^E]$ $[K_i/(K^E + K_i)]$, because the increase in the clamped capacity causes more energy to be stored electrically. The reduction in converted mechanical energy lowers the coupling coefficient.

The effects of three different types of inactive components have been discussed separately. Now their combined effects will be considered, since transducers often contain all three and sometimes more. When a stress rod and glue bonds are considered together the bond stiffness, K_i, is mechanically in series with the ceramic, while the rod stiffness, K_s, is mechanically in parallel with the ceramic and K_i. As shown in the equivalent circuit of Fig. 4.7, the stress rod compliance, $C_s = 1/K_s$, becomes a series element and the glue bond or insulator compliance, $C_i = 1/K_i$, becomes a shunt element.

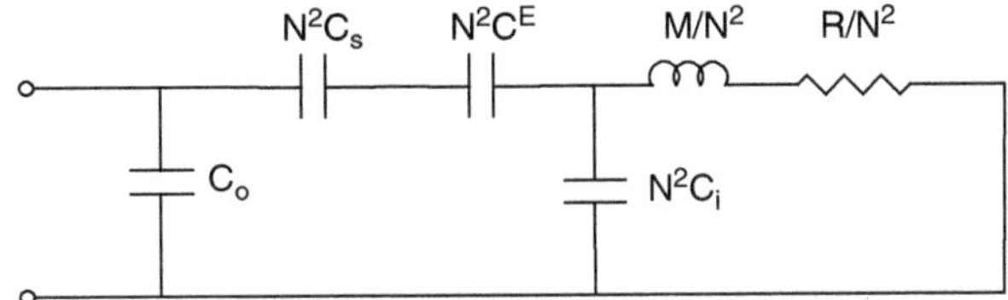

Fig. 4.7 Circuit with both stress rod and glue bond

The motion at the end of the ceramic is an internal degree of freedom that must be eliminated as was done for the glue bond alone. The resulting static equations for the variables at the two transducer ports are:

$$F_b = \left[\frac{K^E K_i}{K^E + K_i} + K_s\right]x - \frac{NK_i}{K^E + K_i}V = K_e^E x - N_e V,$$

$$Q = \frac{NK_i}{K^E + K_i}x + \left[C_0 + \frac{N^2}{K^E + K_i}\right]V = N_e x + C_{0e}V.$$

Thus N_e and C_{0e} are unchanged from the case of the glue bond alone, but the stress rod modifies the effective stiffness. Using these results for the effective circuit parameters in Eq. (4.27) gives

$$k_e^2 = \frac{k^2}{\left[1 + \frac{K_s}{K_i} + (1 - k^2)\frac{K_s}{K^E}\right]\left[1 + \frac{K^E}{K_i}\right]} = \frac{k^2}{\left[1 + \frac{K_s}{K_i} + \frac{K_s}{K^D}\right]\left[1 + (1 - k^2)\frac{K^D}{K_i}\right]}. \quad (4.31)$$

This expression reduces to the results for the stress rod alone when $K_i \to \infty$ and for the glue bonds alone when $K_s \to 0$. Note that the two effects interact in such a way that the combined effect is not equal to the product of the individual effects.

The effect of cable capacitance, C_c, can also be included if C_{0e} is determined from the above equation for Q after C_0 is replaced by $C_0 + C_c$. Then the effective coupling coefficient resulting from all three inactive components acting together is given by Eq. (4.27) as

$$\begin{aligned} k_e^2 &= \frac{k^2}{\left[1 + \frac{K_s}{K_i} + (1 - k^2)\left(\frac{K_s}{K^E} + \frac{C_c}{C_0}\right)\right]\left[1 + \frac{K^E}{K_i}\right]} \\ &= \frac{k^2}{\left[1 + \frac{K_s}{K_i} + \frac{K_s}{K^D} + \frac{C_c}{C_f}\right]\left[1 + (1 - k^2)\frac{K^D}{K_i}\right]}, \end{aligned} \quad (4.32)$$

which reduces to Eq. (4.31) when $C_c = 0$, and to Eq. (4.28) when $K_i \to \infty$ and $K_s \to 0$.

Numerical examples are given in Table 4.2 using typical values of the inactive components specified in terms of C_f and K^D for three active material coupling

Table 4.2 The effective coupling coefficient for various inactive components

Inactive component		$k = .90$	.70	.50
Cable	Eq. (4.28), $C_c/C_f = 0.2$	$k_e = .82$	.64	.46
Stress rod	Eq. (4.29), $K_s/K^D = 0.2$	$k_e = .82$	.64	.46
Glue bond	Eq. (4.30), $K_i/K^D = 5$	$k_e = .88$	.67	.47
The three above	Eq. (4.32)	$k_e = .74$	.56	.39
Eddy currents	Eq. (4.34), $\chi_r = 0.8$	$k_e = .88$	.66	.46

values, k. Note that the combined effect of three inactive components is significantly less than the product of the individual effects.

Magnetic field transducers contain analogous or equivalent inactive components that reduce the coupling coefficient. For example, leakage inductance reduces the coupling coefficient in a way similar to cable capacitance. But magnetic transducers are also subject to the effect of eddy current shielding (see Sect. 3.1.5), which changes the inductance and the electromechanical turns ratio as shown in the circuit of Fig. 3.17. This circuit corresponds to the following expression for k_e^2, analogous to Eq. (4.27):

$$k_e^2 = \frac{N_{me}^2 C_e^H}{N_{me}^2 C_e^H + L_{0e}}, \tag{4.33}$$

where N_{me}, C_e^H, and L_{0e} are the effective turns ratio, open circuit compliance, and clamped inductance. Eddy current shielding gives $N_{me} = \chi_r N_m$, $L_{0e} = \chi_r L_0$, and $C_e^H = C^H$, where χ_r is the real part of the eddy current factor. Using these values Eq. (4.33) gives

$$k_e^2 = \frac{\chi_r k^2}{1 + (\chi_r - 1)k^2} \approx \chi_r k^2, \tag{4.34}$$

where k^2 is the material coupling coefficient when eddy currents are insignificant and $\chi_r = 1$. The approximate form of Eq. (4.34) holds when eddy currents are small and χ_r is near unity. A numerical example is given in Table 4.2.

4.4.3 *The Effect of Dynamic Conditions on the Coupling Coefficient*

The calculations in Sect. 4.3 of the elastic potential energy, the kinetic energy, and the effective dynamic stiffness and mass of a vibrating bar can be used to evaluate the effect of standing waves on the electromechanical coupling coefficient [3, 11, 15, 16]. Equation (4.12) shows that the dynamic stiffness of a fixed-end bar of length L with no mass loading remains close to the static stiffness until the frequency approaches the fundamental mode resonance where L equals one quarter wavelength. At that frequency the stiffness exceeds the static stiffness by the factor $\pi^2/8 = 1.234$. Although the bar is a distributed system it can be treated as a single degree of freedom system with an effective dynamic coupling coefficient defined at the resonance frequency and applicable near resonance. Recall that all the expressions for k^2 in Sect. 1.4.1 show that an increase in stiffness results in a decrease in coupling if the other parameters remain the same. Equation (4.27) is most convenient for calculating the dynamic coupling coefficient as long as we make the reasonable assumption that N and C_0 do not depend on frequency, for it then contains only one frequency-dependent parameter, K^E, given by Eq. (4.12).

For a piezoelectric bar the effective coupling coefficient is reduced by the increased dynamic stiffness in a way that depends on the electrode arrangement. Consider first the segmented bar electrode arrangement that was treated in Sect. 3.2.3.1. Equation (3.46) gives the expressions for N_e and C_{0e} and shows that the standing waves in the bar depend on s_{33}^E. Thus the static stiffness is $K_b^E = A_0/nhs_{33}^E$, and the effective dynamic stiffness of the fundamental mode is $K_d^E = \pi^2 K_b^E/8$ from Eq. (4.12). These values of N_e, C_{0e}, and K_d^E when put in Eq. (4.27) give the effective coupling coefficient, but the result can also be found, and perhaps some insight gained, by noting that the dynamic increase in stiffness is mathematically equivalent to adding a stress rod of stiffness $K_s = [(\pi^2/8) - 1]K_b^E$. When this value of K_s is used in Eq. (4.29), with $k = k_{33}$, we find that the effective dynamic coupling coefficient for the fundamental mode of the bar with no mass loading is

$$k_{ed1}^2 = \frac{k_{33}^2}{k_{33}^2 + (\pi^2/8)(1 - k_{33}^2)}, \tag{4.35a}$$

which may also be written as

$$\frac{k_{ed1}^2}{1 - k_{ed1}^2} = \frac{8}{\pi^2} \frac{k_{33}^2}{1 - k_{33}^2}. \tag{4.35b}$$

When k_{33} is small $k_{ed1}^2 \approx 8k_{33}^2/\pi^2$; when k_{33} approaches unity $k_{ed1} \approx k_{33}$, and, when k_{33} has a typical value of 0.707, $k_{ed1} = 0.669$. Thus in this mode the dynamic reduction of coupling is only about 6 % for the segmented bar, but in a real transducer this reduction would be combined with other reductions caused by inactive components.

A different estimate of the dynamic effect on coupling can be made by calculating the resonance and antiresonance frequencies for a specific mode of the bar as was done following Eq. (4.3a). For the fundamental mode we found $\omega_{r1}/\omega_{a1} = 0.773$, which gives $k_{eff1} = \left[1 - (\omega_{r1}/\omega_{a1})^2\right]^{1/2} = 0.634$, to be compared with 0.669 found above. This is a real difference caused by the dependence of ω_r and ω_a on both the dynamic stiffness and the dynamic mass at two different frequencies; if there were no dynamic effects both calculations would give

$k_{ed1} = k_{eff1} = 0.707$. Also note that if the mass was the same at ω_r and ω_a the expression for k_{eff1} would become $\left[1 - K_d^E(\omega_{r1})/K_d^I(\omega_{a1})\right]^{1/2}$, which is similar to Eq. (1.16) but does not follow from it, because $K_d^E(\omega_{r1})$ and $K_d^I(\omega_{a1})$ differ, not only due to the short circuit/open circuit conditions, but also due to the frequency dependence of the stiffness. Equation (4.27), with the effective dynamic stiffness evaluated at ω_{r1}, gives an effective dynamic coupling coefficient that does not depend on the frequency-dependent part of the stiffness difference between ω_r and ω_a, and also does not depend on mass.

The effective coupling can be calculated for any other odd mode of the bar that could also be excited by this electrode arrangement. For the next mode with no mass loading, where $k_r L = 3\pi/2$, Eq. (4.12) shows that $K_d^E = (9\pi^2/8)K_b^E$, and

$$k_{ed3}^2 = \frac{k_{33}^2}{k_{33}^2 + (9\pi^2/8)\left(1 - k_{33}^2\right)}. \tag{4.36}$$

In this mode for $k_{33} = 0.707$, $k_{ed3} \approx 0.29$, showing that, for the higher modes, the effective coupling is decreased strongly by dynamic effects. For the nth mode ($n = 1, 3, 5\ldots$)

$$\frac{k_{edn}^2}{1 - k_{edn}^2} = \frac{8}{n^2\pi^2}\frac{k_{33}^2}{1 - k_{33}^2}. \tag{4.37a}$$

Note that

$$\sum_{n,\,odd}^{\infty} \frac{k_{edn}^2}{1 - k_{edn}^2} = \frac{8}{\pi^2}\frac{k_{33}^2}{1 - k_{33}^2}\sum_{n,\,odd}^{\infty}\frac{1}{n^2} = \frac{k_{33}^2}{1 - k_{33}^2}, \tag{4.37b}$$

since the value of the infinite series is $\pi^2/8$ [21]. This interesting result came from informal technical memoranda of the Clevite Corporation. An interpretation of this sum over modes will be given following Eq. (4.39b).

The effective coupling coefficients above apply to the modes of a bar with no mass loading. The same procedure that led to Eq. (4.35b) shows that the general result for the fixed-end segmented bar, with arbitrary mass loading, is

$$\frac{k_e^2}{1 - k_e^2} = \frac{K_b^E}{K_d^E}\frac{k_{33}^2}{1 - k_{33}^2}, \tag{4.38}$$

where Eq. (4.12) gives

$$\frac{K_d^E}{K_b^E} = \frac{1}{2}\left(\frac{k_r^E L}{\sin k_r^E L}\right)^2\left[1 + \frac{\sin 2k_r^E L}{2k_r^E L}\right],$$

and k_r^E is a solution of

$$k_r^E L \tan k_r^E L = M_b/M.$$

For example, for $M_b/M = \pi/4$, $k_r^E L = \pi/4$, $K_d^E/K_b^E \approx 1.01$ and $k_e \approx k_{33}$, as expected in this case where the bar is only 1/8 of a wavelength long and the dynamic stiffness only slightly exceeds the static stiffness.

The effective coupling coefficient for the other bar electrode arrangements can be found in a similar way. The length expander bar also has the electric field parallel to the bar (see Sect. 3.2.3.3), but in this case the electric field is not constant along the bar. The controlling elastic constant is s_{33}^D rather than s_{33}^E, and the dynamic stiffness from Eq. (4.12) is $K_d^D = \pi^2 A_0/8 s_{33}^D L$ for the fundamental mode with no mass loading. The values of N_e and C_{0e} are given in Eq. (3.52), and Eq. (4.27) for k_e can be expressed in terms of K_d^D by use of $K_d^E = K_d^D(1 - k_e^2)$. Then Eq. (4.27) gives

$$k_e^2 = \frac{N_e^2}{K_d^I C_{0e}} = \frac{8}{\pi^2} k_{33}^2. \tag{4.39a}$$

The relationship between k_e and k_{33} for the length expander bar differs from that for the segmented bar because K_d^D rather than K_d^E is the effective stiffness.

For the nth mode $k_{en}^2 = 8k_{33}^2/n^2\pi^2$ and

$$\sum_{n,\,\text{odd}}^{\infty} k_{en}^2 = \frac{8k_{33}^2}{\pi^2} \sum_{n,\,\text{odd}}^{\infty} \frac{1}{n^2} = k_{33}^2, \tag{4.39b}$$

using the value of the infinite series given above. This result emphasizes the poor energy conversion capability of the high-order modes. If all the modes were driven together at their modal frequencies with the same energy input, U_e, the total converted energy would be $\sum_{n,\,\text{odd}}^{\infty} k_{en}^2 U_e = k_{33}^2 U_e$, which is the same converted energy as if the bar was driven at a single frequency well below the fundamental mode with the energy U_e.

For the 31 bar with electrodes on the sides (see Sect. 3.2.3.2) N_e and C_{0e} are given in Eq. (3.47), and the controlling elastic constant is s_{11}^E. Thus $K_d^E = \pi^2 A_0/8\, s_{11}^E L$ and Eq. (4.27) gives

$$\frac{k_e^2}{1 - k_e^2} = \frac{8}{\pi^2} \frac{k_{31}^2}{1 - k_{31}^2}. \tag{4.40}$$

The relation between the effective coupling and the material coupling is the same for the segmented bar and the 31 bar because the effective stiffness is K^E in both cases.

The effective coupling coefficients calculated above for the three different bar electrode arrangements are consistent with the energy definition in which

$$k_{\mathrm{e}}^2 = \frac{U_{\mathrm{p}}}{U_{\mathrm{p}} + C_{0\mathrm{e}}V^2/2}.$$

Here $U_{\mathrm{p}} = ½K_{\mathrm{d}}\zeta_0^2$ is the converted mechanical energy with K_{d} given by Eq. (4.12). The rapid decrease of effective coupling with mode number is directly related to the increasing complexity of the standing wave strain distribution which varies as cos $(m\pi z/2L)$ along the bar with m odd. For $m = 1, L = \lambda/4$, the strain is maximum at the fixed end, zero at the free end, and has the same sign throughout the bar. For $m = 3$, $L = 3\lambda/4$, the strain has opposite signs in 2/3 of the bar, leaving only 1/3 in which piezoelectric excitation is effective. Since the elastic energy is proportional to strain squared the converted energy is 1/9 the value for $n = 1$, consistent with the $1/n^2$ decrease of k_{en}^2.

The three different bar electrode arrangements analyzed above were also discussed by Woollett who determined the effective coupling coefficients for the modes of bars free on both ends [3]. The results are the same as Eqs. (4.35b), (4.39a), and (4.40) for the fixed-free bar. Aronov has presented a means for optimization of the effective coupling by changing the electrode shape [22].

The coupling coefficient determined from measured resonance/antiresonance frequencies, as in Eqs. (3.82) and (9.5), is an effective coupling coefficient in the broadest sense, because it includes all the effects such as inactive components, dynamic conditions, and energy loss mechanisms that apply to the particular transducer being measured. Measurement of the coupling coefficient of a transduction material alone requires elimination of all these effects.

4.5 Parameter Based Figure of Merit (FOM)

In addition to the effective coupling coefficient, power factor, bandwidth, and the maximum energy density, the projector Figure of Merit (FOM) is used as a metric for evaluating the maximum power handling capability of a transducer as a projector of underwater sound power. This transducer projector Figure of Merit is based on the operational Figure of Merit definition

$$\mathrm{FOM}_{\nu} = W/(V_0 f_{\mathrm{r}} Q_{\mathrm{m}}) \tag{4.41}$$

where W is the maximum output power (based on electrical drive, stress, or thermal limits, whichever applies) at resonance, f_{r} is the resonance frequency, Q_{m} is the mechanical Q of the transducer, and the subscript "ν" indicates that the figure of merit is referenced to the total volume of the transducer, V_0. The units for FOM_{ν} are in watts/Hz m^3. Values from 60 to 130 are typical for flextensional-type transducer designs. An extensive list has been given by Jones and Lindberg [23] and by Delany [24]. A single crystal cylinder transducer achieved a measured $\mathrm{FOM}_{\nu} = 2174$ [25]. In this operational definition FOM_{ν} values are often obtained from measured values of W, V_0, f_r, and Q_m. Although Eq. (4.41) implies that the FOM_ν is a function of f_r and Q_m, we will show that this is not the case and that the FOM_ν may be calculated from other common transducer parameters. Moreover, it will be shown that the FOM_ν is also not dependent on the total volume, V_0, alone but on the ratio of the volume of active transducer piezoelectric material, V_p, to the total volume, V_0; that is, V_p/V_0.

Equation (4.41) may be referenced to the mass, M, of the transducer by multiplying Eq. (4.41) by V_0/M and since the units for this measure, FOM_m, are traditionally [23, 26] watts/kHz kg, the result must also be multiplied by 1000. It can be shown that $FOM_m \approx B \times FOM_\nu$ with units of W/kHz kg and B the buoyancy. A large FOM_ν and large B are needed for a large FOM_m. Additional discussion on the FOM is given in Sect. 5.1.1 of Chap. 5. The figure of merit for hydrophones, which is proportional to the product of the appropriate "d" and "g" piezoelectric constants and the volume of the material, V_p, is developed in Sect. 6.1.2 of Chap. 6

A different, but equivalent, representation of Eq. (4.41) for evaluating the Figure of Merit for piezoelectric projector transducers is developed in Chap. 5 in a form which is based on the electrical energy, E_e, stored in the transducer. This result may be written as

$$FOM_\nu = 2\pi\eta_{ea}k_e^2E_e/V_0 \tag{4.42}$$

where k_e is the effective coupling coefficient and η_{ea} is the electro-acoustical efficiency of the transducer. We extend that method here, in Chap. 4, and develop an alternative form [27] which provides a more direct evaluation through the electric field intensity, ξ, and the volume of the piezoelectric material, V_p.

Well below resonance, the maximum electrical energy density stored in the piezoelectric material may be simply written as $U_e = \varepsilon^T\xi^2/2$, where ε^T is the free dielectric constant of the material and ξ is the maximum applied electric field in peak Volts/m. The electric energy is then $E_e = V_pU_e = V_p\ \varepsilon^T\xi^2/2$, where V_p is the volume of the piezoelectric material. And since the effective coupling coefficient square may be written as the ratio of the energy converted to mechanical form, E_m, to the electrical energy stored; that is, $k_e^2 = E_m/E_e$, we may write

$$E_m = k_e^2V_p\varepsilon^T\xi^2/2 \tag{4.43}$$

The input power at resonance, $W_i = 2\pi f_r\ Q_mE_m$, may then be written as

$$W_i = 2\pi f_rQ_mV_pk_e^2\varepsilon^T\xi^2/2 \tag{4.44}$$

with the output power given by $W = \eta_{ea}W_i$, where η_{ea} is the efficiency of the transducer. Thus, on using the rms value of the electric field, $\xi_r = \xi/\surd 2$, we get an expression for the output power as

$$W = \eta_{ea}V_p2\pi f_rQ_mk_e^2\varepsilon^T\xi_r^2. \tag{4.45}$$

which is, on its own, a useful statement on the output power of a transducer. Finally, from Eq. (4.45) and the operational definition of FOM_ν, given by Eq. (4.41), we get

$$FOM_\nu = 2\pi\eta_{ea}\varepsilon^T k_e^2 \xi_r^2 V_p/V_0 \tag{4.46a}$$

Equation (4.46a) demonstrates that the FOM_ν has a strong quadratic dependence on the effective coupling coefficient and the maximum electric field intensity. It also has a linear dependence on the efficiency and the dielectric constant, as well as the ratio of the piezoelectric material volume to the total transducer volume. And it is seen to be independent of f_r and Q_m. All the parameters of Eq. (4.46a) are readily available and provide an empirical or theoretical foundation for evaluating and designing transducers for maximum power performance. Equation (4.46a) shows that the FOM_ν is a constant for scaled transducer designs if V_p/V_0 remains a constant, as it should for a scaled design.

The above expression may be used for magnetostrictive transducers by replacing the piezoelectric volume V_p with the magnetostrictive volume, V_m, the free dielectric constant, ε^T, by the free permeability, μ^T, and the rms electric field intensity, ξ_r, by the rms magnetic field intensity, H_r. For magnetostrictive transducers the formula is then

$$FOM_\nu = 2\pi\eta_{ea}\mu^T k_e^2 H_r^2 V_m/V_0 \tag{4.46b}$$

Equation (4.46a) may also be used for evaluating piezoelectric material alone by setting $\eta_{ea} = 1$, $V_0 = V_p$, and $k_e = k$. In this active material case, alone, we may write

$$FOM_\nu(\text{piezoelectric}) = 2\pi\varepsilon^T k^2 \xi_r^2 = 2\pi\xi_r^2 d^2/s^E = 2\pi\xi_r^2 ed \tag{4.47a}$$

where $k^2 = d^2/\varepsilon^T s^E$, s^E is the short circuit elastic modulus, d is the piezoelectric "d" constant, and the piezoelectric constant $e = d/s^E$. As seen, the FOM is directly related to the piezoelectric "ed" product, just as the figure of merit for a hydrophone is directly related to the "gd" product. The choice of which case to use from Eq. (4.47a) may depend on the accuracy or availability of the piezoelectric constants. Since the 33 mode coupling coefficient is approximately twice the 31 mode value in PZT ceramics, operation in the 33 mode should yield a factor of approximately four improvement in the FOM_ν. The FOM_ν for the magnetostrictive case may be readily obtained from Eq. (4.47a) with s^H the open circuit elastic modulus and d the magnetostrictive "d" constant, and written as

$$FOM_\nu(\text{magnetostrictive}) = 2\pi\mu^T k^2 H_r^2 = 2\pi H_r^2 d^2/s^H \tag{4.47b}$$

Sample results [27] are given in Table 4.3 for piezoceramic PZT-4, PZT-8, undoped single crystal PIN-PMN-PT, and doped textured ceramic PMN-PT (see Sect. 13.5). (The lower PIN-PMN-PT FOM is due to the lower, 200 kV/m, maximum electric field of the undoped material.) The maximum electric field, ξ_r

Table 4.3 FOM_v (W/Hz m^3) for piezoelectric materials in the 33 mode

Material	d_{33}(pC/N)	s_{33}^E(pm^2/N)	ξ_r (kV/m)	FOM_v
PZT-4	289	15.5	200	1350
PZT-8	225	13.5	400	3770
PIN-PMN-PT	1320	57.3	200	7640
Textured PMN-PT	517	23.5	400	11,430

values are not exact and depend on the range in which the material parameters become highly dissipative or nonlinear. Although some of the field intensity values have been reduced, the results below do not fully address dissipative heating limitations due to high duty cycle operation.

As may be seen from the transducer results given in Chap. 5, the material FOM_v values are more than ten times higher than typical flextensional transducers, which are less due to inefficiencies, lower effective coupling coefficient, and reduced proportion of active material given through the ratio V_p/V_0. And this is often necessary in the design of a practical transducer for a specific application. Different types of projector designs, such as those given in Chap. 5, are necessary for the wide band of frequencies that are in use. The FOM value should help in choosing the best projector design out of a group of projector transducers which appear to satisfy certain desired operational requirements such as frequency, bandwidth, size, shape, hydrostatic pressure, and, possibly, weight.

The figure of merit is evaluated at the fundamental resonant frequency of the transducer and the value of it does not depended on the frequency or the Q_m. As we will see in Chap. 5, certain types of transducer designs are more appropriate for particular operating frequencies because of the achievement of a desired resonant frequency within the limits of a reasonable size. The FOM value would be helpful at this point in choosing the best design variation within these transducers.

4.6 Summary

Important transducer characteristics such as resonance frequency, f_r, mechanical quality factor, Q_m, characteristic impedance, Z_c, effective electromechanical coupling coefficient, k_e, and figure of merit, FOM, were discussed in more detail in this chapter. It was shown the value of the resonance frequency of a simple vibrator is lower for the displacement response and higher for the acceleration response than the resonance for the velocity response. It was also shown that at low frequencies the frequency-dependent radiation resistance lowers the resonance. The mechanical Q is typically given by $Q_m = \omega_r M/R = f_r/(f_2 - f_1)$ where f_2 and f_1 are frequencies at the -3 dB half-power points of the power curve while M is the mass and R is the mechanical resistance from the equivalent circuit representation. For a vibrating bar the mass is distributed and is less than the static mass of the bar, M. For a quarter wavelength resonator bar, the mass is $M/2$, but at frequencies well below this quarter wavelength resonance the dynamic mass is $M/3$. The quantity $(KM)^{1/2}$ is called the characteristic mechanical impedance, Z_c, of the transducer and it is a measure of the Q_m which may be written as $Q_m = (KM)^{1/2}/R = Z_c/R$.

A number of alternative definitions for the coupling coefficient were discussed in this chapter. One that is the most useful and physically revealing is that the square of the coupling coefficient is equal to the ratio of the energy converted in one form to the total energy stored in another form. This definition was used as a basis to evaluate the effective coupling coefficient, k_e, and the reduction thereof, when inactive materials are added to the transducer. For example, if a stress bolt of stiffness, K_s, is used to compress the piezoelectric drive element of a transducer, it was shown that $k_e^2 = k^2/(1 + K_s/K^D)$ where K^D is the open circuit stiffness of the piezoelectric element.

The volume figure of merit for a piezoelectric projector may be written, in an operational form, as $\text{FOM}_v = W/(V_0 f_r Q_m)$ where W is the power out and V_0 is the total volume of the transducer. It was shown in this chapter that we may also write it in a theoretical form as $\text{FOM}_v = 2\pi\eta_{ea}\varepsilon^T k_e^2 \xi_r^2 V_p/V_0$ where V_p is the volume of the piezoelectric active material, ξ_r is the rms electric field intensity, η_{ea} is the electro-acoustic efficiency, and ε^T is the free dielectric constant. This form shows that the FOM does not fundamentally depend on the resonance frequency f_r or the Q_m but has a quadratic dependence on the electric field and the effective coupling coefficient and is linearly dependent on the efficiency, free dielectric constant, and the fraction of piezoelectric material.

Exercises (Degree of Difficulty: *Lowest, **Moderate, ***Highest)

4.1.* Consider the single mass resonator of Fig. 4.2. Calculate the resonance frequency for bar stiffness $K = 4 \times 10^{10}$ N/m and head mass $M = 1$ kg by first ignoring the bar mass and then considering that the bar contributes an additional 0.3 kg. Also, calculate the mechanical Q_m with $R_2 = 6 \times 10^4$ Ns/m for the two cases.

4.2.* Show that the square of the coupling coefficient $k^2 = N^2C^E/C_f$ where the free capacity $C_f = N^2C^E + C_0$, C_0 is the clamped capacity and the mechanical short circuit compliance $C^E = 1/K^E$ is consistent with Eq. (4.22) and $C_0 = C_f(1 - k^2)$ is consistent with Eq. (4.23).

4.3.** Determine the percentage change in the coupling coefficient if a stress rod with a stiffness that is one-tenth the stiffness of the piezoelectric ceramic short circuit stiffness is used with original coupling coefficient values of 0.9, 0.7, 0.5, and 0.3. What is the percentage change in the resonance and antiresonance frequencies? Assume a simple lumped mass system.

4.4.** Determine the percentage change in the coupling coefficient if an electrical insulator with a stiffness that is ten times greater than the stiffness of the piezoelectric ceramic is used between the head mass and the piezoelectric ceramic. Evaluate with original coupling coefficient values of 0.9, 0.7, 0.5, and 0.3. What is the percentage change in the resonance and antiresonance frequencies? Assume a simple lumped mass system. Tabulate results and compare with Exercise 4.3.

4.5.** Determine the percentage change in the coupling coefficient and the resonance and antiresonance frequencies when both the stress rod and the electrical insulator of Exercises 4.3 and 4.4 are used. Tabulate results and compare with individual effects.

4.6.** Calculate the effective dynamic coupling coefficient for the fundamental mode of a 33 mode segmented bar for k_{33} values of 0.9, 0.7, 0.5, and 0.3. Tabulate the percentage reduction. Compare results for the case of the length expander bar with electrodes only on the ends.

4.7.* Calculate the effective dynamic coupling coefficients for the second and third modes of the end electroded bar of Exercise 4.6.

4.8.** Derive an expression for the dynamic stiffness of the *n*th mode of the fixed-free bar from Eq. (4.12). Then derive expressions for the effective coupling coefficient of the *n*th mode for both the length expander bar and the segmented bar following the procedure in Sect. 4.4.3.

4.9.*** Derive an expression for the dynamic stiffness of the *n*th mode of the 31 mode ring based on the expression for the short circuit resonance frequencies, $f_n = f_0(1 + n^2)^{1/2}$. Then derive an expression for the effective coupling coefficient following the procedure in Sect. 4.4.3. Calculate the effective coupling coefficient, $k_{\text{ed}n}$, for the $n = 0$, 1, 2, 3 modes and for $k_{31} = 0.33$.

References

1. L.E. Kinsler, A.R. Frey, A.B. Coppens, J.V. Sanders, *Fundamentals of Acoustics*, 4th edn. (Wiley, New York, 2000)
2. D. Stansfield, *Underwater Electroacoustic Transducers* (Bath University Press, Bath, UK, 1991)
3. R.S. Woollett, *Sonar Transducer Fundamentals* (Naval Undersea Warfare Center, Newport, Rhode Island, Undated)
4. P.M. Morse, *Vibration and Sound*, 2nd edn. (McGraw-Hill, New York, 1948)
5. F.V. Hunt, *Electroacoustics* (Wiley, New York, 1954)
6. J.F. Hersh, *Coupling Coefficients*. Harvard University Acoustics Research Laboratory Technical Memorandum No. 40, 15 Nov 1957
7. R.S. Woollett, *Transducer Comparison Methods Based on the Electromechanical Coupling Coefficient Concept* (USNUSL, New London, Connecticut)
8. M.B. Moffett, W.J. Marshall, in *The Importance of Coupling Factor for Underwater Acoustic Projectors*. 127th ASA Meeting, June 1994; also Naval Undersea Warfare Center Technical Document 10,691, 7 June 1994
9. Private Communication, Wenwu Cao, Materials Research Institute, Pennsylvania State University (2004)
10. W.P. Mason, *Electromechanical Transducers and Wave Filters*, 2nd edn. (Van Nostrand, New York, 1948)
11. A. Caronti, R. Carotenuto, M. Pappalardo, Electromechanical coupling factor of capacitive micromachined ultrasonic transducers. J. Acoust. Soc. Am. **113**, 279–288 (2003)
12. D.A. Berlincourt, D.R. Curran, H. Jaffe, in *Piezoelectric and Piezomagnetic Materials and Their Function in Transducers*, ed. by W.P. Mason. Physical Acoustics, vol 1-Part A (Academic, New York, 1964)

13. J.C. Piquette, Quasistatic coupling coefficients for electrostrictive ceramics. J. Acoust. Soc. Am. **110**, 197–207 (2001)
14. O.B. Wilson, *An Introduction to the Theory and Design of Sonar Transducers* (US Government Printing Office, Washington, DC, 1985)
15. W.J. Marshall, *Dynamic Coupling Coefficients for Distributed Parameter Piezoelectric Transducers*. USNUSL Report No. 622, 16 Oct 1964
16. N. Lamberti, A. Iula, M. Pappalardo, The electromechanical coupling factor in static and dynamic conditions. Acustica **85**, 39–46 (1999)
17. K. Uchino, *Piezoelectric Actuators and Ultrasonic Motors* (Kluwer Academic, Boston, 1997)
18. IEEE: Standard on Piezoelectricity, ANSI/IEEE Std 176-1987
19. H.G. Baerwald, IRE Intern. Conv. Record, vol 8, Part 6 (1960)
20. R.S. Woollett, Effective coupling factor of single-degree-of-freedom transducers. J. Acoust. Soc. Am. **40**, 1112–1123 (1966)
21. I.S. Gradshteyn, I.M. Ryzhik, *Table of Integrals, Series and Products* (Academic, New York, 1980), p. 7, 0.234, 7
22. B. Aronov, On the optimization of the effective electromechanical coupling coefficients of a piezoelectric body. J. Acoust. Soc. Am. **114**, 792–800 (2003)
23. D.F. Jones, J.F. Lindberg, in *Recent Transduction Developments in Canada and the United States*. Proceedings of the Institute of Acoustics, vol 17, Part 3 (Institute of Acoustics, St. Alabans, UK, 1995), pp. 15–33
24. J.L. Delany, Bender transducer design and operations. J. Acoust. Soc. Am. **109**, 554–562 (2001)
25. H. Robinson, in *High Pressure Characterization of Single Crystal Cylinder Transducers*. 2007 U.S. Navy Workshop on Acoustic Transduction Materials and Devices, State College, PA
26. J.F. Lindberg, The application of high energy density transducer materials to smart systems. Mater. Res. Soc. Symp. Proc. **459**, 509–518 (1997)
27. J.L. Butler, Transducer figure of merit. J. Acoust. Soc. Am. **132**, 2158–2160 (2012)

Chapter 5
Transducers as Projectors

Active sonar and acoustic communication systems rely on electroacoustic transducers which "project" sound that is subsequently detected by hydrophones through a direct path or reflection from a target. Our focus in this chapter is on the projector which is significantly larger and more complex than the hydrophone because of the need to generate high acoustic intensity. Because of the reciprocal nature of transducers, the underlying concepts presented in this chapter will also apply to hydrophones, to be discussed in Chap. 6, although the details may differ considerably. The fundamentals of Chaps. 1 through 4 form the basis for our discussion of both projectors and hydrophones. The foundation and details, for modeling and analyzing projector transducers using equivalent circuits, matrix representations, and finite element models were presented in Chap. 3. Arrays of projectors will be discussed in Chap. 7 and acoustic radiation from transducers will be discussed in Chaps. 10 and 11.

Naval applications present the principal needs for underwater sound transducers and the main motivation for new transducer developments. Surface ships use hull-mounted arrays containing hundreds of projectors for medium frequency active search sonar and high frequency mine hunting sonar. Similar arrays are also mounted in towed bodies. Other projectors are used on surface ships for acoustic communications and depth sounding. Submarines, which depend on underwater sound more than other ships, use all the above types of sonar, but also need sonar for obstacle avoidance and under ice navigation. Submarine transducers must also be capable of operating over a wide range of hydrostatic pressure. Long range active sonar surveillance presents the greatest challenge to transducer development because of the need for high power at low frequency. Underwater communication systems require ship mounted transducers, but also involve networks of fixed transducers with special requirements. In addition there are many non-military applications for underwater sound transducers such as depth sounding, bottom mapping, fish finding, investigation of ship and aircraft wrecks, oil exploration, and various research projects.

J.L. Butler, C.H. Sherman, *Transducers and Arrays for Underwater Sound*,
Modern Acoustics and Signal Processing, DOI 10.1007/978-3-319-39044-4_5

Depending on source level, bandwidth, and system requirements, projectors can take various geometrical and mechanical forms such as spheres, cylinders, rings, piston radiators (see Figs. 1.5, 1.6, 1.7, and 1.8), benders, and amplified motion devices such as flextensional transducers (see Figs. 1.13 and 1.14). Piezoelectric ceramics are the most commonly used material for generating underwater sound, because of the many geometrical shapes in which they can be fabricated, their excellent electromechanical properties, their low electrical losses, and their ability to generate high forces. The desired performance from a projector is typically high power or intensity, high efficiency and broad bandwidth, usually with limitations on size and weight. Greater intensity and narrower acoustic beams require arrays (see Figs. 1.9, 1.10, and 1.11) of projectors that will be discussed in Chap. 7.

Projectors typically have 3D omnidirectionality, 2D omnidirectionality, or are unidirectional. Projectors that are small compared to the wavelength of sound usually have nearly 3D omnidirectionality, which is often the case for flextensional transducers. Spherical transducers are truly omnidirectional for all frequencies while cylindrical transducers have 2D omnidirectionality in a plane perpendicular to the axis, with directionality in any plane through the axis. The cylinder or ring is a common piezoelectric ceramic element which when stacked axially and cemented together with a piston head mass and inertial rear tail mass forms the common Tonpilz piston transducer which approaches a unidirectional beam pattern when the piston size is greater than one-half wavelength. Large arrays of closely packed piston transducers can be used to form a highly directional, unidirectional beam. Communication transducers are usually omnidirectional in the horizontal plane, but have reduced vertical radiation to limit unwanted reflections from the surface and bottom.

The operating depth affects the performance of a projector since acoustic isolation materials generally lose their compressibility under high hydrostatic pressure. Acoustic isolation is often needed to reduce radiation from surfaces that may be out of phase or prevent vibration from communicating with the transducer housing. The properties of piezoelectric ceramic can also be changed under high hydrostatic pressure, since the "permanent" bias can change under the pressure cycling that occurs in submarines. One solution to the deep submergence problem is the free-flooded ring, where the hydrostatic pressure is the same on the inside and the outside, but its radiation pattern is not convenient for forming arrays.

The operating frequency band for a particular application has a strong impact on the type of transducer required. If the transducer is to operate at low frequencies, designs which have low resonant frequencies and manageable sizes, such as benders or flextensional transducers, are most suitable. At the other extreme of high frequency operation the transducers must be small and are usually the piezoelectric ceramic–metal sandwich type. The most common transducer for mid-frequency bands is the Tonpilz, composed of a stack of piezoelectric ceramic elements with a larger piston radiating head mass and a heavy inertial tail mass. These are the transducers most often used to produce intense directional beams. We will discuss a variety of transducer designs with specific advantages for particular operating bands and other requirements. Finally, modal and low profile transducers will be presented and discussed.

High output power and efficiency are of paramount importance for any projector of sound, which is difficult to achieve if wideband operation is also required. This leads to designs that have multiple resonances or operation in the mass controlled region above resonance. The need for higher output has led to the development of the magnetostrictive material Terfenol-D, textured ceramic, and the single crystal materials PMN-PT and PIN-PMN-PT which have increased bandwidth and higher power factor performance.

5.1 Principles of Operation

High power projectors are often operated in the vicinity of resonance where the output motion is magnified by the mechanical Q of the transducer. At this frequency the mechanical mass and stiffness reactance cancel and, if the transducer is also electrically tuned, the input electrical impedance is resistive at resonance. At frequencies off resonance the impedance becomes partially reactive, reducing the power factor and increasing the volt-ampere product required for a given power output. Transducers with a high effective coupling coefficient require fewer volt-amperes and consequently smaller power amplifier capacity for a given power output over a given band.

Below resonance the electrical energy density stored is $U_e = \varepsilon^T E^2/2$, and Eq. (1.19) shows that the energy density converted to mechanical form is $U_m = k_e^2 \varepsilon^T E^2/2$ where ε^T is the free dielectric constant, E is the electric field intensity, and k_e is the transducer effective coupling coefficient. For magnetostrictive transducers the mechanical energy density is $U_m = k_e^2 \mu^T H^2/2$ where μ^T is the free permeability and H is the magnetic field intensity. As discussed in Chap. 2, the mechanical power density at resonance is $P = \omega_r Q_m U_m$ and for piezoelectric ceramics

$$P = \omega_r Q_m k_e^2 \varepsilon^T E^2/2,$$

where Q_m is the mechanical Q and ω_r is the angular mechanical resonance frequency. If the ceramic is sufficiently compressed, so that it is not dynamically stress limited, then the power output is limited by the electric field intensity E, which is typically limited to 10,000 V/in. or approximately 4 kV/cm (see Sect. 13.14). If the volume of active material is AL, the total mechanical power radiated is

$$W = \eta_{ea} \omega_r Q_m k_e^2 C_f V^2/2,$$

where η_{ea} is the electro acoustic efficiency, V is the voltage and $C_f = \varepsilon^T A/L$ is the free capacity.

5.1.1 Projector Figure of Merit

Two figures of merit are often used for projectors. One that depends on the total transducer volume, V_0, is defined as

$$\mathrm{FOM_v} = W/(V_0, f_r, Q_m),$$

which is the power per unit volume divided by Q_m and the resonance frequency, f_r, which may also be written as $\mathrm{FOM_v} = \eta_{ea}\pi k_e^2 V^2 C_f/V_0$, on substituting the expression for W. When weight rather than size is important, the alternative figure of merit involving the total transducer mass, M, is defined as

$$\mathrm{FOM_m} = W/(M, f_r, Q_m),$$

The units for $\mathrm{FOM_v}$ are watts/Hz m^3 while the units for $\mathrm{FOM_m}$ are, traditionally, watts/kHz kg. Transducers which produce the greatest power at a given frequency in the smallest volume or for the least weight have the highest projector figure of merit. See Chap. 4, Sect. 4.5 for more discussion and a parameter-based version of the Projector Figure of Merit (FOM). These figures of merit allow a direct comparison of different projector designs as long as the operating conditions are similar. Figure of merit values [1] are listed in Table 5.1 for some of the flextensional transducers (see Fig. 1.13) which will be discussed in this chapter.

Since the electrical energy stored in the transduction material, E_e, is $C_f V^2/2$, the figure of merit expressions may also be written as, with V_p the volume of the piezoelectric material,

$$\mathrm{FOM_v} = 2\pi\eta_{ea}k_e^2 E_e V_p/V_0 \quad \text{and} \quad \mathrm{FOM_m} = 2\pi\eta_{ea}k_e^2 E_e V_p/M.$$

which are consistent with Eqs. (4.46a) and (4.46b) of Chap. 4. We can see from these expressions that the FOM is $2\pi\eta_{ea}$ times the fraction of stored electrical energy converted to mechanical energy by the effective coupling coefficient per unit total volume or mass. Thus a high transducer figure of merit is obtained by use of a transduction material with high electrical energy storage capacity and high material coupling coefficient, incorporated into an efficient transducer mechanism with a correspondingly high effective coupling coefficient and a small total transducer volume or mass. It should be noted that the FOM, as defined above, includes

Table 5.1 Figures of merit for various PZT driven flextensional transducers

Projector	$\mathrm{FOM_m}$(watts/kHz kg)	$\mathrm{FOM_v}$ (watts/Hz m^3)
Class I barrel stave	26	63
Class IV flextensional	30	64
Class V ring-shell	55	81
Class VI ring-shell	25	71
Astroid	28	128

nothing about the transducer operating conditions. Designing to operate under unusually high static pressure or to withstand high shock or high temperature would tend to increase the mass or volume or reduce the effective coupling coefficient and yield a lower figure of merit. FOM comparisons can be misleading unless the operating conditions are considered.

The acoustic pressure, p, from a projector, as shown in Chap. 10, is proportional to the area, A, and the average normal velocity, u_n; the product Au_n is called the volume velocity or source strength. In the far field this pressure yields the intensity $I = |p|^2/2\rho_0 c_0$. Thus, high intensity requires large area and high normal velocity. For small radiators of limited radiating area, large velocities are needed and the accompanying required displacement, $x_n = u_n/j\omega$, becomes greater as the frequency decreases. This is the basic difficulty in obtaining large intensities from small sources at low frequencies.

Another difficulty results from the low radiation resistance associated with small low frequency sources as this leads to a low mechanical efficiency and low output power. At low frequencies the radiation resistance $R_r = A^2\omega^2\rho_0/4\pi c_0$ (see Chap. 10) where A is the radiating area, and, in this chapter, c_0 is the speed of sound and ρ_0 is the density of the medium. (We will use c and ρ for various other sound speeds and densities, to be identified in each case.) The radiated power is, therefore,

$$W = u_n{}^2 R_r/2 = u_n{}^2 A^2\omega^2\rho_0/c_0 4\pi = (x_n A)^2\omega^4\rho_0/c_0 4\pi.$$

Thus, the displacement area product must increase quadratically to compensate for a linear decrease in frequency, which seriously limits low frequency output power. Magnification, by lever arms (see Sect. 5.5) or quarter wave sections (see Sect. 5.4), can be used to increase the displacement from displacement limited driver materials, while area transformation with pistons (see Sect. 5.3) can be used to increase the radiating area.

Since the power output is proportional to the radiation resistance for a given velocity, it is important that the transducer be designed for maximum radiation resistance. A measure of this is the so-called radiation "efficiency" $\eta_{ra} = R_r/A\rho_0 c_0$ which, for acoustically large uniform radiators, approaches unity as though the transducer radiated plane waves. Although curved surfaces do not produce plane waves, they can approach unity radiation efficiency if the radius of curvature is large compared to the wavelength of interest. It is important that the amplitude and phase do not vary over the surface of the radiator as reductions in amplitude reduce the output and phase variations reduce the radiation resistance and the radiation efficiency. Out-of-phase motion can result from excitation of higher order modes of vibration or from the out-of-phase radiation from the back of the main radiator as in the case of a vibrating piston without a back enclosure. The design of the radiation section of the transducer is as important as the design of the transduction drive section.

Careful choice of the transduction method and material is of paramount importance. If piezoelectric ceramic is used, Navy Type III (PZT-8) yields the least electrical loss under high drive fields since the electrical dissipation does not increase as much under high drive as it does in other piezoelectric ceramics (see

Sects. 13.5 and 13.14). On the other hand, Type I (PZT-4) material is a better choice under lower drive conditions as it has a larger d constant and coupling coefficient. Textured ceramics promise an even greater d constant and higher coupling coefficient. In some applications the magnetostrictive material Terfenol-D may be a better choice than piezoelectric ceramics because of its greater power density and lower impedance. The newer magnetostrictive material, Galfenol, offers no greater power density than piezoelectric ceramics, but has more than an order of magnitude greater tensile strength and may be welded or threaded to other parts of the transducer. Newer single crystal material such as PIN-PMN-PT offers a greater d constant and coupling coefficient, with approximately 6–10 dB greater output potential and a better power factor without an external bias. Single crystal PMN-PT material requires an additional bias field to achieve its full output potential. Bias fields add to the complexity of the design by requiring additional power sources and wiring or permanent magnets. Magnetostrictive materials Terfenol-D and Galfenol require a bias and must also be laminated to reduce eddy current losses. The properties of these transduction materials, as well as other materials used in transducers, are given in Sects. 13.5 through 13.7.

5.2 Ring and Spherical Transducers

The ring, or short thin-walled cylinder, is one of the most common forms of underwater transducer which is used for both projectors and hydrophones. It may be stacked and formed into a line array, end-capped and air-backed or free-flooded and operated in either the piezoelectric 31 or 33 mode. The shape provides omnidirectionality in the plane perpendicular to the axis and near omnidirectionality in planes through the axis, if the height is small compared to the acoustic wavelength in the medium. A small height compared to the wavelength in the active material also prevents the excitation of length extensional and bending modes of vibration. In the fundamental circumferential mode the ring vibrates in a radial direction as a single mass as a result of the expansion and contraction of the circumference induced by an electric or magnetic field and can be represented by a simple lumped equivalent circuit.

5.2.1 *Piezoelectric 31 Mode Ring*

The 31 mode piezoelectric ring is the most commonly used ring transducer. It is usually inexpensive when compared to other transducers and achieves an effective coupling coefficient that is nearly equal to k_{31}. Usually, adding two leads, two end caps and waterproof encapsulation to the ceramic ring is all that is needed to fabricate this transducer. In addition, the equivalent circuit takes on the simplest form of any transducer. A sketch of a 31 mode ring with electrodes on the inside and outside cylindrical surfaces and piezoelectric coordinates 1, 2, 3 is shown in Fig. 5.1.

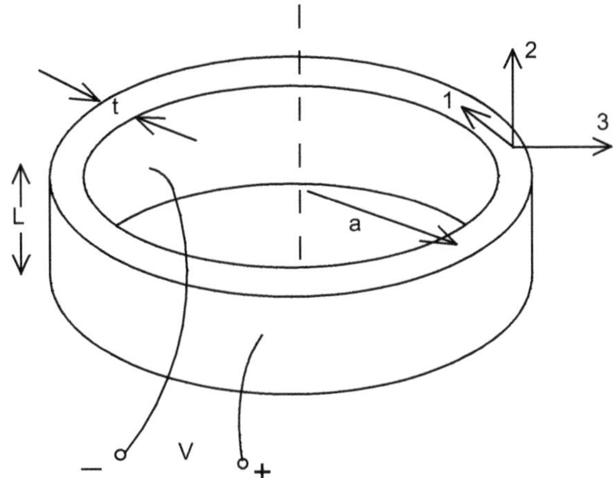

Fig. 5.1 31-Mode piezoelectric ring of mean radius *a*, thickness *t*, and length *L*

The 1 direction is along the circumference, 2 is along the length, and 3 is radial. In the 31 mode we excite cylindrical vibration in a mode where the primary stress and strain are in the circumferential direction, by means of an electric field which is spatially orthogonal to that direction. Although this is not the optimum way to excite circumferential motion, it is commonly used because it permits simple construction. It is assumed that the wall thickness, t, is small compared to the mean radius, a, and the length L is small compared to the mean diameter, $D = 2a$. It is also assumed that the ends are free to move with no loading so that $T_2 = 0$ and that the interior of the cylinder is filled with air and protected from the outside fluid medium by acoustically isolated end caps. Because the electrodes are equipotential surfaces, $E_1 = E_2 = 0$ on the inside and outside surfaces and throughout the cylinder since t is small. These conditions lead to the piezoelectric equation pair

$$S_1 = s_{11}^E T_1 + \mathrm{d}_{31} E_3, \tag{5.1}$$

$$D_3 = \mathrm{d}_{31} T_1 + \varepsilon_{33}^T E_3. \tag{5.2}$$

The radial equation of motion may be developed from Eq. (5.1) by rewriting the circumferential strain, S_1, as ξ/a yielding

$$\xi/a = s_{11}^E F/tL + \mathrm{d}_{31} V/t, \tag{5.3}$$

where ξ is the displacement in the radial direction as the circumference changes from $2\pi a$ to $2\pi(a + \xi)$, as illustrated in Fig. 5.2, and F is the circumferential force $T_1 tL$. The force, F, may be related to the radial force, F_r, as shown in Fig. 5.3.

For $\delta\theta$ small, $F_r = 2F\sin\,(\delta\theta/2) \approx F\delta\theta$ and, since this applies at each point around the ring, the total radial force is $F_r = 2\pi F$. For small displacements the

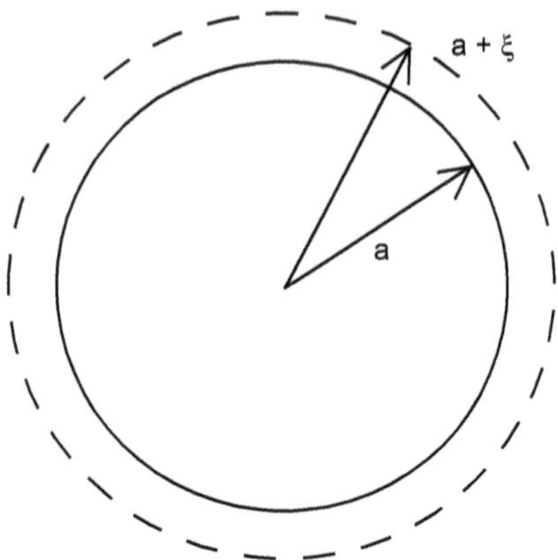

Fig. 5.2 Expansion of the ring circumference

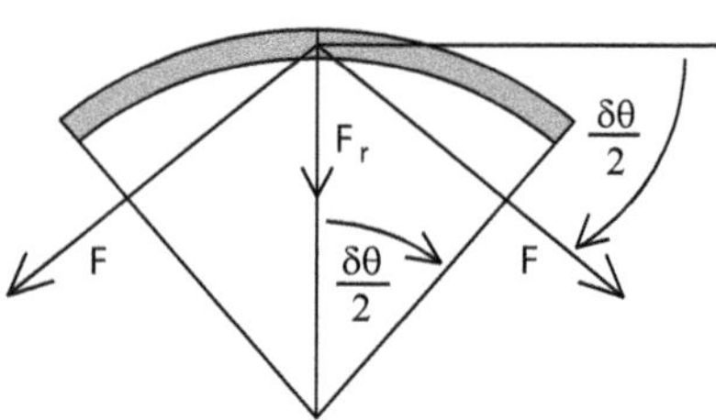

Fig. 5.3 Radial force, F_r

entire ring, of density ρ, moves as a mass, $M = \rho 2\pi atL$, and the radial equation of motion is

$$M\ddot{\xi} = F_0 - F_r = F_0 - 2\pi F, \tag{5.4}$$

where F_0 is any additional force such as the radiation force on the radial moving surface or the force due to an incoming acoustic wave. Substitution of the circumferential force, F, from Eq. (5.3) then leads to

$$M\ddot{\xi} + K^E\xi = NV + F_0, \tag{5.5}$$

where the short circuit radial stiffness $K^E = 2\pi tL/{s_{11}}^E a$ and the electromechanical turns ratio $N = 2\pi L d_{31}/{s_{11}}^E$. Under sinusoidal conditions Eq. (5.5) becomes

$$j\omega M\dot{\xi} + \left(K^E/j\omega\right)\dot{\xi} = NV + F_0. \tag{5.6}$$

With F_0 including the mechanical damping resistance, R_m, and radiation impedance, Z_r, the solution for the velocity $u = \dot{\xi}$ is

$$u = NV/\left(j\omega M + K^E/j\omega + R_m + Z_r\right), \tag{5.7a}$$

where NV is the piezoelectric driving force for a radial one degree of freedom mass/spring system.

Under in-air loading conditions, as opposed to in-water loading, Z_r is negligible and the mechanical resonance frequency occurs when $\omega M = K^E/\omega$ leading to $\omega_r^2 = K^E/M = 1/s_{11}^E \rho a^2$. Since the bar wave speed $c = \left(1/s_{11}^E \rho\right)^{1/2}$, resonance occurs at the frequency

$$f_r = c/2\pi a \quad \text{and} \quad f_r D = c/\pi = \text{constant for each material.} \tag{5.7b}$$

For the case of Navy Type I (PZT-4) material the frequency constant (see Sect. 13.6) is $f_r D = 41$ kHz-inches, affording a simple means for estimating the mean diameter for a desired resonance frequency. Since the wave speed $c = f\lambda$, resonance occurs when the wavelength equals the circumference. Although this is a useful relationship for remembering the condition for resonance, there are no standing waves along the circumference of the cylinder, since the continuous circumference presents no boundary condition. However, one might interpret the wavelength condition as two half-wavelength resonant bars curved to form half-circles and joined at their ends, forming a full wavelength structure. Circumferential waves in the ring satisfy the periodic condition $2\pi a/\lambda = 1, 2, 3, \ldots$, which is the condition for the vibration to optimally reinforce itself.

Equation (5.2) may be used to obtain the input electrical admittance of the ring by eliminating T_1 from Eqs. (5.1) and (5.2) giving

$$D_3 = \left(\mathrm{d}_{31}/s_{11}^E\right)S_1 + \varepsilon_{33}^S E_3, \tag{5.8}$$

where, as in Chap. 2, the clamped dielectric constant $\varepsilon_{33}^S = \varepsilon_{33}^T\left(1 - k_{31}^2\right)$ and $k_{31}^2 = \mathrm{d}_{31}^2/s_{11}^E \varepsilon_{33}^T$. With the dielectric displacement, D_3, given by the charge, Q, per unit area, Eq. (5.8) becomes

$$Q/(2\pi a L) = \left(\mathrm{d}_{31}/s_{11}^E a\right)\xi + \varepsilon_{33}^S V/t, \tag{5.9}$$

and the current is

$$I = \mathrm{d}Q/\mathrm{d}t = \left(2\pi a L \mathrm{d}_{31}/s_{11}^E\right)u + \varepsilon_{33}^S(2\pi a L/t)\mathrm{d}V/\mathrm{d}t. \tag{5.10}$$

Under sinusoidal conditions and with the clamped capacity $C_0 = (2\pi a L/t)\varepsilon_{33}^S$, Eq. (5.10) gives the input electrical admittance

$$Y = I/V = Nu/V + j\omega C_0. \tag{5.11}$$

Substitution of u/V from Eq. (5.7a) then leads to the final form of the electrical admittance as

$$Y = G_0 + j\omega C_0 + N^2/\left(j\omega M + 1/j\omega C^E + R_m + Z_r\right), \tag{5.12}$$

where we have added the electrical loss conductance $G_0 = \omega C_f \tan\delta$ (see Chaps. 2 and 3) and introduced the compliance $C^E = 1/K^E$ for the analogous capacitance component in the electrical equivalent circuit shown in Fig. 5.4.

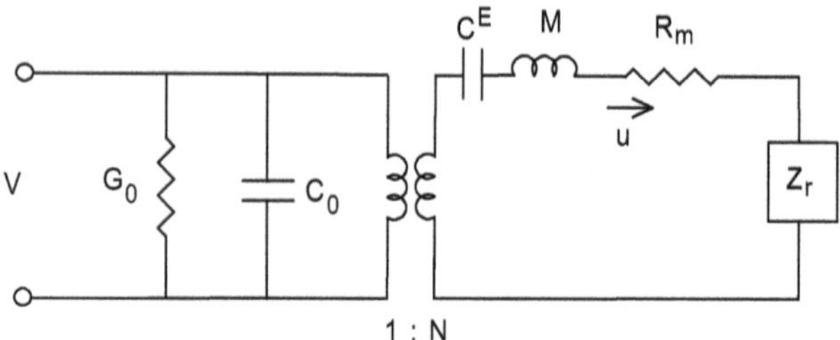

Fig. 5.4 Equivalent circuit for piezoelectric ring

In the model above we assumed polarization along the radial direction with electrodes on the inner and outer cylindrical surfaces. If, however, the electrodes are placed on the top and bottom annular surfaces in Fig. 5.1, and the ring is polarized along the 2 direction, $C_0 = (2\pi at/L)\varepsilon^S$ and $N = 2\pi t d_{31}/s_{11}^E$, rather than the values above for the radially polarized case.

The circuit of Fig. 5.4 or Eq. (5.7a) may be used to obtain the ring radial velocity; however, there is no exact closed form solution for the radiation impedance or for the relationship between the far-field pressure and the ring radial velocity. On the other hand, the Laird-Cohen model [2], which is based on a vibrating cylinder of finite length with infinite rigid cylindrical extensions (see Chap. 10), gives a good approximation to the far-field pressure. A Fourier series approach [3] has also been used to obtain the radiation impedance for the same model (see Sect. 11.1.2). The radiation impedance for the ring transducer may also be approximated by an equivalent sphere of the same radiating area with radius $a_s = (aL/2)^{1/2}$ and impedance (see Sect. 10.4.1),

$$Z_r = R_r + j\omega M_r = A\rho_0 c_0 \left[(ka_s)^2 + jka_s\right] / \left[1 + (ka_s)^2\right], \tag{5.13}$$

where the area $A = 4\pi a_s^2 = 2\pi aL$, ρ_0 and c_0 are the density and sound speed in the water, and the wave number $k = \omega/c_0$.

With Eq. (5.13) as an approximation for the radiation impedance the mechanical Q is

$$\begin{aligned} Q_m &= \omega_r(M + M_r)/(R_m + R_r) = \eta_{ma}\omega_r(M + M_r)/R_r = \eta_{ma}[\omega_r M/R_r + \omega_r M_r/R_r] \\ &\approx \eta_{ma}[(k_r a_s + 1/k_r a_s)(t\rho/a_s\rho_0) + 1/k_r a_s], \end{aligned} \tag{5.14}$$

where $k_r = \omega_r/c_0$. As seen, the first term depends on the wall thickness, t. The remaining term, $1/k_r a_s$, is a result of the ratio $\omega_r M_r/R_r$, which is a measure of the radiation Q and is dependent on the contribution from radiation loading alone. This term would be predominant if the wall thickness was extremely small. For large $k_r a_s$, where the radiation mass is negligible and Eq. (5.7b) holds, Eq. (5.14) simplifies to

$$Q_{\text{m}} \approx \eta_{\text{ma}}(t/a)(\rho c/\rho_0 c_0), \tag{5.15}$$

which shows that cylindrical transducers have broad bandwidth if $t \ll a$.

For example: Typical values might be mechanoacoustical efficiency $\eta_{\text{ma}} = 0.8$, $t/a = 0.2$, $\rho c/\rho_0 c_0 = 22.4 \times 10^6/1.5 \times 10^6 = 14.9$ resulting in a low mechanical $Q_{\text{m}} = 2.4$. In practice the ring resonance typically occurs for $k_r a_s \approx 2$ increasing Q_{m} to ≈ 3.5, as can be seen from Eq. (5.14) with $\rho/\rho_0 = 7.8$.

The radiation mass always (an exception is noted in Sect. 5.8.1) lowers the resonance frequency and in this low ka_s case the mass approximation is $4\pi a_s^3 \rho_0$, and the resonance frequency is

$$f_r \approx (c/2\pi a)\left(1 + 4\pi a_s^3 \rho_0/M\right)^{-1/2} = (c/2\pi a)(1 + \rho_0 a_s/\rho t)^{-1/2}, \tag{5.16}$$

which can result in a significant reduction in the resonance frequency for $t \ll a_s$. For a small ring we may use a spherical radiation model for the far-field acoustic pressure in terms of the radial velocity u given by Eq. (5.7a) or the equivalent circuit of Fig. 5.4:

$$p(r) = j\omega\rho_0 A u e^{-jkr}/4\pi r(1 + jk_r a_s). \tag{5.17}$$

Typical designs of the piezoelectric ring transducer include metal plate or curved-ceramic acoustically isolated end caps to prevent water entering the interior. The end caps are sometimes separated by a stiff rod along the ring axis which maintains the end caps at a small distance from the ends of the ring and prevents clamping of the ring radial motion. Soft acoustic isolation material such as rubber impregnated cork (corprene) or paper (see Sect. 13.2) is also used to isolate the caps. The cylinders may be wrapped with fiberglass to prevent operation in tension under high drive and booted with rubber or potted with polyurethane or both. Under deep submergence the hydrostatic pressure, P_0, compresses the ceramic ring with a circumferential stress $T = P_0 a/t$ allowing operation at higher dynamic stress.

For example: At 1000 ft depth the ambient pressure is 444 psi and for $a/t = 10$ the compressive circumferential stress on the ring is 4440 psi. On the other hand, if the ambient induced stress reaches levels greater than 10,000–15,000 psi, deterioration of the piezoelectric parameters can occur and lead to a reduction in performance (see Sect. 13.14).

Multiple rings acoustically isolated from each other may be stacked to form line arrays which are steerable from broadside to end-fire while maintaining an omnidirectional pattern in the plane perpendicular to the axis of the array (see Chap. 7).

5.2.2 *Piezoelectric 33 Mode Ring*

Although the 31 mode ring transducer is simple and effective, greater coupling, power output, and electrical efficiency can be obtained by operation in the 33 mode where the electric field is directed along the circumference, i.e., in the direction of the primary stress and strain. This is usually achieved by piezoelectric ceramic elements cemented together in mechanical series with electrodes between them as illustrated in Fig. 5.5a. An alternative method uses electrode striping with circumferential polarization, illustrated in Fig. 5.5b, and, as in the segmented ring of Fig. 5.5a, is typically used with all elements wired in parallel.

The segmented construction of Fig. 5.5a allows the addition of inactive materials to modify the transducer performance. A small portion of the ceramic located under the electrodes in the striped case of Fig. 5.5b is unpolarized since the ring is polarized by use of the same electrodes. The elastic modulus of this portion is approximately equal to the open circuit modulus of the polarized material.

The model for the 33 mode ring transducer with inactive sections has been given by Butler [4] and is summarized below using the circuit of Fig. 5.4. With w_{i} the width of the inactive material of elastic modulus s_{i} and w_{a} the width of the active piezoelectric material of elastic modulus s_{33}^E the effective coupling coefficient (see Eq. (4.30)) is

$$k_{\mathrm{e}}^2 = k_{33}^2/\left(1 + s_{\mathrm{i}}w_{\mathrm{i}}/s_{33}^E w_{\mathrm{a}}\right), \tag{5.18}$$

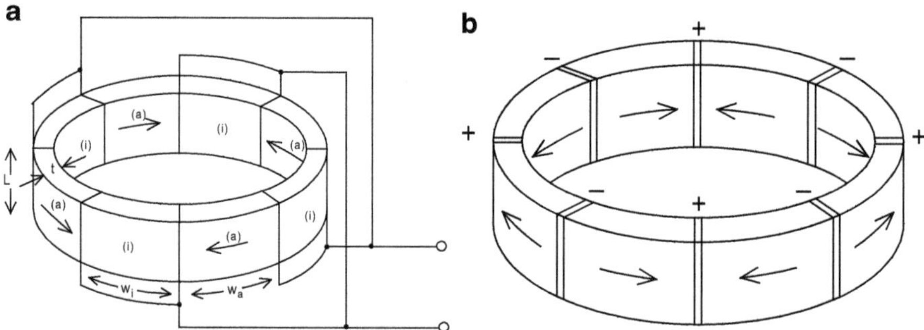

Fig. 5.5 (**a**) Parallel wired segmented ring with inactive segments (*i*) interlaced with active piezoelectric segments. The arrows show the direction of remnant polarization [4]. (**b**) Striped 33-mode piezoelectric ceramic cylinder

where $k_{33}^2 = d_{33}^2/s_{33}^E\varepsilon_{33}^T$. The inactive material may be used to represent cement joints, the portions of a solid ring shielded by striped electrodes or additional material such as aluminum used to reduce the Q_m and optimize the performance of the transducer for a particular application.

With cross-sectional area $A_c = tL$, n active and n inactive segments and total mean circumference $2\pi a = n(w_i + w_a)$, the mass of the ring is $M = nA_c(\rho_i w_i + \rho w_a)$ where ρ_i is the density of the inactive section and ρ is the density of the active section, and the remaining parameters for the equivalent circuit of Fig. 5.4 are:

$$C_0 = nA_c\varepsilon_{33}^T(1 - k_e^2)/w_a, \tag{5.19a}$$

$$N = 2\pi A_c d_{33}/\left(s_i w_i + s_{33}^E w_a\right), \tag{5.19b}$$

$$C^E = n\left(s_i w_i + s_{33}^E w_a\right)/4\pi^2 A_c. \tag{5.19c}$$

The resulting in-air resonance frequency is

$$f_r = (w_i + w_a)/2\pi a\left[(\rho_i w_i + \rho w_a)\left(s_i w_i + s_{33}^E w_a\right)\right]^{1/2}, \tag{5.20}$$

and the mechanical Q is

$$Q_m = [2\pi A_c/(R_m + R_r)]\left[(\rho_i w_i + \rho w_a)/\left(s_i w_i + s_{33}^E w_a\right)\right]^{1/2}. \tag{5.21}$$

Equations (5.20) and (5.21) show that inactive materials with density less than the ceramic density yield a lower Q_m and materials with a higher elastic compliance $s_i > s_{33}^E$ yield a lower resonance frequency.

For the case of no inactive material, $w_i = 0$ and $f_r = (1/2\pi a)\left(1/\rho s_{33}^E\right)^{1/2} = c/2\pi a$ leading to a frequency constant $f_r D = 37$ kHz-inches which is 10 % less than the 31 mode case of Sect. 5.2.1.

More importantly the 33 mode ring has a coupling coefficient and a "d" constant which are approximately double those for the 31 mode ring giving a significantly improved power factor and greater bandwidth. The approximate radiation impedance in Eq. (5.13) and far-field pressure in Eq. (5.17) may also be used to estimate performance for the 33 mode ring.

5.2.3 *The Spherical Transducer*

The thin shell spherical transducer provides a near-ideal omnidirectional beam pattern. It is usually fabricated from two hemispherical shells which are cemented together and operated with a radial electric field applied between the inner and outer electrode surfaces. The two outer electrodes as well as the two inner electrodes of the hemispheres are connected together. A wire from the inner electrodes is passed

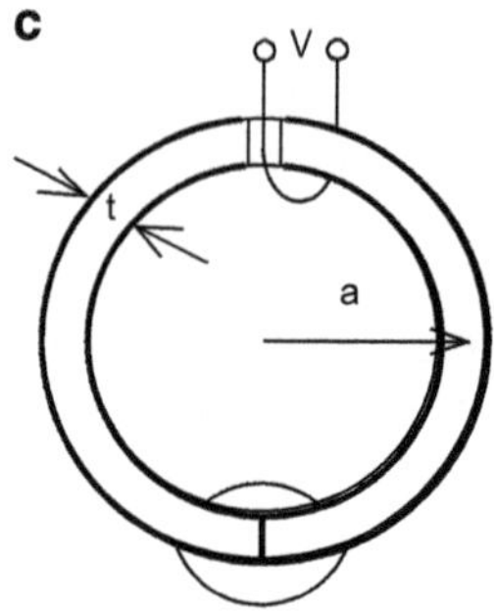

Fig. 5.5 (**c**) Cross section of spherical piezoelectric transducer fabricated from two hemispheres with electrodes on the inner and outer surfaces

through an insulated connector fitted into a small hole in the sphere at the junction of the two hemispheres as illustrated in Fig. 5.5c.

The size of the sphere is usually limited by the fabrication process to about a 6 in. diameter, although larger sizes have been fabricated from spherical triangle segments [5]. The transducer is especially useful where an omnidirectional pattern is desired even at frequencies where the wavelength is small compared to the size of the sphere. Because of the spherical shape, this transducer may be operated at considerable depths, and because of its planar mode of operation it has a planar mode coupling coefficient, k_p, (see Sect. 5.4.3) that is about half way between that of the 31 and 33 mode rings. It is considerably more costly than the 31 mode ring transducer described in Sect. 5.2.1.

The spherical transducer is unique in that it is the only transducer in which the radiation load can be accurately represented by a simple closed form expression allowing an exact algebraic formula for the resonance frequency and mechanical Q under water-loaded conditions. As in the case of a ring transducer, the fundamental omnidirectional radial (or so-called *breathing*) mode of vibration can be represented by a simple lumped equivalent circuit model for the common case where the wall thickness, $t \ll 2a$ with, a, the mean radius. As a radial electric field E is applied between the electrode surfaces a circumferential stress is developed causing the spherical wall to expand in the radial direction with lumped mass $M = 4\pi a^2 t\rho$, where ρ is the density of the material. This mass oscillates radially at a resonance frequency given by $1/2\pi$ times the square root of the ratio of the effective radial stiffness and the mass M.

The analysis is similar to the development of the 31 mode ring in Sect. 5.2.1, except there are now stresses and strains in both the 1 and 2 directions leading to a planar expansion of the spherical shell circumference. The results of this analysis have been given by Berlincourt [6] and are repeated below for the equivalent circuit of Fig. 5.4. With the area $A = 4\pi a^2$ and $a \gg t$ the equivalent circuit parameters are

$$C_0 = A\varepsilon_{33}^T\left(1 - k_p^2\right)/t,\ \ N = 4\pi a d_{31}/s_c^E,\ \ M = 4\pi a^2 t\rho,\ \ C^E = s_c^E/4\pi t \quad (5.22)$$

where the planar coupling coefficient is given by $k_p^2 = d_{31}^2/\varepsilon_{33}^T s_c^E$ and $s_c^E = \left(s_{11}^E + s_{12}^E\right)/2$. At in-air resonance, $\omega_0^2 = 1/MC^E$ and $\omega_0 = c/a$ where the wave speed $c = \left(1/\rho s_c^E\right)^{1/2}$.

The radiation impedance of Eq. (5.13) applies exactly with a_s replaced by the outer radius of the sphere, or, to an excellent approximation for a thin-walled sphere, with a_s replaced by the mean radius a, giving the radiation resistance, R_r, and radiation mass, M_r, as

$$R_r = \rho_0 c_0 A (ka)^2/\left[1 + (ka)^2\right] \text{ and } M_r = M_0/\left[1 + (ka)^2\right], \tag{5.23}$$

where $M_0 = 4\pi a^3 \rho_0$. At resonance, $\omega_r M + \omega_r M_0/\left[1 + (\omega_r a/c_0)^2\right] = 1/\omega_r C^E$ leading to the quadratic equation in $(\omega_r/\omega_0)^2$

$$(\omega_r/\omega_0)^4 + \left[(1 + a\rho_0/t\rho)(c_0/c)^2 - 1\right](\omega_r/\omega_0)^2 - (c_0/c)^2 = 0,$$

with solution

$$(\omega_r/\omega_0)^2 = \left[1 - a\rho_0 c_0{}^2/t\rho c^2 - (c_0/c)^2\right]/2 + \left\{\left[1 - a\rho_0 c_0{}^2/t\rho c^2 - (c_0/c)^2\right]^2/4 + (c_0/c)^2\right\}^{1/2}. \tag{5.24}$$

For the special case where $a/t = \left[(c/c_0)^2 - 1\right]\rho/\rho_0$, Eq. (5.24) simplifies and yields the result $\omega_r = \omega_0 (c_0/c)^{1/2}$. This case corresponds to an extremely thin shell since it gives $a/t = 101$ for PZT-4 and $\omega_r = 0.51\omega_0$; a practical value of a/t, such as 10, gives $\omega_r = 0.95\omega_0$.

The general solution for the mechanical Q_m under water loading is,

$$Q_m = \eta_{ma}\omega_r (M + M_r)/R_r = \eta_{ma}\left[k_r t\rho/\rho_0 + (1 + t\rho/\rho_0 a)/k_r a\right], \tag{5.25}$$

where $k_r = \omega_r/c_0 = 2\pi/\lambda_r$. As seen, Q_m is controlled by the first term in Eq. (5.25) when the sphere is large compared to the wavelength in the medium and by the second term when the sphere is small compared to the wavelength as a result of the heavy mass loading.

A sphere is a strong structure and may be submerged to twice the depth of a cylinder. For the sphere the circumferential stress caused by ambient pressure P_0 is $T = P_0 a/2t$ to be compared with a cylinder value of $P_0 a/t$. Portions of the sphere may be covered or shielded to control the radiation and form beam patterns other than the natural omnidirectional case [7]. A hemispherical section alone may be used to approximate a 180° beam pattern in the vicinity of resonance where, in this case the in-air resonance frequency is 24 % higher [5] than a full sphere. Because of the well-defined mechanical and electrical structure, a spherical transducer was chosen to show that nearly 10 dB greater output may be obtained if a large DC bias is applied to normal pre-polarized piezoelectric ceramic material [8].

5.2.4 The Magnetostrictive Ring

The magnetostrictive ring transducer may also be operated in the radial mode through excitation by a coil of n turns around the ring as illustrated in Fig. 5.6. The equivalent circuit representation was originally developed by Butterworth and Smith [9]. A mobility form of the mechanical portion of the equivalent circuit is shown in Fig. 3.20 where L_0 is the clamped inductance, χ is the eddy current factor, N is the electromechanical turns ratio, C^H is the open circuit effective radial compliance, M is the mass of the ring, $R_{\rm m}$ is the mechanical loss resistance, and $Z_{\rm r}$ is the radiation impedance.

The equivalent circuit components are

$$L_0 = \mu_{33}^S n^2 A_{\rm c}/2\pi a, \quad N = {\rm d}_{33} A_{\rm c} n / a s_{33}^H, \quad C^H = a s_{33}^H / 2\pi A_{\rm c}, \quad M = \rho 2\pi a A_{\rm c}, \quad (5.26)$$

where the transduction parameters are discussed in Chaps. 1, 2, and 3 and $A_{\rm c}$ is the cross-sectional area of the ring which is operated in the 33 mode. As discussed in Chap. 3, a dual circuit and a direct impedance equivalent representation of Fig. 5.7 may also be used to describe the transducer.

Terfenol-D rods may be used to fabricate an octagonal, as shown in Fig. 1.15 [10], or square [11], ring-type structure that supports pistons at the intersection of the rods forming an outer cylindrical structure. A magnetic DC bias field may be applied through the same AC coil or an additional coil. Magnetostrictive ring transducers have also been fabricated from long strips of nickel or permendur (iron-cobalt 50 % alloy) by winding the strip into a scroll. This design is rugged, comparatively simple and can be made very large. It has been used as a low frequency free-flooded projector in deep submergence applications.

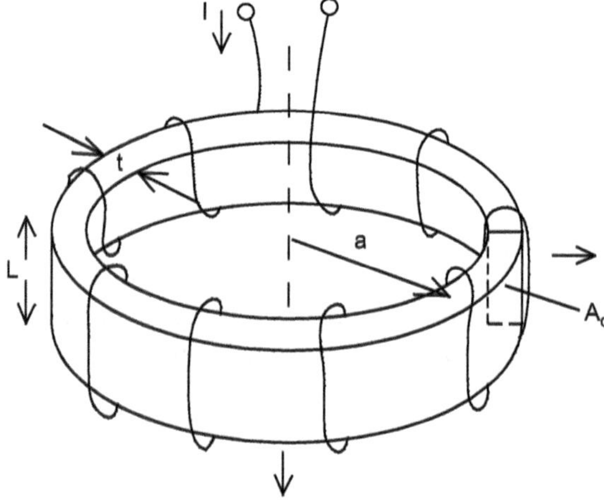

Fig. 5.6 Magnetostrictive ring transducer

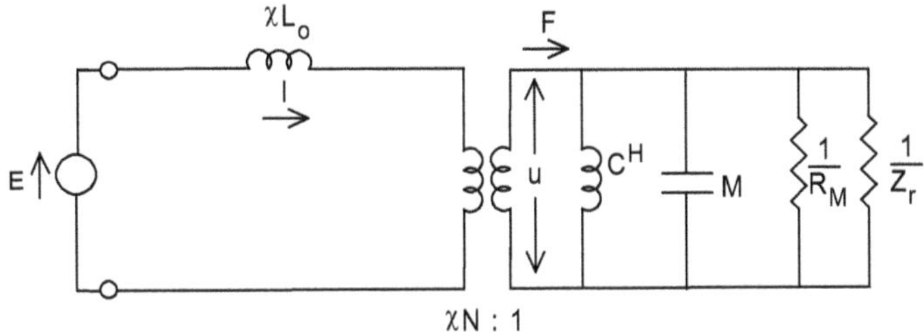

Fig. 5.7 Mobility circuit representation of the mechanical portion

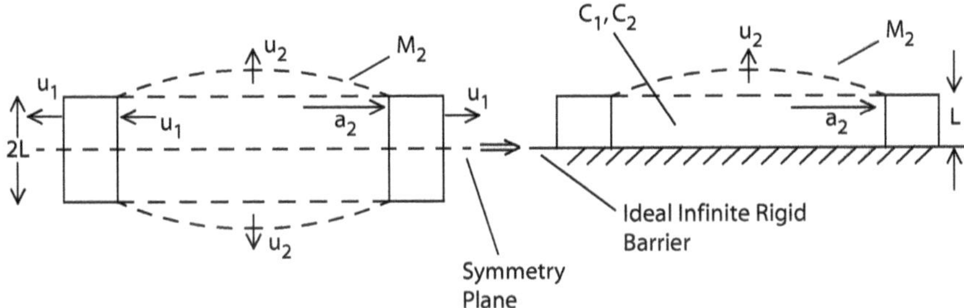

Fig. 5.8 Free-flooded ring with half plane symmetry

5.2.5 *Free-Flooded Rings*

Underwater projector ring transducers are normally operated with isolated end caps and an air-filled interior. Occasionally the interior is filled with a compliant fluid for deep submergence operation; however, the fluid acts as an additional radial stiffness that raises the resonance frequency and reduces the effective coupling coefficient. On the other hand, when the interior is free-flooded, the motion of the inner surface and compression of the interior fluid can be used to advantage if it is allowed to radiate and constructively combine with the radiation from the outer surface. Free-flooded rings have been commonly used for deep submergence applications at depths that other transducer types could not withstand. McMahon [12] and others have fabricated and tested arrays of piezoelectric free-flooded rings.

The cross section of a free-flooded ring of length $2L$, and a simpler half space model with one fluid port, that follows from symmetry, are shown in Fig. 5.8.

The sound field of a free-flooded ring is more difficult to calculate than that of most transducers, because not only the outside, but also the inside, the top and bottom sides and both ports, are radiating surfaces. Toroidal coordinates, in which a constant coordinate surface is a torus, offer a possible approach that gives useful results in the far field when the ring height and thickness are about equal [13]. Other approaches are more practical, such as use of a piston in a rigid baffle to model the radiation from the ports since L is typically much shorter than a wavelength.

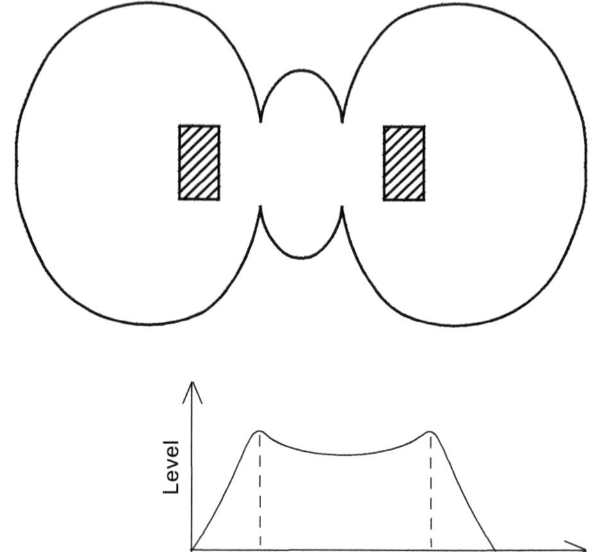

Fig. 5.9 Typical beam pattern for a free-flooded ring

Fig. 5.10 Typical free-flooded ring response

As the ring expands, the outer surface compresses the surrounding outer fluid while the inner surface lets the entrained inner fluid expand resulting in an approximate dipole ring with 180° out-of-phase pressures leading to partial cancellation along the axis of the ring at low frequency. But there is usually no null on the axis because of the different inner and outer areas and the small, but detectable, thickness and height mode radiation from the ring. At the Helmholtz resonance, ω_{h}, the interior fluid (half-length) axial compliance, $C_2 = L/\beta\pi a_2^2$, and the radiation mass, M_2, at the port give $\omega_{\mathrm{h}}^2 = 1/C_2M_2$ where $\beta = \rho_0 c_0^2$ is the bulk modulus of the fluid. Here there is an additional 90° phase shift in the pressure leading to partial addition on the axis resulting in a lobe as shown in Fig. 5.9. Above the Helmholtz resonance, there is further reduction in the phase shift and the radiation mass reactance, ωM_2, becomes significant reducing the output from the inner chamber. The band-pass type response for the free-flooded ring is shown in Fig. 5.10, where f_{r} and f_{h} are the ring and Helmholtz resonance frequencies, respectively.

The Helmholtz (or squirting) resonance is typically at or below ring resonance. If the two resonance frequencies, f_{h} and f_{r}, are closer together, there is less reduction at mid-band. The fall off in level below f_{h} is more rapid than usual because of out-of-phase cancellation.

The acoustic coupling of the inside and outside surfaces is similar to that of the bass reflex loudspeaker system [14] where the back of the speaker cone is ported near the front of the cone through a tube which, with the interior volume, acts as a Helmholtz resonating system. The mutual radiation impedance coupling between the Helmholtz port and speaker cone has been given by Lyon [15]. A reasonably complete circuit model for the half length ring with interior fluid coupling emerges if we consider only the radial mode of vibration and ignore radiation from the

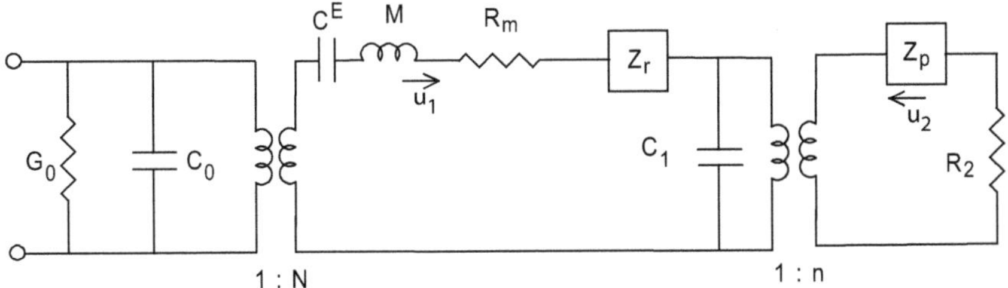

Fig. 5.11 Free-flooded ring equivalent circuit with ring and port radiation impedances Z_r and Z_p

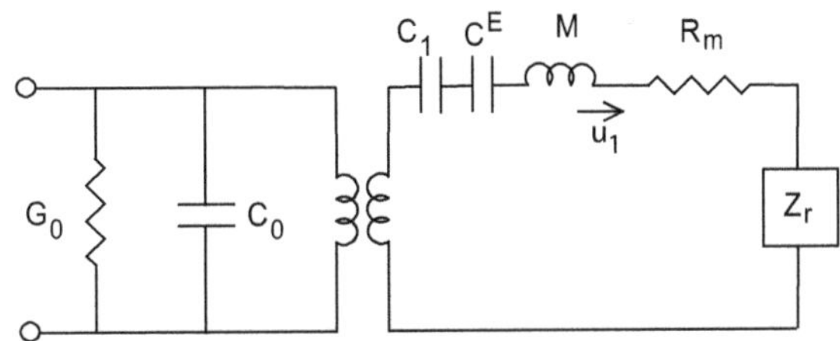

Fig. 5.12 Equivalent circuit for fluid-filled end-capped ring

length and wall thickness modes of the ring. This comparatively simple circuit model, without mutual radiation impedance coupling, is shown in Fig. 5.11.

The quantity Z_r is the self-radiation impedance of the outer surface of the ring of area A_1, Z_p is the self-radiation impedance of the port of area A_2, R_2 is the mechanical loss resistance at the ports due to the viscosity of the fluid oscillating within the ring, and $C_1 = 1/\beta 4\pi L$ is the radial compliance of the inner volume for rigid end caps ($u_2 = 0$).

The ring exterior and interior velocity is u_1 and the velocity at the ports is u_2. For equal volume velocity, well below the Helmholtz resonance, $A_2 u_2 = -A_1 u_1$ so that $u_2 = -u_1 A_1/A_2 = -u_1 2\pi a_2 L/\pi a_2^2 = -u_1 2L/a_2$. Thus, the acoustical transformer ratio, n, in Fig. 5.11 which connects the velocity at the port to the velocity at the ring is $n = -u_1/u_2 = a_2/2L$. Note that the axial compliance C_1/n^2 equals $L/\beta\pi a_2^2$, as it should, since the transformer connects the axial motion to the radial motion. Because of the half-plane symmetry, and with L small enough, the impedance of a piston in a rigid baffle may be used to approximate Z_p. (If one port of the ring is blocked with a heavy mass there is no infinite rigid baffle, the port radiation resistive load is reduced by about one-half and this single-sided free-flooded ring is called a "squirter"). If both ports are blocked, $u_2 = 0$, and the circuit of Fig. 5.11 simplifies further to the circuit of Fig. 5.12 for a fluid-filled end-capped ring.

Here $C_1 = 1/\beta_1 4\pi L$, where β_1 is the bulk modulus of the enclosed fluid. This fluid reduces the coupling coefficient of the ring, k, to the effective coupling coefficient, k_e, given by (see Eq. 4.29)

$$k_e^2 = k^2/\left[1 + (1 - k^2)C^E/C_1\right] = k^2/\left(1 + s^D\beta_1 2a/t\right), \tag{5.28}$$

where s^D is the open circuit elastic modulus of the cylinder for the mode of interest. As can be seen, the reduction in coupling can be significant if t is small and β_1 is high.

The bulk modulus of an interior enclosed fluid not only lowers the effective coupling coefficient but also raises the resonance frequency. These effects can be minimized by insertion of a more compliant fluid, such as silicone oil, or a more compliant structures, such as flatten air-filed cylindrical tubes [16], so-called *squashed tubed.* These tubes have been used inside of fluid-filled flextensional transducers (Sect. 5.5) allowing operation at greater depths.

Under ideal radial operation the ring is short enough to make the length extensional and bending modes well above the fundamental ring resonance. Rings may be stacked, with isolation between them, on top of each other to increase the length and produce a half-wavelength fluid resonator, directly excited by the radial motion of the piezoelectric rings. This extended length tube may be modeled by the addition of the impedance $Z = -1/j\omega C_2 - j\rho_0 c_0 \pi a_2^2 \cot\ kL$ in series with the port radiation impedance Z_p. Note that at low frequencies or for small L this added impedance $Z \approx -1/j\omega C_2 + \rho_0 c_0^2 \pi a_2^2/j\omega L = 0$. Since the tube is no longer short and the ends are not close to the mid-plane, the model for a piston in a rigid baffle is not accurate. The model for the radiation from the end of a tube [17] or an approximate model with an equivalent sphere at a distance L from a rigid baffle may be more appropriate.

The multi-port ring transducer [18, 19] is another type of free-flooded ring where concentric tubes are used to create a band-pass filter at frequencies well below the ring resonance. The transducer is illustrated in cross section and end views in Fig. 5.13 which shows a piezoelectric ring (typically 33 mode) exciting an inner tube resonance through its inner surface and an outer tube resonance through its outer surface.

The inner and outer stiff tubes form two half wavelength fluid resonators with the outer shorter tube set at about an octave above the inner tube. A smooth additive response, similar to the curve of Fig. 5.10, is obtained between the two resonators as a result of two 180° phase shifts which brings the far-field pressures from the inner and outer tubes back in phase. The first 180° shift results from excitation of the two

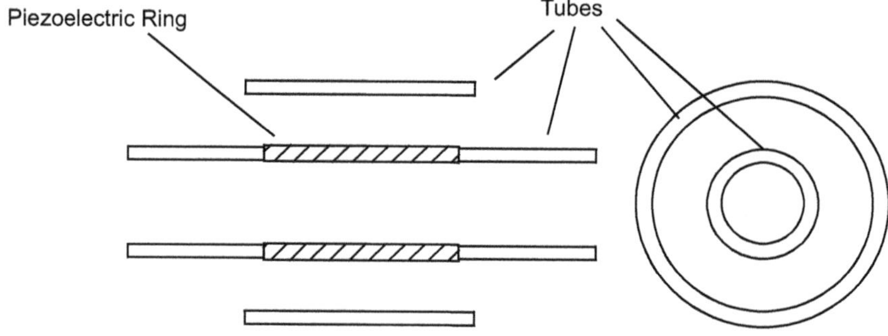

Fig. 5.13 Multiport transducer [19]

tubes by opposite sides of the piezoelectric driver; the second shift occurs in the frequency range between the two tube resonances where one tube is in the mass controlled region while the other is in the stiffness controlled region.

5.2.6 *Multimode Rings*

In the previous sections we considered ring transducers operated in the fundamental "breathing" mode with uniform radial extension excited by a uniform radial or circumferential electric field. Higher order extensional modes of a ring may be excited by a circumferential electric field with azimuthal dependence cos $(n\varphi)$. The resonance frequency, f_n, of these modes is given by [20]

$$f_n = f_0(1 + n^2)^{1/2}, \tag{5.29}$$

where, as before, the fundamental in-air ring resonance $f_0 = c/\pi D$ with c the bar sound speed in the ring and D the mean diameter. The radial surface displacements of the omni, dipole, and quadrupole modes are illustrated in Fig. 5.14. The dipole mode resonance is $f_1 = f_0(2)^{1/2}$, the quadrupole resonance is $f_2 = f_0(5)^{1/2}$, and the higher order resonances $f_n \approx nf_0$ for $n \gg 1$. A cos $(n\varphi)$ voltage distribution is ideal and excites only the nth mode; however, a square wave approximation to this is often sufficient to excite the desired modes, although other conforming modes will be excited at a lower level. The voltage may be applied to electrodes, separated by gaps, with appropriate phase reversals or simply wired directly as shown in Fig. 5.15 for the dipole and quadrupole modes of a 31 mode ring with two and four gaps, respectively.

Ehrlich [21] initially used the ring dipole mode for directional detection, and resolved the directional ambiguity by using the omni mode as a reference to determine which lobe (±) of the dipole pattern had received a signal (see Chap. 6). A transducer model for the ring dipole mode has been given by Gordon et al. [22] and extended by Butler et al. [23] to include the quadrupole mode as a projector. The quadrupole mode can be combined with the omni and dipole modes to obtain the normalized beam pattern function:

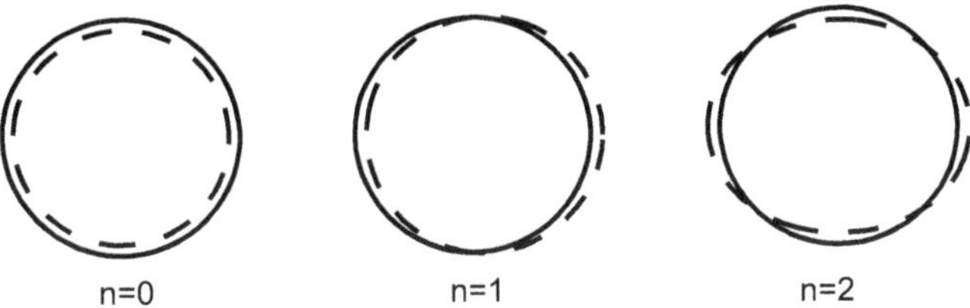

Fig. 5.14 Omni, dipole, and quadrupole ring modes of vibration

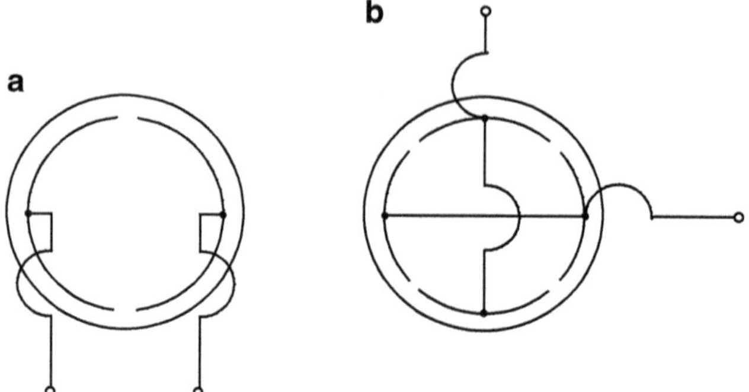

Fig. 5.15 Wiring scheme excitation of the dipole (**a**) and quadrupole (**b**) modes

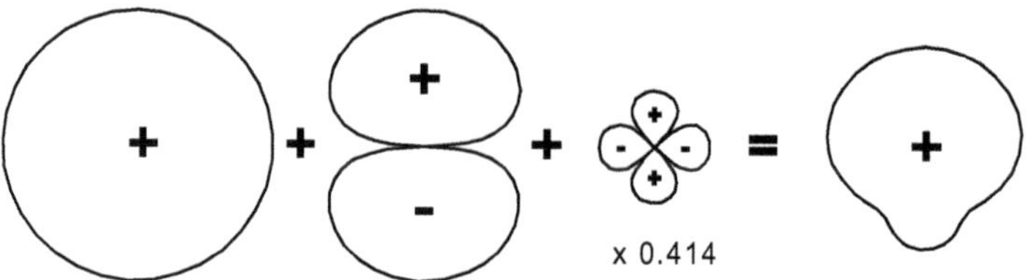

Fig. 5.16 Synthesis of 90° beam [23]

$$F(\varphi) = (1 + A\cos\varphi + B\cos 2\varphi)/(1 + A + B). \tag{5.30}$$

The specific case of $A = 1$ and $B = 0.414$ yields a beam width of 90° which can be steered in 45° increments by changing the voltage distribution on eight electrodes of the ring transducer. The synthesis of this beam is illustrated in Fig. 5.16.

The weighting factors A and B are usually maintained as real constants by the choice of the operating band [23] or by electrical equalization.

In general, the pressure modal contributions, p_n, for each ring mode may be determined from the desired far-field pressure function $p(\varphi)$ through the Fourier series

$$p_n = (\delta_n/\pi)\int_0^{\pi} p(\varphi)\cos(n\varphi)\mathrm{d}\varphi, \tag{5.31}$$

where $\delta_0 = 1$ and $\delta_n = 2$ for $n > 0$. A solution based on the Laird-Cohen [2] cylinder model with rigid axial extensions [see Eq. (10.34)] then leads to the modal velocities, u_n, of the ring as

$$u_n/u_0 = (p_n/p_0)e^{-in\pi/2}H_n^{'}(ka)/H_0^{'}(ka), \tag{5.32}$$

where $H'_n(ka)$ is the derivative of the cylindrical Hankel function of the second kind of order n. The velocities may be related to the applied electric field E_n through an equation similar to Eq. (5.7a):

$$u_n = j\omega\omega_0^2 a d_{31} E_n / \left[\left(1 + n^2\right)\omega_0^2 - \omega^2 + \left(1 - n^2\omega_0^2/\omega^2\right) j\omega Z_n / \rho t \right], \qquad (5.33)$$

where ω_0 is the fundamental angular resonance frequency, a is the mean radius, ρ is the density, and t is the wall thickness of the cylinder and d_{31} is the piezoelectric coefficient [22]. The quantity Z_n is the specific acoustic modal impedance, p_n/u_n, where p_n is the modal pressure, which for small rings may be approximated by the modal impedance of a sphere (See Sect. 10.4.1). Otherwise, the modal impedance may be obtained through the Fourier series approach [3] (see Sect. 11.1.2). Beams formed by the addition of modes, as in Eq. (5.30), will have a beam pattern independent of frequency provided that the coefficients are maintained as real constants over the frequency band of interest.

The same radiation modes can be obtained from a cylindrical array of discrete transducers with appropriate distributions of voltage drive (see Sect. 7.1.4).

5.3 Piston Transducers

While spherical and ring sources are omnidirectional in at least one plane, the piston type transducer generally projects sound into one direction with a directionality that depends on its size compared to the wavelength. Large arrays of these transducers are used to project highly directional, high intensity beams of sound into a particular area, as will be discussed in Chap. 7. Because they are well suited for large close-packed arrays, piston transducers are more commonly used than any other type in underwater acoustics. Moving coil and variable reluctance piston transducers are typically used in the frequency range below 600 Hz where piezoelectric or magnetostrictive transducers would be too long for most applications. However, mechanically leveraged piezoelectric X-spring piston transducers have been operated as low as 300 Hz and as high as 50 kHz. "Tonpilz" lumped mode transducers are typically operated in the range from 1 to 50 kHz, sandwich type transmission line transducers cover the range from 10 to 500 kHz, and piezoelectric plates, diced plates or piezoelectric composites, are used from about 50 kHz to beyond 2 MHz. The most common sonar projector is the Tonpilz and we begin the discussion of piston transducers with this design.

5.3.1 The Tonpilz Projector

The Tonpilz transducer (see Figs. 1.7 and 1.8) takes its name from the German "sound mushroom," or "singing mushroom" [24] presumably because of a large piston head mass driven by a slender drive section which gives it the

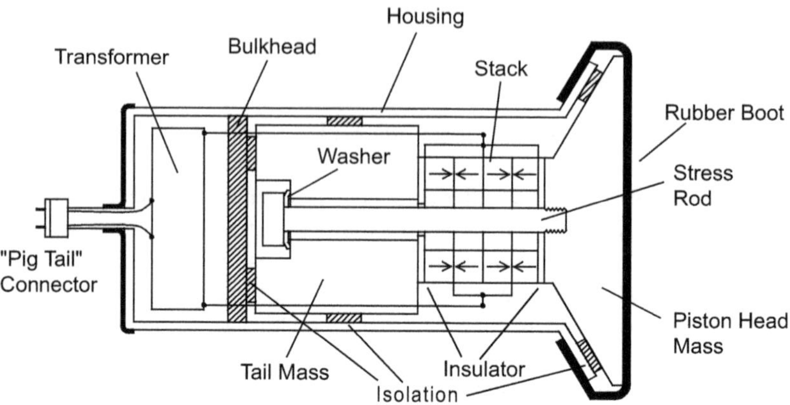

Fig. 5.17 Cross section of typical Tonpilz transducer

cross-sectional appearance of a mushroom. This concept along with a tail mass allows a compact means for obtaining high output at midrange frequencies without the need for an excessively long piezoelectric ceramic or magnetostrictive drive stack. A cross section sketch of a typical Tonpilz transducer is illustrated in Fig. 5.17 showing a 33 mode driven ring-stack [25] of four parallel wired PZT rings driving a relatively light but stiff piston mass with a comparatively heavy tail mass on the other end.

Other parts shown include mechanical isolation, a housing, transformer with tuning network, rubber enclosure around the head, and electrical underwater connector. The stack is held under compression by a stress rod [26] and, sometimes, a compliant conical disc or Bellville washer [27] that decouples the stress rod and maintains compressive stress under thermal expansion. In some designs the circumference of the PZT stack is also fiberglass wrapped for added strength under shock. Typically, the head is aluminum, the tail is steel, the stress rod is high strength steel, and the piezoelectric rings are Navy Type I or III piezoelectric ceramic. The housing is usually steel, and the water tight boot is neoprene or butyl rubber and occasionally polyurethane for short term immersion. The rubber boot is vulcanized to the head to ensure good bonding with no air pockets, which would unload the transducer and reduce the radiation resistance.

The desire is to attain the greatest possible motion of the head piston and radiate as much power as possible near and above mechanical resonance. Although operation below resonance is possible, the Transmitting Voltage Response, TVR, falls off at typically 12 dB per octave below resonance. Moreover, operation below resonance can introduce significantly magnified harmonic distortion when the harmonics occur at or above resonance (see Chap. 12). A number of lumped and distributed modeling techniques and specific formulas that are directly applicable to the Tonpilz transducer are developed in Chaps. 3 and 4. In this section these models will be the basis for a more detailed discussion of the design and analysis of Tonpilz transducers.

A simplified lumped model may be used as an aid in understanding and implementing an initial design. Normally, this model would be followed by more accurate distributed and finite element models. The lumped mode representation is a reasonably good model for a Tonpilz transducer as the size and shape of the respective parts favor such a reduction to lumped masses and spring. As discussed in Chap. 3, high power piezoelectric ceramic or magnetostrictive drive sections of thick-walled cylinders or bars have significant mass because of their densities of 7600 (for piezoelectric ceramic) and 9250 kg/m^3 (for Terfenol-D). Consequently, the mass of the drive section, M_s, should be included and distributed with the head mass, M_h, and tail mass, M_t, in even the simplest lumped models.

A lumped mechanical model of the Tonpilz of Fig. 5.17, with minimum essential parts, is illustrated in Fig. 5.18 and may be used as a basis for the equivalent circuit of Fig. 5.19.

As shown and developed in Chap. 3, the circuit parameters are

$$G_0 = \omega C_f \tan\delta,\ \ C_f = n\varepsilon_{33}^T A_0/t,\ \ C_0 = C_f\left(1 - k_{33}^2\right),$$
$$N = d_{33}A_0/ts_{33}^E,\ \ C^E = nt/A_0 s_{33}^E, \tag{5.34}$$

where $R_0 = 1/G_0$, n is the number of rings in the drive stack, t is the thickness of each ring, $k_{33}^2 = d_{33}^2/s_{33}^E\varepsilon_{33}^T$, C^E is the short circuit compliance of the drive stack, C_{tr} is the compliance of the stress rod assembly, R_m is the mechanical loss resistance, R_r is the radiation resistance, M_r is the radiation mass, u_h is the velocity of the head, u_t is the velocity of the tail, and the relative velocity between the two is $u_r = u_h - u_t$.

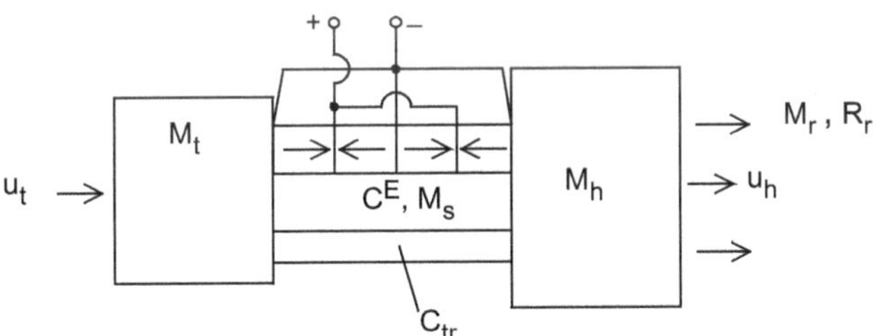

Fig. 5.18 Basic mechanical lumped model

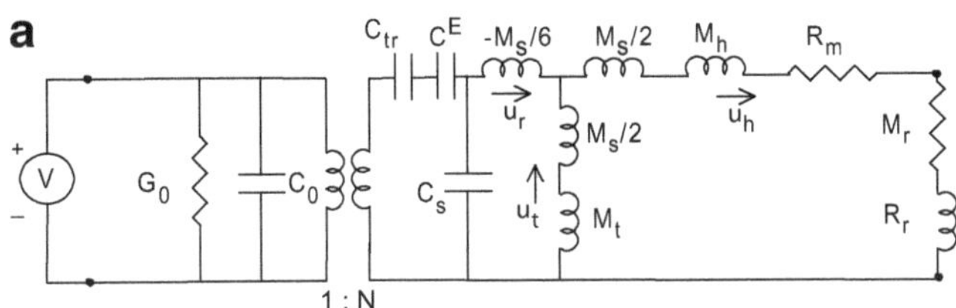

Fig. 5.19 (**a**) Lumped equivalent circuit for a Tonpilz transducer

As the stack expands the head and tail move in opposite directions. The head, tail, and relative velocity magnitudes are related by

$$|u_h/u_t| = M_t/M_h, \quad |u_h/u_r| = 1/(1+M_h/M_t) \quad \text{and}$$
$$|u_t/u_r| = 1/(1+M_t/M_h). \tag{5.35}$$

A large tail to head mass ratio is desirable as it yields a large head velocity, radiating the most power. If $M_t = M_h$ then $|u_t| = |u_h|$ and $u_h = u_r/2$ which is 6 dB less in source level than if $u_t = 0$, which is approached if $M_t \gg M_h$. Typical Tonpilz designs use tail to head mass ratios from 2 to 4, as larger values lead to too much weight. For $M_t/M_h = 4$, $u_h = 0.8u_r$ which is a reduction of only 2 dB from the ideal infinite tail mass case. A design with a low mass piston head allows a large M_t/M_h ratio without as much weight burden.

If a Bellville spring of compliance C_B is used in mechanical series with the stress rod of compliance, C_{sr}, the total stress rod assembly compliance becomes $C_B + C_{sr} = C_{tr}$, which is approximately equal to C_B, since C_B is usually much greater than C_{sr}. In this case the stress rod is decoupled from the tail and acts as a mass rather than a spring that can be added to the mass of the head. Also, since $C_B \gg C^E$, the total electromechanical compliance $C_m \approx C^E$ and there is no loss in coupling coefficient. However, as discussed in Chap. 4, there still is a reduction in coupling due to the electrical insulators (often GRP, Macor, alumina, or un-polarized piezoelectric ceramic) at the ends of the stack (see Fig. 5.17) as well as the cement joints between the rings. The cement joints have elastic properties similar to plastic, such as Lucite, with a thickness of approximately 0.003 in. and may be modeled as the shunt compliance, C_s, connected between the junction of C^E and $-M_s/6$ and ground in Fig. 5.19a. The compliance of the insulators has the same effect as the cement joint compliance and the two may be lumped together. Occasionally, the electrical insulator at the head mass is also used as a mechanical tuning adjustment on the resonance frequency of the transducer. The electrodes are cemented in place between the rings and are typically thin strips of expanded metal, or approximately 0.003 in. thick beryllium copper with punched holes for the cement and dimples or ridges to ensure contact to the electrodes on the piezoelectric ceramic.

If the drive stack is composed of a thin-walled cylinder or thin-walled rings, the stack mass, M_s, may be ignored and the approximate resonance frequency and mechanical Q are

$$\omega_r^2 = (1+M_1/M_t)/M_1C^E \quad \text{and} \quad Q_m = \eta_{ma}\omega_r M_1(1+M_1/M_t)/R_r, \tag{5.36}$$

where $M_1 \approx M_h + M_r$ and the mechanical-acoustical efficiency $\eta_{ma} = R_r/(R_r + R_m)$. These equations show that a lower resonance frequency may be obtained by increasing M_1 through an increase in M_h, but resulting in a higher Q_m and lower output above resonance. A comparatively heavy tail mass reduces the resonance frequency and reduces the Q_m, but leads to a heavier transducer.

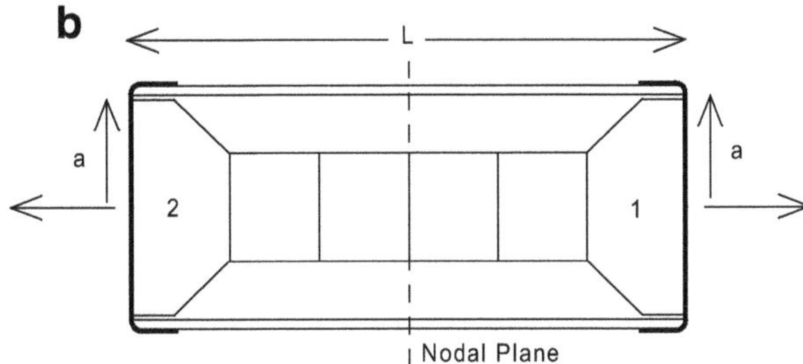

Fig. 5.19 (**b**) Dual piston transducer

The special case of a dual piston transducer, illustrated in Fig. 5.19b, achieves the ideal condition of an infinite tail mass without the added weight of a tail mass, benefiting from the symmetrical motion and resulting in a vibrational node at mid-plane.

In addition to this, the symmetry gives an equivalent rigid baffle condition at the mid-plane (see Chap. 10) yielding a greater radiation loading on each piston, if the distance to the mid-plane is less than a quarter-wavelength in the surrounding medium. The greater radiation loading can also be understood as a result of the mutual radiation impedance Z_{12} between the two pistons. The radiation impedance on piston 1 is $Z_1 = Z_{11} + Z_{12}$ and for equal sized small sources ($ka \ll 1$) the radiation resistance is (see Chap. 7)

$$R_1 \approx R_{11} + R_{11} \sin(kL)/kL = R_{11}[1 + \sin(kL)/kL], \tag{5.37}$$

where R_{11} is the self-radiation resistance of one piston and L is the center-to-center separation between the small pistons. Thus for $kL \ll 1$, the total radiation resistance on piston 1 (and similarly on piston 2) is $R_1 \approx 2R_{11}$, which can significantly reduce Q_m compared to the usual single piston Tonpilz. The dual piston transducer has an additional reduction in Q_m as M_t approaches ∞ in Eq. (5.36), since there is no energy stored in a moving tail mass, as there is in a typical Tonpilz transducer.

The discussion of the transducer in Fig. 5.19b shows that the radiation resistance for a piston with a large rigid baffle is approximately twice that for one without a rigid baffle. This point is sometimes overlooked when the response of a single piston transducer is calculated assuming a rigid baffle, while in the intended application there may be nothing that approximates such a baffle. The un-baffled condition leads to both a far-field pressure and radiation resistance of about half the rigid baffle values when ka is small.

The two degree of freedom circuit of Fig. 5.19a can be further simplified to a one degree of freedom system if $\omega(M_h + M_t + M_s + M_r) \gg R_r + R_m$, which is often the case in practice. The resulting one degree of freedom circuit, with $M_1 = M_r + M_h + M_s/2$, and $M_2 = M_t + M_s/2$, is shown in Fig. 5.20 where

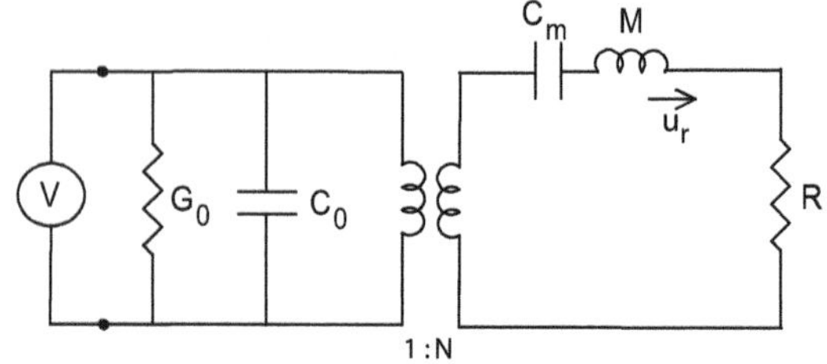

Fig. 5.20 Reduced equivalent circuit of Tonpilz transducer

$$M = M_1/(1 + M_1/M_2) - M_s/6, \;\; C_m = C^E/(1 + C^E/C_{tr}),$$
$$R = (R_r + R_m)/(1 + M_1/M_2)^2, \tag{5.38}$$

which is similar in appearance to the circuit of Fig. 5.4 for the ring transducer. When transformed by N^2 to a fully electrical circuit, it is then similar to the Van Dyke circuit of Fig. 3.16 and the measurement circuit of Fig. 9.1. The circuit of Fig. 5.20 may also be used to represent the dual piston transducer of Fig. 5.19b since here, because of symmetry, the tail mass M_2 for both sides of the transducer is located at the nodal plane and acts like a rigid wall with, in effect, M_2 approaching infinity.

A transducer with a low Q_m is closely matched to the medium as discussed in Chap. 4. With the stiffness $K_m = 1/C_m$ in the simplified circuit of Fig. 5.20 we have

$$\omega_r = (K_m/M)^{1/2} \quad \text{and} \quad Q_m = (K_m M)^{1/2}/R. \tag{5.39}$$

Both a lower resonance frequency and lower Q_m are obtained for a low value of the stiffness K_m. Since $K_m = Y_m A_0/L$ where Y_m is the effective Young's modulus, a low K_m implies a long stack length, L, or small cross-sectional area A_0 or both. Since the power output is proportional to the volume of drive material, A_0L, a reduction in A_0 must be accompanied by a proportional increase in L for a constant A_0L. Low Q_m or nearly matched impedance is achieved through an adjustment of the ratio of the piston head area, A, to the area, A_0, of the drive stack.

For matched impedance under full array loading conditions at resonance the quantity $A_0(\rho c)_t$ must equal $A(\rho c)_w$, where t and w indicate transducer and water. This leads to a head to drive stack ratio $A/A_0 = (\rho c)_t/(\rho c)_w \approx 22.4 \times 10^6/1.5 \times 10^6 \approx 15$ for PZT material. This ratio can be difficult to attain in practice as a large diameter piston head can lead to a flexural resonance near the longitudinal resonance and a thin-walled piezoelectric drive stack can limit the output power. (Even greater area ratios may be needed for single-element matched-loading when the piston diameter is one half wavelength or less). On the other hand, transducers are commonly operated with a $Q_m \approx 3$ and this requires an area ratio of only a factor of 5, which is usually attainable in practice. If power output at a given resonance frequency, rather than Q_m, is a more important issue, the area of the stack is increased, increasing the force, and the length of the stack is proportionally increased, increasing the displacement, while maintaining the same stiffness and nearly the same resonance frequency.

The design of the piston head is almost as important as the design of the drive stack as it is the main interface between the drive stack and the medium. It needs to be large enough to provide a good match to the medium and yet must also provide uniform longitudinal motion. The first flexural resonance of the head should be significantly above the operating band. The first free flexural resonance of a disc of diameter D and thickness t is

$$f_{\mathrm{r}} = 1.65ct/D^2\left(1-\sigma^2\right)^{1/2} = 2.09Mc/\rho D^4\left(1-\sigma^2\right)^{1/2}, \tag{5.40a}$$

where c is the bar sound speed and σ is Poisson's ratio and the mass of the disc is $M = \rho\pi a^2 t$. Equation (5.40a) shows that materials with high sound speed such as alumina with $c = 8500$ m/s and beryllium with $c = 9760$ m/s can yield a high flexural resonance. The density of the head material is also important since low density materials allow a thicker head, raising f_{r} for the same head mass. Thus, the beryllium alloy AlBmet (see Sect. 13.2) is better than alumina with its density of 2100 kg/m^3 compared to 3760 kg/m^3. The sound speeds for steel, aluminum, and magnesium are all roughly the same at 5130 m/s, 5150 m/s, and 5030 m/s, respectively; however, their densities are 7860 kg/m^3, 2700 kg/m^3, and 1770 kg/m^3, respectively, making magnesium the best choice and steel the worst of these three.

A circular piston of constant thickness is not optimum as increasing the thickness to raise the flexural resonance above the operating band causes an increase in mass and lowers the output above resonance. A tapering from the center to the edge removes mass from the outer part which raises the flexural resonance and decreases the total mass. The flexural resonances for rectangular and square plates are lower than for discs and, therefore, pose a greater problem. The first free flexure resonance of a square plate where the four corners move in opposition to the central part may be written as

$$f_{\mathrm{r}} = 1.12ct/L^2\left(1-\sigma^2\right)^{1/2}, \tag{5.40b}$$

where L is the side length of the square plate. A head flexural resonance can cause increased or reduced output at the upper part of the operating band depending on its location relative to the fundamental resonance and the transducer bandwidth. Butler, Cipolla, and Brown [28] showed that although piston flexure can cause adverse effects above the Tonpilz resonance, a small improvement in performance in the lower frequency range can be obtained due to a lowered Tonpilz resonance as shown in Fig. 5.21.

There is a null in the response in the vicinity of flexural resonance since roughly half the area of the piston is out of phase with the other half.

The design of the tail mass is the simplest task. The tail is usually less than a quarter-wavelength long with a diameter slightly greater than the diameter of the

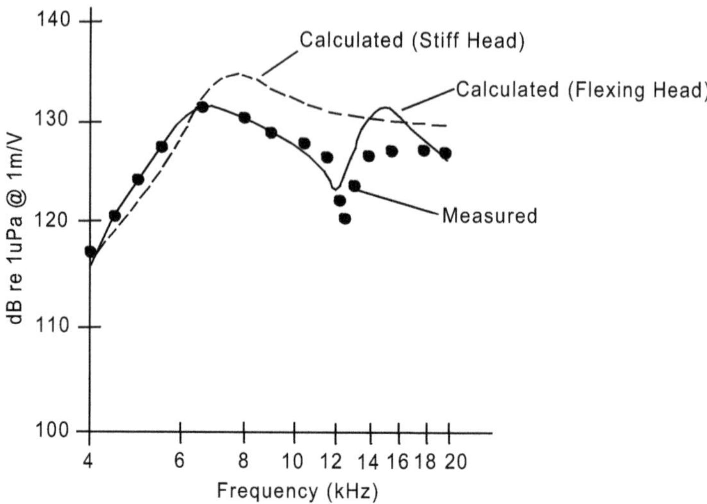

Fig. 5.21 Comparison of the stiff (*dashed line*) and flexing (*solid line*) head calculated responses with the measured (*bulleted line*) constant voltage transmitting response of flexing head Tonpilz transducer [28]

driving stack. It is typically in the form of a solid steel cylinder, occasionally of tungsten in smaller higher frequency designs, with a hole through the center to accommodate the stress rod. The stress rod exerts a compressive stress T_0 on the ceramic to prevent the drive stack from operating in tension under high drive conditions. The corresponding tensile stress in the stress rod of cross-sectional area A_t is $T_t = T_0A_0/A_t$. Because of the typically large area ratio, the stress rod must be made of high strength steel, such as tool steel or Titanium. The stiffness of the stress rod is usually made about 10 % or less of the drive stack stiffness to prevent significant reduction in the effective coupling coefficient [see Eq. (4.29)].

The transducer is mounted within the housing with isolation material such as corprene for stresses up to 200–300 psi and with paper (see Sect. 13.2) for stresses up to 1000 psi. The total stiffness of the isolation material, K_i, should be low enough to place its resonance with the total transducer mass, M_{tot}, well below the lowest frequency, f_1, of the operating band; that is $(K_i/M_{tot})^{1/2}/2\pi < f_1$. The transducer may be mounted from the head, the tail, or a central nodal plane. However, nodal mounting (see Fig. 1.5) replaces the most stressed piece of piezoelectric ceramic with inert material, lowering the effective coupling coefficient. Tail and nodal mounting can yield a higher efficiency because of the lower velocity at those points with less motion transmitted to the housing. However, head mounting with rubber or syntactic foam can be most effective under higher pressure as no ambient pressure is transmitted to the drive stack. The transducer housing is an essential part of the Tonpilz transducer as it not only protects the electrical components from the water, but also prevents an acoustical short circuit between the front of the piston and the out-of-phase rear of the piston. It also isolates radiation from the tail mass and lateral vibrations from the drive stack. Sonar transducers are often designed to withstand high explosive shock which can excite the fundamental resonance causing the tail and head mass to move oppositely creating a great tensile stress on the drive stack. The stress rod and fiberglass wrapping around the stack help to prevent damage to the stack under explosive shock conditions. See Woollett [29], Stansfield [30], and Wilson [24] for additional information on the Tonpilz transducer design.

The Tonpilz design, illustrated in Fig. 5.17, shows a stack of four 33 axial mode Navy Type I or III rings wired in parallel. Typically, the axial thickness of the rings is 0.25–0.50 in. and the number of rings is from two to twelve depending on the frequency of operation and power required. Greater output can be obtained using biased prepolarized piezoelectric ceramics [8], biased PMN, biased single crystal PMN-PT, prepolarized PIN-PMN-PT, textured ceramic (see Sect. 13.5), or magnetically biased Terfenol-D magnetostrictive material (see Sect. 13.7). As discussed in Chaps. 2 and 3, Terfenol-D can attain large displacements before magnetic saturation. However, it has additional losses due to ohmic loss in the surrounding coil as well as eddy current loss in the material, and it also has effective coupling coefficient reduction due to the magnetic circuit [31] (see Sect. 13.9). However, with these problems under control (e.g., laminations to reduce eddy currents and a well-designed magnetic circuit) a Terfenol-D transducer can produce greater output than an equivalent size Navy Type I or III piezoelectric ceramic transducer and do so with lower input electrical impedance [32, 33]. Two Terfenol-D magnetostrictive Tonpilz designs are shown in Fig. 5.22, illustrating two different permanent magnet (typically rare earth) biasing schemes.

The dual drive case needs the same number of turns in each leg as the single drive case for the same performance.

The permanent magnets may be eliminated if a DC current is added to the coil, but at a cost of lower electrical efficiency and heating due to the additional DC ohmic loss in the coil. The magnets may also be eliminated by replacing one of the Terfenol-D legs, with positive magnetostriction, by a leg with negative magnetostriction, such as $SmDyFe_2$, and fitting each leg with separate coils with oppositely directed diodes [34]. With this arrangement one half of a cycle would be directed to one leg, while the second half of the cycle would be directed to the other leg. This results in a half cycle of expansion followed by a half cycle of contraction with the transducer oscillating at the same frequency as the drive frequency without a bias field. This method also provides a means for doubling the ultimate strain of the transducer, since there is no bias field and the strain goes from the maximum negative value to the maximum positive value. On the other hand, the effective coupling coefficient of the transducer is reduced by approximately 30 % since one magnetostrictive leg is not activated while the other is driven [see Eq. (4.29), Sect. 4.4.2].

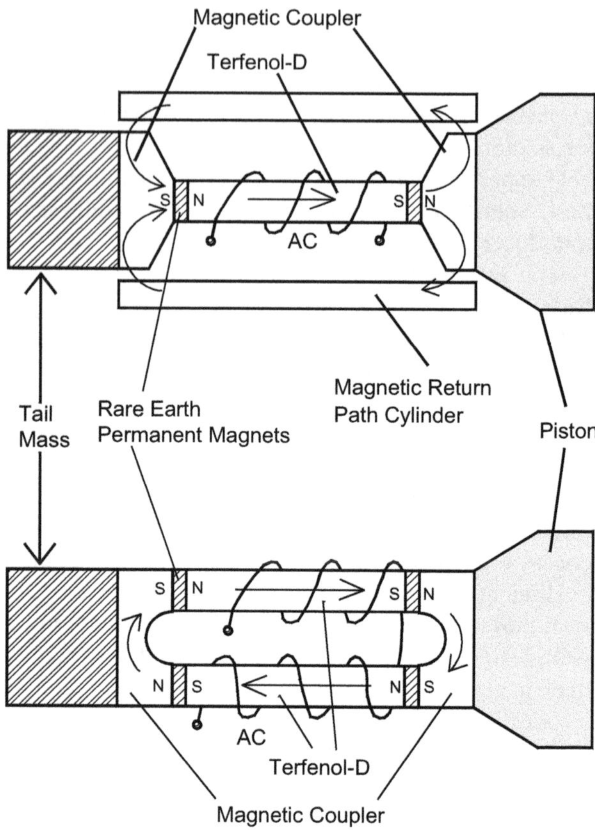

Fig. 5.22 Magnet bias designs for Tonpilz transducer. *Arrows* indicate magnetic flux directions

5.3.2 The Hybrid Transducer

The hybrid transducer [35] is a piston transducer with a drive stack composed of a mechanically coupled magnetostrictive and piezoelectric ceramic drive assembly. It has unique properties such as self-tuning, enhanced motion at one end with cancelled motion at the opposite end, and wideband doubly resonant response. This unique performance is a result of an inherent 90° phase shift between piezoelectric ceramic and magnetostrictive materials under constant voltage drive. This phase shift can be understood by considering the constitutive equations (see Chaps. 2 and 3) at low frequencies where the radiation impedance load is usually negligible. Here the 33 mode piezoelectric strain is $S_3 = d_{33}E_3$, while the 33 mode magnetostrictive strain is $S_3{}' = d_{33}{}'H_3$, where the prime is used to distinguish the magnetostrictive and piezoelectric strain and d constants. For bars of the materials each of length L and cross-sectional area, A_0, and for the same sinusoidal voltage, V, applied to each, the piezoelectric voltage is given by $V = \mathrm{EL}$ and the magnetostrictive voltage is given by $V = j\omega nA_0\mu^T H$ leading to the strains

$$S_3 = d_{33}V/L \quad \text{and} \quad S_3' = \left(d_{33}'V/L\right)(n/j\omega L_f), \tag{5.41}$$

where n is the number of coil turns and the free inductance $L_f = \mu^T n^2 A_0/L$. The j in the magnetostrictive part of Eq. (5.41) shows that this strain, S_3', and voltage are 90° out of phase and the two strains, S_3 and S_3', differ in phase by 90° for the same voltage, V. Terfenol-D and piezoelectric ceramic are a unique combination in that they have nearly the same coupling coefficient, mechanical impedance, and short circuit wave speed. (Although the magnetostrictive drive section of the hybrid transducer automatically provides the necessary 90° phase shift, it is possible to replace this section with a piezoelectric section if an electrically imposed 90° phase shift is introduced [36]).

The 90° phase difference with comparable impedance and coupling provides a favorable condition for addition of waves at one end and cancellation at the opposite end. This condition can be implemented in a transmission line or Tonpilz transducer structure. The transmission line case [37] allows a physical understanding of the process and is illustrated in Fig. 5.23 showing two quarter wavelength sections and the sequence of events from time $t = 0$ to $t = T/2$ where T is the period of the vibration.

Initially the piezoelectric section expands causing a stress wave to travel to the front through a distance of one quarter wavelength arriving at the time that the magnetostrictive section expands (as a result of the 90° delay in the magnetostrictive section), adding to the magnetostrictive motion and launching a wave into the medium. In the next quarter cycle the magnetostrictive expansion moves to the left and arrives just when the piezoelectric section is contracting resulting in a cancellation at the rear of the transducer. The result is a transducer with large motion at one end and no motion at the other end. If the phase of one of the drive sections is reversed, motion at the ends will be interchanged.

This transducer does not strictly obey reciprocity in the sense that if we transmit as a projector from the front with a null in the rear the receiving hydrophone response will be from the rear with a null in the front. Bobber [38] obtained similar nonreciprocal transmit receive results with a checkerboard array of independent magnetostrictive and piezoelectric transducer elements. The reciprocity formulas of Chap. 9 hold provided that one set of leads are reversed during the receive mode. An array of hybrid symmetrical Tonpilz transducers, of the design illustrated in Fig. 5.24, has been developed with a 15 dB front to back ratio in the vicinity of resonance at 4.25 kHz [39].

The electrical input to the two sections was wired in parallel applying the same voltage to each section and providing tuning between the inductance of the magnetostrictive section and the capacitance of the piezoelectric section. The transducer was modeled using a pair of "*T*" networks coupled through the center mass M.

An evaluation of the effective coupling coefficient of the hybrid transducer [39] poses special consideration since this transducer is composed of both magnetostrictive magnetic field and piezoelectric electric field sections. However, if we extend Mason's energy definition and consider the effective coupling at electrically tuned

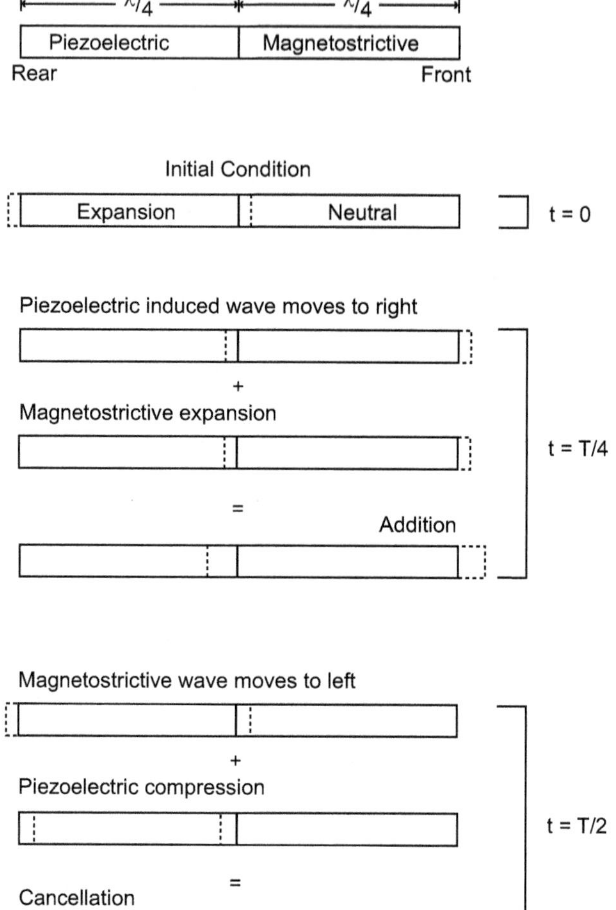

Fig. 5.23 Sequence of events leading to an enhancement at one end and cancellation at the opposite end of the hybrid transducer. Each section is one-quarter-wavelength long. T is the period of one cycle, and λ is the wavelength measured in the respective materials [37]

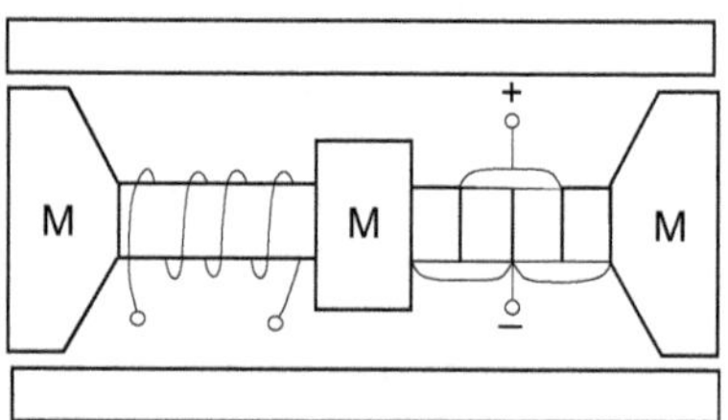

Fig. 5.24 Directional hybrid transducer

resonance, an interesting aspect of the hybrid design becomes apparent. Consider first the case where the magnetostrictive section is replaced by a piezoelectric section so that the transducer is composed of two identical piezoelectric sections. The coupling coefficient of each section is written as $k^2 = E_m/(E_m + E_e)$, where E_m is the mechanical energy converted and E_e is the electrical energy stored. The total effective coupling coefficient for both identical piezoelectric sections connected is then given by $k_e^2 = 2E_m/(2E_m + 2E_e) = k^2$ with no change in coupling, as might be expected. Consider now the hybrid case where we replace one piezoelectric section with an equivalent magnetostrictive section with the same coupling coefficient. Here the total mechanical energy is $2E_m$, as before; however, as a result of the shared exchange of magnetic and electrical energy of the two sections at electrical resonance, the total electrical energy is only E_e with a resulting effective coupling coefficient given by

$$k_e^2 = 2E_m/(2E_m + E_e) = 2k^2/\left(1 + k^2\right).$$

Thus, for $k = 0.5$ the effective coupling is increased to 0.63 while for $k = 0.7$ the effective coupling is increased to 0.81. To evaluate this effect, the electrical tuning should be set at the same frequency as the mechanical resonance and the effective coupling determined through the dynamic representation by $k_e^2 = 1 - (f_r/f_a)^2$. The increase in effective coupling is a result of the inherent electrical tuning that arises from the electrical connection of the capacitive-based piezoelectric section and the inductive-based magnetostrictive section with a resulting reduction of stored electrical energy.

The hybrid transducer design can take on another form and provide a wideband response with a smooth transition between two distinct resonance frequencies associated with the magnetostrictive and piezoelectric sections. A sketch of this broadband transducer is shown in Fig. 5.25 with piezoelectric and magnetostrictive stiffness, K_1^E and K_2^E, and masses, m_1 and m_2, along with head, central, and tail masses M_1, M_2, and M_3, respectively.

At low frequencies the stiffness K_1^E leads to a high impedance and couples the masses, M_1 and M_2, together so that the front section acts as one mass $M_f = M_1 + M_2 + m_1$. This mass, M_f, along with the tail mass M_3, resonates with the magnetostrictive short circuit stiffness, K_2^E, leading to the resonance frequency $\omega_1 = \left[K_2^E(1 + M_f/M_3)/M_f\right]^{1/2}$. At high frequencies the tail mass M_3 is decoupled

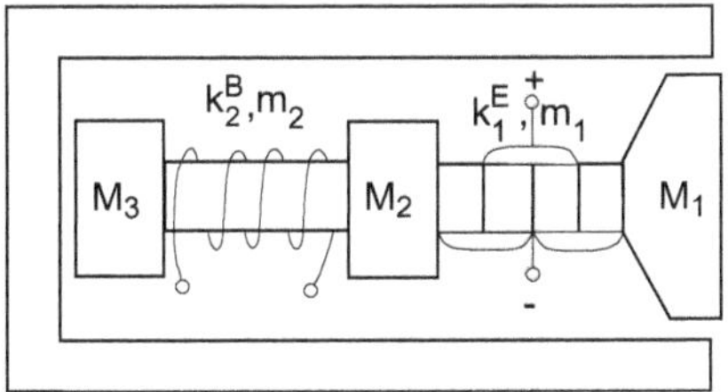

Fig. 5.25 Broadband hybrid transducer

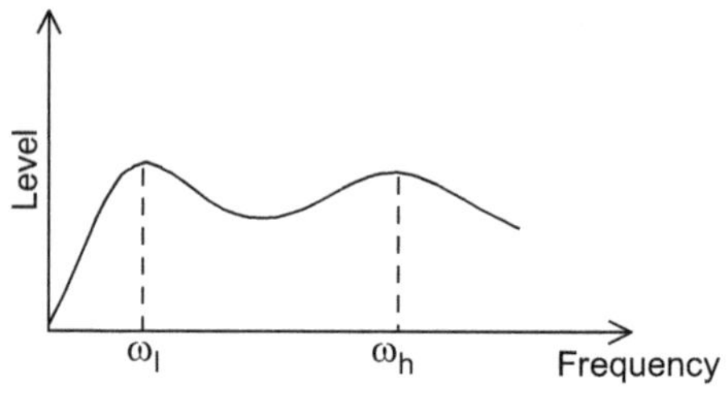

Fig. 5.26 Typical hybrid wideband response

through K_2^E leading to the resonance frequency $\omega_\mathrm{h} = \left[K_1^E(1 + M_1/M_2)/M_1\right]^{1/2}$. Because of the 90° phase shift, there is little reduction in the response between the resonance frequencies and, moreover, the electrical inductance and capacitance cancel providing electrical self-tuning. In practice the Terfenol-D section is positioned as shown in Fig. 5.25 and operates most effectively in the lower portion of the band because of eddy current considerations and the need for greater power output at the lower frequencies. A typical TVR is illustrated in Fig. 5.26.

Both high frequency [40] and low frequency [41] versions (see Fig. 1.16) of the wideband hybrid designs have been developed. A broadband doubly resonant response [36] can also be obtained with two piezoelectric stacks if one is reverse wired allowing addition between the resonance frequencies. This gives a wideband response, as in the case of the Hybrid transducer, but the response falls off faster below the fundamental resonance because of the phase reversal between the two piezoelectric stacks.

5.4 Transmission Line Transducers

While Tonpilz transducers are usually designed for the range from 1 to 50 kHz, transmission line sandwich type transducers are used mostly in the range from 10 to 500 kHz and piezoelectric ceramic bars or plates in the range from 50 kHz to beyond 2 MHz. In the frequency range above 10 kHz, lumped elements become very small and in this range of small wavelengths the wave nature of the transducer becomes more influential. Also, the area ratio between the piston head and the drive stack must be smaller because of more pronounced effects of head flexure with extensive overhang. We consider in this section sandwich and plate designs as well as composite material designs for transducers operating in the frequency range above 10 kHz.

5.4.1 Sandwich Transducers

The sandwich transmission line transducer was first introduced by Langevin [42] as a layered symmetrical structure of metal, quartz, and metal (see Fig. 1.3) that lowered the resonance frequency of a thin quartz thickness mode plate and provided a practical underwater sonar projector. A number of other sandwich transducer designs using piezoelectric ceramics have been given by Liddiard [43]. The front section of the sandwich is approximately matched to the medium by use of a matching material with intermediate impedance, such as aluminum, magnesium, or glass reinforced plastic (GRP), e.g., G10, rather than through a change in area, as in the case of the Tonpilz transducer. The material of the rear section of the sandwich, often steel or tungsten, is used to block or reduce the motion of the rear surface. The blocking and matching is usually accomplished through quarter wavelength layers. These procedures lead to a transducer with both a lower resonance frequency and lower Q_m than the piezoelectric ceramic drive plate alone. The modeling methods presented in Chaps. 3 and 4 can be readily applied to the transmission line transducer.

The quarter wavelength resonator section is an important concept for the transmission line transducer. In Sect. 3.2.1 it was shown that the impedance, Z_0, at one end of a transmission line, of length L and bar impedance $Z_b = \rho c A_0$, with load impedance Z_L at the other end may be written as

$$Z_0 = Z_b[Z_L + jZ_b \tan kL]/[Z_b + jZ_L \tan kL], \tag{5.42}$$

where $k = 2\pi/\lambda$ is the wave number in the bar (see Fig. 5.27).

If now the bar is one-quarter wavelength long, $kL = \pi/2$, Eq. (5.42) becomes $Z_0 = Z_b^2/Z_L$ which shows that Z_b acts as a transformer that converts the load admittance, $Y_L = 1/Z_L$, to Z_0. Thus, a small load impedance becomes a large input impedance and a large load impedance becomes a small input impedance. It also shows that if we wish to match a transducer of impedance Z_0 to a particular load impedance, Z_L, the impedance of the quarter wavelength matching layer must be

$$Z_b = (Z_0 Z_L)^{1/2}, \tag{5.43}$$

the geometric mean of the load and transducer impedance.

For example: Consider the case of matching the plane wave mechanical impedance of PZT, which is approximately 22.4×10^6 kg/m^{2}s, to the plane wave water load, which is 1.5×10^6 kg/m^{2}s. If the matching quarter wavelength layer is of the same area as the transducer, we need a material with $\rho c = 5.8 \times 10^6$ kg/m^{2}s, which is approximated by glass reinforced plastic (GRP) with $\rho c = 4.9 \times 10^6$ kg/m^{2}s (see Sect. 13.2).

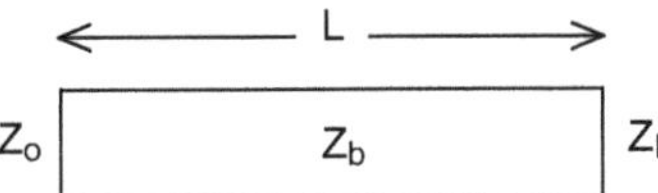

Fig. 5.27 Bar of length L with load impedance Z_L and input impedance Z_0

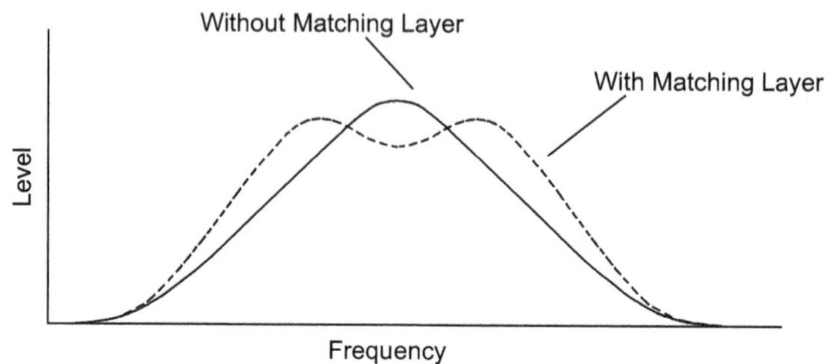

Fig. 5.28 Transmission line transducer response without (*solid line*) and with (*dashed line*) matching layer

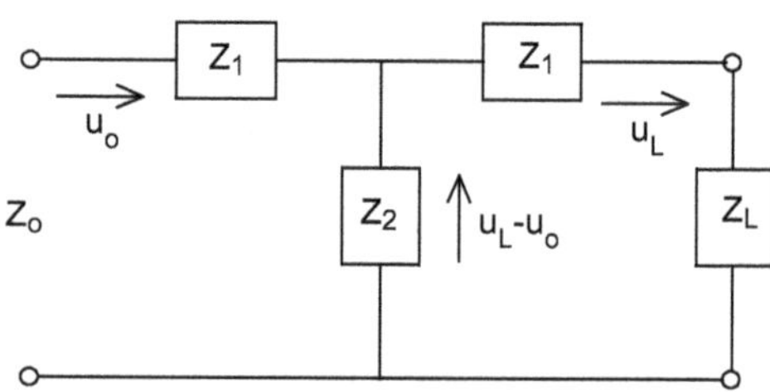

Fig. 5.29 Transmission line *T* network representation

Under ideal matching conditions there is no reflection and all the input power is passed on to the load. However, this only happens precisely at the frequencies where the section is a quarter wavelength and odd multiples thereof. At this frequency there is greater loading reducing the otherwise higher Q resonance of the device. With sufficient loading the main resonance is reduced leading to a double humped response curve illustrated in Fig. 5.28.

The bar in Fig. 5.27 may also be represented by the transmission line T network of Fig. 5.29 where $Z_1 = jZ_b \tan(kL/2)$ and $Z_2 = -jZ_b/\sin(kL)$, as was developed in Chap. 3. With the velocity at the input u_0 and at the load u_L, the circuit equations give $Z_2(u_0 - u_L) = (Z_1 + Z_L)u_L$ and use of $\tan kL/2 = (1 - \cos kL)/\sin kL$ leads to the velocity ratio

$$u_L/u_0 = 1/[\cos kL + j(Z_L/Z_b)\sin kL]. \tag{5.44}$$

For example: For $L = \lambda/2$, $kL = \pi$ and $u_L/u_0 = -1$; however, for $L = \lambda/4$, $kL = \pi/2$ and $u_L/u_0 = -jZ_b/Z_L$ giving a 90° phase shift, but more importantly, an output velocity increase by the ratio Z_b/Z_L. For matched conditions the magnification ratio $Z_b/Z_L = (Z_0/Z_L)^{1/2} = (22.4 \times 10^6/1.5 \times 10^6)^{1/2}$, for a velocity increase of approximately 3.9 times as a result of the quarter wavelength section between PZT of impedance $Z_0 = 22.4 \times 10^6$ kg/m^{2}s and the water load of impedance $Z_L = 1.5 \times 10^6$ kg/m^{2}s.

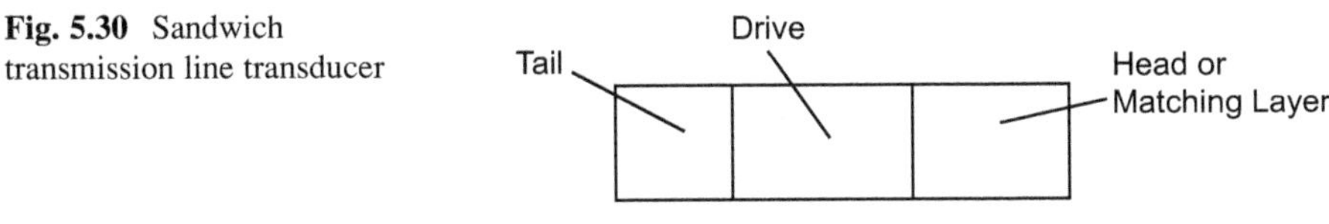

Fig. 5.30 Sandwich transmission line transducer

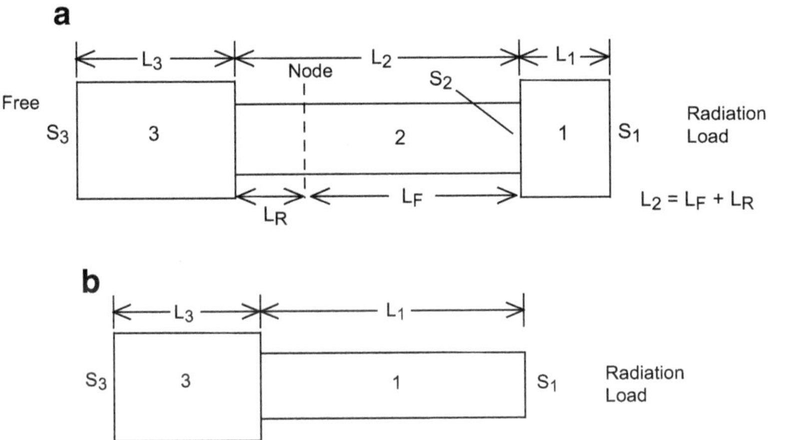

Fig. 5.31 (**a**) Three part transmission line transducer. (**b**) Two part transmission line transducer

The use of an additional quarter wavelength section on the tail end can provide a nearly rigid boundary condition on the rear of the drive section forcing greater motion on the front. Then with a free condition behind the tail section, $Z_L = 0$, Eq. (5.44) gives $u_0 = u_L \cos kL$ and at quarter wave resonance, $kL = \pi/2$, yields the rigid condition $u_0 = 0$. From Eq. (5.42) the impedance for $Z_L = 0$ is $Z_0 = jZ_b \tan kL$ yielding an infinite impedance at $kL = \pi/2$ and an impedance proportional to Z_b at frequencies above and below quarter wave resonance. Consequently materials of high impedance such as steel or tungsten are used for the tail section. A sketch of a transmission line transducer sandwich design with equal area is illustrated in Fig. 5.30. One design option is to make the length of the tail and drive sections each one quarter wavelength long, creating a half wavelength resonator with an additional quarter wavelength head section as a matching layer. A typical response curve with and without the matching layer is illustrated in Fig. 5.28.

A specific three layer model for the transducer shown in Fig. 5.31a may be developed by the methods presented in Chaps. 3 and 4, such as cascading three "*T*" networks or matching three solutions of the wave equation at the interfaces and setting the impedance or stress at the ends to zero to obtain the in-air resonance condition. With $Z_i = \rho_i c_i S_i$, $k_i = \omega / c_i$ and S_i the area of the *i*th section, resonance occurs when the total reactive impedance is zero, i.e.,

$$(Z_1/Z_2)\tan k_1L_1 + \left[1 - \left(Z_1Z_2/Z_3^2\right)\tan k_1L_1 \tan k_3L_3\right]$$
$$\tan k_2L_2 + (Z_3/Z_2)\tan k_3L_3 = 0. \quad (5.45)$$

With $L_2 = L_F + L_R$, where L_F and L_R are the distances to the node from front section 1 and rear section 3, respectively, the location of the node may be found from

$$\tan k_1L_1 \tan k_2L_F = S_2\rho_2c_2/S_1\rho_1c_1 \text{ or } \tan k_3L_3 \tan k_2L_R = S_2\rho_2c_2/S_3\rho_3c_3. \quad (5.46)$$

Equation (5.46) may also be obtained from the methods used to obtain Eq. (5.45), but using only two sections with a rigid condition at one end and a free condition at the other end.

Knowledge of the nodal position is important as it is the plane of no motion and, therefore, a good place to mount a transducer (see, for example, Fig. 1.5). However, since this location varies with additional radiation mass loading, a rubber interface between the nodal area and the housing is usually necessary. The plane of the node is also the location of the greatest stress and thus a poor location for a cement joint, unless a compressive stress rod is used. The mechanical Q_m may be written as (see Sect. 3.1.2)

$$Q_m = \eta_{ma}(1 + m/M)\,\omega_r m/R,$$

where R is the radiation resistance and η_{ma} is the mechanoacoustic efficiency. The effective head mass, m, and tail mass, M, may be determined from the kinetic energy (see Sect. 4.2.2) and further simplified at resonance by use of Eq. (5.46), resulting in

$$m = (\rho_1S_1L_1/2)[1 + \text{Sinc}\, 2k_1L_1/\, \text{Sinc}\, 2k_2L_F] + m_r,$$

$$M = (\rho_3S_3L_3/2)[1 + \text{Sinc}\, 2k_3L_3/\, \text{Sinc}\, 2k_2L_R],$$

where m_r is the radiation mass and the effective masses are referred to the velocities of the two ends.

For the special case where there is no section 2, as illustrated in Fig. 5.31b, quarter wavelength resonance occurs at $k_1L_1 = \pi/2$ and $k_3L_3 = \pi/2$ and with plane wave loading, ρ_0c_0 the specific acoustic impedance of the medium and $\eta_{ma} = 1$, we get

$$Q_m = (\pi/4)(\rho_1c_1/\rho_0c_0)(1 + S_1\rho_1c_1/S_3\rho_3c_3). \quad (5.47)$$

Using Eq. (5.47) with $S_1 = S_3$, we may construct Table 5.2 for various common material combinations (see Sect. 13.2) for this two part transmission line transducer.

Table 5.2 Transmission line transducer combinations

Tail	($\rho_3 c_3 \times 10^{-6}$)	Head	($\rho_1 c_1 \times 10^{-6}$)	Q_m
PZT	22.4	PZT	22.4	23.5
Steel	40.4	PZT	22.4	18.2
Tungsten	83.5	PZT	22.4	14.9
PZT	22.4	Aluminum	13.9	11.8
PZT	22.4	Magnesium	8.9	6.5
PZT	22.4	GRP (G-10)	4.9	3.1

Table 5.2 shows the comparatively high mechanical Q_m for an all-PZT transducer and how it may be lowered, to some extent, using a high impedance quarter wavelength backing tail and, to a greater extent, using a quarter wavelength matching head.

5.4.2 *Wideband Transmission Line Transducers*

Normally electroacoustic underwater transducers are operated in the vicinity of the fundamental resonance frequency where maximum output is obtained. Wideband performance can be obtained above resonance to some extent but, it is often limited by the next overtone resonance. Because of phase differences, the presence of the overtone resonance generally creates a cancellation at some frequency between the two resonant frequencies, typically resulting in a significant notch in the response that limits the bandwidth.

The problem is also often caused by unwanted lateral modes of vibration of the driving stack or flexural modes of vibration of the piston radiator. With these modes under control, however, there is still a major wideband impediment caused by the overtones of the fundamental that cannot be eliminated. The 180° phase difference between the mass controlled region of the fundamental and the stiffness controlled region of the first overtone causes cancellation and a notch in the response. The notch may be eliminated through the introduction of an additional phase shifted resonance between these modes [44].

A specific method for eliminating the notch is explained by reference to Figs. 5.32a, b, c, d and 5.33a, b which illustrate the calculated resulting acoustic pressure amplitude.

Figure 5.32a shows a piezoelectric longitudinal bar resonator operating in the 33 mode and composed of four separate piezoelectric elements wired in parallel with polarization directions, as shown by the arrows, for additive motion in the longitudinal direction. The dashed lines illustrate the symmetrical displacement of the bar for a voltage $+V$. The fundamental resonance occurs when the bar is one-half wavelength long, and the second harmonic when the bar is one wavelength long, but the second harmonic cannot be excited by the voltage arrangement of Fig. 5.32a. Because of the electrical symmetry, only the fundamental resonance

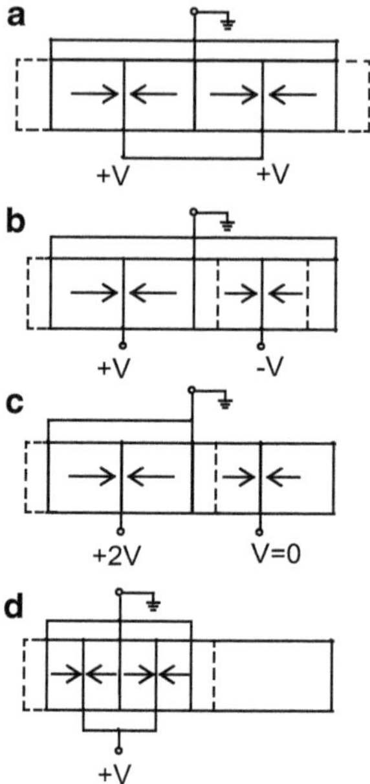

Fig. 5.32 Physical models illustrating the operation of the transducer [44]

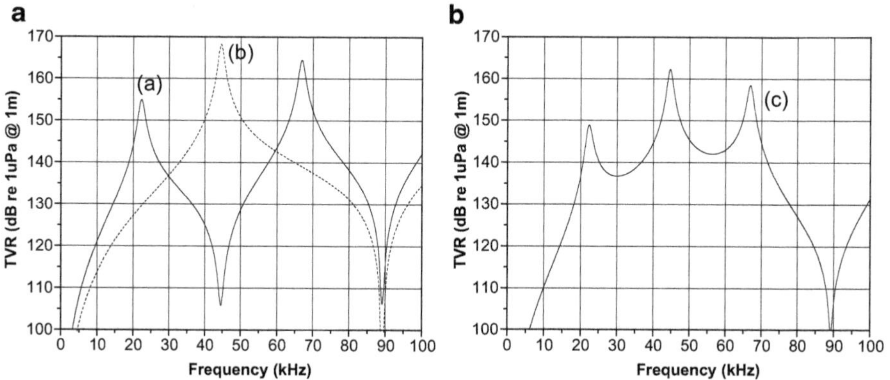

Fig. 5.33 Even (**a**), odd (**b**), and combined (**c**) mode excitations [44]

and all the odd harmonics are excited. If f_1 is the fundamental resonance, then the odd harmonic frequencies are $f_{2n+1} = (2n + 1)f_1$ for $n = 1, 2, 3, \ldots$. The amplitude response of the acoustic pressure to the right of the bar is shown in Fig. 3.33 by the curve labeled (a) with a fundamental resonance at 22.5 kHz, a third harmonic at 67.5 kHz and a null at 45 kHz. The null occurs at the frequency of the second harmonic that cannot be excited by this electrode arrangement. The calculated null at this frequency is deep because the vibration of the mass controlled region of the fundamental is exactly 180° out of phase with the vibration of the stiffness controlled region of the third harmonic. The existence of the unexcited even modes provides a means for constructively adding another resonant response at these nulls.

The even harmonics (but not the odd) are excited by the arrangement of Fig. 5.32b where the polarity of the voltage, V, on the right-hand pair of elements is reversed relative to that applied to the left pair. This causes a contraction on the right element pair while the left element pair expands as illustrated by the dashed lines. The excited even harmonic resonances are given by $f_{2n} = 2nf_1$ for $n = 1, 2, 3, \ldots$. The second harmonic acoustic pressure amplitude response is plotted as curve (b) in Fig. 5.33 with resonance at approximately 45 kHz which is just the location of the null for the wiring arrangement of Fig. 5.32a. The second harmonic motion on the right side of the bar is 180° out of phase with the fundamental motion as may be seen by comparing the displacements of Fig. 5.32a and b. The additional phase shift shown in Fig. 5.32a and b yields the in-phase condition at mid-band and allows the constructive addition of the second harmonic of Fig. 5.32b to the fundamental of Fig. 5.32a when the two systems are added.

The sum of the voltage conditions of Fig. 5.32a and b leads to the condition illustrated in Fig. 5.32c showing 2 V volts on the left piezoelectric pair and 0 V on the right piezoelectric pair. Since the $V = 0$ voltage drive section is no longer active in generating a displacement it may be replaced by an inactive piezoelectric section as shown in Fig. 5.32d. The wideband acoustic pressure amplitude response for the cases of Fig. 5.32c or d are given in Fig. 5.33, curve (c), showing the addition of the second harmonic resonance at 45 kHz filling in the original null. The harmonic frequencies for this case are $f_n = nf_1$ for $n = 1, 2, 3, \ldots$. The first null now appears in the vicinity of 90 kHz at twice the frequency of the 45 kHz null for the original case (a) of Fig. 5.33 and thus doubling the bandwidth. This null occurs when the piezoelectric pair on the left side is one wavelength long. The transducer now utilizes the fundamental and the second and third harmonic of the composite bar. The bandwidth can be further increased by reducing the length of the active piezoelectric section relative to the total length and allowing the excitation of higher harmonic modes such as the fourth, fifth, and sixth modes.

The mechanical Q_m at each of the multiple resonance frequencies can be quite different for some wideband transducer designs. In these cases feedback may be used to control the response. A sketch of a wideband transmission line transducer [44] with resonance frequencies 12, 25, and 38 kHz and with feedback is shown in Fig. 5.34a. The piezoelectric stack is one wavelength long at approximately 42 kHz limiting the bandwidth to 40 kHz in this design. Calculated transmission line circuit analysis results without and with feedback are shown in Fig. 5.34b.

As seen, the multiple resonant frequencies have no nulls between them and the feedback yields a smoothed response over a band of two octaves. Feedback causes the voltage to maintain a constant velocity of the piezoelectric element in the vicinity of the junction between the transmission line and the piezoelectric stack. The implementation of a feedback system with an integrator, wideband transducer, and sensor is also illustrated in Fig. 5.34a. Either integration or differentiation is necessary to provide the required 90° phase shift and lossless damping at the major resonance frequency. High frequency oscillations are minimized through the use of

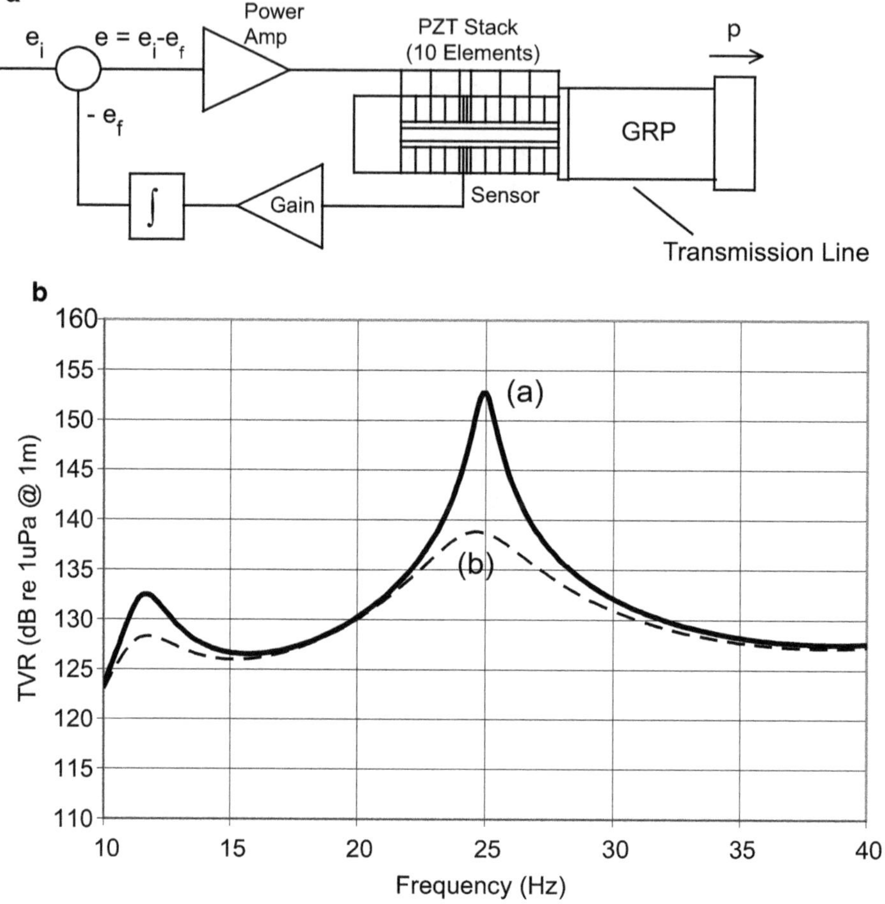

Fig. 5.34 (**a**) Multiply resonant transducer with feedback. (**b**) Calculated transmitting voltage response without (*a*) and with (*b*) feedback [44]

integration rather than differentiation and the strategic location of the sensor along the length of the piezoelectric ceramic stack.

Rodrigo [45] described a doubly resonant system with an additional spring-mass between a Tonpilz transducer and the medium, as shown in Fig. 5.35a, achieving a bandwidth on the order of one octave. Figure 5.35b shows a lumped equivalent circuit of the sketch of Fig. 5.35a.

At the parallel resonance of M' and C' a high impedance load is presented to the Tonpilz section which, if resonant at the Tonpilz resonance, creates a double humped response curve as in Fig. 5.28. This loading effect is similar to the loading effect of free-flooded rings and quarter wave matching layers and also similar to the Helmholtz loading used in audio band-pass woofers [46] and the air bubble-loaded transducer of Sims [47]. These doubly resonant transducers may be operated with

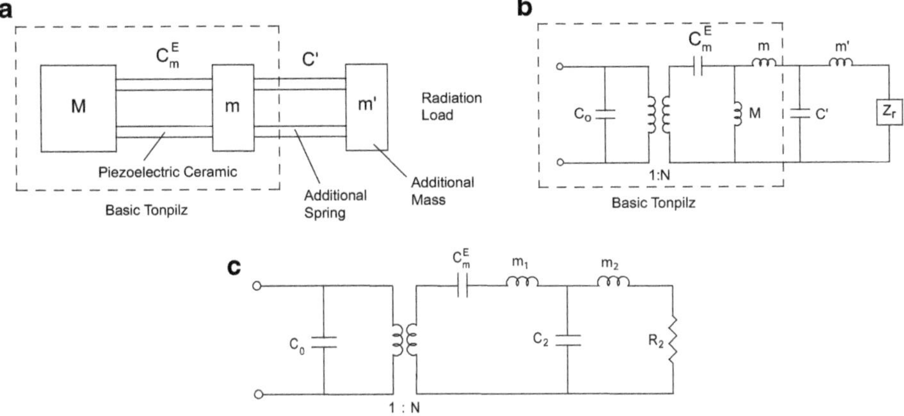

Fig. 5.35 (**a**) Rodrigo broadband design [45]. (**b**) Equivalent circuit for Rodrigo design. (**c**) Approximate circuit representation of Fig. 5.35

the resonance frequencies set at the same frequency or at different frequencies. The equivalent circuit of Fig. 5.35b may be reduced to the simpler circuit of Fig. 5.35c through use of the second Norton transform of Sect. 13.10 along with the typical condition where the total resistance is considerably less than the mass reactance, specifically when $R_r + R_l \ll \omega[(M + m + m' + m_r)(m' + m_r)]^{1/2}$. As may be seen, a large tail mass, M, can promote this condition. In this representation we have introduced the mechanical loss resistance, R_l, and let $Z_r = R_r + j\omega m_r$, where R_r is the radiation resistance and m_r is the radiation mass. In the resulting simplified circuit of Fig. 5.35c

$$m_1 = m/(1 + m/M),\;\; m_2 = \left(m' + m_r\right)/\left[1 + \left(m + m' + m_r\right)/M\right][1 + m/M],$$

$$C_2 = C'(1 + m/M)^2,\;\; R_2 = (R_r + R_l)/\left[1 + \left(m + m' + m_r\right)/M\right]^2.$$

This circuit removes the tail mass, M, as a separate component of the original circuit and accordingly allows a simpler representation of doubly resonant transducers and a simpler algebraic solution of the two resonance frequencies.

An improved alternative wideband method using multiple resonant sections between the driver and the medium has also been given by S. C. Butler [48] resulting in a broadband transducer with three resonance frequencies and a bandwidth of two octaves. A sketch of this transducer is shown in Fig. 5.36 showing two additional passive resonator sections 1–2 and 3–4 in front of the main active resonator 5–6–7, where 6 is the piezoelectric driver section, 7 is the tail mass, and 5 is a front mass.

At the lowest resonant frequency, f_1, sections 1 through 4 act as a single mass which combines with 5 and resonates with sections 6 and 7. At the next resonance

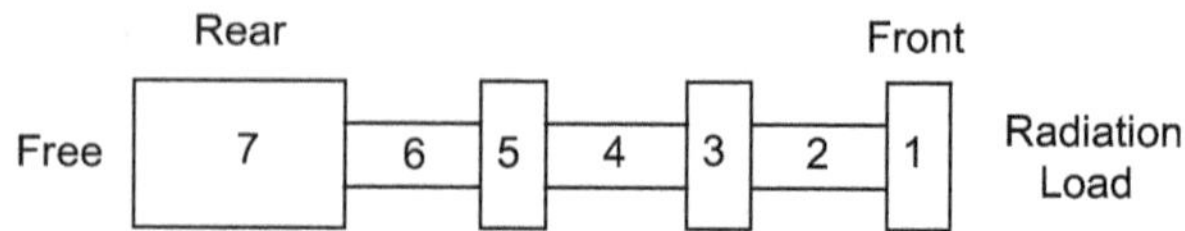

Fig. 5.36 Triply resonant transducer

frequency, f_2, sections 1 through 3 act as a mass and resonate with sections 4 and 5, being decoupled by compliant section 6 from section 7. Finally, at the third resonant frequency, f_3, sections 1 through 3 act as a resonator on being decoupled from the remaining part of the transducer through the compliance of section 4.

In some applications, such as finite amplitude nonlinear parametric systems (see Sect. 7.7), widely separated dual band transducer systems are desired rather than one wideband transducer. Here the upper band is used for nonlinear generation of the difference frequency and the lower band is used for receiving the difference frequency. Lindberg [49] has described a two band transducer system with an array of the high band transducers attached at their nodes to the piston of the lower band transducer. In this arrangement each of the smaller transducers moves as part of the piston head in the low frequency band and decouples from the larger piston in the higher frequency band thereby allowing independent operation of the two transducer systems.

5.4.3 Large Plate Transducers

Large diameter plates of piezoelectric ceramics with electrodes on the two major surfaces are often used to obtain high intensity narrow beams in one comparatively simple design usually without other components. These designs are limited to a plate thickness of approximately 0.5 in. and diameters of about 7 in., unless multiple plates are cemented together in a mosaic to create a larger diameter. The plates are usually back-isolated by a pressure release material such as corprene or, for deeper depth operation, paper (see Sect. 13.2), supported by a steel back plate. The piezoelectric ceramic plates, often in the form of a disc, are operated in the thickness mode and because of lateral clamping due to the large dimensions of the plate, the plate thickness resonance is higher than a longitudinal bar resonance by approximately 20 %. The 0.5 in. maximum thickness limits the lowest half wavelength resonance frequency to about 150 kHz for typical piezoelectric ceramic materials. The equivalent circuit for the thickness mode of a plate transducer is shown in Figure 3.39 and discussed in Sect. 3.2.3.4. The in-air half wavelength antiresonance frequency (electrical open circuit) occurs at $f_a = c/2L$, where $c = \left(c_{33}^D/\rho\right)^{1/2}$ and L is the thickness, while the electrical short circuit resonance occurs when $Z_a/2 + Z_b - Z_0N^2 = 0$ (see Fig. 3.40). This condition along with the square of the thickness coupling coefficient $k_t^2 = h_{33}^2\varepsilon_{33}^S/c_{33}^D$ and the trigonometric relation, $\sin 2x = 2\sin x \cos x$, with $x = \omega_r L/2c$ leads to the equation

$$k_t^2 = (\omega_r L/2c)\ \cot(\omega_r L/2c) \quad \text{or} \quad k_t^2 = (\pi f_r/2f_a)\ \cot(\pi f_r/2f_a).$$

For $k_t = 0.50$, $\omega_r L/2c = 1.393$ and the in-air short circuit resonance frequency $f_r = 0.887c/2L$. The second form of the equation for k_t^2 may be used to obtain k_t from measured values of f_r and f_a. Under water-loaded conditions the size of the plate or disc is often much greater than the wavelength of sound in water, in which case, the mechanical radiation impedance is resistive and equal to $A_0\rho_0c_0$.

Depth sounders with 2 in. diameter discs operating at around 200 kHz are commonplace. These units often have a hole in the center to shift a radial resonance outside the band of interest [50]. Larger diameter units with very narrow beams, on the order of 1°, have been used for greater depths. These 200 kHz large diameter transducer designs have also been used for finite amplitude generation of difference frequencies by simultaneously transmitting two frequencies, for example, 195 and 205 kHz, yielding an exceptionally narrow beam at the difference frequency of 10 kHz (see Sect. 7.7).

Implementation of the thickness mode of disc transducers is complicated by the fundamental radial mode and its lower order overtones that can interfere with the thickness mode. An equivalent circuit model for the radial mode can be developed from the impedance expression given by Nelson and Royster [51]. After a little algebraic manipulation the expression for the electrical input admittance for a thin disc of radius a and thickness L may be written as

$$Y_0 = j\omega C_0 + N^2/(Z_r + Z_m^E),$$

where

$$C_0 = C_f\left(1 - k_p^2\right),\ C_f = \pi a^2\varepsilon_{33}^T/L,\ N = 2\pi a d_{31}^2/s_{11}^E(1-\sigma^E),$$

$$k_p^2 = 2d_{31}^2/\varepsilon_{33}^T s_{11}^E(1-\sigma^E),$$

and where Z_r is the radial load impedance on the edges of the disc of area $2\pi aL$. The short circuit mechanical impedance of the disc is

$$Z_m^E = -j\ 2\pi aL\rho c\left[J_0(ka)/J_1(ka) - \left(1-\sigma^E\right)/ka\right],$$

where J_0 and J_1 are the Bessel functions of the first kind of order 0 and 1, $ka = \omega a/c$, the sound speed $c = \left[1/s_{11}^E\rho\left\{1-(\sigma^E)^2\right\}\right]^{1/2}$, Poisson's ratio $\sigma^E = -\ s_{12}^E/s_{11}^E$, and ρ is the density of the disk. The equivalent circuit for a ring, shown in Fig. 5.4, may be used to represent a radial disc mode, if the ring mechanical impedance, $j\omega M + 1/j\omega C^E$, is replaced by the disc radial mechanical impedance, Z_m^E.

At very low frequencies $Z_m^E \approx 2\pi L/j\omega s_{11}^E(1-\sigma^E)$, where $2\pi L/s_{11}^E(1-\sigma^E)$ is the short circuit stiffness. At mechanical resonance, $Z_m^E = 0$ and with free edges ($Z_r = 0$), the radial resonance frequencies of the disc are given by the roots of the equation

$$J_0(ka)/J_1(ka) = \left(1 - \sigma^E\right)/ka.$$

The lowest root [6] for $\sigma^E = 0.31$ is $ka = 2.05$ leading to the fundamental short circuit resonance frequency $f_r = 2.05c/2\pi a = 0.65c/D$, where the diameter $D = 2a$. The higher order radial modes have little effect on the thickness mode performance for $D \gg L$. A 3D model of the piezoelectric disc as the driver of a transmission line Langevin transducer, including radial and thickness modes, has been given by A. Iula et al. [52].

Another problem area results from surface waves such as shear or Lamb waves generated at the edges of the disc. These waves travel at a slow speed comparable to the speed of sound in water and launch a wave at the coincident angle given by sin $\alpha = c_0/c_s$, where c_0 is the speed in the medium and c_s is the speed of the surface waves. The result can be an apparent lobe in the beam pattern centered at an angle, α, that does not change significantly with the diameter or thickness of the plate. The effect is more pronounced for rectangular plates with straight edges than for circular plates. The surface wave lobe can be reduced by a quarter wavelength margin between the edge of the plate and the electrode.

Lateral resonances and surface wave effects may also be mitigated by dicing the transducer into smaller elements. The dicing is usually done with very fine wire saws and cross cut at 90° yielding an array of separate small square bar or post-like structures. The cutting tool is adjusted to leave a thin interconnecting ceramic structure on the back side to hold the structure together with the array of small square radiators on the front side. This also leaves one electrode on the back surface as a common connection but necessitates a separate connection to each square radiator on the front.

5.4.4 Composite Transducers

Composite piezoelectric transducers ("piezocomposite") consist of piezoelectric ceramics such as PZT embedded in an inert polymer matrix such as polyurethane, silicone or other types of rubber, polyethylene, or an epoxy. The composite designation follows that of Newnham [53] which describes the connectivity of the piezoelectric material and the connectivity of the polymer material illustrated in Fig. 3.37a. Thus, 1-3 connectivity refers to the piezoelectric ceramic connected only along a single (1) direction while the polymer is connected along three (3) directions. In the case of 0-3 connectivity small piezoelectric ceramic particles are suspended in a polymer and are not connected (0) to each other while the polymer material is connected in three directions (3). In the case of 2-2 connectivity both the piezoelectric ceramic and the polymer are connected in two directions (2) (as alternating rows of ceramic bars and polymer material). The 2-2 composites have been used in medical ultrasound arrays and can be manufactured by dicing piezoelectric material and filling the cuts with polymer. Although 0-3 composites are difficult to polarize, they have the advantage of high flexibility. Because of their low coupling coefficient, 0-3 composites are more suitable as hydrophones than as projectors. We will concentrate on the more commonly used 1-3 connectivity which has found use in projectors.

The 1-3 composites consist of arrays of piezoelectric rods arranged in a polymer matrix as illustrated in Fig. 5.37a and b. The piezoelectric rods are not connected by piezoelectric ceramic material along the x and y direction while the polymer material is connected along the x and y directions as well as the z direction. These 1-3 composites can function as transmitters and as receivers and can be proportioned for high performance operation up to approximately 1000 psi. The composites can be tailored for high coupling coefficient and broad-band performance. The performance of the composite depends on the volume fraction of the rods, the specific material and aspect ratio of the piezoelectric rods, the composition of the polymer, the electrode or cover plate stiffness, the spatial array period compared to the wavelength in the composite and the overall size compared to the wavelength in the surrounding medium. Because of advances in injection molding technology [54], 1-3 composites can be manufactured in high volume at a cost comparable to the cost of a corresponding solid sheet of piezoelectric material. The most common configurations are the regular arrays of round and square piezoelectric rods ranging in size from 20 μm to 5 mm in width.

A comparison between a 1-3 composite with 15 % PZT-5H and 85 % polymer matrix, and solid piezoelectric ceramic PZT-5H is shown in Table 5.3 [55].

Note that the thickness mode coupling coefficient is greater in the composite than in the PZT-5H plate and that the mechanical impedance is much lower.

A simple but effective thickness mode model for piezocomposites has been developed by Smith and Auld [56] which allows the calculation of the essential parameters of the composite piezoelectric as a function of the volume fraction, ν, of the piezoelectric ceramic material. A tensor model has also been given by Avellaneda and Swart [57].

The Smith-Auld model uses the piezoelectric constitutive equation set in the 3 direction with stress and electric displacement as functions of strain and electric field:

$$T_3 = c_{33}^E S_3 - e_{33}^t E_3,$$

$$D_3 = e_{33} S_3 + \varepsilon_{33}^S E_3.$$

After applying boundary conditions and making assumptions and approximations the model gives piezocomposite effective values for the short circuit elastic modulus c^E, the clamped dielectric constant ε^S, and the piezoelectric "e" constant. The thickness (3) mode operation of large diameter plates often assumes clamped conditions along the lateral (1) and (2) directions (see Sect. 5.4.3). Usually the disc or plate is large compared to the wavelength, and each piezoelectric ceramic element of the composite and its adjacent polymer responds as though it were within a cell with rigid boundaries (see Fig. 5.37b). This is so since each cell acts on

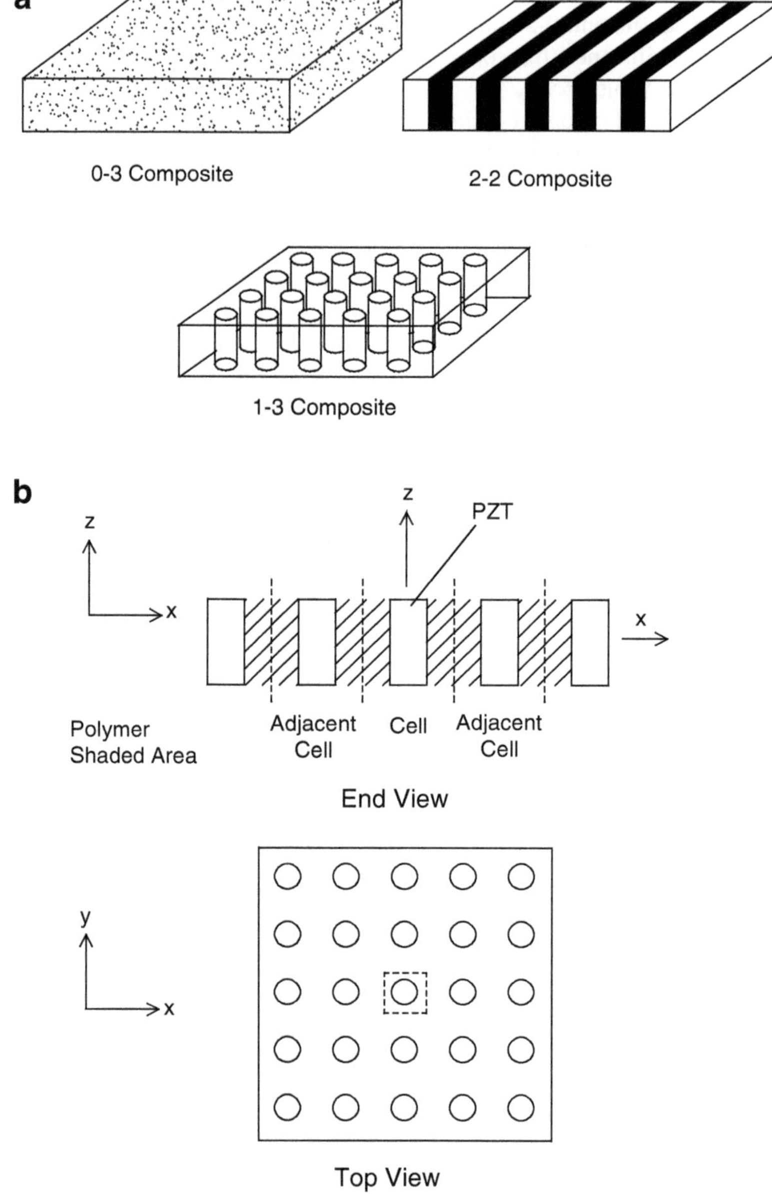

Fig. 5.37 (**a**) Three piezocomposite configurations in current use for transducers. (**b**) Model for 1-3 composite

adjacent cells with equal force (except for the smaller number of cells on the periphery which are free to move and do not satisfy the clamped assumption). Thus in this model, it is assumed that the effective strains $\mathbf{S_1}$ and $\mathbf{S_2}$ at the cell surfaces are zero, and therefore are zero throughout the cell, giving

Table 5.3 Comparison of 1-3 composite and PZT-5H

Property	Units	1-3 Composite	PZT-5H
Relative dielectric constant		460	3200
Dissipation factor		0.02	0.02
Coupling coefficient k_t		0.63	0.51
Mechanical Q		20	65
d_{33}	(pC/N)	550	650
Impedance	(Mrayls)	6	30
Density	(kg/m^3)	1800	7500

$$\mathbf{S_1} = \nu S_1^c + (1-\nu)S_1^p = 0 \quad \text{so that} \quad S_1^p = -S_1^c \nu/(1-\nu),$$

where the superscripts c and p refer to piezoelectric ceramic and polymer, respectively, and the same relationship holds for $\mathbf{S_2}$. The effective density is given by

$$\boldsymbol{\rho} = \nu\rho^c + (1-\nu)\rho^p,$$

where **bold** signifies effective composite value.

Additional assumptions are: The lateral stresses in the 1 and 2 directions are the same in the polymer and the piezoelectric ceramic. The strain and electric field are independent of the 1 and 2 directions throughout the structure. The electric field components E_1 and E_2 are zero because the electrode surfaces are equipotentials. The electric field E_3 is the same in the ceramic and polymer. The ceramic and polymer move together along the 3 direction so that S_3 is the same for both. This model leads to a set of effective values which may be written as

$$\mathbf{c_{33}^E} = \nu\left[c_{33}^E - 2\left(c_{13}^E - c_{12}\right)^2/\mathbf{c}\right] + (1-\nu)c_{11},$$

$$\mathbf{e_{33}} = \nu\left[e_{33} - 2e_{31}\left(c_{13}^E - c_{12}\right)/\mathbf{c}\right],$$

$$\boldsymbol{\varepsilon}_{\mathbf{33}}^{\mathbf{S}} = \nu\left[\varepsilon_{33}^S + 2(e_{31})^2/\mathbf{c}\right] + (1-\nu)\varepsilon_{11},$$

$$\mathbf{c} \equiv c_{11}^E + c_{12}^E + \nu(c_{11} + c_{12})/(1-\nu).$$

The c_{11} and c_{12} elastic modulus and ε_{11} dielectric constant without superscripts refer to the polymer matrix.

The results may be used in an additional set of constitutive equations in which D and S are the independent variables. This set is most useful in representing the resonant operation of a thickness mode composite transducer with large lateral dimensions. The 33 mode version of this set may be written as

$$T_3 = \mathbf{c_{33}^D} S_3 - \mathbf{h_{33}} D_3,$$

$$E_3 = -\mathbf{h}_{33} S_3 + \left(1/\boldsymbol{\varepsilon}_{33}^{\mathbf{S}}\right) D_3,$$

where $\mathbf{h}_{33} = \mathbf{e}_{33}/\boldsymbol{\varepsilon}_{33}^{\mathbf{S}}$

$$\mathbf{c}_{33}^{\mathbf{D}} = \mathbf{c}_{33}^{\mathbf{E}} + (\mathbf{e}_{33})^2/\boldsymbol{\varepsilon}_{33}^{\mathbf{S}}.$$

This set applies to the equivalent circuit of Fig. 3.40 and Eq. (3.54) where the effective thickness coupling coefficient, k_t, sound speed, υ_t, and impedance, Z_t, may be written as

$$\mathbf{k}_t = \mathbf{h}_{33}\left(\boldsymbol{\varepsilon}_{33}^{\mathbf{S}}/\mathbf{c}_{33}^{\mathbf{D}}\right)^{1/2}, \quad \boldsymbol{\upsilon}_t = \left(\mathbf{c}_{33}^{\mathbf{D}}/\boldsymbol{\rho}\right)^{1/2}, \quad \mathbf{Z}_t = \left(\mathbf{c}_{33}^{\mathbf{D}}\boldsymbol{\rho}\right)^{1/2} A_0,$$

with

$$Z_0 = 1/j\omega C_0, \;\; C_0 = A_0\boldsymbol{\varepsilon}_{33}^{\mathbf{S}}/t, \;\; Z_a = jZ_t \tan \omega t/2\boldsymbol{\upsilon}_t \;\text{ and }\; Z_b = -jZ_t/\sin \omega t/\boldsymbol{\upsilon}_t.$$

In the equivalent circuit A_0 is the cross-sectional area of the composite transducer. With t the thickness, the half wavelength fundamental thickness mode resonance occurs when $t=\lambda/2$. Since the sound speed $\boldsymbol{\upsilon}_t = f\lambda$, the fundamental resonance frequency $f_1 = \boldsymbol{\upsilon}_t/2t$. This is often expressed as the thickness mode frequency constant $f_1 t = \boldsymbol{\upsilon}_t/2$. The excited harmonic resonance frequencies are odd integer multiples of f_1.

Smith and Auld [58] have numerically evaluated the model for various compositions as a function of the volume fraction of the ceramic, ν. As may be expected, the effective density, $\boldsymbol{\rho}$, and effective relative dielectric constant $\boldsymbol{\varepsilon}_{33}^{\mathbf{S}}/\varepsilon_0$ vary linearly with ν. However, the effective elastic constant $\mathbf{c}_{33}^{\mathbf{D}}$, piezoelectric constant $\mathbf{e}_{33}$, and specific acoustic impedance, $\mathbf{Z}_t/A_0$, depart from linearity for volume fractions of PZT greater than 75 % as a result of the greater lateral clamping of the individual PZT rods by adjacent PZT rods. At very low PZT volume fraction the effective thickness mode sound speed, $\boldsymbol{\upsilon}_t$, approaches the sound speed in the polymer matrix material while at high PZT volume fractions the speed approaches that of the PZT material.

The most interesting change occurs in the effective thickness coupling coefficient, $\mathbf{k}_t$. There is a reduced effective coupling coefficient at low PZT volume fractions, less than 5 %, where the composite acts more like a piezoelectric polymer. At high volume fractions, greater than 99 %, the effective coupling, $\mathbf{k}_t$, approaches k_t, as might be expected as, here, the composite acts as a solid disk (see Sect. 4.4.1.3). On the other hand, for volume fractions in the range from 15 to 95 %, the effective coupling coefficient of the 1-3 composite, $\mathbf{k}_t$, significantly exceeds k_t and, because of reduced lateral clamping, approaches the value of k_{33} with a maximum value at a volume fraction of approximately 75 %. For hydrostatic mode hydrophone operation, with free rather than clamped lateral boundary conditions, the optimum volume fraction for maximum coupling is approximately 10 % (see Sect. 6.3.2).

5.5 Flextensional Transducers

Flextensional transducers (see Fig. 1.13) are generally used as low to medium frequency high power projectors radiating sound by the flexure of a metal or GRP shell excited by a drive stack operated in an extensional mode. The most common design is a Class IV type (see Fig. 1.14) which is an oval or elliptical shell driven by a piezoelectric ceramic stack along the major axis of the shell with amplified motion along the direction of the minor axis. Since the inactive shell makes up a significant portion of the transducer stiffness, it causes a significant reduction in the coupling coefficient, resulting in an effective value of $k_e \approx k_{33}/2$ for a 33 mode driven system.

The first flextensional transducer has been attributed [59] to Hayes [60] (see Fig. 1.4); however, it was the later work and patent of Toulis [61] that led to the Class IV design, which was modeled by Brigham [62] and later encoded for computer design and analysis by Butler [63]. Other early flextensional designs by Merchant [64] and Abbott [65], as well as the modeling by Royster [66], laid the foundation for more recent designs by Jones and McMahon [67] and Nelson and Royster [51]. Dogan and Newnham [68] developed a very compact design, and Butler [69–72] extended the flextensional concept in various ways. Since flextensional transducers make use of flexural modes of an elastic shell the analysis and modeling required to determine equivalent circuit parameters is considerably more complicated than the longitudinal mode cases discussed earlier, and it is usually necessary to make simplifying approximations. The results of such modeling will be used in this section, and the details will be found in the papers referred to above. Approximate models are essential for preliminary design, but final design should be based on finite element numerical methods (see Sect. 3.4).

The original types of flextensional transducers have been given a Class designation, illustrated in Fig. 1.13, that follows the historical order and distinguishes the models by the type of shell and drive system. In this section the flextensional designs which we believe are of most technical significance or are most commonly used will be discussed.

5.5.1 The Class IV and VII Flextensional Transducers

The Class IV flextensional transducer [61] is illustrated in Figs. 1.14 and 5.38.

The oval shell is typically high strength aluminum (steel and GRP have also been used) which is fitted with mechanically isolated end caps and booted with rubber. The drive stack is usually Type I (PZT-4) or Type III (PZT-8) piezoelectric ceramic, although electrostrictive PMN and laminated magnetostrictive Terfenol-D have been used for greater output power. In operation the driver oscillates along the x direction causing a small oscillating motion at the shell ends and symmetrical amplified motion of the shell in the y direction. The peak amplified motion is given approximately by the ratio of the semi-major to semi-minor axes, b/a. As the stack

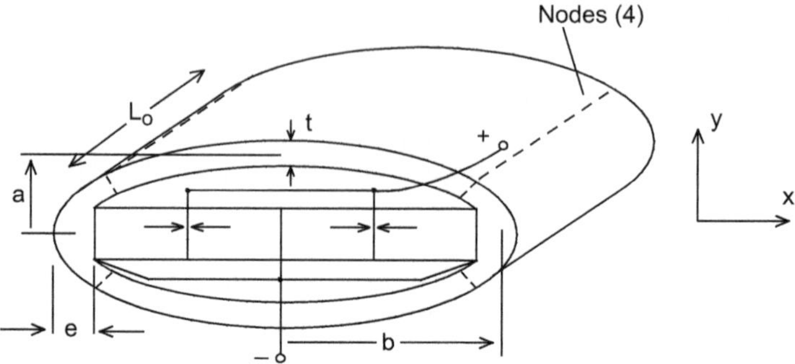

Fig. 5.38 Class IV flextensional transducer [61]

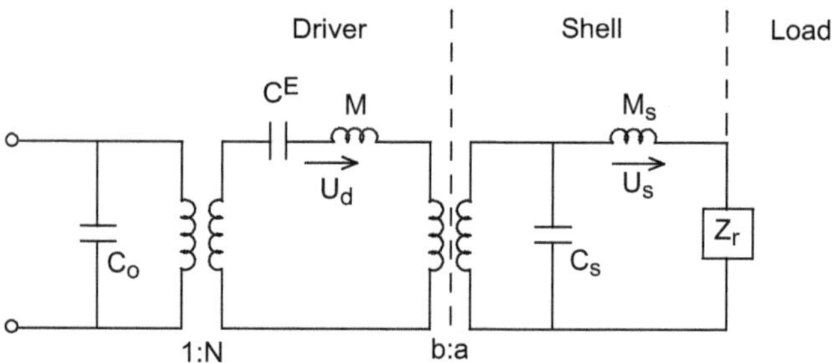

Fig. 5.39 Simplified flextensional equivalent circuit

expands, the ends move outward causing a major portion of the shell to move inward resulting in four nodes approximately located by the dashed lines in Fig. 5.38. Since the motion in the y direction is much larger, and has much greater area, than the motion in the x direction, the out-of-phase radiation from the ends is negligible. The drive stack excites the shell in its fundamental quadrupole bending mode of vibration which is modified by the boundary conditions at the shell ends imposed by the drive stack. In addition to this mode, the higher frequency octopole bending mode and fundamental extensional shell mode may also be excited. The octopole mode produces a sharp poorly loaded (because of multiple phase reversals) resonance while the shell extensional mode produces a well-loaded strong resonance.

The simplified fundamental quadrupole mode equivalent circuit, shown in Fig. 5.39, is useful as an aid in understanding the essential features of the transducer.

Here we have taken advantage of the symmetry in both the x and y directions, to reduce the number of circuit elements and output ports. The piezoelectric driver moves with velocity u_d at its ends and the shell moves with average velocity u_s which is loaded by the radiation impedance and radiates the acoustic pressure. The velocity at the shell midpoint is magnified by the ratio b/a. The mass M_s is the dynamic mass of the shell and C_s is the effective compliance of the shell. The quantities M and C^E represent the mass and short circuit compliance of the piezoelectric drive stack of clamped capacity C_0 and electromechanical turns ratio N.

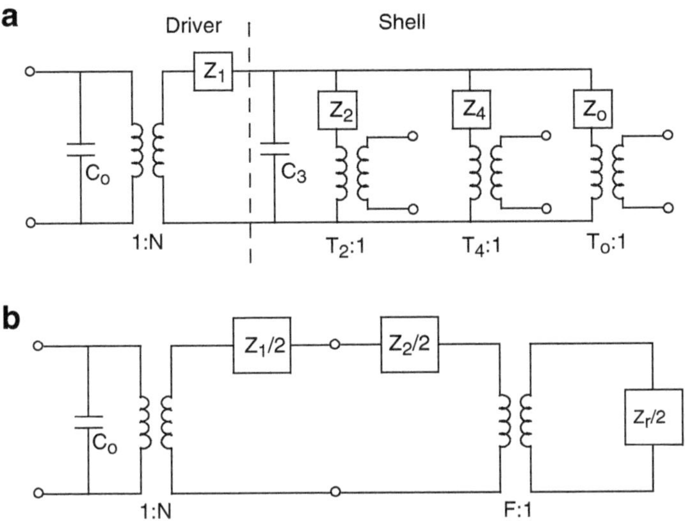

Fig. 5.40 (**a**) Brigham's flextensional equivalent circuit. (**b**) Quadrupole mode equivalent circuit

A more detailed equivalent circuit which includes the quadrupole and octopole bending modes and extensional omni mode has been given by Brigham [62] and is illustrated in Fig. 5.40a. Here the driver is represented by a transmission line of impedance

$$Z_1 = -j\rho_1 c_1 A_1 \cot(\omega L_1/2c_1),$$

where ρ_1 is the density, c_1 is the longitudinal short circuit sound speed, A_1 is the cross-sectional area, and L_1 is the full length of the drive stack. Element C_3 is the compliance of metal end pieces or electrical insulating pieces between the stack and the shell, Z_2, Z_4, and Z_0 are the impedances of the quadrupole, octopole, and omni extensional modes while T_2, T_4, and T_0 are the magnification transformer ratios for the respective modes. This model assumes symmetry and does not include odd modes. However, in the presence of a baffle or in an array with other transducers the pressure distribution on the two major surfaces will usually not be the same which may excite odd modes of vibration. This flextensional transducer model, as well as a model for an interactive array of such transducers, has been encoded [63] for computer analysis.

Although the omni extensional shell mode does affect the quadrupole mode, a simplified version of the Brigham model is obtained if only the quadrupole mode is retained with some other simplifying assumptions. This approximate representation is shown in Fig. 5.40b. The impedances are divided by two to account for the horizontal plane of symmetry yielding a one port model. The stack mechanical impedance Z_1 arises from the left right symmetry of Fig. 5.38, avoiding the T network representation of Sect. 3.2.3.1. The quadrupole impedance is, with L_0 the length of the shell and $S_0 = (\pi/2)\left(a^2 + b^2\right)^{1/2}/\sqrt{2}$,

$$Z_2 = -j(K_2/\omega)\left(1 - \omega^2/\omega_2{}^2\right),$$

where

$$K_2 \approx Y\ t^3 L_0 / \left[5.14 S_0 a^2 (a/b)^{1/2} (t/e)^{3/2}\right] \quad \text{and}$$
$$\omega_2 \approx \pi^2 tc[2.5 - 0.25 \ \cos(\pi a/b)]/8 S_0^2 \sqrt{3}\,.$$

The parameters e and L_0 are defined in Fig. 5.38, and the transformation factor, F, of Fig. 5.40b is given by

$$F^2 \approx \left[(1 + a/b)^4 - 15(a/b)^2\right] / \left[30(a/b)^3 (1 + a/4b)\right].$$

In the circuit of Fig. 5.40b, C_0 is the clamped capacitance for one element of the drive stack, the electromechanical turns ratio may be written as $N = g_{33} C_0 Y_{33}^D$ and Z_r is the radiation impedance. A simple computer model based on these approximate equations and similar equations for the octopole and omni extensional shell mode have been encoded [63] and may be used as a flextensional design tool for initial analysis of individual transducers and arrays (before a more extensive finite element model is constructed).

Since the transducer operates in the vicinity of the quadrupole mode, with negligible motion from its ends, the radiation is approximately omnidirectional and the radiation load may be approximated by an equivalent sphere. The beam pattern does, however, show a reduced output in the y direction (see Fig. 5.38) which is a result of the time delay between waves arriving from the two sides. The assembly of the transducer requires precision machining of the interface shanks and the flats on the inner ends of the shell. The stack is usually made oversize, so that during assembly, with the shell compressed on the major surfaces, the stack can be inserted and receive a compressive bias when the compression of the shell is relieved. Significant over pre-compression is necessary as the compression on the stack is reduced as the hydrostatic pressure is increased.

As mentioned earlier, only even modes are normally excited because of the symmetry of the design. However, there is a significant odd mode, in the vicinity of the quadrupole mode, that may be excited by unequal pressure on the two major surfaces or by asymmetrical excitation of the drive stack (see Fig. 1.14). The odd mode is essentially a rigid body motion of the shell as a reaction to the fundamental bending mode of the drive stack. Butler [69, 70] has shown that, by driving the stack of a Class IV flextensional transducer into a bending mode (see Sect. 5.6), this odd dipole mode can be excited, and, when combined with the quadrupole mode yields a directional flextensional transducer as illustrated in Fig. 5.41. Here the +/+ motion of the quadrupole (nearly omni) mode is combined with the ± motion of the dipole mode to create reduced motion of the shell and far-field pressure on one side [69], giving directional far-field pressure [70] as illustrated in Fig. 5.42. This method of obtaining a directional far-field

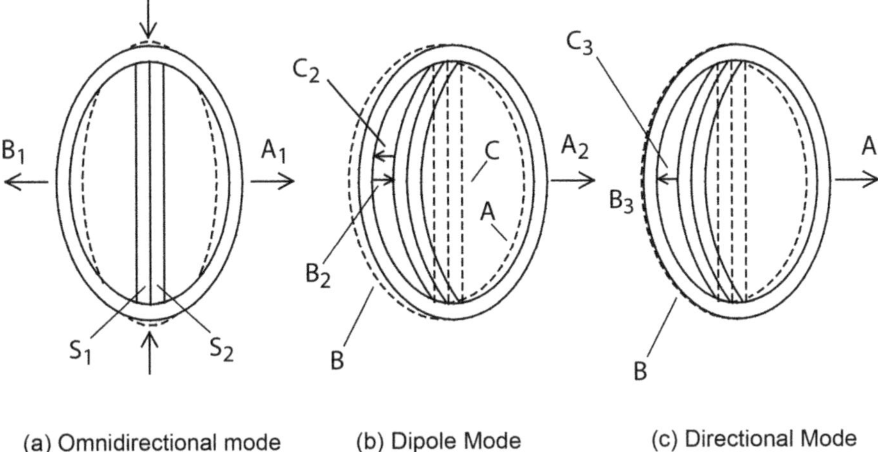

Fig. 5.41 Schematic illustration of a flextensional transducer showing the sequence of events leading to directionality. The neutral state is represented by the dashed lines

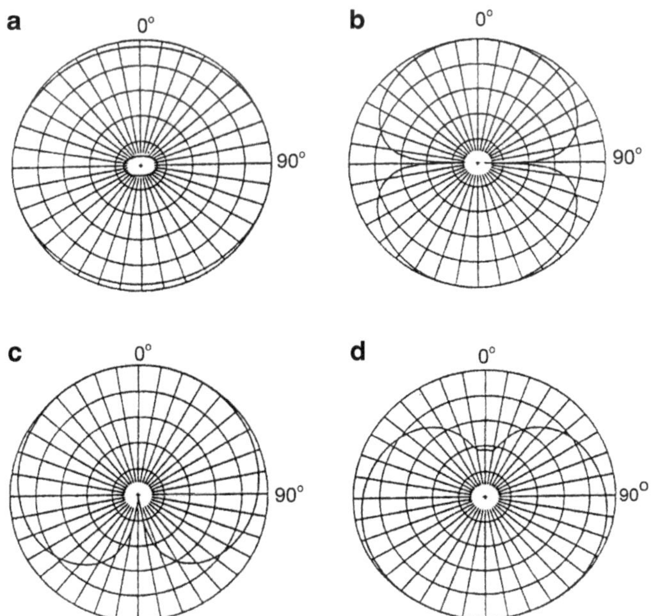

Fig. 5.42 Measured single element 900 Hz radiation patterns operating in the (**a**) quadrupole mode, (**b**) dipole mode, (**c**) directional mode, and (**d**) directional mode drive leads reversed (10 dB/division) [70]

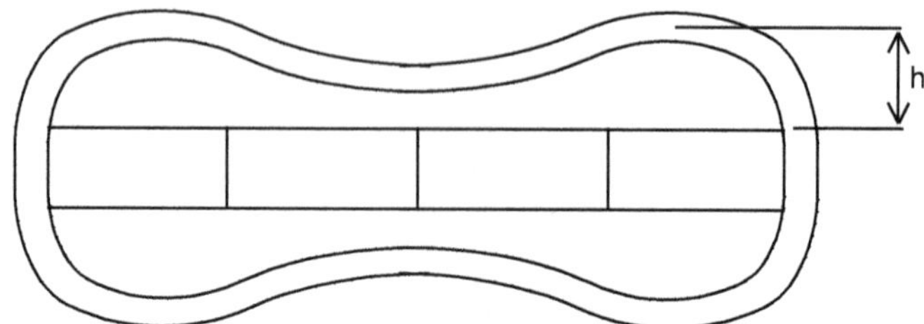

Fig. 5.43 Class VII "dog bone" flextensional transducer

pattern is basically the same as that for the multimode transducer described in Sect. 5.2.6.

The Class VII, "dog bone" [64], is illustrated in Fig. 5.43 and is seen to be the same as the Class IV except that the shell is inverted. In this case the entire shell moves outward as the piezoelectric stack moves outward, and both are in-phase below resonance. Hydrostatic pressure on the major surfaces causes further compression of the piezoelectric stack rather than a release of the initial compression, as in the Class IV transducer. Although this Class VII design provides opportunity for operation at a greater depth, the portion of the shell of height h causes a greater compliance than in the Class IV design resulting in additional bending and a lower effective coupling coefficient.

5.5.2 *The Class I Barrel Stave Flextensional*

The Class I convex flextensional transducer, shown in cross section in Fig. 5.44, was one of the first flextensional transducer to be fully modeled [66]. The shell has slots along the axial, z, direction to reduce the axial stiffness that would be imposed by lateral Poisson's coupling. The design is axi-symmetric about the z-axis, except for the slots. The concave version, shown in the same figure, also has slots and may be fabricated by separate metal staves and in this is called the "barrel stave" flextensional transducer [67]. As in the case of the dog bone flextensional transducer, compression rather than tension is experienced by the drive stack under hydrostatic pressure on the barrel stave transducer. The cylindrical shape makes this transducer an attractive choice for underwater towed lines and sonobuoy applications.

A model for the barrel stave transducer has been given by Moffett et al. [73], and the influence of the magnetic circuit on the effective coupling coefficient of a magnetostrictive Terfenol-D driven barrel stave transducer has been given by Butler et al. [31]. The transducer is generally rubber booted to prevent water ingression; however, this can lead to problems under pressure when the rubber is forced into the slots causing reduced output. A boot-free design, with axial pleats in an otherwise continuous metal shell to reduce circumferential stiffening, has been developed by Purcell [74].

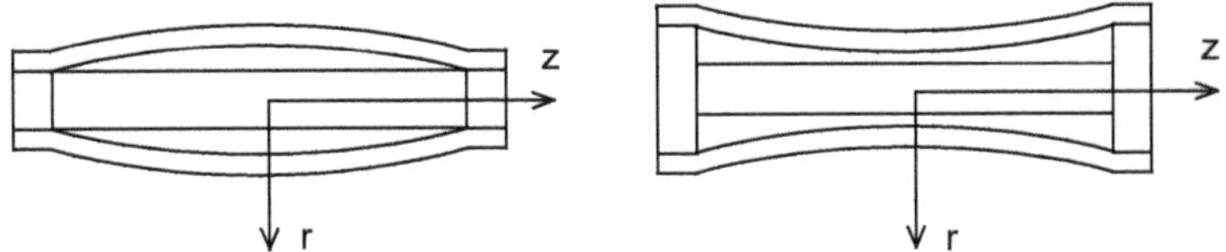

Fig. 5.44 The Class I convex and concave barrel stave flextensional transducer

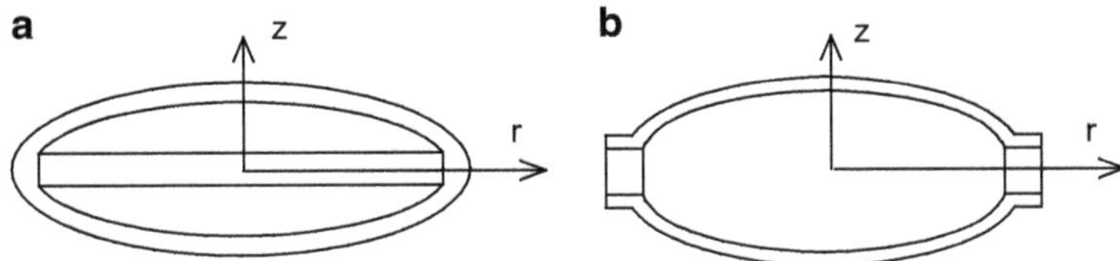

Fig. 5.45 The Class V disc-driven (**a**) and ring drive (**b**) flextensional transducer

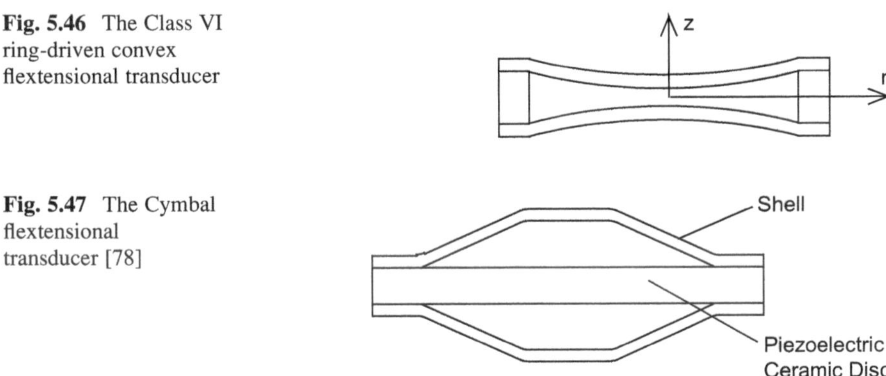

Fig. 5.46 The Class VI ring-driven convex flextensional transducer

Fig. 5.47 The Cymbal flextensional transducer [78]

5.5.3 The Class V and VI Flextensional Transducers

The Class V planar piezoelectric disc driven and ring-driven convex-shell flextensional transducers are shown in cross section in Fig. 5.45 while the concave [65] Class VI ring-driven shell design is shown in cross section in Fig. 5.46.

The disc-driven version has perfect axial symmetry about the z axis while the ring-driven designs are only approximately axially symmetric because of slots in the shell and metal sections in the piezoelectric 33 mode drive ring for shell attachment. A 600 Hz Class V ring shell, based on a patent by McMahon and Armstrong [75], has been fabricated and successfully tested [76]. This transducer was used in a comparison with a theoretical model by Butler [77]. The convex Class V ring shell design requires an interior compressed air bladder for deep operation as the convex shell produces circumferential tension in the piezoelectric ring under ambient pressure. The concave Class VI ring shell produces circumferential compression on the ring and can generally operate without an air bladder at greater depths than the Class V ring shell.

The disc-driven Class V has a stiffer drive structure and can withstand greater stress than the ring shell designs. A model for this design was first given by Nelson and Royster [51] and has since been incorporated into the computer model FIRST by Butler [77], based on the original piezoelectric disc model by Mason [78]. Newnham and Dogan [68] have developed a miniature Class V flextensional transducer called the "Cymbal" because of the shape of the shell shown in cross section in Fig. 5.47. Note that the shell is attached to the disc surface rather than the disc end, as in the case of Fig. 5.45. This allows a simple fabrication procedure for small transducers. These small transducers can achieve a low resonance frequency and have been used as both hydrophones and projectors.

5.5.4 Astroid, Trioid, and X-Spring Transducers

An internally and externally driven Astroid (hypocycloid of four cusps [79]) transducer [71] is illustrated in Fig. 5.48. The internal drive case is illustrated with four piezoelectric stacks with a steel center piece and metallic shell, while the external drive case is illustrated with four magnetostrictive rods forming a square. As the rods and stacks expand, the four shells bend outward producing amplified motion as in the case of a Class IV flextensional. This is added, in phase, to the extensional motion of the drivers yielding an even greater motion of the shell. The Astroid may be modeled as a pair of Class IV flextensional transducers with sections I and III acting as one transducer and sections II and IV acting as the other, both driven by stacks with length $\sqrt{2}$ times the length of one of the stacks shown in Fig. 5.48.

An X-spring ("transducer-spring") [72] version of the Astroid is illustrated in Fig. 5.49a showing four pistons attached to the point of greatest motion of the shell

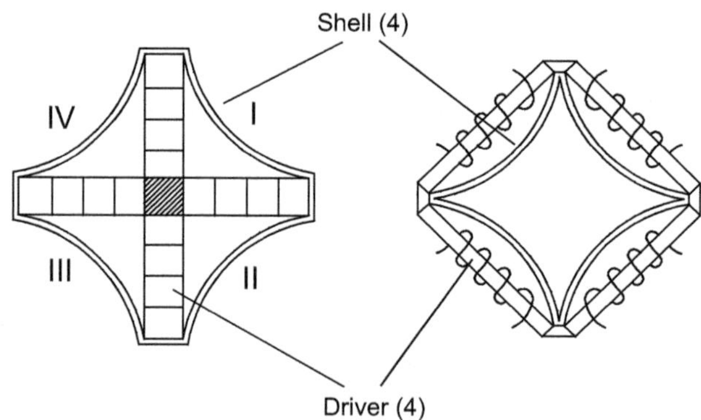

Fig. 5.48 Internally and externally driven Astroid flextensional transducer [71]

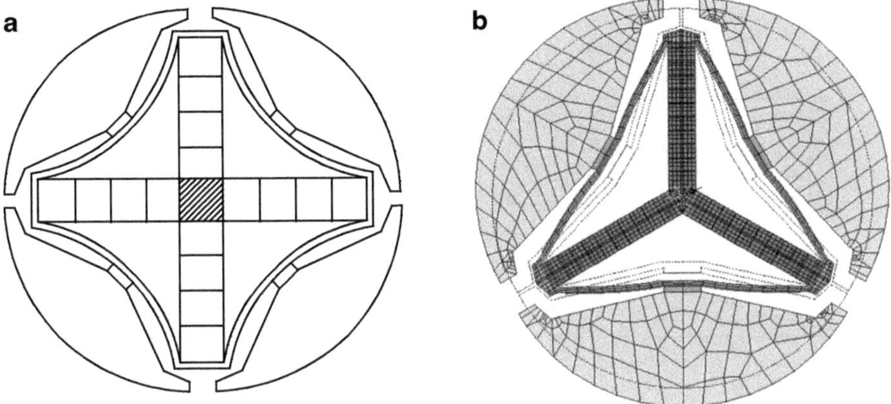

Fig. 5.49 (**a**) X-spring version of the Astroid flextensional transducer. (**b**) FEA dynamic in-air motion of the Trioid transducer

lever arm amplifier. A variation on the Astroid transducer is the Trioid transducer [80] shown in Fig. 5.49b, illustrating the (FEA exaggerated) dynamic motion with three pistons, three piezoelectric stacks in an expansion mode along with the high strength leveraging shell between the piston and stacks. This version produces greater magnification and resonates at a lower frequency as a result of the longer length of the three leveraging arms between the ends of the stacks.

Transducer designs like these provide a means for low Q_m wideband performance, high output, low frequency operation from cylindrical shaped implementations of limited diameter. The low resonance along with low Q_m is due, in part to the magnified piston mass and radiation load impedance onto the piezoelectric stacks. For an output displacement magnification factor of 3, the load magnification onto the piezoelectric stack is nine times greater ($3^2 = 9$).

Two other X-spring versions of Class IV and Class VI (ring shell) flextensional transducers are shown in Fig. 5.50a and b. The magnified motion in the direction of a may be understood by noting that the lever arm of length, H, in Fig. 5.50a is given by $H^2 = a^2 + b^2$, where b is the half length of the piezoelectric driver. The derivative yields $2H\mathrm{d}H = 2a\mathrm{d}a + 2b\mathrm{d}b$, where $\mathrm{d}a$ is the change in length of a, and $\mathrm{d}b$ is the change in length of the piezoelectric stack. Under ideal conditions the lever arm pivots at its intersections and is stiff enough so that $\mathrm{d}H$ is negligible yielding the magnification factor

$$M_f = \mathrm{d}a/\mathrm{d}b = -b/a.$$

The negative sign shows that an extension of b causes a contraction of a in the convex case of Fig. 5.50a. The concave case of Fig. 5.50b displays an in-phase positive sign.

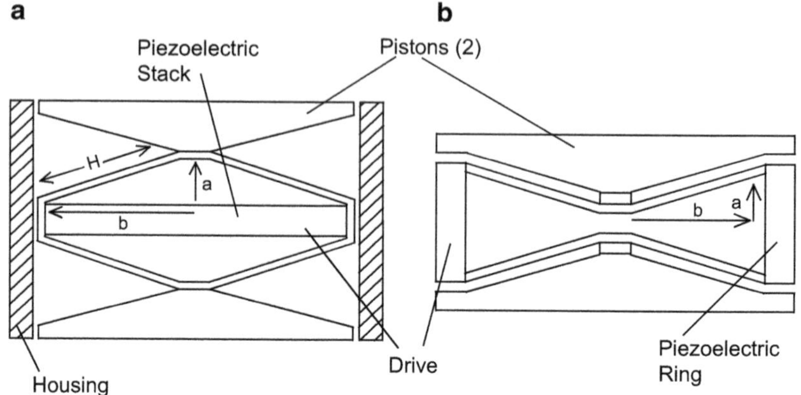

Fig. 5.50 X-spring version of the Class IV (**a**) and Class VI (**b**) flextensional transducer [72]

The magnification factor causes the piston displacement to be M_f times greater than the stack displacement or equivalently, the stack velocity is $1/M_f$ times the piston velocity. In addition to the displacement transfer, the forces on the piston are transferred to the driver stack through the lever arm and are magnified by the lever arm action, resulting in a force M_f times greater on the drive stack. Since the stack velocity is $1/M_f$ times the piston velocity, this increase in force and reduction in velocity results in an effective radiation load magnification of M_f^2 on the drive stack, yielding a better match to the medium and a lower Q_m. In practice values of $M_f = 3$ are readily obtained yielding a ninefold increase in the effective loading on the active driver. However, since the lever arm exhibits bending as well as extensional compliance there is a reduction in the effective coupling coefficient.

The equivalent circuit of Fig. 5.39 serves as a model for the X-spring transducer where M_s is the mass of the piston (plus the dynamic mass of the lever arms), C_s is the compliance of the lever arm, and the magnification transformer ratio is b/a. The X-spring is capable of greater output than a conventional flextensional because the piston is mounted at the point of maximum motion. A single piston version of the X-spring [81] is illustrated in Fig. 5.51a where the piezoelectric ring drives concave and convex shells, connected by a stiff rod yielding amplified motion at the point of contact of the piston.

As the ring moves outward in the radial, r, direction the piston of mass, M_s, moves upward in the z direction relative to the $-z$ motion of the ring of mass M. Thus the ring moves not only in the radial direction but also in the axial direction and, therefore, also serves as a reaction mass. This transducer can be modeled with the circuit of Fig. 5.39 with the addition of the ring mass in parallel with the compliance C_s of the lever arms where u_s is the velocity of the piston in the axial direction and u_d is the velocity of the piezoelectric ring in the radial direction. The X-spring may also be used as an actuator [72].

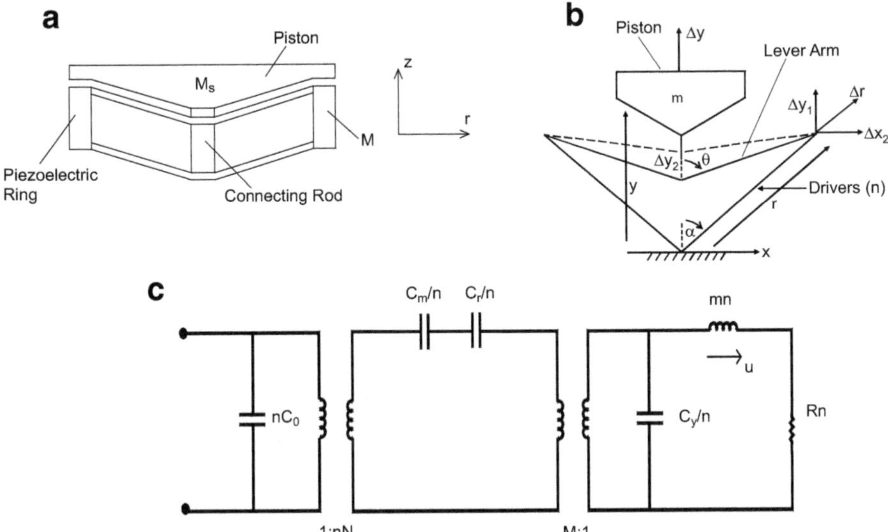

Fig. 5.51 (**a**) Single piston X-spring transducer. (**b**) Mechanical model of a symmetry section of an Astroid type Transducer. (**c**) Equivalent circuit of Astroid ($n = 4$) type transducers

5.5.5 Lumped Mode Equivalent Circuit

A simple lumped mode equivalent circuit for the X-spring, Astroid, Trioid, and Octoid (see Sect. 5.7.2) flextensional type transducers may be developed from the symmetrical section of the mechanical model shown in Fig. 5.51b. In this model the symmetry planes are through the mid-section of the two piezoelectric drivers. The lever arms, typically high strength steel, are connected between the piston and the drivers and provide magnification of the displacement from the piezoelectric drivers.

The number of symmetry sections, n, is related to the angle $\alpha = 180/n$ of the piezoelectric driving stave. For example, for the Astroid $n = 4$ and $\alpha = 45°$ and the subtended angle, 2α, of the two piezoelectric drivers of Fig. 5.51b is 90°. (For the X-spring $n = 2$, $2\alpha = 180°$, the Trioid $n = 3$, $2\alpha = 120°$, and the octoid $n = 4$, $2\alpha = 45°$.) The angle θ in Fig. 5.51b determines the partial magnification factor $M_f = \tan\theta$ given by the ratio b/a as identified in Fig. 5.50 for the lever arms. As the piezoelectric drivers expand with a displacement Δr, both of the components Δx_2 and Δy_1 contribute to the displacement, $\Delta y = \Delta y_1 + \Delta y_2$, of the piston of mass m where the magnified displacement $\Delta y_2 = \Delta x_2 \tan\,\theta$ and direct motion $\Delta y_1 = \Delta r \cos\alpha$. The total magnified motion of the piston is then

$$M = \Delta y/\Delta r = \cos\alpha + \sin\alpha\ \tan\theta$$

The lever arm of stiffness K adds a stiffness $K_r = K\cos\theta$ to the drive stack and a compliance between the piston and the lever arms of stiffness $K_y = K\cos(\theta\text{–}\alpha)$. An equivalent circuit can be developed with m the mass of the piston and water radiation mass, R the radiation resistance and mechanical loss resistance, compliances $C_r = 1/K_r$ and $C_y = 1/K_y$ along with the piezoelectric stack short circuit compliance, C_m, and the clamped capacitance, C_0. Finally with N the electromechanical turns ratio we get for each stack, the equivalent circuit of Fig. 5.51c where n is the number of symmetry sections shown in Fig. 5.51b.

As may be seen in the equivalent circuit of Fig. 5.51c, the series lever arm compliance component, C_r, stiffens up the drive system while the other shunt lever arm component, C_y, adds a compliance between the drive and the mass. The series compliance, C_r, raises the resonance while the shunt compliance, C_y, lowers the resonance and both components lower the coupling coefficient (see Sect. 4.4.2). The magnification factor is represented as a transformer with step up ratio of $1/M$. The transform increases the output velocity, u, by the factor M (with a typical factor of 3) and increases the impedance load on the drive stacks by a factor M^2 (typically 9) magnifying both the mass and resistive load on the drive yielding a low resonant frequency and low mechanical Q. Because the lever arms can flex and, moreover, may not act as ideal hinges at the junction between the lever arm and drive stacks, this simple equivalent circuit should be considered as an initial analytical model to be followed by an FEA numerical model.

5.6 Flexural Transducers

Except for Sect. 5.5.1 where we discussed the directional flextensional transducer, we have considered only piezoelectric or magnetostrictive drive systems where the motion is extensional as shown in Fig. 5.52a. In this section we will discuss flexural transducers which operate in in-extensional bending modes where the neutral plane length does not change as the driver bends as shown in Fig. 5.52b.

As the bar bends the part above the neutral plane expands while the part below the neutral plane contracts leading to no net extension. Structures are generally more compliant in bending than in tension which leads to lower resonance frequencies for a given size. The fundamental longitudinal extensional-resonance

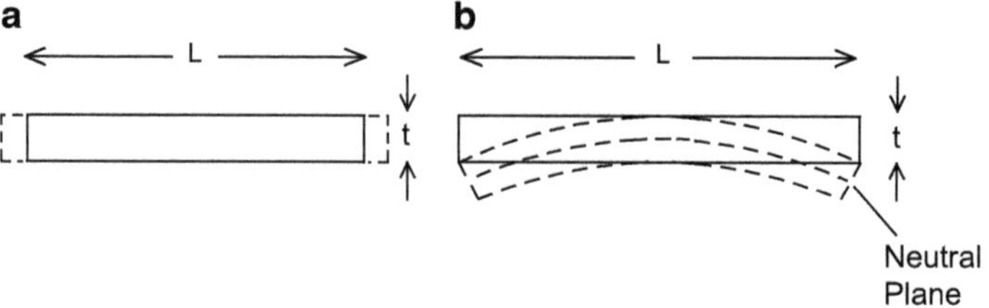

Fig. 5.52 Extensional (**a**) and inextensional (**b**) modes of vibration

frequency, f_r, and the bending inextensional-resonance frequency, f_i, for the free-free bars of Fig. 5.52 are

$$f_r = c/2L \text{ and } f_i \approx tc/L^2 \text{ leading to } f_i \approx 2f_r t/L,$$

where c is the bar sound speed and t is the thickness. Both resonance frequencies are approximately the same for a thick bar with $t = L/2$, but for a thin bar with $t = L/20$, $f_i \approx f_r/10$, and the flexural resonance frequency is a decade below the length mode extensional resonance frequency. Thus bender mode transducers are well suited to those low frequency applications where large transducers would be impractical. The excitation of bending modes requires a reversal in the drive system making one portion experience extension while the other experiences contraction about the neutral plane. In this section we will discuss bender bar and disc transducers as well as the slotted cylinder (bender) transducer and a bender mode drive X-spring transducer.

As mentioned at the beginning of Sect. 5.5 the analysis and modeling of transducers based on flexural modes is more complicated than those based on longitudinal modes. Although the bender bar and bender disc transducers have relatively simple geometries, which makes analytical modeling more feasible, only the results will be given here with the details to be found in the references.

5.6.1 Bender Bar Transducer

Figure 5.52b illustrates a bar vibrating in its fundamental mode for free-free boundary conditions with fundamental resonance frequency $f_1 = 1.028\, tc/L^2$ and overtones $f_2 = 2.756 f_1$ and $f_3 = 5.404 f_1$. It turns out that the same resonance frequencies are obtained if the bar is rigidly clamped at its ends [82]. On the other hand, if the bar is simply supported at its ends, the fundamental resonance frequency is considerably lower, by roughly a factor of 2, and is given by $f_1 = 0.453\, tc/L^2$ with overtones at $f_2 = 4f_1$ and $f_3 = 9f_1$. The first few modes of vibration [83] are illustrated in Fig. 5.53.

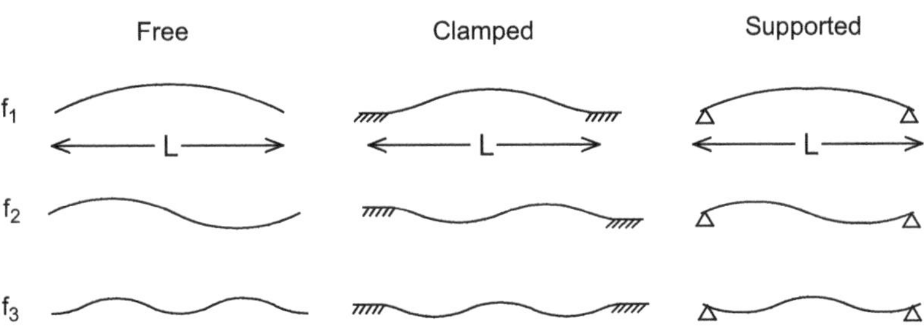

Fig. 5.53 Free, clamped, and simply supported flexural modes of vibration

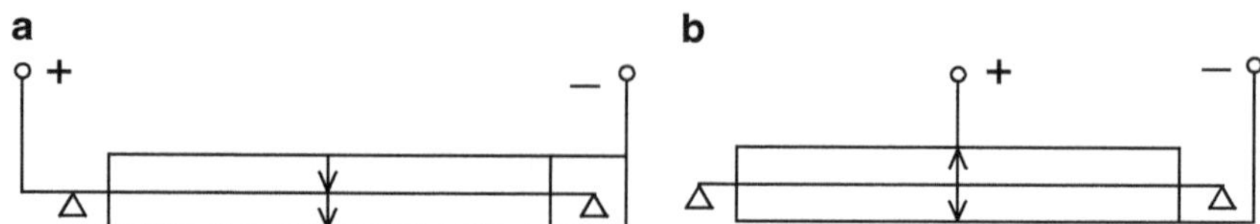

Fig. 5.54 Parallel (**a**) and series (**b**) wired bilaminar piezoelectric bars

The clamped resonance frequency is higher because the zero slope boundary condition at the rigid clamps effectively shortens the active vibration length. Because of the reduced active length, the effective coupling coefficient is less for the clamped case than it is for the simply supported case, which usually makes the latter more suitable for transducer applications. For the same resonance frequency a simply supported bender bar can be made approximately twice as thick as a clamped bender bar allowing greater strength under hydrostatic pressure and greater power capability. Woollett [84] has analyzed the bender bar transducer and has developed a number of useful models. In many cases the fundamental mode equivalent circuit representations may be reduced to the simple Van Dyke form.

Two common configurations for 31 mode excitation of the bender bar are illustrated in Fig. 5.54a and b, which shows the simple supports at the optimum position, i.e., at the nodal plane. In case (a) two identical and like oriented piezoelectric 31 mode bars are cemented together with an electrode contact between them and wired in parallel while in case (b) the polarization directions are reversed and the two are wired in series. In both cases a positive connection is made to the tip of the polarization direction arrow in the top piece and a negative connection is made to the tip of the arrow in the bottom piece. As the top piece expands laterally the bottom piece contracts laterally, causing upward bending with a bending reversal on the next half cycle. The normal velocity distribution for the simple support case of Fig. 5.53, with x measured in the lateral direction from the midpoint between the two supports, is

$$u(x) = u(0)\cos\ \pi x/L,$$

leading to an rms average velocity of 0.707 $u(0)$, where $u(0)$ is the peak velocity at the midpoint of the bar at $x = 0$.

A major drawback with the bender mechanism is the variation of stress through the thickness, with zero stress at the neutral plane, which lowers the coupling coefficient since all the material does not operate at its peak potential. This can be mitigated to some extent by replacing part of the ceramic with inactive material such as aluminum or brass as shown in Fig. 5.55a. The central portion of the piezoelectric bender where the stress is small is replaced with inactive material, which causes an increase in the effective coupling coefficient by placing piezoelectric material where the stress is high. In Fig. 5.55b the lower piezoelectric piece is replaced with inactive material to give greater ability to withstand hydrostatic stress, although it reduces the effective coupling coefficient.

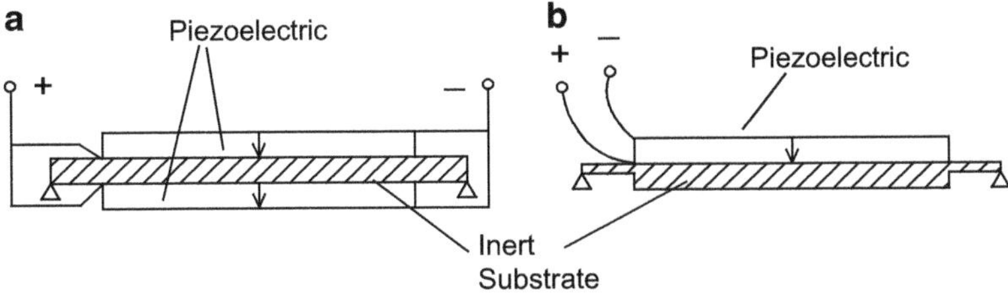

Fig. 5.55 Bilaminar and trilaminar benders

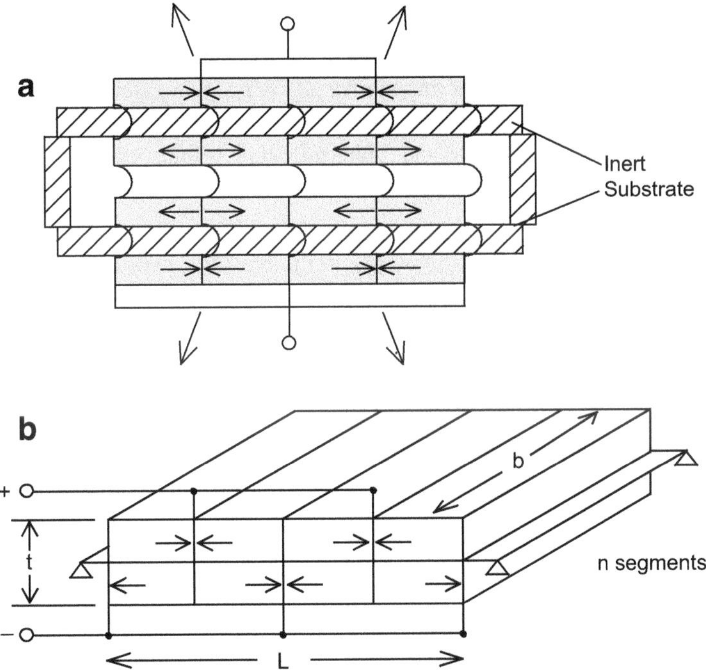

Fig. 5.56 (**a**) A dual trilaminar bender 33-mode piezoelectric bar transducer. (**b**) The 33-mode bender with length L, thickness t, width b, and n segments. *Arrows* show direction of polarization

In this design only the metal layer, with greater tensile strength than ceramic, goes into tension for static pressure on the piezoelectric layer. Although we have illustrated the mechanism using the 31 mode, the 33 mode is more suitable for high power operation. A dual 33 mode bender transducer is illustrated in Fig. 5.56a, which expands and contracts in the directions of the arrows. Arrays of these transducers are capable of producing high power at low frequencies.

Woollett [84] has developed an equivalent circuit model for the simply supported 33 mode segmented bender bar, illustrated in Fig. 5.56b with four segments. The top and bottom of each segment is reverse polarized to excite the bending motion. The transducer is ideally mounted with simple supports at the mid-plane. We may use the equivalent circuit of Fig. 5.4 if we let the circuit velocity, u_r, be Woollett's rms reference velocity which is equal to $0.707u_p$, where u_p is the peak velocity. With this reference velocity, the dynamic mass is equal to the static mass. The circuit components are

$$C_0 = n^2bt\varepsilon_{33}^S/L,\ \ C^E = 12s_{33}^E L^3/\pi^4bt^3,\ \ M = \rho tbL,\ \ N = \left(\pi/\sqrt{2}\right)d_{33}nbt^2/s_{33}^E L^2,$$

and the dynamic effective coupling coefficient, k_e, is given by

$$k_e^2/\left(1 - k_e^2\right) = \left(6/\pi^2\right) k_{33}^2/\left(1 - k_{33}^2\right).$$

The circuit of Fig. 5.4 may also be used over an extended frequency range if we replace the lumped mechanical impedance $Z_m^E = j\left[\omega m - 1/\omega C^E\right]$ by

$$Z_m^E = -j\left(2\pi bt^3/3L^2c_f\right)/[\tan\left(\omega L/2c_f\right) + \tanh(\omega L/2c_f)],$$

where the flexural wave velocity $c_f = \left(\omega^2t^2/12\rho s_{33}^E\right)^{1/4}$.

The maximum mechanical stress, T_m, in the bar for a hydrostatic pressure P is

$$T_m = (3/4)(L/t)^2P,$$

which can limit the maximum operating depth of the transducer, especially if the ratio *L*/*t* is large. Achieving a simple support mounting is also an important part of the design, and a number of hinge mounts have been considered for this purpose [84]. Another type of flexural bar mounting uses a bar free on both ends (see Fig. 5.52b), but with the end portions shielded to prevent out-of-phase radiation [85].

Although we have concentrated on piezoelectric ceramic material as the active bender material, magnetostrictive material may also be used, but with considerable difficulty, because of the mechanical loading of the driving coil which inhibits the bending motion and the need to reverse the direction of the magnetic bias field. This last problem could be solved by use of two different materials, one with positive magnetostriction, the other with negative magnetostriction, but otherwise similar properties [84]. The rare earth magnetostrictive material composed of Samarium, Dysprosium, and Iron has negative magnetostriction [86] and, used with Terfenol-D, (Terbium, Dysprosium, Iron, $Tb_{.27}\ Dy_{.73}\ Fe_{1.9}$), might be a candidate for a magnetostrictive bender bar transducer. The biasing problem in benders could also be solved if a magnetostrictive material with significant remanent bias was available. The need for bias in other types of magnetic field transducers could also be eliminated by use of materials with negative magnetostriction (see end of Sect. 5.3.1).

5.6.2 Bender Disc Transducer

The bender disc flexural transducer is excited through the planar radial mode of a disc, which has a basic coupling coefficient, k_p, a value between k_{31} and k_{33}. The fundamental resonance frequency for a clamped edge disc [83] of diameter, D, and thickness, t, is $f_r = \left[1.868c/(1-\sigma^2)^{1/2}\right]t/D^2 \approx 2ct/D^2$ (for Poisson's ratio $\sigma \approx 0.33$). The simply supported edge fundamental resonance frequency is $f_r = \left[0.932c/(1-\sigma^2)^{1/2}\right]t/D^2 \approx ct/D^2$, which is again seen to be half that of the clamped edge case. The disc has a fundamental resonance that is approximately twice as high as a bar of length equal to the disc diameter. Under water loading the radiation mass causes a reduction in the resonance frequency which may be estimated from the formula [87]

$$f_w \approx f_r/[1 + 0.75(a/t)(\rho_0/\rho)]^{1/2},$$

where a and t are the radius and thickness of the disc and ρ and ρ_0 are the density of the disc and water medium, respectively. It can be seen that the resonance frequency of a thin low density disc can be significantly reduced by water mass loading. A simplified equivalent circuit may be obtained from the mechanical compliance of the disc and the dynamic mass of a simply supported disc which may be written as

$$C_m^E = \left(2s_{11}^E/3\pi\right)\left(1-\sigma^2\right)a^2/t^3 \quad \text{and} \quad M = 2\pi a^2 t\rho/3.$$

The resulting resonance, based on these lumped parameters, is $f_r = (1/2\pi)\left(C_m^E\rho\right)^{-1/2} = \left[0.955c/(1-\sigma^2)^{1/2}\right]t/D^2$ and is within 2.5 % of the in-air resonance value given above. The clamped capacity $C_0 = C_f\left(1-k_e^2\right)$ and the electromechanical turns ratio $N = k_e\left(C_f/C_m^E\right)^{1/2}$ may be determined from the calculated free capacity and the effective coupling coefficient, k_e. Woollett [87] has shown that $k_e \approx .75k_p$ for a bilaminar disc.

Although the disc does not provide a resonance frequency as low as a bar, it is a more commonly used transducer because of its high planar coupling coefficient compared to a 31 mode bar and also because of its simplicity compared to a 33 mode bar. The flexural disc transducer is limited to about 7 in. diameter, the largest piezoelectric disc that can be fabricated. Woollett [87] has analyzed the bender disc transducer and has developed a detailed model and equivalent circuit. He has also shown that a trilaminar disc transducer has maximum coupling when the outer piezoelectric discs and the inactive brass layer have equal thickness, in which case the coupling coefficient is about 9 % greater than an all piezoelectric ceramic bender disc. A simply supported trilaminar design and dual bilaminar design are illustrated in Fig. 5.57a and b.

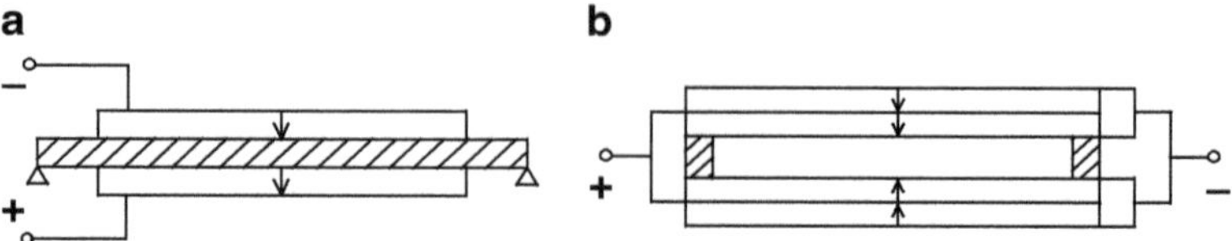

Fig. 5.57 Trilaminar (**a**) and dual bilaminar (**b**) bender disc designs

The inner support ring of the dual bilaminar is designed to be stiff in the axial direction and compliant in the radial direction to approximate a simple support condition [87]. Various methods have been considered for approximating an ideal simple support condition around the edge of the disc, and in some designs the measured resonance frequency occurs between the theoretical rigid and simple support resonance frequencies.

Woollett has also developed a number of useful design formulas for estimating the performance of the double bilaminar bender disc design of Fig. 5.57b with four PZT-4 (Type I) discs of radius a and a two-disc thickness of t. In this simply supported case the in-air resonance, f_r, and antiresonance, f_a, frequencies are

$$f_r = 705\,t/a^2 \quad \text{and} \quad f_a = 771\,t/a^2,$$

with an effective coupling coefficient of $k_e = 0.41$. The ratio of the resonant frequency in air and in water, f_{rw}, and the mechanical Q_m are

$$f_r/f_{rw} = (1 + 0.10a/t)^{1/2}, \;\; Q_m = 7.0(f_r/f_{rw})^3 \eta_{ma},$$

where η_{ma} is the mechanoacoustic efficiency. He also shows that the electromechanical turns ratio is $N = 106a$ and the hydrophone sensitivity below resonance is $M = 0.01a^2/h$ (Vm^2/N). The above results are for a diameter, D, small compared to the wavelength of sound in the medium and for a thickness, t, considerably less than the radius, a. This case is usually met in practice since, typically, $t \approx a/10$ yielding a low resonance frequency for a small size. In this case the in-air frequency constant $f_r D = 0.141$ kHz m which is approximately one-tenth the frequency constant of an extensional 33 mode bar (see Sect. 13.6).

A major drawback to benders is the low tensile stress limit in the piezoelectric material under hydrostatic pressure. The induced radial stress, T_s and T_c, for simple support and clamped edge conditions, respectively, in a disc of radius a and thickness t, for a hydrostatic pressure P are approximately

$$T_s \approx 1.25(a/t)^2 P \quad \text{and} \quad T_c \approx (a/t)^2 P.$$

These formulas set the limit on the thickness-to-radius ratio, t/a, and consequently limit the whole transducer design, particularly if a 2000 psi tensile strength limit of piezoelectric ceramic is considered. The high pressure design illustrated in Fig. 5.58

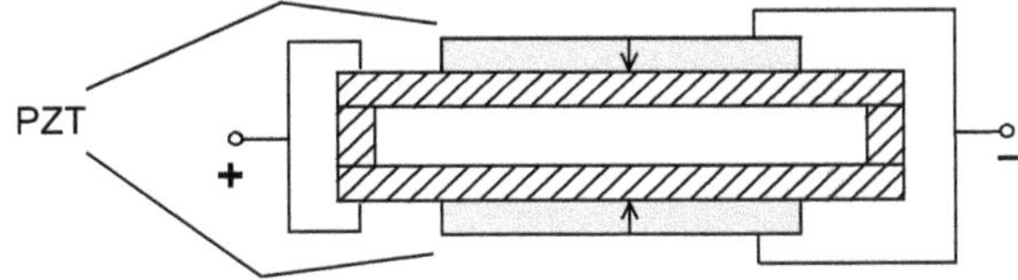

Fig. 5.58 High pressure dual bender design with inactive metal inner discs

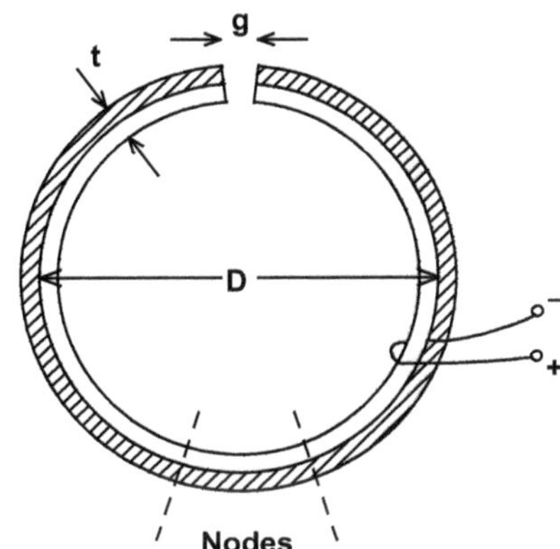

Fig. 5.59 A 31-mode bending slotted cylinder

replaces the inner piezoelectric ceramic layers with a metal disc substrate which can withstand 25 times the tensile strength of piezoelectric ceramic.

However, because of the inactive substrate the effective coupling coefficient is reduced by approximately 30 %. An alternative bender disc transducer design avoids the piezoelectric tensile limit problem by driving metal bender discs (or bars) by piezoelectric ceramic stacks located in the vicinity of the simply support peripheral region of the discs [88, 89].

5.6.3 Slotted Cylinder Transducer

The slotted cylinder transducer, illustrated in Fig. 5.59, is excited into its bending mode through the action of the inner piezoelectric cylinder on the outer metal substrate.

It is an original invention of W. T. Harris [90] that had been dormant until H. Kompanak [91] made improvements and began using these transducers in oil well applications. They have now found other applications in underwater sound because of their compact shape and low frequency performance. In one possible fabrication method, a slotted metal aluminum tube, with gap width, *g*, is slipped over and compression cemented to a 31 mode piezoelectric cylinder, which is then slotted as illustrated in Fig. 5.59. The gap is usually small and the unit is typically capped (with isolation) on its ends, air filled and rubber booted for underwater

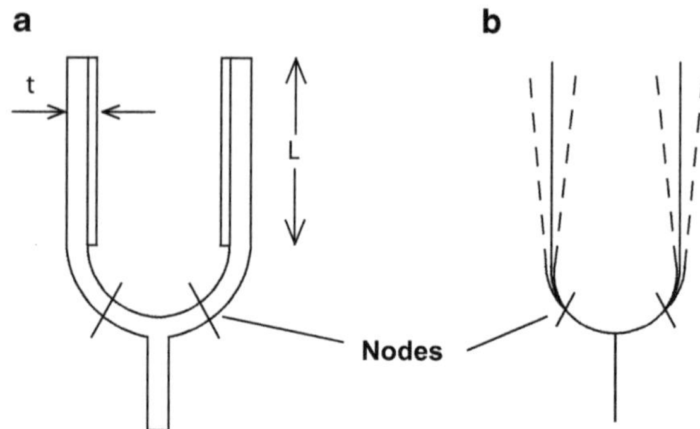

Fig. 5.60 Piezoelectric tuning fork (**a**) and fundamental mode of vibration (**b**)

operation. Metal is used on the outside to make the piezoelectric ceramic experience compression, rather than tension, under hydrostatic pressure.

The transducer operates in a bending mode in a manner analogous to piezoelectric excitation of a tuning fork, illustrated in Fig. 5.60a and b.

As the piezoelectric material shrinks, the tines of the tuning fork and the sides of the cylinder shrink, causing an inward bending motion followed by outward motion as the piezoelectric material expands. The vibration is similar to cantilevers clamped on one end and free on the other end. The fundamental resonance frequency of the tuning fork cantilever model for thickness t, length L, and bar speed c is $f_1 = 0.1615ct/L^2$ with overtones $f_2 = 6.267f_1$ and $f_3 = 17.55f_1$. If we now relate this to the slotted cylinder of Fig. 5.59 and assume the distance between the nodes is small and approximately equal to the gap size we get the slotted cylinder fundamental resonance frequency

$$f_{\mathrm{i}} \approx 0.0655(1 + 4g/\pi D)ct/D^2 \approx 0.0655ct/D^2.$$

If we compare this with the fundamental extensional ring mode resonance $f_{\mathrm{e}} = c/\pi D$ we get, for the same diameters, $f_{\mathrm{i}} = 0.206f_{\mathrm{e}}(t/D)$. Thus, for $t = D/10$, the fundamental resonance frequency of the slotted cylinder is $f_{\mathrm{i}} \approx f_{\mathrm{e}}/50$, much lower than the resonance frequency of a complete ring of the same diameter.

An equivalent circuit model [92] has been developed based on the kinetic and potential energies of a slotted cylinder of length L, effective density ρ, effective Young's modulus Y, with the following expressions for the dynamic mass, M, and stiffness, K^E:

$$M = 5.40\rho LtD \quad \text{and} \quad K^E = 0.99\ YL(t/D)^3.$$

This model gives the resonance frequency $f_{\mathrm{r}} = 0.0682ct/D^2$, only slightly different from the tuning fork model. This one layer model may be decomposed into a bilaminar model, by a method given by Roark and Young [93], which accounts for separate layers of the piezoelectric ceramic and metal. The effective coupling coefficient, k_{e}, may be determined by the modulus substitution method [94] or the

FEA methods of Sect. 3.4.3 and used to determine the electromechanical turns ratio $N = k_e\left(C_f K^E\right)$ (see Sect. 1.41) with C_f the measured free capacity. The clamped capacity is then $C_0 = C_f\left(1 - k_e^2\right)$ to complete the lumped representation.

The radial motion of the cylinder is approximately a cosine distribution which reduces the effective source strength and the radiation loading. This can be approximately taken into account by use of an equivalent sphere model for the radiation impedance with factors α, β multiplying the resistive and reactive terms, respectively. The radiation mass loading can cause a significant reduction in the resonance frequency if the cylinder wall $t \ll D$. The radius of an equivalent sphere with the same area as the cylinder is $(\mathrm{LD})^{1/2}/2$ leading to the water mass loading $M_w \approx \beta(\pi/2)\rho_0(\mathrm{LD})^{3/2}$ with the resulting in-water resonance frequency

$$f_w = f_r/(1 + M_w/M)^{1/2} \approx f_r/\left[1 + 0.29\beta(\rho_0/\rho)(\mathrm{LD})^{1/2}/t\right]^{1/2}.$$

Using the same equivalent sphere, the in-water radiation resistance, R_w, gives the mechanical Q_m

$$Q_m = 2\pi f_w M(1 + M_w/M)/R_w \approx 16(\rho c_0 D/\alpha\rho_0 cL)\left[1 + 0.29\beta(\rho_0/\rho)(LD)^{1/2}/t\right].$$

According to these two expressions, the resonance frequency and mechanical Q_m are most effected by the water loading if the shell wall thickness and density are small. Notice also that the Q_m is reduced as the length, L, is increased and the density of the shell is decreased. Because of the velocity distribution, the effective radiating area is only about half the actual area and reasonable agreement is obtained for $\alpha \approx \beta \approx 1/2$. A Fourier transform model for a finite cylinder with rigid extensions and with a cosine velocity distribution [3] is a more accurate model for the radiation load. As with any transducer design that includes bending and other complications, a finite element model or more extensive analytical model should be implemented before fabrication is undertaken. B. S. Aronov has provided an analytical model that allows optimization of the effective coupling coefficient of the slotted ring transducer [95].

The original slotted cylinder design has been modified for 33 mode operation and greater output. A 33 mode version of the tapered slotted cylinder [90] is illustrated in Fig. 5.61 and can be seen to be framed from the geometry of a smaller circle inside and tangent to a larger circle.

With the 33 mode drive the mechanical stress is increased with peak values normally in the region opposite the gap in the vicinity of the nodes. The tapered design minimizes the stress by having the greatest thickness in this region. Because of the tapering, the dynamic mass of the tines is reduced in the region near the gap which raises the fundamental resonance frequency for the same outer diameter. This raises the radiation resistance near resonance and increases the output and mechanoacoustic efficiency, albeit at a greater size for a given resonance.

Fig. 5.61 A 33-mode tapered slotted cylinder

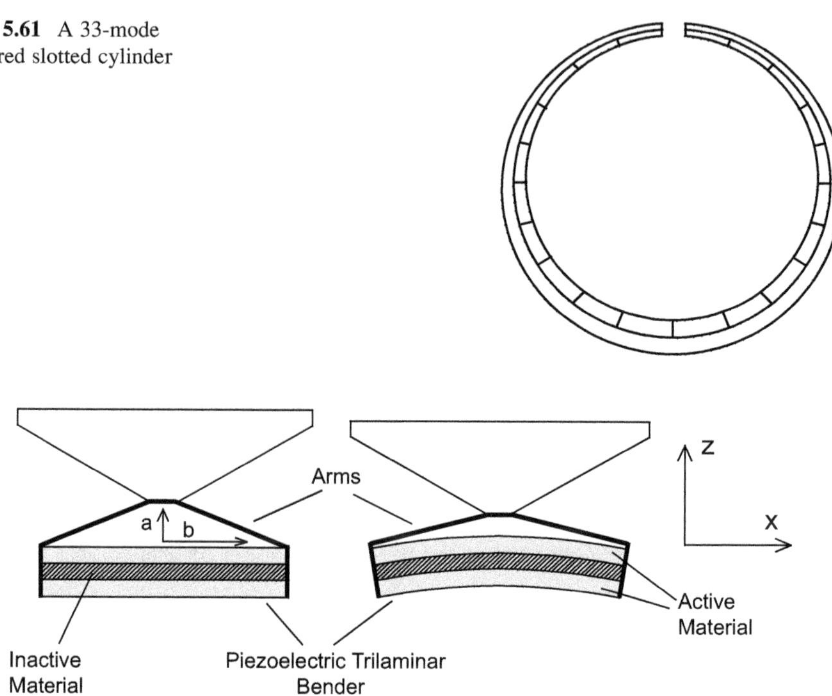

Fig. 5.62 Bender mode X-spring transducer showing bending action and the piston motion in the z direction [81]

This tapered slotted cylinder transducer design has been modeled and programmed for computation under the name TSCAT (Tapered Slotted Cylinder Acoustic Transducer) [96].

5.6.4 Bender Mode X-Spring Transducer

The bender mode X-spring transducer [81] combines a number of the principles we have discussed in Sect. 5.6. It is based on the X-spring [72] discussed in Sect. 5.5.4 but with a drive stack operating in its bending mode as diagrammed in Fig. 5.62 and illustrated in Fig. 1.17.

This transducer combines the drive action of a bender bar and the magnification of an X-spring flextensional transducer. As the piezoelectric trilaminar section bends upwards the levers are extended outward causing the connected piston mass to move in the $-z$ direction with a magnification ratio b/a. As in the case of the ring-driven single X-spring piston, illustrated in Fig. 5.51, the piezoelectric bender acts as an inertial reaction mass moving in the $+z$ direction. The resultant motion of the piston in the z direction is reduced by the inertial motion; however, this reduction is minimal if the mass of the driver is considerably greater than the mass of the piston. This transducer can operate at frequencies much lower than other flextensional or X-spring transducers because of the higher compliance of the bender bar drive. It may also be driven by a bender disc instead of a bender bar.

5.7 Modal Transducers

In the section we present three piezoelectric-based modal projectors: The modal power wheel projector uses eight high power Tonpilz transducers with a common tail mass to create steerable modal synthesized beam patterns. The octoid flextensional transducer is similar to the power wheel projector but utilizes magnified motion and achieves a lower resonant frequency. The leveraged cylindrical transducer is similar to the octoid transducer but is driven by a piezoelectric ring instead of eight piezoelectric stacks.

5.7.1 Power Wheel Transducer

The multimodal rings (or cylinders) described in Sects. 5.2.6 and 6.5.5 obtain their directionality from the sum of the first three distributed omni, dipole, and quadrupole modes of vibration of a piezoelectric cylinder. These modes are simultaneously excited and added together with the proper weighting functions to obtain various beam structures, allowing 360° coverage in eight incremented beams. The increased directivity index of the transducer allows the use of less power and less heating of the transducer. This section adds the modal power projector [97] to the piezoelectric cylinder designs by considering the case of acoustic mode generation through the use of a discrete set of eight high power Tonpilz transducers coupled through a common tail mass. The common tail mass is not necessary for the modal projector, but it does provide a desirable lower Q, lower resonance, and greater output.

The concept is to generate beam patterns by exciting acoustical modal beam patterns in the surrounding fluid simultaneously with specific weighting functions to obtain a desired beam pattern. Many projector transducer configurations can generate dipole modes in addition to the basic omnidirectional monopole mode. And a smaller number can also generate a quadrupole mode which can improve the directionality without serious reduction in the efficiency, when combined together with the first two modes. If we limit the beam pattern to the first three modes, as in Sect. 5.2.6, the three mode beam pattern function may be rewritten as

$$p(\theta)/p(0) = [1 + A_1 \cos\theta + A_2 \cos 2\theta]/[1 + A_1 + A_2] \tag{5.48}$$

where the weighting function for the dipole mode is A_1, the quadrupole mode is A_2, and the monopole mode is unity. Note that the individual beam pattern functions are

Table 5.4 Modal beam characteristics from Eq. (5.48)

Curve	A_1	A_2	BW (deg)	DI (dB)	BL (dB)	90 L (dB)
(a)	1	0	131	4.8	Null	−6
(b)	1	0.414	90	7.1	−15	−12
(c)	1.6	0.80	78	8.0	−25	−25

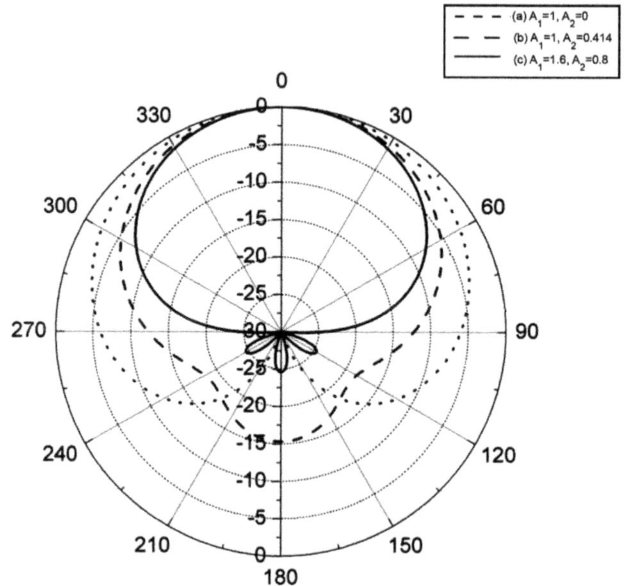

Fig. 5.63 Three modal projector beam patterns with weighting factors: (**a**) $A_1 = 1$, $A_2 = 0$ (**b**) $A_1 = 1$, $A_2 = 0.414$; and (**c**) $A_1 = 1.6$, $A_2 = 0.8$

unity for the monopole mode, cos θ for the dipole mode, and cos 2θ for the quadrupole mode. The dipole mode has two lobes of alternate phase and two nulls while the quadrupole mode has four lobes with alternating phase on adjacent lobes along with four nulls.

A number of interesting and useful beam patterns (see Sect. 6.5.6) may be generated from Eq. (5.48). Some of these beam pattern characteristics are given in Table 5.4, where BW is the (3 dB down) beam width, DI is for the case of axial symmetry, BL is the level of the back lobe radiation, and $90L$ is the level at $\pm 90°$. The corresponding beam patterns are shown in Fig. 5.63.

Curve (a) is the case of the classic two-mode true cardioid, curve (b) is the tri-modal ring case used by Butler et al. [5] and curve (c) is the case used in the present modal projector discussed here. Curve (c) has a DI gain increase of more than 3 dB over the true cardioid of curve (a). The beam pattern of case (c) has been optimized so that the three back lobes, as well as the level at $\pm 90°$, are all at a level of -25 dB. In the case of a cylindrical array of rings, as in a modal projector array, the DI increase of case (c) is 6 dB compared with cylindrical omnidirectionality

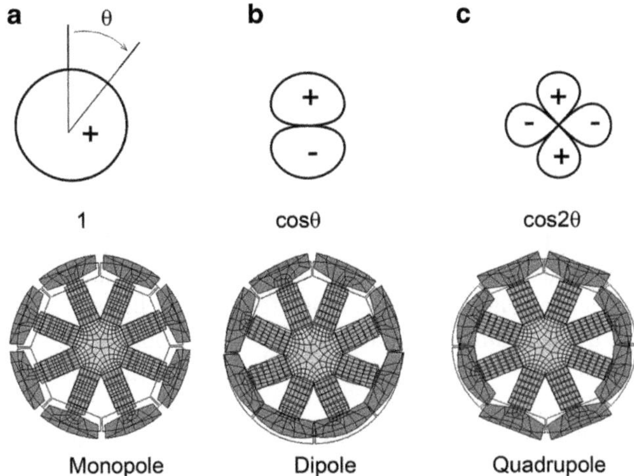

Fig. 5.64 Monopole, dipole, and quadrupole modes with corresponding beam patterns

from an extended length cylinder. This 6 dB means that the power required to achieve a given source level is down to only 25 % of the power needed if the ring were operated in the cylindrical omnidirectional mode alone.

The structure and motion of the transducer in the omnidirectional monopole mode is shown in the finite element illustration of Fig. 5.64a. Here eight elements are shown, with each element having a stack of six piezoelectric ceramic pieces, driving pistons from a common center mass and all being driven in-phase. Because of the in-phase drive, there is no motion of the center mass and there is a displacement null at the center. Consider now opposing elements of the transducer ring array. With m the piston mass, K the short circuit stiffness of the stack, and R the resistive load, the angular resonance frequency, ω_m, and mechanical Q for the monopole mode, Q_m, may be written as

$$\omega_m = (K/m)^{1/2} \quad \text{and} \quad Q_m = \omega_m m/R$$

Operation in the dipole mode is illustrated in Fig. 5.64b where the bottom four stacks are driven out of phase from the top four. The result is motion of translation in the vertical direction and there is a net motion of the center mass in opposition to the motion of the pistons and opposite elements act like Tonpilz elements each with a tail mass, M, equal to one-half the center mass. This motion ultimately yields a beam pattern null in the horizontal plane and oppositely phased lobes in the vertical direction with a beam pattern function of the form $\cos\theta$. If we consider opposite pairs in this dipole mode, with $R \ll \omega_d M$, the angular resonance frequency, ω_d, and mechanical Q for the dipole mode, Q_d, may be written as

$$\omega_d \approx (K/m)^{1/2}(1 + m/M)^{1/2} \quad \text{and} \quad Q_d \approx (\omega_d m/R)(1 + m/M)$$

And we can see that, in the case of the dipole mode, the resonance frequency is increased by $(1 + m/M)^{1/2}$ while the Q_d is increased by $(1 + m/M)$ and it also turns out that, well below resonance, the velocity is decreased by $1/(1 + m/M)$, in this ideal lumped model representation. For $M = 3$ m the frequency increase would be 15 %, the Q_d increase would be 33 %.

A finite element illustration of the quadrupole mode with its cos 2θ beam pattern function is shown in Fig. 5.64c. The excitation of the quadrupole mode is accomplished by alternating the phase of adjacent pairs. As seen, opposite elements are in-phase and, therefore, we would expect the resonance frequency of this mode to be the same as the monopole mode for an ideal lumped mode model. The classic true cardioid beam pattern Fig. 5.63a can be obtained from the first two modes while all three modes are necessary for the improved beam patterns Fig. 5.63b and c.

The excitation of the combined modes of Eq. (5.48) requires adjusting the voltage drive so that the complex on-axis acoustic pressure amplitude and phase of the dipole and quadrupole modes become identical to the amplitude and phase of the monopole mode at each frequency. Once this voltage adjustment is accomplished, the weighting coefficients may be applied to each mode and then summed to obtain the desired beam pattern. If the required voltage distribution is implemented at each frequency across the frequency band of interest, the beam pattern will be the same at all frequencies providing that the individual monopole, dipole, and quadrupole beam patterns do not change in structure. These modal transducers have been driven with PZT-8 ceramic, PMM-PT single crystal, and Terfenol-D magnetostrictive drive materials [98].

5.7.2 *Octoid Transducer*

The performance of the modal power wheel may be enhanced by the addition of surrounding leveraging shell between the eight drive stacks and the eight pistons. This high strength shell inclusion is similar to one used in the Astroid transducer discussed and shown as Fig. 5.49 in Sect. 5.5.4. And the results are similar in that the shell provides magnified motion to the pistons resulting in a wider-band, lower-resonant frequency performance. The structure and (exaggerated) dynamics of the first three modes of vibration of this octoid transducer [99] are illustrated in Fig. 5.65 with dashed lines indicating the FEA static condition.

As the eight piezoelectric stacks expand in unison, the pistons move outward with a monopole/omnidirectional magnified motion (of approximately three). The dipole mode is obtained by driving the top four stacks outward and the bottom four stacks inward. As may be seen, the top three pistons move outward, while the bottom three move inward, creating a null in the horizontal plane. The two other pistons move in a rocking motion with the top part in phase with the top pistons and the bottom part in phase with the bottom pistons yielding a dipole pattern. In the quadrupole mode the top and bottom pistons are driven outward while the left and

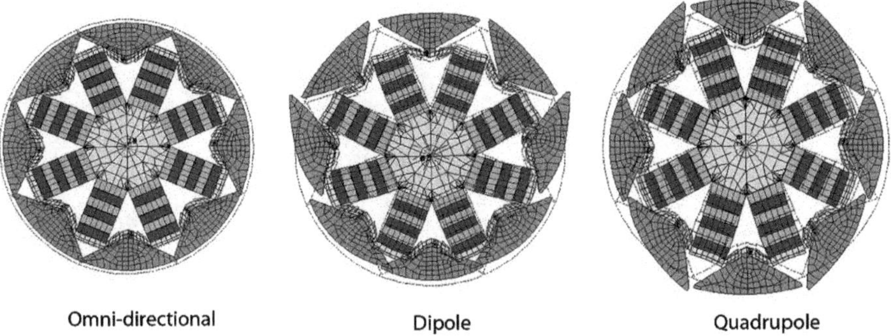

Fig. 5.65 FEA of first three modes of vibration for 6 in. octoid

right pistons are driven inward. The remaining four pistons rock in adjacent-phased favorable manner resulting in nulls in two orthogonal planes forming a quadrupole beam pattern.

With a 6 in. diameter structure the in-water resonances are, approximately: 8 kHz for the omni mode, 9.5 kHz for the quadrupole mode, and 12 kHz for the dipole mode. These resonant frequencies are approximately 40, 20, and 25 % lower than the comparable modal power projector resonance values, without a leveraging shell.

5.7.3 Leveraged Cylindrical Transducer

The leveraged cylindrical transducer is similar to the octoid transducer, of Sect. 5.7.2, in that it uses a cylindrical leveraging shell connected to eight conformal pistons except it is driven by a piezoelectric cylinder instead of eight radial stacks [100]. The piezoelectric cylinder or ring can be in the form of a 31 mode cylinder, a striped 33 mode cylinder or eight-piezoelectric-stack cylindrical-structure with wedges connecting to the leveraging shell and pistons as shown in Fig. 5.66. The lever arms, pistons and the upper portion of the wedges could be one complete assembly, EDM cut from steel stock in one automated pass or as titanium structure using an additive manufacturing electron beam melting process. In this form the transducer has interior space in the center and, for the same size, it resonates at a frequency 30 % lower than the octoid power wheel.

The modal motion of the transducer operating in the omnidirectional (monopole), dipole, and quadrupole modes is shown in Fig. 5.67. This motion is more uniform and does not exhibit rocking modes as in the case of the octoid, as shown in Fig. 5.65. The close proximity of the modes allows wider bandwidth coverage. The dynamic dipole resonance is 1.2 times the omni and the quadrupole resonance is 1.3 times the omni. The omni mode is excited by driving all eight ring staves in phase, the dipole with the four upper staves out of phase with the four lower staves and the

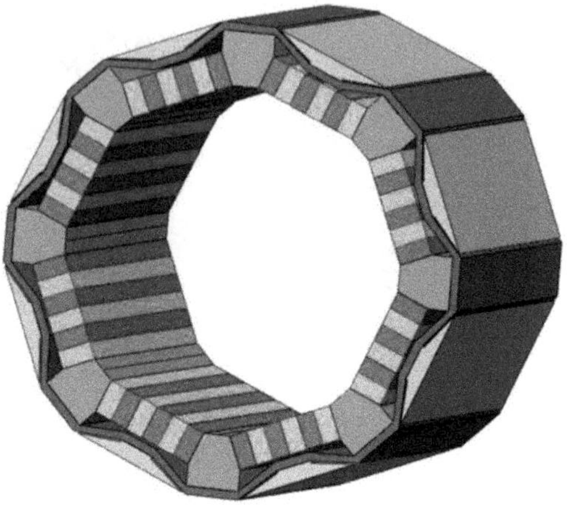

Fig. 5.66 Leveraged cylindrical transducer

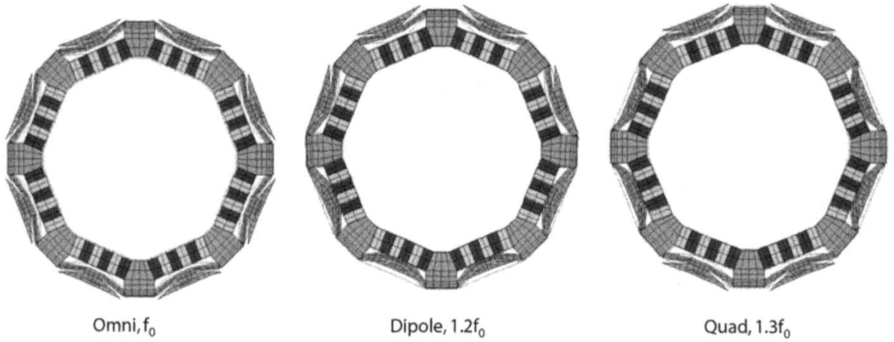

Omni, f_0
Dipole, $1.2f_0$
Quad, $1.3f_0$

Fig. 5.67 Monopole (omni), dipole, and quadrupole modes of vibration

c
b
Symmetry
Symmetry
a

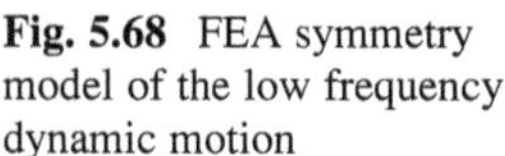

Fig. 5.68 FEA symmetry model of the low frequency dynamic motion

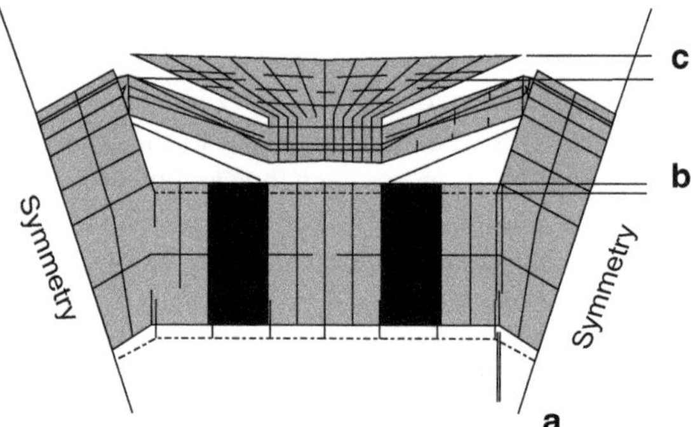

quadrupole mode with the two upper and two lower staves out of phase with the two right and the two left staves.

The dynamic modal response of the monopole mode, well below resonance, of a 1/8th symmetry model is shown in Fig. 5.68. Here the ratio of the radial displacement of the stack, b, relative to the piezoelectric activated stack displacement, a, is $b/a = 2.25$, the ratio of the piston motion, c, to the radial displacement of the stack, b, is $c/b = 2.0$ leading to a piston magnified displacement of $c/a = 4.5$. These motions are the result of the circumferential extension of the stacks and the leveraged amplification of the piston as a result of the circumferential displacement.

The transducer has the distinct advantage of a steerable transducer, allowing improved performance for a steered array of transducers [101, 102]. The single self-contained modal transducer may be combined with other modal transducer elements to form such an array. Because of its lower resonance it can be used in arrays that require half wavelength center-to-center spacing and as elements of a doubly steered array (see Sect. 7.8).

A less expensive and less powerful version of this transducer is a simple 31 mode modal ring with a steel shell and no pistons. In-air resonance of the monopole mode of a 4.4 in. mean diameter ring alone is 9.3 kHz while the resonance of the ring with a leveraging shell is 5.8 kHz, providing a significant 38 % reduction in resonance.

5.8 Low Profile Piston Transducers

Low profile transducers allow sonar arrays to be mounted on the hull of a vessel without significant reconstruction. We present here two novel low profile transducers that yield nearly the same low resonance frequency from the same small volume but from different piezoelectric drive systems. The first transducer is based on piston motion derived from dual cantilever piezoelectric drives while the second transducer achieves piston motion from four shear mode d_{36} single crystal drivers.

5.8.1 *Cantilever Mode Piston Transducer*

The cantilever mode transducer (CMX) produces a low frequency response from a short "piston type" transducer [103, 104]. It is a compliant cantilever-driven transducer which uses a translator arm (lever arm with magnification unity) that is driven by two piezoelectric cantilevers at opposite sides of the translator arm and mounted on an inertial tail mass. This transducer, coupled to a piston, yields a low frequency wideband response from low-profile housings. Under closely packed array-loaded conditions the transducer also yields a resonance lower than normally attained under array conditions, along with a desirable fixed velocity output, reducing array interaction problems at this resonance.

This transducer uses a pair of cantilever bending mode drivers attached to the radiating piston through a translator which converts the horizontal motion of the cantilevers to the vertical motion of the piston. A finite element illustration of the dynamic motion is shown in Fig. 5.69. Notice that the depth of this low profile transducer is less than the size of the piston head. The cantilever motion is

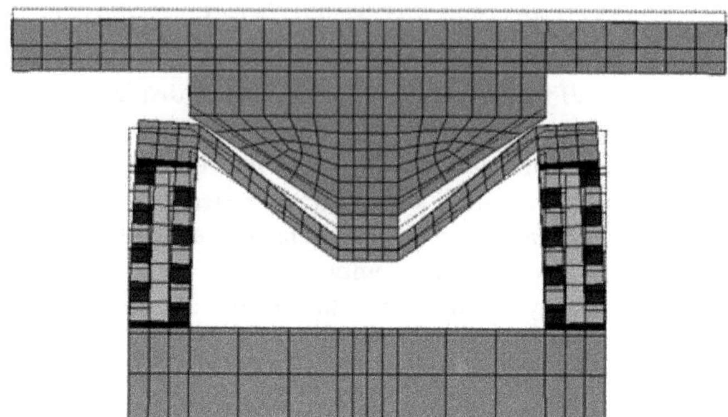

Fig. 5.69 FEA animation of piston for inward bending cantilever

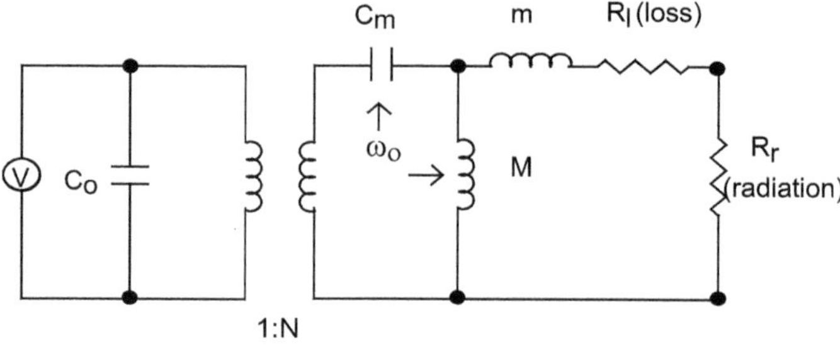

Fig. 5.70 Equivalent circuit representation of a cantilever mode transducer

excited through the action of phased reversed piezoelectric 31 mode plates on either side of a central substrate or 33 mode bender piezoelectric stacks. As seen, the tail mounted piezoelectric cantilever motion is in the horizontal direction and the piston motion is in the vertical direction as a result of the translated motion by the translator arm. As the cantilevers bend outward, the translator arm (shell) moves upward causing the piston to move uniformly at the maximum displacement of the translator arm. This combination gives maximum source level which is generated by the source strength product AU, where A is the area of the piston and U is the uniform velocity of the piston.

The low resonance frequency and flat wideband response can be understood through the simplified piezoelectric equivalent circuit shown in Fig. 5.70. This circuit illustrates the input drive voltage, V, the clamped capacity, C_0, the electromechanical turns ratio, N, the mechanical compliance, C_m, of the cantilever benders, the tail mass M, the piston head mass, m, the mechanical loss resistance, R_l, and the radiation resistance R_r under fully array-loaded conditions. The motion translator lever arms are set to a magnification ratio of unity, as the cantilever drivers provide sufficient displacement on their own.

The special low-frequency wideband response is a result of the low resonance frequency of the CMX transducer where the piston reactance, $\omega_o m$, is small compared to the resistive load, $(R_l + R_r)$, leading to $Q_m \ll 1$ resulting in a lower parallel resonance, $f_0 = \omega_o/2\pi$, between M and C_m compared to the typical Tonpilz resonance, f_r, between m and M in parallel and C_m. This parallel resonance at f_0 is not only lower in frequency than the typical value f_r, but it is of much higher mechanical impedance than the typically low impedance common series resonance at f_r. Because of its low frequency and low piston mass, m, the transducer achieves an even lower resonant frequency under full array loading. This advantageous condition is unusual for transducers, since most transducers experience an increase in resonance as the array size increases and achieve the higher in-air resonance under full array loading.

This effect can be explained by first considering the radiation resistance and reactance Chap. 10 curves of Fig. 10.19 for a piston type transducer as a function of the size parameter, ka. Here "a" is the effective radius of the piston or the effective radius of an array, the wave number "k" $= \omega/c$ where c is the sound speed in the medium and ω is the angular frequency, $2\pi f$. We see that as the piston size or array size increases the radiation resistance, R, increases and approaches the fully loaded condition $\rho c A$, where ρ is the density of the medium and A is the area of the piston or array. We also see that the radiation reactance, $X = \omega m_r$, where m_r is the radiation mass, initially increases linearly with a constant value of the radiation mass, m_r but then after $ka \approx 1.5$, decreases and approaches zero as the radiation resistance approaches the full radiation loading $\rho c A$.

Thus, at low frequencies, where ka is small, the radiation mass loading, m_r, should add to the piston mass, m, and produce a resonant frequency lower than the in-air resonance. However, as discussed, as the effective radius, a, increases, X and m_r approach zero and the resonance frequency should return to the in-air value as it does for most transducers in large arrays. However, in the case of the CMX transducer, the resonance frequency decreases as the array size increases. This is a result of another lower resonance, which comes into play because of the unusually high transducer compliance and small piston mass reactance along with the comparatively large radiation resistance loading leading to $\omega m \ll \rho c A$. This new resonance, ω_0, is controlled by the tail mass and is significantly lower than the conventional resonance frequency, ω_r.

This new resonance, ω_0, may be understood with reference to the typical piston type transducer equivalent circuit of Fig. 5.70 where N is the electromechanical turns ratio, C_0 the clamped capacitance, C_m the mechanical short circuit compliance, M is the tail mass, m is the piston mass, R_l the mechanical loss resistance, V is the input voltage and we now call R_r the radiation resistance and let the total resistance be $R_t = R_r + R_l$ and also let the total mass be $m_t = m + m_r$. Under in-air or in-water full array rho-c loading conditions, $m_r = 0$ and $m_t = m$. The normal short circuit angular resonance frequency is given by

$$\omega_r^2 = (1/C_m m_t)(1 + m_t/M) \tag{5.49}$$

while the new tail mass resonance is given by

$$\omega_0^2 = 1/C_m M \tag{5.50}$$

After collecting terms, the pressure response for a piston of area A in a rigid baffle may be shown to be

$$p(\omega) = [j\omega\rho ANV/2\pi]/[j\omega m_t(1 - \omega_r^2/\omega^2) + R_t(1 - \omega_0^2/\omega^2)] \tag{5.51}$$

which, when written in this form, displays two resonance conditions: when $\omega = \omega_r$ and when $\omega = \omega_0$ along with two coefficients ωm_t and R_t, respectively. It is worth restating that in this discussion ωm is small as a result of the low frequency operation and because cantilever compliance, C_m, is high, only a small value of m is needed to obtain a low resonance frequency.

Consider now the Eq. (5.51) coefficients ωm_t and R_t of the resonant conditions which occurs for $\omega = \omega_r$ and $\omega = \omega_0$, respectively. For a small piston or array with $R_t \ll \omega m_t$, the first term in the denominator of Eq. (5.51), which contains ω_r, controls the response. However, for a large array where $R_t \gg \omega m_t \approx \omega m$ the second term in the denominator, which contains, ω_0, now controls the response. These mathematical statements can be visualized from the equivalent circuit of Fig. 5.70. For a small array where R_t is small compared to the mass reactance ωm_t the mass is essentially in parallel with M resulting in Eq. (5.49) for $\omega_{r,}$ while for a large array where R_t is large compared to ωm, the mass m is no longer in parallel with M and we get the lower resonance given by Eq. (5.50) for ω_0. The two resonances may be shown to be related under full array loading as

$$\omega_0^2 = \omega_r^2/(1 + M/m) \tag{5.52}$$

and it is seen that $\omega_0 < \omega_r$. Note that for $M = m$ that $\omega_0 = 0.707\omega_r$, for $M = 2\,m$ we get $\omega_0 = 0.577\omega_r$, and that for $M = 3\,m$ we get $\omega_0 = 0.5\omega_r$.

The voltage source of Fig. 5.70 is normally a low impedance source and under these conditions, with $V = 0$, the circuit of one transducer in an array may be represented as shown in Fig. 5.71, where R_0 represents the mechanical losses in the compliance C_m and may be written as $R_0 = Q_0/\omega C_m$. The Q_0 values can range from 100 up to 1000 (for piezoelectric materials) and consequently, R_0 is typically a large quantity. On the other hand, R_l represents the mounting loss and, by design, is

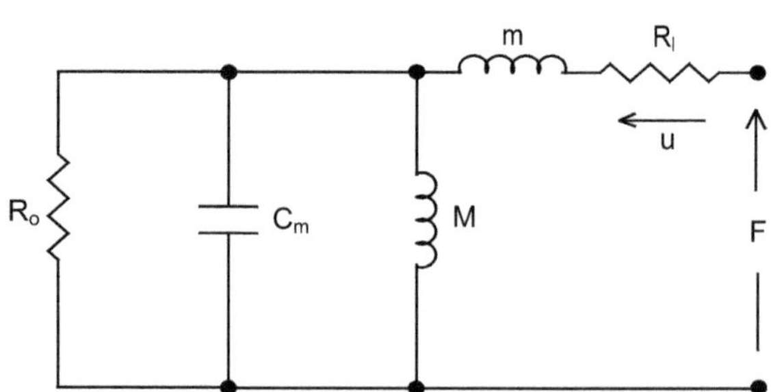

Fig. 5.71 Mechanical impedance of cantilever mode transducer

typically small compared to the radiation resistance R_r. The force, F, is the force on this transducer due to surrounding transducers in the array (minus the self-radiation impedance). The input mechanical impedance may then be written as

$$Z(\omega) = R_l + j\omega m + 1/[j\omega C_m + 1/j\omega M + 1/R_0] \quad (5.53)$$

At the new lower resonance, ω_0, and with R_0 very large

$$Z(\omega_0) = R_l + j\omega_0 m + R_0 \approx R_0 \quad (5.54)$$

and the velocity $u(\omega_0) \approx F/R_0$. At the normal resonance, ω_r, and with R_0 very large

$$Z(\omega_r) \approx R_l \quad (5.55)$$

and the velocity $u(\omega_r) \approx F/R_l$.

Consequently since $R_0 \gg R_l$, the velocity of $u(\omega_0) \ll u(\omega_r)$ and the piston impedance is high and hardly moves at this new lower resonance acting like a rigid piston. On the other hand, the piston moves significantly at ω_r and acts like a compliant surface causing large motions and possibly high Q_m and high stresses in the transducer. Accordingly, one might expect the interaction issues at ω_0 to be considerably less than at the typical Tonpilz resonance ω_r.

Finite element calculated TVR response results are shown in Fig. 5.72 for various arrays of closely packed PZT-8 driven CMX transducers with 3×3 in.

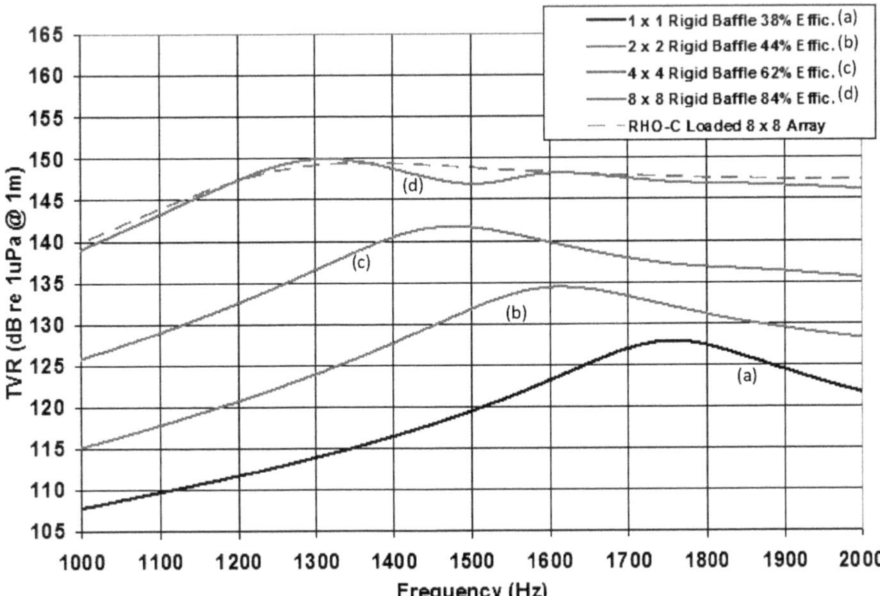

Fig. 5.72 FEA calculated TVR array results for various array configurations of the cantilever mode transducer

pistons, with an in-air resonance of 2.10 kHz. The five water-loaded cases range from a 1×1 single element array, with resonance at 1.75 kHz, to an 8×8 array of 64 elements, with a resonance of 1.30 kHz along with an ideal ρc fully array-loaded case shown as the dashed line. These results demonstrate that there is a decrease in the resonance frequency as the array size increases. They also show that the 8×8 array performance is very close to the performance of an ideally loaded array suggesting that there are no significant array interaction issues. Here the electro-mechanical efficiency is 84 % and more than twice the single element value of 38 %, where the radiation resistance is considerably lower.

5.8.2 Shear Mode Piston Transducer

Piezoelectric single crystal material with a high shear mode d_{36} (2200 pm/V) value, high coupling k_{36} values greater than 0.80, and high elastic compliance 190×10^{-12} m^2/N are available. This crystal cut [105, 106] provides shear mode operation with the electric field in the same direction as the polarizing field allowing drive fields exceeding 10 V/mil, providing this material as a candidate for a low-frequency, low-profile shear mode projector. Previously it was necessary to electrically drive piezoelectric material in a direction other than the polarization direction, in order to excite the shear mode [107]. And this had limited the shear mode application to accelerometers or hydrophones where high drive fields are not needed.

A shear mode d_{36} tonpilz transducer, with a three inch diameter piston and depth of 2.5 in. along with an in-water resonant frequency of approximately 2 kHz, has been developed [108, 109]. Although the operation of this shear mode transducer is quite different, the size and resonant frequency are quite similar to the cantilever mode piston transducer of Sect. 5.8.1. A cross section of the four bar shear mode element design is shown in the FEA model of Fig. 5.73. The d_{36} bars are attached to the center stem of the piston and to the tail mass through four pieces attached to the tail mass. In operation the shear mode bars cause the piston to move relative to the tail mass and achieve a low resonant frequency through the high compliance of

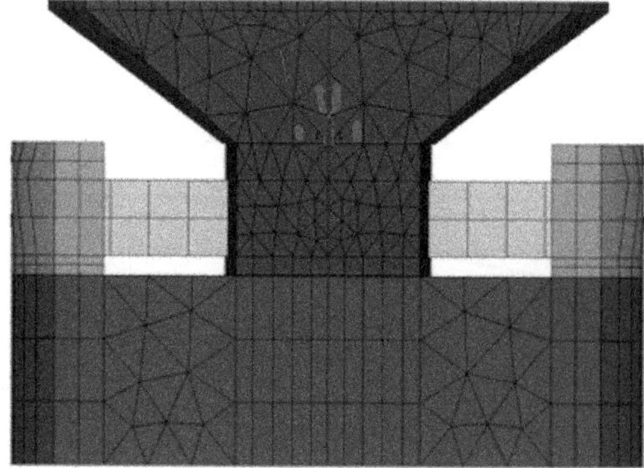

Fig. 5.73 Cross-sectional view of an FEA model of the shear mode Tonpilz transducer (Note that although the stem appears to be in contact with the tail mass, it is not in contact in the FEA model)

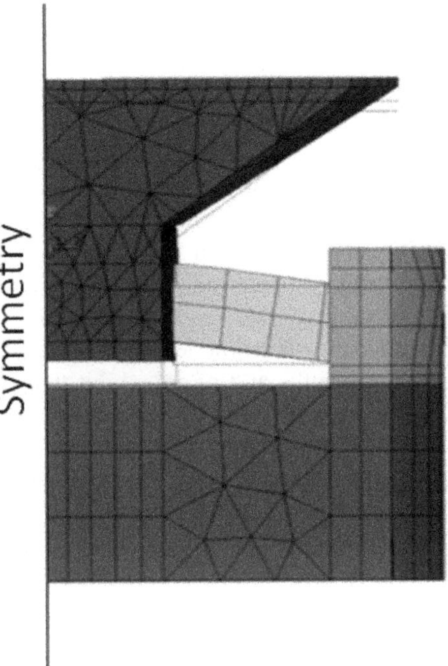

Fig. 5.74 In-air exaggerated dynamic motion of a one-eighth symmetry model of the shear mode Tonpilz transducer

the shear mode operation. The FEA in-air 2.7 kHz exaggerated dynamic motion of the fundamental shear mode of the piston transducer is shown in Fig. 5.74. In this design the tail mass is considerably greater than the mass of the piston and stem resulting in greater motion of the piston than the tail, which is typical of most tonpilz designs. Large displacement values can be obtained through the high value of the d_{36} parameter.

If the transducer were to be isolation-mounted from the tail to the housing, the hydrostatic pressure on the piston would compress the piston and stem causing great hydrostatic induced shear stress on the single crystal elements; thus, limiting its operation to only modest depths. This can be avoided by using compressed air within the housing, or by isolation mounting from the head to the housing. In the latter case, the resonance frequency of the mounting system must be significantly below the transducer resonance or operating band. Compliant oil backing techniques maybe possible but this could lead to large housing requirements which could negate the low profile size.

The performance and size of the shear mode low profile transducer is similar to the low profile cantilever mode piston transducer of Sect. 5.8.1. Both have low resonant frequencies with low mass piston heads as a result of the very compliant piezoelectric drive systems. Because of this, it may be possible that a shear mode array of transducers could also exhibit a reduction in frequency as the array size increases rather than exhibit the conventional increase, as would normally be expected.

5.9 Summary

In this chapter we considered spherical, cylindrical, ring, piston (Tonpilz), transmission line, composite, flextensional, flexural, modal, and low profile transducers. Piezoelectric ceramics are currently the most commonly used for underwater sound, because of the many geometrical shapes in which they can be fabricated, their excellent electromechanical properties, their low electrical losses, and their ability to generate high forces and high source levels. After a brief discussion of the principles of operation, a detailed presentation was given for the commonly used 31 mode ring followed by a discussion of the 33 mode ring, the magnetostrictive ring and the planar mode spherical transducer. Free-flooded transducers can operate at great depths and often exhibit two resonances providing a wide bandwidth of operation as presented by free-flooded rings and multiport transducers. Multimodal leveraged transducer cylinders were shown to be a means for generating various steerable-directional beam patterns with the possibility of improving array response through steering the element as well as the array.

The Tonpilz piston transducer is the most commonly used transducer in arrays and Sect. 5.3.1 is devoted to this transducer, performance and construction along with a simplified lumped equivalent circuit model used to represent it. A novel hybrid transducer with combined magnetostriction and piezoelectric drives was shown to yield a wide bandwidth, partial self-tuning and an improved coupling coefficient. Transmission line transducers have lengths comparable to one-half wavelength and are usually used in higher frequency applications. Their representation is based on the solution to the wave equation and equivalent circuit components are in the form of trigonometric functions. They usually are composed of three sections plus a possible matching layer (which may also be the third layer). One of the best water-matched combinations is a magnesium head section, tungsten tail section, and piezoelectric drive center section. Other designs include multiply and triply resonant forms with additive output between these resonances. Large plate and composite transducer were also described for higher frequency operation.

Flextensional and flexural transducers are often the choice for low frequency operations since they can produce low frequency resonances from a reasonable size. Flextensional transducers also provide amplified motion from a shell attached to the piezoelectric drive section. Various flextentional legacy design classes were discussed as well as newer astroid, trioid, and X-spring designs. Flexural transducer designs can provide even lower resonances as the resonance frequency is inversely proportional to the square of the length. Flexural bender bar and disc along with slotted cylinder and X-spring designs were presented and discussed. New modal projecting transducers (power wheel, octoid, leveraged cylinder) which form directional beams through the addition of the monopole (omni), dipole, and quadrupole modes were also presented. Finally, novel low profile low frequency cantilever mode piston and shear mode piston transducers were presented and discussed.

Although we have described a number of electroacoustic projector designs, there are some which were not discussed such as the historic moving coil Fessenden

Oscillator [110] and several low frequency variable reluctance transducer designs by Massa [111]. Also, we have not covered other powerful sources of low frequency acoustic energy that do not fit into the six electroacoustic types described in this book. These omitted designs include the hydro-acoustic source by Bouyoucus [112], spark sources [113–115] and the discharge excited Edgerton "Boomer" [116] and the FSI "Buble Gun" [117].

Exercises (Degree of Difficulty: *Lowest, **Moderate, ***Highest)

5.1.* Calculate the in-air ring mode short circuit resonance frequencies of a thin-walled, 0.508 cm (0.2 in.) Type 1 (PZT-4) piezoelectric ceramic ring transducer of mean diameter 10.16 cm (4 in.) and height 5.08 cm (2 in.) operating in the 31 mode. Compare the result with the resonance frequency of a thin-walled spherical transducer operating in the planar mode with the same diameter and wall thickness. Use the frequency constants in Sect. 13.6.

5.2.** Calculate the in-air short circuit resonance frequencies for the ring transducer of Exercise 3.1 operating in the 33 mode. Calculate the change in resonance frequency and coupling coefficient if the 33 mode is obtained by electrode striping the ring, replacing 10 % of the active circumferential length of short circuit elastic modulus, s_{33}^E, by Type I material, but with open circuit modulus s_{33}^D. Use equations in Sect. 5.2.2 and Sect. 13.5.

5.3.* Calculate the approximate in-water resonance short-circuit frequencies and mechanical Q_m's for the ring and spherical transducers of Exercise 5.1. This resonance is the frequency of maximum output for a constant voltage transmitting response. Assume the ring has end caps that are mechanically isolated from the ring with air on the inside and approximate spherical loading to obtain the water mass loading for the ring.

5.4.*** Evaluate the ring parameters of Exercise 5.1 for the equivalent circuit of Fig. 5.4 and compute the TVR. Compare the results with a finite element model if available.

5.5.* Calculate the volume and mass figures of merit (FOM_v and FOM_m) for the transducer of Exercise 5.1.

5.6.** Determine the Helmholtz resonance frequency (often called cavity resonance) of the ring of Exercise 5.1. Assume the ring is small compared to the wavelength at this frequency and use the piston in a rigid baffle low frequency radiation mass loading $M_r = 8\rho a^3/3$ or use the water field added tube length $\Delta L = 8a/3\pi$ (see Exercise 5.7). Assume symmetry about a plane midway through the ring; thus the effective cavity depth is $L/2$ not L.

5.7.* Show that the water load added length, $\Delta L = 8a/3\pi$, is equivalent to the low frequency radiation mass loading $M_r = 8\rho a^3/3$. Justify the Exercise 5.6 statement "We assume symmetry about a plane midway through the ring; thus the effective cavity depth is $L/2$ not L."

5.8.*** Consider the Tonpilz projector illustrated in Fig. 5.17 with piston area $0.0162\ \mathrm{m}^2$, head mass $M_\mathrm{h} = 10$ kg, and tail mass $M_\mathrm{t} = 40$ kg and a drive stack consisting of four Type 1 (PZT-4) piezoelectric ceramic rings of mean diameter 10.16 cm (4 in.), height 0.508 cm (0.2 in.), and wall thickness 1.016 cm (0.4 in.) operating in the 33 mode and wired as illustrated in Fig. 5.17. Also assume a tie rod with one-tenth the stiffness of the stack and insulators plus glue joints with a total of 10 times the stiffness of the stack. Calculate the in-air short circuit resonance frequency, the turns ratio N, the free capacitance C_f, and the effective coupling coefficient.

(a) Ignore the effects of the tie rod, insulator, glue joints, and mass of the stack.
(b) Ignore the effects of the tie rod, insulator, glue joints, but include stack mass.
(c) Include effects of the tie rod, insulator, glue joints, and stack mass.

5.9.*** Use the equivalent circuit of Fig. 5.20 to calculate the response of the projector of Exercise 5.8, case A. Then use the equivalent circuit of Fig. 5.19 to calculate the TVR and maximum SPL response of the projector of Exercise 5.8. Assume equivalent sphere radiation loading and piston in a rigid baffle radiation loading. Compare results. Assume 100 % and then 80 % mechanoacoustic efficiency. Determine k_eff and Q_m. Compare results from circuit model with transmission line, matrix or FEA models. Calculate the maximum power output and the $\mathrm{FOM_v}$ for maximum electric field voltage drive.

5.10.* Determine the in-air frequency diameter constants, fD, for thin-walled Terfenol-D and Galfenol magnetostrictive ring transducers under open, f^H, and short, f^B, circuit conditions (see Sect. 13.7).

5.11.* Calculate the in-air ring mode open circuit and short circuit resonance frequencies and free inductances of thin-walled, 0.508 (0.2 in.) ideally biased magnetostrictive Terfenol-D and Galfenol ring transducers of mean diameter 10.16 cm (4 in.) and height 5.08 cm (2 in.) operating in the 33 mode with a coil of 100 turns. Compare the resonance frequency results with the PZT-4 ring of Exercise 5.1.

5.12.*** In Sect. 5.3.2 an equation for estimating the effective coupling coefficient, $k_\mathrm{e}^2 = 2k^2/\left(1 + k^2\right)$ was given for the hybrid transducer for the case where the piezoelectric ceramic and magnetostrictive drivers have the same coupling coefficient. This approach evaluated the coupling coefficient from the total mechanical energy stored and the shared electrical energy at resonance for the two sections wired in parallel. Using the same approach develop an expression for k_e^2 where the drivers have two different coupling coefficients, k_1 and k_2, and show that it reduces to the original expression for $k_1 = k_2$.

References

1. J.F. Lindberg, The application of high energy density transducer material to smart systems, Mat. Res. Soc. Symp. Proc. **459**, 509–519 (1997). See also D.F. Jones, J.F. Lindberg, Recent transduction developments in Canada and the United States, Proceedings of the Institute of Acoustics, 17, Part 3, 15–33 (1995)
2. D.T. Laird, H. Cohen, Directionality patterns from acoustic radiation from a source on a right cylinder. J. Acoust. Soc. Am. **24**, 46–49 (1952)
3. J.L. Butler, A.L. Butler, A Fourier series solution for the radiation impedance of a finite cylinder. J. Acoust. Soc. Am. **104**, 2773–2778 (1998)
4. J.L. Butler, Model for a ring transducer with inactive segments. J. Acoust. Soc. Am. **59**, 480–482 (1976)
5. Channel Industries, Inc., Santa Barbara, CA 93111
6. D.A. Berlincourt, D.R. Curran, H. Jaffe, Piezoelectric and piezomagnetic materials and their function in transducers, in *Physical Acoustics*, ed. by W.P. Mason, vol. 1 (Academic, New York, 1964)
7. J.L. Butler, Solution of acoustical-radiation problems by boundary collocation. J. Acoust. Soc. Am. **48**, 325–336 (1970)
8. M.B. Moffet, M.D. Jevnager, S.S. Gilardi, J.M. Powers, Biased lead zirconate titanate as a high-power transduction material. J. Acoust. Soc. Am. **105**, 2248–2251 (1999)
9. S. Butterworth, F.D. Smith, Equivalent circuit of a magnetostrictive oscillator. Proc. Phys. Soc. **43**, 166–185 (1931)
10. J.L. Butler, S.J. Ciosek, Rare earth iron octagonal transducer. J. Acoust. Soc. Am. **67**, 1809–1811 (1980)
11. S.M. Cohick, J.L. Butler, Rare-earth iron "square ring" dipole transducer. J. Acoust. Soc. Am. **72**, 313–315 (1982)
12. G.W. McMahon, Performance of open ferroelectric ceramic cylinders in underwater transducers. J. Acoust. Soc. Am. **36**, 528–533 (1964)
13. C.H. Sherman, N.G. Parke, Acoustic radiation from a thin torus, with application to the free-flooding ring transducer. J. Acoust. Soc. Am. **38**, 715–722 (1965)
14. A.L. Thuras, *Translating Device*, U.S. Patent 1,869,178 (26 July, 1932)
15. R.H. Lyon, On the low-frequency radiation load of a bass-reflex speaker (L). J. Acoust. Soc. Am. **29**, 654 (1957)
16. A.J. Shashaty, The elastic problem of the flattened cylinder type of underwater acoustical compliance element. J. Acoust. Soc. Am. **66**(6), 1818 (1979)
17. H. Levine, J. Schwinger, On the radiation of sound from an unflanged circular pipe. Phys. Rev. **73**, 383–406 (1948)
18. J.L. Butler, *Multiport Underwater Sound Transducer*, U.S. Patent 5,184,332 (2 February, 1993)
19. A.L. Butler, J.L. Butler, *A Deep-Submergence, Very Low-Frequency, Broadband, Multiport Transducer*, Oceans 2002 Conference, Biloxi, MS, see also Sea Technology, pp. 31–34 (November 2003)
20. A.E.H. Love, *Mathematical Theory of Elasticity*, 4th edn. (Cambridge University Press, London, 1934), p. 452
21. S.L. Ehrlich, P.D. Frelich, *Sonar Transducer*, U.S. Patent 3,290,646, (6 December, 1966)
22. R.S. Gordon, L. Parad, J.L. Butler, Equivalent circuit of a ring transducer operated in the dipole mode. J. Acoust. Soc. Am. **58**, 1311–1314 (1975)
23. J.L. Butler, A.L. Butler, J.A. Rice, A tri-modal directional transducer, J. Acoust. Soc. Am. **115**, 658–665 (2004). J.L. Butler, A.L. Butler, *Multimode Synthesized Beam Transducer Apparatus*, U.S. Patent 6,734,604 B2, (11 May, 2004)
24. O.B. Wilson, *Introduction to Theory and Design of Sonar Transducers*, (Peninsula Publishing, Los Altos, 1988)

25. H.B. Miller, Origin of the 33-driven ceramic ring-stack transducer. J. Acoust. Soc. Am. **86**, 1602–1603 (1989)
26. H.B. Miller, Origin of mechanical bias for transducers, J. Acoust. Soc. Am. **35**, 1455 (1963). H.B. Miller, U.S. Patent 2,930,912, (March 1960)
27. W.C. Young, *Roark's Formulas for Stress and Strain*, 6th edn. (McGraw-Hill, New York, 1989), pp. 452–454
28. J.L. Butler, J.R. Cipolla, W.D. Brown, Radiating head flexure and its effect on transducer performance. J. Acoust. Soc. Am. **70**, 500–503 (1981)
29. R.S. Woollett, *Sonar Transducer Fundamentals*, Section II, (Naval Underwater Systems Center, Newport) (n.d.)
30. D. Stansfield, *Underwater Electroacoustic Transducers* (Bath University Press, Bath, 1990)
31. J.L. Butler, M.B. Moffett, K.D. Rolt, A finite element method for estimating the effective coupling coefficient of magnetostrictive transducers. J. Acoust. Soc. Am. **95**, 2533–2535 (1994)
32. M.B. Moffett, A.E. Clark, M. Wun-Fogle, J.F. Lindberg, J.P. Teter, E.A. McLaughlin, Characterization of Terfenol-D for magnetostrictive transducers, Hawaii. J. Acoust. Soc. Am. **89**, 1448–1455 (1991)
33. S.C. Butler, *A 2.5 kHz Magnetostrictive Tonpilz Sonar Transducer Design*, SPIE 9th Symposium on Smart Structures and Materials, Conference Proceedings, Session 11, (March 2002), San Diego. Also, of historical interest, J.L. Butler, S.J. Ciosek, Development of two rare-earth transducers, U.S. Navy J. Underw. Acoust. **27**, 165–174 (1977)
34. W.M. Pozzo, J.L. Butler, *Elimination of Magnetic Biasing Using Magnetostrictive Materials of Opposite Strain*, U.S. Patent 4,642,802, (10 February 1987)
35. J.L. Butler, A.E. Clark, *Hybrid Piezoelectric and Magnetostrictive Acoustic Wave Transducer*, U.S. Patent 4,443,731, (17 April 1984). *Hybrid Transducer*, U.S. Patent 5,047,683, (10 September 1991)
36. S.C. Thompson, *Broadband Multi-Resonant Longitudinal Vibrator Transducer*, U.S. Patent 4,633,114, (1987). See also S.C. Thompson, M.P. Johnson, E.A. Mclaughlin, J.F. Lindberg, Performance and recent developments with doubly resonant wideband transducers, in *Transducers for Sonics and Ultrasonics*, ed. by M.D. McCollum, B.F. Hamonic, O.B. Wilson (Technomic Publishing, Lancaster, 1992). S.C. Butler, Development of a high power broadband doubly resonant transducer (DRT), UDT Conference Proceedings, (November 2001), Waikiki, Hawaii
37. J.L. Butler, S.C. Butler, A.E. Clark, Unidirectional magnetostrictive/piezoelectric hybrid transducer. J. Acoust. Soc. Am. **88**, 7–11 (1990)
38. R.J. Bobber, A linear, passive, nonreciprocal transducer. J. Acoust. Soc. Am. **26**, 98 (1954)
39. J.L. Butler, A.L. Butler, S.C. Butler, Hybrid magnetostrictive/piezoelectric Tonpilz transducer, J. Acoust. Soc. Am. **94**, 636–641 (1993). S.C. Butler, J.F. Lindberg, A.E. Clark, Hybrid magnetostrictive/piezoelectric Tonpilz transducer, Ferroelectrics, **187**, 163–174 (1996)
40. J.L. Butler, *Design of a 10 kHz Wideband Hybrid transducer, Image Acoustics*, (31 December 1993), and *Design of a 20 kHz Wideband Hybrid transducer, Image Acoustics*, (31 May 1994) with S.C. Butler and in collaboration with W.J. Hughes, Applied Research Laboratory, Penn State University
41. S.C. Butler, F.A. Tito, A broadband hybrid magnetostrictive/piezoelectric transducer array, Oceans 2000 MTS/IEEE Conference Proceedings, Vol. 3 (September, 2000)
42. P. Langevin, British Patent 145,691, (28 July 1921)
43. G.E. Liddiard, Ceramic sandwich electroacoustic transducers for sonic frequencies, in *Acoustic Transducers, Benchmark Papers in Acoustics*, ed. by I.D. Groves, vol. 14 (Hutchinson Ross, Stroudsburg, 1981)
44. J.L. Butler, A.L. Butler, *Ultra Wideband Multiply Resonant Transducer*, MTS/IEEE Oceans 2003, San Diego, (September 2003). *Multiply Resonant Wideband Transducer Apparatus*, U.S. Patent 6,950,373 (27 September 2005)

45. G.C. Rodrigo, *Analysis and Design of Piezoelectric Sonar Transducers*, Ph.D. Thesis, London, (1970). J.R. Dunn, B.V. Smith, Problems in the realization of transducers with octave bandwidths, Proceedings of The Institute of Acoustics, Vol. 9, Part 2 (1987)
46. H.C. Lang, *Sound Reproducing System*, U.S. Patent 2,689,016, (14 September 1954)
47. C.C. Sims, Bubble transducer for radiating high-power low-frequency sound in water, J. Acoust. Soc. Am. **32**, 1305–1308 (1960). *Underwater Resonant Gas Bubble*, U.S. Patent 3,219,970 (1965). T.H. Ensign, D.C. Webb, Electroacoustic Performance Modeling of the Gas-Filled Bubble Projector, in *Transducers for Sonics and Ultrasonics*, ed. by M.D. McCollum, B.F. Hamonic, O.B. Wilson (Technomic, Lancaster, 1992)
48. S.C. Butler, *Triply Resonant Transducer*, MTS/IEEE Oceans 2003, San Diego (September, 2003). S.C. Butler, Triple-resonant transducers, IEEE Trans. Ultrason. Ferroelectr. Freq. Control. **59**, 1292–1300 (2012). *Broadband Triply Resonant Transducer*, U.S. Patent 6,822,373B1 (23 November 2004)
49. J.F. Lindberg, *Parametric Dual Mode Transducer*, U.S. Patent 4,373,143 (9 February 1983)
50. G.W. Renner, Private communication with J.L. Butler.
51. R.A. Nelson, L.H. Royster, On the vibration of a thin piezoelectric disk with an arbitrary impedance on the boundary, J. Acoust. Soc. Am. **46**, 828–830 (1969). Development of a mathematical model for the Class V flextensional underwater acoustic transducer, J. Acoust. Soc. Am. **49**, 1609–1620 (1971)
52. A. Iula, R. Carotenuto, M. Pappalardo, Än approximate 3-D model of the Langevin transducer and its experimental validation. J. Acoust. Soc. Am. **111**, 2675–2680 (2002)
53. R. Newnham, L. Bowen, K. Klicker, L. Cross, Composite piezoelectric transducers. Mater. Eng. **2**, 93–106 (1980)
54. L.J. Bowen, U.S. Patent 5,340,510, (23 August 1984)
55. L.J. Bowen, R. Gentilman, D. Fiore et al., Design, fabrication and properties of SonoPanel 1-3 piezocomposite transducers. Ferroelectrics **187**(1), 109–120 (1996)
56. W.A. Smith, B.A. Auld, Modeling 1-3 composite piezoelectrics: thickness-mode oscillations. IEEE Trans. Ultrason. Ferroelectr. Freq. Control **38**, 40–47 (1991)
57. M. Avellaneda, P.J. Swart, Calculating the performance of 1-3 piezoelectric composites for hydrophone applications: an effective medium approach. J. Acoust. Soc. Am. **103**, 1449–1467 (1998)
58. W.A. Smith, B.A. Auld, Modeling 1-3 composite piezoelectrics: thickness mode oscillations. IEEE Trans. Ultrason. Ferroelectr. Freq. Control. **38**, 40–47 (1991)
59. K.D. Rolt, The history of the flextensional electroacoustic transducer. J. Acoust. Soc. Am. **87**, 1340–1349 (1990)
60. H.C. Hayes, *Design and Construction of Magnetostrictive Transducers*, Summary Technical Report of Division 6, Vol. 13, National Defense Research Committee (1946). H.C. Hayes, *Sound Generating and Directing Apparatus*, U.S. Patent 2,064,911, (22 December 1936)
61. W.J. Toulis, *Flexural-Extensional Electromechanical Transducer Apparatus*, U.S. Patent 3,277,433 (4 October 1966)
62. G.A. Brigham, Lumped parameter analysis of the class IV (oval) flextensional transducer, Technical Report, TR 4463, NUWC, Newport, (15 August 1973). G.A. Brigham, Analysis of the class IV flextensional transducer by use of wave mechanics, J. Acoust. Soc. Am. **56**, 31–39 (1974). G.A. Brigham, B.Glass, Present status in flextensional transducer technology, J. Acoust. Soc. Am. **68**, 1046–1052 (1980)
63. J.L. Butler, FLEXT (Flextensional Transducer Program), Contract N66604-87-M-B328 to NUWC, Newport, Image Acoustics, Cohasset, MA
64. H.C. Merchant, *Underwater Transducer Apparatus*, U.S. patent 3,258,738 (28 June 1966)
65. F.R. Abbott, *Broad Band Electroacoustic Transducer*, U.S. Patent 2,895,062, (14 July 1959)
66. L.H. Royster, Flextensional underwater acoustic transducer. J. Acoust. Soc. Am. **45**, 671–682 (1969)
67. G.W. McMahon, D.F. Jones, *Barrel Stave Projector*, U.S. Patent 4,922,470 (1 May 1990)

68. R.E. Newnham, A. Dogan, *Metal-Electroactive Ceramic Composite Transducer*, U.S. Patent 5,729,007, (17 March 1998). See also, A. Dogan, *Flextensional "Moonie and Cymbal" Actuators*, Ph.D. thesis, The Pennsylvania State University (1994). A. Dogan, K. Uchino, R.E. Newnham, Composite piezoelectric transducer with truncated conical endcaps "Cymbal", IEEE Trans. Ultrason. Ferroelectr. Freq. Control. **44**, 597–605 (1997). J.F. Tressler, R.E. Newnham, W.J. Hughes, Capped ceramic underwater sound projector: The "cymbal" transducer, J. Acoust. Soc. Am. **105**, 591–600 (1999)
69. J.L. Butler, *Directional Flextensional Transducer*, U.S. Patent 4,754,441 (28 June 1988). S.C. Butler, A.L. Butler, J.L. Butler, Directional flextensional transducer, J. Acoust. Soc. Am. **92**, 2977–2979 (1992)
70. S.C. Butler, J.L. Butler, A.L. Butler, G.H. Cavanagh, A low-frequency directional flextensional transducer and line array. J. Acoust. Soc. Am. **102**, 308–314 (1997)
71. J.L. Butler, *Flextensional Transducer*, U.S. Patent 4,846,548 (5 September 1989). See also H.C. Hayes, *Sound Generating and Directing Apparatus*, U.S. Patent 2,064,911 (22 December 1936). J.L. Butler, K.D. Rolt, A four-sided flextensional transducer, J. Acoust. Soc. Am. **83**, 338–349 (1988)
72. J.L. Butler, *Electro-Mechanical Transduction Apparatus*, U.S. Patent 4,845,688 (4 July 1989)
73. M.B. Moffett, J.F. Lindberg, E.A. McLaughlin, J.M. Powers, An equivalent circuit model for barrel stave flextensional transducers, in *Transducers for Sonics and Ultrasonics*, ed. by M.D. McCollum, B.F. Hamonic, O.B. Wilson (Technomic, Lancaster, 1993)
74. C.J.A. Purcell, *Folded Shell Projector*, U.S. Patent 5,805,529 (8 September 1998)
75. G.W. McMahon, B.A. Armstrong, U.S. Patent 4, 524,693 (25 June 1985)
76. G.W. McMahon, B.A. Armstrong, A 10 kw ring-shell projector, in *Progress in Underwater Acoustics*, ed. by H.M. Merklinger (Plenum press, New York, 1987)
77. J.L. Butler, An electro-acoustic model for a flextensional ring shell transducer, (The program FIRST), Contract N66604-88-M-B155, to NUWC, Newport, RI, Image Acoustics, Inc., Cohasset, MA (31 March 1988)
78. W.P. Mason, *Piezoelectric Crystals and Their Application to Ultrasonics* (D. Van Nostrand, New York, 1950)
79. C. Hodgman, *C.R.C. Standard Mathematical Tables*, 12th edn. (Chemical Rubber Company, Cleveland, 1961), p. 421
80. J.L. Butler, A.L. Butler, *Multi Piston Electro-Mechanical Transduction Apparatus*, U.S. Patent 7,292,503 (6 November 2007)
81. J.L. Butler, A.L. Butler, *Single-Sided Electromechanical Transduction Apparatus*, U.S. Patent 6,654,316 B1 (25 November 2003)
82. P.M. Morse, *Vibration and Sound* (McGraw-Hill, New York, 1948)
83. H.F. Olson, *Acoustical Engineering* (D. Van Nostrand, New York, 1957)
84. R.S. Woollett, *The Flexural Bar Transducer* (Naval Undersea Warfare Center, Newport, 1986)
85. J.W. Fitzgerald, *Underwater Electroacoustic Transducer*, U.S. Patent 5,099,461 (24 March 1992)
86. A.E. Clark, private communication
87. R.S. Woollett, *Theory of the Piezoelectric Flexural Disk Transducer with Applications to Underwater Sound*, USL Research Report No. 490, Naval Undersea Warfare Center, Newport (1960)
88. A. L. Butler, J. L. Butler and V. Curtis, End Driven Bender, 2016 U.S. Navy Workshop on Acoustic Transduction Materials and Devices (The Pennsylvania State University, May 10, 2016)
89. D.J. Erickson, *Moment Bender Transducer*, U.S. Patent 5,204,844 (20 April 1993)
90. W.T. Harris, U.S. Patent 2,812,452 (November 1957)
91. H.W. Kompanek, U.S. Patents: 4,220,887 (September 1980), 4,257,482 (March 1981), 4,651,044 (March 1987)

92. J.L. Butler, *An Approximate Electro-Acoustic Model for the Slotted High Output Projector Transducer*, Contract N62269-87-M-3792, NAVAIR, MD, Image Acoustics, Inc., (30 December 1987)
93. W.C. Young, *Roark's Formulas for Stress and Strain*, 6th edn. (McGraw-Hill, New York, 1989), pp. 117–120
94. K.D. Rolt, J.L. Butler, Finite element modulus substitution method for sonar transducer effective coupling coefficient, in *Transducers for Sonics and Ultrasonics*, ed. by M.D. McCollum, B.F. Hamonic, O.B. Wilson (Technomic, Lancaster, 1992)
95. B.S. Aronov, Piezoelectric slotted ring transducer. J. Acoust. Soc. Am. **133**, 3875–3884 (2013)
96. J.L. Butler, TSCAT, *The Computer Program TSCAT for a Tapered Slotted Cylinder Transducer*, Contract N66604-93-D-0583, NUWC, Newport, RI, Image Acoustics, Inc., MA (30 June 1994)
97. A.L. Butler, J.L. Butler, *Modal Acoustic Array Transduction Apparatus*, US Patent 7,372,776 B2 (13 May 2008)
98. J.L. Butler, A.L. Butler, S.C. Butler, The modal projector. J. Acoust. Soc. Am. **129**, 1881–1889 (2011)
99. A.L. Butler, J.L. Butler, The octoid modal vector projector. J. Acoust. Soc. Am. **130**, 2505 (2011)
100. J.L. Butler, Älternative tonpilz and bender transducer designs. J. Acoust. Soc. Am. **134**, 4092 (2013)
101. J.L. Butler, A. Butler, M.J. Ciufo, Doubly steered array of modal transducers. J. Acoust. Soc. Am. **132**, 1985 (2012)
102. J.L. Butler, A.L. Butler, *Doubly Steered Acoustic Array*, US Patent 8,599,648 B2 (3 December 2013)
103. J.L. Butler, A.L. Butler, Cantilever mode piston transducer array. J. Acoust. Soc. Am. **133**, 3360 (2013)
104. A.L. Butler, J.L. Butler, *Cantilever Driven Transduction Apparatus*, US Patent 7,453,186 B1 (18 November 2008)
105. Pengdi Han, U.S. Patent Application Number 2006/0012270 A
106. S. Zhang, F. Li, W. Jiang, J. Luo, R.J. Meyer, W. Cao, T.R. Shrout, Face shear piezoelectric properties of relaxor-$PbTiO_3$ single crystal. J. Appl. Phys. **98**, 182903 (2011)
107. W. Cao, S. Zhu, B. Jiang, Analysis of shear modes in a piezoelectric vibrator. J. Appl. Phys. **83**, 4415–4420 (1998)
108. D.J. Van Tol, R.J. Meyer, *Acoustic Transducer*, US Patent 7,615,912 B2 (10 November 2009)
109. R.J. Meyer, T.M. Tremper, D.C. Markley, D.J. Van Tol, P. Han, J. Tian, *Low Profile, Broad Bandwidth Projector Design Using d_{36} Shear Mode*, Navy Workshop on Transduction Materials and Devices, Penn State (11–13 May 2010)
110. G.W. Stewart, R.B. Lindsay, *Acoustics* (D. Van Nostrand, New York, 1930) pp. 248–250. H.J.W. Fay, in *Acoustic Transducers, Benchmark Papers in Acoustics*, 14, ed. by I. Groves (Hutchinson Ross, Stroudsberg, 1981). K.D. Rolt, *The Fessenden Oscillator: History, Electroacoustic Model, and Performance Estimate*, 127th Meeting of the Acoustical Society of America (June 1994)
111. D.P. Massa, High-power electromagnetic transducer array for Project Artemis, J. Acoust. Soc. Am. **98**(5), 2901–2902 (1995). F.W. Massa, F. Massa, Electromagnetic transducers for high-power low-frequency, deep-water applications, J. Underw. Acoust. **20**(3), 621–629 (July 1970). F.W. Massa, *Electromagnetic Transducers for Underwater Low-frequency High-power Use*, U.S. Patent 4,736,350, (5 April 1988)
112. J.V. Bouyoucos, Hydroacoustic transduction, J. Acoust. Soc. Am. **57**, 1341–1351 (1975). *Self-Excited Hydrodynamic Oscillators*, Acoustic Research Laboratory, Harvard University, TM No. 36 (31 July 1955)
113. D.D. Caulfield, *Predicting Sonic Pulse Shapes of Underwater Spark Discharges*, WHOI Report 62-12 (March 1962)

114. J.L. Butler, K.D. Rolt, *Feasibility of High Power, Low Frequency, High Efficiency Plasma Spark Gap Projector*, Final Report, SBIR Topic N92-088 (23 June 1993), Image Acoustics, Inc. and Massa Products Corporation
115. R.B. Schaefer, D. Flynn, *The Development of a Sonobuoy Using Sparker Acoustic Sources as an Alternative Explosive SUS Devices*, Oceans "99 (IEEE, Seattle, 16 September 1999)
116. Originally manufactured by EG&G, currently manufactured by Applied Acoustic Engineering, Ltd., Great Yarmouth, Norfolk
117. Falmouth Scientific, Inc. Cataumet, MA

Chapter 6
Transducers as Hydrophones

All the applications of underwater projectors described at the beginning of Chap. 5 also require the use of hydrophones. In most active sonar systems the same transducers serve as both projectors and hydrophones, but there are good reasons in some cases to use separate hydrophones for reception, e.g., hydrophones in towed line arrays can be well removed from the self-noise of the ship. In addition passive search and surveillance sonar, as well as passive ranging sonar, use only hydrophones. Passive sonobuoys and various noise monitoring functions also require only hydrophones.

Hydrophones detect the pressure variations of acoustic signals and noise in the water and produce an output voltage proportional to the pressure. In addition they generate a noise voltage due to thermal agitation in any internal resistances. Thus the performance criteria for hydrophones are quite different from that for projectors. While projectors are usually operated in the vicinity of resonance, with power output as the major concern, hydrophones are usually operated over a wide band below resonance, and the open circuit output voltage and signal-to-noise ratio are of most concern. The smallest signal detectable by a hydrophone is equal to, or slightly less than, the ambient sea noise unless the internal hydrophone noise plus the preamplifier input noise exceeds the sea noise. The noise voltage generated by a hydrophone and its preamplifier noise may be compared with the sea noise by relating it to an equivalent noise pressure in the water using the hydrophone sensitivity.

Hydrophones are usually smaller and simpler than projectors, but, since reciprocity holds in most cases, the transducer models developed in Chap. 3 and the projector designs discussed in Chap. 5 may also be used for hydrophones. The commonly used Tonpilz transducer design discussed in Chaps. 3 and 5 is very effective as a high power projector, but it is usually not the most suitable design for a single hydrophone or for hydrophone arrays intended only for passive sonar. Tonpilz transducers do serve very well as both projectors and hydrophones in closely packed active arrays, although they may not be suitable if such arrays are used passively outside the active band where unwanted resonances may exist.

J.L. Butler, C.H. Sherman, *Transducers and Arrays for Underwater Sound*,
Modern Acoustics and Signal Processing, DOI 10.1007/978-3-319-39044-4_6

All the transducer mechanisms described in Chap. 2 could be used as hydrophones, but the advantages of piezoelectric ceramics and single crystal piezoelectric materials are so great that they dominate hydrophone applications and will accordingly be the main concern of this chapter. Piezoelectric ceramic can be formed in many shapes and sizes, any of which, when provided with suitable electrodes, electrical cables, and waterproofing, becomes a hydrophone. For example, hollow piezoelectric ceramic cylinders are a very common type of hydrophone that can be made in several variations by applying different electrodes and enclosing the interior in different ways. With single crystal piezoelectric material smaller, wider band hydrophones may be developed.

In this chapter we will consider hydrophones that are sensitive to the pressure and the pressure gradient (or velocity) as well as those sensitive to higher order modes of vibration of the hydrophone structure that might be excited by the acoustic signal. Pressure sensitive hydrophones are the most common; they are normally operated below resonance where their response is frequency independent and omnidirectional. Velocity sensitive hydrophones have a figure eight beam pattern with a 6 dB per octave rise in response with increasing frequency. Designs and special considerations for scalar (pressure) and vector (pressure gradient and particle velocity) sensors as well as intensity probes and small multielement designs will be given. The relationship of hydrophone internal thermal noise to other hydrophone parameters will be discussed in detail as well as the relationship of thermal noise to radiation resistance.

A brief introduction to hydrophone equivalent circuits, as well as definitions of important parameters, was given in Sect. 2.8.7. Arrays of hydrophones will be discussed in Chap. 8, including associated noise sources. We begin with fundamental principles of hydrophone operation before describing various hydrophone designs.

6.1 Principles of Operation

The free field voltage sensitivity of a hydrophone is defined as the ratio of the open circuit voltage amplitude to the free field pressure amplitude of an incident plane sound wave, and is usually symbolized by M, for microphone. (The symbol M_0 has also been used for open circuit voltage sensitivity to differentiate this sensitivity from the short circuit sensitivity [1] represented by the symbol M_s.) In general the sensitivity depends on the frequency, the direction of the incident plane wave, the properties of the active material, such as piezoelectric ceramic, and the geometry of the hydrophone. For directional hydrophones the sensitivity is usually defined for a plane wave arriving on the maximum response axis (MRA). The sensitivity will be derived for many different types of hydrophones starting with the simplest case of a rectangular piezoelectric ceramic plate as shown in Fig. 6.1.

The acoustical part of the analysis in this first section will be kept simple by assuming that the frequency is low enough, and consequently the wavelength large enough, that the acoustic pressure amplitude is uniform over the entire hydrophone.

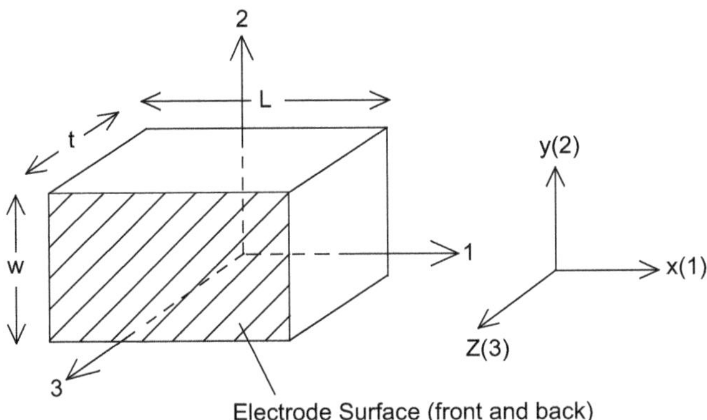

Fig. 6.1 Piezoelectric ceramic with two electrodes perpendicular to the direction of polarization 3 axis

Under these conditions the sensitivity is sometimes called the pressure sensitivity. It is also assumed that the hydrophone is operated below its fundamental resonance where the sensitivity is independent of frequency. In practice there is a reduction in the output voltage at very low frequencies where the large but finite input impedance of the preamplifier becomes comparable to the impedance of the hydrophone. Piezoelectric materials are most often used for pressure sensitive hydrophones, in part because of the flat response below resonance. A more complete discussion of hydrophone sensitivity, including directionality, diffraction, and noise, will be given in Sects. 6.6 and 6.7.

6.1.1 Sensitivity

Piezoelectric hydrophones are sensitive to pressure, but it is interesting to consider their sensitivity in terms of the acoustic particle displacements that accompany the small pressures they are designed to detect [1].

For example: The pressure spectral density of Sea State Zero (SS0) ambient noise is about 44 dB//(μPa)^2/Hz at 1 kHz (see Fig. 6.37) which is equivalent to a pressure of $p \approx 160$ μPa. In a plane wave signal of the same pressure the particle displacement is $x = p/\rho c\omega = 1.7 \times 10^{-14}$m = 0.00017 Angstroms. The displacement of the sensitive surface of a hydrophone caused by this plane wave signal is less than the displacement in the water, since piezoelectric materials are considerably stiffer than water. This displacement is the order of 10,000 times smaller than the crystal lattice dimensions of the piezoelectric material, which is in motion itself because of thermal lattice vibrations.

Small displacements are detectable because the particle displacement of the acoustic wave is coherent over the sensitive surface of the hydrophone, while the random thermal motion in the hydrophone material is incoherent. The thermal vibrations in the hydrophone material cause internal hydrophone noise, which will be discussed in Sect. 6.7.

Since the objective of a hydrophone sensitivity calculation is determination of the voltage caused by stresses induced by the acoustic pressure, equations of state in the form shown in Eq. (2.4) giving electric field in terms of stress and electric displacement are the most direct. Consider the case where the plate has electrodes on the surfaces shown in Fig. 6.1, which establish the polar axis parallel to the thickness along the 3 or z direction. Then the component of electric field sensed by the electrodes is E_3, given by Eq. (2.4) as,

$$E_3 = -g_{31}T_1 - g_{32}T_2 - g_{33}T_3 + \beta_{33}^T D_3. \tag{6.1a}$$

The electric field intensity $E_3 = -\partial V/\partial z$ and the voltage

$$V = -\int_{-t/2}^{t/2} E_3 \mathrm{d}z = -E_3 t, \tag{6.1b}$$

since E_3 is constant and t is the plate thickness. Under open circuit conditions, where $D_3 = 0$, Eqs. (6.1a and 6.1b) yield

$$V = g_{31}tT_1 + g_{32}tT_2 + g_{33}tT_3, \tag{6.2a}$$

and the open circuit output voltage is proportional to the thickness between the electrodes, the $g_{3\mathrm{i}}$ constants, and the stresses, T_i.

If, as in Fig. 6.2a, the surfaces normal to the 1 and 2 directions are free to move but shielded from the incident acoustic pressure by a stiff structure and gap, or pressure release material, then $T_1 = T_2 = 0$.

Since only the surface normal to the 3 direction is exposed to the pressure, only T_3 is non-zero and equal to the pressure amplitude of an incident plane wave, p_i. In this case Eq. (6.2a) gives the hydrophone 33 mode sensitivity

$$M_{33} = V/p_\mathrm{i} = g_{33}t, \tag{6.2b}$$

On the other hand, if only the surfaces normal to the 1 axis are exposed as in Fig. 6.2b, then $T_2 = T_3 = 0$ and the 31 mode receiving sensitivity is

$$M_{31} = V/p_\mathrm{i} = g_{31}t, \tag{6.2c}$$

The 31 mode sensitivity is approximately one-half that of the 33 mode since in typical lead zirconate titanate (PZT) materials, $g_{31} \approx -g_{33}/2$. If both the surfaces normal to the 1 and 2 axes are exposed to the acoustic pressure, while the surface

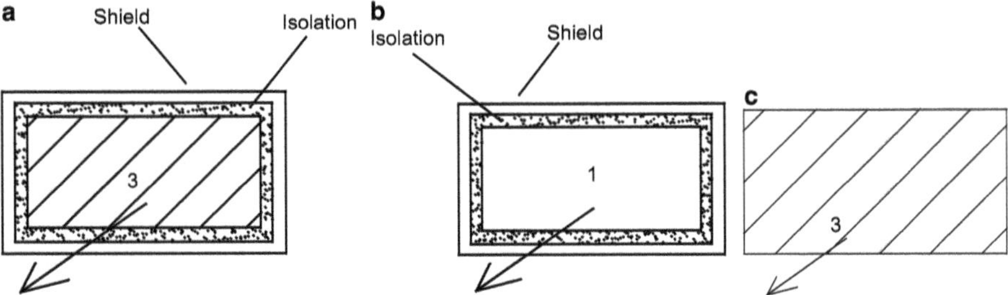

Fig. 6.2 (**a**) 1 and 2 directions shielded from p_i. (**b**) 2 and 3 directions shielded from p_i. (**c**) 1, 2, and 3 directions exposed to p_i

normal to the 3 axis is shielded, then $M = g_{31}t + g_{32}t = 2g_{31}t$ since $g_{32} = g_{31}$ for the PZT materials. In the final example, illustrated in Fig. 6.2c, all surfaces are exposed to the incident pressure, and this case, called the hydrostatic sensitivity, is given by

$$M_h = (g_{33} + 2g_{31})t = g_h t, \tag{6.2d}$$

where $g_h = g_{33} + 2g_{31}$ is the hydrostatic g constant.

For example: For PZT-4, $g_h = 0.0249 + 2(-0.0106) = 0.0037$ V m/N which is an order of magnitude below g_{33} (see Sect. 13.5). However, other materials are available that are better suited for the hydrostatic mode such as lead metaniobate or lead titanate where $g_h = 0.0320 + 2(-0.0017) = 0.0286$ which is very close to g_{33}. Although the resulting hydrostatic sensitivity is high for this material, the dielectric constant is approximately one-fifth of that for PZT-4 yielding a higher impedance and lower figure of merit.

Numerical values of hydrophone sensitivity are referenced to 1 V per μPa, and the formulas of Eqs. (6.2b, 6.2c, and 6.2d) are accordingly multiplied by 10^{-6}. The resulting receiving voltage sensitivity, on a logarithmic scale, is RVS = 20 log $|M|$ in dB//1 V/μPa. For example, a 0.01 m thick PZT-4 plate hydrophone operating in the 33 mode has RVS = −192 dB//V/μPa. (The free field voltage sensitivity notation, FFVS, is sometimes used instead of the receiving voltage sensitivity notation, RVS, for hydrophone response.)

6.1.2 *Figure of Merit*

The figure of merit of a hydrophone is usually considered to be $M^2/|Z_h|$ where Z_h is the electrical input impedance of the hydrophone. The impedance is important because the internal hydrophone noise is generated in the resistive part of the impedance. This figure of merit is related to the signal-to-noise ratio and is independent of passive schemes one might use to increase the sensitivity such as re-wiring hydrophones from a parallel to a series configuration or use of a transformer. If a transformer of turns ratio N were used to raise the output voltage sensitivity, M, to NM, the impedance would be increased to N^2Z_h and both the numerator and the denominator of M^2/Z_h would be increased by N^2 yielding the same figure of merit.

If, on the other hand, there are N hydrophones instead of one, the figure of merit is increased by the factor, N. For example, if $N = 4$ and the hydrophones are wired in a series-parallel configuration, with parallel pairs wired in series or series pairs wired in parallel, the impedance of the four hydrophones is still Z_h, but because of the series pair the output voltage doubles and the sensitivity increases from M to $2M$. The figure of merit is now $(2M)^2/Z_h$ showing an improvement factor of 4, equal to the number of hydrophones. Thus, increasing the number of hydrophones or, equivalently, increasing the active volume of the hydrophone, yields an improved figure of merit. On the other hand, rewiring a given number of hydrophones yields no improvement in the figure of merit, as might be expected.

At low frequencies, well below resonance, $Z_h = 1/j\omega C_f$ and $M^2/Z_h = M^2 j\omega C_f$. Since M is constant at low frequency we can define a frequency-independent hydrophone figure of merit $\mathrm{FOM_h}$ as M^2C_f. For the low frequency piezoelectric ceramic hydrophone of Fig. 6.1 the free capacitance $C_f = \varepsilon_{33}^T LW/t$ and, with the shielding condition of Fig. 6.2a, the sensitivity is given by Eq. (6.2b) leading to

$$\mathrm{FOM_h} = M^2C_f = (g_{33})^2\varepsilon_{33}^T LWt = g_{33}d_{33}V_p, \tag{6.3}$$

where V_p is the volume of the piezoelectric material and $g_{33} = d_{33}/\varepsilon_{33}^T$. As seen, the greater the volume of piezoelectric ceramic material, the greater the figure of merit. The better hydrophone materials have a high "*gd*" product, as in the case of single crystal material (see Sect. 13.5). The "*gd*" product is often referred to as the figure of merit of the material, but it also depends on the mode in which the material is used. (The projector figure of merit for the material is proportional to the "*ed*" product, see Sect. 4.5) For example, PZT materials operating in the 33 mode have a *gd* product 5.5 times greater than when operating in the 31 mode. An alternative figure of merit incorporates the noise generating loss factor, tan δ, and uses $C_f/\tan\delta$ rather than C_f. In this case Eq. (6.3) becomes

$$M^2C_f/\tan\delta = g_{33}d_{33}V_0/\tan\delta, \tag{6.4}$$

emphasizing the importance of a low dissipation factor. More will be said about the figure of merit and its relationship to noise in Sect. 6.7.

6.1.3 *Simplified Equivalent Circuit*

The introductory equivalent circuit of Sect. 2.8.1 and the more extensive circuits developed in Chap. 3 may be readily applied to the receiving case by considering the total force on the transducer as $F = Z_r u + F_b$ where Z_r is the radiation impedance and u is the velocity. The clamped (or blocked) force, F_b, is the integral of the incident pressure over the active area of the hydrophone, since we can ignore the scattered wave from this assumed small hydrophone (see Sect. 6.6). For example, the one-degree of freedom lumped model of Fig. 3.14 may be represented by the circuit of Fig. 3.15 with the addition of the acoustic input force source, F_b, in series with the radiation impedance and the voltage source removed from the electrical terminals. At frequencies well below the resonance frequency where the impedance $1/j\omega C^E$ dominates, the hydrophone circuit representation of Fig. 3.15 reduces to the circuit of Fig. 6.3, where V is the open circuit output voltage for the incident force F_b and $G_0 = \omega C_f \tan\delta$. The electromechanical transformer may be removed from the circuit if the mechanical compliance, C^E, is replaced by the electrical capacitance N^2C^E and the mechanical force F_b is replaced by the electrical voltage F_b/N.

Then a Thevenin series circuit representation (see Sect. 13.8) of the form shown in Fig. 6.4 may be developed where

$$V_t = k^2 F_b / [N(1 - j\tan\delta)] \quad \text{and} \quad C_f' = C_f(1 - j\tan\delta), \tag{6.5}$$

showing the simplicity of the circuit and the important effect of the electrical dissipation in the piezoelectric material. Since $k^2 = N^2C^E/C_f$ the voltage source may also be written as $V_t = F_b k(C^E/C_f)^{1/2}/(1 - j\tan\delta)$ and, for the case illustrated in Figs. 6.1 and 6.2a, this can be reduced to $V_t = (F_b/LW)g_{33}t/(1 - j\tan\delta)$ which is equivalent to the earlier result in Eq. (6.2b) for $\tan\delta = 0$. Normally $\tan\delta \ll 1$ and may be ignored for sensitivity evaluation, but it must be included in the impedance

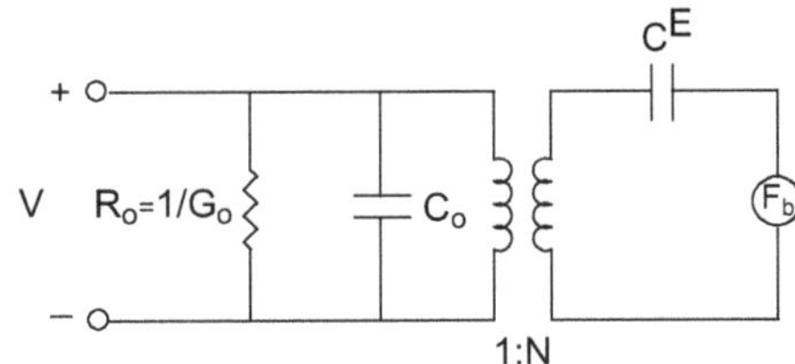

Fig. 6.3 Low frequency hydrophone equivalent circuit representation

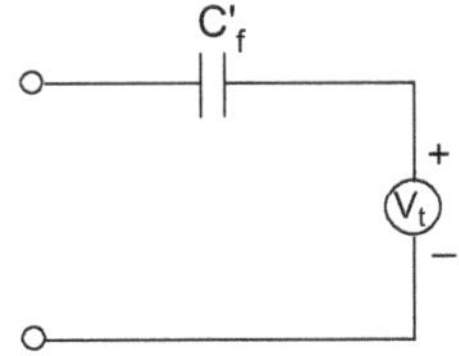

Fig. 6.4 Thevenin circuit representation of Fig. 6.3 where the free capacity C'_f is complex

when the equivalent series resistance is used for evaluating hydrophone noise (see Sect. 6.7).

A hydrophone may also be modeled and analyzed as a projector, and the TVR and impedance may then be used to obtain the RVS through the reciprocity formula (see Sect. 9.5)

$$\mathrm{RVS} = \mathrm{TVR} + 20\log|Z| - 20\log f - 294\,\mathrm{dB}, \tag{6.6}$$

with results in dB//V/μPa. Although the modeling of small hydrophones that operate below resonance can be quite simple, larger hydrophones that operate in the vicinity of resonance are usually more complicated, especially if one considers the scattering of the incident acoustic wave. In this case, modeling as a projector and use of reciprocity may be a simpler way to obtain the receiving voltage sensitivity, and this approach has often been used in finite element modeling.

6.1.4 Other Sensitivity Considerations

The boundary conditions on piezoelectric elements can strongly affect the sensitivity. The simple example of Sect. 6.1.1 assumed the piezoelectric plate to be small with equal pressure on all surfaces. For the case of the 33 mode with other surfaces shielded we obtained a sensitivity $M = g_{33}t$ for a pressure p_i on both exposed surfaces. This is illustrated in Fig. 6.5a showing the physical condition and the corresponding lumped equivalent circuit with the mass of the plate, m, and force $F_b = p_i A$ with A the cross-sectional area of surface 1 or 3.

At low frequencies, where $\omega m \ll 1/\omega C^E$, the force at terminal 2 is F_b, and, because of symmetry, the circuit of Fig. 6.5a becomes equivalent to the circuit of Fig. 6.3. If the hydrophone is blocked or clamped at surface 1, terminal 1 is open circuit (i.e., $u_1 = 0$) as illustrated in Fig. 6.5b and the mass, $m/2$, between terminals 1 and 2 may be deleted. It is seen that at low frequencies the force at node 2 is still F_b and the sensitivity is still $M = g_{33}t$ as in the case of Fig. 6.5a. If, on the other hand, we now remove the block on surface 1 and allow a shielded pressure release condition on surface 1, as in Fig. 6.5c, the force F_b at terminal 1 vanishes and terminal 1 of the equivalent circuits of Fig. 6.5a or b become short circuited. The short circuit at terminal 1 places the associated mass, $m/2$, in parallel with the series combination of $-m/6$, C^E and the secondary of the transformer resulting in the equivalent circuit of Fig. 6.5c. Even though the reactance of the shunt mass, $m/2$, is small at low frequencies, the two equal $m/2$ masses form a voltage divider circuit (see Sect. 13.8) yielding a Thevenin equivalent force equal to $F_b/2$ in series with a mass $m/4$ resulting in a low frequency sensitivity $M = g_{33}t/2$. This illustrates the need for a large inertial mass at terminal 1 to approximate the blocked condition of Fig. 6.5b and avoid a possible 6 dB loss in sensitivity from the pressure release condition illustrated in Fig. 6.5c.

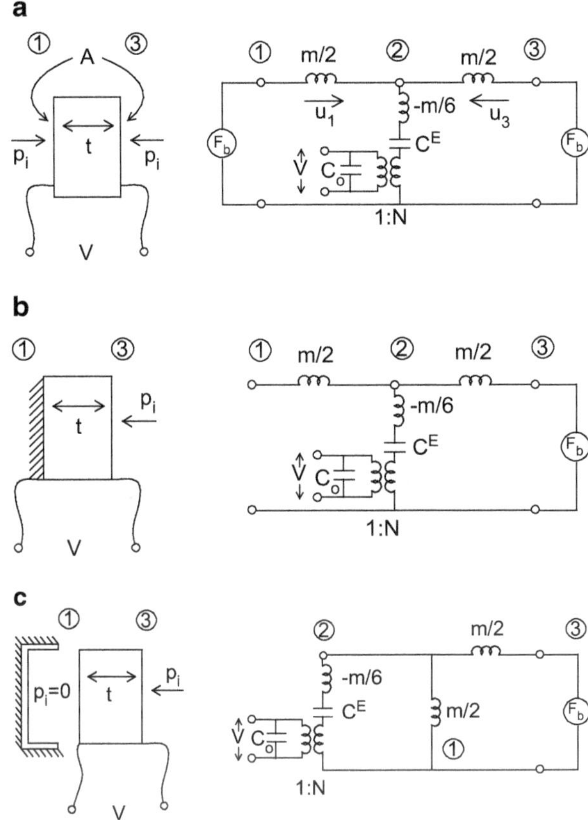

Fig. 6.5 (**a**) Equal pressure hydrophone. (**b**) Single pressure but blocked on side 1. (**c**) Single pressure but free on side 1

The sensitivity at the end of a connecting cable (see Sect. 13.16) depends on the capacitance of the cable relative to the capacitance of the hydrophone. The situation is illustrated in Fig. 6.6, with cable capacitance C_c forming a voltage divider with the hydrophone free capacitance C_f, resulting in an output voltage, at the end of the cable, of

$$V_c = VC_f/(C_f + C_c), \tag{6.7a}$$

where V is the output voltage from the hydrophone alone.

If, for example, the cable capacitance, C_c, were equal to the free capacitance, then $V_c/V = ½$, causing a 6 dB reduction in effective sensitivity. However if the hydrophone capacitance was increased by a factor of 4, by parallel wiring four hydrophones, each the same as the original hydrophone, then the voltage would be the same but the capacitance C_f would be quadrupled yielding a reduction factor of $V_c/V = 0.8$. If the four hydrophones were wired in series the voltage would be quadrupled and the capacitance would be decreased by a factor of 4 also leading to a

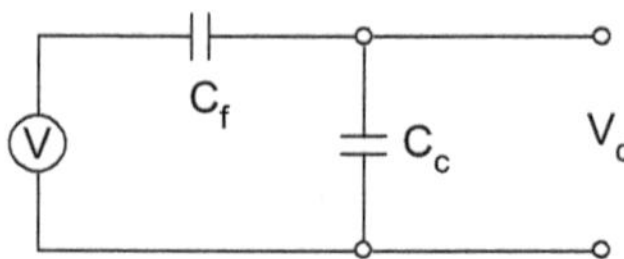

Fig. 6.6 Hydrophone with cable of capacitance, C_c

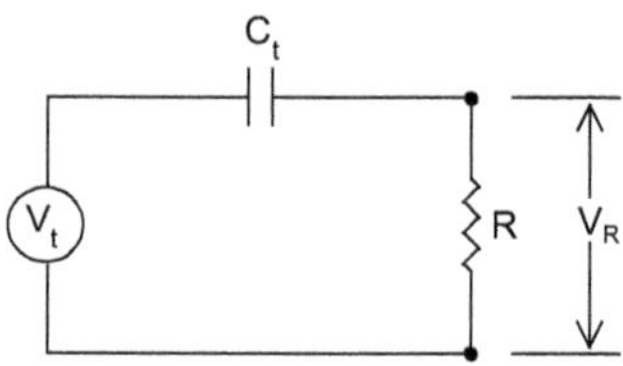

Fig. 6.7 Hydrophone with preamplifier input resistance, R

reduction factor of 0.8. Thus the use of four hydrophones, instead of one, reduces the loss from 6 to 1.9 dB whether wired in series or parallel.

The circuit of Fig. 6.6 may be replaced by its Thevenin equivalent where the voltage source is now the open circuit output voltage and the series impedance is the impedance with $V = 0$. This equivalent Thevenin circuit is shown in Fig. 6.7 where $V_t = V_c = V\, C_f/(C_f + C_c)$, the capacitance is $C_t = C_f + C_c$, and we have added a possible load resistance, R, which could represent the input impedance of a preamplifier.

The resistor, R, and capacitive reactance, $1/j\omega C_t$, act as a voltage divider and the voltage developed across the resistor is $V_R = V_t\, R/(R + 1/j\omega C_t)$. The magnitude may be written as

$$|V_R/V_t| = 1/\left[1 + (1/\omega RC_t)^2\right]^{1/2}. \tag{6.7b}$$

We see that the effect of the resistive load on the output voltage is small if $\omega RC_t \gg 1$. At the frequency where $\omega RC_t = 1$ the voltage level is down 3 dB, and $f = 1/2\pi RC_t$ is called the low frequency cut-off. A value of R can be chosen to roll-off the sensitivity below the band of interest, reducing the reception of external noise. It should be noted that the piezoelectric ceramic dissipative resistance $R_0 = 1/\omega C_f \tan\delta$ alone does not cause a low frequency cut-off as it has the same frequency dependence as the reactance $1/j\omega C_f$. This can be seen by replacing R with R_0 and, for no cable, C_t with C_f in Eq. (6.7b) yielding the small, but frequency independent, voltage reduction

$$|V_R/V_t| = 1/\left[1 + (\tan\delta)^2\right]^{1/2}. \tag{6.7c}$$

In this section we have discussed hydrophones operating at frequencies well below resonance. However, there are important applications where hydrophones are operated at resonance, such as active sonar systems where the transducer operates in the vicinity of resonance as both projector and hydrophone. In this case the full equivalent circuit, transmission line, or matrix model should be used as in the case of the projector (see Chaps. 3 and 5). For electric field transducers the projector resonance is at the short circuit mechanical resonance, f_r, while the hydrophone resonance is at the open circuit antiresonance, f_a. These two resonances differ considerably if the effective coupling coefficient, k_e, is high since $f_r/f_a = \left(1 - k_e^2\right)^{1/2}$, which must be taken into account in the overall system design.

6.2 Cylindrical and Spherical Hydrophones

Circular cylindrical and spherical hydrophones are probably the most commonly used designs because of their high sensitivity, wide-band smooth response up to and possibly through resonance, generally low impedance, good hydrostatic pressure capability, and simplicity. Cylindrical hydrophones need end caps to maintain air backing and encapsulation to prevent water leakage while spherical hydrophones need only encapsulation. Projector equivalent circuits for ring or spherical transducers apply to hydrophones by use of the reciprocity relation in Eq. (6.6) or by inserting a force source in series with the radiation impedance. In general the value of the force is Ap_iD_a, where p_i is the incident free field pressure (often referred to as p_{ff}), D_a is the diffraction constant, and A is the effective capture, or aperture, area equal to the radiating area used in evaluating the radiation impedance or D_a. The diffraction constant $D_a \approx 1$ for pressure sensitive hydrophones that are small compared to the wavelength.

The calculation of hydrophone sensitivity in Sect. 6.1.1 was simplified by assuming that the frequency was low enough to make the pressure uniform over the entire active surface of the hydrophone and equal to the free field pressure (p_{ff}) of the incident wave. A small hydrophone does not disturb the sound field, and thus, measures the pressure that would exist if it were not present. A large hydrophone, or a small hydrophone in a large mounting structure, will change the pressure field. For example, a large rigid wall will cause a doubling of the pressure at the wall since the wave is completely reflected, and the two pressures are in phase and add at the wall. Thus, a small hydrophone in a rigid baffle would measure twice the pressure that would be measured without the baffle. This increase (or, in other cases, decrease) in the pressure is a measure of the diffraction constant D_a. With p_b the average pressure on the surface of the hydrophone when it is clamped, so that its active surface cannot move, D_a is defined as p_b/p_{ff}. The diffraction constant is the ratio of the clamped force to the free field force as defined by Eqs. (1.6) and (1.7) and, in general, it is not a constant, but a function of frequency as well as the particular geometry of the hydrophone; it is discussed in more detail in Sect. 6.6.

6.2.1 Performance with Shielded Ends

Hydrophones are often used at frequencies below their fundamental antiresonance, which occurs under open circuit conditions, for electric field transducers. The introductory discussion for a piezoelectric plate hydrophone in Sect. 6.1.1 will now be extended to a ring or tube hydrophone, illustrated in Fig. 6.8, with mean radius, a, wall thickness, $t < a$, and length $L < 2a$, with electrodes on the inner and outer cylindrical surfaces.

Also shown is a shielded-end piezoelectric tube with two end-caps isolated from the tube with a highly compliant material, such as corprene (see Sect. 13.2), which is suitable for ambient pressures less than 2 kPa ($\approx$300 psi). Our starting point is Eq. (6.1a) with $D_3 = 0$ for open circuit conditions and $T_2 = 0$ because of the pressure release isolation material on the ends of the tube. If we also assume the tube to be air-filled the interior pressure is nearly zero and, since there is no compressive stress across the wall, the stress in the radial 3 direction vanishes and $T_3 = 0$. The result is then $E_3 = -g_{31}T_1$ and, since E_3 is a constant over the thickness, t, as in the plate of Sect. 6.1.1, the output voltage $V = g_{31}tT_1$.

The circumferential stress in the ring, $T_1 = F/tL$, and, as shown in Sect. 5.2.1, the circumferential force, $F = F_r/2\pi$, where F_r is the radial force on the cylinder. If the hydrophone is operating below resonance and the size is small compared to the wavelength, then $D_a \approx 1$ and the radial force $F_r = p_i 2\pi aL$. Successive elimination of T_1, F, and F_r yields the low frequency ring receiving sensitivity

$$M = V/p_i = g_{31}a. \tag{6.8}$$

Note that here the sensitivity is dependent on the mean radius, a, of the ring rather than the thickness, t, as in the case of a plate given by Eq. (6.2b).

In the frequency band which includes the transducer resonance, the hydrophone is not necessarily small compared to the wavelength and the diffraction constant should be included in the force, i.e., $F_b = D_a p_i 2\pi aL$. This force may then be used in series with the mechanical radiation impedance as a clamped (or "blocked" open circuit Thevenin) source and, from the equivalent circuit of Fig. 5.4, may be represented by the circuits of Figs. 6.9 and 6.10 with $Z_m = R_m + j\omega M_m + 1/j\omega C^E$,

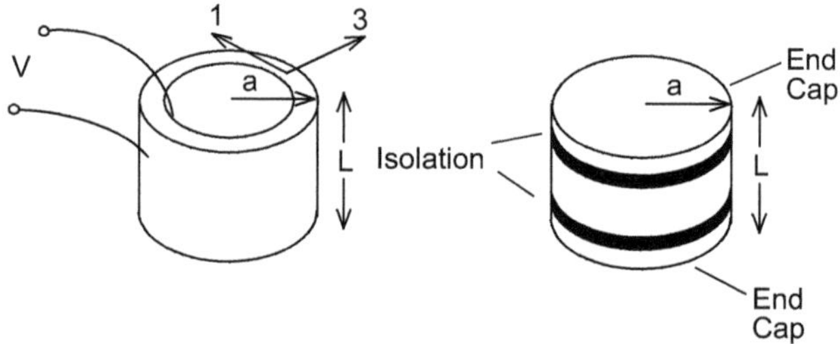

Fig. 6.8 Piezoelectric ring (tube) polarized in the radial (3) direction and operated in the 31-mode with isolated ends caps

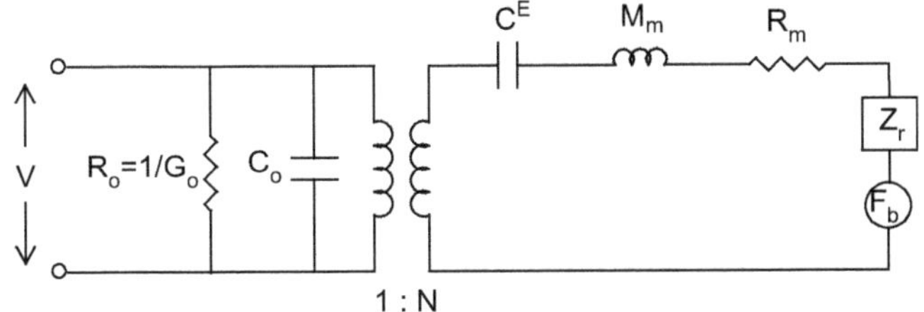

Fig. 6.9 Lumped equivalent circuit of a hydrophone

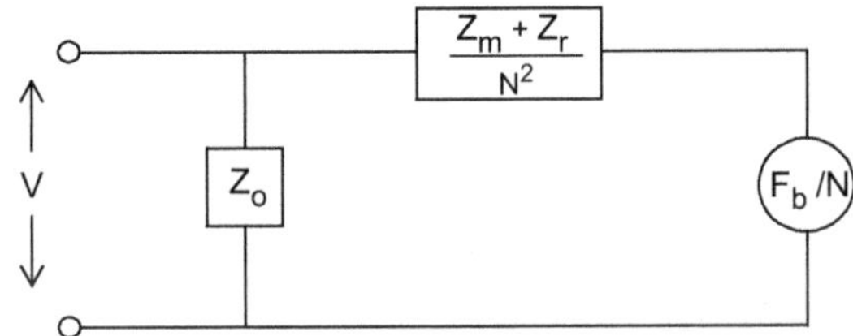

Fig. 6.10 Electrical equivalent circuit for a wide band hydrophone

$Z_r = R_r + j\omega M_r$, $Z_0 = 1/(G_0 + j\omega C_0)$ and F_b all transformed to the electrical side of the electromechanical transformer of turns ratio, N.

The output voltage, V, may then be obtained from the input force, F_b, as in a voltage divider circuit, yielding

$$V = Z_0(F_b/N)/\left[Z_0 + (Z_m + Z_r)/N^2\right]. \tag{6.9}$$

After some algebra, Eq. (6.9) may also be written as

$$V = g_{31}ap_iD_a/\left[1 - (\omega/\omega_a)^2 + j\omega/\omega_aQ_a\right], \tag{6.10}$$

where, for calculating voltage sensitivity, we have treated $G_0 = \omega C_f \tan\delta$ as negligible and used $N = 2\pi Ld_{31}/s_{11}^E$, $C^E = s_{11}^E a/2\pi tL$, from Sect. 5.2.1 as well as

$$\omega_a^2 = 1/\left[(M_m + M_r)C^D\right] \quad \text{and} \quad Q_a = 1/\omega_a C^D(R_m + R_r), \tag{6.11}$$

where $C^D = C^E\left(1 - k_{31}^2\right)$; ω_a is the antiresonance frequency and Q_a is evaluated at antiresonance.

At frequencies well below resonance $\omega \ll \omega_a$, $D_a \approx 1$ and Eq. (6.10) becomes $V = g_{31}ap_i$ with a flat response as expected from Eq. (6.8). At antiresonance Eq. (6.10) gives $V = -jg_{31}ap_iD_aQ_a$ showing a 90° phase shift and a sensitivity increase by the factor Q_a. Under air-loaded conditions the ring resonates at a frequency where the mean circumference, $2\pi a$, is one dilatational wavelength in the material, i.e., $c^D = f_a 2\pi a$, where c^D is the open circuit sound speed. Then the low frequency sensitivity can be written as $|V/p_i| = g_{31}D_ac^D/2\pi f_a$. If we consider the bandwidth to be from nearly zero frequency up to the first antiresonance, we get, the sensitivity bandwidth product,

$$Mf_{\rm a} = D_{\rm a} g_{31} c^D / 2\pi. \tag{6.12}$$

This shows that for a given material and design concept, such as a ring, the product of the sensitivity and bandwidth is proportional to the product of the "g" constant, the sound speed, c^D, and the diffraction constant. Thus, increasing the bandwidth by raising the resonance causes the sensitivity to decrease. For a ring the $Mf_{\rm a}$ product can be increased by a factor of approximately 2 by operation in the 33 mode. A bandwidth sensitivity product is a useful general design concept that can also be developed for other kinds of hydrophones. It shows that while using more material increases the sensitivity, it also lowers the resonance frequency and decreases the bandwidth. Or, equivalently, increasing the bandwidth lowers the sensitivity.

The diffraction constant for short cylinders may be approximated by that of a sphere (see Sect. 6.6), with radius such that its area equals the sensitive area of the cylinder, $2\pi aL$. Values of $D_{\rm a}$ have been given by Trott [2] for capped and shielded-end cylindrical hydrophones. The input impedance may be obtained from the admittance

$$1/Z = Y = j\omega C_0 + \omega C_{\rm f} \tan\delta + N^2 / \left(Z_{\rm m}^E + Z_{\rm r}\right). \tag{6.13a}$$

The sensitivity for cylindrical hydrophones may be improved through polarization along the circumferential direction as discussed in Sect. 5.2.2 using Eqs. (5.18–5.21). The equivalent circuit for this case [3] may be used to calculate the transmitting response, and then the receiving response through reciprocity. The sensitivity for an end-shielded tangentially poled 33-mode segmented or striped-cylinder, illustrated in Fig. 5.5b, operating below resonance may be written as

$$M = g_{33} a[2\pi a/tn] = g_{33} a[w/t] = g_{33} w[a/t],$$

where $a \gg t$. In the first expression [4] the integer, n, is even and is the number of striped electrodes which increases in proportion to the circumference of the ring. Since the length of a segment $w = 2\pi a/n$, the factor in the brackets in the first expression may be written as w/t as shown above. This case yields a higher sensitivity than that of the 31-mode cylinder of Eq. (6.8) since $g_{33} > g_{31}$ and $w > t$. Another interpretation is also given above in the third expression where the sensitivity is shown to depend on the distance between the electrodes, w, magnified by the radius to wall thickness ratio a/t. The 33 mode case is best operated under end-shielded conditions as the free or end-capped conditions yield a lower sensitivity because of cancellation from 31 mode reception.

6.2.2 Spherical Hydrophones

Equations (6.9) through (6.13a) and the equivalent circuit of Figs. 6.9 and 6.10 may also be used to represent a hollow spherical hydrophone. The spherical transducer is described as a projector in Sect. 5.2.3 with

$$C_0 = 4\pi a^2 \varepsilon_{33}^T \left(1 - k_p^2\right)/t, \quad N = 4\pi a d_{31}/s_c^E, \quad C^E = s_c^E/4\pi t \text{ and } M_m = 4\pi a^2 t\rho, \tag{6.13b}$$

where ρ is the density of the piezoelectric ceramic, $s_c^E = \left(s_{11}^E + s_{12}^E\right)/2$ and $k_p^2 = d_{31}^2/\varepsilon_{33}^T s_c^E$. The force $F_b = 4\pi a^2 p_i D_a$, and the expression for D_a given by Eq. (6.53) is exact for the spherical hydrophone. Equations (6.10) and (6.11), which describe the receiving response of a cylindrical hydrophone, may then be used for a spherical hydrophone with $C^D = s_c^D/4\pi t$ where $s_c^D = s_c^E\left(1 - k_p^2\right)$, giving higher values for ω_a and Q_a.

The receiving responses of spherical and cylindrical hydrophones of equal wall thickness, with the diameter and height of the cylinder equal to the diameter of the sphere, are compared in Fig. 6.11 for a radius of 0.0222 m [5]. As seen, although they both have the same sensitivity at low frequencies because of equal radii, the spherical antiresonance frequency is higher. The higher resonance yields a greater bandwidth which, on the other hand, is reduced, somewhat, in sensitivity by the value of the diffraction constant in the vicinity below the antiresonance. The diffraction constant has less effect on the response of the cylindrical hydrophone because of its lower antiresonance.

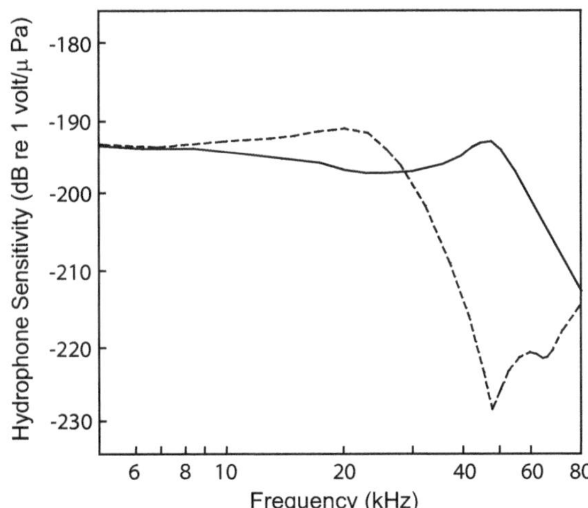

Fig. 6.11 Comparison of theoretical RVS results for a sphere (*solid line*) and a cylinder (*dashed line*) of diameter and height equal to the diameter of the sphere with radius $a = 0.0222$ m [5]

6.2.3 Performance with End Caps

The shielded end condition of Sect. 6.2.1 assumed that the caps were rigid but isolated from the ends of the piezoelectric tube. There are other end conditions which can improve (or reduce) the sensitivity of a tubular hydrophone. These conditions are most easily evaluated well below resonance. Of particular interest is the commonly used case of rigid end caps attached to the ends of an air-filled 31 mode piezoelectric tube illustrated in Fig. 6.12a. The end caps stiffen the piezoelectric ceramic tube which raises the resonance frequency and would lower the sensitivity, if it were not for the in-phase addition to sensitivity as a consequence of the end caps acting as pistons.

Since $g_{32} = g_{31}$, radial-polarized, end-exposed tubes exhibit, in phase, additive 31 and 32 mode sensitivity from both axial stress and radial induced circumferential stress. If the end caps each have area, A, the force on either end of the cylinder is Ap_i and this causes an axial stress $T_2 = p_i A/A_0$ where A_0 is the axial cross-sectional area of the tube where we assume $D_a \approx 1$. For a thin-walled tube of thickness t and mean radius, a, the ratio $A/A_0 \approx \pi a^2/2\pi a t = a/2t$ resulting in a significant stress increase from the end caps. The development for the circumferential sensitivity of Sect. 6.2.1 applies if we let $T_2 = p_i a/2t$ instead of $T_2 = 0$ yielding

$$-E = g_{31}T_1 + g_{32}T_2 = V/t = g_{31}ap_i/t + g_{32}ap_i/2t, \tag{6.14a}$$

leading to

$$M = V/p_i = (g_{31} + g_{32}/2)a = 3g_{31}a/2. \tag{6.14b}$$

The resulting increase in sensitivity is approximately 3.5 dB compared to the shielded end condition of Fig. 6.8 in Sect. 6.2.1.

An alternative end cap designed 31 mode cylindrical hydrophone uses thin concave end caps, as illustrated in Fig. 6.12b, to further improve the sensitivity through additional magnified circumferential stress as a result of the bending of the concave end caps. This hydrophone operates in the same manner as the ring shell (Class VI) flextensional transducer discussed in Sect. 5.5.3, but using the 31 mode instead of the 33 mode.

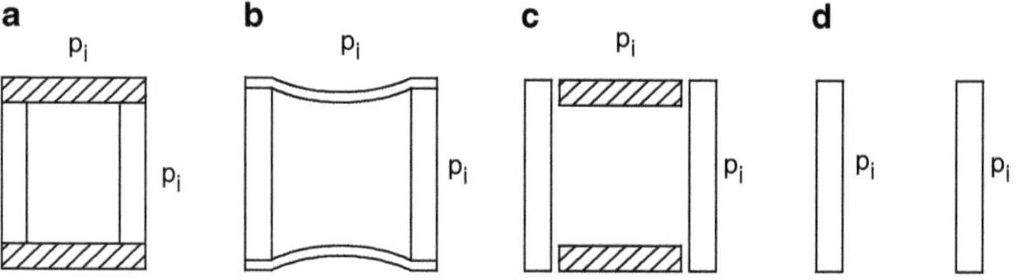

Fig. 6.12 Cylindrical hydrophone with (**a**) rigid end caps, (**b**) concave caps, (**c**) exposed ends, and (**d**) free-flooding

If end caps are not used but the ends of the piezoelectric tube are exposed and the interior is shielded, as illustrated in Fig. 6.12c, then Eq. (6.14a) becomes

$$-E = g_{31}T_1 + g_{32}T_2 = V/t = g_{31}ap_i/t + g_{32}p_i, \quad (6.15a)$$

leading to

$$M = V/p_i = g_{31}a + g_{32}t = g_{31}a(1 + t/a). \quad (6.15b)$$

Thus there is only a small improvement for a thin-walled cylinder, e.g., 0.8 dB if $t = a/10$. We have assumed the tube is air-filled with no fluid-borne acoustic pressure on the inside. If the pressure was the same on the inside and outside of the tube, as illustrated in Fig. 6.12d, there would be no net circumferential compression and $T_1 = 0$. But there would be thickness compression with $T_3 = p_i$, as well as axial compression, yielding

$$-E_3 = V/t = g_{33}T_3 + g_{32}T_2 = g_{33}p_i + g_{32}p_i = (g_{33} + g_{32})p_i. \quad (6.16a)$$

Since for PZT piezoelectric materials $g_{32} = g_{31}$ and $g_{31} \approx -g_{33}/2$, we then get

$$M = V/p_i = -g_{31}t \approx g_{33}t/2, \quad (6.16b)$$

which is considerably less than the other cases since the wall thickness, t, is usually much smaller than the mean radius, a. This case yields the sensitivity for a 31 mode free-flooded ring well below the cavity resonance. McMahon [6] has evaluated the case of a fluid-filled tube with rigid and also with flexing end caps. Because of the stiffening action of the contained inner fluid, there is considerable reduction in the sensitivity if the tube wall is thin and comparatively compliant.

We have considered in this section the end-capped thin-walled 31 mode cylinder because of the improved performance obtained from the caps. Langevin [4] and Wilder [7] have evaluated the low frequency response of cylindrical hydrophones under these end conditions for both 31 and 33 modes and included more details on the effect of a finite wall thickness. Although 33 mode cylinder operation yields a higher sensitivity, because of the higher g_{33} value, it does not benefit from rigid end caps and actually shows reduced sensitivity because of the opposite signs associated with the 33 and 31 modes. The 33 mode ring does, however, benefit from concave ring shell (Class VI) flextensional end cap operation, as this increases the radial stress which adds to the radial stress from the direct pressure on the cylindrical surface.

6.3 Planar Hydrophones

Planar hydrophones are typically used in the closely packed sonar arrays discussed in Chaps. 7 and 8. They may be designed and analyzed by the projector methods of Chap. 5, using reciprocity [see Eq. (6.6)] to obtain the receiving response. In active arrays where the projector is also used as the hydrophone the transducer is usually designed as a projector to achieve maximum source level and the resulting hydrophone response is usually found to be adequate in the active band. However, it may not be satisfactory if used as a passive array outside the active band. Some sonar scanning systems require a narrower receive beam than the projector beam, leading to separate projector and hydrophone arrays, which allows the hydrophone design to be quite different from the projector design. The hydrophones may be operated below resonance providing less phase and amplitude variation from one to another in the active band, and the amount of piezoelectric ceramic material required may be much less than that for the projectors. Moreover, a separate receiving array can be designed to reduce noise more effectively than can be done in a projector array. Finally, the broadband capability of a separate receiving array also makes it useful for passive search operation.

6.3.1 Tonpilz Hydrophones

Figure 6.13 shows a Tonpilz hydrophone design with parallel wired ceramic sections enclosed in an air-filled housing. The 33 mode piezoelectric ceramic is normally sandwiched between a large light piston head and a heavier (approximately three times) tail mass. A lumped mass equivalent circuit is shown in Fig. 6.14 where R_m is the mechanical loss resistance, R_r is the radiation resistance, m_r is the radiation mass, m_c is the mass of the ceramic of length L, F_b is the force on the face of the piston of area A and the piezoelectric section thickness $t = L/n$ where n is the number of piezoelectric ceramic sections.

The circuit of Fig. 6.14 can be reduced to a simpler form under the common condition that the total resistance $R = R_m + R_r \ll \omega M_t$. The procedure is to develop a Thevenin (see Sect. 13.8) circuit to the right of terminals *A–B* and invoke the above condition. The resulting circuit, shown in Fig. 6.15 where $\alpha = M_t/(M_t + m)$ and M_t and m are defined in Fig. 6.14, leads to the output voltage

$$V \approx [g_{33} t p_i D_a (A/A_0) M_t/(M_t + m)] / \left[1 - (\omega/\omega_a)^2 + j\omega/\omega_a Q_a\right]. \tag{6.17}$$

Under open circuit conditions the transducer resonates at the antiresonance frequency, ω_a, where $\omega_a^2 = 1/\left[C^D(m\alpha - m_c/6)\right]$, $Q_a = 1/\omega_a R C^D$, and $C^D = C^E\left(1 - k_{33}^2\right)$. Well below this frequency the response is flat and the output voltage,

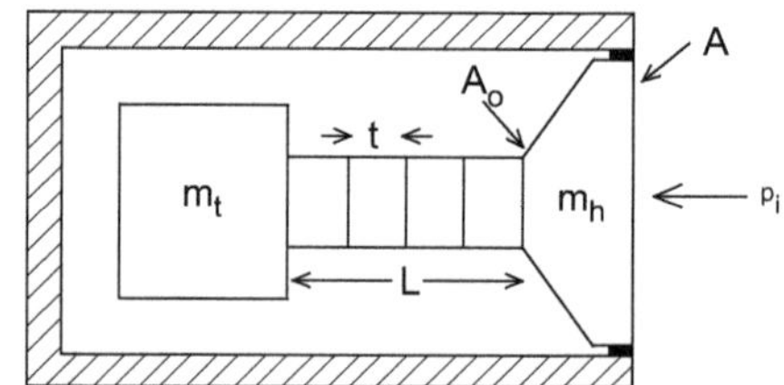

Fig. 6.13 Tonpilz piston hydrophone with piezoelectric stack of length, L, cross-sectional area, A_o, piston head of mass, m_h, and area, A, and tail mass, m_t

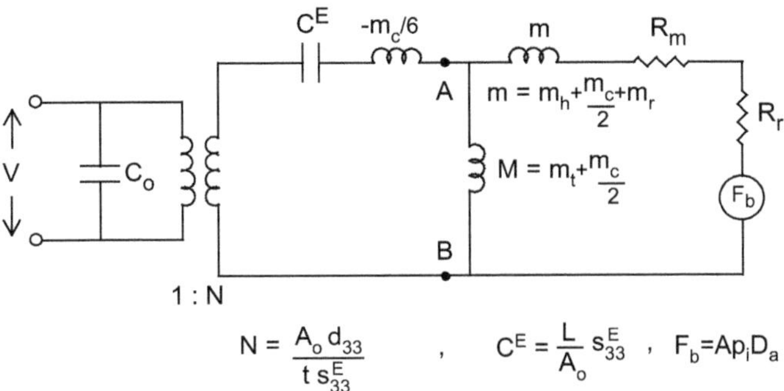

Fig. 6.14 Lumped mass equivalent circuit of Tonpilz hydrophone with $t = L/n$

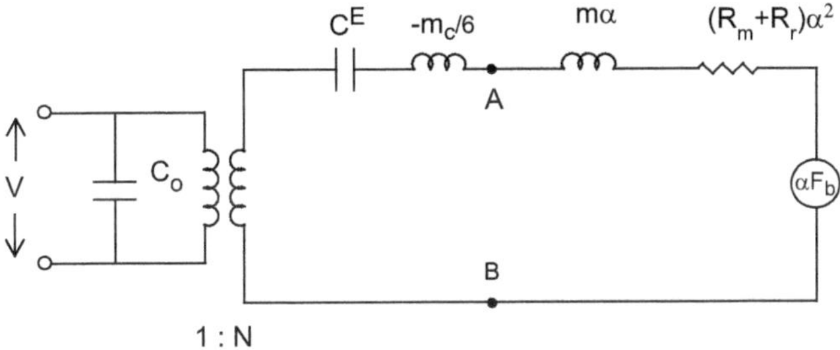

Fig. 6.15 Equivalent circuit of Fig. 6.14 for $(R_r + R_m) \ll \omega M_t$ and $\alpha = M_t/(m + M_t)$

V, is given by the numerator of Eq. (6.17). We see that the output voltage for a hydrophone is proportional to the thickness, t, of each piezoelectric section, which is contrary to the operation of a projector in which the output pressure is inversely proportional to the thickness. The output is also increased by the ratio of the piston area, A, to piezoelectric ceramic area, A_0. We also see the need for a large tail mass to make the factor $M_t/(M_t + m)$ approach unity, e.g., there is a 6 dB reduction in sensitivity if $M_t = m$. There is a 6 dB gain in sensitivity for this Tonpilz type of hydrophone in a rigid baffle, where $D_a = 2$, compared to the individual hydrophone with no baffle. The rigid baffle condition is nearly achieved for the central elements of a large close-packed array.

As in the case of projectors, higher frequency piston hydrophones have the characteristics of a transmission line transducer in the form of a sandwich or a simple plate of piezoelectric material. The principles discussed in Sect. 5.4 apply but with somewhat different interpretations. For example, a quarter wavelength section causes an incoming wave of pressure p_i to be amplified to a pressure Qp_i where Q is the quality factor of the quarter wavelength section, given by the ratio of the effective mass reactance (see Sect. 4.2.2) to resistive load at resonance. In some applications large diameter half wavelength thick plates are used for both transmit and receive.

6.3.2 The 1-3 Composite Hydrophones

The 1-3 composite models for a projector, discussed in Sect. 5.44, may be used to analyze 1-3 composite hydrophones for the case where the sound pressure impinges only on the front face of the transducer. The case where the sound also impinges on the edge can cause a reduction in sensitivity, since it is similar to the hydrostatic mode discussed in Sect. 6.1.1. The so-called hydrostatic case typically arises under the condition of the composite transducer immersed in an acoustic pressure field with wavelength large enough that the pressure is nearly the same over all surfaces (see Fig. 6.16).

As discussed earlier for an individual PZT hydrophone, this leads to an output that includes the 31, 32, and 33 modes. Since the 31 and 32 modes have identical g constants that are opposite in sign and approximately half g_{33}, the sum gives nearly complete cancellation of the output voltage. However, the arrangement of ceramic and polymer in the 1-3 composite reduces the influence of the 31 and 32 modes and produces a significant improvement in hydrostatic voltage sensitivity. The improvement occurs through a greater reduction (approximately 40 %) in the effective d_{31} value than in the effective d_{33} value (approximately 20 %), yielding an improved d_{h}. This result is based on a hydrostatic model by Smith [8].

In this hydrostatic model effective parameters for the composite will be indicated in **bold**. Thus, the hydrostatic low frequency sensitivity below resonance may be written as

$$M = V/p = (\mathbf{g_{33}} + 2\mathbf{g_{31}})t = \mathbf{g_h}t, \tag{6.18a}$$

where t is the thickness along the 3 direction (polarization direction), p is the acoustic pressure, and V is the open circuit voltage output. The effective hydrostatic g constant is

$$\mathbf{g_h} = \mathbf{d_h}/\boldsymbol{\varepsilon}_{33}^{\mathbf{T}} \text{ where } \mathbf{d_h} = \mathbf{d_{33}} + 2\mathbf{d_{31}}. \tag{6.18b}$$

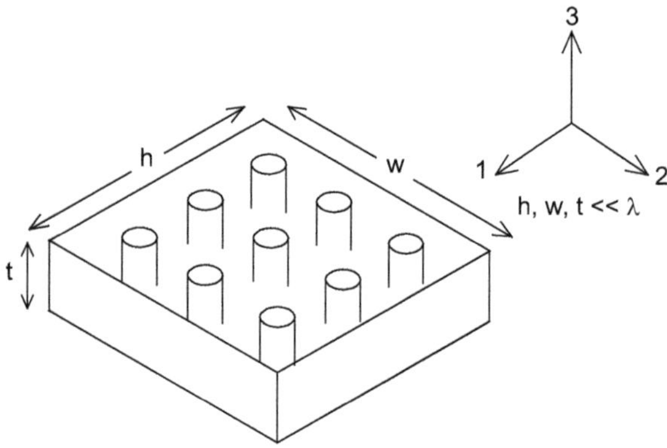

Fig. 6.16 Hydrostatic 1-3 composite model

The hydrostatic model by Smith [8] is based on the piezoelectric constitutive equation set:

$$S = s^E T + \mathrm{d}^t E, \tag{6.19a}$$

$$D = \mathrm{d}T + \varepsilon^T E, \tag{6.19b}$$

where s^E is the short circuit elastic modulus, ε^T is the free dielectric constant, and d is the piezoelectric charge constant. A number of approximations are made that are similar to those in the thickness mode case in Sect. 5.4.4. A major difference, however, is in the treatment of the effective lateral strain, $\mathbf{S_1}$. Clamped end conditions were assumed in the thickness mode model and $\mathbf{S_1}$ was set equal to zero. However, for the hydrostatic model it is assumed that $\mathbf{S_1}$ is the sum of the strains in the ceramic, S^C, and the polymer, S^P, proportioned by the volume fraction, ν, of the piezoelectric material; that is,

$$\mathbf{S_1} = \nu S_1^c + (1-\nu) S_1^p. \tag{6.20}$$

The Smith hydrostatic model results are:

$$\mathbf{d_{33}} = \nu s_{11}^E d_{33}/\mathbf{s}, \tag{6.21}$$

$$\mathbf{d_{31}} = \nu d_{31} + \nu(1-\nu)\left(s_{12}^E - s_{13}^E\right) d_{33}/\mathbf{s}, \tag{6.22}$$

$$\boldsymbol{\varepsilon}_{\mathbf{33}}^{\mathbf{T}} = (1-\nu)\varepsilon_{11}^T + \nu\varepsilon_{33}^T - \nu(1-\nu) d_{33}^2/\mathbf{s}, \tag{6.23}$$

$$\mathbf{s}_{\mathbf{33}}^{\mathbf{E}} = s_{33}^E s_{11}^E/\mathbf{s}, \tag{6.24}$$

$$\mathbf{s}_{\mathbf{13}}^{\mathbf{E}} = \left[\nu s_{13}^E s_{11}^E + (1-\nu) s_{33}^E s_{12}^E\right]/\mathbf{s}, \tag{6.25}$$

$$\begin{aligned}\mathbf{s}_{\mathbf{11}}^{\mathbf{E}} + \mathbf{s}_{\mathbf{12}}^{\mathbf{E}} = {} & \nu\left[s_{11}^E + s_{12}^E - 2\left(s_{13}^E\right)^2/s_{33}^E\right] + (1-\nu)\left[s_{11}^E + s_{12}^E - 2(s_{12})^2/s_{11}^E\right] \\ & + 2\left(\nu s_{13}^E/s_{33}^E + (1-\nu) s_{12}^E/s_{11}^E\right)\mathbf{s}_{\mathbf{13}}^{\mathbf{E}},\end{aligned} \tag{6.26}$$

$$\mathbf{s}_{\mathbf{h}}^{\mathbf{E}} = \mathbf{s}_{\mathbf{33}}^{\mathbf{E}} + 2\left(\mathbf{s}_{\mathbf{11}}^{\mathbf{E}} + \mathbf{s}_{\mathbf{12}}^{\mathbf{E}}\right) + 4\mathbf{s}_{\mathbf{13}}^{\mathbf{E}}, \tag{6.27}$$

where

$$\mathbf{s} = (1-\nu) s_{33}^E + \nu s_{11}^E. \tag{6.28}$$

The effective hydrostatic coupling coefficient and material figure of merit may then be written as

$$\mathbf{k_h} = \mathbf{d_h}/\left(\boldsymbol{\varepsilon}_{\mathbf{33}}^{\mathbf{T}}\mathbf{s}_{\mathbf{h}}^{\mathbf{E}}\right)^{1/2} \text{ and } \mathbf{d_h g_h} = \mathbf{d}_{\mathbf{h}}^2/\boldsymbol{\varepsilon}_{\mathbf{33}}^{\mathbf{T}}. \tag{6.29}$$

Results based on this model for the case of PZT-5H and Stycast, a stiff polymer, show an optimum hydrostatic hydrophone performance for a PZT-5H volume fraction of 10 % where the coupling coefficient and figure of merit are maximum. Hayward, Bennett, and Hamilton [9] have compared the Smith [8] hydrostatic model with finite element results. Although the volume fraction for maximum performance is in agreement with the Smith model, the magnitude of the maximum figure of merit, $\mathbf{d_h g_h}$, appears to be somewhat overestimated by the Smith model.

Avellaneda and Swart [10] have developed a similar but more extensive tensor model for the composite hydrostatic mode of operation. They have also studied the effect of the polymer Poisson's ratio, σ, on the hydrophone performance. Their conclusion is that best results are obtained with $\sigma < -d_{31}/d_{33}$ which causes the most significant decoupling and reduction of the sensitivity of the lateral 31 modes, thus increasing the effective g_h. A PZT-5A/Stycast composite satisfies this condition. Another method for reducing the lateral mode is to use a polymer with embedded air-filled voids.

A simplified equivalent circuit for the composite hydrophone is shown in Fig. 6.17.

Here we assume that the hydrophone is operating well below the first resonance of the piezoelectric composite slab. The voltage is given by the product of the appropriate effective **g** constant ($\mathbf{g_{33}}$ or $\mathbf{g_h}$), the piezoelectric thickness, t, along the direction of polarization and the incoming acoustic pressure, p. If the entire front (silvered or copper clad) surface area, A, is large compared to the wavelength of sound in the medium, then there is pressure doubling at the surface, $D_a = 2$ and p is replaced by $2p$. The electrical impedance is given by the free capacity, C_f, and a shunt dissipation resistor, R_0, where tan δ is the dissipation factor (typically, in this case, 0.01–0.02) and ω is the angular frequency. The 1-3 composite material containing piezoelectric ceramic rods may be configured as a piston hydrophone with a front piston or matching layer and tail mass covering the rods. The rod lengths vary from 0.32 to 2.54 cm [11].

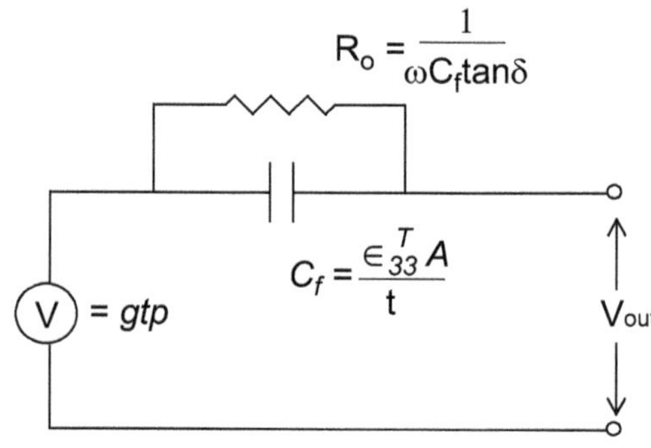

Fig. 6.17 Simplified low frequency hydrophone equivalent circuit

6.3.3 *Flexible Hydrophones*

Although 1-3 composites can be configured to conform to surfaces of moderate curvature, the 0-3 composite is as flexible as a slab of rubber. This occurs since the piezoelectric elements are particles that are not connected to each other (0) but are suspended in a rubber-like matrix which is connected to itself in three directions (3); thus, the "0-3" terminology. An important application of flexible hydrophones arises from the need to reduce flow noise generated as a ship moves through the water. For example, an array designed for low frequency reception might have hydrophone center-to-center spacing of 0.5 m. Use of large area hydrophones that almost fill these spaces with material that has continuous sensitivity would provide spatial averaging and reduced flow noise, as will be discussed in Sect. 8.4.3. The 0-3 composites offer a more continuous distribution of sensitivity than the 1-3 composites, while the flexible piezoelectric polymer [12] is continuous on an even smaller scale.

The polymer materials poly(vinylidene fluoride), PVDF or PVF_2, and 0-3 composites are possible candidates for large ship mounted or towed [13] hydrophone arrays. These materials are available in thin sheets and offer low characteristic impedance with a better match to the water than piezoelectric ceramics. The polymer sheets are stretched along the 1 direction during poling which produces different transverse g constants of $g_{31} = 180 \times 10^{-3}$ and $g_{32} = 20 \times 10^{-3}$ with $g_{33} = -290 \times 10^{-3}$. Because of their low specific acoustic impedance and very small thickness, these materials may be used in front of and, in some cases, mounted on the front face of projectors. In this position, however, the projector will act as a resonant baffle and the PVDF material will detect these vibrations. Since the 31 mode sensitivity of the piezoelectric polymer is comparable to the 33 mode sensitivity, significant and usually detrimental output can be obtained through reception at the ends of the sheet; however, increasing the thickness of the electrodes or bonding the polymer to a thin stiff plate [14] reduces this effect. This increases the stiffness of the hydrophone in the 1 and 2 directions, which reduces the sensitivity in these directions, and has only a minimal effect on the 33 direction [15]. Increasing the stiffness improves the g_h and increases the length mode resonance frequency which increases both the sensitivity and bandwidth.

The hydrostatic hydrophone mode is especially useful in applications where high ambient pressure makes it difficult to shield the ends of the material. Typical approximate measured values of the hydrostatic hydrophone properties are compared in Table 6.1 for 1-3 and 0-3 composites, PVDF and PZT-5H.

Table 6.1 Hydrostatic hydrophone material properties

Property	Units	1-3 Composite	0-3 Composite	PVDF	PZT-5H
g_h	(Vm/N) $\times 10^{-3}$	66	55	90	1.5
$g_h d_h$	(m^2/N) $\times 10^{-12}$	18	1	1	0.1
Rel. dielectric constant		460	40	13	3200

It can be seen that both the 1-3 and 0-3 composites and PVDF have a significantly higher hydrostatic mode sensitivity and figure of merit than PZT-5H, but they all have significantly reduced relative dielectric constant resulting in a higher input impedance.

6.4 Bender Hydrophones

The bender hydrophone may be analyzed using a projector model (see Sect. 5.6) and reciprocity. As in the case of the projector, a trilaminar hydrophone design with simply supported boundary conditions provides the best sensitivity, but may not be the best design for deep submergence or low cost. Circular bender hydrophones may be analyzed using the models in Chap. 5 or those by Woollett [16], Antonyak and Vassergisre [17], and Aronov [18]. The bender bar has also been modeled by Woollett [19].

Probably the simplest and most commonly used design is the dual thin piezoelectric disc on a brass plate shown in Fig. 6.18. A simple approximate equivalent circuit for estimating the performance is given in Fig. 6.19.

Because the piezoelectric ceramic elements are wired in parallel and there is symmetry about the plane through the ring, the equivalent circuit may be applied to one of the dual elements. With A the effective aperture area (on one side), Y_e^E the effective short circuit Young's modulus, and ρ the effective density, the simply supported compliance and mass are

$$C^E = \left[2/3\pi Y_e^E\right]\left[a^2/t^3\right] \text{ and } M_m = 2\pi a^2 t\rho/3. \tag{6.30}$$

The mechanical mass M_m has the subscript m to distinguish it from the sensitivity M, and the effective parameters are for the composite piezoelectric ceramic

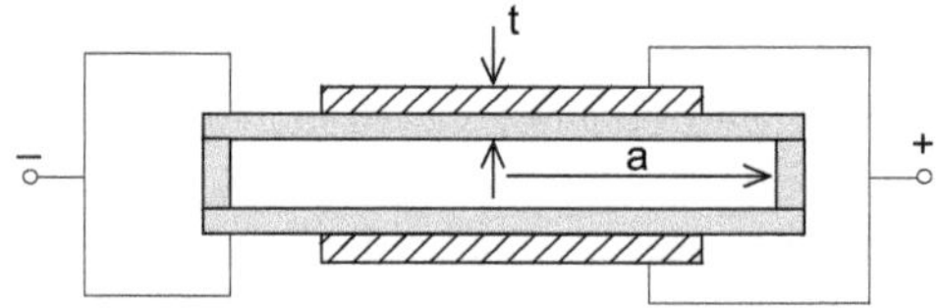

Fig. 6.18 Simplified dual bender piezoelectric disc hydrophone supported on brass plates mounted on a brass ring

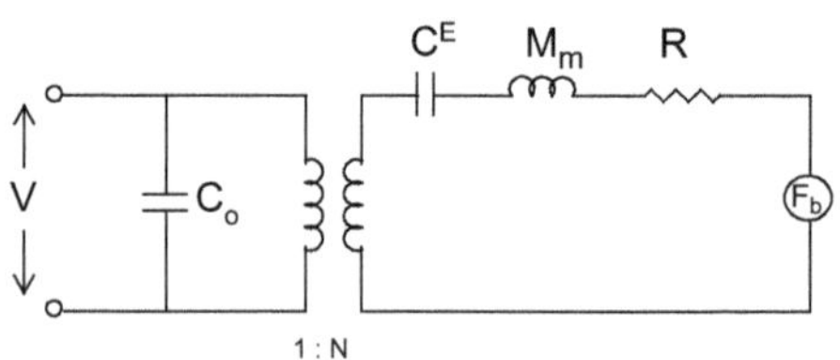

Fig. 6.19 Equivalent circuit for dual bender disc hydrophone

disc and the inactive brass substrate illustrated in Fig. 6.18. The resulting resonance frequency is $\omega_r = (3c^E/2)t/a^2$ where $c^E = \left(Y_e^E/\rho\right)^{1/2}$ is the short circuit effective sound speed and the antiresonance frequency $\omega_a = (3c^D/2)t/a^2$ where $c^D = \left(Y_e^D/\rho\right)^{1/2}$ is the open circuit effective sound speed of the composite disc. The clamped capacitance and electromechanical turns ratio are given by

$$C_0 = C_f\left(1 - k_e^2\right) \text{ and } N = k_e\left(C_f/C^E\right)^{1/2}, \tag{6.31}$$

where C_f is the free capacitance and k_e is the effective coupling coefficient. In Fig. 6.19 the force $F_b = Ap_iD_a$, R is the sum of the mechanical resistance and the radiation resistance. (The radiation mass may be added to the mechanical mass M_m.) $D_a \approx 1$ for most benders, since they are generally small compared to the wavelength up through antiresonance, ω_a.

The receiving sensitivity may be obtained from the equivalent circuit of Fig. 6.19 and written as

$$M = V/p_i = k_e A\left(C^E/C_f\right)^{1/2} D_a/\left[1 - (\omega/\omega_a)^2 + j\omega/\omega_a Q_a\right]. \tag{6.32}$$

where the numerator gives the low frequency sensitivity. For the case where the ceramic disc extends to the inner edge of the ring support and has a thickness $t/2$, Eq. (6.32) becomes

$$M = V/p_i = g_e a(a/t)\left(D_a/\sqrt{3}\right)/\left[1 - (\omega/\omega_a)^2 + j\omega/\omega_a Q_a\right], \tag{6.33}$$

where the quantities ω_a and Q_a are evaluated under water loading conditions. Equation (6.33) shows that the sensitivity is proportional to the effective g constant, g_e, with $g_e^2 = k_e^2/Y_e^E\varepsilon_e^T$, where $\varepsilon_e{}^T$ is the effective free dielectric constant of the transducer assembly and a/t is the radius-to-thickness ratio of the composite plate. The numerator may be written in terms of the antiresonance frequency as $g_e(3c^D/2\omega_a)(D_a/\sqrt{3})$ showing a sensitivity bandwidth product given by $Mf_a = g_e c^D 3D_a/\left(4\pi\sqrt{3}\right)$ for the bender hydrophone.

The maximum stress in a simply supported plate is given by $T = 1.25(a/t)^2P$ where P is the ambient hydrostatic pressure. Thus the numerator in Eq. (6.33) may also be written as $g_e a(T/P)^{1/2}D_a/(3.75)^{1/2}$, showing the sensitivity for a given radius and maximum acceptable T/P ratio. It should be mentioned that bender discs, mounted as shown in Fig. 6.18, only approximate the ideal simply supported condition with a resulting resonance that falls between simply supported and fixed boundary conditions.

6.5 Vector Hydrophones

Typical piezoelectric ceramic hydrophones detect the acoustic pressure, a scalar quantity, and convert this pressure into a proportional output voltage. These pressure or scalar sensors, which were the subject of the previous sections, show no directional sensitivity when small compared to the wavelength. In this section we will discuss acoustic sensors which are sensitive to both the magnitude and the direction of the acoustic wave and are accordingly referred to as vector sensors. The main reason for interest in vector sensors is the ability of these sensors to yield directional information from compact single or dual element hydrophone designs. Arrays of vector sensors are discussed in Chap. 8.

The typical vector sensor responds to a component of the acoustic particle velocity with a "figure eight," or cosine, directivity pattern. This pattern has a 90° beam width and a DI of 4.8 dB. The response of piezoelectric ceramic vector sensors below resonance decreases with decreasing frequency at a rate of 6 dB/octave and with a 90° phase shift relative to the response of a pressure hydrophone. This reduced sensitivity at the lower frequencies can lead to a large equivalent self-noise and reduced signal-to-noise ratio. Some vector sensors detect the gradient of the pressure in a particular direction. For an x directed wave, the gradient of the pressure, p, is directly related to the acceleration, a, and velocity, u, through one of the basic acoustic equations, Eq. (10.1a), with $a = -(1/\rho)\partial p/\partial x$ where ρ is the density of the medium. For sinusoidal motion

$$u = -\,(1/j\omega\rho)\partial p/\partial x, \tag{6.34}$$

and the particle velocity is seen to be proportional to the gradient of the pressure.

Three-dimensional coverage can be obtained by three vector sensors oriented in three orthogonal directions (see Sect. 8.5.1). Vector sensor performance can be derived from two omni sensors connected as a dipole, from the pressure differential across a sensitive surface or enclosed volume or from excitation of the dipole mode of vibration of a buoyant body. Use of directional dipole microphones is commonplace and has been well documented [20, 21]. Higher order mode sensors that use the quadrupole mode are discussed in Sect. 6.5.5.

It is convenient and customary to evaluate vector sensors by measuring their plane wave pressure sensitivity using the same apparatus used to evaluate pressure sensors. In this section we will call this sensitivity M_p, to distinguish it from M, the sensitivity of a pressure sensor. The plane wave velocity sensitivity, M_u, and the plane wave acceleration sensitivity, M_a, of vector sensors can then be determined from M_p as follows:

$$M_p = V/p,$$

$$M_u = V/u = \rho c V/p = \rho c M_p,$$

$$M_a = V/a = V/j\omega u = (\rho c/j\omega)M_p,$$

where V is the output voltage and c is the sound speed in the medium. The acceleration response of a piezoelectric ceramic velocity sensor is flat below resonance, because M_p is proportional to frequency. Since velocity hydrophones often use accelerometers as the transduction mechanism the sensitivity is often referenced to 1 g or 1 mg where $g = 9.81\ \mathrm{m/s^2}$ is the acceleration of gravity. Section 6.7.5 includes an example where $M_a = -17$ dB//V/g, which converts to $M_p = -205$ dB//V/μPa.

6.5.1 Dipole Vector Sensors, Baffles, and Images

Two hydrophones that are very small compared to the wavelength, with their outputs connected together, may be used to create a dipole vector sensor. Consider Fig. 6.20a which shows two small hydrophones intercepting a plane wave of amplitude p_i, arriving at the angle θ from the axis of the two sensors.

With s the distance between the two hydrophones the distance between the wave front at the origin and the wave fronts at the hydrophones is $d = (s/2)\cos\theta$ and, with pressure sensitivities M_1 and M_2, Fig. 6.20a shows that the summed voltage output is

$$\begin{aligned} V &= M_1 p_i e^{-jk(r-d)} + M_2 p_i e^{-jk(r+d)} \\ &= p_i e^{-jkr}\left[M_1 e^{jk(s/2)\cos\theta} + M_2 e^{-jk(s/2)\cos\theta}\right], \end{aligned} \tag{6.35}$$

which with $ks/2 = \pi s/\lambda$ may be written as

$$V = p_i e^{-jkr}[(M_1 + M_2)\cos\{(\pi s/\lambda)\cos\theta\} + j(M_1 - M_2)\sin\{(\pi s/\lambda)\cos\theta\}]. \tag{6.36}$$

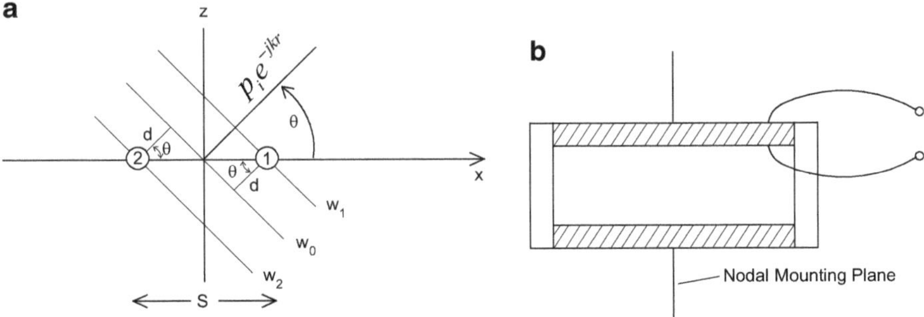

Fig. 6.20 (**a**) Two omni scalar sensors, 1 and 2, receiving plane wave fronts w_1 and w_2. (**b**) End-capped 31 mode cylinder pressure sensor mounted at nodal plane

If the sensitivities in Eq. (6.36) are the same, i.e., $M_2 = M_1$, only the first term remains resulting in an omnidirectional response for $s/\lambda \ll 1$. In this case Eq. (6.36) becomes

$$V = p_{\mathrm{i}}\mathrm{e}^{-jkr}2M_1 \cos\{(\pi s/\lambda)\cos\theta\} \approx 2M_1 p_{\mathrm{i}}\mathrm{e}^{-jkr},$$

resulting in a flat response below resonance if the pressure sensitive hydrophones, 1 and 2, are piezoelectric with M_1 and M_2 constant. On the other hand, if the phase is reversed on the second hydrophone, the outputs are subtracted, and only the second term remains leading to

$$V = 2jM_1 p_{\mathrm{i}}\mathrm{e}^{-jkr}\sin\{(\pi s/\lambda)\cos\theta\} \approx 2jM_1 p_{\mathrm{i}}\mathrm{e}^{-jkr}(\pi s/\lambda)\cos\theta, \tag{6.37a}$$

where the approximate result holds for $s/\lambda \ll 1$. Equation (6.37a) shows the cosine directivity pattern, the 90° phase shift, and the 6 dB/octave slope in the frequency response since $1/\lambda = f/c$. The separation between the hydrophones must be less than $\lambda/4$ to achieve a dipole pattern. This system has the advantage that it may be used as either a scalar pressure sensor by summing the outputs or as a vector dipole sensor by differencing the outputs.

A basic disadvantage of dipole sensors, excluding the multimode sensor (see Sect. 6.5.5), is that only small sensors may be used since the center-to-center spacing or overall length must be small. The dipole does, however, have the advantage of insensitivity to acceleration caused by mechanical vibrations since each of the two small omnidirectional pressure sensors can be symmetrically mounted at a nodal plane as illustrated in the cross section of Fig. 6.20b for an end-capped 31 mode hydrophone. Then, when accelerated, each half of each sensor yields voltage of opposite phase from each side of the nodal plane yielding zero voltage for the summed output. The hydrophone pair of Fig. 6.20a can also be considered a pressure gradient sensor since it detects the difference in pressure over a small distance.

We now consider single pressure sensors and dipole sensors near a soft baffle as they might be used in external hull-mounted arrays (see Sect. 8.5). It can be seen from Figs. 6.20a and 6.21a, using the method of images, that a single pressure sensor at a distance $s/2$ from a large plane soft baffle is equivalent to a dipole sensor with separation s. Thus with sensitivity M_1 its voltage output is the same as the dipole result in Eq. (6.37a).

Next consider a dipole sensor near a soft baffle. Figure 6.21b shows a dipole sensor with small separation distance L, at a midpoint distance $s_1/2$ in front of a soft baffle, creating a reversed image behind the baffle also at a distance $s_1/2$. The result is equivalent to a two element array of aligned identical dipole sensors with center-to-center distance s_1. The voltage output, obtained by use of the Product Theorem (see Sect. 7.1.1), and Eq. (6.37a) is

$$\begin{aligned} V &= [2\cos\{(\pi s_1/\lambda)\cos\theta\}]\left[2jM_1 p_{\mathrm{i}}\mathrm{e}^{-jkr}\sin\{(\pi L/\lambda)\cos\theta\}\right] \\ &\approx 4jM_1 p_{\mathrm{i}}\mathrm{e}^{-jkr}(\pi L/\lambda)\cos\theta, \end{aligned} \tag{6.37b}$$

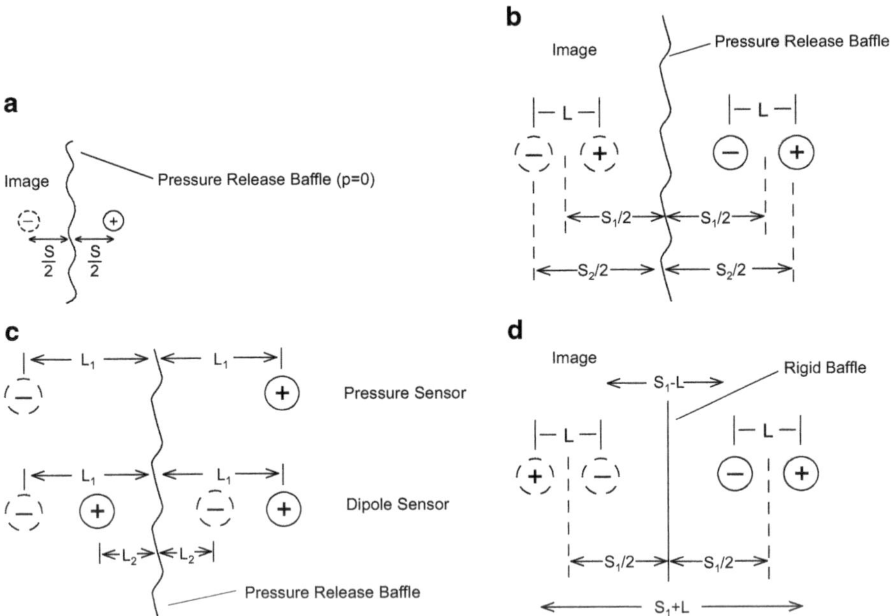

Fig. 6.21 (**a**) Hydrophone in front of a large planar soft baffle. (**b**) Dipole sensor near a large planar soft baffle. (**c**) Comparison of a pressure sensor and a dipole with same stand-off from a soft baffle. (**d**) Dipole sensor with separation L at a distance $S_1/2$ in front of a large planar rigid baffle

where both $\pi s_1/\lambda \ll 1$ and $\pi L/\lambda \ll 1$. For a dipole of separation L, with no baffle the voltage output from Eq. (6.37a) is

$$V_{\mathrm{f}} \approx 2jM_1 p_{\mathrm{i}} \mathrm{e}^{-jkr}(\pi L/\lambda)\cos\theta. \tag{6.37c}$$

Thus, the soft baffle case of Eq. (6.37b) shows an improvement in output by a factor of 2 (6 dB) over the no baffle case of Eq. (6.37c) demonstrating that a soft baffle provides a favorable additive condition for a nearby dipole sensor.

Since the size of a sensor and the distance it projects from a baffle are important for hull-mounted arrays, we will compare the output of a dipole and a single pressure sensor when both have the same maximum projection from a soft baffle. Let the center of the single sensor be at distance L_1 from the baffle as shown in Fig. 6.21c.

As we've seen above this case is equivalent to a dipole with separation $s = 2L_1$, and Eq. (6.37a) gives the output as

$$V_1 = 4jM_1 p_{\mathrm{i}} \mathrm{e}^{-jkr}(\pi L_1/\lambda)\cos\theta. \tag{6.37d}$$

Let the center of the outer sensor of the dipole in Fig. 6.21c also be at distance L_1 from the baffle and the inner sensor be at L_2. Equation (6.37b) with $L = L_1 - L_2$ then gives the dipole output as

$$V_2 = 4jM_1p_ie^{-jkr}(\pi/\lambda)(L_1 - L_2)\cos\theta. \tag{6.37e}$$

The output of a dipole is maximum when the separation between the two poles is as large as possible, in this case when L_2 is as small as possible. But L_2 cannot be less than the radius of an individual sensor, and therefore

$$V_2 = V_1(L_1 - L_2)/L_1 < V_1.$$

Thus the output of the dipole is less than the output of the single pressure sensor when both have the same maximum projection from the soft baffle. This conclusion may also be obtained from a more intuitive approach by considering the sum of outer (real and image) and inner (real and image) pole pairs with separations $2L_1$ and $2L_2$, illustrated by the dipole sensor in the lower part of Fig. 6.21c. As seen, the outer pair is the same as the pressure sensor arrangement in the upper part of Fig. 6.21c; and thus, yields a voltage given by Eq. (6.37d). The output voltage for the inner pair has the same form as Eq. (6.37d) but with separation $2L_2$ and reversed polarity, with negative sign, yielding

$$V_3 = -4jM_1p_ie^{-jkr}(\pi L_2/\lambda)\cos\theta. \tag{6.37f}$$

The total result by superposition is the sum $V_1 + V_3$ which again leads to Eq. (6.37e). It can now be seen that the two inner (real and image) poles produce a dipole of opposite sign compared to the outer pair; reducing the output voltage of the outer pair. Thus the output for the dipole sensor arrangement, with four poles, is less than the outer pair alone, as in the pressure sensor arrangement in the upper part of Fig. (6.21c).

If a similar result holds for other types of vector sensors, such as those shown in Figs. 6.24 and 6.27, it has important implications for use of vector sensors in hull mounted arrays where a compliant baffle is necessary for noise reduction. This subject will be discussed further in Sect. 8.5.3 where vector and pressure sensors will be compared using a somewhat more realistic model of a compliant noise reduction baffle.

The case of a dipole near a rigid baffle is shown in Fig. 6.21d. The image is not reversed in the way it is for the soft baffle in Fig. 6.21b, making it equivalent to an array of two out-of-phase dipoles, with a reduced output given by

$$V = -4M_1p_ie^{-jkr}(\pi s_1/\lambda)(\pi L/\lambda)\cos^2\theta.$$

This response has characteristics similar to a quadrupole sensor with a 180° phase shift, a slope of 12 dB/octave with frequency and a $\cos^2\theta$ directivity pattern.

In some applications there may be other considerations for using vector sensors near soft baffles. In these cases the separation between the surfaces or elements of the vector sensor should be made as large as possible, consistent with dipole conditions, and the inner surface or inner element should be positioned on the soft baffle for maximum overall sensitivity. The standoff distance from the soft baffle to the outer surface or element should be less than $\lambda/8$ so that the distance to the image is less than $\lambda/4$ for dipole performance.

6.5.2 Pressure Gradient Vector Sensor

A body submerged in a fluid experiences compression forces due to uniform in-phase acoustic pressure. The body also experiences translational forces when the pressure varies across it. Consider Fig. 6.22 which shows an incident wave on an end-capped cylindrical tube of small cross-sectional area A and length L with a plane wave arriving at an angle θ, much like Fig. 6.20, but with distance $d = (L/2)$ $\cos\theta$. At end 1 the pressure is $p_i e^{-jk(r-d)}$ while at end 2 it is $p_i e^{-jk(r+d)}$, and the net force in the x direction is given by the difference as

$$F = Ap_i e^{-jkr} 2j \ \sin\{(\pi L/\lambda) \ \cos\theta\} \approx 2jAp_i e^{-jkr} (\pi L/\lambda) \ \cos\theta, \tag{6.38}$$

where $\pi L/\lambda = kL/2$ and the approximation holds for $L/\lambda \ll 1$. The similarity to Eq. (6.37) is evident showing the cos θ dependence, the 90° phase shift and the 6 dB/octave response. However, in Sect. 6.5.1 we used small omni pressure sensors and detected the difference in the output voltages, while in the present case a net force F is developed on the tube along the x direction which must be detected and converted to a voltage.

One method of detecting the force is to use the voltage output from a pair of accelerometers (see Sect. 6.5.4) attached to a rigid center plate as illustrated in Fig. 6.23a.

The accelerometers may be fabricated from piezoelectric rings or discs attached to a mass M with reversed polarity for the same polarization, P_0, direction. As the

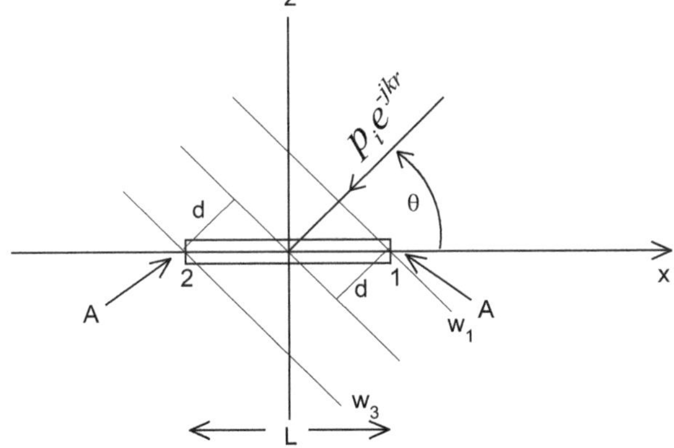

Fig. 6.22 Force on tube of length L and cross-sectional area A

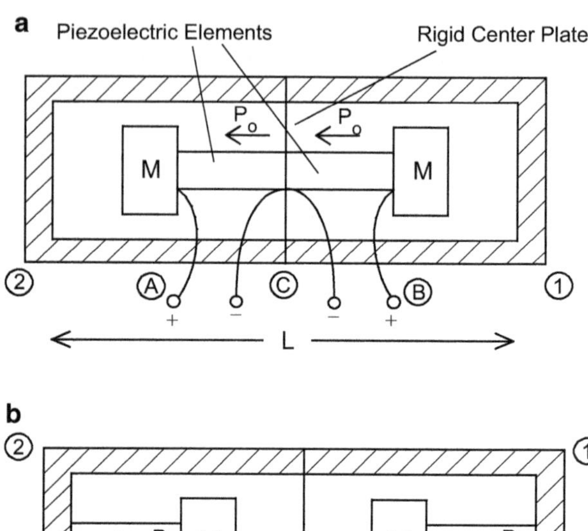

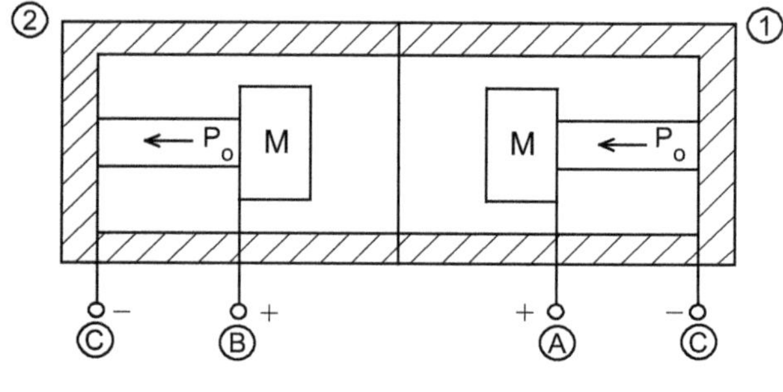

Fig. 6.23 (**a**) Central accelerometers for net force detection. (**b**) End mounted accelerometers

tube of length L moves to the left, the piezoelectric piece on the left experiences compression producing a positive voltage at terminal A. The piece on the right experiences expansion, but because of the reversed polarity, also produces a positive voltage at terminal B, both relative to the common terminal C. If these voltages are summed, the net force and resulting motion will be detected. The accelerometers may also be moved to the ends of the tube, as shown in Fig. 6.23b, where if added they would also detect the force. For small lengths, L, the net force is proportional to the x-component of the pressure gradient; i.e., $F = -AL\partial p_i/\partial x$, consistent with Eq. (6.38). Since these sensors detect acceleration, they are sensitive to acceleration from mechanical vibrations and consequently require mechanical isolation from these sources of noise. Figure 6.23a, b shows pairs of accelerometers which provide a significant reduction of unwanted pressure sensitivity through the cancellation of compression waves or forces by the subtraction process. However, in some applications one accelerometer, as illustrated in Fig. 6.24, may be sufficient.

The accelerometer sensor differs from the dipole sensor of Sect. 6.5.1 in that it detects the net force on a body due to a pressure differential while the dipole sensor detects the difference in pressure.

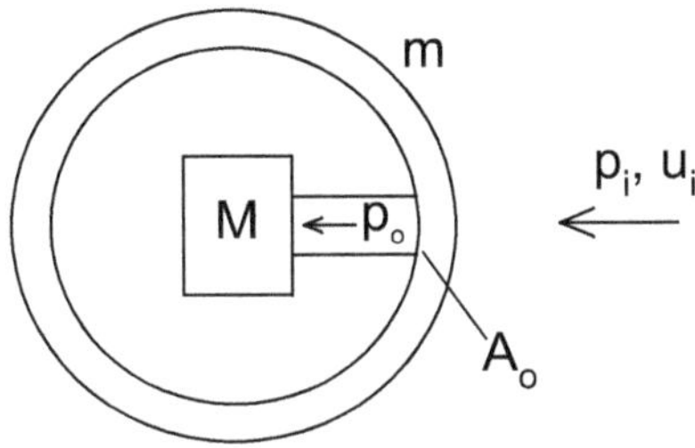

Fig. 6.24 Electric field accelerometer vector sensor with housing mass, *m*

6.5.3 Velocity Vector Sensor

The same system may be interpreted as a velocity sensor. If the tube and the accelerometers of Fig. 6.23a and b are considered as one mass m then the velocity, u, of the tube in the x direction is $u = F/j\omega m$ and Eq. (6.38) becomes

$$u = 2jAp_i e^{-jkr}(\pi L/j\omega m\lambda)\cos\theta = p_i e^{-jkr}(AL/mc)\ \cos\theta. \tag{6.39}$$

Since the buoyancy of the system is the ratio of the water mass displaced by the tube to the mass of the tube sensor system, the buoyancy is $B = \rho AL/m$ and Eq. (6.39) becomes

$$u = B\left(p_i e^{-jkr}/\rho c\right)\cos\theta = Bu_i\cos\theta, \tag{6.40}$$

where u_i is the particle velocity of the incident plane wave arriving at angle θ with the x axis.

Equation (6.40) shows that the velocity of the tube is proportional to the component of the incident acoustic particle velocity vector parallel to the tube with a proportionality constant equal to the buoyancy of the tube sensor system. Thus the tube and particle velocities are in-phase with no frequency dependence between them for incident plane waves. If an electric field accelerometer is used to convert this motion to a voltage the voltage will be frequency dependent, but if a moving coil accelerometer is used there will be no frequency dependence above the fundamental resonance up to the first unwanted parasitic resonance.

The spherical velocity sensor, which uses an accelerometer inside a buoyant sphere, has been analyzed by Leslie, Kendall, and Jones [22] for a wide range of frequencies. They also have shown that for small spheres Eq. (6.40) becomes

$$u = [3B/(2 + B)]u_i\cos\theta. \tag{6.41}$$

Note that $u = u_i\cos\theta$ for neutral buoyancy where $B = 1$. A similar design, but using a buoyant foam, has been given by Gabrielson et al. [23], and an analysis of suspension systems has been given by McConnell [24]. Accelerometers, as illustrated in Figs. 6.23 and 6.24, may be used to detect the motion as discussed in Sect. 6.5.4.

6.5.4 Accelerometer Sensitivity

Accelerometers are the major transduction device for vector sensors as discussed in Sects. 6.5.2 and 6.5.3. Specific accelerometer arrangements which are insensitive to pressure but sensitive to motion as a result of net force or particle motion are illustrated in Fig. 6.23a, b. The sensitivity may be obtained from the Tonpilz hydrophone model of Figs. 6.13, 6.14, and 6.15 and the resulting sensitivity equation, Eq. (6.17). The device may be simplified as shown in Fig. 6.24 along with the corresponding equivalent circuits of Fig. 6.25.

We note that in Eq. (6.17) the force $p_i D_a A$ is equal to $j\omega u(m+M_t)$ where $j\omega u$ is the acceleration and u the velocity of the sensor. Substitution of $j\omega u(m+M_t)$ for $p_i D_a A$ in Eq. 6.17 and use of Eq. (6.41), for a spherical structure, leads to

$$M_u \approx [g_{33}tj\omega \ \cos\theta \ M_t/A_0][3B/(2+B)]/\left[1-(\omega/\omega_a)^2+j\omega/\omega_a Q_a\right], \quad (6.42a)$$

where M_t is the mass attached to the end of the piezoelectric stack plus the dynamic mass of the stack. The accelerometer sensitivity below resonance is given by the numerator of Eq. (6.42a). Note that there is a 6 dB/octave slope for a given particle velocity but a flat response for a given particle acceleration as shown by

$$M_a = M_u/j\omega \approx [g_{33}t \ \cos\theta M_t/A_0][3B/(2+B)]/\left[1-(\omega/\omega_a)^2+j\omega/\omega_a Q_a\right].$$

If the acceleration sensitivity is derived or measured, one may easily obtain the velocity sensitivity through $M_u = j\omega M_a$ and the useful pressure sensitivity through

$$M_p = j\omega M_a/\rho c.$$

If the sensor is not a small sphere, a buoyancy factor other than $[3B/(2+B)]$ would be needed. In some cases the small sphere factor or simply B may be a useful approximation. These sensors are typically designed for maximum sensitivity with $B \approx 1$.

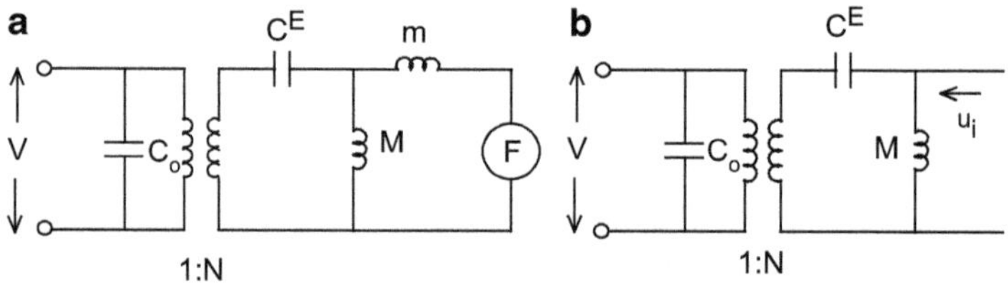

Fig. 6.25 Equivalent circuit of accelerometer for input force (**a**) and velocity (**b**)

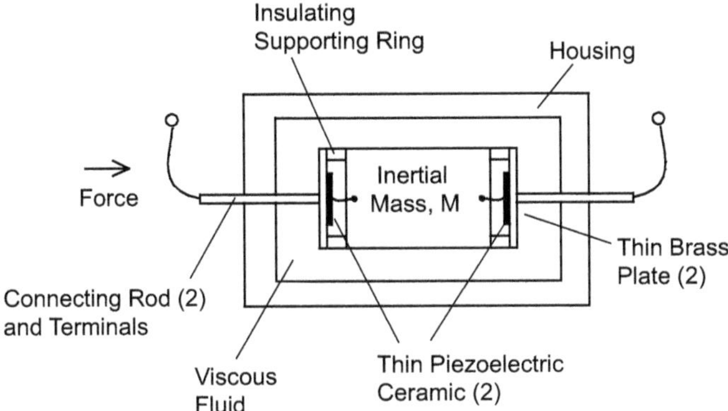

Fig. 6.26 Directional accelerometer based hydrophone with viscous damping for resonance operation [26]

The moving coil transduction mechanism, which is basically a velocity sensor, is also frequently used as an accelerometer [25]. These "geophones" are operated above their very low resonance where the output voltage is

$$V \approx B_f l_c u_i \cos\theta[3B/(2+B)], \tag{6.42b}$$

with B_f the flux density and l_c the coil length. Note that for this transduction mechanism, there is no frequency dependence between the particle velocity and the voltage output. These accelerometers are often limited at the high end of the band by resonances in the coil suspension system.

A piezoelectric accelerometer with a higher operating frequency band is illustrated in Fig. 6.26.

This accelerometer [26] consists of two flexural disc transducers mounted on an inertial mass. The thin brass plates are edge supported on the inertial mass and driven at the center by small connecting rods which translates the motion of the housing to the center of the disc. The piezoelectric discs are wired out of phase giving additive outputs as the housing moves causing one disc to move inward and the other outward. This differs from Fig. 6.23b in that there is a common inertial mass and that bender piezoelectric ceramic discs on a metal substrate are used to convert the motion to a voltage. The resonance is damped by the viscosity of the interior fluid. The size of the accelerometer is approximately one cubic inch. Another design which takes advantage of the high sensitivity of piezoelectric ceramic benders is illustrated in Fig. 6.27 with overall dimensions of approximately one inch diameter and one-half inch thickness. In this design [27] a trilaminar bender with equal plate thickness is used for maximum performance along with syntactic foam to achieve buoyancy and maximum sensitivity.

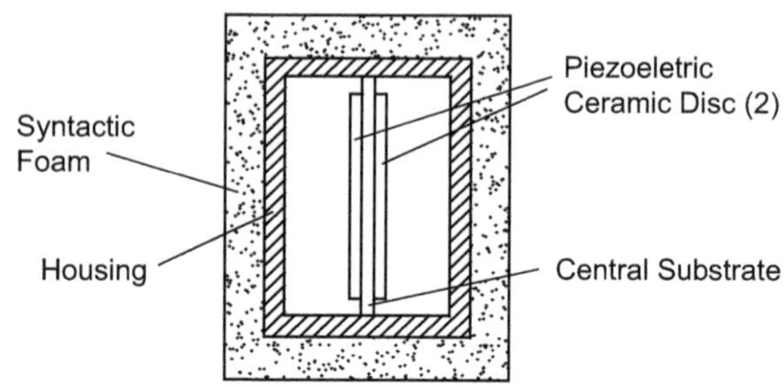

Fig. 6.27 Directional accelerometer based hydrophone with trilaminar bender and syntactic foam buoyancy

6.5.5 *Multimode Vector Sensor*

In Sect. 6.2 we considered piezoelectric ceramic rings and hollow spheres operating in the fundamental extensional "breathing" mode resulting in omnidirectional beam patterns and flat response below antiresonance, f_a. Other modes of vibration of these hydrophones may be used to obtain directional response as discussed for projectors in Sect. 5.2.6. There we focused on the first three ring extensional modes, the omni ($n = 0$), dipole ($n = 1$) and quadrupole ($n = 2$) modes, illustrated in Figs. 5.14 and 5.15, with hydrophone antiresonances given by $f_1 = f_a\sqrt{2}$ and $f_2 = f_a\sqrt{5}$, respectively. The first mode is sensitive to the pressure and the hydrophone operates as a scalar sensor. The second (dipole) mode is sensitive to the pressure gradient and the hydrophone acts as a vector sensor. The third (quadrupole) mode is sensitive to the cross gradient of the dipole mode and has been termed a dyadic sensor [24]. A model for ring modes has been given by Gordon, Parad, and Butler [28] and a model for spherical modes has been given by *Ko*, Brigham, and Butler [5]. This transducer may be analyzed as a hydrophone by the projector model given in Sect. 5.2.6 and the use of reciprocity.

The response and phase shift may be determined for each mode from the force exerted on the cylinder by a plane wave. Consider the plane wave, $p = p_i e^{-jkx}$, shown in Fig. 6.28, incident perpendicular to the axis of a cylindrical transducer of radius a.

As the wave progresses along the x axis, the pressure along the circumference follows a spatial dependence $x = a \cos\theta$ and the incident plane wave pressure may be written in the form $p = p_i e^{-jka\cos\theta}$. The incident pressure is scattered off the cylinder and in general the diffraction constant, as given by Eq. (6.51), is dependent on the sum of the incident wave and the scattered wave pressures and the velocity distribution around the cylinder. If the cylinder is small compared to the wavelength, the incident wave field is not significantly disturbed and the scattered wave is negligible. Thus, for the n'th modal velocity distribution, $\cos n\theta$, on the cylinder with active area $A = 2\pi aL$, Eq. (6.51) with $dS = La d\theta$ and L the length of the cylinder gives the force

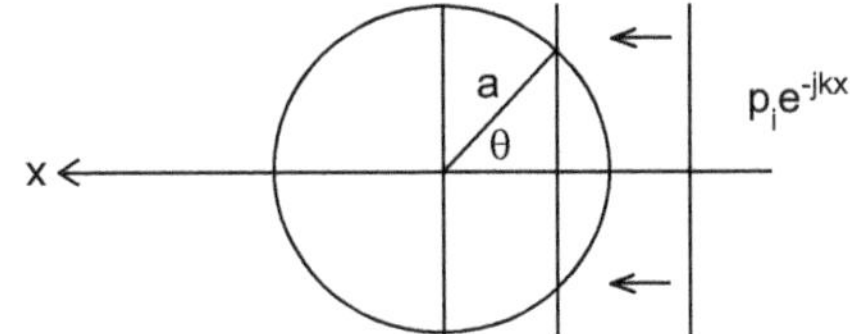

Fig. 6.28 Plane wave incident on a cylinder of radius a and length L

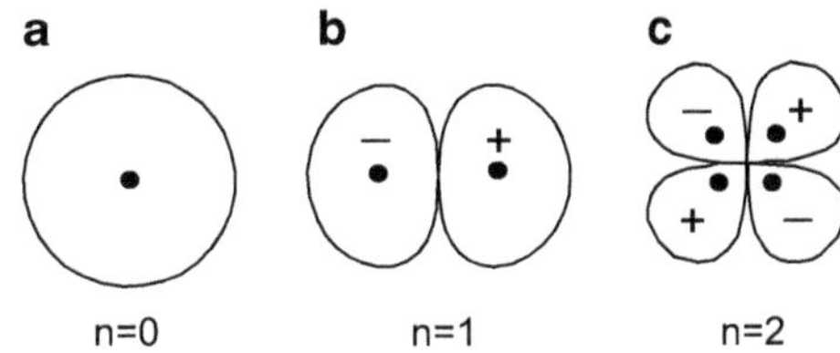

Fig. 6.29 (**a**) omni, (**b**) dipole, and (**c**) quadrupole point sensors

$$F_n \approx aLp_{\text{i}} \int_{-\pi}^{+\pi} \mathrm{e}^{-jka\cos\theta} \cos n\theta \mathrm{d}\theta = 2\pi aLp_{\text{i}} J_n(ka)/j^n$$
$$\approx Ap_i e^{-jn\pi/2}(ka/2)^n/n! \tag{6.43}$$

where we have used an integral representation [29] of the Bessel function of order n, and the last approximate form of Eq. (6.43) is valid for low frequencies where $ka \ll 1$. The corresponding diffraction constant is $D_{\text{a}}(n) = F_n/Ap_{\text{i}}$. This representation shows the $n\pi/2$ phase-dependence (90° phase shift for each mode) and, since $k = \omega/c$, also shows the ω^n dependence of the amplitude resulting in a rise of $6n$ dB/octave in the receiving frequency response as the frequency is increased. The modal beam patterns may be simulated by point sensors: a single sensor for the omni mode, two out-of-phase omni sensors for the dipole mode, and parallel reversed dipole pairs for the quadrupole mode as illustrated in Fig. 6.29.

The most commonly used higher order mode is the dipole mode which has a figure eight directional response, and in this case the piezoelectric ceramic cylinder or sphere may be considered a vector sensor. The dipole mode has a constant beam width of 90° and a DI = 4.8 dB. The dipole generated from a cylindrical or spherical mode maintains the dipole form for a wide range of frequencies and, in principle, is not limited to small size hydrophones. Ehrlich and Frelich [30] used orthogonal pairs of dipoles along with the omni mode to obtain directional information from a small multimode piezoelectric cylinder. The inner electrode of the 31 mode piezoelectric ceramic cylinder is separated into four sections for two orthogonal dipole excitations with beam pattern voltage outputs, V_{N} and V_{E}, proportional to $\sin\theta$ and $\cos\theta$, respectively. Since the phase of each lobe of a dipole alternates, the general direction of the incoming signal is determined by comparing the lobe phase with the phase of the omni mode. The bearing can be obtained by dividing one channel by the other giving $V_{\text{N}}/V_{\text{E}} = \tan\theta$ with the bearing angle $\theta = \mathit{arc}\tan(V_{\text{N}}/V_{\text{E}})$.

An alternate approach is to phase shift one of the channels by 90° yielding the quadrature sum $V_{\text{E}} + j\,V_{\text{N}} = V_1(\cos\theta + j\sin\theta) = V_1 \mathrm{e}^{j\theta}$ with the bearing angle given by the phase angle which can then be compared to the omni mode phase angle to eliminate directional ambiguity. This quadrature sum case is interesting in that it creates a toroidal beam pattern with uniform magnitude V_1 in the horizontal X, Y plane and a null on the vertical Z axis [31]. The null in the axial direction results from cancellation of the oppositely phased segments of the cylinder that are used to create the dipole. This null is maintained even though the north south and east west dipoles are separately phase shifted to attain the voltage function $V_1e^{j\theta}$.

6.5.6 Summed Scalar and Vector Sensors

Addition of the output from scalar and vector sensors can yield useful hydrophone beam pattern characteristics. The most well-known example is the cardioid beam pattern formed from the sum of scalar omni and vector dipole hydrophones, with voltage amplitude and phase adjusted for equal far-field pressure amplitude and phase, yielding the normalized true cardioid beam pattern function,

$$P(\theta) = (1 + \cos\theta)/2, \tag{6.44a}$$

illustrated in Fig. 6.30. This pattern has a null at 180°, is 6 dB down at 90°, has a 3 dB down beam width of 131° and a DI of 4.8 dB, the same DI as the dipole, since the omni adds no directionality.

We note that the cardioid beam width is considerably wider than the 90° beam of the dipole. We call the function of Eq. (4.44a) a true cardioid since it satisfies the mathematical description of a cardioid curve. There are improvements on the equal amplitude summation such as the super-cardioid with the pattern function

$$P(\theta) = (1 + 1.7\ \cos\theta)/2.7. \tag{6.44b}$$

This case has a front-to-back ratio of 11.7 dB, is 8.6 dB down at 90°, has a beam width of 115° and a DI of 5.7 dB, as illustrated in Fig. 6.31a.

The hypercardioid with a pattern function

$$P(\theta) = (1 + 3\ \cos\theta)/4, \tag{6.45}$$

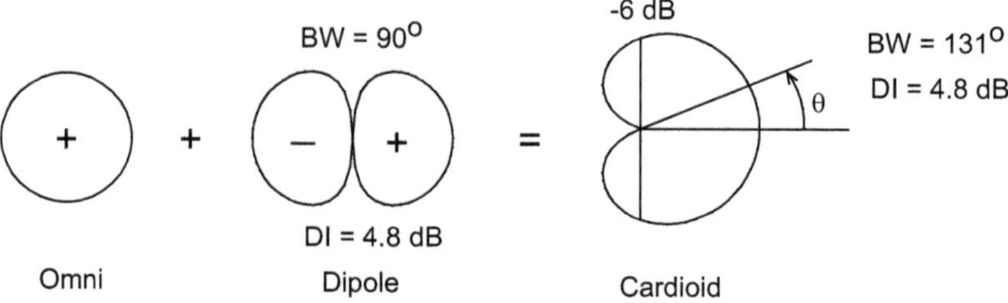

Fig. 6.30 Synthesis of a cardioid pattern

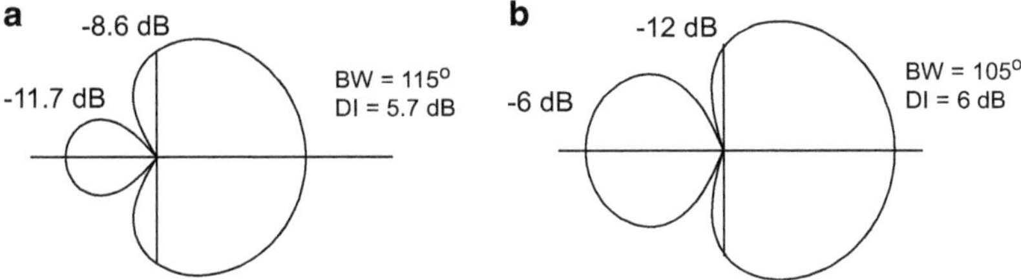

Fig. 6.31 Super cardioid (**a**) and hyper cardioid (**b**) beam patterns

is shown in Fig. 6.31b. It has the highest DI possible with one omni and one dipole sensor yielding a DI of 6.0 dB, is 12 dB down at 90°, has a beam width of only 105° but a front-to-back ratio of only 6 dB. As seen, there is a trade-off between the front-to-back ratio and the beam width, with the smallest beam width of 90° achieved for the dipole alone with 0 dB front-to-back ratio. As the dipole component is increased relative to the omni component both the front-to-back ratio and beam width are decreased. Other means must be used to increase the front-to-back ratio while also achieving a narrower beam.

As discussed in Sect. 5.2.6, the addition of the quadrupole mode can be used to achieve both a narrower beam and a better front-to-back ratio. The normalized beam pattern function for this case may be generally written as

$$P(\theta) = (1 + A \cos\theta + B \cos 2\theta)/(1 + A + B).$$

The corresponding axi-symmetric DI [see Eq. (1.20) and Sect. 13.13] is given by $\mathrm{DI} = 10 \log D_{\mathrm{f}}$ where

$$D_f = 15(1 + A + B)^2/\left(15 + 5A^2 + 7B^2 - 10B\right).$$

The case of the true cardioid, Eq. (6.44a), is given by $A = 1$ and $B = 0$. Adding a quadrupole with weighting factor $B = 0.414$ gives the beam pattern [32]

$$P(\theta) = (1 + \cos\theta + 0.414 \cos 2\theta)/2.414, \tag{6.46a}$$

which has a $\mathrm{DI} = 7.1$ dB, a 90° beam width, is 12 dB down at 90° and has a front-to-back ratio of 15 dB as illustrated in Fig. 6.32a.

The quadrupole mode may also be used to achieve a wider beam cardioid type pattern of Fig. 6.30 with a null at 180° through the addition of a bidirectional pattern to the cardioid pattern [33]. The bidirectional pattern may be formed through the addition of an omni mode of amplitude 1 and a quadrupole mode of amplitude −1, both weighted by a factor 0.5, shown added to the cardioid pattern function in Eq. (6.46b),

$$P(\theta) = [(1 + \cos\theta) + 0.5(1 - \cos 2\theta)]/2. \tag{6.46b}$$

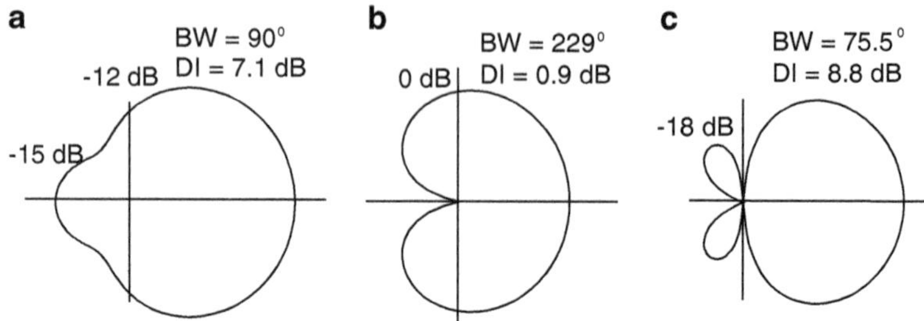

Fig. 6.32 Addition of the quadrupole mode with omni, dipole, and quadrupole ratios (**a**) 1:1:0.414, (**b**) 1:2/3:−1/3, and (**c**) 1:2:1

This beam pattern function, illustrated in Fig. 6.32b, yields a pattern that looks like the true cardioid of Fig. 6.30 with a null at 180°. However, the beam width is 229°, rather than 131°, and the level is the same at ±90° as the level at 0°, rather than −6 dB, which is a direct result of the bidirectional contribution at ±90°. This type of spatial coverage provides a very broad beam with a narrow deep null which may be directed toward a specific unwanted noise source. Equation (6.46b) may also be written as

$$P(\theta) = 3[1 + (2/3)\ \cos\theta - (1/3)\ \cos 2\theta]/4, \tag{6.46c}$$

where it is seen that the relative dipole pressure is 2/3 the omni pressure and the relative quadrupole pressure is −1/3 the omni pressure.

A narrower beam with a null at 180° can also be achieved with stronger contributions from the dipole and quadrupole modes, such as in the pattern function,

$$P(\theta) = (1 + 2\ \cos\theta + \cos 2\theta)/4, \tag{6.47a}$$

illustrated in Fig. 6.32c with nulls at ±90° and 180°. This same function may also be obtained by multiplying dipole and cardioid patterns yielding,

$$P(\theta) = \cos\theta(1 + \cos\theta)/2, \tag{6.47b}$$

which is identical to the three term expansion in Eq. (6.47a).

Equation (6.47b) may also be written as

$$P(\theta) = \left(\cos\theta + \cos^2\theta\right)/2, \tag{6.47c}$$

and interpreted as the pattern formed by a certain combination of the difference of two dipole vector sensors and the sum of the same two dipole vector sensors as illustrated in Fig. 6.33.

Conceptually this expression may be considered as an array of two vector sensors with small separation, S, compared to the wavelength. By the product

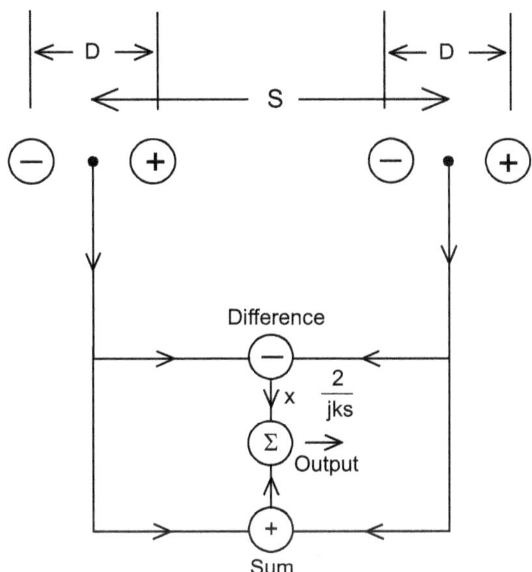

Fig. 6.33 Array difference and sum of two dipole vector sensors

theorem (see Sect. 7.1.1) the array beam pattern is the product of the dipole beam pattern, $jkD\cos\theta$, and the beam pattern of the array. In the case of the sum the array pattern is omni for small separation, S, and the net result is the pattern of a dipole, $2jkD\cos\theta$, giving the first term in Eq. (6.47c). In the case of the difference, the array pattern is also a dipole for small separations, $jkS\cos\theta$, which multiplies the individual dipole pattern to give $(jkS)(jkD)\cos^2\theta$. Multiplying this term by $2/jks$ and summing with the first term then yields the directional function in Eq. (6.47c). A similar result may be obtained from the time delay of the two dipoles as described by Olson [20].

The three cases described by Eqs. (6.47a, 6.47b, and 6.47c) are mathematically identical and demonstrate that the beam pattern function can be obtained from the addition of the omni, dipole, and quadrupole modes, or the product of a dipole and cardioid, or, finally, the difference of two dipoles and the addition of the sum of two dipoles as discussed. The first case is a straight addition of higher order modes and the latter case is equivalent to the addition and subtraction of a two element array, while the second case is the result of a multiplicative processing scheme. Multiplicative processing generally yields a poorer signal-to-noise ratio than additive processing [34] and is usually only suitable in systems where the noise level is very low. Thompson [35] has shown that raising a pattern function of a symmetric linear array of point sources to some integer power greater than unity can be identified as the pattern function of a larger symmetric array with the same individual transducer separations.

The beam pattern given by Eq. (6.47a) and illustrated in Fig. 6.32c results in a beam width of 75.5° and back lobes at ±120° with a level of −18 dB. These back lobes have a negative phase compared to the main lobe and may be decreased to approximately −25 dB through additional contribution from the positive omni mode with the resulting normalized beam pattern function

$$P(\theta) = (1.25 + 2\ \cos\theta + \cos 2\theta)/4.25.$$

This optimized beam pattern function yields a slightly wider 78° beam width with an additional positive back lobe at 180°, and main beam contributions at ±90°, all equal to the −25 dB approximate level of the other two back lobes at ±120° [33].

An alternative approach for attaining a beam pattern similar to a cardioid is to combine the outputs of two scalar sensors with a phase shift between them. The normalized beam pattern function for two scalar sensors with a 90° phase shift may be written as

$$P(\theta) = \cos\{(\pi/4)[1 - (4s/\lambda)\cos(\theta)]\}, \tag{6.47d}$$

where θ is the angle measured from the axis, λ is the wavelength of sound in the water, and s is the center-to-center separation between the sensors. For quarter wavelength separation, $4s/\lambda = 1$, the pressure adds in one direction and cancels in the other yielding a beam pattern that looks similar to the cardioid pattern of Fig. 6.30, but is not a true cardioid. The beam pattern generated by quarter wave spacing and 90° phase shift is broader than that of a cardioid and is 3 dB down at ±90°, with a 180° beam width and a DI = 3 dB, rather than 6 dB down at ±90°, a 131° beam width and a DI = 4.77 dB ≈ 4.8 dB for a true cardioid described by the mathematical function $(1 + \cos\theta)/2$. This particular spacing and phasing combination is equivalent to end-fire steering (see Sect. 8.1.2).

We have seen in Eq. (6.37a) that the dipole pattern can be obtained from the difference of two scalar sensors with a small separation. The cardioid and other patterns can also be obtained as variations on the dipole with an additional small phase shift between the two scalar outputs. The result is

$$P(\theta) = 2\sin[(\pi s/\lambda)\cos\theta + \delta/2], \tag{6.47e}$$

where s is the scalar sensor separation and δ is the additional phase shift. When $s/\lambda \ll 1$ and $\delta = 2\pi s/\lambda$ the cardioid is obtained. The value $\delta = 2\pi s/3\lambda$ maximizes the DI at 6 dB, giving the same pattern as Eq. (6.45) with a null at 105° and a small back lobe. Different values of δ form a series of directional patterns called limacons [1, 20].

Summary results for "cardioid type" beam pattern functions of this section are given in Table 6.2.

The various beam pattern functions above can also be obtained from a cylindrical array of discrete transducers with appropriate distributions of voltage drive [33] (see Sect. 7.14). Narrower beams may be obtained using higher order radiation modes.

Table 6.2 Summary of "cardioid type" beam pattern function characteristics

Function	Figure	BW	Level at ±90°	Level at 180°	DI
By mode summation					
Eq. (6.44a)	6.30	131 (deg)	−6 dB	−∞ dB	4.8 dB
Eq. (6.44b)	6.31a	115	−8.6	−11.7	5.7
Eq. (6.45)	6.31b	105	−12	−6	6.0
Eq. (6.46a)	6.32a	90	−12	−15	7.1
Eq. (6.46b)	6.32b	229	0	−∞	0.9
Eq. (6.47a)	6.32c	75.5	−∞	−∞	8.8
By phase shifting					
Eq. (6.47d)	–	180	−3	−∞	3.0
Eq. (6.47e)	4.31b	105	−12	−6	6.0

Notes: Equation (6.47d) evaluated for 90° phase shift and $4s/\lambda=1$. Equation (6.47e) evaluated for $\delta=2\pi s/3\lambda$

6.5.7 *Intensity Sensors*

The product of the output of a scalar sensor, which measures the acoustic pressure, and a vector sensor, which measures the acoustic particle velocity, yields a measure of the acoustic intensity. The intensity is a vector defined in Sect. 10.1 and given by the product of the pressure and particle velocity. The time averaged value is given by $\langle I\rangle = ½\mathrm{Re}(pu^*)$ as shown in Sect. 13.3, and for plane waves is $pp^*/2\rho c = \rho cu^*/2$ (where * means the complex conjugate). Integration of the intensity over all directions gives the power. Smith et al. [36] investigated the use of such a probe for transducer measurements in a small room or water-filled tank. The intensity may be determined, for example, by two small scalar sensors with small separation, s. The summed voltage for this case, illustrated in Fig. 6.20a and given by Eq. (6.36), is repeated here for $s \ll \lambda$ as

$$V = p_{\mathrm{i}}\mathrm{e}^{-jkr}[(M_1 + M_2) + j(M_1 - M_2)(\pi s/\lambda)\cos\theta], \tag{6.48}$$

where θ is the angle between the direction of an incident plane wave and the axis of the sensor pair. If the two sensors are summed with $M_2 = M_1$, the output voltage is

$$V_0 = 2M_1 p_{\mathrm{i}}\mathrm{e}^{-jkr}, \tag{6.49a}$$

while the differenced output voltage is

$$V_{\mathrm{d}} = j2M_1 p_{\mathrm{i}}\mathrm{e}^{-jkr}(\pi s/\lambda)\cos\theta. \tag{6.49b}$$

The magnitude of the conjugate product of the omni and dipole voltages,

$$|V_0 V_d^*| = 4|p_{\mathrm{i}}|^2|M_1|^2(\pi s/\lambda)\cos\theta = 8\rho c I_{\mathrm{i}}|M_1|^2(\pi s/\lambda)\cos\theta, \tag{6.50}$$

is proportional to the component of the incident intensity parallel to the axis of the sensor pair. Turning the sensor for maximum output identifies the direction of the source.

Intensity sensors are often referred to as intensity probes. They can take a variety of forms [24, 37], such as the pressure/velocity probe by Gabrielson et al. [38], the pressure/pressure probe by Ng [39], the pressure/acceleration probes by Sykes [40] and Schloss [41], and the velocity/velocity probe by McConnell et al. [42]. G. C. Lauchle has investigated the feasibility of detecting and using the imaginary component of intensity [43], which is called the reactive intensity to distinguish it from the active intensity or time average intensity. The reactive intensity can be defined as Im (pu^*) as shown in Sect. 13.3. Stanton and Beyer [44] developed a probe for measuring the real and imaginary parts of the complex intensity.

A magnetostrictive/piezoelectric hybrid form of intensity sensor [45] is illustrated in Fig. 6.34 showing a cantilever transducer composed of a piezoelectric ceramic 31 mode bar with output voltage V_p and two magnetostrictive thin bars on either side with one coil and voltage output V_m. The two magnetostrictive strips are oppositely poled (by magnets or, if possible, through remanence) so that they do not produce an output from a compressive scalar component but do produce an output for a pressure differential $p_1 - p_2$ since it causes a net bending force. On the other hand, the piezoelectric bar is symmetrical and produces no output on bending, but does produce an output from the pressure p_1, a scalar. The product of the two outputs is proportional to the intensity. If the bending resonance is set below the operating band, the voltage output from the magnetostrictive sensor is flat above that resonance without the usual 6 dB/octave rise associated with piezoelectric vector sensors.

If an intensity probe is used near a soft surface the pressure will be small and the product of the pressure and velocity will also be small. Likewise, if the intensity probe is near a rigid surface the velocity will be small and again the output will be small. In such cases Lubman [46] has suggested that the pressure and velocity be separately measured to achieve a sensor system with less sensitivity to interference from nearby structures or to fading in an oceanographic environment.

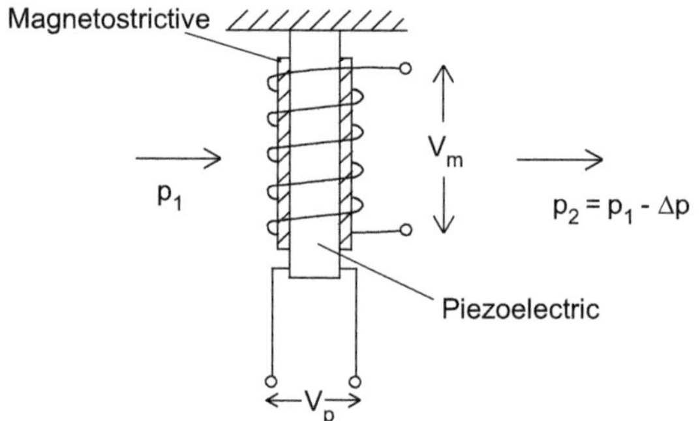

Fig. 6.34 Hybrid intensity probe [45]

6.6 The Plane Wave Diffraction Constant

A small hydrophone does not disturb the sound field, and, therefore, measures the free field plane wave pressure amplitude, p_i, that would exist if it was not present. A large hydrophone, or a small hydrophone in a large mounting structure, does disturb the pressure field, and the diffraction constant, D_a, is a measure of this disturbance. D_a is defined as p_b/p_i, where p_b is the spatially averaged pressure on the surface of the hydrophone when it is clamped, so that its active surface cannot move. For a hydrophone, small compared to the wavelength, set in a large rigid baffle $D_a = 2$, while the same hydrophone, with no baffle, does not disturb the field and $D_a = 1$. The total force on the hydrophone surface results from the free field pressure of the incident wave plus the pressure of the scattered wave and the pressure caused by the motion of the hydrophone surface (related to the radiation impedance). In Eq. (1.5) the force is divided into the part caused by the motion, u_0, of the surface, $F_r = -Z_r u_0$, and the part that is independent of the motion, the clamped force, F_b, as shown schematically in Fig. 6.35a.

In the general case where the motion of the hydrophone surface is not uniform the clamped force is defined by Eq. (1.6) as the sum of the incident pressure and the scattered pressure, as illustrated in Fig. 6.35b, weighted by the velocity distribution function and integrated over the surface. The diffraction constant as defined above is the ratio of the clamped force to the free field force and given by Eqs. (1.6) and (1.7) as

$$D_a = \frac{F_b}{Ap_i} = \frac{1}{Ap_i}\iint_A \left[p_i\left(\vec{r}\right) + p_s\left(\vec{r}\right)\right]\frac{u^*\left(\vec{r}\right)}{u_0{}^*}\mathrm{d}S, \tag{6.51}$$

where F_b is the clamped force, A is the area of the moveable surface of the hydrophone, Ap_i is the free field force caused by the incident pressure amplitude,

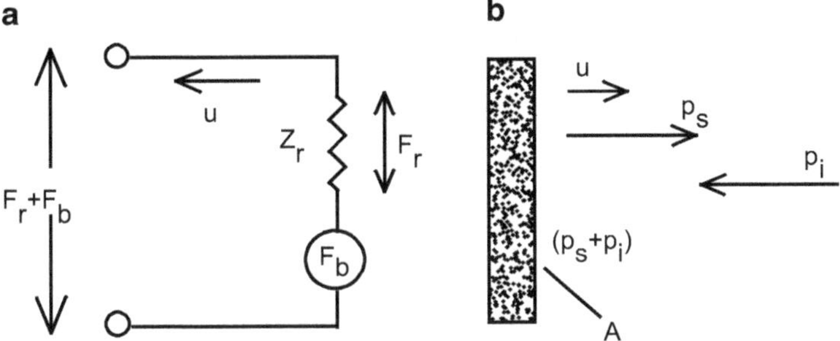

Fig. 6.35 (**a**) Clamped (blocked) force F_b and radiation force F_r. (**b**) Incident pressure, p_i, and scattered pressure, p_s

p_i, p_s is the scattered pressure, $[p_i(\vec{r}) + p_s(\vec{r})]$ is the total clamped pressure on the surface and $u^*\left(\vec{r}\right)/u_0^*$ is the complex conjugate of the velocity distribution function where u_0 is a reference velocity. Woollett [47] has applied this formulation for the determination of diffraction constants for vector sensors. The diffraction constant is a function of the hydrophone size relative to the wavelength, its shape, its velocity distribution, and the arrival direction of the plane wave. It is customary to define the diffraction constant for a plane wave arriving on the Maximum Response Axis (MRA) [48], which is consistent with the usual definition of the sensitivity. We will define in Sect. 11.3.1 a more general diffraction constant which depends on the direction of arrival of the plane wave.

The free field voltage sensitivity of a hydrophone, M, was defined earlier as the ratio of the open circuit voltage to the free field pressure of a plane sound wave arriving on the MRA. However, it is the clamped pressure, $F_b/A = p_b = D_a p_i$, that appears in an equivalent circuit or a pair of transducer equations, and the sensitivity must be calculated from

$$M = V/p_i = VD_a A/F_b. \tag{6.52}$$

The results in Sect. 6.1.1 are correct because in those cases the hydrophone dimensions were assumed to be small compared to the wavelength making $D_a = 1$ and $F_b/A = p_i$.

The most illuminating discussion of the diffraction constant has been given by Bobber [48] who points out that the name is a misnomer since in many cases it is reflection or interference that causes the clamped force to differ from the free field force. Thus, for example, in the case of a flat hydrophone in a plane baffle, p_s in Eq. (6.51) is a reflected wave rather than a diffracted wave. In other cases, such as a thin line or ring hydrophone, diffraction, reflection, and scattering are all insignificant, but partial cancellation of the incident wave between different parts of the hydrophone surface reduces the clamped force and makes the diffraction constant less than one.

Henriquez [49] has calculated the diffraction constant for several specific idealized hydrophone shapes that can be used to estimate values for actual hydrophones. For a spherical hydrophone of radius a,

$$D_a = \left(1 + k^2 a^2\right)^{-1/2}, \tag{6.53}$$

where, because of symmetry, the direction of the wave is irrelevant. In this case the response is omnidirectional at all frequencies, but the diffraction constant introduces a strong frequency dependence; for large ka the frequency dependence of D_a causes a 6 dB/octave reduction in the sensitivity with increasing frequency. Equation (6.53) is derived in Sect. 11.3.1.

For a long cylindrical hydrophone of radius a with the wave direction perpendicular to the axis,

$$D_{\mathrm{a}} = (2/\pi ka)\left[J_1^2(ka) + N_1^2(ka)\right]^{-1/2}, \tag{6.54}$$

where J_1 and N_1 are first order Bessel and Neumann functions. For a thin ring hydrophone of radius a with the wave direction in the plane of the ring,

$$D_{\mathrm{a}} = J_0(ka). \tag{6.55}$$

For a circular piston of radius a at the end of a long tube with the wave direction perpendicular to the face of the piston the results of Levine and Schwinger [50] were used to determine D_{a}. All these cases are compared in Fig. 6.36 [49].

These results show that in all cases $D_a \approx 1$ for small ka, decreases as ka increases for the sphere, cylinder, and ring but increases with ka to a maximum value of 2 for the piston at the end of a tube. The different behavior in the last case occurs since, as the piston gets large, the wave scattered from it becomes the same as the wave reflected from an infinite rigid plane for which the pressure on the surface is doubled. Thus the diffraction constant for hydrophones mounted close to a hard surface is 2. The value 2 is also obtained for a piston in a plane rigid baffle or a piston which is large compared to the wavelength. On the other hand, the diffraction constant for small hydrophones on a soft surface is approximately zero because the pressure is near zero on a soft surface.

Weighting the pressure on the surface by the velocity distribution function in Eq. (6.51) can be significant in any case where the velocity is nonuniform, e.g., bender (Sect. 6.4) or multimode hydrophones (Sect. 6.5.5). When a hydrophone is capable of vibrating in more than one mode it may have more than one diffraction constant [48].

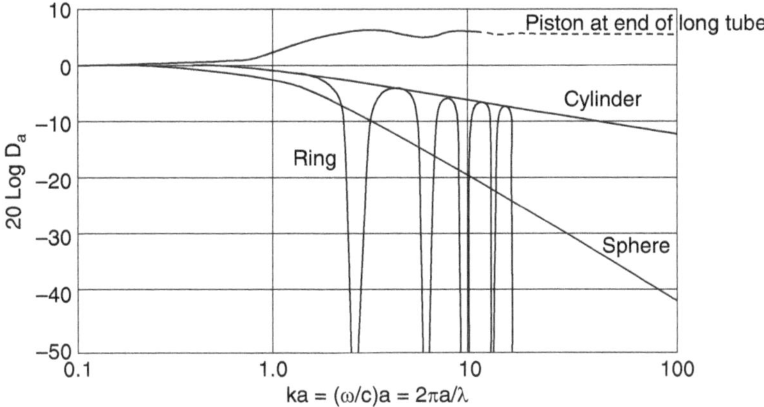

Fig. 6.36 Diffraction constant D_{a} for a piston, sphere, cylinder, and ring [49]. *Note*: The cylinder and ring curves should decrease as 20 log D_{a} and not 10 log D_{a}, as shown. See Z. Milosic [51]. Thus, the dB values for the ring and cylinder curves should be multiplied by 2

However, if the transducer was designed such that a certain mode was not driven electrically, that mode would produce no voltage output when excited acoustically. But it could still influence the voltage output of the hydrophone via acoustic coupling between modes.

Three important acoustical parameters, the diffraction constant, D_a, the directivity factor, D_f, and the radiation resistance, R_r, are related, in general, by [48]

$$D_a^2 = \frac{4\pi R_r D_f}{\rho c k^2 A^2} = \frac{4\pi c R_r D_f}{\rho \omega^2 A^2}. \qquad (6.56)$$

As seen from Eq. (6.56), the diffraction constant may alternatively be calculated from the radiation resistance and the directivity factor, or, in general, when two of the three parameters are known the third can be found from this relationship. For example, an un-baffled transducer of any shape, but small compared to the wavelength, has $D_a = D_f = 1$ showing that $R_r = \rho c k^2 A^2 / 4\pi = \omega^2 A^2 \rho / 4\pi c$, which will be derived in Chap. 10. See Sect. 11.3.1 for a development of Eq. (6.56).

The diffraction constant is defined in Eq. (6.51) in terms of the weighted average pressure on the sensitive surface of a hydrophone, but diffraction can have important effects at a point on a hydrophone surface. Such effects often occur when measuring the response of hydrophones and are discussed briefly in Sect. 9.7.5.

6.7 Hydrophone Thermal Noise

A basic limitation on hydrophone performance is the electrical noise generated internally by thermal agitation in its components and in the water; this noise must not exceed the total sea noise for good performance. The internal noise of a hydrophone is caused by its energy dissipation mechanisms, i.e., the electrical and mechanical dissipation factors of the transduction material, the mechanical resistance in any other moving hydrophone components, including mounting, and the radiation resistance. Electrical dissipation and mechanical resistance allow electrical and mechanical energy in the hydrophone to be lost as internal thermal energy, and they also allow thermal energy in the hydrophone materials to cause electrical noise. Radiation resistance differs in that it is the means by which mechanical energy in the transducer generates acoustic energy in the water, but it also allows thermal energy in the water to cause electrical noise. It follows from the equivalent circuit concept that both the electrical and mechanical loss mechanisms can be represented by a Thevenin equivalent electrical series resistance, R_h. This resistance then determines the equivalent total Johnson thermal mean squared noise voltage, $\langle V_n^2 \rangle$, by

$$\langle V_n^2 \rangle = 4KTR_\mathrm{h}\Delta f, \tag{6.57}$$

where K is Boltzman's constant (1.381×10^{-23} J/K), T is the absolute temperature, and Δf is the bandwidth. At 20°C Eq. (6.57) leads to

$$10\log\langle V_n^2 \rangle = -198\,\mathrm{dB} + 10\log R_\mathrm{h} + 10\log \Delta f. \tag{6.58}$$

Equation (6.58) emphasizes the fact that the noise increases with the bandwidth. The value −198 dB is the noise level below 1 V in a 1 Hz band for a 1 Ω resistor at 20 °C. The resistance R_h is the resistive part of the total electrical input impedance of the hydrophone, including the motional impedance transformed by the electro-mechanical turns ratio, N. (See Sects. 13.15 and 6.7.7 for more discussion of hydrophone internal noise.)

The value of the noise voltage in Eq. (6.58) becomes more useful when converted to an equivalent mean squared noise pressure, $\langle p_n^2 \rangle = \langle V_n^2 \rangle / M^2$ by means of the hydrophone sensitivity M. The resulting level of the equivalent noise pressure is then

$$\begin{aligned} 10\log\langle p_n^2 \rangle &= -198\ \mathrm{dB} + 10\log R_\mathrm{h} - 20\log M + 10\log \Delta f \\ &= -198\ \mathrm{dB} - 10\log M^2/R_\mathrm{h} + 10\log \Delta f \end{aligned} \tag{6.59}$$

The equivalent noise pressure level is normally evaluated in a 1 Hz band ($\Delta f = 1$) and referred to as the noise pressure spectral density. In the following spectral densities will be used and the term $10 \log \Delta f$ will be omitted. Equation (6.59) shows that for noise considerations the most meaningful form of the hydrophone figure of merit discussed in Sect. 6.1.2 is M^2/R_h.

The noise pressure obtained from Eq. (6.59) by use of the plane wave sensitivity is equal to the pressure of a plane wave incident on the MRA of the hydrophone that would give the same voltage as the internal noise and is usually called the equivalent plane wave noise pressure. The combination of Sea State Zero (SS0) noise and thermal noise shown in Fig. 6.37 is usually considered to be approximately the minimum ambient noise in the sea, and therefore suitable for comparison with the hydrophone internal noise. Obviously, a good hydrophone for a particular application should have internal noise lower than the expected ambient sea noise.

Since hydrophones often have a directional response, the sensitivity averaged over all directions is appropriate for use in Eq. (6.59) to convert the noise voltage to a noise pressure suitable for comparison with sea noise (see Sects. 6.7.1 and 11.3.1). This determination of the equivalent noise pressure was called by Woollett the isotropic acoustic equivalent of electrical noise. Woollett also described a procedure for comparing hydrophone noise with ambient noise, including the noise at the input stage of a preamplifier [52].

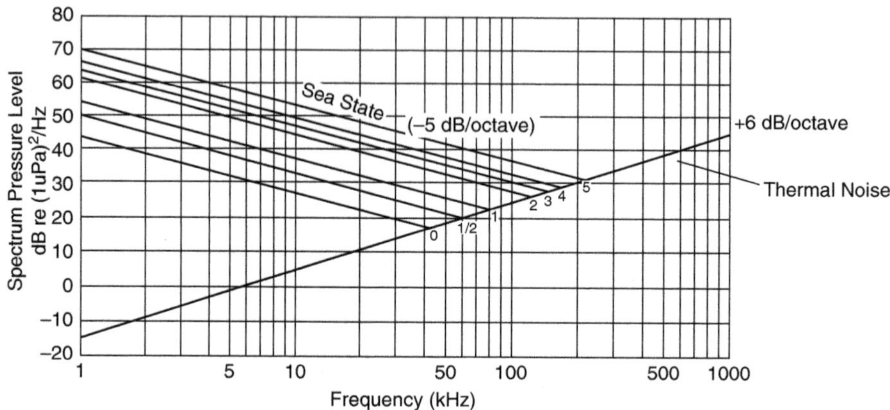

Fig. 6.37 Sea state and thermal noise [54]

6.7.1 Directivity and Noise

The relationship of hydrophone noise to the directivity factor and the electroacoustic efficiency may be shown by writing Eq. (6.59) in a form independent of the sensitivity M by use of the directivity factor, D_f, the transmitting current response, S, and the reciprocity factor, J. We start with the directivity factor which may be written, in terms of a projector, as

$$D_f = I_0/I_a = I_0 4\pi d^2/W_0 = 4\pi d^2 p_0^2/\rho c W_0, \tag{6.60}$$

where, using rms values, I_0 and p_0 are the on-axis intensity and pressure at a distance d from the transducer, and I_a, is the spatial average of the intensity at the same distance. The output power, W_0, is given by $\eta_{ea}\ i^2 R_h$ where η_{ea} is the electroacoustic efficiency and i is the input current. Using these relationships in Eq. (6.60) and solving for p_0^2 yields

$$p_0^2 = \eta_{ea} D_f \rho c\ i^2 R_h/4\pi d^2, \tag{6.61}$$

which may be written in terms of the transmitting current response (at distance d, rather than 1 m)

$$S^2 = p_0^2/i^2 = \eta_{ea} D_f R_h \rho c/4\pi d^2. \tag{6.62}$$

The transmitting current response is related to the receiving response by the reciprocity factor $M/S = J = 2d/\rho f$ (see Sect. 9.5) leading to

$$M^2 = D_f \eta_{ea} R_h c/\rho \pi f^2, \tag{6.63}$$

and the resulting RVS is

$$\text{RVS} = 20\log M = 10\log R_{\text{h}} - 20\log f + 10\log \eta_{\text{ea}} + \text{DI} - 123.2, \qquad (6.64)$$

showing the relationship between the sensitivity, the input electrical resistance, frequency, the electroacoustic efficiency, and the DI.

By eliminating M between Eqs. (6.64) and (6.59) we have the desired result

$$10\log\langle p_n^2\rangle = 20\log f - 74.8 - 10\log \eta_{\text{ea}} - \text{DI}. \qquad (6.65)$$

It should be noted that the DI does not increase the sensitivity of a hydrophone, but it does decrease the isotropic acoustic equivalent of internal noise [52] for hydrophones with directivity, thus increasing the signal-to-noise ratio.

Equation (6.64) may be further manipulated by obtaining the output voltage V for a 1 μPa reference acoustic pressure by writing $M = V/p_{\text{ff}} = V \times 10^6$, and eliminating 10 log R_{h} by use of Eq. (6.58). The result is the signal-to-noise voltage ratio, SNR, for a 1 μPa acoustic signal when only internal hydrophone noise is present:

$$\text{SNR} = 10\log V^2/\langle V_n^2\rangle = \text{DI} + 10\log \eta_{\text{ea}} - 20\log f - 45.2\,\text{dB}. \qquad (6.66)$$

As seen, the signal-to-noise increases with the DI and efficiency but decreases with the frequency. As the frequency increases the directivity factor D_{f} approaches $4\pi Af^2/c^2$ for hydrophones in a plane baffle; then the term 20 log f cancels in Eqs. (6.65) and (6.66) leaving a dependence on the area, A. Thus, the noise decreases and the signal-to-noise increases as A increases, as will also be shown in Sect. 6.7.4.

6.7.2 *Low Frequency Hydrophone Noise*

At frequencies well below resonance the mechanical, R_{m}, and radiation, R_{r}, resistances are small compared to the mechanical reactance $1/j\omega C^E$. However, the electrical dissipation follows the frequency dependence of the mechanical reactance and may often be the main loss mechanism, and therefore the main internal noise mechanism, at low frequencies. In this case the equivalent electrical resistance is due to the electrical dissipative part alone and using Eq. (6.5) we get

$$\begin{aligned} R_{\text{h}} &= \text{Real}\{1/j\omega C_{\text{f}}(1 - j\tan\delta)\} = (\tan\delta/\omega C_{\text{f}})/(1 + \tan^2\delta) \\ &\approx \tan\delta/\omega C_{\text{f}}, \end{aligned} \qquad (6.67\text{a})$$

since, typically, $\tan^2\delta \ll 1$. Although the dissipation is usually small and the capacitance is typically high, the $1/\omega$ dependence of R_{h} can cause significant noise at very low frequencies. This resistance may be used in Eq. (6.59) to obtain the low frequency equivalent noise pressure level:

$$10\log\langle p_n^2\rangle \approx -206\,\mathrm{dB} + 10\log(\tan\delta/C_\mathrm{f}) - 20\log M - 10\log f, \qquad (6.67b)$$

which increases by 3 dB/octave with decreasing frequency. The dissipation factor, tan δ, is a constant for small signals over a wide range of frequencies for typical piezoelectric ceramic materials and is usually measured at 1 kHz. Note that the voltage sensitivity term in Eq. (6.67b) does not change with frequency for pressure sensitive piezoelectric ceramic hydrophones operating well below resonance.

For example: Consider a PZT-4 hollow spherical hydrophone with 0.02 m radius and 0.002 m wall thickness which has a low frequency sensitivity of −193 dB and a capacitance of 29 nF (see Sect. 6.2.2). Using $\tan\delta = 0.004$, from Sect. 13.5, Eq. (6.67b) shows the internal noise to be about 8 dB//(μPa)^2/Hz or 37 dB below SS0 at 1 kHz (see Fig. 6.37), with a greater difference at lower frequencies.

6.7.3 More General Description of Hydrophone Noise

Some insight into the relative importance of electrical dissipation, mechanical resistance, and radiation resistance in determining hydrophone noise may be obtained by examining the real part of the electrical input impedance, Z_h. For a lumped mode equivalent circuit such as that illustrated in Fig. 6.38, with short circuit mechanical impedance $Z_\mathrm{m}^E = R_\mathrm{m} + j\left[\omega M_\mathrm{m} - 1/\omega C^E\right]$, radiation impedance $Z_\mathrm{r} = R_\mathrm{r} + j\omega M_\mathrm{r}$ and clamped admittance $Y_0 = G_0 + j\omega C_0$, the electrical input impedance is given by

$$Z_\mathrm{h} = 1/Y_\mathrm{h} = 1/\left[G_0 + j\omega C_0 + N^2/\left(Z_\mathrm{m}^E + Z_\mathrm{r}\right)\right]. \qquad (6.68)$$

The radiation mass M_r is a constant for low frequencies but is a decreasing function of frequency at higher frequencies, usually in the vicinity of and above resonance. The input electrical resistance, $R_\mathrm{h} = \mathrm{Real}\,(Z_\mathrm{h})$, when used in Eqs. (6.58) and (6.59), gives the complete Johnson thermal noise voltage and the corresponding equivalent noise pressure. The quantity R_h can be written compactly as

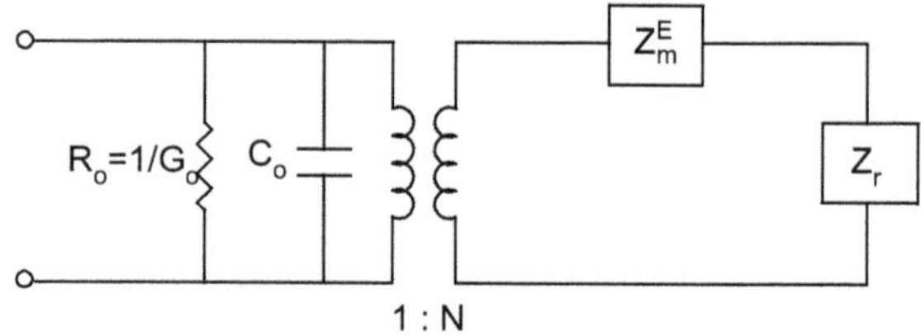

Fig. 6.38 Simplified equivalent circuit for hydrophone electrical input impedance, Z_h

$$R_h = \frac{R/N^2 + G_0|Z/N^2|^2}{|1 + Y_0 Z/N^2|^2}, \tag{6.69}$$

where $Z = Z_r + Z_m$, $R = R_r + R_m$, $X = \omega M' - 1/\omega C^E$, and $M' = M_m + M_r$. To consider the form R_h takes in limited frequency regions, it is convenient to express it in terms of familiar transducer parameters as follows:

$$R_h = \frac{k^2\left\{\frac{R}{N^2}\left[1 - \frac{\omega^2}{\omega_a^2} + \frac{\omega \tan\delta}{\omega_r Q_m}\right] + \frac{X}{N^2}\left[\frac{\omega}{\omega_a Q_a} + \tan\delta\left(\frac{\omega^2}{\omega_r^2} - 1\right)\right]\right\}}{\left[1 - \frac{\omega^2}{\omega_a^2} + \frac{\omega \tan\delta}{\omega_r Q_m}\right]^2 + \left[\frac{\omega}{\omega_a Q_a} + \tan\delta\left(\frac{\omega^2}{\omega_r^2} - 1\right)\right]^2}, \tag{6.70}$$

where the resonance and antiresonance frequencies are $\omega_r = (M'C^E)^{-1/2}$ and $\omega_a = (M'C^D)^{-1/2}$, respectively, $Q_m = \omega_r M'/R$ and $Q_a = \omega_a M'/R$, and the resistances are to be evaluated at ω_r and ω_a.

It is evident from this expression for R_h that the three loss mechanisms are coupled in a complicated way and cannot be clearly separated, except as approximations in limited frequency regions. However, Sect. 13.15 gives a different approach in which the three losses are treated separately and Eq. (6.78b) gives an approximation for R_h that holds when electrical dissipation is negligible. At frequencies well below ω_r or ω_a Eq. (6.70) reduces to

$$R_h = \frac{\left[\tan\delta/\omega C_f + k^4 R_r/N^2 \eta_{ma}\right]}{(1 + \tan^2\delta)}, \tag{6.71}$$

and in this form the loss mechanisms are clearly separated. At still lower frequencies the first term dominates, and R_h becomes equal to the value in Eq. (6.67a) where only electrical dissipation is important as will be shown in detail by the following example.

The calculations made following Eq. (6.67a and 6.67b) for a PZT-4 spherical hydrophone of radius 0.02 m can be extended here to find the frequency at which the mechanical resistance term in Eq. (6.71), $k^4 R_r/N^2 \eta_{ma} = k^2 R_r C^E/\eta_{ma} C_f$, becomes significant compared to the electrical dissipation term, $\tan\delta/\omega C_f$. For the spherical hydrophone the radiation resistance and the diffraction constant are given by

$$R_r = 4\pi a^2 \rho c (ka)^2 / \left[1 + (ka)^2\right],$$

$$D_a = \left[1 + (ka)^2\right]^{-1/2},$$

the free capacity $C_f = 29$ nF, the short circuit compliance $C^E = s_c^E/4\pi t = 1.6 \times 10^{-10}$m/N, $k^2 = k_p^2 = 0.34$ and $\tan\delta = 0.004$ (see Sects. 6.2.2 and 13.5). Assuming that $\eta_{ma} = 0.8$ we find the results for R_h in the first three lines of Table 6.3, then add 20 log D_a to the previously calculated low frequency value of M to get the

Table 6.3 Hydrophone noise $\langle p_n^2 \rangle$, sea state 0 & thermal sea noise in dB//(μPa)²/Hz

f(kHz)	R_r (kg/s)	$\frac{k^2 R_r C^E}{\eta_{ma} C_f}$	$\frac{\tan\delta}{\omega C_f}$	R_h	D_a	MD_a^2	$\langle p_n^2 \rangle$	SS0	Thermal sea
1	53	0.12	22	22	0 dB	−193 dB	8 dB	45 dB	−15 dB
10	3100	7.3	2.2	9.5	−2	−195	7	27	5
20	5500	13	1.1	14	−6	−199	12	23	11
44(f_r)	7000	16	–	49[a]	−12	−199	18	17	18
54(f_a)	7200	17	–	167[b]	−15	−198	22	14	22

[a]From Eq. (6.72b)
[b]From Eq. (6.72c)

effective sensitivity (MD_a^2) and finally determine $\langle p_n^2 \rangle$ from Eq. (6.59). Note that $D_f = 1$ for the spherical hydrophone at all frequencies.

Table 6.3 shows that at 1 kHz the hydrophone thermal noise is caused almost entirely by tan δ, while at 10 kHz the influence of ($R_r + R_m$) exceeds that of tan δ, and at 20 kHz ($R_r + R_m$) has become dominant. We have also assumed that tan δ and η_{ma} are frequency independent, which means that we have assumed (for simplicity) that R_m has the same frequency dependence as R_r. The fundamental resonance frequency of this hydrophone is about 44 kHz (see below), and the low frequency approximation for both R_h and M is reasonable up to 20 kHz as long as the diffraction constant, D_a, is included in the sensitivity.

At mechanical resonance, ω_r, where $X = 0$, Eq. (6.70) simplifies to

$$R_h = \frac{(k^2 R/N^2)\left[k^2 + \tan\delta/Q_m\right]}{\left[k^2 + \tan\delta/Q_m\right]^2 + (1 - k^2)/Q_a^2}, \tag{6.72a}$$

but in most cases $k^2 \gg \tan\delta/Q_m$ and

$$R_h \approx \frac{k^4 R/N^2}{k^4 + (1 - k^2)/Q_a^2} = \frac{k^2 R_r C^E/\eta_{ma} C_f}{k^4 + (1 - k^2)/Q_a^2}. \tag{6.72b}$$

Thus, in the vicinity of resonance, the noise is strongly determined by the radiation resistance and mechanoacoustic efficiency, and very little by electrical dissipation as long as k^2 is high.

For example: Continuing the spherical hydrophone example at resonance we have the hydrophone mass = 0.075 kg, the radiation mass ≈ 0.008 kg, $f_r = 44$ kHz, $f_a = 54$ kHz, $Q_m = 2.6$, and $Q_a = 3.1$. The sensitivity is raised to −187 dB by the resonance [see Eq. (4.10)] and lowered to −199 dB by the diffraction constant. This spherical hydrophone is similar to the spherical hydrophone in Fig. 6.11, where the radius is 10 % greater. The final result of these effects, given in the fourth line of Table 6.3, shows the internal thermal noise at resonance to be about the same as the thermal noise in the sea.

At antiresonance, ω_a, $X = \omega_a M' k^2$, and if $1/Q_a$ is large compared to both $k^2 \tan \delta/\left(1-k^2\right)$ and $\omega_a \tan \delta/\omega_r Q_m$, the hydrophone resistance can be approximated by

$$R_h = k^4 Q_a^2 R/N^2 = Q_a^2 k^2 R_r C^E/\eta_{ma} C_f. \tag{6.72c}$$

Equation (6.10) shows that the sensitivity is increased above the low frequency value by the factor Q_a at $\omega = \omega_a$, making it -183 dB before the D_a^2 of -15 dB is applied, which then gives the results in the fifth line of Table 6.3 where the internal noise is again about the same as the thermal sea noise.

The values of $\langle p_n^2 \rangle$ in Table 6.3, when used in Eq. (6.65) with DI = 0, give reasonable values of electroacoustic efficiency, i.e., very low at 1 kHz and high, up to 80 %, at the other frequencies. The thermal sea noise exists in the sea and also in the hydrophone, as $\langle p_n^2 \rangle$ is defined by Eq. (6.59) where R_h includes the radiation resistance. Therefore, at high frequencies where tan δ has little effect on the noise $\langle p_n^2 \rangle$ is approximately equal to the thermal sea noise increased by the factor $1/\eta_{ma}$ and decreased by the factor D_f as will be seen more generally in the next section. In Table 6.3 with $\eta_{ma} = 0.8$ and $D_f = 1$ the expected difference is about 1 dB, which is within the accuracy of the calculations.

6.7.4 Comprehensive Hydrophone Noise Model

A comprehensive hydrophone noise model is developed, in detail, in Sect. 13.15 where a different approach is taken by extending the definition [53] of Johnson thermal noise to include mechanical components such as the mechanical resistance and radiation resistance, thereby developing a noise force for these components (also see Sect. 6.7.7). This noise force is then directly converted to an equivalent noise pressure through the capture area and diffraction constant for the hydrophone. The model is shown to agree with the model by Mellen [54] and with the results in Sect. 6.7.3. The approach leads to an equivalent hydrophone noise pressure given by

$$\begin{aligned} \langle p_n^2 \rangle = &\left[4\pi KT(\rho/c) f^2/D_f \eta_{ma}\right] \\ &\times \left[1 + \left(\tan \delta/k^2\right)\left\{\omega/\omega_r Q_m + (Q_m \omega_r/\omega)\left(1-\omega^2/\omega_r^2\right)^2\right\}\right], \end{aligned} \tag{6.73}$$

which can be reduced to

$$\langle p_n^2 \rangle = 4\pi KT(\rho/c) f^2/D_f \eta_{ea}. \tag{6.74}$$

Equation (6.74) may be rewritten, at 20 °C in dB//(μPa)^2 Hz, as

$$10\log\langle p_n^2\rangle = 20\log f - 74.8 - 10\log\eta_{\mathrm{ea}} - \mathrm{DI}, \tag{6.75}$$

which is identical to Eq. (6.65) which was developed through reciprocity in Sect. 6.7.1. The incident plane wave signal pressure squared, $|p_{\mathrm{i}}|^2$, must be greater than the value given by Eq. (6.74) to obtain a signal-to-noise ratio greater than unity.

The total electrical noise voltage, $\langle V_n^2\rangle$, may be obtained from Eq. (6.74) and an expression for the hydrophone voltage sensitivity which may be developed from the equivalent circuit, of Fig. 6.19. This circuit is not restricted to bender hydrophones and the resulting sensitivity, given by Eq. (6.32), is a good approximation for most pressure sensitive hydrophones. Accordingly, we may express the hydrophone sensitivity as

$$|V/p|^2 = k^2A^2D_{\mathrm{a}}^2C^E/C_{\mathrm{f}}\left[(\omega/\omega_{\mathrm{a}}Q_{\mathrm{a}})^2 + \left(1-\omega^2/\omega_{\mathrm{a}}^2\right)^2\right]. \tag{6.76}$$

and on replacing p^2 by $\langle p_n^2\rangle$ the mean squared noise voltage may be written as

$$\langle V_n^2\rangle = \langle p_n^2\rangle k^2A^2D_{\mathrm{a}}^2C^E/C_{\mathrm{f}}\left[(\omega/\omega_{\mathrm{a}}Q_{\mathrm{a}})^2 + \left(1-\omega^2/\omega_{\mathrm{a}}^2\right)^2\right]. \tag{6.77}$$

Substitution of Eq. (6.74) and the second form of Eq. (6.56) into Eq. (6.77) yields

$$\langle V_n^2\rangle = 4\mathrm{KTR_r}k^2C^E/C_{\mathrm{f}}\eta_{\mathrm{ea}}\left[(\omega/\omega_{\mathrm{a}}Q_{\mathrm{a}})^2 + \left(1-\omega^2/\omega_{\mathrm{a}}^2\right)^2\right]. \tag{6.78a}$$

The preamplifier electrical noise should be less than this quantity and the hydrophone should be properly matched to the preamplifier for best performance [52, 55–57].

The approximate hydrophone input resistance, R_{h}, at any frequency, may be obtained from Eq. (6.78a) by substitution of Eq. (6.57) (with $\Delta f = 1$) yielding

$$R_{\mathrm{h}} = R_{\mathrm{r}}k^2C^E/C_{\mathrm{f}}\eta_{\mathrm{ea}}\left[(\omega/\omega_{\mathrm{a}}Q_{\mathrm{a}})^2 + \left(1-\omega^2/\omega_{\mathrm{a}}^2\right)^2\right]. \tag{6.78b}$$

Equation (6.78b) gives the same results as Eq. (6.70) under the usual condition of the electrical dissipation factor $\tan\delta \ll 1$.

6.7.5 *Vector Sensor Internal Noise*

The internal thermal noise in vector sensors may be analyzed by the methods described above. However, in the vector sensor case the output is usually derived from a pressure or force differential which falls at a rate of 6 dB per octave with decreasing frequency yielding a smaller sensitivity than a scalar sensor at the lower frequencies. The smaller sensitivity results in a larger pressure equivalent of the electrical dissipative noise in the low frequency range than that given by Eq. (6.67b) with M constant.

Consider the case of two small closely spaced identical omnidirectional piezoelectric hydrophones, as discussed in Sects. 6.5.1 and 6.5.2 and illustrated in Fig. 6.20a. The series wired, on-axis, summed scalar omni and differenced vector directional voltage outputs may be obtained from Eqs. (6.49a, 6.49b) and written as

$$V_0 = 2M_1 p_i, \text{ and } V_d = 2M_1(\pi sf/c_0)p_i, \tag{6.79}$$

where p_i is the free field incident pressure, M_1 is the sensitivity of each hydrophone, s is the separation between the small hydrophones and $\pi sf/c = \pi s/\lambda \ll 1$. The equivalent thermal scalar, p_{0n}, and vector, p_{dn}, noise can then be determined from Eq. (6.59) with $M = 2M_1$ for the scalar hydrophone and $M = 2M_1(\pi sf/c)$ for the vector hydrophone and written as

$$10\log\langle p_{0n}^2\rangle = -198\,\text{dB} + 10\log 2R_h - 20\log 2M_1, \tag{6.80a}$$

or

$$10\log\langle p_{dn}^2\rangle = -198\,\text{dB} + 10\log 2R_h - 20\log 2M_1 - 20\log f\pi s/c_0. \tag{6.80b}$$

Since $f\pi s/c < 1$ the additional term, $-20 \log f\pi s/c$, is positive, increasing the noise of the vector hydrophone at a rate of 6 dB per octave, as the frequency decreases. The low frequency scalar piezoelectric ceramic hydrophone noise equation, Eq. (6.67b), may also be used for a vector hydrophone with the addition of the same extra noise term, $-20\log(f\pi s/c)$, and written as

$$10\log\langle p_{dn}^2\rangle \approx -206\,\text{dB} + 10\log 2\tan\delta/C_f - 20\log 2M_1 - 30\log f \\ - 20\log \pi s/c, \tag{6.80c}$$

showing a 9 dB/octave internal noise increase as the frequency decreases, which could be a serious limitation on low frequency signal detection.

In Table 6.3 of this chapter the equivalent thermal hydrophone noise was compared with SS0 ambient noise for the specific case of a 0.04 m diameter piezoelectric ceramic spherical hydrophone. At 1 kHz it was found that the hydrophone noise was 33 dB below the SS0 noise of 45 dB//μPa²/Hz. If a dipole was made from two of the same hydrophones, the smallest separation possible would be $s = 0.04$ m, giving $\pi s/c = 8.5 \times 10^{-5}$. Using Eq. (6.90a), with $\tan\delta = 0.01$ and $C_f = 29$ nF for the spherical hydrophone, we find the dipole equivalent noise at 1 kHz to be 31 dB//μPa²/Hz, which is 19 dB higher than the single hydrophone noise, but still 14 dB below SS0 noise. This result holds for the incident wave parallel to the dipole axis; if the wave was incident at the angle θ, the separation, s, would be replaced by $s\cos\theta$, and for $\theta = 60°$ the dipole noise would be 6 dB higher, but still 8 dB below SS0 noise. At lower frequencies the dipole noise would increase at 9 dB per octave, while the sea state noise increases at about 5 dB per octave, but shipping noise becomes more important than sea state noise in this low frequency region. This example suggests that internal thermal noise may not be a serious limitation on use of dipole sensors in the sonar frequency range as long as piezoelectric materials with low dielectric loss are used.

The internal noise in accelerometers is also important since they are commonly used in velocity sensors; measured results on two accelerometers designed for use with a velocity sensor will provide examples. In one case [58] the piezoelectric ceramic flexural disc accelerometer has $C_f = 3.4$ nF and $\tan\delta = 0.014$, giving a low frequency input resistance of $R_h = 656\,\Omega$ at 1 kHz. The low frequency acceleration sensitivity is -17 dB//V/g which corresponds to a pressure sensitivity of -205 dB//V/μPa at 1 kHz (see conversion relations at the beginning of Sect. 6.5). Equation (6.59) then gives the equivalent noise pressure of the accelerometer at 1 kHz as

$$10\log\langle p_n^2\rangle = -198\,\mathrm{dB} + 205\,\mathrm{dB} + 10\log R_h = 35\,\mathrm{dB}//\mu\mathrm{Pa}^2/\mathrm{Hz}, \qquad (6.81)$$

which is close to the measured value of 38 dB in Figure 13 of reference [58]. The measured accelerometer noise in the other example is almost the same, about 37 dB//μPa^2/Hz in Figure 5 of reference [59]. Thus it appears that the internal noise in vector sensors is about the same whether based on accelerometers or on dipole hydrophones when the same piezoelectric ceramic is used.

It was pointed out by Lo and Junger [60] that vector intensity sensors do not respond to isotropic noise. This result may be useful in applications when ambient noise is the limitation, if it has a significant isotropic component [61, 62].

6.7.6 Vector Sensor Susceptibility to Local Noise

Vector sensors also show a strong dependence on local external noise sources at the lower frequencies. Consider a vector sensor consisting of two small scalar sensors separated by s, with a voltage at each sensor given by $V_1 = M_1 p_1$ and $V_2 = M_1 p_2$, and a voltage difference of $V_d = M_1(p_1 - p_2) = M_1 \Delta p$. As a simple example, also consider a nearby small spherical noise source situated at a distance, r, on a line passing through the two small scalar sensors. In practice the noise sources are not this simple and usually consist of a spatial distribution of sources resulting from turbulence and structural vibrations. The omni case is used here only to show the strong sensitivity of vector sensors to nearby noise sources and to inhomogeneous noise fields (also see Sect. 8.5.3).

Since the noise source is nearby, the wave front is spherical with a pressure gradient that depends on the distance from the source. Let the pressure from this source be written as $p = A\mathrm{e}^{-jkr}/r$, where A is proportional to the source strength, with the gradient given by

$$\partial p/\partial r = -jk(1 + 1/jkr)p. \tag{6.82}$$

With $\partial p \approx \Delta p$ and $\partial r \approx s$ we can approximate Eq. (6.82) as

$$\Delta p = -jks(1 + 1/jkr)p, \tag{6.83}$$

yielding the vector sensor voltage response to the noise source as

$$V_{\mathrm{d}} = -jM_1 ks(1 + 1/jkr)p. \tag{6.84}$$

For a noise source which produces the same pressure at the vector sensor but is located at a large distance, R, such that $kR \gg 1$, we may write $V_{\mathrm{dp}} = -jksM_1 p$. The ratio of the mean squared voltage outputs of the nearby and the distant noise sources is then

$$\langle V_d^2 \rangle / \langle V_{\mathrm{dp}}^2 \rangle = 1 + 1/(kr)^2, \tag{6.85}$$

which shows the increased sensitivity of vector sensors to nearby noise sources as compared to distant noise sources. This effect is particularly strong at low frequencies where $kr < 1$. The sensitivity of vector hydrophones is dependent on the curvature of the wave field. Accordingly, one would expect them to be even more sensitive to nearby dipole and quadrupole sources. It is important to use a screen or dome to move noise sources as far away as possible. Since the pressure differential may be considered as a force on a body and since the velocity $u = -(1/j\omega\rho)\partial p/\partial r$ these effects apply to pressure gradient and particle velocity hydrophones.

As discussed before, vector sensors can also suffer from mechanically generated noises and should therefore be mounted with a suspension resonance well below the operating band. This is particularly important for vector sensors based on accelerometers which, by their very nature, detect mechanical vibrations as well as accelerations derived from acoustic signals. In the case of an isotropic noise field, vector sensors can have a signal-to-noise advantage over small scalar sensors by 4.8 dB, due to the DI of the cosine directional pattern. Moreover, in some cases the nulls in their directivity pattern may be directed towards a strong noise source. Noise in vector sensors has been discussed in detail by Gabrielson [63]. Arrays of vector sensors will be discussed in Chap. 8.

6.7.7 Thermal Noise from Radiation Resistance

Thermal electrical noise at the output of a hydrophone consists of direct electrical noise plus electrical noise converted from mechanical noise. The direct electrical noise is ordinary Johnson noise in the electrical resistance associated with the electrical components of the transducer; in a piezoelectric transducer it depends on the electrical dissipation factor, tan δ. The converted mechanical noise begins as random thermal motion inside or on the surfaces of mechanical components such as the radiating head mass, plus random thermal motion of molecules in the medium which bombard the head mass. These internal and external thermal fluctuations result in a random noise force on the head mass that is converted to a random noise voltage by the transduction mechanism. For comparison with noise from other sources it is convenient to convert the thermal noise voltage to an equivalent plane wave noise pressure as in Sect. 6.7.4. Of all the internal noises, only the thermal noise in the medium originates as an actual pressure.

R. H. Mellen [64] calculated the thermal noise pressure in the medium originally and also demonstrated its relationship to the radiation resistance of the transducer. His expression for the noise pressure on the vibrating surface of a hydrophone is based on expressing the thermal motion in the medium in terms of normal modes of vibration. Each mode represents a degree of freedom, and, at thermal equilibrium, equipartition of energy gives each mode the energy KT where K is Boltzmann's constant and T is the absolute temperature. The mean squared noise pressure per unit band Δf is then shown to be

$$\langle p_{\mathrm{r}}^2 \rangle = 4\pi f^2 KT\rho/c. \tag{6.86}$$

This is the thermal noise pressure of the medium that would be detected by a small omnidirectional hydrophone and is shown in comparison with sea state noise in Fig. 6.37. This thermal noise is considered to be ideally isotropic and, in the case of a directional hydrophone, would be reduced by the directivity factor D_{f} before being added to noise from other sources.

The most direct way to see the relationship of radiation resistance to thermal sea noise requires looking inside the hydrophone and considering the force and velocity that apply to the radiation resistance alone as was done in Sect. 13.15. This can be done by converting the mean squared pressure in Eq. (6.86) to a mean squared force using

$$\langle F_{\mathrm{r}}^2 \rangle = \langle p_{\mathrm{r}}^2 \rangle A^2 D_{\mathrm{a}}^2 / D_{\mathrm{f}}. \tag{6.87}$$

Here we are converting the actual isotropic noise pressure in the medium to the force on the hydrophone by using the diffraction constant averaged over all directions, $\overline{D} = D_{\mathrm{a}}/D_{\mathrm{f}}^{1/2}$, which is derived in Sect. 11.3.1, where $D(\theta, \phi)$ is defined as the directional plane wave diffraction constant. Note that this pressure to force conversion differs from the one used in Sect. 13.15 to convert force to the equivalent plane wave pressure ($F = p_{\mathrm{ff}} A D_{\mathrm{a}}$). Substituting Eq. (6.86) into Eq. (6.87) gives

$$\langle F_{\mathrm{r}}^2 \rangle = 4KT\left(\pi\rho f^2 A^2 D_{\mathrm{a}}^2\right)/cD_{\mathrm{f}} \tag{6.88}$$

Use of the general relation for radiation resistance in Eq. (6.56), $R_{\mathrm{r}} = \omega^2 \rho A^2 D_{\mathrm{a}}^2 / 4\pi c D_{\mathrm{f}}$, then gives

$$\langle F_r^2 \rangle = 4\text{KTR}_r \tag{6.89}$$

per unit band Δf. Equation (6.89) shows the direct relationship between the noise force, F_r, and the radiation resistance, R_r, consistent with the Nyquist Theorem. Thus, the radiation resistance may be treated as the source of the thermal noise as in Sects. 6.7.4 and 13.15.

The thermal noise contribution can be expressed as a noise force, F_r in series with a noise-less resistance R_r as shown in Fig. 13.10. The electromechanical transformer, of turns ratio N, converts the mechanical quantities to its right to electrical quantities by dividing forces by N and impedances by N^2. Thus with $V_r = F_r/N$, we can also write

$$\langle V_r^2 \rangle = 4\text{KTR}_r/N^2 \tag{6.90}$$

which represents a noise voltage source at the mechanical input to the hydrophone where the head moves as a result of the thermal fluctuations in the medium.

The output noise voltage at the electrical terminals of the hydrophone is modified by the transducer which acts as a filter on the input noise. The hydrophone output noise voltage is discussed in Sect. 6.7.4 and shown in Sect. 13.15 to be

$$\langle V_n^2 \rangle = 4\text{KTR}_\text{h} = 4\text{KTR}_r k^4/N^2\eta_{ea}\left[(\omega/\omega_a Q_a)^2 + \left(1 - \omega^2/\omega_a^2\right)^2\right] \tag{6.91}$$

where R_h is the hydrophone series electrical resistance, η_{ea} is the electroacoustical efficiency and $k^2 = N^2C^E/C_f$. Equation (6.91) shows that the hydrophone noise output voltage is proportional to the radiation resistance, is inversely proportional to the efficiency, and also depends on frequency as well as transducer parameters. It can be seen that if the transducer was 100 % efficient, there would still be noise from the medium at the hydrophone terminals, which sets a fundamental limit on the detectable signal of a hydrophone. On the other hand, since $\eta_{ea} = \eta_{em}\eta_{ma,}$ where the mechanoacoustical efficiency is $\eta_{ma} = R_r/(R_r + R_m)$ we may replace R_r/η_{ea} in Eq. (6.91) by $(R_r + R_m)/\eta_{em}$ where R_m is the mechanical loss resistance and η_{em} is the electromechanical efficiency. Thus, under vacuum loading conditions, where $R_r = 0$, there is a noise contribution from R_m/η_{em} which is dependent on the mechanical and electrical losses alone. This condition is approximated under air loading conditions where $R_r \ll R_m$.

6.8 Summary

In this chapter various hydrophones have been described including omnidirectional pressure sensitive and dipole velocity sensitive vector sensors. The chapter covers principles of operation and hydrophone types: cylindrical and spherical, planar, bender, vector as well as the diffraction constant and thermal noise. Hydrophones typically operate below the open circuit resonance and sometime up to this resonance. The below-resonance hydrophone sensitivity $M = V/p_i = gt$, where V is the open circuit voltage, p_i is the acoustic pressure, t is the thickness between the electrodes, and g is the piezoelectric "g" constant. If the mechanical stress is in the polarized 3 direction and the output electrodes are normal to that then g is g_{33} and maximum output is obtained. If the stress is in an orthogonal direction then g is g_{31} with a value of approximately $-g_{33}/2$. For the case of a 31 mode cylinder $M = g_{31}a$, where a is the mean radius.

For a block of material and equal pressure on all surfaces, the hydrostatic sensitivity $M = (g_{33} + 2g_{31})t$ which can be nearly zero for most piezoelectric materials. The challenge is to design hydrophones where cancellation from other surfaces is mitigated by shielding them from the acoustic pressure. The other challenge is to design for a high value hydrophone figure of merit $\text{FOM}_h = M^2 C_f = gdV_p$ where C_f is the free capacity, V_p is the volume of piezoelectric material, and d is the piezoelectric "d" constant. It was shown that a spherical hydrophone of the same diameter as a cylinder hydrophone of the same height has a wider frequency response. The high sensitivity of planar Tonpilz hydrophones and the flexibility and hydrostatic sensitivity of 1-3 composite hydrophones and models for them were also presented. Bender hydrophones are compliant and provide good sensitivity for a small size but are typically limited in depth capability. The most common ones are dual bender disc and an equivalent circuit for this design was given.

Vector hydrophones are sensitive to a pressure differential or velocity or, in some designs, the acceleration rather than the acoustic pressure. Because of this, they are not omnidirectional and have a dipole beam pattern with a DI of 4.8 dB. Various designs and the response near baffles were described in Sect. 6.5. The response for pressure gradient, velocity, and accelerometer vector hydrophones was analyzed. Multimodal vector sensors which include monopole, dipole, and quadrupole performance were also discussed in addition to intensity sensors.

The plane wave diffraction constant D_a was shown as a means for obtaining the receiving sensitivity of a hydrophone that is not small compared to wavelength. With p_i the incident acoustic pressure and V the output voltage $M = V/p_i = VD_aA/F_b$, where A is the area and F_b is the blocked force which is used at the mechanical terminals of the equivalent circuit. If the transducer is small compared to wavelength, $D_a = 1$ and if set in or just in front of a rigid baffle $D_a = 2$. For a sphere of radius, a, and wave number k, $D_a = (1 + k^2a^2)^{-1/2}$. An important general equation which relates the D_a to the directivity factor, D_f, and radiation resistance, R_r, is $D_a^2 = 4\pi R_r D_f/\rho ck^2A^2$.

The limitation on hydrophones is the noise in the medium and the electrical noise generated in the hydrophone. The Johnson electrical thermal noise voltage V_n level may be written as $20 \log V_n = -198 \text{ dB} + 10 \log R_h + 10 \log \Delta f$ and seen to be dependent on the bandwidth Δf and the resistive, R_h, part of the total electrical input impedance of the hydrophone. This resistive part is due to the electrical and mechanical losses in the transducer as well as the radiation resistance. Models for transducer noise were given in this chapter. Other noises include sea state noises and flow and ship noises. Models and measured results were given for these noise sources, as well as means for mitigation. Increasing the directivity index of a hydrophone reduces the noise and is one of the advantages of a hydrophone array. An additional discussion on noise sources and mitigation is given in Chap. 8. Also, see Sect. 13.15 for a Comprehensive Noise Model.

Exercises (Degree of Difficulty: *Lowest, **Moderate, ***Highest)

6.1.** Calculate the in-air ring-mode open-circuit resonance frequency and the low frequency receiving response and free capacitance of a thin-walled, 0.508 cm (0.2 in.) Type 1 (PZT-4) piezoelectric ceramic ring hydrophone of mean diameter 10.16 cm (4 in.) and height 5.08 cm (2 in.) operating in the 31 mode. Compare results with a thin-walled spherical hydrophone operating in the planar mode, with the same wall thickness and diameter. What is the length mode resonance frequency of the ring? Assume air backing with isolated end caps on the ends of the ring.

6.2.*** Determine the receiving response through the antiresonance of the ring hydrophone of Exercise 6.1. Assume equivalent sphere radiation loading and spherical diffraction constant. Calculate the electrical impedance and use reciprocity to determine the TVR. Compare with the TVR obtained directly in Exercise 6.4.

6.3.*** Consider two small ring Type I hydrophones one-tenth the size of the one in Exercise 6.1 and separated by 5.08 cm (2 in.). First, calculate antiresonance and the low frequency RVS value for each alone. Are these results scalable from the results of Exercise 6.1? Next, determine the low frequency summed sensitivity and the low frequency differenced sensitivity for a plane wave incident in the axial direction. Calculate the frequency at which you would expect this vector sensor to differ from an expected cosine beam pattern. Calculate the frequencies at which the summed and differenced modes yield a null in the axial direction.

6.4.*** Determine the low frequency electrical noise voltage, equivalent noise pressure, and signal-to-noise ratio for a 1 μPa plane wave signal on the MRA for the ring hydrophone of Exercise 6.1. Repeat for the wide band noise response through antiresonance. Assume an electrical dissipation factor of 0.004, equivalent sphere radiation loading and diffraction constant.

6.5.** Determine the low frequency electrical noise voltage and equivalent noise pressure for the summed and differenced dual hydrophones of Exercise 6.3. Assume an electrical dissipation factor of 0.004.

6.6.** The Tonpilz of Exercise 5.8 (case A), as shown in Fig. 5.17, is to be also used as a hydrophone. Calculate the low frequency signal sensitivity and equivalent noise pressure for a dissipation factor of 0.004. Assume array baffle conditions and that $D_a = 2$. Sketch and explain the operation of the TR switch for achieving transmit and receive operation from a single transducer.

6.7.** Calculate the first four short circuit extensional ring mode resonance frequencies of the hydrophone of Exercise 6.1 for Type I material operating in the 31 mode. How many electrodes are needed to detect (or excite) each of these four modes? What is the highest mode that can be excited below the length mode resonance?

6.8.*** The internal thermal noise for a 100 % efficient hydrophone can be represented by the mechanical radiation resistance of the hydrophone, although it is caused by a thermal noise pressure in the medium acting on the surface of the hydrophone, as described by Mellen [54]. Radiation resistance is usually thought of as a measure of how a transducer radiates, but in this case, it is also a measure of how the transducer receives. What is the basic physical reason for this dual role of radiation resistance?

6.9.* As an example of the usefulness of Eq. (6.56) use the known radiation resistance in Eq. (10.44) and directivity factor for a monopole sphere to find the diffraction constant for a spherical hydrophone.

References

1. C.L. LeBlanc, *Handbook of Hydrophone Element Design Technology*, NUSC Technical Document 5813 (NUWC, Newport, 11 October 1978)
2. W.J. Trott, Sensitivity of piezoceramic tubes, with capped or shielded ends, above the omnidirectional frequency range. J. Acoust. Soc. Am. **62**, 565–570 (1977)
3. J.L. Butler, Model for a ring transducer with inactive segments. J. Acoust. Soc. Am. **59**, 480–482 (1976)
4. R.A. Langevin, Electro-acoustic sensitivity of cylindrical ceramic tubes, J. Acoust. Soc. Am. **26**, 421–427 (1954). An example of a tangentially poled tube hydrophone is given by T.A. Henriquez, An extended-range hydrophone for measuring ocean noise, J. Acoust. Soc. Am. **52**, 1450–1455 (1972)
5. S. Ko, G.A. Brigham, J.L. Butler, Multimode spherical hydrophone, J. Acoust. Soc. Am. **56**, 1890–1898 (1974). See also, S. Ko, H.L. Pond, Improved design of spherical multimode hydrophone, J. Acoust. Soc. Am. **64**, 1270–1277 (1978). J.L. Butler, S.L. Ehrlich, Superdirective spherical radiator, J. Acoust. Soc. Am. **61**, 1427–1431 (1977)
6. G.W. McMahon, Sensitivity of liquid-filled, end-capped, cylindrical, ceramic hydrophones. J. Acoust. Soc. Am. **36**, 695–696 (1964)
7. W.D. Wilder, Electroacoustic sensitivity of ceramic cylinders. J. Acoust. Soc. Am. **62**, 769–771 (1977)
8. W.A. Smith, Modeling 1-3 composite piezoelectrics: Hydrostatic response. IEEE Trans. Ultrason. Ferroelectr. Freq. Control **40**, 41–49 (1993)
9. G. Hayward, J. Bennett, R. Hamilton, A theoretical study on the influence of some constituent material properties on the behavior of 1-3 connectivity composite transducers. J. Acoust. Soc. Am. **98**, 2187–2196 (1995)
10. M. Avellaneda, P.J. Swart, Calculating the performance of 1-3 piezoelectric composites for hydrophone applications: an effective medium approach. J. Acoust. Soc. Am. **103**, 1449–1467 (1998)
11. Private communication with Brian Pazol, MSI, Littleton, MA 01460
12. G.M. Sessler, Piezoelectricity in polyvinylidenefluoride. J. Acoust. Soc. Am. **70**, 1596–1608 (1981)

13. T.D. Sullivan, J.M. Powers, Piezoelectric polymer flexural disk hydrophone. J. Acoust. Soc. Am. **63**, 1396–1401 (1978)
14. J.L. Butler, *Analysis of a Polymer Sandwich Hydrophone* (Image Acoustics, Cohasset, 26 October 1983)
15. M.B. Moffett, D. Ricketts, J.L. Butler, The effect of electrode stiffness on the piezoelectric and elastic constants of a piezoelectric bar. J. Acoust. Soc. Am. **83**, 805–811 (1988)
16. R.S. Woollett, *Theory of the Piezoelectric Flexural Disk Transducer with Applications to Underwater Sound*, USL Research Report No. 490 (Naval Undersea Warfare Center, Newport, 1960)
17. Y.T. Antonyak, M.E. Vassergisre, Calculation of the characteristics of a membranes-type flexural-mode piezoelectric transducer. Sov. Phys. Acoust. **28**, 176–180 (1982)
18. B.S. Aronov, Energy analysis of a piezoelectric body under nonuniform deformation. J. Acoust. Soc. Am. **113**, 2638–2646 (2003)
19. R.S. Woollett, *The Flexural Bar Transducer* (Naval Undersea Warfare Center, Newport, n.d.)
20. H.F. Olson, *Acoustical Engineering* (D. Van Nostrand, New York, 1957)
21. J. Eargle, *Sound Recording* (D. Van Nostrand, New York, 1976)
22. C.B. Leslie, J.M. Kendall, J.L. Jones, Hydrophone for measuring particle velocity. J. Acoust. Soc. Am. **28**, 711–715 (1956)
23. T.B. Gabrielson, D.L. Gardner, S.L. Garrett, A simple neutrally buoyant sensor for direct measurement of particle velocity and intensity in water. J. Acoust. Soc. Am. **97**, 2227–2237 (1995)
24. J.A. McConnell, Analysis of a compliantly suspended acoustic velocity sensor, J. Acoust. Soc. Am. **113**(3), 1395–1405 (2003). See also Ph. D. Thesis, *Development and Application of Inertial Type Underwater Acoustic Intensity Probes*, Penn State University, State College, December 2004
25. P. Murphy, Geophone design evolution related to non-geophysical applications, in *Acoustic Particle Velocity Sensors*, ed. by M.J. Berliner, J.F. Lindberg (American Institute of Physic, AIP Conference Proceedings 368, Woodbury, 1995) pp. 49–56
26. J.L. Butler, *Directional Transducer*, U.S. Patent 4,326,275 (20 April 1982)
27. M.B. Moffett, J.M. Powers, A bimorph flexural-disc accelerometer for underwater use, in *Acoustic Particle Velocity Sensors*, ed. by M.J. Berliner, J.F. Lindberg (American Institute of Physic, AIP Conference Proceedings 368, Woodbury, 1995), pp. 69–83. See also, M.B. Moffett, D.H. Trivett, P.J. Klippel, P.D. Baird, A piezoelectric, flexural-disk, neutrally buoyant, underwater accelerometer, IEEE-T-UFFC, 45, 1341–1346 (1998). J.L. Butler, A low profile motion sensor, SBIR N96-165, NUWC, Contract N66604-97-M-0109 (16 July 1997)
28. R.S. Gordon, L. Parad, J.L. Butler, Equivalent circuit of a ring transducer operated in the dipole mode. J. Acoust. Soc. Am. **58**, 1311–1314 (1975)
29. M. Abramowitz, I.A. Stegun, *Handbook of Mathematical Functions, Eq. 9.12.1* (Wiley, New York, 1972)
30. S.L. Ehrlich, P.D. Frelich, *Sonar Transducer*, U.S. Patent 3,290,646 (6 December 1966)
31. S.L. Ehrlich, Private communication to J. L. B.
32. A.L. Butler, J.L. Butler, J.A. Rice, A tri-modal directional transducer, J. Acoust. Soc. Am. **115**, 658–665 (2004). Also see Oceans 2003 Proceeding (September 2003). J.L. Butler, A.L. Butler, *Multimode Synthesized Beam Pattern Transducer Apparatus*, U.S. Patent 6,730,604 B2 (11 May 2004)
33. J.L. Butler, A.L. Butler, *A Directional Power Wheel Cylindrical Array*, ONR 321 Maritime sensing (MS) Program Review (NUWC, Newport, 18 August 2005)
34. R.J. Urick, *Principles of Underwater Sound*, 3rd edn. (Peninsula, Los Altos Hills, 1983)
35. W. Thompson Jr., Higher powers of pattern functions—a beam pattern synthesis technique. J. Acoust. Soc. Am. **49**, 1686–1687 (1971)
36. P.W. Smith, T.J. Schultz, *On Measuring Transducer Characteristics in a Water Tank*, BBN Report No. 876 (September 1961). P.W. Smith, T.J. Schultz, C.I. Malme, *Intensity Measurement in Near Field and Reverberant Spaces*, BBN Report No. 1135 (July 1964). See also,

T.J. Schultz, P.W. Smith, C.I. Malme, Measurement of Acoustic Intensity in Reactive Sound Field, J. Acoust. Soc. Am. **57**, 1263–1268 (1975)
37. K.J. Bastyr, G.C. Lauchle, J.A. McConnell, Development of a velocity gradient underwater acoustic intensity sensor. J. Acoust. Soc. Am. **106**, 3178–3188 (1999)
38. T.B. Gabrielson, J.F. McEachern, G.C. Lauchle, *Underwater Acoustic Intensity Probe*, U.S. Patent No. 5,392,258 (1995)
39. K.W. Ng, *Acoustic Intensity Probe*, U.S. Patent No. 4,982,375 (1991)
40. A.O. Sykes, *Transducer for Simultaneous Measurement of Physical Phenomena of Sound Wave*, U.S. Patent No. 3,274,539 (1966)
41. F. Schloss, *Intensity Meter Particle Acceleration Type*, U.S. Patent No. 3,311,873 (1967)
42. J.A. McConnell, G.C. Lauchle, T.B. Gabrielson, *Two Geophone Underwater Acoustic Intensity Probe*, U.S. Patent No. 6,172,940 (2001)
43. G.C. Lauchle, K. Kim, Acoustic intensity scattered from an elliptic cylinder, in *Proceedings of Workshop on Directional Acoustic Sensors* (Newport, April 2001). G.C. Lauchle, Intensity Measurements of DIFAR Signals, ONR Maritime Sensing Program Review (Newport, May 2004)
44. T.K. Stanton, R.T. Beyer, Complex measurement in a reactive acoustic field, J. Acoust. Soc. Am. **65**, 249–252 (1979). See also, P.J. Westervelt, Acoustic impedance in terms of energy functions, J. Acoust. Soc. Am. **23**, 347–348 (1951)
45. J.L. Butler, A.E. Clark, *Hybrid Piezoelectric and Magnetostrictive Acoustic Wave Transducer*, U.S. Patent 4,443,731 (17 April 1984)
46. D. Lubman, Benefits of acoustical field diversity sonar, J. Acoust. Am. **70**(S1), S101 (A) (1981). D. Lubman, Antifade sonar employs acoustic field diversity to recover signals from multipath fading, in *Acoustic Particle Velocity Sensors*, ed. by M.J. Berliner, J.F. Lindberg (American Institute of Physics, AIP Conference Proceedings 368, Woodbury, 1995), pp. 335–344
47. R.S. Woollett, Diffraction constants for pressure gradient transducers. J. Acoust. Soc. Am. **72**, 1105–1113 (1982)
48. R.J. Bobber, Diffraction constants of transducers. J. Acoust. Soc. Am. **37**, 591–595 (1965)
49. T.A. Henriquez, Diffraction constants of acoustic transducers. J. Acoust. Soc. Am. **36**, 267–269 (1964)
50. H. Levine, J. Schwinger, On the radiation of sound from an unflanged circular pipe. Phys. Rev. **73**, 383–406 (1948)
51. Z. Milosic, J. Acoust. Soc. Am. **93**, 1202 (1993)
52. R.S. Woollett, Procedures for comparing hydrophone noise with minimum water noise. J. Acoust. Soc. Am. **54**, 1376–1380 (1973)
53. C. Kittel, *Elementary Statistical Physics* (Wiley, New York, 1958), pp. 148–149
54. R.H. Mellen, Thermal-noise limit in the detection of underwater acoustic signals. J. Acoust. Soc. Am. **24**, 478–480 (1952)
55. J.W. Young, Optimization of acoustic receiver noise performance. J. Acoust. Soc. Am. **61**, 1471–1476 (1977)
56. R.S. Woollett, Hydrophone design for a receiving system in which amplifier noise is dominant. J. Acoust. Soc. Am. **34**, 522–523 (1962)
57. T.B. Straw, *Noise Prediction for Hydrophone/Preamplifier Systems*, NUWC-NPT Technical Report 10369, NUWC, Newport, 3 June 1993)
58. P.D. Baird, EDO directional acoustic sensor technology, in *Proceedings of Workshop on Directional Acoustic Sensors*, Newport (April 2001)
59. P.A. Wlodkowski, F. Schloss, Advances in acoustic particle velocity sensors, in *Proceedings of Workshop on Directional Acoustic Sensors*, Newport (April 2001)
60. Y.E. Lo, M.C. Junger, Signal-to-noise enhancement by underwater intensity measurements. J. Acoust. Soc. Am. **82**, 1450–1454 (1987)

61. V.I. Ilyichev, V.A. Shchurov, The properties of the vertical and horizontal power flows of the underwater ambient noise, in *Natural Physical Sources of Underwater Sound*, ed. by B.R. Kerman (Kluwer Academic, The Netherlands, 1993), pp. 93–109
62. D. Haung, R.C. Elswick, J.F. McEachren, Acoustic pressure-vector sensor array. J. Acoust. Soc. Am. **115**, 2620 (2004)
63. T.B. Gabrielson, Modeling and measuring self-noise in velocity and acceleration sensors, in *Acoustic Particle Velocity Sensors*, ed. by M.J. Berliner, J.F. Lindberg (AIP Conference Proceedings 368, Woodbury, 1995), pp. 1–48
64. R.H. Mellen, Thermal-noise limit in the detection of underwater acoustic signals. J. Acoust. Soc. Am. **24**, 478–480 (1952)

Chapter 7
Projector Arrays

Naval applications are the main motivation for the development of large, innovative sonar systems. Therefore, the development of large acoustic arrays is closely related to new ship construction, and especially to new submarines since they depend so strongly on acoustics [1, 2]. The main function of active sonar on submarines is searching for surface ships and other submarines, but avoidance of mines and sea mounts, as well as underwater communications, are also very important. Active search requires large projector arrays operating in the 2–10 kHz region for medium range performance, while obstacle avoidance uses smaller, higher frequency arrays. All submarine applications require transducers capable of withstanding hundreds of pounds per square inch of hydrostatic pressure without significant change in performance. On surface ships active sonar is used mainly for searching for submarines, with transducers similar to those in submarine arrays except for the hydrostatic pressure requirement. Extremely long range active sonar requires lower frequency and higher power (see Fig. 1.10), which leads to many problems in transducer and array design as well as possible environmental effects.

Active sonar arrays often contain hundreds of individual projectors in order to radiate sufficient acoustic power in well-defined directions. The transducers in these arrays are usually mounted on a plane, cylindrical, or spherical surface as shown in Fig. 7.1 and Figs. 1.10, 1.11, and 1.12 and enclosed behind an acoustic window in the hull of a ship, a submarine, or a towed body.

The transducers are packed closely together as shown in the figures to maximize acoustic loading, and to form and steer acoustic beams. However, the close-packed transducers also influence each other by acoustic coupling through the water, causing the vibration of each transducer to be determined, not only by its electrical driver, but also by the vibration of the other transducers. The importance of acoustic coupling became apparent during testing of large arrays of efficient transducers after WWII, when it was found that coupling can degrade array performance, and even cause transducer failure in some cases. The effects of acoustic coupling can be predicted by including it in the array design process in terms of mutual radiation impedances.

J.L. Butler, C.H. Sherman, *Transducers and Arrays for Underwater Sound*,
Modern Acoustics and Signal Processing, DOI 10.1007/978-3-319-39044-4_7

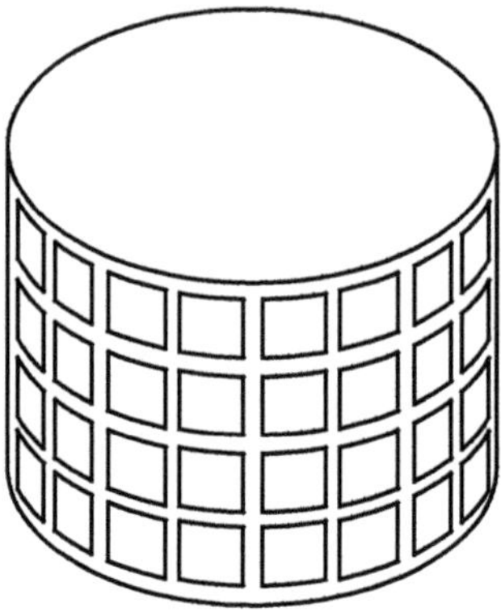

Fig. 7.1 Diagram of a cylindrical array used in active sonar

An active sonar system design begins with a goal of detecting certain targets at specified ranges from specific platforms. The goal determines the appropriate frequency band and the required acoustic performance characteristics, such as the maximum source level, maximum acoustic power, directivity index, number of steered beams, and beam widths. The platform and other operating conditions usually impose constraints on array size, weight, shape, and available electrical power, which, in turn, constrain the acoustic characteristics. Sonar system design will not be discussed here; Urick [3] gives a simple description of the whole process in his Chaps. 2 and 13, Horton [4] discusses all aspects of sonar and Bell [2] describes the 20 year development of a specific surface ship active sonar system.

Only the array design will be discussed here, and it begins with the relationships between frequency, dimensions, and directivity which are given in Sect. 7.1. These relations are used to determine array size and shape and the number and arrangement of transducers consistent with the constraints. These factors then determine the individual transducer size, shape, and weight. As a first step in array analysis the acoustic coupling can be neglected, which makes the transducer velocities proportional to the applied voltages, and allows easy calculation of preliminary directivity patterns and DIs over the full range of frequencies and steering angles. The DI and the specified source level give a preliminary estimate of the average power required from each transducer. The transducer size, frequency, and average power indicate the type of transducer that might be suitable and whether it will operate near the limits of transduction capabilities. But the question of transducer limits cannot be answered without including acoustic coupling in the array analysis, because the coupling makes the transducer power vary from one transducer to another.

Array analysis including acoustic coupling determines the velocities of each transducer using assumed transducer parameters and calculated mutual impedances with specified voltage amplitudes and phases as will be described in Sect. 7.2. Calculation of mutual impedance will be discussed in Sect. 7.3, with examples that can be used in array analysis. Solution for the velocities requires inversion of a complex matrix, and then the total radiation impedance of each transducer can be found, as well as transducer currents and electrical impedances. Because of acoustic coupling the velocities, radiation impedances, resonance frequencies, and radiated power vary from one transducer to another. Furthermore, the transducer velocities do not have the same phasing and shading as the voltages, which means that the beam patterns calculated from these velocities usually have lower DIs than those calculated without considering mutual coupling.

The total acoustic power radiated by the array, for specified voltages, can be determined from the calculated velocities and radiation resistances. Comparing this power with the maximum power required by the originally specified source level and the recalculated DIs gives the required maximum transducer voltage amplitudes and maximum velocity amplitudes. This is the critical transducer information. The velocities determine the stresses in the transducers, which must be low enough to avoid mechanical failure, while the voltages determine the electric fields and currents which must be low enough to avoid electrical breakdown (see Sect. 2.8.5).

The information provided by the array analysis gives a basis for determining whether an existing transducer design is adequate, and, when a new design is needed, whether it is feasible. The transducer problems caused by the coupling arise mainly from the variations of velocity and current, which change with frequency and steering. Since the electrical amplifiers are usually approximate voltage sources, the voltage magnitudes and phases are under control, but the variations in the velocities and currents are not. Each transducer and its power amplifier must be able to withstand the worst case of high velocity or high current. Although only a small fraction of the transducers will be subjected to these high values for a given frequency and steering, other transducers will be subjected to similar high values as the beam is steered and the frequency changed. Thus the array output is limited by the failures that might result from the velocity and current variations caused by the coupling.

When array analysis indicates that the assumed transducer characteristics are not adequate the electrical and mechanical limits might be raised or brought closer together (see Sects. 2.8.5 and 2.9) by improving construction techniques or considering a different transduction material or mechanism. With a revised transducer design and a new set of transducer parameters, another array design iteration can be initiated by solving the array equations again and recalculating the acoustic performance. When this integrated transducer/array design process leads to satisfactory results a partial array can be built and tested for comparison with the results of array analysis. The experimental testing must also include a complete array and consider all the system components [1, 2].

In echo ranging systems it is usually possible to use the projector array for the receiving array, although when high performance against noise is required, and space is available, it may be advantageous to use a separate receiving array [1]. Hydrophone arrays will be discussed in Chap. 8.

There are several levels of complexity in array analysis depending on the transducer characteristics and the array geometry. The simplest case is the surface array of fixed velocity distribution (FVD) transducers, which can also be called a baffled array since each transducer is part of a common surface formed by the other transducers as shown in Fig. 7.2a. Arrays of non-FVD transducers present a much more complex situation because these transducers may vibrate in more than one

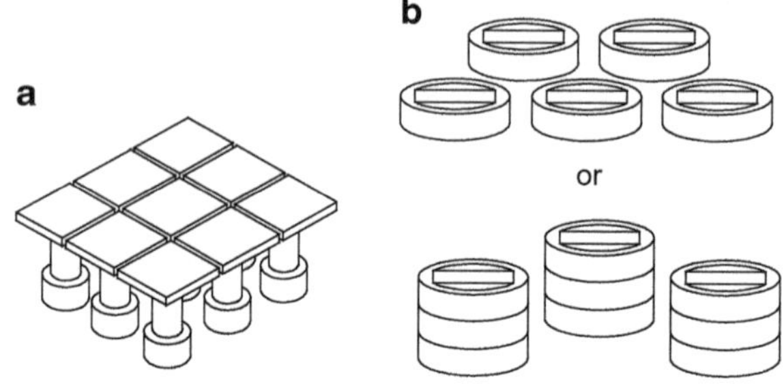

Fig. 7.2 Diagrams of surface and volume arrays. (**a**) Surface array of close-packed longitudinal resonator transducers forming a plane surface. (**b**) Volume array of oval flextensional transducers

mode with a modal composition that depends on location in the array. Then the mutual radiation impedance concept must be extended to include coupling between different modes (see Sect. 7.4).

Another level of complexity occurs when the transducers in the array do not form part of a closed surface but instead are arranged throughout a volume as illustrated in Fig. 7.2b. Then, with no common baffle surrounding the transducers, the coupling between transducers includes scattering in addition to the direct effect of one transducer on another (see Sect. 7.5). In large volume arrays classical methods of analysis are not practical, but finite element modeling is applicable.

Acoustic pressure variations in the near field of arrays may also be a limitation since the pressure at a point on or near the array surface may be high enough to initiate cavitation (see Sect. 10.3.2). In some cases the hydrostatic pressure inside sonar domes on surface ships is raised above the outside pressure to avoid cavitation. The acoustic pressure variations on the array surface are related to the total radiation impedance variations among the transducers, since the radiation impedance is the spatial average of the acoustic pressure over a transducer surface. Therefore, array analysis that shows large variations of total radiation impedance may indicate the potential for cavitation.

Section 7.6 will illustrate the detailed information that can be obtained by finite element modeling of arrays, including an example of the pressure and displacement distribution on an array surface. In Sect. 7.7 a different type of projector array, the nonlinear parametric array, will be presented. An array concept in which the array transducer elements, as well as the array, are steered into the same direction will be developed in Sect. 7.8.

7.1 Array Directivity Functions

7.1.1 The Product Theorem

The far-field directivity function of an array is the sum of the far fields of all the individual transducers (see Chap. 10). Arrays of identical transducers forming a plane surface are the easiest to analyze if it is assumed that a plane baffle extends beyond the edges of the array for several wavelengths which makes the far fields of all the individual transducers have the same directivity function in the array. It will be assumed that the baffle is rigid and that the radiating faces have arbitrary shape with area A. The pressure produced by the *i*th transducer in the array is given by the Rayleigh integral in Eq. (10.25a), where R is the distance from a point on the transducer surface to a field point. In the far field R is much greater than the transducer dimensions and can be approximated by r_i in the denominator of Eq. (10.25a) and by $R = r_i - r_0 \sin\theta \cos(\phi - \phi_0)$ in the phase factor, where r_i is the distance from the center to a point in the far field and r_0 is the distance from the center to a point on the transducer surface (see Fig. 7.3). The result for transducers with uniform velocity is

$$p_i(r_i, \theta, \phi) = \frac{j\rho cku_i A}{2\pi} \frac{e^{-jkr_i}}{r_i A} \iint e^{jkr_0 \sin\theta \cos(\phi-\phi_0)} dS_0. \tag{7.1}$$

The integral in Eq. (7.1), divided by A, is the dimensionless individual transducer directivity function which depends only on θ and ϕ; it can be written as $f(\theta, \phi)$ and is the same for every transducer in the array as long as they are oriented in the same way. Then Eq. (7.1) can be written as

$$p_i(r_i, \theta, \phi) = \frac{j\rho cku_i A}{2\pi} f(\theta, \phi) \frac{e^{-jkr_i}}{r_i}. \tag{7.2}$$

Equation (7.2) shows that the far-field pressure produced by each transducer is proportional to its frequency, velocity, and area or to its acceleration and area since $jcku = j\omega u$ is the acceleration.

An array of N such transducers arranged in any way on the plane, but all having the same orientation with respect to the array coordinate system, has a far field given by

$$p(r, \theta, \phi) = \frac{j\rho ckA}{2\pi} \frac{f(\theta, \phi)}{r} \sum_{i=1}^{N} u_i e^{-jkr_i}, \tag{7.3}$$

where all the r_i are approximated by r in the amplitude but not in the phase, with r the distance from the center of the array coordinate system to a point in the far

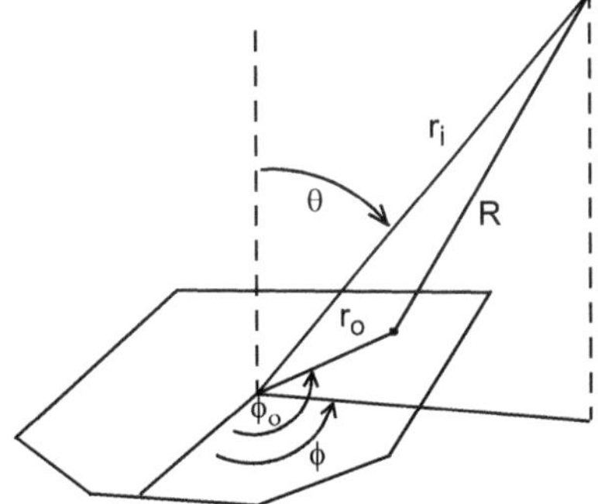

Fig. 7.3 Coordinates used in Eq. (7.1) for the field of a piston transducer with arbitrary shape

field. Equation (7.3) can be used to calculate the array directivity function when the transducers have different velocities and are arranged arbitrarily on a plane. It is evident that the array directivity function is the product of the individual transducer directivity function, $f(\theta, \phi)$, and the directivity function of an array consisting of point sources located at the center of each transducer and given by the summation in Eq. (7.3). This result is known as the Product Theorem [5]. The Product Theorem does not apply to arrays on curved surfaces where the transducer axes do not all point in the same direction.

7.1.2 Line, Rectangular, and Circular Arrays

Equation (7.3) can be used to calculate the far-field directivity function of a plane array with any geometrical arrangement of the transducers, as long as they are oriented in the same way, with any distribution of velocities among the transducers. When the transducers all vibrate with the same velocity and are also arranged on a uniform rectangular grid, as shown in Fig. 7.4, Eq. (7.3) can be evaluated in a simple form.

Let the plane of the array be the xy plane, with one corner of the array at the origin, with N transducers parallel to the x-axis with spacing D, and M transducers parallel to the y-axis with spacing L. The distance r_i in Eq. (7.3) is the distance from the center of a transducer designated by n, m to a far-field point where $n=0, 1, 2, \ldots N-1$ and $m=0, 1, 2, \ldots M-1$. For example, consider the transducer at $n=0$, $m=2$ and a far-field point in the yz plane ($\phi = 90°$); it can be seen from Fig. 7.4 that $r_i = r_{02} = r - 2L \sin\theta$, and in general that

$$r_i = r_{\mathrm{nm}} = r - \sin\theta(nD\cos\phi + mL\sin\phi). \tag{7.4}$$

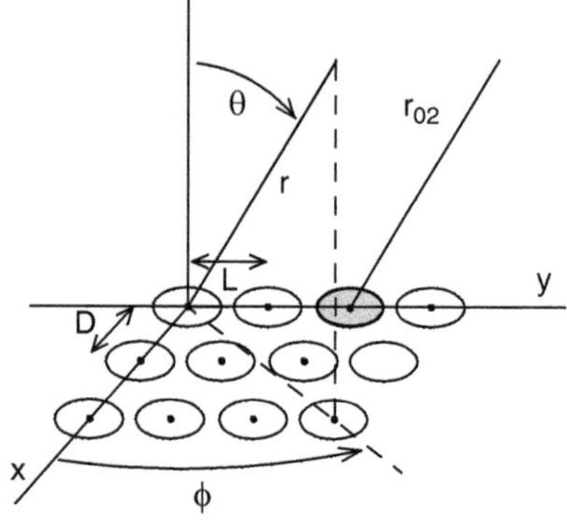

Fig. 7.4 Coordinates for calculating the far field of a rectangular array of identical piston transducers. The *shaded* transducer is located at $n=0$, $m=2$

Equation (7.3) can then be written as

$$p(r, \theta, \phi) = \frac{j\rho ckuA}{2\pi} f(\theta, \phi) \frac{\mathrm{e}^{-jkr}}{r} \sum_{n=0}^{N-1} \mathrm{e}^{jknD \sin\theta \cos\phi} \sum_{m=0}^{M-1} \mathrm{e}^{jkmL \sin\theta \sin\phi} \tag{7.5}$$

where u is the transducer velocity. Since each of these summations is a geometric series, the far-field pressure of the array can be written as

$$\begin{aligned} p(r, \theta, \phi) = & \left[\frac{jNM\rho ckuAf(\theta, \phi)\mathrm{e}^{-jkR}}{2\pi r}\right] \frac{\sin\left(\frac{1}{2}NkD \sin\theta \cos\phi\right)}{N \sin\left(\frac{1}{2}kD \sin\theta \cos\phi\right)} \\ & \times \frac{\sin\left(\frac{1}{2}MkL \sin\theta \sin\phi\right)}{M \sin\left(\frac{1}{2}kL \sin\theta \sin\phi\right)} \end{aligned} \tag{7.6a}$$

where $R = r - (N-1)D \sin\theta \cos\phi/2 - (M-1)L \sin\theta \sin\phi/2$, and the added phase factors are used to refer the phase of the array output to a point at the center of the array. The last two factors in Eq. (7.6a) have each been normalized to unity at $\theta = 0$.

For small, omnidirectional transducers, where $f(\theta, \phi)$ is a constant equal to unity, the array directivity function depends on the product of the second and third factors in Eq. (7.6a), and it has a simple form in two planes. For $\phi = 0°$ and 180° (the *xz* plane) the second factor in Eq. (7.6a) is the directivity function of a line array function, $\sin[N\pi(D/\lambda)\sin\theta]/N\sin[\pi(D/\lambda)\sin\theta]$, of omnidirectional transducers and the third factor is unity. Similarly, for $\phi = 90°$ and 270° (the *yz* plane) the third factor is a line array function, $\sin[M\pi(L/\lambda)\sin\theta]/M\sin[\pi(L/\lambda)\sin\theta]$, and the second factor is unity. Equation (7.6a) reduces to the field of a line array by setting $M = 1$ and $\phi = 0$. It reduces further to the field of a continuous line by letting the number of transducers increase ($N \to \infty$) while the spacing decreases ($D \to 0$) such that $\mathrm{ND} = L_0$, the length of the line, and NAu equals the source strength, Q_0, which gives the far-field pressure,

$$p(r, \theta, 0) = \left[\frac{j\rho ckQ_0\mathrm{e}^{-jkR}}{2\pi r}\right] \frac{\sin\left(\frac{1}{2}kL_0 \sin\theta\right)}{\left(\frac{1}{2}kL_0 \sin\theta\right)}. \tag{7.6b}$$

This differs from the far field of the continuous line source in Eq. (10.22) by a factor of 2 because the line array is in an infinite plane baffle, while the line source is in free space. The approximate results for beam width, directivity index, and first side lobe levels given in Chap. 10 for the continuous line and rectangle also hold for a line or rectangular array of discrete transducers if the separation does not exceed about one half wavelength. The results for the directivity factor are:

$D_f \approx 2ND/\lambda$, for a line of N transducers,
$D_f \approx 4\pi NMDL/\lambda^2$, for a rectangle of NM transducers.

An interesting special case of Eq. (7.6a) is the directivity pattern in the perpendicular plane on the diagonal of a square array ($\phi = 45°$, $M = N$, $D = L$); this pattern is proportional to the square of a line array pattern of N transducers with spacing $D/\sqrt{2}$:

$$p(r, \theta, \pi/4) = \left[\frac{jN^2 \rho ckuA e^{-jkR}}{2\pi r}\right]\left[\frac{\sin\left[\frac{1}{2}Nk(D/\sqrt{2})\sin\theta\right]}{N\sin\left[\frac{1}{2}k(D/\sqrt{2})\sin\theta\right]}\right]^2. \tag{7.6c}$$

This pattern is interesting because it does not change sign from one lobe to the next. Furthermore, the main beam width is greater than it is for the pattern in a plane parallel to a side of the array, despite the diagonal being longer than the side, and the side lobes are lower [6]. This occurs because, as viewed from a diagonal, the diamond shape of the array is equivalent to shading to suppress side lobes (see Sect. 7.1.4).

Another useful array geometry that can be formulated in a simple way is a uniformly spaced circular array of radius a in a plane as shown in Fig. 7.5.

Consider N transducers (where N is a multiple of 4) all vibrating with the same velocity, where $\phi_0 = 360/N$ is the angular separation in degrees between adjacent transducers. In this case the pattern is the same in any plane perpendicular to the array that passes through the center of the circle and two of the transducers. For the pattern in the xz plane the distance from the centers of the two transducers on the x-axis to a far-field point is $r \pm a \sin\theta$. For the two transducers on the y-axis the distance is r, and for each four transducers displaced from the x-axis by the angles $\pm n\phi_0$ and $\pm(180° - n\phi_0)$ the distance is $r \pm a \sin\theta \cos n\phi_0$. When the contributions from all these transducers are combined in pairs the exponentials reduce to cosines and Eq. (7.3) gives for the pattern in any perpendicular plane that passes through the center of the circle and two of the transducers:

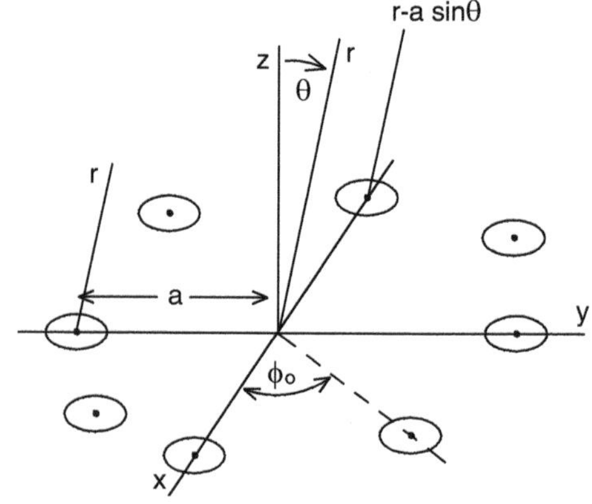

Fig. 7.5 Coordinating for calculating the far field in the xz plane of a circular array of identical transducers. The far field is the same in the xz, yz and other planes passing through two transducers and the center of the array

$$p(r,\theta) = \frac{j\rho ckuA}{2\pi} f(\theta) \frac{e^{-jkr}}{r} \left[2\cos(ka\sin\theta) + 2 + 4\sum_{n=1}^{\frac{N}{4}-1} \cos(ka\sin\theta\cos n\phi_0) \right]. \tag{7.7a}$$

When $N \to \infty$ and $\phi_0 \to 0$ with $N\phi_0 = 2\pi$, Eq. (7.7a) gives the far field of a thin ring, which can be derived in a different way and is given in Eq. (10.36a). Equation (7.7a) when normalized to unity at $\theta = 0°$ can be written as

$$\frac{p(\theta)}{p(0)} = \frac{f(\theta)}{Nf(0)} \left[2\cos(ka\sin\theta) + 2 + 4\sum_{n=1}^{\frac{N}{4}-1} \cos(ka\sin\theta\cos n\phi_0) \right]. \tag{7.7b}$$

See Chap. 8, Sect. 8.6 for a discussion of steered planar circular arrays.

A circular array with empty interior has higher side lobes than a line array since it corresponds to shading a line array more strongly near the ends. Circular arrays are of interest because they have smaller grating lobes than a line array, as will be discussed below. Phased circular arrays were discussed by Thompson [7]. Another type of circular array consisting of a circular piston surrounded by concentric annular pistons has been discussed by Zielinski and Wu [8]. Such an array can be shaded to produce a searchlight beam pattern with arbitrarily suppressed equal side lobes.

7.1.3 Grating Lobes

Equation (7.6b) for the continuous line source is the familiar sin *x*/*x* function with side lobes that decrease in amplitude as θ increases. However, actual arrays are usually not continuous, because they consist of discrete transducers that can be individually phased to steer the beam. At low frequency, where the spacing between transducers is small in terms of wavelengths, the directivity function of discrete arrays is similar to that for the continuous line, but at higher frequencies these directivity functions become quite different. The difference can be illustrated by the first of the line array functions in Eq. (7.6a) with $\phi = 0$. Then for $\theta = \pm 90°$ (negative values of θ are in the half plane $\phi = 270°$) and D equal to one half wavelength ($kD = \pi$) the pressure is zero for N even, and proportional to $1/N$ for N odd, since most of the individual contributions cancel. However, for D equal to a full wavelength ($kD = 2\pi$) the contributions add in phase at $\theta = \pm 90°$ making the normalized pressure unity and equal to the main lobe pressure as shown in Fig. 7.6.

These large lobes at $\pm 90°$ are called grating (or aliasing) lobes because they duplicate the height of the main lobe except as modified by the individual hydrophone directivity function, $f(\theta, \phi)$. As kD increases above 2π, the first grating lobes appear at smaller angles and other grating lobes also appear. For example, when $kD = 4\pi$ the first grating lobes appear at $\pm 30°$ and the second at $\pm 90°$.

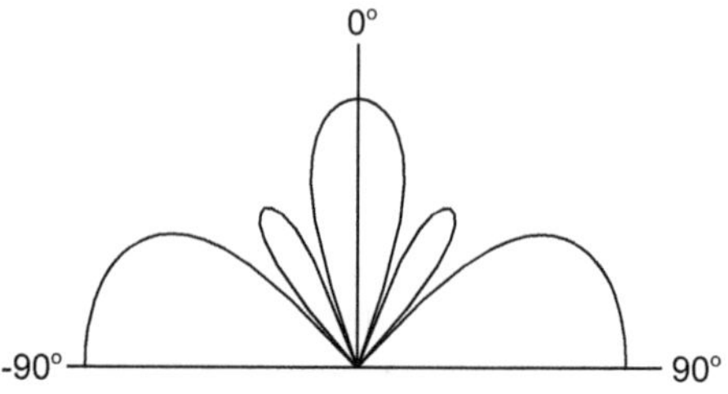

Fig. 7.6 Example of grating lobes at ±90° for a line array of three transducers with $kD = 2\pi$

Since grating lobes present a serious problem in a sonar system the spacing between transducers must be significantly less than one wavelength at the top of the frequency band (this criterion is changed by steering the beam, see Sect. 7.1.4). It is also usually the case in projector arrays that the transducer radiating faces must almost fill the spaces between transducers to maximize radiation resistance and power, which makes the directivity function of the individual transducers have a significant effect. If the transducer radiating faces are square, with side length d, the function $f(\theta, \phi)$ for an individual transducer in the xz plane is given by Eq. (7.6b) with $L_0 = d$. Then the normalized array pattern of a line array of N square transducers with spacing D is given by the Product Theorem, or specifically by Eq. (7.6a) with $M = 1$, as

$$p(\theta) = \frac{\sin\left(\frac{1}{2}kd\sin\theta\right)}{\left(\frac{1}{2}kd\sin\theta\right)}\ \frac{\sin\left(\frac{1}{2}NkD\sin\theta\right)}{N\sin\left(\frac{1}{2}kD\sin\theta\right)}. \tag{7.8a}$$

The transducer directivity function in the first factor has a significant effect at 90°. Furthermore, if the transducers completely fill the array ($d = D$ and packing factor = 1) Eq. (7.8a) reduces to

$$p(\theta) = \frac{\sin\left(\frac{1}{2}NkD\sin\theta\right)}{\frac{N}{2}kD\sin\theta,} \tag{7.8b}$$

which is the directivity function of a continuous line of length ND with no grating lobes. Thus the directivity of the individual square transducers cancels the grating lobes when $d = D$, because it has nulls at the angles where the grating lobes occur.

Such complete cancellation of the grating lobes is not feasible in practice because it requires each transducer to completely fill its allotted space, while a steerable array must have at least small gaps between transducers. If the line array contained circular transducers packed as closely as possible (radius equal to $D/2$) the first factor in Eq. (7.8a) would be $2J_1(kD\sin\theta/2)/(kD\sin\theta/2)$(see Sect. 10.2.2) which would reduce, but not cancel, the grating lobes. For example, the grating lobe at $\theta = 90°$ for $kD = 2\pi$ would be reduced by 7.5 dB. It will be seen below that when the beam is steered the grating lobes are also steered, and the directivity of the transducers then has much less effect in reducing the grating lobes.

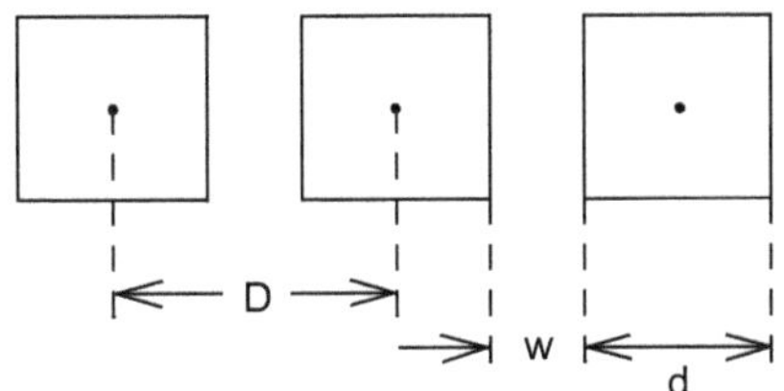

Fig. 7.7 Array of square transducers with spacing D and gaps of width w between them

Square pistons with small gaps between them do provide some control over grating lobes. With a gap width of $w = (D-d)$, as shown in Fig. 7.7, the first factor in Eq. (7.8a) (the transducer directivity function) can be written in terms of D and w and, for $w \ll D$, approximated by

$$\frac{\left[\sin\left(\frac{1}{2}kD\sin\theta\right) - \frac{1}{2}kw\sin\theta\cos\left(\frac{1}{2}kD\sin\theta\right)\right]}{\left(\frac{1}{2}kD\sin\theta\right)}.$$

Thus for the critical case of $\theta = 90°$ and $kD = 2\pi$ the transducer pattern reduces the grating lobe by the factor $(kw/2\pi)$; e.g., this factor is equal to 20 dB for gap widths that are 1/10 the transducer spacing. Results of this kind are also given by Stansfield [9]. As mentioned before, this type of grating lobe control only applies to unsteered beams.

Nonuniform spacing between transducers is another way of controlling grating lobes [10, 11]. The effect of nonuniform spacing is quite different from the effect of varying the drives of the individual transducers (shading). In a line array of omnidirectional transducers with $kD = 2\pi$ shading changes the sum of the outputs at $\theta = 0°$ and at 90° equally and, therefore, does not change the grating lobe relative to the main lobe. It is a general result that shading does not reduce grating lobes relative to the main lobe, since both occur when all transducer outputs add in phase. On the other hand, in a line array with nonuniform spacing all the outputs add in phase at $\theta = 0°$, but they do not add in phase at $\theta = 90°$ which reduces the grating lobe relative to the main lobe. The same type of grating lobe reduction occurs in the circular array since it is similar to a line array with nonuniform spacing. Since steering the beam also steers the grating lobes, their control by use of nonuniform spacing will be discussed in more detail in the next section. Grating lobes can also be controlled by use of staggered arrays of transducers with special shapes [12]. Element orientation and positioning can also affect grating lobes and will be discussed in Sect. 7.1.5.

7.1.4 Beam Steering and Shaping

The directivity functions discussed so far resulted from plane arrays of transducers all vibrating in phase with the same velocity amplitude. These functions can be modified in various ways by adjusting the amplitudes and phases of the transducer voltages, which adjusts the amplitudes and phases of the transducer velocities. For example, adjusting the amplitudes (called shading) can be used to reduce the side lobes relative to the main lobe. Shading will be discussed in Chap. 8 on receiving arrays where reduction of side lobes is especially important to discriminate against noise coming from directions other than the direction of the main beam.

The main reason for adjusting the velocity phases is to steer the beam, an essential capability for an active sonar search system. As an example consider adjusting the phases of the transducer velocities of the rectangular array in Fig. 7.2, while keeping the velocity amplitudes equal. When the velocities are all in phase the beam points in the direction $\theta = 0$, perpendicular to the plane of the array, called the broadside direction. Steering the beam to an arbitrary direction, specified by $\theta = \theta_0$ and $\phi = \phi_0$, requires application of phase shifts that make the far fields of all the transducers in phase in that direction. It can be seen from Eq. (7.5) that this occurs if the velocity phases are shifted progressively such that the transducer at location n, m has its velocity phase shifted by

$$\mu_{\mathrm{nm}} = -nkD \sin \theta_0 \cos \phi_0 - mkL \sin \theta_0 \sin \phi_0. \tag{7.9}$$

When these phase shifts are included in Eq. (7.5), each summation is a geometric series as before, and the result can be written in the same form as Eq. (7.6a):

$$p(r, \theta, \phi) = \left[\frac{jNM\rho ckuAf(\theta, \phi)\mathrm{e}^{-jkR}}{2\pi r}\right] \frac{\sin NX}{N \sin X} \frac{\sin MY}{M \sin Y}, \tag{7.10a}$$

where

$$X = \frac{kD}{2}[\sin \theta \cos \phi - \sin \theta_0 \cos \phi_0],$$

and

$$Y = \frac{kL}{2}[\sin \theta \sin \phi - \sin \theta_0 \sin \phi_0].$$

Equation (7.10a) differs from Eq. (7.6a) in that the maximum value, corresponding to the peak of the main lobe, occurs at $\theta = \theta_0$ and $\phi = \phi_0$, showing that the main beam has been steered to the desired direction. This does not mean that the entire pattern has been exactly rotated into that direction. The resulting pattern is still the product of the array pattern and the transducer pattern, but the main beam is broadened and the side lobes are raised relative to the unsteered pattern as the steering angle increases toward end-fire (see Glossary) at $\theta_0 = 90°$. A specific example for a ten element line array with approximate half wavelength spacing shows that the beam width approximately doubles when steered into the region of $\theta_0 = 60 - 90°$ [13].

To discuss the effect of steering on the grating lobes it will be convenient to consider a line array of uniformly spaced omnidirectional transducers. Equation (7.10a) gives this case for $M = 1$, with f equal to a constant and $\phi = \phi_0 = 0$; then the normalized pattern can be written as

$$\frac{p(\theta, \theta_0)}{p(0,0)} = \frac{\sin NX}{N \sin X}, \tag{7.10b}$$

where $X = \frac{kD}{2}(\sin\theta - \sin\theta_0)$ and θ_0 is the steering angle. Grating lobes occur when $X = \pm\pi, \pm 2\pi, \ldots$; with the first ($X = \pm\pi$) appearing at angles which satisfy

$$\sin\theta = \sin\theta_0 \pm 2\pi/kD = \sin\theta_0 \pm \lambda/D. \tag{7.10c}$$

For $kD = 2\pi$ with no steering grating lobes appear at $\theta = \pm 90°$, but if the beam is steered to 30°, the grating lobe at −90° is steered to −30°, and, as the beam is steered to 90°, it moves from −30° to 0°. Equation (7.10c) also shows how the spacing must be reduced, or the frequency lowered, to eliminate the grating lobe when the beam is steered. For example, when $kD = \pi$ and the beam is steered to 90° there is a full grating lobe at −90°, but for $kD < \pi$ the grating lobe is reduced and nearly eliminated at $kD = \pi/2$.

It was shown qualitatively in the previous section that a line array with nonuniform spacing between the transducers reduces the grating lobes relative to the main lobe. We now consider this method of grating lobe control quantitatively [10, 11]. Figure 7.8 shows a line array of N transducers with spacings that are nonuniform but symmetric about the center; $2D_1$ is the spacing between the two center transducers, D_2 is the spacing between a center transducer and the next transducer on each side, etc.

Thus the distance from the center of the array to the nth transducer on each side is $x_n = \sum_{i=1}^{n} D_i$, and the distances from the nth transducers on each side to a wavefront determined by r and θ are $r \pm x_n \sin\theta$. Combining the contributions

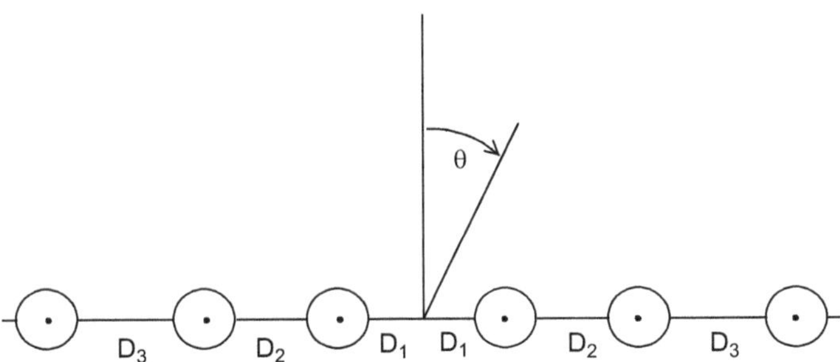

Fig. 7.8 Line array with non-uniform spacing. Geometric spacing with $R = 1.5$

from all the transducers in pairs, similar to the derivation of Eq. (7.7a), gives the far field as

$$p(\theta) = 2\left[\frac{j\rho ckuAe^{-jkr}}{2\pi r}\right]\sum_{n=1}^{N/2} \cos\left(kx_n \sin\theta\right). \tag{7.11a}$$

Note that for uniform spacing D, $x_n = nD$. The normalized directivity pattern is

$$\frac{p(\theta)}{p(0)} = \frac{2}{N}\sum_{n=1}^{N/2} \cos\left(kx_n \sin\theta\right). \tag{7.11b}$$

A special case of nonuniform spacing makes x_n a geometric progression by having a fixed ratio, R, between the spacings of adjacent transducers; thus $D_2/D_1 = R$, $D_3/D_2 = R$, and $D_3/D_1 = R^2$. Then $D_i/D_1 = R^{(i-1)}$ and

$$x_n = D_1\sum_{i=1}^{n} R^{(i-1)} = D_1(R^n - 1)/(R - 1).$$

The expression for the directivity pattern then becomes

$$\frac{p(\theta)}{p(0)} = \frac{2}{N}\sum_{n=1}^{N/2} \cos\left[kD_1 \sin\theta(R^n - 1)/(R - 1)\right]. \tag{7.11c}$$

Since this sum of cosines is less than $N/2$, except for $\theta = 0$ when it equals $N/2$, Eq. (7.11c) shows that the grating lobes are reduced relative to the main lobe. An array with this type of nonuniform spacing is sometimes called a logarithmic array because $\log D_i/D_1 = (i-1)\log R$.

As we have seen uniform spacing with $kD = 2\pi$ results in a grating lobe at $\theta = 90°$ with peak value equal to the main lobe peak. For $\theta = 90°$ and $kD_1 = 2\pi$ Eq. (7.11c) gives

$$\frac{p(90°)}{p(0)} = \frac{2}{N}\sum_{n=1}^{N/2} \cos\left[2\pi(R^n - 1)/(R - 1)\right]. \tag{7.11d}$$

As an example, for $R = 1.05$ Eq. (7.11d) gives a grating lobe 1.4 dB below the main lobe for $N = 6$ and 5.2 dB below for $N = 8$. Chow [10] showed that the relative grating lobe level is given approximately by $\log\left[1/N(N-1)(\ln R)\right]$ which gives 4.4 dB for $R = 1.05$ and $N = 8$, as compared to 5.2 dB found from Eq. (7.11d).

As discussed before, the beam can be steered to the direction θ_0 with phase shifts of $kx_n \sin\theta_0$ which replaces $\sin\theta$ in Eqs. (7.11b and 7.11c) by $(\sin\theta - \sin\theta_0)$. The effectiveness of a specific value of R can then be evaluated by calculating $p(\theta)/p(0)$ at specific values of θ for comparison with a uniformly spaced array.

Phasing may be used to form narrow beams from sections of a curved array that otherwise would radiate broader beams than desired. For example, the phases of the transducer staves in a section of the cylindrical array shown in Fig. 7.1 can be adjusted to put their outputs in phase on a plane tangential to the cylinder. The required phase shifts can be approximated by $ka(1 - \cos\beta_i)$ as can be seen from Fig. 7.9a.

The phased array has a pattern similar to the pattern of a plane array with a length equal to the projected length of the cylindrical array, $2a\sin\alpha$. However, unintended shading results from the variable spacing of the projected transducer locations and from the individual transducer directivity. These two effects partially compensate each other, but often the directivity is less important, and the array is inverse shaded (higher amplitude near the ends than at the center) which increases the side lobes relative to the main lobe. To limit this effect the angular sector of a cylinder that can be used at one time is about 120°, unless additional shading is used to eliminate the inverse shading. In the case of cylindrical arrays the pattern can be steered in azimuth without further distortion by switching the set of phases to a different set of transducers. These patterns can also be steered to depression angles by additional phasing as described above for plane arrays. Spherical arrays can also be phased to a plane and then steered in azimuth, and to some extent in depression angle, by switching to other sets of transducers. Sometimes it is necessary to defocus, or broaden, a beam, e.g., a plane array can be defocused by phasing the staves to a curved surface such as a cylinder using phase shifts that decrease from the center toward the ends.

Beam formation and steering of a cylindrical array, such as shown in Fig. 7.1, may also be accomplished by the modal method discussed in Sects. 5.2.6 and 6.5.6 for cylindrical transducers. The procedure is the same except the modal distributions are approximated by the discrete amplitude and phase of the Tonpilz piston transducers rather than by use of the continuously distributed modes of an elastic cylinder. The omni radiation mode is approximated by driving all the transducers with the same amplitude and phase. The dipole radiation mode is approximated by driving the transducers with a discrete fit to the dipole function $\cos\theta$ with a phase reversal for the transducers in the back half. The quadrupole radiation mode is

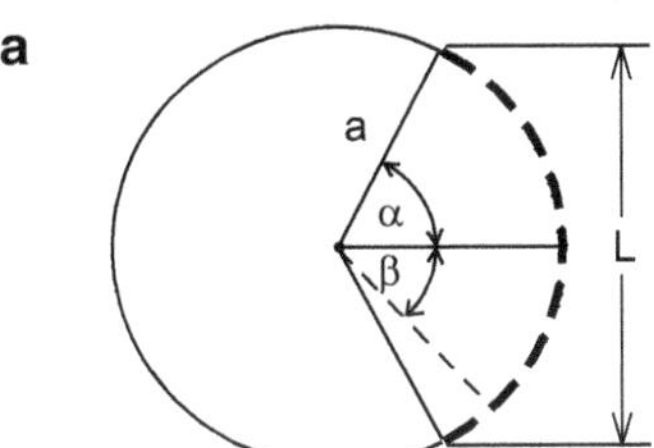

Fig. 7.9 (**a**) Phasing a cylindrical array of radius a to a plane. $L = 2a\sin\alpha$ is the projected length of the array. A transducer located at angle β has the phase $ka(1-\cos\beta)$ added

approximated by driving the transducers with a discrete fit to the quadrupole function cos 2θ with phase reversals in successive quadrants. Higher order modes and a sufficient number of transducers are necessary for narrow beams. The voltage amplitude and phase of each transducer is initially adjusted for equal far-field pressure from each mode and then added with weighting factors to obtain a particular beam pattern. This superimposed resultant voltage amplitude and phase for each frequency is then distributed around the transducers to obtain the same beam pattern at each frequency. Butler and Butler [14] implemented this method using a cylindrical array of 40 piston transducers arranged in eight staves, each composed of five parallel wired transducers, and obtained the normalized beam pattern,

$$P(\theta) = 3[1 + (2/3)\cos\theta - (1/3)\cos 2\theta]/4,$$

with dipole and quadrupole weighting factors of 2/3 and −1/3, respectively. This distribution produced a constant width hemi-cylindrical beam pattern with a null at 180°, and nearly uniform pressure from −90° to +90° over a one octave band.

Transient effects, such as the time to reach steady state, can be significantly extended in large arrays. For the transducers alone Q_m cycles are required to reach 96 % of steady state. For a broadside beam the rise time would not be increased, but for a beam steered to the angle θ parallel to a side of an array of length L it would be increased by $L \sin\theta/\lambda$ periods. For example, if transducers with $Q_m = 5$ were used in an array with a length of 5 wavelengths the rise time for an end-fire beam would be doubled. This effect would sometimes be important in determining the appropriate pulse length for an operational system or in making measurements to evaluate an array. Note that the transducer Q_m is determined by internal loss mechanisms and radiation loading, while the increase in rise time caused by the array is a completely different mechanism that depends on array size.

Most of the discussion in this section and all the expressions derived for the beam patterns of specific arrays are based on arrays in a large, rigid plane baffle. Since such baffles do not exist, it is important to note that the sound field in directions close to the plane (θ near or equal to 90°) will be less than that given by these expressions. In many cases the difference at $\theta = 90°$ will be 6 dB. This can be seen (see Sect. 10.3.3) by considering that the rigid plane case is equivalent to a plane array that is exactly the same on both sides of the plane. The same array in a perfectly soft plane is equivalent to an array in which the normal velocity distribution on one side is exactly the negative of that on the other side. Note that the latter case has zero field at 90°. Superposition of the two cases shows that the field of a plane array with no baffle is ½ the sum of the fields with rigid and soft baffles. Therefore, at 90° the field of an array with no baffle is ½ the field of the same array in a large rigid baffle. It will be shown in Chap. 11 that at other angles the factor is approximately $(1 + \cos\theta)/2$. The factor $(1 + \cos\theta)/2$ could have an important effect in predicting the performance of arrays when the beam is steered close to end-fire. The factor also has a beneficial effect, similar to the effect of the individual transducer directivity, in reducing grating lobes near 90° when the beam is not steered far from broadside.

Phasing to steer the far-field beam also causes changes in the near field which might be undesirable. Changing the phases of the transducer velocities changes the near-field pressure distribution which changes the mutual impedances and the total radiation impedance of each transducer. Thus steering the array, especially steering to end-fire, which increases the pressure at one end of the array, might cause cavitation and increase problems caused by acoustic coupling.

7.1.5 *Staggered Arrays*

We now present a staggering means for reducing grating lobes and side lobe levels. Consider a four element line array along with a second staggered four element line array to improve the beam structure. The model for a single line array of N elements of length, L, with center-to-center spacing, S, is illustrated in Fig. 7.9c for $N=4$.

The normalized beam pattern, P_1, as a function of angle, θ, with steering angle, θ_0, measured from the array normal, and with λ the wave length, may be written as

$$P_1(\theta) = [(\sin y)/y][(\sin Nx)/(N \sin x)]$$

where

$$y = (\pi L/\lambda)(\sin \theta) \quad \text{and} \quad x = (\pi S/\lambda)(\sin \theta - \sin \theta_0)$$

The first factor for $P_1(\theta)$ is the beam pattern for the elements and the second factor is the beam pattern for an array of N point source elements spaced at a distance S. If there were two lines stacked vertically and perfectly aligned to form a rectangular array, the horizontal beam pattern would be the same.

However, if the lines were staggered as illustrated in Fig. 7.9b with one line displaced by the distance $S/2$, the staggered line horizontal beam pattern function, P_2 for both lines would now be given by

$$P_2(\theta) = [(\sin y)/y][(\sin Nx)/(2N \sin x/2)]$$

It can then be shown from the above three equations that

$$P_2(\theta) = P_1(\theta) \cos [(\pi S/2\lambda)(\sin \theta - \sin \theta_0)].$$

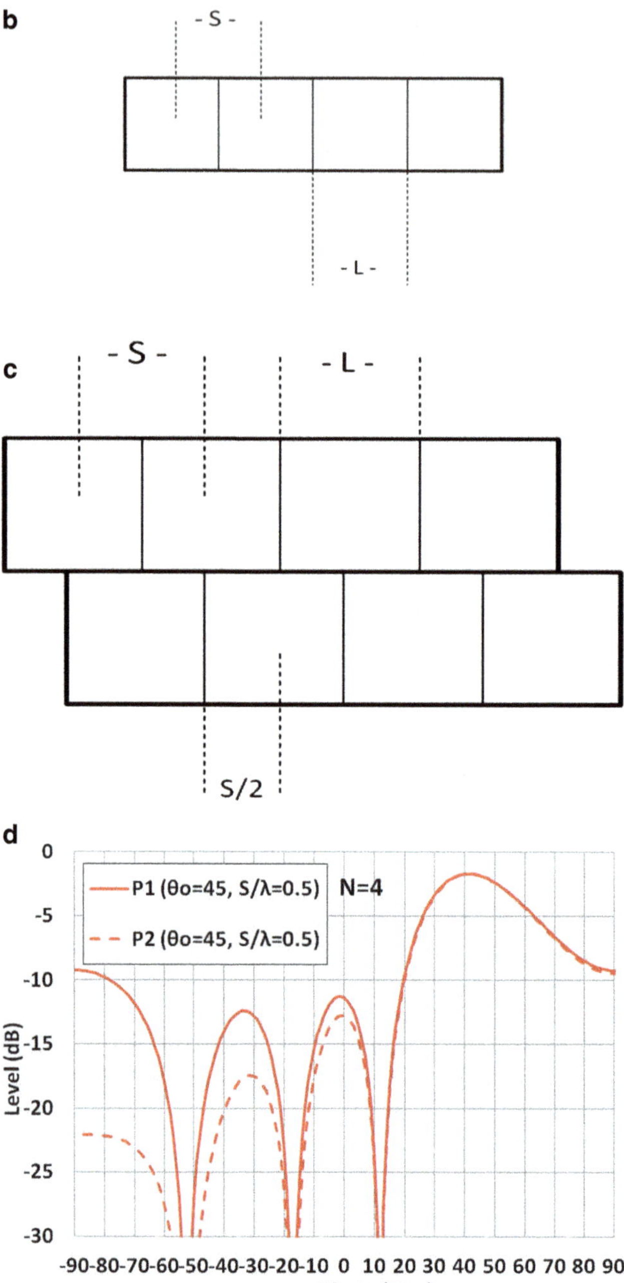

Fig. 7.9 (**b**) Four element line array, with center-to-center separation S. (**c**) Staggered four element line array, with center-to-center separation $S/2$. (**d**) Side lobe comparison of in-line, P_1, and staggered, P_2, results for 45° steering.

This equation shows that the value of staggered beam pattern function, P_2, equals the un-staggered beam pattern function, P_1, on the main lobe for $\theta = \theta_0$, since cos (0)=1; however, at other angles we see that $P_2 < P_1$. This improved condition is illustrated in Fig. 7.9d for the case of steering angle $\theta_0 = 45°$ for an array of two staggered lines of four close-packed elements with center-to-center spacing $S = L = \lambda/2$. As seen for the staggered case of P_2 and $\theta < 0$, the side lobe level is reduced by more than 10 dB in the region from $-70°$ to $-90°$ and approximately 5 dB in the regions from $-20°$ to $-50°$. Staggering virtually injects elements between the actual adjacent elements and allows array operation up to twice the frequency through the effective reduction in element center-to-center separation by a factor of two of the array illustrated in Fig. 7.9c.

Staggering may also be used in arrays where steering is in the vertical plane as well as in the horizontal plane. The simplest cases is for a square array of close-packed elements of length L with center-to-center spacing S. If the array is twisted 45° so that it appears as a diamond rather than as a square then the array lines will appear as end-to-end diamond elements with other elements filling in the spaces. The conventional and the diamond-shaped arrangements are illustrated in Fig. 7.9e and f, respectively.

In this diamond-shaped case the effective element center-to-center spacing is $S/\sqrt{2}$ in both the vertical and horizontal planes allowing an improved steering beam structure in both planes. In this arrangement there is an added bonus of area shading since the radiating area decrease from the center line to the edge of the array.

Equation (7.10a) of Sect. 7.1.4 may be used to compare normalized beam patterns from square- and diamond-shaped arrays by setting $M = N$, $D = L$ and replacing the quantity in brackets, [ ], by the normalized square element beam pattern function by $[(\sin x)(\sin y)/xy]$ where $x = (kL/2)\sin\theta\cos\varphi$ and $y = (kL/2)\sin\theta$ $\sin\varphi$. Un-steered beam patterns along the horizontal, $\varphi = 0°$, and diagonal, $\varphi = 45°$, directions (corresponding to the orientations shown in Fig. 7.9e and f) for a 2×2 array with $kL = kS = 7.28$, are shown in Fig. 7.9g.

The dashed curve, along the diagonal of the array, where $\varphi = 45°$, shows a 15 dB first side lobe reduction when compared to the beam pattern along the length, where $\varphi = 0°$. We also can see that the locations of the side lobes have been moved further out in angle. The orientation along the diagonal provides an effective center-to-center staggered spacing of 0.707 s, instead of s, allowing a 40 % increase in operating frequency with improved suppression of grating lobe effects.

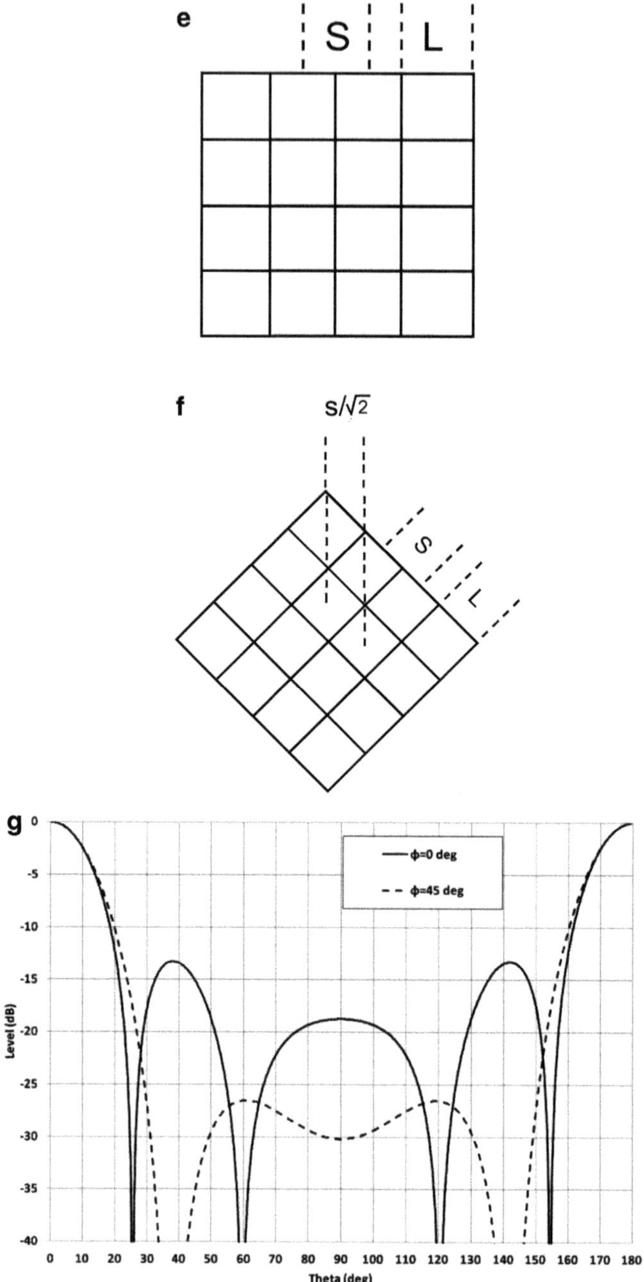

Fig. 7.9 (**e**) Square array with center-to-center separation S. (**f**) Diamond array with center-to-center separation $S/\sqrt{2}$. (**g**) Normalized patterns for $\varphi = 0°$ (*solid line*) and (*dashed line*) $\varphi = 45°$

7.1.6 Effects of Random Variations

There are several reasons why the ideal directivity patterns discussed above are not achieved in practice, the main ones being transducer manufacturing tolerances, position tolerances in the array, and different radiation reactances in an array caused by acoustic coupling. Any of these can cause the velocity phase of each transducer to vary from the desired value, and such variations always reduce the on-axis far-field pressure since it depends on all the individual contributions adding up in phase.

Radiation reactance variations and manufacturing tolerances combine to give variations in mechanical impedance and in resonance frequency that can be related to variations in velocity phase. The velocity for any mechanical vibrator is given by F/Z where $Z = R + jX = |Z|e^{j\phi}$ is the mechanical impedance and F, the driving force, is proportional to voltage or current. Since $X = |Z| \sin \phi$, variations in ϕ can be related to variations in X and Z by

$$\frac{\mathrm{d}X}{\mathrm{d}\omega} = |Z| \cos \phi \frac{\mathrm{d}\phi}{\mathrm{d}\omega} + \sin \phi \frac{\mathrm{d}}{\mathrm{d}\omega} |Z|.$$

Near resonance $|Z| \approx R$, $\cos \phi \approx 1$, and $\sin \phi \approx 0$ which gives $\mathrm{d}X/\mathrm{d}\omega = R\mathrm{d}\phi/\mathrm{d}\omega$. Using the definition of Q_m in Eq. (4.6), and since $|\mathrm{d}Z/\mathrm{d}\omega| = \mathrm{d}X/\mathrm{d}\omega$ for R independent of frequency, we have

$$\left|\frac{\mathrm{d}Z}{\mathrm{d}\omega}\right|_{\omega_r} \approx \frac{2RQ_m}{\omega_r} = \left.\frac{\mathrm{d}X}{\mathrm{d}\omega}\right|_{\omega_r} \approx R\left.\frac{\mathrm{d}\phi}{\mathrm{d}\omega}\right|_{\omega_r},$$

which can be written as

$$\mathrm{d}\phi|_{\omega_r} \approx 2Q_m \frac{\mathrm{d}\omega_r}{\omega_r}. \tag{7.12}$$

Thus a relative change in resonance frequency gives a change in phase amplified by the factor $2Q_m$. Since Q_m values are typically 3–10 these variations may have a significant effect in arrays. For example, an array of transducers with Q_m of 10 and ω_r variations of 5 %, i.e., $\mathrm{d}\omega_r/\omega_r = 0.05$, gives phase variations across the array up to 57.5°.

A large array with random variations among the transducers can be considered a partially coherent array that will generate a component of the radiated field that is nearly omnidirectional [15]. The incoherent omnidirectional component sets a lower limit on side lobe levels. We can define the random fraction, F, as the ratio of incoherent radiated power to coherent radiated power,

$$F \equiv W_r/W = I_r/I_a = I_r D_f/I_0,$$

where I_0 is the on-axis coherent intensity, I_a is the average coherent intensity, and I_r is the incoherent omnidirectional intensity. Then $I_r/I_0 = F/D_f$ is the lower limit on side lobes and can be written in dB as

$$10\log(I_r/I_0) = 10\log F - DI. \tag{7.13}$$

As an example: If $F = 5\ \% = 0.05$, the lowest average relative side lobe level possible is –13 dB – DI. The greater the DI, the lower the side lobe level limit for a given random fraction F.

7.2 Mutual Radiation Impedance and the Array Equations

7.2.1 *Solving the Array Equations*

In the previous section array directivity functions were discussed assuming that the transducer velocities were known. In this section the acoustic coupling between transducers will be included and its effect on the velocities will be analyzed. Consider an array of transducers forming part of a closed surface, e.g., the cylindrical array shown in Fig. 7.1. Each transducer produces a sound field, and the total sound field at every point in the water is the superposition of all the individual sound fields. The far-field beam pattern is the result of all these fields combining at a large distance from the array. Similarly, on the surface of each transducer all the individual acoustic pressures combine to give the total pressure, which, at a point designated by $\vec{r_i}$ on the surface of the ith transducer in an array of N transducers, can be written as (see Fig. 7.10)

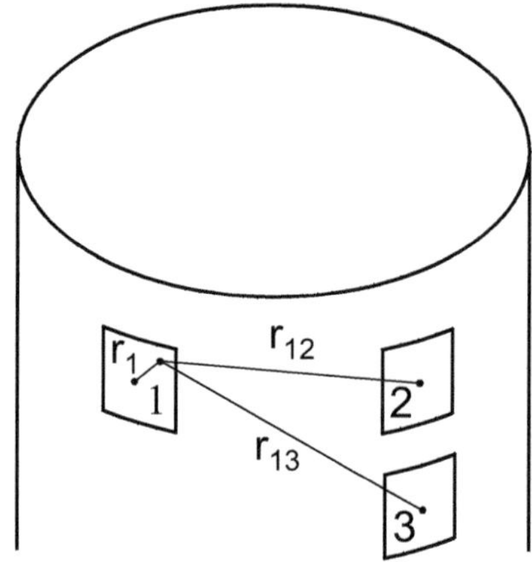

Fig. 7.10 Notation for Eq. (7.14) showing the pressures from transducers 2 and 3 in a cylindrical array contributing to the total pressure at a point on transducer 1

$$p\left(\vec{r_i}\right) = \sum_{j=1}^{N} p_j\left(\vec{r_{ij}}\right). \tag{7.14}$$

Note that the sum includes the ith transducer itself, that $p_j\left(\vec{r_{ij}}\right)$ is the pressure produced by the jth transducer at the designated point on the ith transducer, and that r_{ij} is the distance from the center of the jth transducer to the designated point. The force exerted on the ith transducer by the pressures from all the transducers is

$$F_i = \sum_{j=1}^{N} \iint p_j\left(\vec{r_{ij}}\right) \mathrm{d}S_i. \tag{7.15}$$

For transducers with uniform velocity distributions this force can be expressed in terms of the radiation impedance of the ith transducer by dividing by the velocity of that transducer, u_i, as was done for a single transducer in Eq. (1.4):

$$Z_i = R_i + jX_i = \frac{F_i}{u_i} = \frac{1}{u_i} \sum_{j=1}^{N} \iint p_j\left(\vec{r_{ij}}\right) \mathrm{d}S_i = \sum_{j=1}^{N} \frac{u_j}{u_i} Z_{ij}, \tag{7.16}$$

where

$$Z_{ij} = R_{ij} + jX_{ij} = \frac{1}{u_j} \iint p_j\left(\vec{r_{ij}}\right) \mathrm{d}S_i, \tag{7.17}$$

is defined as the mutual radiation impedance, or interaction impedance, between the ith and jth transducers. This definition applies to any pair of transducers that form part of the closed surface, including those that may have different sizes and shapes. The values of Z_{ij} depend on parameters, such as the separation and relative orientation, of a given pair of transducers on a given surface, but they do not depend on the other transducers in the array. It follows from the acoustical reciprocity theorem that $Z_{ji} = Z_{ij}$ (see Sect. 11.2.2). The practical difficulties that arise in calculating values of Z_{ij} will be discussed at the beginning of the next section.

Z_{ii} is the self-radiation impedance of the ith transducer, i.e., the radiation impedance when all the other transducers in the array are clamped or when it is the only transducer on the same surface. If the transducers are identical the self-impedance is usually considered to be the same for all the transducers in an array, but often that is not the case because self-impedance depends on the surroundings. For example, in a truncated cylindrical array, as in Fig. 7.1, the self-impedance of transducers near the top and bottom of a vertical column differs from the self-impedance of transducers near the center of the column. The impedance Z_i, defined in Eq. (7.16), is the total radiation impedance of the ith transducer; it depends on the self-impedance, all the mutual impedances, and the velocities of all the other transducers in the array. The resistive part of Z_i also has the usual relationship to the power radiated by the ith transducer, i.e., the power equals $\frac{1}{2}R_i u_i u_i^*$. Large variations of the total radiation impedance are most likely at low frequencies where the transducer radiating faces are small compared to the wavelength, and have the most effect at resonance where the transducer mechanical impedance is small. However, the interactions have the beneficial effect of increasing the total radiation resistance of most transducers in an array, such that R_i usually exceeds the self-radiation resistance, R_{ii}. In some cases the increase in radiation resistance approaches the ideal of full array loading, with the average $R_i \approx \rho c A_i$ where A_i is the transducer area.

Analysis of an array requires inclusion of all the acoustic interaction forces in Eq. (7.15) in the equations that describe each transducer [16]. This can be accomplished by replacing the self-radiation impedance in the equations for a transducer alone by the total radiation impedance. Thus, in an array of electric field transducers the total mechanical impedance of the ith transducer is $Z_{\text{m}}^E + Z_i$. Using Eqs. (2.72a and 2.72b), with $F_{\text{b}} = 0$ for projectors, and assuming that all the transducers in the array have the same mechanical impedance Z_{m}^E, as well as the same turns ratio N_{em} and clamped admittance Y_0, the set of $2N$ array equations for electric field transducers is:

$$0 = \left(Z_{\text{m}}^E + \sum_{j=i}^{N} \frac{u_j}{u_i} Z_{\text{ij}} \right) u_i - N_{\text{em}} V_i, \tag{7.18}$$

$$I_i = N_{\text{em}} u_i + Y_0 V_i, \tag{7.19}$$

for $i = 1, 2, \ldots N$. Similar equations hold for magnetic field transducers. The rest of this chapter will be written in terms of electric field transducers, but the discussion is easily modified for magnetic field transducers.

If the array geometry and the sizes and shapes of the transducer faces are specified, the Z_{ij} can be calculated in principle, but are usually approximated in practice. Then, if the transducer parameters are known, Eq. (7.18) can be used to calculate the voltages, if the velocities, u_j, of each transducer are specified, simply by adding all the terms. On the other hand, if the voltages are specified Eq. (7.18) is a set of N simultaneous equations which can be solved by matrix inversion for the velocities. It would seem reasonable to specify the velocities since they determine the beam pattern, but there are various reasons why that is usually not practical as will be discussed in Sect. 7.2.2. In practice a beamformer provides the voltage for each transducer with appropriate phase and amplitude which then is applied to a power amplifier [17]. The amplifier impedance and tuning elements or transformers affect the voltage that is applied to the transducer and must be included in the analysis. These components can be combined with the transducer components, in ABCD matrix form (see Sect. 3.3.2), and represented by a Thevenin equivalent mechanical impedance, Z_{mi}^T, and force, F_i^T [17] (see Sect. 13.8). Then Eq. (7.18) is replaced by

$$F_i^T = Z_{\text{mi}}^T u_i + F_i = Z_{\text{mi}}^T u_i + \sum_{j=1}^{N} Z_{\text{ij}} u_j. \tag{7.20a}$$

When each transducer is represented by

$$V_i = AF_i + Bu_i, \tag{7.20b}$$

as in Sect. 3.32, and ignoring amplifier and tuning impedances, comparison with Eq. (7.20a) shows that $F_i^T = V_i/A$ and $Z_{\mathrm{mi}}^T = B/A$, which are consistent with a Thevenin equivalent circuit representation.

It is convenient in performing array calculations to form the impedance matrix, Z^T, with the elements:

$$Z_{\mathrm{ij}}^T = Z_{\mathrm{ij}} \text{ for } i \neq j \quad \text{and} \quad Z_{\mathrm{ii}} + Z_{\mathrm{mi}}^T \quad \text{for} \quad i = j.$$

In this impedance matrix the diagonal elements are the self-radiation impedances plus the Thevenin mechanical impedance, while the off-diagonal elements are the mutual radiation impedances. Eq. (7.20a) can then be written as the matrix equation $F^T = Z^T u$ with the solution for the velocities given by

$$u = \left(Z^T\right)^{-1} F^T. \tag{7.20c}$$

The matrix inversion must be done at numerous frequencies to determine the array performance. The frequency-dependent ABCD parameters, which depend only on transducer and electrical parameters, and not on radiation impedances, can be conveniently transferred at each frequency from a transducer design program to an array analysis program such as the Transducer Design and Array Analysis Program TRN [18]. The off-diagonal elements generally decrease with distance from the diagonal, because they represent more distant interactions. However, near resonance the diagonal elements also become quite small when the reactance components cancel.

When the velocities are determined the total acoustic force and the total radiation impedance on each transducer are given by Eq. (7.16) as

$$F_i = u_i \sum_{j=1}^{N} \frac{u_j}{u_i} Z_{\mathrm{ij}},$$

$$Z_i = F_i/u_i.$$

Then, using the transducer ABCD parameters, the voltages, currents, and electrical impedances for each transducer can be found from

$$V_i = AF_i + Bu_i, \tag{7.21a}$$

$$I_i = CF_i + Du_i, \tag{7.21b}$$

$$(Z_e)_i = \frac{AZ_i + B}{CZ_i + D}, \tag{7.21c}$$

where the transducer parameters, A, B, C, and D, are determined as in Sect. 3.3.2. The total radiation resistances, R_i, and the velocities give the total acoustic power radiated by the array as

$$W = \frac{1}{2}\sum_{j=1}^{N} R_j |u_j|^2, \tag{7.21d}$$

and the intensity averaged over all directions, referred to a distance of one meter, is $W/4\pi$. Using the velocities and the far fields of the individual transducers the intensity in the steered direction (the MRA) can be calculated and used with the average intensity to determine the directivity factor and the DI. With W and DI the source level is given by Eq. (1.25). From the velocities the far-field beam pattern of the array can also be found to complete the description of the array performance.

Thus determination of the acoustic performance of a specific projector array with specified parameters (including mutual impedances) is straightforward, but the design of an array to achieve a specified acoustic performance may not be straightforward. The array design and the transducer design are not separable, the number of variables is large, and our knowledge of some parameters, such as mutual impedances and loss factors, is far from perfect. Design iterations, partial array measurements, and full array evaluations are necessary, especially when the specified performance requires use of materials under conditions that approach their physical limitations.

7.2.2 Velocity Control

The discussion in the previous section shows that, because of the acoustic interactions, a direct proportionality does not exist between the individual transducer velocities and the applied voltages. Since the transducer velocities determine the array directivity patterns the most direct way to get the desired patterns would be to specify the velocity amplitudes and phases and use the array equations to find the voltage amplitudes and phases required to give those velocities. This approach would not eliminate the coupling, but it would control the velocities. However, it is impractical for the following reasons:

(1) Although unwanted velocity variations are eliminated it causes unwanted voltage variations.
(2) The beamformer design is more complicated because the voltage phases would not be simple progressive shifts.

(3) The mutual impedances are often not known with sufficient accuracy to justify a more complicated beamformer design.
(4) Active sonar often uses FM and noise pulses with considerable bandwidth which would further complicate this approach.

Carson [19] proposed a method of velocity control to deal with the variations in transducer velocity caused by the acoustic interactions. His approach makes the velocities approximately proportional to the voltages by adding electrical components to each transducer such that the input electrical impedance is high compared to the variations in the radiation impedance. The frequency dependence of both the added electrical impedance and the radiation impedance limit the effectiveness of this approach to a rather narrow band, but it could be useful in the part of the band where interactions are most severe. There are two such frequency regions: The first is near the transducer resonance frequency where $Z^E_{\mathrm{m}} + Z_i$ is small and approximately equal to $R_{\mathrm{m}} + R_i$; the second is at the low end of the band where the distances between transducer centers are small in terms of wavelengths and the mutual impedances are most significant. Such velocity control may be achieved inadvertently to some degree in the resonance region since addition of electrical tuning elements is normal practice to improve efficiency and bandwidth. As Woollett has said [20], "—tuning reactances may contribute a desirable isolation effect between transducers in an array." The optimization of this type of velocity control would require detailed analysis, including accurate values for the mutual impedances, and consideration of efficiency and bandwidth. Stansfield [9] gives a practical discussion of velocity control.

In some cases it may be possible to use negative velocity feedback to control the transducer velocities in an array. The feedback signal must be designed to reduce the voltage on each transducer by an amount proportional to the velocity magnitude of that transducer. The array equation, Eq. (7.18), can then be written with a feedback voltage $V_{\mathrm{fi}} = Gu_{\mathrm{i}}$ subtracted from the applied voltage V_i:

$$\left(Z^E_{\mathrm{m}} + \sum_{j=1}^{N} \frac{u_j}{u_i} Z_{\mathrm{ij}}\right) u_i = N_{\mathrm{em}}(V_i - V_{\mathrm{fi}}) = N_{\mathrm{em}}(V_i - Gu_i),$$

where G is the feedback gain. Solving for u_i, and separating the self-impedance from the interactions, the velocity of the ith transducer can be written as

$$u_i = N_{\mathrm{em}} V_i \Big/ \left[Z^E_{\mathrm{m}} + Z_{\mathrm{ii}} + \sum_{j \neq i}^{N} \frac{u_j}{u_i} Z_{\mathrm{ij}} + N_{\mathrm{em}} G\right].$$

If there were no interactions (i.e., all $Z_{\mathrm{ij}} = 0$ for $i \neq j$) all the transducer velocities would be proportional to the applied voltages with the same proportionality constant $\left[N_{\mathrm{em}} / \left(Z^E_{\mathrm{m}} + Z_{\mathrm{ii}} + N_{\mathrm{em}} G\right)\right]$, as long as all the transducers in the array were the same. The sum over the interactions, which differs from one transducer to another, interferes with this simple proportionality. But it can be restored approximately if the feedback gain can be increased enough to make $\left[Z_{\rm m}^E + Z_{\rm ii} + N_{\rm em}G\right]$ considerably greater than the variations in $\sum_{j\neq i}^{N} \frac{u_j}{u_i} Z_{\rm ij}$. One approach to acquiring a feedback signal is discussed briefly in Sect. 5.4.2, where one of the piezoelectric ceramic rings in the ring stack of a Tonpilz transducer is used as the feedback sensor [21]. Another approach requires attaching a small accelerometer to the inside of the head mass of a Tonpilz transducer [22].

7.2.3 Negative Radiation Resistance

Large arrays often contain some inoperative transducers because of water leakage, broken electrode connections, etc. The array equations can be used to study the effects of such undriven transducers in specified locations simply by setting the appropriate voltages, V_i, equal to zero in Eq. (7.18). Those transducers that are not driven electrically will be driven acoustically by the other transducers. Equation (7.18) shows, when no voltage is applied to the *i*th transducer in an array and the velocity of that transducer is not zero, that its total radiation impedance, Z_i, equals $-Z_{\rm m}^E$, and its radiation resistance is negative. Negative radiation resistance means that the transducer is absorbing acoustic power from the sound field radiated by the other transducers. This causes a reduction of total power radiated by the array and distortion of the beam pattern.

Negative radiation resistance can also occur in other situations. Just as electrically undriven transducers in an array have negative radiation resistance, so do electrically undriven modes of non-FVD transducers in an array when the pressure distribution is such that those modes are excited (see Sect. 7.4). Negative radiation resistance can also occur in certain array configurations even when all the transducers are electrically driven. The simplest case of this kind was studied by Pritchard [23] using an array with one transducer at the center of a circle of transducers where the radiation resistance of the central transducer varies from negative to positive values as the frequency is varied. This would seem to be an impractical case, although it might occur in arrays of nested concentric circles.

7.3 Calculation of Mutual Radiation Impedance

7.3.1 Planar Arrays of Piston Transducers

An approximate calculation of the mutual radiation impedance between two small piston sources of arbitrary shape in a rigid plane (see Fig. 7.11) will be made to illustrate the general procedure.

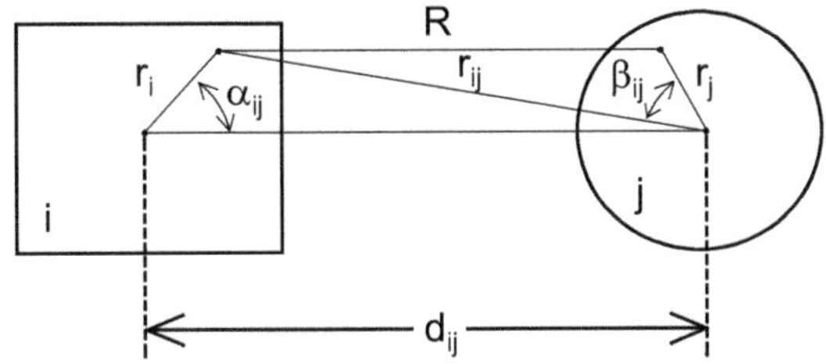

Fig. 7.11 Notation for calculating the mutual radiation impedance between two piston transducers in a plane from Eq. (7.23)

The pressure produced by the *j*th piston at a point on the surface of the *i*th piston is given by Eq. (10.25) as:

$$p_j(r_{ij}) = \frac{j\rho c k u_j}{2\pi} \iint \frac{e^{-jkR}}{R} dS_j, \tag{7.22}$$

where $R^2 = r_{ij}^2 + r_j^2 - 2r_j r_{ij} \cos\beta_j$, $r_{ij}^2 = d_{ij}^2 + r_i^2 - 2r_i d_{ij} \cos\alpha_{ij}$, d_{ij} is the distance between the centers of the pistons, and r_j, β_j, r_i, and α_{ij} are defined by Fig. 7.11. Inserting Eq. (7.22) in Eq. (7.17) gives an expression for the mutual impedance:

$$Z_{ij} = \frac{1}{u_j} \iint p_j(r_{ij})\, dS_i = \frac{j\rho c k}{2\pi} \iiiint \frac{e^{-jkR}}{R} dS_j dS_i. \tag{7.23}$$

Equation (7.23) illustrates the difficulty of calculating mutual impedance since, with the expression for R above, these integrals are impossible to evaluate analytically except in a few special cases. However, we can obtain a simple but useful result from Eq. (7.23) by making extreme approximations. Assume that both pistons are much smaller than their separation, and also sufficiently smaller than the wavelength, that R can be approximated by d_{ij} in both the denominator and the exponent of the integrand. Then the integrand is constant, the integral yields $A_j\, A_i$, the product of the areas of the two pistons, and the mutual resistance and reactance can be written as

$$R_{ij} + jX_{ij} = \frac{\rho c k^2 A_j A_i}{2\pi} \frac{je^{-jkd_{ij}}}{kd_{ij}} = \frac{\rho c k^2 A_j A_i}{2\pi} \left[\frac{\sin kd_{ij}}{kd_{ij}} + j \frac{\cos kd_{ij}}{kd_{ij}} \right]. \tag{7.24a}$$

The dependence on kd_{ij} in Eq (7.24a) also holds for the mutual impedance between small monopole spheres when scattering is neglected (see Sect. 7.5). For circular pistons of the same size with areas equal to πa^2 Eq. (7.24a) becomes

$$\frac{R_{ij} + jX_{ij}}{\rho c A} = \frac{1}{2}(ka)^2 \left[\frac{\sin kd_{ij}}{kd_{ij}} + j \frac{\cos kd_{ij}}{kd_{ij}} \right]. \tag{7.24b}$$

It is evident that Eq. (7.24a) is only a good approximation when the pistons are so small that there is no significant phase variation across their surfaces, i.e., when $ka \ll 1$; however, the simplicity of Eq. (7.24a) and its generality regarding piston

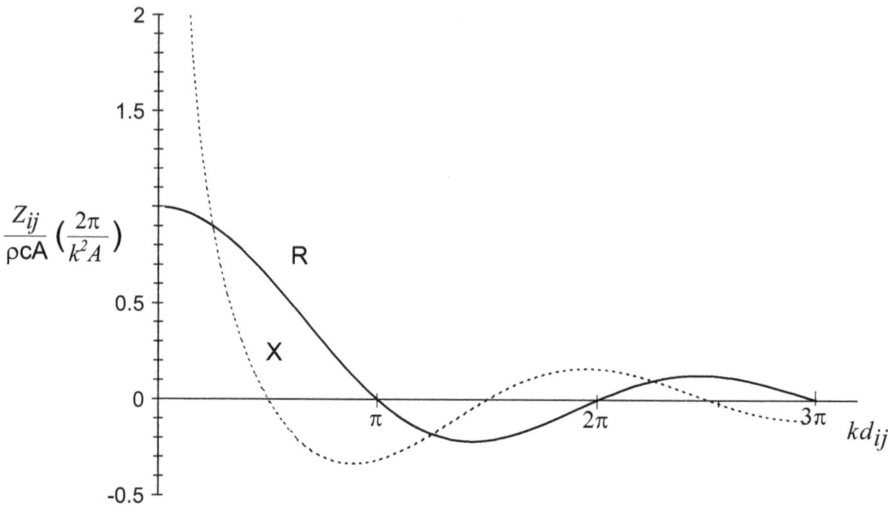

Fig. 7.12 Mutual radiation resistance and reactance for two small piston transducers of area, *A*, in a plane from Eq. (7.24b)

shape make it a useful approximation. It also correctly illustrates the general behavior of mutual impedance in that the resistive and reactive parts oscillate about zero with a magnitude decreasing inversely with the separation between the two pistons as shown in Fig. 7.12.

The real part of Eq. (7.24b) for circular pistons reduces to the correct value of the self-resistance for small ka, i.e., $R_{11} = ½\rho cA(ka)^2$ (see Sect. 10.4.2), when $kd_{\rm ij}$ goes to zero. Another useful form of Eq. (7.24a) that holds for small, baffled pistons of any shape can be written as

$$R_{\rm ij} + jX_{\rm ij} = \rho cA\frac{k^2A}{2\pi}\frac{je^{-jkd_{\rm ij}}}{kd_{\rm ij}} = R_{11}\frac{je^{-jkd_{\rm ij}}}{kd_{\rm ij}}. \tag{7.24c}$$

For two small pistons close together ($kd_{12} \ll 1$) with equal velocities the total impedance of each is $Z_1 = Z_2 = Z_{11} + Z_{12}$, and the total resistance is $R_1 = R_2 = R_{11} + R_{12} = R_{11}(1 + \sin kd_{12}/kd_{12}) \approx 2R_{11}$, showing the increase in acoustic loading caused by the coupling.

The mutual radiation impedance for circular pistons of arbitrary size in a rigid, plane baffle was studied first by Wolff and Malter [24], then by Klapman [25] by integrating Eq. (10.25a) after expanding the integrand, giving results in the form of infinite series. Stenzel [26] also made calculations in this way. Pritchard [23, 27] took the approach used by Bouwkamp [28], in which the far-field intensity directivity function of two piston sources is integrated over complex angles. Thompson [29] also used Bouwkamp's approach to compute self and mutual impedances for annular and elliptical pistons. This approach has the advantage of avoiding the complexities of the near field, and it is clear physically that integrating the intensity

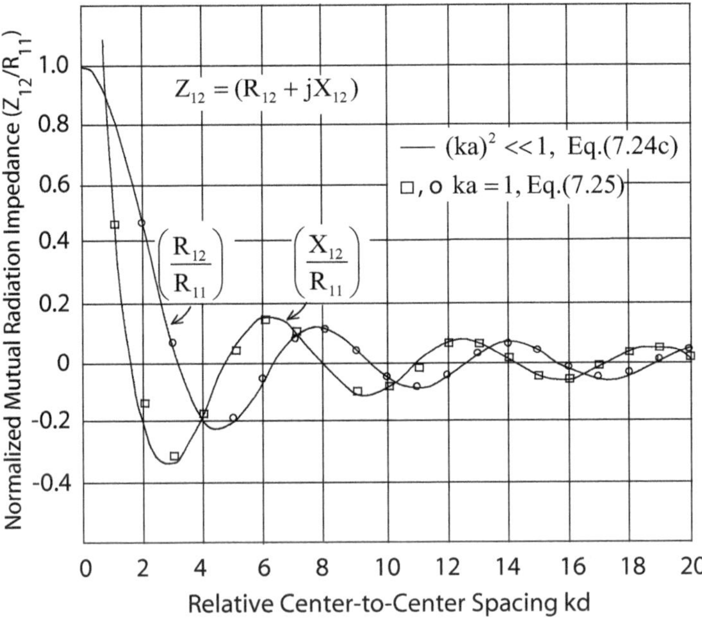

Fig. 7.13 Normalized mutual radiation impedance as a function of relative separation for two rigid circular disks in an infinite rigid plane. Approximate results for small disks and exact results for $ka = 1$ [23]

over all directions gives the radiated power and thus the radiation resistance. The fact that integration over complex angles also gives the radiation reactance is not obvious physically, because it is a mathematical result that follows from representing the physical variables in complex form. This method which involves the Hilbert Transform will be discussed in Chap. 11.

Pritchard [23] expressed the mutual impedance between circular pistons of arbitrary radius, a, in an infinite rigid plane in terms of an infinite series, which was extended by Porter [17] to include the case of pistons with unequal radii, a and b. The final result is

$$Z_{\mathrm{ij}} = 2\rho c\pi a^2 \sum_{p=0}^{\infty} \frac{1}{\pi^{1/2}} \Gamma(p+1/2) \left(\frac{a}{d_{\mathrm{ij}}}\right)^p h_{\mathrm{p}}^{(2)}(kd_{\mathrm{ij}}) \sum_{n=0}^{p} \left(\frac{b}{a}\right)^{n+1} \times \left(\frac{J_{p-n+1}(ka) J_{n+1}(kb)}{n!(p-n)!}\right). \tag{7.25}$$

In Eq. (7.25) Γ is the gamma function, $h_{\mathrm{p}}^{(2)}$ is the spherical Hankel function of the second kind, and J is the Bessel function. Figure 7.13 gives an example of Z_{ij} calculated from Eq. (7.25) taken from Pritchard's paper [23] and compared with Eq. (7.24a).

The results are almost identical for $ka = 1$, but the differences are significant for larger ka, and Eq. (7.25) has become an integral part of several array analysis programs.

Note that Z_{ij} is defined as the integral of the pressure caused by one vibrating transducer over the surface of another which is not vibrating, as was done in Eq. (7.23). If both transducers were vibrating with the same velocity and the pressure fields of both were included in the integration over the surface of one the result would be $Z_{ii} + Z_{ij}$. In the Bouwkamp method both transducers must be vibrating because the far field of both is used in the calculation, and thus the result obtained is $2(Z_{ii} + Z_{ij})$.

Mutual radiation impedance has also been calculated for square and rectangular pistons in an infinite, plane, rigid baffle [30, 31]. When normalized by dividing by $\rho c A$ the results are very similar for squares and circles if the areas are equal. For squares and rectangles the orientation with respect to each other influences the mutual impedance as well as the distance between centers, but for squares with $ka \leq 2$ the orientation effect is negligible [30].

Pritchard [23] also derived a useful result, sometimes called the hydraulic impedance transformation, which had been given earlier by Toulis [32]. Consider a plane array of N small, square piston transducers with separations between adjacent transducer centers small compared to the wavelength as shown in Fig. 7.14.

If all transducers have the same velocity the total radiation resistance of the ith transducer is given by Eqs. (7.16) and (7.24a) as

$$R_i = \sum_{j=1}^{N} R_{ij} = \rho c \frac{(kA_1)^2}{2\pi} \sum_{j=1}^{N} \frac{\sin kd_{ij}}{kd_{ij}}, \tag{7.26}$$

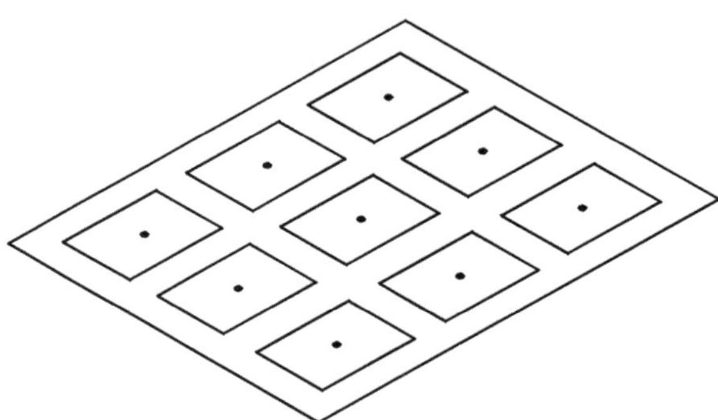

Fig. 7.14 Array of small square piston transducers for development of the hydraulic impedance transformation given in Eq. (7.29a). The argument starts with very small pistons and a packing factor $\ll 1$. The piston size is then increased until the array area is completely filled and the packing factor $= 1$

where A_1 is the area of one transducer. The total radiation resistance of the array is

$$R_A = \sum_{i=1}^{N} R_i = \rho c \frac{(kA_1)^2}{2\pi} \sum_{i=1}^{N} \sum_{j=1}^{N} \frac{\sin kd_{ij}}{kd_{ij}}. \tag{7.27}$$

Now imagine increasing the size of each piston while keeping the centers fixed until the entire array area is vibrating (although this violates one of the conditions on Eq. (7.24a) it appears to have approximate validity for the resistive part of the impedance for small pistons). At a given frequency the change of piston size with fixed centers does not change the kd_{ij} values or the value of the double sum. The area of each piston is then $A_1 = A_A/N$ where A_A is the total array area, and the total radiation resistance of the array is given by Eq. (7.27) as

$$R_A' = \rho c \frac{(kA_A)^2}{2\pi N^2} \sum_{i=1}^{N} \sum_{j=1}^{N} \frac{\sin kd_{ij}}{kd_{ij}} = \rho c A_A, \tag{7.28}$$

where the last step holds if the array is large enough to be fully radiation loaded. Equation (7.28) can be solved for the double sum, which can then be substituted into Eq. (7.27). This removes the dependence on all the kd_{ij}, and the total radiation resistance of the array relative to the active area of the array can be written as

$$\frac{R_A}{\rho c N A_1} = \frac{N A_1}{A_A} = \frac{\text{Active area}}{\text{Total area}} = \text{Packing factor} = \text{pf}, \tag{7.29a}$$

while the total radiation resistance of the array relative to the total area of the array is

$$\frac{R_A}{\rho c A_A} = (\text{pf})^2. \tag{7.29b}$$

Since R_A/N is the average radiation resistance of one piston we also have

$$\overline{R}_i = (R_A/N) = \rho c A_1 \text{pf}. \tag{7.29c}$$

The significance of these results can be seen by applying Eq. (7.29c) to a large array of small, circular, close-packed pistons. The packing factor, and thus the average radiation resistance of one piston relative to its own area, is $\pi/4$. For one piston alone, e.g., with $ka = ½$, the radiation resistance relative to its own area would be $½(ka)^2 = 1/8$. Thus, for $ka = ½$, the combined interactions in a large array increase the average resistive loading of each piston by a factor of 2π, from 1/8 to $\pi/4$. Recall that for two small pistons close together the interactions result in an increase by a factor of 2.

Equations (7.29b and 7.29c) have approximate validity only for small transducers under conditions where the velocities are nearly equal, and it can then be applied to arrays of transducers of other shapes. The ideal of full array loading can only be approached for transducer shapes, such as squares, rectangles, triangles, or hexagons, which permit the packing factor to be nearly unity. Note that Eq. (7.29c) also does not reveal the variations of radiation resistance among the transducers, which are often important.

7.3.2 Nonplanar Arrays, Nonuniform Velocities

Mutual radiation impedance has been calculated for circular piston sources on a rigid sphere [33] and for rectangular piston sources [34] and rings [35] on an infinite rigid cylinder. The mathematical details are given in Sects. 11.1.1 and 11.1.2; here some general features of the results will be discussed. Comparison of mutual impedance for different baffle shapes provides some useful insights. When the separation exceeds about two wavelengths the rate at which the mutual impedance magnitude decreases depends on the curvature of the baffle. This is shown in Fig. 7.15 where mutual resistances for a plane baffle and for cylindrical and spherical baffles are compared over a wide range of separations. The lack of dependence on curvature for small separations is shown again in more detail in Fig. 7.16 for square pistons on a cylindrical baffle.

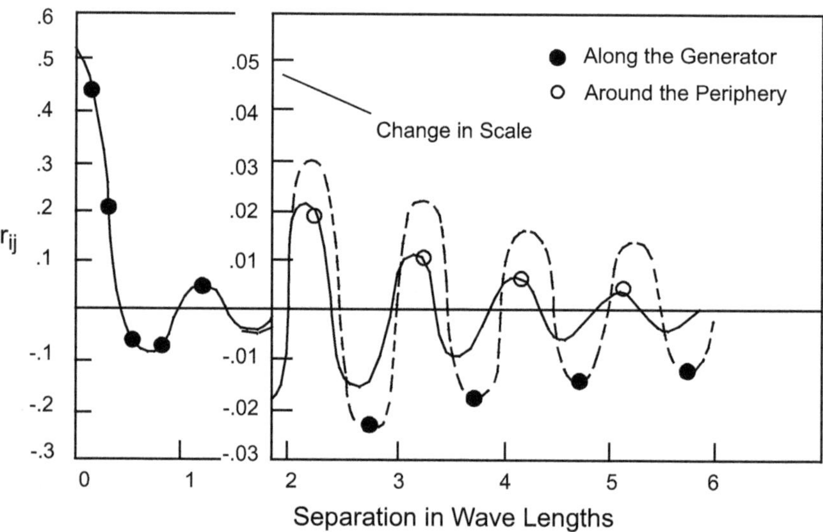

Fig. 7.15 Illustration of dependence of mutual radiation resistance on curvature of the baffle. The *dashed curve* is for pistons in a plane, the *solid dots* for square pistons with the same angular position on a cylinder. The *solid curve* is for circular pistons with the same axial position on a cylinder. The cylinder radius is 6 wavelengths, $r_{ij} = R_{ij}/\rho cA$ [34]

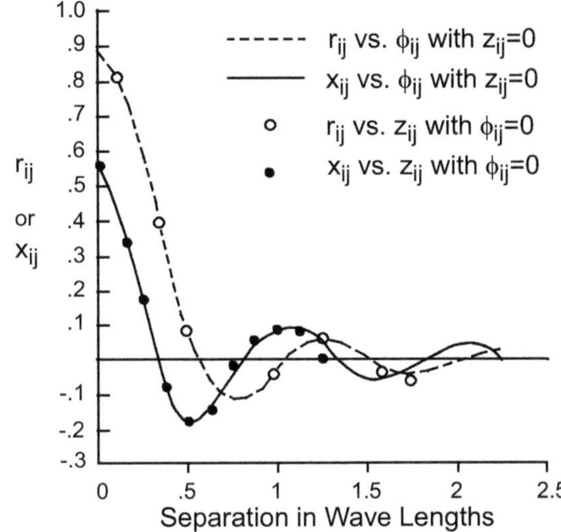

Fig. 7.16 Mutual radiation resistance and reactance for square piston transducers of $\lambda/2$ sides on an infinitely long, rigid cylinder of about 6λ radius. $R_{ij} = R_{ij}/\rho cA$, $x_{ij} = X_{ij}/\rho cA$, φ_{ij} = angular separation, z_{ij} = axial separation of the pistons [34]

The impedance values are about the same for small separation along the length where there is no curvature and separation around the circumference where the radius of curvature is 5.5 wavelengths [34]. It is evident from these results that interaction effects are more important in large plane arrays than in large curved arrays. The combined interactions from distant transducers might be comparable to the interactions from nearest neighbor transducers in a large plane array, while it might be negligible in a curved array. The calculations also show that for transducer separations exceeding about ½ wavelength the mutual impedance is small compared to the self-impedance, and thus that the most serious effects of the interactions occur at low frequency. In addition, it can be seen from Eq. (7.18) that the interactions have the most effect at frequencies where the internal mechanical impedance of the transducers, Z^E_m, is relatively low, i.e., near resonance.

The rate at which the mutual impedance decreases with separation between transducers also depends on the acoustic properties of the baffle surface. An acoustically soft surface causes the interactions to decrease more rapidly, because the pressure on one transducer exerted by another is lowered by the soft surface [36–39]. However, in close-packed arrays the surface impedance is determined mainly by the radiating faces of the transducers, each of which is approximately rigid, although moveable. Therefore the mutual impedance determined for transducers on a rigid surface correctly accounts for the motions of the other transducers which form the surface surrounding any particular transducer in close-packed arrays.

The definition of mutual radiation impedance in Eq. (7.17) holds when the velocity distribution on the transducers is uniform. When it is not uniform, but fixed, the definition must be generalized as was done for the self-impedance in

Sect. 1.3. The generalization starts by rewriting Eq. (1.4b) as the total radiation impedance of the ith transducer in an array of N transducers:

$$Z_i = \frac{1}{U_i U_i^*} \iint p\left(\vec{r_i}\right) u_i^*\left(\vec{r_i}\right) \mathrm{d}S_i, \tag{7.30}$$

where U_i is the reference velocity, $u_i\left(\vec{r_i}\right)$ is the fixed velocity distribution, and $p\left(\vec{r_i}\right)$ is the total pressure caused by all the transducers in the array acting on the surface of the ith transducer, given by Eq. (7.14). Thus Z_i can be written as

$$Z_i = R_i + jX_i = \frac{1}{U_i U_i^*} \iint \sum_{j=i}^{N} p_j\left(\vec{r_{ij}}\right) u_i^*\left(\vec{r_i}\right) \mathrm{d}S_i = \sum_{j=1}^{N} Z_{ij} \frac{U_j}{U_i}, \tag{7.31}$$

where

$$Z_{ij} \equiv \frac{1}{U_j U_i^*} \iint p_j\left(\vec{r_{ij}}\right) u_i^*\left(\vec{r_i}\right) \mathrm{d}S_i, \tag{7.32}$$

is defined as the mutual radiation impedance for FVD transducers with nonuniform velocity distributions. This definition retains the usual relationship to the power, i.e., the time average power radiated by the ith transducer is $\frac{1}{2} R_i U_i U_i^*$, and it reduces to the uniform velocity case in Eq. (7.17) when $u_i\left(\vec{r_i}\right)$ is a constant equal to U_i. The useful reciprocal relation $Z_{ji} = Z_{ij}$ also holds for nonuniform velocity distributions, but only when all points of each transducer surface vibrate in phase or when the surface is divided into regions that are 180° out of phase [16].

Very few numerical calculations of self or mutual radiation impedance have been made for transducers with nonuniform velocity distributions. Porter [40] made calculations for circular disks in an infinite, rigid plane with nonuniform velocity distributions that represented flexural vibrations of plates with clamped and supported edges. This work extended Pritchard's results based on the Bouwkamp method. Chan [41] discussed the same flexural mode cases using a different mathematical approach. Porter's results show that when referred to the average surface velocity the self-resistances for pistons, clamped edge plates, and supported edge plates differ significantly when ka exceeds ~1.5, while the self-reactances differ significantly when ka exceeds ~0.5. Numerical results for the mutual impedances were given only for $ka = 1$ where the differences between the two flexural vibrators and the piston are small.

7.4 Arrays of Non-FVD Transducers

7.4.1 Modal Analysis of Radiation Impedance

The fixed velocity distribution (FVD), defined in Sect. 1.3, is not always adequate, since in general transducers vibrate in more than one mode with the amplitude and phase of each mode determined by both the electrical drive and acoustic forces. Since each mode is a FVD in a given medium, and since transducers are often operated in relatively narrow frequency bands where one mode is dominant, the single mode FVD assumption is usually adequate. But in some situations, especially in arrays, the pressure variations across the surface result in nonuniform forces on individual transducers that excite more than one mode. Then each transducer velocity distribution can vary with frequency and steering angle.

We will begin by considering the radiation impedance of an individual transducer vibrating in more than one mode. Its sound field is the superposition of the fields of all the modes, and its radiation impedance depends on the modal impedances and the acoustic coupling between modes. The normal velocity on the surface of a transducer can be written as a sum of modes

$$u\left(\vec{r}\right) = \sum_{n} U_n \eta_n\left(\vec{r}\right), \tag{7.33}$$

where the coefficients, U_n, are the complex modal velocity amplitudes, the $\eta_n\left(\vec{r}\right)$ are the modal normal velocity functions, and $\vec{r}$ is the position vector on the surface. In principle the summation index goes to infinity, but for most practical problems only a few modes would be considered. Modal normal velocity functions are available in analytical form only in the simplest cases, but their use here permits a clear definition of concepts such as modal radiation impedance.

Each modal velocity function produces a modal pressure field given by $U_n p_n\left(\vec{r}\right)$, where $p_n(\vec{r})$ is the pressure for unit velocity. Since each mode is a FVD the self-radiation impedance of the nth mode referred to U_n is given by Eq. (1.4b), as

$$Z_{\mathrm{nn}} = \frac{1}{U_n U_n^*} \iint_{S_n} U_n p_n\left(\vec{r}\right) U_n^* \eta_n\left(\vec{r}\right) \mathrm{d}S. \tag{7.34}$$

The total radiation impedance of the nth mode, referred to U_n, includes the pressures from all the other modes, and is given by

$$Z_n = \frac{1}{U_n U_n^*} \iint_{S_n} \sum_{\mathrm{m}} U_{\mathrm{m}} p_{\mathrm{m}}\left(\vec{r}\right) U_n^* \eta_n\left(\vec{r}\right) \mathrm{d}S = \sum_{\mathrm{m}} \frac{U_m}{U_n} Z_{\mathrm{nm}}, \tag{7.35}$$

where Z_{nm} is the mutual radiation impedance between the nth and mth modes, defined by

$$Z_{nm} = \frac{1}{U_m U_n^*} \iint_{S_n} U_m p_m(\vec{r}) U_n^* \eta_n(\vec{r}) dS. \tag{7.36}$$

Note that $Z_{mn} = Z_{nm}$ because of acoustic reciprocity (see Sect. 11.2.2) and $Z_{nm} = 0$ when n and m refer to orthogonal modes such as the dipole and monopole modes of a sphere.

The acoustic power radiated by the mth mode is

$$P_m = ½ U_m U_m^* \mathrm{Re} Z_m = ½ \mathrm{Re} \sum_n U_n U_m^* Z_{mn} \tag{7.37}$$

The important difference from the FVD case is that the vibration of the transducer is not characterized by one reference velocity. However, the total radiation impedance of the transducer, Z_r, must be defined with respect to some reference velocity such that its real part has the usual relationship to the total radiated power. In most cases the lowest mode ($n = 0$) is dominant and therefore U_0 is the most suitable choice for the reference velocity. The total power radiated by the transducer is then written in terms of Re Z_r and equated to the sum of the modal powers in Eq. (7.37)

$$½ U_0 U_0^* \mathrm{Re} Z_r = \sum_m P_m = \sum_m \frac{1}{2} \mathrm{Re} \sum_n U_n U_m^* Z_{mn}. \tag{7.38}$$

The total radiation impedance of the transducer referred to U_0 is then defined such that its real part is consistent with Eq. (7.38), i.e.,

$$Z_r = \sum_m \sum_n \frac{U_n U_m^*}{U_0 U_0^*} Z_{mn} = \sum_m \frac{U_m U_m^*}{U_0 U_0^*} Z_m. \tag{7.39}$$

7.4.2 *Modal Analysis of Arrays*

The pressure distribution on the face of an individual transducer in an array is not uniform, and not symmetrical with respect to that transducer, except for certain array locations (see Fig. 7.17a). If the radiating faces of the transducers are capable of vibrating in more than one mode the pressure distribution will excite other modes, and the relative amount of each mode will vary with location in the array and with frequency and steering angle (see Fig. 7.17b). The modes of each transducer, discussed in Sect. 7.4.1, can be treated as separate transducers at the same array location, with each mode of each transducer acoustically coupled to all the modes of all the transducers [42].

For transducers in which M modes are to be considered the velocity profile of the ith transducer in an array can be written, as in Eq. (7.33):

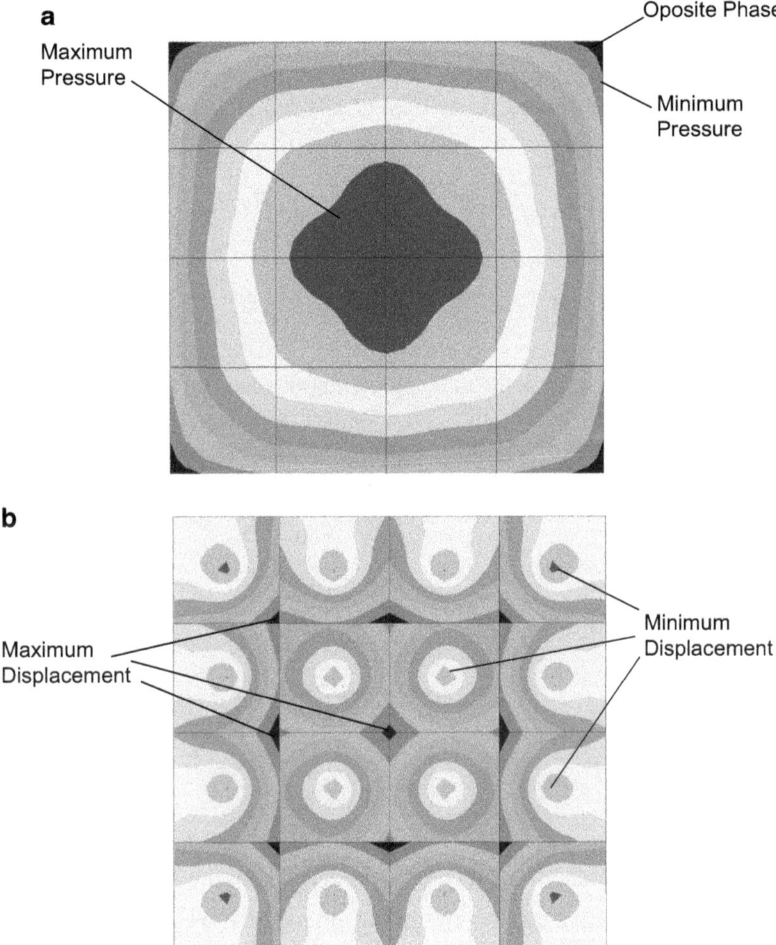

Fig. 7.17 (**a**) Pressure contours on a 16 element array. (**b**) Displacement contours on the 16 element array

$$u_i\left(\vec{r_i}\right) = \sum_{m=1}^{M} U_{\mathrm{mi}}\,\eta_m\left(\vec{r_i}\right), \tag{7.40}$$

where the U_{mi} are the modal velocity amplitudes and the mode functions are designated by $\eta_{\mathrm{m}}\left(\vec{r_i}\right)$ assuming that the transducers in the array are identical.

Equations can be written for each mode analogous to the transducer equations that were derived in Chap. 2 for the lumped-parameter models of the various transducer types. These equations can be expanded into array equations by replacing the self-radiation impedance by the total radiation impedance of each mode, including the acoustic couplings between all the modes of all the transducers. If the transducers in question have M significant modes and there are N transducers in the array, $2NM$ equations will be required to describe the array performance. Array equations in the same form as Eqs. (7.18) and (7.19) apply to each mode of each transducer:

$$0 = \left(Z_m^E + Z_{\mathrm{mi}}\right)U_{\mathrm{mi}} + N_m V_i, \quad m = 1, 2, \ldots M, \ i = 1, 2 \ldots N, \tag{7.41}$$

$$I_i = \sum_{m=1}^{M} N_m U_{\mathrm{mi}} + Y_0 V_i. \tag{7.42}$$

In these array equations Z_{m}^E is the mechanical impedance and N_{m} is the transduction coefficient of the mth mode; these parameters are usually the same for each transducer in an array, but they differ from mode to mode. The quantity Z_{mi} is the total radiation impedance of the mth mode of the ith transducer which can be written as an extension of Eq. (7.35) by summing over all the transducers in the array:

$$Z_{\mathrm{mi}} = \sum_{n=1}^{M}\sum_{j=1}^{N}\frac{U_{nj}}{U_{\mathrm{mi}}}Z_{\mathrm{nmij}}, \tag{7.43}$$

where

$$Z_{\mathrm{nmij}} = \frac{1}{U_{\mathrm{nj}}U_{\mathrm{mi}}^*}\iint p_{\mathrm{nj}}\left(\vec{r_i}\right)U_{\mathrm{mi}}^*\eta_{\mathrm{m}}\left(\vec{r_i}\right)\mathrm{d}S_i, \tag{7.44}$$

and $p_{\mathrm{nj}}\left(\vec{r_i}\right)$ is the pressure produced by the nth mode of the jth transducer. Z_{nmij} is the mutual radiation impedance between the nth mode of the jth transducer and the mth mode of the ith transducer. For example, Z_{nnij} is the mutual impedance between the ith and jth transducers, both vibrating in the nth mode, Z_{nmii} is the mutual impedance between the nth and mth mode of the ith transducer, and Z_{nnii} is the self-impedance of the ith transducer vibrating in the nth mode. The following reciprocal relations hold:

$$Z_{\mathrm{nnij}} = Z_{\mathrm{nnji}}, \ Z_{\mathrm{nmii}} = Z_{\mathrm{mnii}}, \ Z_{\mathrm{mnji}} = Z_{\mathrm{nmij}},$$

and when the nth mode is symmetric and the mth mode is antisymmetric, $Z_{\mathrm{nmii}} = 0$, but $Z_{\mathrm{nmij}} \neq 0$, because modes of opposite symmetry cannot excite each other on the same transducer, but they can excite each other on different transducers.

A specific example may be the best way to clarify these concepts. Consider an array of two identical transducers with square radiating faces as shown in Fig. 7.18.

Both are driven electrically to excite only the piston modes, but unsymmetrical acoustic forces can also excite rocking modes in which the square surfaces rotate about a nodal line that bisects the surface. It can be seen from Fig. 7.18 that the

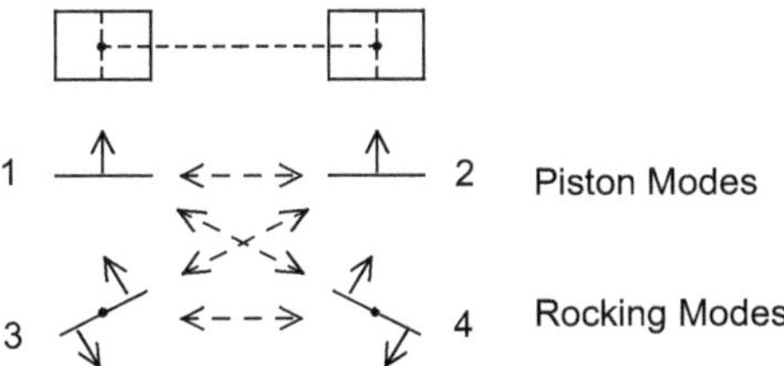

Fig. 7.18 Two transducers each vibrating in a piston mode and a rocking mode. For analysis the two can be considered as four transducers with acoustic interactions between them as shown by the *dashed arrows*

pressure distribution across the surface of each transducer would excite the rocking mode with a node perpendicular to the line between the centers of the transducers but would not excite the other rocking mode with a node parallel to that line. In this array of two transducers, each with two modes, there are 16 self and mutual impedances given by Eq. (7.44); some are equal to each other and some are zero. Numbering the piston and rocking modes 1 and 3 for the first transducer, and 2 and 4 for the second transducer permits these Z_{nmij} to be written with only two subscripts, i.e., Z_{ij} with $i, j = 1, 2, 3, 4$. (see Fig. 7.18). For example, Z_{1211}, the mutual impedance between the first mode of the first transducer and the second mode of the first transducer, becomes Z_{13}. Using this simplified notation all the radiation impedances for this array of two transducers are:

$Z_{11} = Z_{22}$; the self-impedances of the first and second transducers in the piston mode.

$Z_{33} = Z_{44}$; the self-impedances of the first and second transducers in the rocking mode.

$Z_{13} = Z_{31} = Z_{24} = Z_{42} = 0$, the mutual impedances between the piston and rocking modes of the same transducer (the piston mode function is a constant which is symmetric, while the rocking mode function is a linear function of position on the square radiating face which is antisymmetric about the center which makes these impedances vanish).

$Z_{12} = Z_{21}$; the mutual impedances between the piston modes of the two transducers.

$Z_{34} = Z_{43}$; the mutual impedances between the rocking modes of the two transducers.

$Z_{14} = Z_{41} = Z_{23} = Z_{32}$; the mutual impedances between the piston mode of one transducer and the rocking mode of the other transducer.

The symmetry of this array makes the velocity and the total radiation impedance for the piston and rocking modes the same for each transducer. Designating the piston mode velocity amplitude by U_1 and the rocking mode velocity amplitude by U_3, the total radiation impedances are

$$Z_1 = Z_2 = Z_{11} + Z_{12} + \frac{U_3}{U_1} Z_{14}, \tag{7.45a}$$

$$Z_3 = Z_4 = Z_{33} + Z_{43} + \frac{U_1}{U_3} Z_{14}. \tag{7.45b}$$

These impedances depend on the relative velocities of the two modes which is determined by the array equations. In this case Eq. (7.41) becomes for the two modes

$$0 = \left(Z_1^E + Z_{11} + Z_{12}\right) U_1 + Z_{14} U_3 + N_1 V_1, \tag{7.46a}$$

$$0 = \left(Z_3^E + Z_{33} + Z_{43}\right) U_3 + Z_{14} U_1, \tag{7.46b}$$

where Z_1^E and Z_3^E are the internal mechanical impedances of the piston mode and the rocking mode. The quantity N_1 is the transduction coefficient that drives the piston mode when the voltage V_1 is applied, but $N_3 = 0$ because the transducer is designed to make V_1 drive only the piston mode. The rocking mode is driven only by acoustic coupling, as can be seen from Eq. (7.46b), which gives

$$U_3 = \frac{-Z_{14} U_1}{Z_3^E + Z_{33} + Z_{43}}, \tag{7.47}$$

showing that Z_{14} is the mutual impedance that excites the rocking mode.

Combining Eqs. (7.45b) and (7.47) gives

$$Z_3 = R_3 + jX_3 = -\,Z_3^E = -\,R_3^E - jX_3^E,$$

which shows that the radiation resistance of the rocking mode is negative as expected for an electrically undriven mode. The power associated with each rocking mode is also negative:

$$W_3 = W_4 = ½ U_3 U_3^* R_3 = -½ |U_3|^2 R_3^E,$$

showing that the rocking modes absorb power from the sound field produced by the electrically driven piston modes.

The acoustic power radiated by the piston modes for a fixed voltage drive is affected by the rocking modes in two ways. The piston mode velocity is changed as determined by Eqs. (7.46a and 7.46b), and the total radiation resistance of the piston modes is changed as shown by Eq. (7.45a). Thus the power radiated by each piston mode in the presence of the rocking mode is

$$W_1 = W_2 = ½ U_1 U_1^* R_1 = ½ U_1 U_1^* \left[R_{11} + R_{12} + \mathrm{Re}\left(\frac{U_3}{U_1} Z_{14} \right) \right].$$

In a larger array where the velocities were not all the same the changes in piston mode velocities caused by the rocking modes would result in beam pattern distortion as well as reduction of power.

The formal analysis outlined above is a systematic way of including the effects of non-FVD transducers in arrays. Very little numerical work has been done on this type of array problem, and finite element modeling (FEM or FEA) might be the most practical approach (see Sect. 7.6).

7.5 Volume Arrays

The arrays considered so far contain transducers with radiating surfaces that conform approximately to the array surface, such as an array of Tonpilz transducers whose radiating faces form a portion of a cylindrical surface as shown in Figs. 7.1 and 1.10. These are called surface arrays, and, although the transducers influence each other acoustically, there is no scattering among them. But some transducers, e.g., flextensional transducers, cannot form a surface array because they radiate from two or more sides. When these transducers are mounted in an open structure they form an array that will be called a volume array. Even when an array is constructed by mounting such transducers on a closed surface it has characteristics that are more like a volume array than a surface array, because of the scattering among the transducers. Figure 7.2 illustrates surface and volume arrays.

In volume arrays the interactions are more complicated than in surface arrays because of scattering. The field of an individual transducer in a surface array can be calculated assuming that the other transducers are clamped and form part of the surface; thus there is no scattering from the other transducers. But in a volume array the field of an individual transducer is strongly affected by the presence of other transducers even when they are clamped, because the other transducers scatter the field of the first. For transducers close together the scattered fields can contribute significantly to the far field and exert asymmetrical forces on each transducer that can excite modes that are not excited by the electrical drive. Thus transducers that can be considered to have FVD when used alone or in surface arrays may show their non-FVD characteristics when used in volume arrays.

The analysis of such arrays is probably best done by FEA, and even that might be difficult for large arrays. However, the principles involved can be illustrated analytically by considering small spherical sources driven electrically in the monopole mode. Scattering of the radiated spherical wave field by another sphere can be approximated as scattering of a plane wave by a rigid sphere if the two spheres are not too close together. Consider two pulsating spheres of radius a with centers separated by distance d as shown in Fig. 7.19, and let p_1 be the pressure radiated by the first sphere which is given by Eq. (10.15b) with $Q = 4\pi a^2 u_1$:

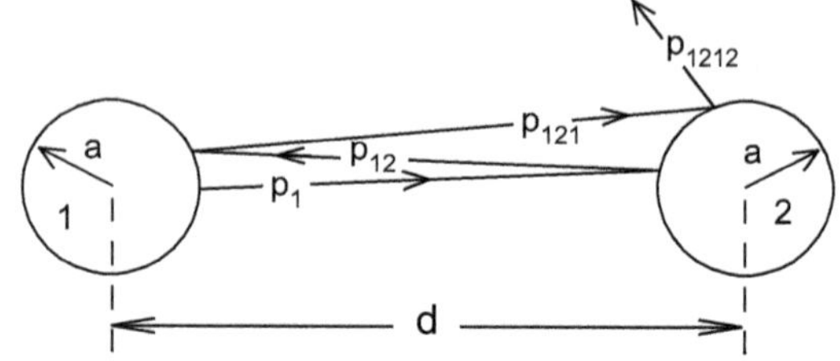

Fig. 7.19 Scattering of the field of one pulsating sphere by another sphere in a volume array. Each p_1, p_{12}, etc. represents a spherical wave

$$p_1(r_1) = j\rho cka^2 u_1 \frac{e^{-jkr_1}}{r_1}, \tag{7.48}$$

where r_1 is the distance from the center of the first sphere and u_1 is its normal velocity. When this spherical wave reaches the second sphere only a small part of the spherical wavefront is intercepted, and, if d is large enough, this part can be approximated by a plane wave. Scattering of a plane wave by a rigid sphere is a problem with a known solution [43], and, for $ka \ll 1$, the scattered pressure can be put in the form

$$p_{12}(r_2) = p_1(d)\delta_0 \frac{e^{-jkr_2}}{kr_2}, \tag{7.49}$$

where r_2 is the distance from the center of the second sphere and $\delta_0 = (ka)^3/3$. Such multiple scattering continues indefinitely with reduced scattered pressure at each step. For example, p_{121} is p_{12} scattered from the first sphere, p_2 is the pressure radiated by the second sphere, p_{21} is the scattering by the first sphere of the wave radiated by the second sphere, etc. (see Fig. 7.19).

The total radiation impedance of the first sphere is the integral of the total pressure over the surface of that sphere divided by its normal velocity:

$$Z_1 = \frac{1}{u_1} \iint_{S_1} [(p_1 + p_{12} + p_{121} + \cdots) + (p_2 + p_{21} + \cdots)] dS_1 = Z_{11} + \frac{u_2}{u_1} Z_{12}, \tag{7.50}$$

where

$$Z_{11} = \frac{1}{u_1} \iint_{S_1} [p_1(r_1 = a) + p_{12}(r_2 = d) + p_{121}(r_1 = a) + \cdots] dS_1, \tag{7.51}$$

which shows explicitly that the pressure components are to be evaluated on the surface of the first sphere. For $ka \ll 1$ the integrals are approximated by multiplying by the spherical area which gives, after using Eqs. (7.48) and (7.49),

$$Z_{11} = 4\pi a^2 j\rho c(ka)^2 \left[\frac{\mathrm{e}^{-jka}}{ka} - \delta_0 \frac{\mathrm{e}^{-2jkd}}{(kd)^2} + \delta_0^2 \frac{\mathrm{e}^{-jk(2d+a)}}{k^3 d^2 a} \right]. \tag{7.52}$$

The first term in Eq. (7.52) is the self-radiation impedance for a pulsating sphere with $ka \ll 1$. The second term is a small correction to the self-impedance caused by scattering from the second sphere, and the third term is a still smaller correction caused by re-scattering from the first sphere. In a similar way the mutual radiation impedance between the two spheres is

$$\begin{aligned} Z_{12} &= \frac{1}{u_2} \iint_{S_1} [p_2(r_2 = d) + p_{21}(r_1 = a) + \cdots] \mathrm{d}S_1 \\ &= 4\pi a^2 j\rho c(ka)^2 \left[\frac{\mathrm{e}^{-jkd}}{kd} - \delta_0 \frac{\mathrm{e}^{-jk(d+a)}}{k^2 da} \right]. \end{aligned} \tag{7.53}$$

The first term in Eq. (7.53) is the mutual radiation impedance neglecting scattering. It has the same dependence on separation between the spheres as was found in Eq. (7.24a) for two small pistons of arbitrary shape in a rigid plane. The second term is a small correction to the mutual impedance caused by scattering of the radiated pressure from the second sphere by the first sphere. The assignment of scattering terms to self or mutual impedance is such that all the terms in Z_{11} originate from the first sphere, while all the terms in Z_{12} originate from the second sphere. In larger arrays the impedances would contain additional terms arising from direct radiation and scattering from other spheres.

It should be pointed out that the analysis in this section assumed that the spheres were FVD transducers capable of vibrating only in the monopole mode. Transducers of practical size in a realistic volume array would probably be vibrating in the dipole mode as well as the monopole mode, as a result of the nonuniform pressure on their surfaces. Volume arrays of non-FVD transducers probably present the ultimate challenge for array analysis.

This example of scattering effects on radiation impedance serves mainly to illustrate qualitatively the physical nature of the problem. In most volume arrays the transducer separations would be small, and the simplified scattering calculation above would be quantitatively useless. Thompson has done a more exact analysis of scattering and acoustic coupling between spherical radiators [44, 45]. All the scattering events, and excitation of other modes, can be included by finite element numerical modeling.

7.6 Near Field of a Projector Array

Most array structures are complicated enough that analytical modeling cannot be expected to give detailed results accurately, but it can provide approximate quantitative guidance as to how well a proposed array concept might perform. An analytical approach might also miss important features, such as excitation of unexpected modes, that a finite element approach would reveal. Since finite element modeling can be applied to both the transducer structure and the sound field, at least for small arrays, it is probably the best way to obtain reliable quantitative data for an array design. Such FEA information based on small arrays can also guide the formulation of analytical approximations for application to large arrays that would be too time consuming by FEA.

The basics of the FEA approach to transducer and array analysis was given in Sect. 3.4; here only an example of results for a small array of 16 close-packed piezoelectric Tonpilz transducers with square radiating faces vibrating in water is given in Fig. 7.17a and b.

The contour plots in Fig. 7.17a show the pressure magnitude, while those in Fig. 7.17b show the normal displacement magnitude, on the surface of the array. Both plots are symmetrical about the two lines that bisect the array and about the two diagonals because all the transducers are driven in phase with the same voltage magnitude. This symmetry was used to reduce the size of the computation to one quadrant of the array with appropriate boundary conditions on the sides of the quadrant that join the rest of the array. Although each transducer has the same electrical drive, each is affected differently by the sound pressure as can be seen from the pressure distribution in Fig. 7.17a. As a result, the displacement distributions differ from each other and depart considerably from the uniform piston motion that is usually desired. The individual transducer faces can bend and rock independently of each other under the influence of the piezoelectric drivers and the different pressure distributions, which causes discontinuities in the displacement at the small gaps where the edges of the transducer faces meet.

Figure 7.17a can be described quantitatively quite easily because the pressure is a smooth function of position on the array. The pressure is maximum at the center, falls slowly to 80 % about halfway toward the center of an edge, falls to 60 % at three quarters of the way and to about 10 % at the edge. Near the corners the pressure falls to zero, and at the corners it is out of phase with about 20 % of the maximum pressure amplitude. The displacement contours in Fig. 7.17b are not as easy to summarize quantitatively, but the displacement is in phase over the entire array and the variations are not as great as the pressure variations. The maximum displacement occurs at the center of the array and also on the inner corners of the outside transducers. A displacement minimum occurs near the center of each transducer. For the inside transducers the minimum value is about 65 % of the value at the center of the array, for the outside transducers at the corners the minimum is about 55 %, and for the others on the outside it is about 60 %.

It can be seen from Fig. 7.17a that the pressure distribution on the corner transducers is likely to excite a rocking mode with a diagonal node, while on the side transducers a rocking mode with a node parallel to the side of the array is likely to be excited. The displacement distributions in Fig. 7.17b are approximately consistent with this interpretation, since a mixture of piston mode and rocking mode can be seen. But this situation is more complicated than the case of piston and rocking modes analyzed in Sect. 7.4.2, because a bending mode is also involved in the motion of these transducers.

7.7 The Nonlinear Parametric Array

In Chap. 2 we mentioned some of the mechanical and electrical nonlinear mechanisms that exist in electroacoustic transducers. The process of sound propagation through water also has nonlinear characteristics that have significant effects if the sound pressure amplitude is high enough. We have ignored these characteristics in calculating sound fields, because in the normal operation of transducers the amplitudes are not high enough for acoustic nonlinear effects in the medium to be significant. However, under certain conditions the acoustic nonlinearities do have effects that can be useful and important [46].

In this section we will describe a special type of projector array, the parametric array, which exists only because of the nonlinearities in the medium. The nonlinear mechanisms cause a high amplitude sound wave to generate other sound waves with frequencies at the harmonics of the original wave. However, when two high amplitude sound waves of different frequencies propagate in the same region they each generate harmonics, and they also interact to generate sound waves with frequencies equal to the sum and difference of the primary (i.e., the original) frequencies. These new waves do not come from the transducer that radiates the primary waves; they are generated in the medium, and their amplitude can be significant when the near fields of the primaries form well-collimated, overlapping, high amplitude beams that can interact with each other over a considerable distance. We will give a simplified physical description of difference frequency generation in the near field and an approximate calculation of the far field of the difference frequency component.

Westervelt [47] recognized that the most interesting of the new waves generated by nonlinear effects in the medium might be the difference frequency component, which, because of its low frequency, would be absorbed much less than the other waves and, hence, propagate farther. His analysis showed that the far field of the difference frequency component would have very narrow beams with very low side lobes compared to conventionally radiated beams, characteristics that could have significant practical value. These results can be predicted by starting with an approximate form of the wave equation that includes the most important nonlinear terms. Then an approximate solution of this nonlinear wave equation shows the existence of, and gives the amplitude of, the difference frequency pressure generated in the near field by these nonlinear mechanisms. The near-field difference frequency pressure can then be regarded as an acoustic source, and its far field can be calculated by ordinary linear acoustics. The following analysis is based on Westervelt [47], Beyer [46], and Kinsler, Frey, et al. [13].

After simplification of various nonlinear terms an approximate wave equation for the acoustic pressure can be written in the form

$$c^2\left(1+\tau\frac{\partial}{\partial t}\right)\nabla^2 p-\frac{\partial^2 p}{\partial t^2}=-\frac{\beta}{\rho c^2}\frac{\partial^2}{\partial t^2}\left(p^2\right). \tag{7.54}$$

Equation (7.54) is the linear acoustic wave equation except for the dissipation term proportional to τ and the nonlinear term proportional to β. τ is related to the shear and bulk viscosities of the medium and $\beta = 1 + B/2A$, where B/A is the nonlinear parameter of the medium ($B/A \approx 5$ for sea water [46]). Although τ and β are never zero, the terms containing these parameters are negligible under many conditions, and then this nonlinear wave equation reduces to the lossless linear wave equation. Note also that if only the term containing β is neglected, the equation is linear and has familiar solutions that represent damped plane waves given by

$$p(x, t) = P_0 \mathrm{e}^{-\alpha x} \mathrm{e}^{j(\omega t - kx)}, \tag{7.55}$$

where $k = \omega/c$ and α, the absorption coefficient, is approximately equal to $\omega^2\tau/2c$. It is necessary to include damping in the wave equation when analyzing nonlinear effects, because damping is one of the factors that determines the size of the nonlinear interaction region. Note that the approximate absorption coefficient depends on frequency squared.

Now consider a transducer driven electrically with two frequencies, ω_a and ω_b, called the primary frequencies, where $\omega_a - \omega_b$ is small compared to ω_a or ω_b. Also consider the transducer to be large compared to the primary wavelengths, so that it radiates highly directional overlapping beams at both frequencies. Assume also that the transduction mechanism is perfectly linear and that the two driving voltages produce only two velocities on the radiating surface at the frequencies ω_a and ω_b. When the term containing β is neglected, the radiated pressure in the near field consists of two damped plane waves, which we now write in real form, as will be needed when products occur in the following analysis:

$$p_0 = P_a \mathrm{e}^{-\alpha_a x} \cos\ (\omega_a t - k_a x) + P_b \mathrm{e}^{-\alpha_b x} \cos\ (\omega_b t - k_b x) = p_a + p_b. \tag{7.56}$$

We will now keep the term containing β and find an approximate solution of Eq. (7.54) by the perturbation method, a method of finding approximate solutions of nonlinear equations. In this case, where the nonlinear term in Eq. (7.54) is proportional to β, it is convenient to assume the solution in the form of a perturbation series in powers of the small quantity $\delta = \beta/\rho c^2$ as follows:

$$p = p_0 + \delta p_1 + \delta^2 p_2 + \cdots, \tag{7.57}$$

where p_0 is the zeroth-order solution, δp_1 is the first-order solution, etc. When Eq. (7.57) is substituted into the one-dimensional version of Eq. (7.54), and the terms are separated into groups according to powers of δ, we find that the terms that do not depend on δ satisfy the equation

$$c^2\left(1 + \tau\frac{\partial}{\partial t}\right)\frac{\partial^2 p_0}{\partial x^2} - \frac{\partial^2 p_0}{\partial t^2} = 0, \tag{7.58a}$$

which determines p_0, the zeroth-order solution already given in Eq. (7.56). Similarly, the terms that depend on δ to the first power satisfy

$$c^2\left(1+\tau\frac{\partial}{\partial t}\right)\frac{\partial^2 p_1}{\partial x^2}-\frac{\partial^2 p_1}{\partial t^2}=-\frac{\partial^2 p_0^2}{\partial t^2}, \tag{7.58b}$$

which determines p_1 and the first-order solution, δp_1. The equations that determine the higher order solutions, p_2, etc., can be found in the same way, but will not be needed here.

Use of the perturbation method has replaced the nonlinear Eq. (7.54) by a set of linear equations, Eqs. (7.58a, 7.58b,...) which can be solved one after the other, since the second equation depends on the solution of the first, etc. When p_0 is used in Eq. (5.57b) we get an inhomogeneous equation with seven terms on the right-hand side involving time derivatives of p_a and p_b. These terms are source terms, each of which contributes part of the complete solution of Eq. (7.58b). The source terms involving only p_a or p_b contribute harmonics of ω_a and ω_b to the solution, while the terms involving both p_a and p_b contribute sum and difference frequency components. Here we are only interested in the difference frequency component, and define p_{1d} as the part of $\delta p_1=(\beta/\rho c^2)p_1$ that gives the first-order difference frequency pressure. The parts of Eq. (7.58b) that are related to p_{1d} are:

$$\begin{aligned}\left(1+\tau\frac{\partial}{\partial t}\right)\frac{\partial^2 p_{1d}}{\partial x^2}-\frac{\partial^2 p_{1d}}{c^2\partial t^2}&=-\frac{\beta}{\rho c^4}\frac{\partial^2}{\partial t^2}(2p_a p_b)\\ =\frac{\beta P_a P_b}{\rho c^4}\mathrm{e}^{-(\alpha_a+\alpha_b)x}\omega_d^2\cos\left(\omega_d t-k_d x\right)&=\frac{\beta P_a P_b}{\rho c^4}\mathrm{e}^{-2\alpha x}\omega_d^2\mathrm{e}^{j(\omega_d t-k_d x)},\end{aligned} \tag{7.59}$$

where $\omega_d=(\omega_b-\omega_a)$ is the angular difference frequency, $k_d=\omega_d/c$ is the acoustic wavenumber at the difference frequency, and $\alpha=(\alpha_a+\alpha_b)/2$ is the average absorption coefficient for the primary frequencies. The last form of Eq. (7.59) has been converted back to complex notation for convenience in solving the linear equation in which it now appears.

Equation (7.59) can be solved by considering the inhomogeneous term on the right-hand side to represent a virtual acoustic line source vibrating at the frequency ω_d, phased to end-fire by the factor $\mathrm{e}^{-jk_d x}$ and shaded by the exponential factor $\mathrm{e}^{-2\alpha x}$. The volume of the source is approximately AL, where A is the area of the radiating surface of the projector, and L is the length of the virtual source. Under certain conditions L is of the order $1/\alpha$, i.e., large enough to reduce the primary wave amplitudes sufficiently to make nonlinear interaction insignificant at the outer end of the virtual line source.

The right-hand side of Eq. (7.59) is $j\rho\omega_d$ times the source strength density (volume velocity per unit volume) of the virtual line source. The far field can be calculated from linear acoustics in the same way that the far field of a uniform line source is calculated in Sect. 10.2.1. The differential source strength, $\mathrm{d}Q$, in Eq. (10.21) is equal to the right-hand side of Eq. (7.59) times $A\mathrm{d}x$

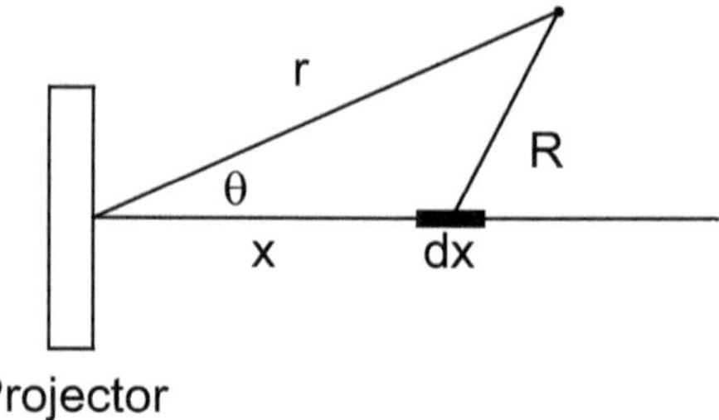

Fig. 7.20 Coordinates for calculating the difference frequency far field of a parametric array

and divided by $j\rho\omega_d$, and the differential contribution to the far-field pressure from the element dx is

$$\mathrm{d}p_{1\mathrm{d}} = \frac{\beta P_\mathrm{a} P_\mathrm{b} \omega_\mathrm{d}^2}{4\pi\rho c^4} \mathrm{e}^{-2\alpha x} \mathrm{e}^{j(\omega_\mathrm{d} t - k_\mathrm{d} x)} \frac{\mathrm{e}^{-jk_\mathrm{d} R}}{R} A\mathrm{d}x, \tag{7.60}$$

where $R = \left[r^2 + x^2 - 2rx\cos\theta\right]^{1/2}$ is the distance from the element to a far-field point at distance r from the center of the projector at the angle θ as shown in Fig. 7.20.

The time factor is retained in this case to make clear that the difference frequency far field is being calculated. Integrating Eq. (7.60) from 0 to L with R approximated by r in the denominator and by $(r - x\cos\theta)$ in the exponent gives

$$\begin{aligned} p_{1\mathrm{d}}(r,\theta,t) &= \frac{\beta P_\mathrm{a} P_\mathrm{b} \omega_\mathrm{d}^2 A}{4\pi\rho c^4} \frac{\mathrm{e}^{j(\omega_\mathrm{d} t - k_\mathrm{d} r)}}{r} \int_0^L \mathrm{e}^{-jx[k_\mathrm{d}(1-\cos\theta) - j2\alpha]} \mathrm{d}x \\ &\approx \frac{\beta P_\mathrm{a} P_\mathrm{b} \omega_\mathrm{d}^2 A}{4\pi\rho c^4 [k_\mathrm{d}(1-\cos\theta) - j2\alpha]} \frac{\mathrm{e}^{j(\omega_\mathrm{d} t - k_\mathrm{d} r)}}{jr} \end{aligned} \tag{7.61}$$

where the approximation of the integral holds for L large. We have also neglected the term $\tau\partial^3 p_{1\mathrm{d}}/\partial t\partial x^2$ in Eq. (7.59), which is equivalent to neglecting the absorption of the difference frequency wave as it propagates to the far field, as is usually done when calculating far fields. It is convenient to write this result as the magnitude

$$|p_{1\mathrm{d}}(r,\theta)| = \frac{\beta P_\mathrm{a} P_\mathrm{b} \omega_\mathrm{d}^2 A}{8\pi\rho c^4 \alpha r} \frac{1}{\left[1 + (k_\mathrm{d}/\alpha)^2 \sin^4(\theta/2)\right]^{1/2}}, \tag{7.62}$$

where the factor $\left[1 + (k_\mathrm{d}/\alpha)^2 \sin^4(\theta/2)\right]^{-1/2}$ is the directivity function normalized to unity at $\theta = 0$. The directivity function shows that the full beam width at -3 dB intensity is $4\sin^{-1}(\alpha/k_d)^{1/2}$, and that there are no side lobes as θ increases from 0° to 90°. Under typical conditions the beam widths are a few degrees [13].

The result in Eq. (7.62) is essentially the same as that from Westervelt's original model in which the nonlinear interaction that generates the difference frequency occurs entirely in the near field of the projector. This model was discussed here as an example of a mechanism that generates a difference frequency, but it does not include other mechanisms that are also important. Later models that include additional nonlinear absorption of primary energy as well as nonlinear generation of difference frequency in the far field of the projector are necessary for accurate determination of the difference frequency source level [46, 48–52].

Moffett and Mellen [48] have combined all the important features of these later models into a comprehensive model that has been used by Moffett and Konrad [53] and Moffett and Robinson [54] as the basis for a parametric array design procedure. This procedure provides design curves for determining the difference frequency source level for different downshift ratios, i.e., the ratio of the mean primary frequency to the difference frequency. Figure 7.21 is one set of such curves for a downshift ratio of 5, i.e., the difference frequency is 1/5 the mean primary frequency [53].

The difference frequency rms source level, SL, is found from

$$SL = SL_0 + G,$$

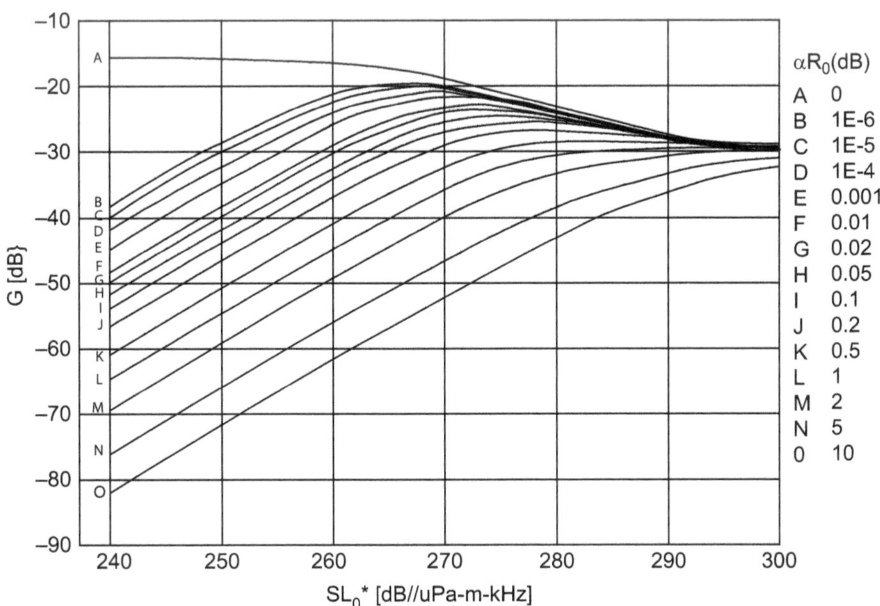

Fig. 7.21 Parametric gain G for scaled primary source level. SL_0* and down shift ratio of 5 [53]

where SL_0 is the rms source level of one primary frequency component (the two primaries are assumed to have the same source level) and G (the parametric gain) is the ordinate of Fig. 7.21. The abscissa of Fig. 7.21 is the scaled primary source level,

$$SL_0^* = SL_0 + 20\log f_0,$$

where f_0 is the mean primary frequency expressed in kHz. The parameter on the curves of Fig. 7.21 is αR_0, where α is the absorption coefficient in dB/m at f_0, and $R_0 = Af_0/c$ is the Rayleigh length, with A the area of the projector and c the sound speed.

As an example, consider a square piston transducer 0.5 m on a side radiating primary frequencies of 90 and 110 kHz with a difference frequency of 20 kHz (down shift ratio of 5). Using $A = 0.25$ m^2 and $\alpha = 0.03$ dB/m in seawater at 100 kHz gives $\alpha R_0 \approx 0.5\,\text{dB}$. Also consider that the primaries have equal rms source levels of $SL_0 = 237\,\text{dB}//\mu\text{Pa}\cdot\text{m}$, giving a scaled source level of $SL_0^* = 277\,\text{dB}//\mu\text{Pa}\cdot\text{m}\cdot\text{kHz}$. From Fig. 7.21 the parametric gain is $G = -28$ dB, and the difference frequency source level is, therefore, $209\,\text{dB}//\mu\text{Pa}\cdot\text{m}$. The source level predicted by the incomplete model in Eq. (7.62) is about 9 dB higher, mainly because it does not include nonlinear absorption of the primary beams.

With the same primary source levels and greater downshift ratios of 10 and 20 the difference frequency source levels would be reduced to 197 and $186\,\text{dB}//\mu\text{Pa}\cdot\text{m}$ at 10 kHz and 5 kHz, respectively [53]. The decrease is 11 dB per octave of difference frequency, whereas Eq. (7.62) predicts 12 dB per octave.

Westervelt [47] also pointed out that a small amplitude plane wave interacting with a high amplitude near-field wave of a different frequency would result in sum and difference frequency components, and that this could be the basis for a parametric receiving array. This type of acoustic receiver has been studied [55, 56], and design procedures have been given [53].

7.8 Doubly Steered Arrays

Line or planar arrays are often electrically steered to directions at an angle θ from the broadside direction and in some cases steered toward the end-fire direction at $\theta_s = 90°$ where θ_s is the steering angle. As the beam approaches 90°, the beam width increases and the source level decreases. In this section we show how the decrease in source level can be mitigated by electrically steering the radiation from the *array elements* into the same direction as the array is steered to; and thus the name, doubly steered array [57].

Consider first a conventional array where the array product theorem states that for a planar or line array of equally sized elements, the overall beam pattern function $F(\theta, \theta_s)$ is the product of the element beam pattern, $F_e(\theta)$, and the array beam pattern, $F_a(\theta, \theta_s)$, with point sources replacing the elements. That is,

$$F(\theta, \theta_s) = F_e(\theta) F_a(\theta, \theta_s) \tag{7.63}$$

This product generally yields a narrower beam pattern and higher broadside un-steered DI than $F_a(\theta, \theta_s)$ alone. For a conventional line element of length L and wavelength λ of un-steered elements

$$F_e(\theta) = \sin\left[(\pi L/\lambda)\sin\theta\right]/\left[(\pi L/\lambda)\sin\theta\right]. \tag{7.64}$$

And for an array of N elements with center-to-center spacing s and steering angle θ_s

$$F_a(\theta, \theta_s) = \sin Nx/N \sin x, \quad \text{where } x = (\pi s/\lambda)(\sin\theta - \sin\theta_s). \tag{7.65}$$

For a close-packed array the element length is equal to the center-to-center spacing, $L = s$.

The point source array beam pattern function, F_a, value is always unity in the direction of steering when $\theta = \theta_s$. However, this is not the case for the un-steered element beam pattern function F_e which is unity only at $\theta = 0$. If the array is steered to end-fire at 90° then $F_a(90, 90) = 1$ but $F_e(90) = 2/\pi$ (≈ -4 dB) which is a considerable loss in source level. Not only that, if the array is typical and the elements are spaced at $\lambda/2$, the function $F(90, 90) = F(-90, 90)$ and the same level of—4 dB is attained at both $\pm 90°$ instead of only +90°. If the element function is replaced by an omnidirectional element function there is no −4 dB reduction but there is a full strength beam at both $\pm 90°$ reducing the DI by 3 dB. This may be avoided by using $\lambda/4$ wavelength spaced elements but would require twice as many half size elements and this could also result in array interaction problems.

If now the elements are electrically steerable, such as the modal transducer elements of Sects. 5.2.6 and 5.7.3, then Eq. (7.63) becomes

$$F(\theta, \theta_s) = F_e(\theta, \theta_s) F_a(\theta, \theta_s) \tag{7.66}$$

where elements are steered in the same direction as the array is steered. In this case the beams of the modal elements would be in the same direction of the array beam.

For the case of modal electrically steered array elements (see Sect. 5.2.6)

$$F_e(\theta, \theta_s) = \left[1 + A_1 \cos(\theta - \theta_s) + A_2 \cos 2(\theta - \theta_s)\right]/\left[1 + A_1 + A_2\right] \tag{7.67}$$

Here beam patterns may be formed which approximate piston or line transducers. The modal transducer elements may be steered in the direction to which the array is steered; thus, maintaining beam pattern coincidence along the maximum response axis and at selected steered angles. For example, if the line array were end-fired, the modal beam pattern would also be incremented (i.e., steered) to this direction adding to the end-fire output. This would mitigate a common end-fire problem where array element beam patterns are often directed broadside, even when arrays are steered to

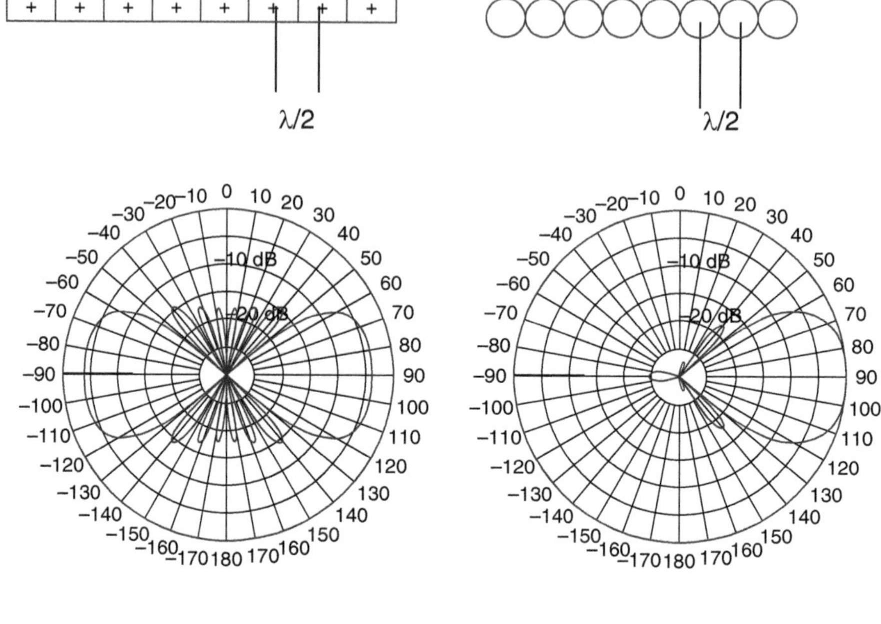

Fig. 7.22 Comparison of leveraged cylindrical transducer end-fire line array beam pattern with conventional line array

end-fire. Moreover, and maybe more importantly, it allows end-fire steering without the need for quarter wavelength half size elements. Implementation of this doubly steered array concept could also eliminate array interaction unloading problems in the rear cancellation direction during end-fire radiation conditions.

We compare end-fire beams of conventional line with a leveraged cylindrical transducer (see Sect. 5.7.3) array in Fig. 7.22 with dipole strength, $A_1 = 2$, and quadrupole strength, $A_2 = 1$. [We could have also chosen the simpler modal cardioid pattern case with $A_1 = 1$ and $A_2 = 0$.] Because of the ring shell and piston loading on the cylindrical structure, these elements resonate at a lower frequency than the cylinder alone and can be spaced at one-half a wavelength. Because these elements can be electrically steered in the same end-fire direction as the array is steered, they yield a 4 dB gain compared to a line array and, moreover, provide a single end-fire beam even with half-wavelength spacing providing an additional 3 dB gain in the DI. There would be an even greater 6 dB improvement when compared with a planar piston array, due to the single sided piston condition [see Sect. 11.2.4, Eq. (11.44b)] and the absence of symmetrical addition which occurs in line arrays.

7.9 Summary

In this chapter on projector arrays the directivity, mutual radiation impedance, array equations, non-fixed velocity distribution, volume arrays, near field of arrays, parametric arrays, and doubly steered arrays were presented, analyzed, and discussed. The array product theorem showed that for planar or line arrays of the same type of transducers the beam patterns and response is equal to the product of the beam pattern of one of the transducer elements and the beam pattern of an array of point source elements representing the central location of the elements. This was followed by an analysis and development of the acoustic pressure for line, rectangular, and circular arrays along with approximations for the directivity factor.

The cause and effect of grating lobes was presented and it was shown that this occurs when the center-to-center separation of elements approaches one wavelength and that these lobes cannot be reduced by shading. The discussion on beam steering included a presentation on shading the array with nonuniform spacing and how this also reduces grating lobes. Staggering array elements was presented as a method for reducing both side and grating lobes. A brief study of array random variations showed that the limit of side lobes depends on the directivity index, DI, and that a high DI is needed for very low side lobes.

Array equations were developed and a matrix inversion solution was given based on the *ABCD* parameter representation of the transducers. The mutual radiation impedance formulation was then developed for a planar array of piston transducers. It was shown that for small pistons or spheres the mutual impedance between elements i and j is $Z_{ij} = R_{ii}[\sin(kd)/kd + j\cos(kd)/kd)]$, where d is the center-to-center separation and R_{ii} is the self-radiation resistance. Discussion and analysis of nonplanar arrays and arrays with non-fixed velocity distributions, volume arrays and near-field effects were also given.

The nonlinear parametric array was shown to yield a narrow beam with low side lobes from a comparatively small array. And this is shown to be a result of the nonlinear interaction of two acoustic beams of slightly different frequencies launching an end-fire array at the difference frequency. Finally, doubly steered linear arrays were presented where the transducer element is electrically steered in the same direction as the array is steered and shown to achieve a significantly improved end-fire response.

Exercises (Degree of Difficulty: *Lowest, **Moderate, ***Highest)

7.1.* Consider two small pistons of radius, a, with $ka \ll 1$, separated by a center-to-center distance, d, in a large rigid baffle. Show that the total radiation resistance of each approaches $2R_{ii}$ as kd becomes small. Why is this case similar to the case of a single piston approaching a rigid wall?

7.2.* Show that Eq. (7.21c) follows from Eqs. (7.21a) to (7.21b).

7.3.* Consider a large array of square pistons of side length L and with spacing $L/10$ between the pistons. What is the packing factor? What is the average radiation resistance of the individual pistons? What is the total radiation resistance of an array of N pistons?

7.4.* Show that Eq. (7.30) reduces to $Z = F/U$ for a uniform velocity distribution.

7.5.* Explain why the product theorem is useful in array beam pattern analysis? Would it be useful in determining the DI or the radiation impedance?

7.6.*** A close-packed array of ideal square pistons shows no grating lobes when it is not steered, because the individual piston patterns null out the array grating lobes. But when the array is steered grating lobes appear at angles such that they are only partially nulled out. Show this by calculating the beam pattern for a line array of ten close-packed, square pistons with side length of one-wavelength and center-to-center spacing of one wavelength steered to 45° (see Eqs. (7.10b and 7.10c)).

7.7.** Show that Sinc $(kd/2) = 2\sin(kd/2)/kd$ approaches 1 for $kd/2 \ll 1$. Likewise show that $J_1(ka)$ goes to $ka/2$ for $ka \ll 1$; and thus, $2J_1(ka)/ka$ also approaches 1 for $ka \ll 1$.

7.8.** Determine the value of α/k_d required for a parametric array full beam width of 4° and calculate the corresponding beam pattern.

7.9.** To see extreme effects of acoustic coupling consider a small transducer located at the center of a circle of radius d of N identical transducers being driven to make all have the same velocity. Use Eqs. (7.16) and (7.24c) and show that the total radiation resistance of the center transducer could become negative for $N > 5$ when $\lambda < 4d/3$ and also in other frequency regions for greater values of N. What does negative radiation resistance mean physically? What is the total radiation reactance of the center transducer when $\lambda = 2d$.

7.10.*** As a simple example of the effects of beam steering on the radiation impedance of transducers in an array consider a line array of two small transducers driven to have the same velocity amplitude but steered by a velocity phase difference of μ, i.e., $u_2 = u_1 e^{j\mu}$. How does the total radiation impedance of the two transducers differ, in general, and specifically for end-fire steering?

References

1. H.H. Schloemer, *Technology Development of Submarine Sonar Hull Arrays*, Naval Undersea Warfare Center Division Newport, Technical Digest, September 1999 [Distribution authorized to DOD components only] Also Presentation at Undersea Defense Technology Conference and Exhibition, Sydney, Australia, 7 February 2000
2. T.G. Bell, *Probing the Ocean for Submarines*, Naval Sea Systems Command, Undersea Warfare Center, Division Newport, 28 March 2003 [Distribution authorized to DOD and US DOD contractors only]
3. R.J. Urick, *Principles of Underwater Sound*, 3rd edn. (Peninsula, Los Altos Hills, 1983)
4. J.W. Horton, *Fundamentals of Sonar*, 2nd edn. (U.S. Naval Institute, Annapolis, 1959)

5. W.O. Pennell, M.H. Hebb, H.A. Brooks et al., *Directivity Patterns of Sound Sources*, NDRC C4 – sr287-089, Harvard Underwater Sound laboratory, April 29, 1942; Reference in Chapter 5 of NDRC, Div 6, Vol. 13, 1946
6. W.S. Burdic, *Underwater Acoustic System Analysis*, 2nd edn. (Prentice Hall, Englewood Cliffs, 1991)
7. W. Thompson Jr., Directivity of a uniform-strength, continuous circular-arc source phased to the spatial position of its diameter. J. Acoust. Soc. Am. **105**, 3078–3082 (1999)
8. A. Zielinski, L. Wu, A novel array of ring radiators. IEEE J. Ocean. Eng. **16**, 136–141 (1991)
9. D. Stansfield, *Underwater Electroacoustic Transducers* (Bath University Press, Bath, 1990). Fig. 6.11
10. Y.L. Chow, On grating plateaux of nonuniformly spaced arrays. IEEE Trans. Antennas Propag. **13**(2), 208–215 (1965)
11. A. Ishimaru, Theory of unequally spaced arrays. IRE Trans. Antennas. Propag. **10**(6), 691–702 (1962)
12. F.J. Pompei, S.C. Wooh, Phased array element shapes for suppressing grating lobes. J. Acoust. Soc. Am. **111**, 2040–2048 (2002)
13. L.E. Kinsler, A.R. Frey, A.B. Coppens, J.V. Sanders, *Fundamentals of Acoustics*, 4th edn. (Wiley, New York, 2000)
14. J.L. Butler, A.L. Butler, *A Directional Power Wheel Cylindrical Array*, ONR 321 Maritime Sensing (MS) Program Rev. (NUWC, Newport, 18 August 2005)
15. J.L. Butler, C.H. Sherman, Acoustic radiation from partially coherent line sources. J. Acoust. Soc. Am. **47**, 1290–1296 (1970)
16. C.H. Sherman, Analysis of acoustic interactions in transducer arrays. IEEE Trans. Sonics Ultrason. **SU-13**, 9–15 (1966)
17. D.T. Porter, *NUSC Train of Computer Programs for Transmitting Array Prediction*, Naval Underwater Systems Center Technical Document 8159 (26 January 1988)
18. TRN, Transducer Design and Array Analysis Program, NUWC, Newport, RI. Developed by M. Simon, K. Farnham with array analysis module based on the program ARRAY, by J.L. Butler (Image Acoustics, Cohasset)
19. D.L. Carson, Diagnosis and cure of erratic velocity distributions in sonar projector arrays. J. Acoust. Soc. Am. **34**, 1191–1196 (1962)
20. R.S. Woollett, *Sonar Transducer Fundamentals* (Naval Underwater Systems Center, Newport, n.d.), p. 147
21. A.L. Butler, J.L. Butler, *Ultra Wideband Active Acoustic Conformal Array Module*, ONR 321MS Program Review (Naval Undersea Warfare Center, Newport, 17–20 May 2004)
22. J. Zimmer, *Submarine Hull Mounted Conformal Array Employing BBPP Technology*, ONR 321MS Program Review (Naval Undersea Warfare Center, Newport, 17–20 May 2004)
23. R.L. Pritchard, Mutual acoustic impedance between radiators in an infinite rigid plane. J. Acoust. Soc Am. **32**, 730–737 (1960)
24. I. Wolff, L. Malter, Sound radiation from a system of vibrating circular diaphragms. Phys. Rev. **33**, 1061 (1929)
25. S.J. Klapman, Interaction impedance of a system of circular pistons. J. Acoust. Soc. Am. **11**, 289–295 (1940)
26. H. Stenzel, *Leitfaden zur Berechnung von Schallvorgangen* (Springer, Berlin, 1939)
27. R.L. Pritchard, Tech. Memo. No. 21, Appendix C, Acoustics Research Laboratory, Harvard Univ., NR-014-903 (15 January 1951)
28. C.J. Bouwkamp, A contribution to the theory of acoustic radiation. Phillips Research Reports **1**, 262–277 (1946)
29. W. Thompson Jr., The computation of self and mutual radiation impedances for annular and elliptical pistons using Bouwkamp's integral. J. Sound Vib. **17**, 221–233 (1971)
30. E.M. Arase, Mutual radiation impedance of square and rectangular pistons in a rigid infinite baffle. J. Acoust. Soc. Am. **36**, 1521–1525 (1964)
31. J.L. Butler, *Self and Mutual Impedance for a Square Piston in a Rigid Baffle* (Image Acoustics, Contract N66604-92-M-BW19, Cohasset, 20 March 1992)
32. W.J. Toulis, Radiation load on arrays of small pistons. J. Acoust. Soc. Am. **29**, 346–348 (1957)

33. C.H. Sherman, Mutual radiation impedance of sources on a sphere. J. Acoust. Soc. Am. **31**, 947–952 (1959)
34. J.E. Greenspon, C.H. Sherman, Mutual radiation impedance and near field pressure for pistons on a cylinder. J. Acoust. Soc. Am. **36**, 149–153 (1964)
35. D.H. Robey, On the radiation impedance of an array of finite cylinders. J. Acoust. Soc. Am. **27**, 706–710 (1955)
36. F.B. Stumpf, F.J. Lukman, Radiation resistance of magnetostrictive-stack transducer in presence of second transducer at air-water surface. J. Acoust. Soc. Am. **32**, 1420–1422 (1960)
37. W.J. Toulis, Mutual coupling with dipoles in arrays. J. Acoust. Soc. Am. **37**, 1062–1063 (1963)
38. F.B. Stumpf, Interaction radiation resistance for a line array of two and three magnetostrictive-stack transducers at an air-water surface. J. Acoust. Soc. Am. **36**, 174–176 (1964)
39. C.H. Sherman, Theoretical model for mutual radiation resistance of small transducers at an air-water surface. J. Acoust. Soc. Am. **37**, 532–533 (1965)
40. D.T. Porter, Self and mutual radiation impedance and beam patterns for flexural disks in a rigid plane. J. Acoust. Soc. Am. **36**, 1154–1161 (1964)
41. K.C. Chan, Mutual acoustic impedance between flexible disks of different sizes in an infinite rigid plane. J. Acoust. Soc. Am. **42**, 1060–1063 (1967)
42. C.H. Sherman, *General Transducer Array Analysis*, Parke Mathematical Laboratory Report No. 6, Contract N00014-67-C-0424 (February 1970)
43. P.M. Morse, K.U. Ingard, *Theoretical Acoustics* (McGraw Hill, New York, 1968)
44. W. Thompson Jr., Acoustic coupling between two finite-sized spherical sources. J. Acoust. Soc. Am. **62**, 8–11 (1977)
45. W. Thompson Jr., Radiation from a spherical acoustic source near a scattering sphere. J. Acoust. Soc. Am. **60**, 781–787 (1976)
46. R.T. Beyer, *Nonlinear Acoustics* (US Government Printing Office, Washington, DC, 1975)
47. P.J. Westervelt, Parametric acoustic array. J. Acoust. Soc. Am. **35**, 535–537 (1963)
48. M.B. Moffett, R.H. Mellen, Model for parametric acoustic sources. J. Acoust. Soc. Am. **61**, 325–337 (1977)
49. H.O. Berktay, D.J. Leahy, Farfield performance of parametric transmitters. J. Acoust. Soc. Am. **55**, 539–546 (1974)
50. M.B. Moffett, R.H. Mellen, On parametric source aperture factors. J. Acoust. Soc. Am. **60**, 581–583 (1976)
51. M.B. Moffett, R.H. Mellen, Nearfield characteristics of parametric acoustic sources. J. Acoust. Soc. Am. **69**, 404–409 (1981)
52. M.B. Moffett, R.H. Mellen, Effective lengths of parametric acoustic sources, J. Acoust. Soc. Am. **70**, 1424–1426 (1981). See also "Erratum", **71**, 1039 (1982)
53. M.B. Moffett, W.L. Konrad, *Nonlinear Sources and Receivers, Encyclopedia of Acoustics*, vol. 1 (Wiley, New York, 1997), pp. 607–617
54. M.B. Moffett, H.C. Robinson, *User's Manual for the CONVOL5 Computer Program*, NUWC-NPT Technical Document 11,577 (25 October 2004)
55. P.H. Rogers, A.L. Van Buren, A.O. Williams Jr., J.M. Barber, Parametric detection of low-frequency acoustic waves in the nearfield of an arbitrary directional pump transducer. J. Acoust. Soc. Am. **56**, 528–534 (1974)
56. M.B. Moffett, W.L. Konrad, J.C. Lockwood, A saturated parametric acoustic receiver. J. Acoust. Soc. Am. **66**, 1842–1847 (1979)
57. J.L. Butler, A.L. Butler, M.J. Ciufo, Doubly steered array of modal transducers. J. Acoust. Soc. Am. **132**, 1985(A) (2012)

Chapter 8
Hydrophone Arrays

The goal of both passive and active sonar systems is reliable long-range detection and ranging capability, but the basic considerations that influence performance of the two types of sonar are quite different. The receiving array in passive systems such as towed arrays or wide aperture ranging arrays must be able to detect signals with unknown frequency content, and therefore must operate over a frequency band much greater than the band of a typical active system. And they must do so in the presence of interfering noise. Chapter 6 shows that there are many ways to design hydrophones with adequate broad band sensitivity that are small, lightweight, and inexpensive compared to the high power projectors needed for active sonar. But the main problem in passive sonar is control of the interfering noise, especially in ship-mounted arrays.

In small active sonar systems, such as depth sounders or fish finders, it is quite feasible to use the projector array as the receiver, since the frequency of the received echoes is known to be the same as the transmitted frequency or only slightly Doppler shifted. But in many naval active sonar systems where high performance is critical there is much to be gained by use of a separate receiving array. A separate receiving array, either hull mounted or towed, can be made larger than the projector array and, therefore, has more directionality to aid in bearing determination and noise discrimination; it can also be designed to include noise reduction measures which may not be compatible with a projector array. When separate passive and active sonar systems are available, the best results are sometimes achieved by using the passive array, rather than the active array, to receive the echoes from active transmissions [1].

Array considerations such as size, geometrical arrangement, phasing, and shading will be discussed in Sects. 8.1 and 8.2. They are important for reducing noise, but adequate noise control often requires broadening the concept of a receiving array to include components other than the hydrophones, such as the inner and outer decouplers. The subject of underwater noise characterization and control is too large [1–5] to be covered thoroughly here, but we will indicate the important

J.L. Butler, C.H. Sherman, *Transducers and Arrays for Underwater Sound*,
Modern Acoustics and Signal Processing, DOI 10.1007/978-3-319-39044-4_8

characteristics of the major types of noise that interfere with reception of underwater sound and briefly discuss the major methods of noise reduction.

Part of the noise that limits receiving array capability is internal noise arising from thermal agitation in the hydrophone materials. This noise is a transducer characteristic that was discussed in Sect. 6.7 as part of hydrophone design; it is incoherent and too small to be important in most ship-mounted arrays, but could be important in fixed or drifting arrays at very low frequency where it could exceed ambient noise. Ship-mounted hydrophone arrays must contend with several other types of noise in addition to the ambient noise: flow noise resulting from motion through the water, structural noise excited by flow and by machinery, and acoustic noise from the propulsor. Ambient noise can be reduced by array directivity, using knowledge of the noise correlation and noise directionality to optimize array design. Flow noise can be reduced by hydrophone features such as size, and by separating the hydrophones from the flow. This separation can be achieved either by placing the hydrophones inside a water-filled dome or covering them on the outside with an acoustically transparent layer (the outer decoupler) analogous to a wind screen on a microphone. Structural noise comes from the inside and can be controlled by layers of material that both absorb noise and reflect it away from the array (the inner decoupler). Structural noise might also be controlled to some extent by ship design, but that approach is usually incompatible with other ship functions.

The problems of noise control and the weight associated with the inner and outer decouplers required for large arrays of pressure sensors have led to consideration of arrays of vector sensors. Since these sensors respond to acoustic particle velocity or acoustic intensity, rather than pressure, the relevant noise characteristics are different and might be easier to control in some cases. Several types of individual vector sensors were described in Sect. 6.5, and in Sect. 8.5 vector sensor arrays will be discussed.

Before beginning quantitative discussion of hydrophone array design and noise control we will qualitatively describe several types of arrays. One way of controlling noise is to remove the array from the vicinity of the noise sources. This is usually done in hull-mounted arrays when they are installed near the bow of a ship as far from the propulsor and the propulsion machinery as possible, but this approach can be carried much further by towing an array behind the ship. Towing is especially feasible for line receiving arrays that can be made thin to reduce drag and flexible to facilitate retrieval and onboard storage. Such arrays can also be made long enough to achieve high directionality at low frequency in the horizontal plane, and the beam can be steered for searching in that plane.

In towed arrays, the hydrophones, cables, preamps, and interior fluid are encased in a flexible hose that maintains the hydrophone spacing and waterproofs the whole assembly. If vector sensors are used, the hose and interior structure must also be able to maintain the individual sensor orientation, otherwise the orientation must be measured. Towed arrays can be completely removed from the flow and structural noise of the towing ship, and, at a sufficient distance, they are also removed from the ship's radiated noise when the acoustic beams are steered away from the ship. But the flow produces turbulent pressure fluctuations on the outside of the hose and excites vibrations in the hose wall and the interior components, especially for small diameter arrays. Some of the hydrophone materials discussed in Chap. 6, such as 0–3 ceramic–polymer composites and piezoelectric polymers such as PVDF, can be used for long, slender, flexible hydrophones to reduce high wavenumber noise in towed arrays by area averaging.

Passive arrays can be used to determine the bearing of a sound source, but they cannot determine the range of the source, unless more than one array is used. Two arrays can determine range by triangulation if their separation is sufficient to give a reliable estimate of the difference in bearing at the two array locations. Three arrays can determine range by measuring phase differences that determine the curvature of an incoming wave front, which can be done with separations small enough to fit three arrays on one submarine. However, there are several critical requirements for successful wave front curvature ranging: precise positioning of the arrays, identical characteristics for each of the three arrays, sufficient distance between the arrays, and good noise control. Maximizing the separation between arrays unavoidably places the aft array near the machinery and propulsion noise sources and also near structural features that differ from those near the mid and forward arrays. Dealing with these conflicting requirements involves detailed theoretical and experimental analysis of the expected noise characteristics and optimum use of all noise control measures [1].

Passive arrays mounted on torpedoes face special problems, because the small size and high speed of the vehicle makes it difficult to achieve sufficient aperture and to separate the array from structural noise sources. Since the range requirements for torpedo operations are modest, higher frequencies that are compatible with the small size of the vehicle can provide effective sonar performance.

Arrays of drifting sensors and arrays for autonomous vehicles are other special cases with individual requirements. Robust array design is especially important for long towed arrays, fixed arrays, and drifting arrays. Means of detecting, locating, and compensating for individual hydrophone failures are important, as well as array configurations that minimize sensitivity to individual failures.

8.1 Hydrophone Array Directional Response

8.1.1 Directivity Functions

Hydrophones are used mainly at frequencies below their fundamental resonance where their dimensions are small compared to the acoustic wavelength and their response is omnidirectional. To obtain the high degree of directionality required for determining the bearing of a received signal, it is usually necessary to use a large array with dimensions of at least several wavelengths. The hydrophone spacing must be less than one half wavelength at the highest frequency to avoid grating lobes in steered arrays, unless unequal spacing is used (see Sects. 7.1.3 and 8.1.2). Thus large arrays must contain a large number of hydrophones.

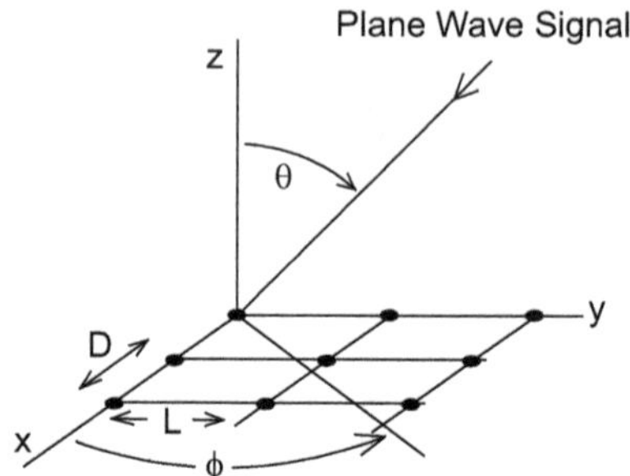

Fig. 8.1 Plane wave signal incident from direction θ, ϕ on a rectangular array of hydrophones lying in the x, y plane

Many of the results in Sect. 7.1 for the far-field directivity patterns of projector arrays also apply to hydrophone arrays because of acoustic reciprocity (see Sect. 11.2.2). In the receiving case a plane sound wave approaches the array from an unknown direction and exerts a force on each hydrophone. The force is converted by the transduction mechanism in the hydrophone to a voltage and detected by the electronic components of the system. The main beam of the directivity pattern must be steered through all directions of interest to determine the presence of a signal and its bearing. To analyze array performance the signal will be assumed to be a plane acoustic wave of amplitude p_i arriving from the direction given by the angles θ and ϕ as shown in Fig. 8.1.

The plane wave can be expressed as

$$p(x, y, z)\mathrm{e}^{j\omega t} = p_\mathrm{i}\mathrm{e}^{j\left(k_x x + k_y y + k_z z\right)}\mathrm{e}^{j\omega t} = p_\mathrm{i}\mathrm{e}^{j\left(\vec{k}\cdot\vec{r} + \omega t\right)},$$

where $\vec{r} = \hat{\imath}x + \hat{\jmath}y + \hat{k}z$ is the position vector and $\vec{k}$ is the wavevector with components

$$k_x = k\sin\theta\cos\phi,\ \ k_y = k\sin\theta\sin\phi,\ \text{ and }\ k_z = k\cos\theta,$$

and magnitude $k = \omega/c = \left[k_x^2 + k_y^2 + k_z^2\right]^{1/2}$, where k is called the wave number and c is the sound speed in water. Later we will encounter waves traveling with speeds that differ from c, e.g., flexural waves in a plate with speed c_p and wave number ω/c_p. We will consider first the directional response to acoustic waves and use the wavevector components above in terms of the angles that define the direction of the plane wave. In Sect. 8.1.4 we will discuss the more general wavevector response in terms of the components k_x, k_y, and k_z.

The acoustic signal is assumed to produce the same pressure amplitude at each hydrophone in an array, but the phases vary depending on the hydrophone location and the direction of arrival of the signal. The pressure on a single hydrophone located at x, y in a planar array lying in the plane $z = 0$, as shown in Fig. 8.1, is

$$p(x, y, 0) = p_i e^{jk \sin\theta(x\cos\phi + y\sin\phi)}. \tag{8.1}$$

The voltage output from the hydrophone at x, y when added to the outputs of all the other hydrophones gives the array output for any geometrical arrangement of any number of hydrophones in the planar array.

For a rectangular array of N by M identical, omnidirectional hydrophones with x- and y-spacings D and L, as shown in Fig. 8.1, the hydrophone designated by n, m is located at $x = nD, y = mL$ where $n = 0, 1, 2, \ldots, N-1$ and $m = 0, 1, 2, \ldots, M-1$, and Eq. (8.1) gives the pressure on that hydrophone as

$$p_{nm} = p_i e^{jk \sin\theta(nD\cos\phi + mL\sin\phi)}. \tag{8.2}$$

The sum of the individual hydrophone outputs is the simplest form of the array output. It gives the unsteered, unshaded beam, and when normalized by dividing by the number of hydrophones and expressed as a voltage, by use of the effective hydrophone sensitivity M_0, it is:

$$V_A(\theta, \phi) = \frac{1}{NM}\sum_{n,m} V_{nm} = \frac{M_0}{NM}\sum_{n,m} p_{nm} = \frac{M_0}{NM} p_i \sum_{n=0}^{N-1} e^{jknD\sin\theta\cos\phi} \sum_{m=0}^{M-1} e^{jkmL\sin\theta\sin\phi}. \tag{8.3}$$

The voltage output of the hydrophone at the location n, m is $V_{nm} = M_0 p_{nm}$, where M_0 is the product of the diffraction constant and the low frequency sensitivity as discussed in Sect. 6.6. The individual voltage outputs usually go to a preamp and to a phase shifter for steering the beam before all the channels are combined (see Sect. 8.1.2). Equation (8.3) can easily be generalized to include individual hydrophone directivity by using the product theorem (see Sect. 7.1.1) and multiplying by the hydrophone directivity function.

The summations in Eq. (8.3) are geometric series identical to those in Eq. (7.5), showing that the receiving array beam pattern is the same as the projector array beam pattern in Eq. (7.6) as required by acoustic reciprocity. The summations combine all the phase information from the individual hydrophones to give the magnitude of the array voltage output as a function of the direction of the incident wave as follows:

$$|V_A(\theta, \phi)| = [p_i M_0][\sin(NX)/N\sin(x)][\sin(MY)/M\sin(Y)], \tag{8.4a}$$

where

$$X = (kD/2)[\sin(\theta)\cos(\phi)] \text{ and } Y = (kL/2)[\sin(\theta)\sin(\phi)].$$

As in Sect. 7.1 the second and third factors in Eq. (8.4a) give the normalized patterns in planes parallel to the sides of the array for $\phi = 0°$ and $90°$. Each of these factors is the pattern of a line array; thus the pattern of the rectangular array is the product of two line array patterns, a result that is sometimes called the second product theorem [6]. Other array geometries are often used, and their directional characteristics can be determined from Eq. (8.1). For example, the results for circular projector arrays are also applicable to hydrophone arrays.

For hydrophone spacing of about one half wavelength ($\lambda/2$), or less, the beam width and directivity factor approximations for large, continuous line and rectangular radiators (see Sect. 10.2) can be used for discrete arrays with ND the length of the line and NMDL the area of the rectangle. The approximate directivity factors are:

$$D_{\mathrm{f}} \approx 2\mathrm{ND}/\lambda, \quad \text{for a line of } N \text{ hydrophones},$$

$$D_{\mathrm{f}} \approx 4\pi \mathrm{NMDL}/\lambda^2, \quad \text{for a rectangle of NM hydrophones},$$

and the directivity indices are $\mathrm{DI} = 10 \log D_{\mathrm{f}}$. Better approximations for these directivity factors are given by Burdic [7] and Horton [8].

Equations (8.3) and (8.4a) assume that the incident signal pressure and the scattered pressure, which is accounted for by the diffraction constant, are the only pressure fields acting on each hydrophone in the array. However, the incident pressure makes each hydrophone vibrate and radiate a pressure field that interacts, via the mutual radiation impedances, with all the other hydrophones in the array as discussed in Chap. 7. But these interactions are not important unless the hydrophones are being used near their resonance frequency.

The array output for a plane array with a continuous, separable distribution of sensitivity per unit area, $m_1(x_0)m_2(y_0)$, can be written in terms of Fourier transforms:

$$V_{\mathrm{A}} = p_i \iint_S m_1(x_0) m_2(y_0) \mathrm{e}^{jk \sin\theta (x_0 \cos\phi + y_0 \sin\phi)} \mathrm{d}x_0 \mathrm{d}y_0, \tag{8.5a}$$

or as

$$V_{\mathrm{A}} = p_i \int_{-\infty}^{\infty} m_1(x_0) \mathrm{e}^{jkx_0 \sin\theta \cos\phi} \mathrm{d}x_0 \int_{-\infty}^{\infty} m_2(y_0) \mathrm{e}^{jky_0 \sin\theta \sin\phi} \mathrm{d}y_0. \tag{8.5b}$$

Thus the directivity pattern of a plane or line array is the Fourier transform of the sensitivity function. Equation (8.5b) shows again that the pattern function of the array is the product of two separate pattern functions of which Eq. (8.4a) is an example. This Fourier transform relationship between pattern functions and sensitivity functions, which was first recognized by Michelson [9], allows patterns to be calculated easily in some cases using known Fourier transforms.

8.1.2 Beam Steering

Progressive phase shifting (or time delaying) of the hydrophone outputs steers a received beam just as phase shifting the applied voltages steers a projected beam (see Sect. 7.1.4). A necessary part of passive sonar search is continually steering the beam, or simultaneously forming multiple beams, to cover all the directions of interest. The major beam of the array can be steered to the direction θ_0, ϕ_0 by phase shifting the output of the *nm*th hydrophone by the amount

$$\mu_{nm} = -nkD \sin\theta_0 \cos\phi_0 - mkL \sin\theta_0 \cos\phi_0.$$

It can be seen from Eq. (8.3) that these phase shifts cause the exponents to vanish when $\theta = \theta_0$ and $\phi = \phi_0$ which makes all the hydrophone outputs add in phase and form the major beam in that direction. With these phase shifts Eq. (8.4a) becomes

$$|V_A(\theta,\phi)| = [p_i M_0] \left|\frac{\sin NX}{N \sin X}\right| \left|\frac{\sin MY}{M \sin Y}\right|, \tag{8.6}$$

where

$$X = \frac{kD}{2}[\sin\theta\cos\phi - \sin\theta_0\cos\phi_0],$$

and

$$Y = \frac{kL}{2}[\sin\theta\sin\phi - \sin\theta_0\sin\phi_0].$$

The main beam and side lobes change significantly as the beam is steered resulting in a different directivity factor for each steering direction.

For an array on a curved surface the output of each hydrophone can be phase shifted to make the array directional response approximate that of a planar array tangent to the curved array and given by Eq. (8.4a). The phase shifts must be proportional to the perpendicular distance from each hydrophone to the plane; e.g., for a cylindrical array of radius a each vertical column (or stave) of hydrophones must be phase shifted by $ka(1 - \cos\beta_i)$, where β_i is the angle from the center of the array to the ith hydrophone (see Fig. 8.2).

The same phasing to a plane can be done for arrays of any shape, such as those that conform to the hull of a ship. Cylindrical arrays intended for passive search usually have staves of hydrophones extending all, or most, of the way around the circumference of the cylinder. Usually about 120° of the cylindrical surface is used for each beam. For more than 120° the inverse shading resulting from the projected hydrophone locations would raise the side lobes and need to be compensated by additional shading. The acoustic beam from one set of staves can be steered in

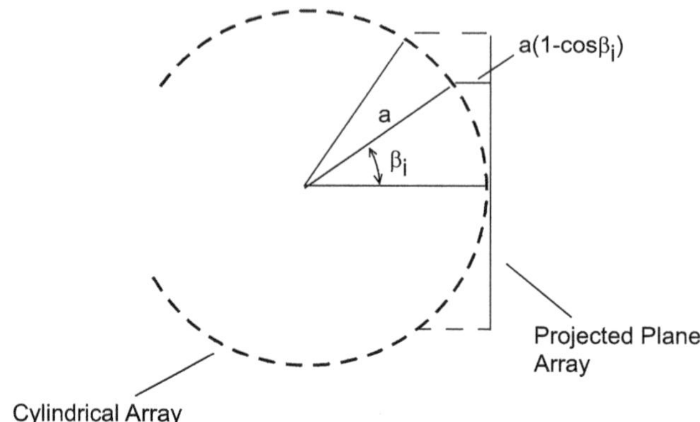

Fig. 8.2 Top view of cylindrical array with 13 staves being phased to a plane to form a beam

azimuth by electrically switching to another set of staves, displaced by one or more staves. The beam can also be steered in depression angle by phase shifting as would be done for a plane array. When a spherical array is phased to a plane, beam steering can be done in both azimuth and depression angle by switching to different sets of hydrophones.

8.1.3 *Shading and Directivity Factor*

Shading a receiving array means adjusting the amplitudes of the hydrophone outputs to achieve a desired change in the directivity pattern. Since the side lobes allow noise and reverberation to mix with the signal from the steered direction, the reduction of side lobes is usually the main objective of shading. Shading an array such that the sensitivity decreases from the center to the sides, results in lower side lobe levels and a wider main beam, compared to a uniformly shaded array. Shading that increases the sensitivity from the center toward the sides, called inverse shading, increases the side lobes and decreases the main beam width. Thus, each application requires consideration of whether main beam width or relative side lobe level is more important [10]. Shading is accomplished by multiplying the magnitude of the signal in each channel by a shading coefficient, a_{nm}. Then the shaded array output, including progressive phasing to steer the beam, is a generalization of Eq. (8.3) which can be written as

$$V_{\mathrm{A}}(\theta,\phi) = \frac{M_0 p_i}{NM}\sum_{n=0}^{N-1}\sum_{m=0}^{M-1} a_{nm} e^{j[knD(\sin\theta\cos\phi - \sin\theta_0\cos\phi_0) + kmL(\sin\theta\sin\phi - \sin\theta_0\sin\phi_0)]}. \tag{8.7a}$$

For a line array on the x-axis, i.e., for $M = 1$, $a_{n0} = a_n$, and $\phi = \phi_0 = 0$, Eq. (8.7a) becomes

$$V_{\mathrm{A}}(\theta) = \frac{M_0 p_{\mathrm{i}}}{N} \sum_{n=0}^{N-1} a_n \mathrm{e}^{jknD(\sin\theta - \sin\theta_0)}. \tag{8.7b}$$

Neither of these expressions can be summed in a simple analytical way as was done for Eq. (8.3), but they can be written in other ways that are more convenient for some purposes. For example, consider a line array containing an odd number of hydrophones and, except for the center one, consider them in pairs consisting of the two adjacent to the center, the next two beyond them, etc. If the array is symmetrically shaded the shading coefficients associated with each pair are the same, and the response of each pair is proportional to $2\cos[mkD(\sin\theta - \sin\theta_0)]$. Then Eq. (8.7b) can be written as

$$V_{\mathrm{A}}(\theta) = \frac{M_0 p_{\mathrm{i}}}{(2M+1)} \sum_{m=0}^{M} \epsilon_m a_m \cos[mkD(\sin\theta - \sin\theta_0)], \tag{8.7c}$$

where the number of hydrophones, N, is now written as $2M + 1$, and $\epsilon_0 = 1$, $\epsilon_m = 2$ for $m > 0$. A similar expression holds for an array containing an even number of hydrophones. Pritchard [11] used these expressions to calculate the directivity factor of line arrays in series form. Thompson [12] used similar expressions to synthesize large array directivity patterns with certain features by squaring, cubing, etc. the directivity patterns of small arrays with the same features.

Pritchard [11] and Davids et al. [10] gave detailed discussions of optimum shading of receiving arrays based mostly on the Dolph method [13] of shading using Chebyshev polynomials. This method is optimum in the sense that it gives the narrowest main lobe possible for a line array of equally spaced point hydrophones when all side lobes are made equal at a specified level relative to the main lobe. Or, equivalently, for a specified main lobe width the method gives the lowest possible side lobes with equal relative levels. Pritchard's results showed the relationships among main lobe width, relative side lobe level, directivity index, and number and spacing of hydrophones. Urick [14] compares the Dolph–Chebyshev shading with several other types of shading. Albers [6] describes the Dolph procedure in detail with numerous illustrations. Other methods of shading are also used; e.g., binomial shading, in which the amplitudes are proportional to the coefficients of a binomial expansion, gives the narrowest main lobe possible with no side lobes at all; Gaussian shading does the same. Taylor shading [15] is a modification of Dolph–Chebyshev shading in which the outermost side lobes roll off smoothly. Wilson [16] gives a clear comparison of the effects of varying the shading, the spacing, the phasing, and the number of hydrophones. Figures 8.3, 8.4, 8.5, 8.6, and 8.7 give examples of steering and shading a 12 hydrophone line array with $\lambda/2$ spacing with self-explanatory captions.

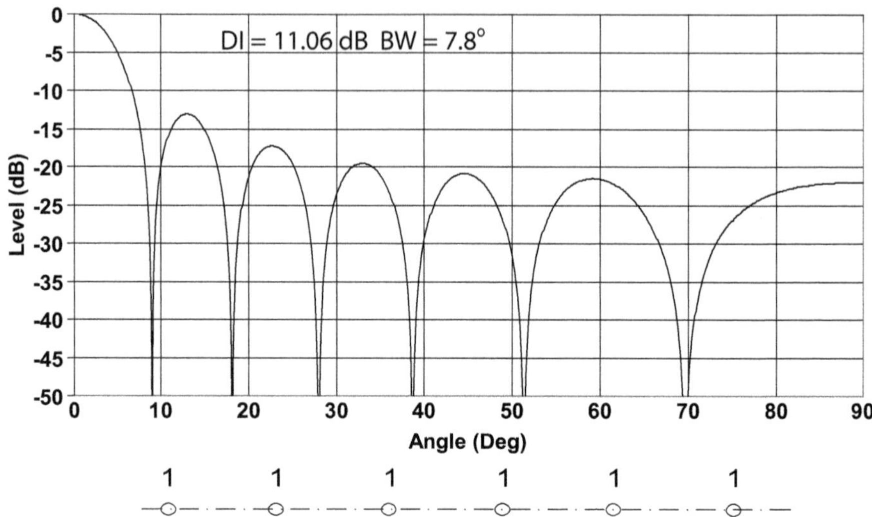

Fig. 8.3 Unshaded 12 element $\lambda/2$ spaced array

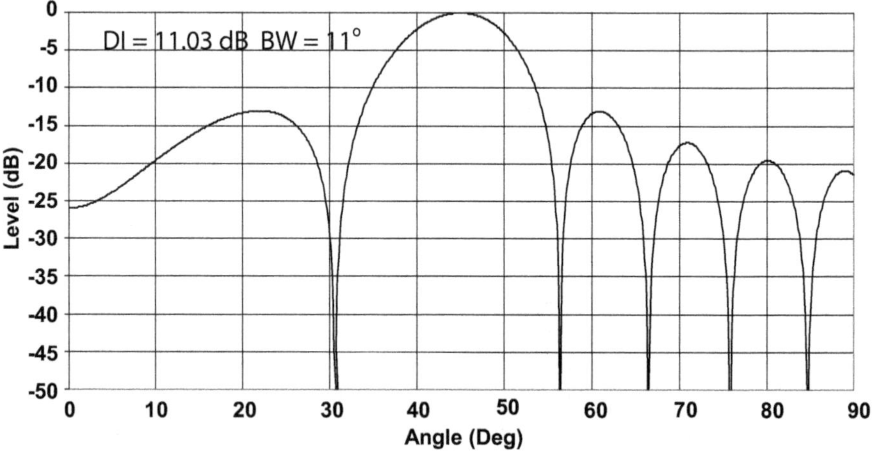

Fig. 8.4 Unshaded 12 element $\lambda/2$ spaced array steered to 45°

Pritchard [17] also gave a simple approximation for the directivity factor of a steered line array of $2M + 1$ uniformly spaced and symmetrically shaded hydrophones based on Eq. (8.7c). It can be written as

$$D_{\mathrm{f}} = (2D/\lambda)/\sum_{m=0}^{M} \in_m b_m^2, \tag{8.8a}$$

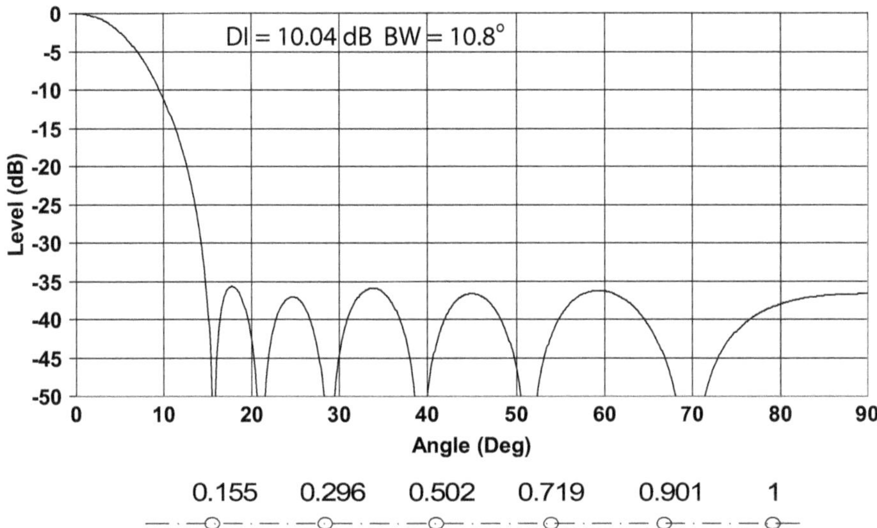

Fig. 8.5 Dolph–Chebyshev shaded 12 element $\lambda/2$ spaced array, for −37 dB side lobes [10]. Coefficients: 1, 0.901, 0.719, 0.502, 0.296, 0.155

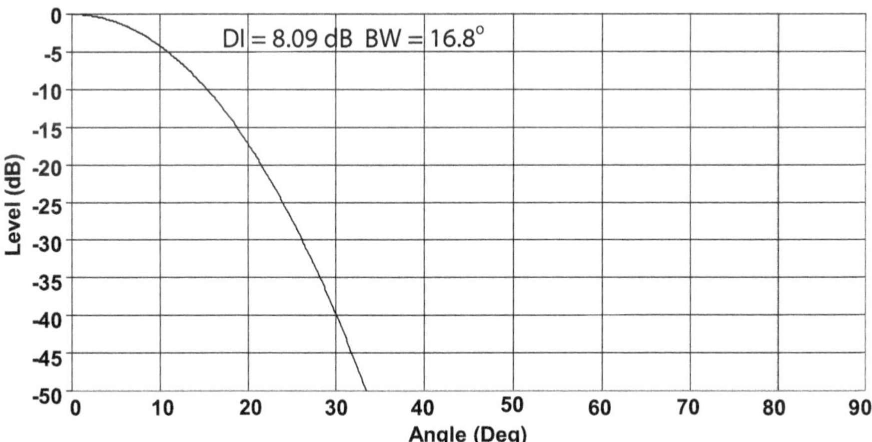

Fig. 8.6 12 element $\lambda/2$ spaced array with binomial shading. Coefficients: 1, 0.714, 0.357, 0.119, 0.024, 0.002

where the b_m are the symmetric shading coefficients normalized such that $\sum_{m=0}^{M} \in_m b_m = 1$. For a uniform array (i.e., no shading) the normalization condition makes all the $b_m = (2M+1)^{-1}$ and $D_{\mathrm{f}} = (2D/\lambda)(2M+1) \approx 2L/\lambda$, where L is the length of the line array. Since the unsteered line array beam has an omnidirectional toroidal shape in the plane perpendicular to the array, the main beam shape changes

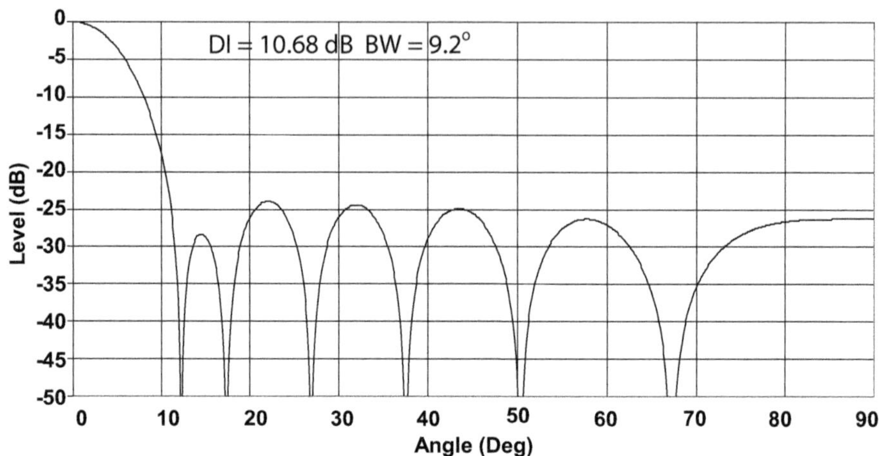

Fig. 8.7 12 element $\lambda/2$ spaced Taylor shading for 25 dB side lobes. Coefficients: 1, 0.91, 0.77, 0.58, 0.44, 0.40

drastically as the beam is steered, from toroidal to conical to a broad searchlight beam at end-fire. For moderate steering D_f does not change much and Eq. (8.8a) is a useful approximation, but at end-fire D_f is approximately twice the value at broadside, i.e., twice the value given by Eq. (8.8a) [7, 17].

The approximation in Eq. (8.8a) is convenient for quick estimates, but with current computers Eqs. (8.7a) and (8.7b) can be used to calculate complete directivity patterns and accurate directivity factors. For example, combining acoustic reciprocity with the definitions of directivity factor in Eq. (1.20) and intensity in Eq. (7.19), and using Eq. (8.7b), the directivity factor of a line array of N shaded omnidirectional hydrophones steered to the direction θ_0 can be obtained in the following form for convenient computation:

$$D_f(\theta_0) = \frac{\sum_{n=0}^{N-1}\sum_{m=0}^{N-1} a_n a_m}{\sum_{n=0}^{N-1}\sum_{m=0}^{N-1} a_n a_m \cos\left[(n-m)kD\sin\theta_0\right]\dfrac{\sin\left[(n-m)kD\right]}{(n-m)kD}}. \tag{8.8b}$$

For half wavelength spacing ($kD = \pi$), Eq. (8.8b) reduces to $\sum_{n=0}^{N-1}\sum_{m=0}^{N-1} a_n a_m / \sum_{n=0}^{N-1} a_n^2$, which does not depend on the steering angle. For no shading (all the a_n equal), Eq. (8.8b) takes the more convenient form with $q = (n - m)$:

$$D_f(\theta_0) = \frac{N}{1 + \frac{2}{N}\sum_{q=1}^{N-1}(N-q)\cos\left(qkD\sin\theta_0\right)\frac{\sin(qkD)}{qkD}}, \tag{8.8c}$$

which reduces to N when $kD = \pi$. This result was given by Burdic [7] as the array gain for isotropic noise (see Sect. 8.2 for discussion of array gain) and by Horton [8] for $\theta_0 = 0$.

The directivity of small hydrophone arrays can be increased significantly by superdirective shading, which requires making some of the shading coefficients negative. Such arrays have a significantly narrower main beam and higher directivity factor than the same array with all positive shading coefficients, but of course the main beam response is reduced. A more general definition of superdirectivity includes any directional radiator that is small compared to the wavelength such as the dipole modes of a sphere or cylinder. Simple examples of superdirectivity are given by a pair of small hydrophones separated by a small distance, $D \ll \lambda$, with the outputs subtracted. Eq. (8.7b) with $\theta_0 = 0, \theta = 90° - \alpha, a_1 = 1, a_2 = -1$ and with the hydrophones located at $D/2$ on each side of the origin gives the effective sensitivity of the pair as $2M_0 \sin[(kD/2)\cos\alpha] \approx M_0 kD \cos\alpha$ for $kD \ll 1$. Thus the directivity pattern is given by $\cos\alpha$, the dipole pattern, which has $D_f = 3$, DI $= 4.8$ dB and sensitivity of M_0kD, which decreases with frequency by 6 dB per octave (see Sect. 6.5.1). If there was a small phase shift of kD between the two hydrophones before subtracting, the result would be $M_0kD(\cos\alpha + 1)$, which is the cardioid pattern (see Fig. 6.30) with $D_f = 3$ and DI $= 6.8$ dB and sensitivity of $2M_0kD$. The same hydrophone pair with the same separation and with outputs added would be nearly omnidirectional with $D_f \approx 1$, DI ≈ 0, and sensitivity of $2M_0$.

Pritchard [11, 18] discussed the maximum directivity of line arrays and showed that applying the Dolph method to the case where the hydrophone spacings are small compared to the wavelength leads to the situation where at least one shading coefficient approaches zero. For spacing less than a quarter wavelength the optimum pattern (in the Dolph sense) requires some of the coefficients to be negative, and such patterns are superdirective. One of Pritchard's specific examples is an array of five hydrophones with $\lambda/8$ spacing, or a total length of $\lambda/2$; the shading coefficients are +5, −6, +24, −16, +5, and the DI is 4.9 dB [11]. Equation (8.8a) applied to a five hydrophone array of total length $\lambda/2$ with uniform shading gives $D_f = 1.25$ and a DI of about 0.9 dB, showing that the superdirective shading increases the DI by about 4 dB. Note the large variation in the superdirective coefficients, and that the sum of all the coefficients is only about 3 % of the value it would have if there were no sign reversals, and the sensitivity is about 30 dB below that for a conventionally shaded array.

The disadvantages of superdirective arrays are a low sensitivity to plane wave signals, high side lobes, narrow band performance, and increased susceptibility to incoherent noise because of the phase reversals. Superdirective arrays have found little use for these reasons. The effect of ambient noise on superdirective arrays will be considered in Sect. 8.4.1.

Grating lobe control is very important in receiving arrays where it is usually desirable to make the bandwidth as broad as possible. Much of the discussion of grating lobe control in Sect. 7.1.3 is also applicable to receiving arrays, but use of the individual hydrophone directivity is quite limited because the hydrophones are usually omnidirectional over most of the band. Nonuniform spacing of the hydrophones, such as the logarithmic array discussed in Sect. 7.1.4, would also be effective in receiving arrays. And it should be reiterated that shading does not reduce grating lobes relative to the main lobe. Grating lobes are also important for high wavenumber noise and will be discussed and illustrated further in the next section from this more general point of view.

8.1.4 Wavevector Response of Arrays

Equation (8.4a) for the response of a plane rectangular array of omnidirectional pressure sensitive hydrophones, with $k \sin\theta \cos\phi$ replaced by k_x and $k \sin\theta \sin\phi$ replaced by k_y, is a more general form of the array response that can also be applied to non-acoustic plane pressure waves with wavevector components k_x and k_y:

$$|V_A(k_x, k_y)| = p_0 M_0(k_x, k_y) \left|\frac{\sin(Nk_x D/2)}{N \sin(k_x D/2)}\right| \left|\frac{\sin(Mk_y L/2)}{M \sin(k_y L/2)}\right|. \tag{8.4b}$$

In Eq. (8.4b) p_0 is the pressure amplitude, and the sensitivity, M_0, depends on k_x and k_y unless the individual hydrophones are small compared to the wavelengths associated with k_x and k_y. Similarly, Eq. (8.6) with $X = (D/2)(k_x - k_{0x})$ and $Y = (L/2)(k_y - k_{0y})$ gives the response when the array is steered to a wavevector with components k_{0x} and k_{0y}. The non-acoustic waves of interest for ship-mounted arrays usually occur as flexural wave noise and flow noise (see Sects. 8.3.2 and 8.3.3) which have lower propagation speeds than acoustic waves. Therefore at a given angular frequency, ω, the noise wavenumbers, ω/speed, are higher than the acoustic wavenumber and the noise wavelengths are smaller.

The wavenumber response of an unsteered line array is illustrated qualitatively in Fig. 8.8a. The major lobe appears at $k_x = 0$, with grating lobes at $k_x = 2\pi n/D$ ($n = \pm 1, \pm 2, \ldots$) i.e., when D is an integral number of wavelengths.

The minor lobes that occur between the grating lobes are omitted for this discussion to simplify the figure. When the major lobe is steered by k_{0x}, all the grating lobes are also shifted by k_{0x}; Fig. 8.8b is a specific example where the steering wavenumber is $k_{0x} = \pi/D$. As the major lobe is steered from $k_{0x} = 0$ to $k_{0x} = \pi/D$, all the grating lobes shift by π/D and may receive noise components that are present with wavenumbers in this range. For example, if noise was present at $k_x = 5\pi/2D$ it would be received on the first grating lobe when the major lobe was steered to $k_{0x} = \pi/2D$. This corresponds to steering the acoustic beam by $\theta_0 = 30^\circ$ at the frequency where D is half the acoustic wavelength, since $k \sin\theta_0 = k_{0x}$. The wavevector response of arrays will be discussed further in Sect. 8.4.2 for two-dimensional arrays and flexural wave noise.

Interest in the wavevector characteristics of noise has led to the development of two special purpose arrays. One is the wavevector filter for measuring the wavevector characteristics of a noise field [5, 19, 20]; the other is a projector

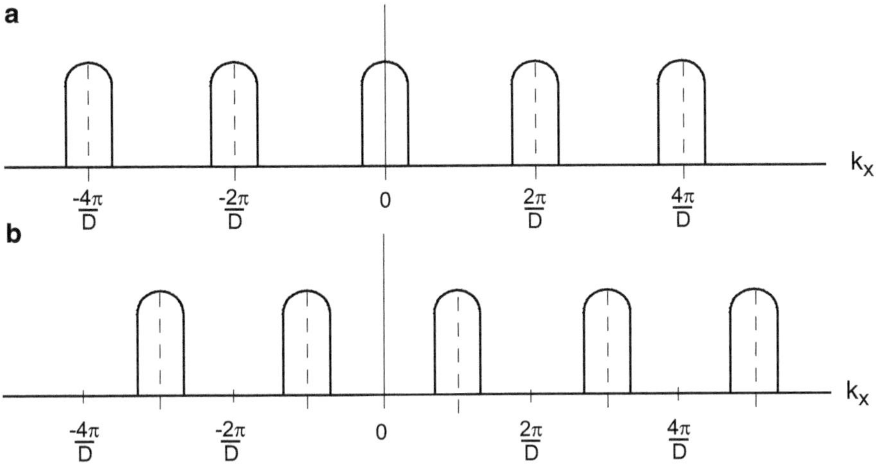

Fig. 8.8 Wave number response of a line array with spacing D between hydrophones. (**a**) unsteered (**b**) steered to acoustic end-fire, $k_{0x} = \pi/D$. The minor lobes between the grating lobes are omitted

array for generating a pressure field with known non-acoustic wave numbers [21]. Wavevector filters have been used to measure the characteristics of the turbulent boundary layer (TBL) [20], to separate TBL noise from acoustic and flexural wave noise [22], and to determine characteristics of flexural waves on experimental hull simulators [1]. The wave number generator has been used to evaluate the properties of baffles and decouplers at known wavenumbers corresponding to noise wavenumbers expected in hull-mounted arrays [1].

8.2 Array Gain

The directivity factor, D_f, and directivity index, $\mathrm{DI} = 10\log D_f$, are defined in Chap. 1 as measures of how well a projector or an array of projectors concentrates acoustic radiation in the direction of the main lobe. Specifically, the directivity factor is the ratio of the radiated intensity in the direction of the main lobe to the radiated intensity averaged over all directions. In receiving arrays similar definitions of directivity factor and directivity index are useful as measures of the array's ability to discriminate against plane waves coming from directions other than the main lobe direction. For receiving arrays the directivity factor that follows from acoustic reciprocity is the ratio of the power response to a plane wave arriving on the main lobe (MRA) to the average power response to plane waves of the same pressure amplitude arriving from all directions (see Sect. 11.2.2). Since the ability of receiving arrays to discriminate against noise is of critical importance, the directivity factor of a receiving array is often expressed differently, as the increase in signal-to-noise ratio of the array over the signal-to-noise ratio of one omnidirectional hydrophone [14]. When the signal is a plane wave arriving on the main lobe and the noise is an isotropic, incoherent mixture of plane waves, this definition is equivalent to the other definition. In the simplest case the array sums the signal coherently and the noise incoherently, resulting in an increase in signal-to-noise ratio. But a more general measure of noise discrimination capability is needed, one that can account for partial coherence in both the noise and the signal. The array gain, AG, defined as 10 times the logarithm of the increase in signal-to-noise ratio of the array over the signal-to-noise ratio of one array hydrophone alone, is such a measure [14]. It can be expressed in terms of the statistical properties of the signal and noise fields, and, like the directivity index, it depends on frequency or on a specified frequency band. The array gain can be written as

$$\mathrm{AG} = 10\log\frac{\langle S^2\rangle/\langle N^2\rangle}{\langle s^2\rangle/\langle n^2\rangle}, \tag{8.9}$$

where $\langle s^2\rangle$ and $\langle n^2\rangle$ are the mean square (i.e., time average of the square) signal and noise outputs of a single hydrophone alone, and $\langle S^2\rangle$ and $\langle N^2\rangle$ are the mean square signal and noise outputs of the array. In this definition it is not necessary for the individual hydrophones to be omnidirectional or to have any other special characteristics, as long as the single hydrophone and the hydrophones in the array are identical. Note that specifying array geometry, steering, and frequency determines the directivity index, but the same array geometry with the same steering and frequency has different array gains depending on the coherence of the signal and noise fields.

The array signal and noise outputs can be written in terms of the single hydrophone outputs as

$$S = \sum_i a_i s_i \quad \text{and} \quad \langle S^2\rangle = \sum_i\sum_j a_i a_j \langle s_i s_j\rangle, \tag{8.10a}$$

$$N = \sum_i a_i n_i \quad \text{and} \quad \langle N^2\rangle = \sum_i\sum_j a_i a_j \langle n_i n_j\rangle, \tag{8.10b}$$

where the a_i are amplitude shading coefficients and the summations extend over all the hydrophones in the array. The quantities $\langle s_i s_j\rangle$ and $\langle n_i n_j\rangle$ are cross correlation functions defined by [23]

$$\langle s_i s_j\rangle = \lim_{T\to\infty}\frac{1}{T}\int_0^T V_i(t)V_j(t-\tau_{ij})\mathrm{d}t,$$

where V_i and V_j are the outputs of the ith and jth hydrophones, which are separated by a distance d_{ij}, and τ_{ij} is a time delay introduced between those outputs. For discussing beam steering of arrays it will be convenient to use phase shifts between

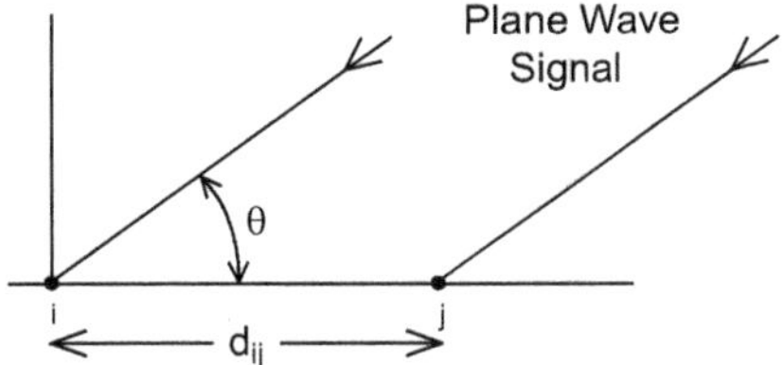

Fig. 8.9 Coordinates for calculation of the cross correlation function for two hydrophones receiving a plane wave signal

the received signals, ϕ_{ij}, rather than time delays. Then the normalized cross correlation function is defined as [14]

$$\rho_{ij}^{s}(d_{ij}, \phi_{ij}) = \frac{\langle s_i s_j \rangle}{\left[\langle s_i^2 \rangle \langle s_j^2 \rangle\right]^{1/2}}. \tag{8.11}$$

For example, when the signal is a plane wave of wavenumber k arriving at the angle θ from the line between the ith and jth hydrophones (see Fig. 8.9) the outputs of the two hydrophones are proportional to $s_i = V_0 e^{j\omega t}$ and $s_j = V_0 e^{j(\omega t + kd_{ij}\cos\theta + \phi_{ij})}$.

Taking the product of the real parts, and time averaging (see Chap. 13.3) gives $\langle s_i s_j \rangle = \frac{1}{2} V_0^2 \cos\left(kd_{ij}\cos\theta + \phi_{ij}\right)$ and $\langle s_i^2 \rangle = \langle s_j^2 \rangle = \frac{1}{2} V_0^2$, which leads to the normalized cross correlation function:

$$\rho_{ij}^{s}(d_{ij}, \phi_{ij}) = \cos\left(kd_{ij}\cos\theta + \phi_{ij}\right). \tag{8.12}$$

When the signal is the sound radiated from a complicated vibrating structure such as a ship, it is often only partially coherent [24]. Partial coherence in the signal can also be caused by random variations in the medium, especially in very large arrays where the transmission path from source to hydrophone is not exactly the same for all hydrophones.

The normalized cross correlation function of the noise received from the ith and jth hydrophones is defined in the same way:

$$\rho_{ij}^{n}(d_{ij}, \phi_{ij}) = \frac{\langle n_i n_j \rangle}{\left[\langle n_i{}^2 \rangle \langle n_j^2 \rangle\right]^{1/2}}. \tag{8.13}$$

Examples of specific noise fields will be given in Sect. 8.3.1. Urick [14] gives cross correlation functions for single frequency and flat bandwidth plane wave signals and isotropic noise. The normalized cross correlation functions will be called spatial correlation functions or just spatial correlations in the following.

The signal and noise fields are usually homogeneous over the surface of an array, and therefore have the same mean square values at the location of every hydrophone in the array; thus we assume that

$$\langle s_i^2 \rangle = \langle s^2 \rangle, \quad i = 1, 2, \ldots, \tag{8.14a}$$

$$\langle n_i^2 \rangle = \langle n^2 \rangle, \quad i = 1, 2, \ldots. \tag{8.14b}$$

With this simplification the array gain can be written in terms of the spatial correlation functions as follows:

$$\mathrm{AG} = 10 \log \frac{\sum_i \sum_j a_i a_j \rho_{ij}^{\mathrm{s}}}{\sum_i \sum_j a_i a_j \rho_{ij}^{\mathrm{n}}}. \tag{8.15}$$

The signal can usually be considered completely coherent, and in some cases the noise can be considered completely incoherent, e.g., the internal hydrophone noise is completely incoherent. Since this is a useful case for comparison with other situations, we consider an unshaded ($a_i = 1$) planar array with no beam steering ($\phi_{ij} = 0$) and a plane wave signal arriving from the broadside direction ($\theta = 90°$); then for the coherent signal and the incoherent noise the spatial correlations are:

$$\rho_{ij}^{\mathrm{s}} = 1 \quad \text{for all} \quad i, j$$

and

$$\rho_{i,i}^{\mathrm{n}} = 1, \quad \rho_{i,j}^{\mathrm{n}} = 0 \quad \text{for} \quad i \neq j.$$

For these spatial correlations, and an array of N hydrophones, the sum of the signal correlations is N^2 and the sum of the noise correlations is N regardless of how the hydrophones are arranged. Then the array gain is given by Eq. (8.15) as

$$\mathrm{AG} = 10 \log N.$$

We have seen from Eq. (8.8c) that the directivity factor for a line of N unshaded, omnidirectional hydrophones is equal to N, making $\mathrm{DI} = 10 \log N$, when $kD = \pi$. Thus the directivity index is numerically equal to the array gain at the frequency where the array spacing is one half wavelength for a line array in incoherent noise. However, in general, the directivity index and the array gain are not the same. The directivity index depends on frequency and array parameters such as the number of hydrophones and their spatial arrangement, while the array gain depends on signal and noise characteristics as well as frequency and array parameters. The directivity index and the array gain are different measures of array performance that may be numerically equal for isotropic, incoherent noise for certain array geometries and certain frequencies.

Another simple comparison case occurs when the signal is coherent and the noise is partially coherent with a constant spatial correlation of $\rho_{\mathrm{n}} < 1$ for $i \neq j$, i.e.,

$$\rho_{ii}^{\mathrm{n}} = 1, \quad \rho_{ij}^{\mathrm{n}} = \rho_{\mathrm{n}}, \quad \text{for } i \neq j.$$

Again the sum of the signal correlations is N^2, while the sum of the noise correlations is $N + \left(N^2 - N\right)\rho_{\mathrm{n}}$, and from Eq. (8.15) the array gain is

$$\mathrm{AG} = 10 \log \frac{N}{1 + (N-1)\rho_{\mathrm{n}}}.$$

This result shows that the array gain is close to zero dB for coherent noise ($\rho_{\mathrm{n}} \approx 1$), close to 10 log$N$ for incoherent noise ($\rho_{\mathrm{n}} \approx 0$), and approximately 10 log($1/\rho_{\mathrm{n}}$) for large N and intermediate values of ρ_{n}. The last case shows that a small degree of noise coherence, if it is independent of the distance between hydrophones, could seriously limit the array gain, e.g., for $\rho_{\mathrm{n}} = 0.1$ the array gain would be limited to 10 dB no matter how large the array is. However, real noise is either incoherent or partially coherent with values of ρ_{ij}^{n} that are close to unity for closely spaced hydrophones, which rapidly diminishes as the hydrophone separation increases. The effect of the partial coherence is to make the array gain less than 10 log$N$ for the usual array spacing of a half wavelength or less. Section 8.3.1 gives examples of spatial correlation for specific types of partially coherent ambient noise, and the corresponding array gains are calculated in Sect. 8.4.1.

8.3 Sources and Properties of Noise in Arrays

8.3.1 Ambient Sea Noise

Dynamic pressure variations in the sea arise from many different mechanisms [2], and pressure sensitive hydrophones respond to all of them. Pressure changes caused by tides and surface waves, oceanic turbulence, and seismic disturbances such as microseisms all contribute to ambient noise in the sea, but the frequency is too low to make these sources important in most sonar applications. The thermal sea noise caused by molecular fluctuations in the water is incoherent and only important for passive arrays that operate above about 40 kHz (see comparison of various types of ambient sea noise in Fig. 8.10).

For most passive arrays the dominant sources of ambient noise are ship traffic and the turbulence, splashing and bubbles caused by surface waves. Both radiate acoustic waves, but with different directional and coherence characteristics. Shipping noise is important from 10 to 500 Hz; it can arise from individual nearby ships and, at the lowest frequencies, from the combined effects of many distant ships. Noise generated by wind depends strongly on frequency and wind speed [25]. Geographical variations in shipping density and changes in wind speed make these major sources of ambient noise quite variable with respect to both time and location, and the curves of Fig. 8.10 should be considered as averages. Intermittent

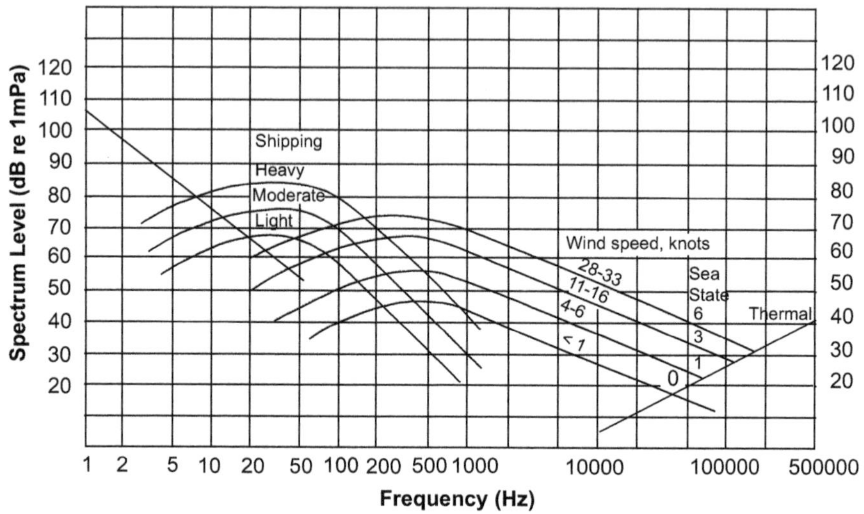

Fig. 8.10 Average deep-water ambient-noise spectra. From [14] (Fig. 7.5) with permission from Peninsula Publishing, Westport, CT 06880. *Note*: mPa is microPa in this figure

noise sources, such as whales, other biological organisms, and storms, increase the variability of ambient noise [2, 14, 26–28].

Ambient noise has been found to be significantly directional in certain frequency ranges. The low frequency noise arrives mainly from horizontal directions as would be expected if it came from distant shipping and storms. Under some conditions higher frequency noise arrives from near vertical directions with an intensity pattern of approximately $\cos^2\theta$, with θ measured from the vertical. This is consistent with the noise originating at the surface [14, 29] and radiating a dipole directional pattern as expected for sources located at the air–water interface. The spatial coherence of ambient noise in deep water has been measured [30], and simple analytical models have been developed that illustrate the combined effects of directionality and coherence [31, 32].

The directionality of the noise, as well as its spatial coherence, influences the noise output of an array. We will compare two simple models of ambient noise fields: an isotropic noise field, where the noise intensity is the same from all directions, and a directional noise field representing sea surface noise where the intensity has a $\cos^2\theta$ directionality. The isotropic case is considered mainly for its simplicity since sea noise is probably never isotropic, although in some cases it has an isotropic component. For isotropic noise the single frequency spatial correlation function depends only on one geometrical variable, the distance between hydrophones, and is given by [31, 32]

$$\rho_{ij}^{\mathrm{n}} = \sin kd_{ij}/kd_{ij}, \tag{8.16}$$

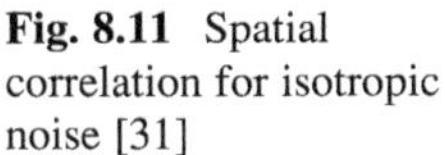
Fig. 8.11 Spatial correlation for isotropic noise [31]

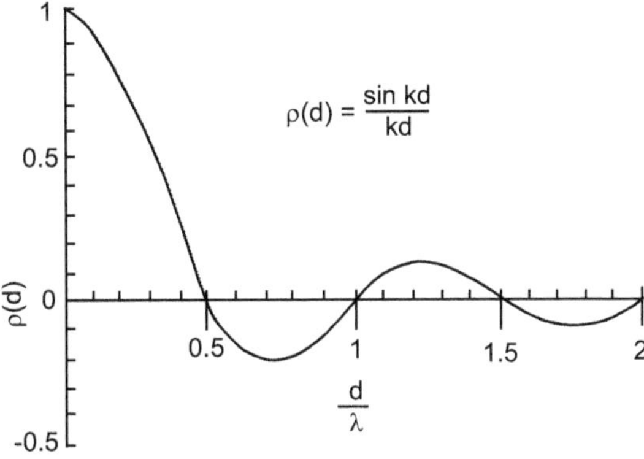

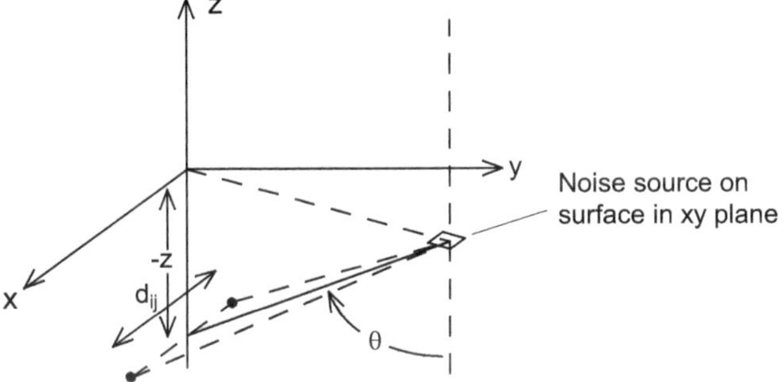

Fig. 8.12 Each small portion of the sea surface in the *xy* plane is a noise source which radiates into the water with a $\cos^2\theta$ intensity pattern. A pair of hydrophones is shown oriented parallel to the surface at depth z

where d_{ij} is the distance between the *i*th and the *j*th hydrophones and $k = 2\pi/\lambda$ (see Fig. 8.11). The factor $\cos\phi_{ij}$ is included in the spatial correlation of the noise when a phase shift of ϕ_{ij} is introduced between the *i*th and *j*th hydrophone outputs [14]. Equation (8.16) and Fig. 8.11 show that for a line array with uniform $\lambda/2$ spacing in isotropic noise all the $\rho_{ij}^{\mathrm{n}} = 0$ for $i \neq j$, and the array gain is $10 \log N$ as it is for completely incoherent noise.

For directional surface noise with $\cos^2\theta$ directionality the single frequency spatial correlation function depends on the orientation of a pair of hydrophones with respect to the surface as well as on their separation. Figure 8.12 shows the geometry and illustrates the case where the pair of hydrophones is oriented parallel to the surface.

The result for the spatial correlation, shown in Fig. 8.13 [31], is

$$\rho_{ij}^{\mathrm{n}} = 2J_1\left(kd_{ij}\right)/kd_{ij}, \tag{8.17}$$

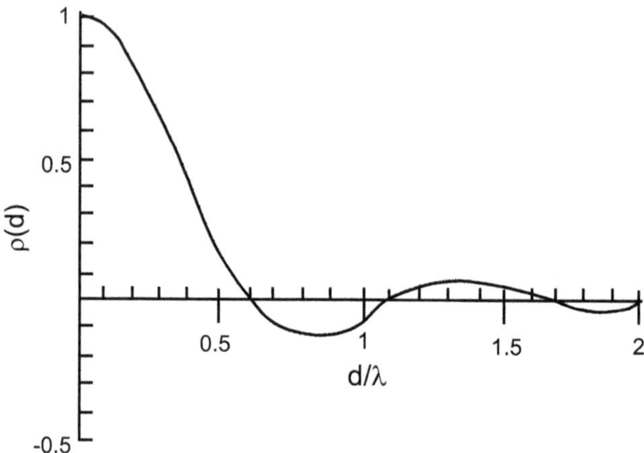

Fig. 8.13 Spatial correlation for surface noise with $\cos^2\theta$ directionality and with the hydrophones oriented parallel to the surface [30]

where J_1 is the first-order Bessel function, and the factor cos φ_{ij} is included when phase shifts are used. For both isotropic and surface noise, the spatial correlation equals unity when $i = j$ and $d_{ij} = 0$, and diminishes rapidly as d_{ij} increases as shown in Figs. 8.11 and 8.13. Note that the surface noise coherence is slightly higher than that for the isotropic noise for small hydrophone separations but diminishes more rapidly for larger separations.

It is interesting to note that the spatial correlation function for isotropic noise in Eq. (8.16) has the same dependence on hydrophone separation as the mutual radiation resistance between small sources given in Eq. (7.24a) [24, 33]. Both are proportional to $\sin kd_{ij}/kd_{ij}$, and, for small kd, where d is the separation of adjacent hydrophones, the relationship between them can be written as

$$R_{ij}/\rho c A_1 = \left(k^2 A_1/2\pi\right)\left(\rho_{ij}^{\mathrm{n}}\right),$$

where A_1 is the area of one hydrophone. This provides a simple approximation for the sum of all the noise correlations for any large rectangular plane array of small hydrophones by using Eq. (7.8) to obtain

$$\sum_i \sum_j \rho_{ij}^{\mathrm{n}} = \sum_i \sum_j \frac{\sin kd_{ij}}{kd_{ij}} = \frac{2\pi N^2}{k^2 A_{\mathrm{A}}},$$

where N is the total number of hydrophones in the array and A_{A} is the total area of the array. Each hydrophone can be considered to occupy an area of d^2, making the total array area equal to Nd^2. Then the array gain is

$$\begin{aligned} \mathrm{AG} &= 10\log\frac{N^2}{2\pi N^2/N(kd)^2} = 10\log\frac{N(kd)^2}{2\pi} \\ &= 10\log N - 10\log\left[2\pi/(kd)^2\right]. \end{aligned} \tag{8.18}$$

This result is approximately valid only for kd less than about $\pi/2$; for $kd = \pi/2$ the reduction from 10 log N is about 4 dB. For smaller kd the reduction is greater since the noise is more correlated.

8.3.2 Structural Noise

The ambient noise discussed in the previous section is acoustic noise radiated from the surface of the sea or from other distant sources; it is the dominant noise for stationary arrays, except for those subject to strong currents. However, for passive arrays mounted on ships or towed by ships, the ambient noise may be important at low ship speeds, but as the speed is increased other sources of noise increase, and usually dominate the ambient noise. Acoustic noise radiated from the turbulence and cavitation caused by the propeller reaches the array by various water paths such as grazing incidence along the hull, surface or bottom reflection, or reverberation. Vibration of the ship structure, excited by both machinery and turbulent flow along the hull, is transmitted to the vicinity of the array by flexural waves that produce decaying pressure waves in the water near the array. The flexural waves may also transmit vibrations to the hydrophones through their mounting structure. Finally, flow along the hull results in turbulent boundary layer pressure fluctuations close to the hydrophones of a hull-mounted array. These latter types of non-acoustic noise present the main noise control problem for hull-mounted arrays at medium to high ship speeds.

The basic properties of the machinery and flow excited structural noise will be discussed first in terms of a simple model based on submerged flat plates [3–5, 34]. Consider first the case of a thin, flat infinite elastic plate with air on one side and water on the other as a very simplified model of a ship's hull. Machinery may excite the plate on one side, and flow may excite it on the other side, causing flexural waves that travel to the array which would typically be mounted in the forward part of the ship. The plate might also represent a sonar dome in which case there would be water on both sides with the interior water stationary with respect to the plate. In either case the vibrating plate produces pressure waves in the water that cause noise in a nearby array. These waves differ from acoustic waves, but all pressure fluctuations cause noise in a hydrophone, and their characteristics must be known in order to control them.

Flexural waves in a thin plate travel with a speed that depends on the mechanical properties of the plate, the frequency, and the media adjacent to the plate. With vacuum on both sides, called the free plate case, the wave number is

$$k_{\mathrm{p}} = \left(\mu\omega^2/D\right)^{1/4} \tag{8.19}$$

and the wave speed is

$$c_{\mathrm{p}} = \omega/k_{\mathrm{p}} = \left(\omega^2 D/\mu\right)^{1/4} = \left[\omega^2 Y h^2/12\rho\left(1-\sigma^2\right)\right]^{1/4}, \tag{8.20}$$

where $D = Yh^3/12(1-\sigma^2)$ is the flexural rigidity, $\mu = \rho h$ is the mass per unit area, Y is Young's modulus, σ is Poisson's ratio, ρ is the density, and h is the thickness of the plate. With water on one, or on both sides, the plate flexural wave speed is reduced by the mass loading of the water, and cannot be expressed in a simple form such as Eq. (8.20), but it still increases with frequency in a similar way. The frequency at which the free plate flexural wave speed, c_{p}, given by Eq. (8.20), equals the speed of sound in a given medium ($c_{\mathrm{p}} = c$) is called the coincidence frequency for that medium. The type of pressure waves produced in the medium depends on whether the frequency is above or below the coincidence frequency, which from Eq. (8.20) is given by

$$\omega_{\mathrm{c}} = c^2\left[12\rho\left(1-\sigma^2\right)/Yh^2\right]^{1/2}. \tag{8.21}$$

In many cases the frequencies of interest to sonar are below the coincidence frequency; e.g., for a 0.0508 m (2 in.) thick steel plate in water $f_{\mathrm{c}} = \omega_{\mathrm{c}}/2\pi$ is about 5 kHz and for a 1 in. thick steel plate it is about 10 kHz.

Below the coincidence frequency flexural waves traveling at speed c_x in a water-loaded plate produce subsonic pressure waves in the water traveling at speed c_x parallel to the plate, and with amplitude decaying exponentially with distance from the plate [3]. The decay factor, at a distance d from the plate, is $\exp\left[-\left(k_x{}^2 - k^2\right)^{1/2} d\right] = \exp\left[-\omega(1/c_x{}^2 - 1/c^2)^{1/2} d\right]$, which corresponds to rapid decay for flexural wave numbers well above the acoustic wave number. The evanescent nature of these pressure waves is critical in determining their contribution to the noise in hydrophones mounted on or near some part of a ship. It is difficult to discuss the effects in a general way, because the mechanisms that excite flexural waves in a ship structure, as well as scattering from structural inhomogeneities such as frames, result in a broad spectrum of wave numbers. However, simple models, such as a point or line force, exciting an infinite plate at a fixed frequency are useful and indicate that the predominant wavenumbers are the free plate wavenumbers in Eq. (8.19). Some simplified calculations of structural noise reduction will be discussed in Sect. 8.4.2.

8.3.3 *Flow Noise*

The wavevector-frequency spectrum of turbulent boundary layer (TBL) pressure fluctuations has been analytically modeled [35–37], and also measured in particular cases. However, it is difficult to avoid contamination by acoustic and structural noise associated with most experimental arrangements. The TBL on the hull of a moving ship can be thought of as a mixture of waves of different frequencies and wavevectors traveling mainly in the direction of flow with a spectrum of wavenumbers, k_x. A small part of the TBL energy also travels transverse to the flow with wavenumbers, k_y, and a smaller part travels out from the hull, including a small amount of acoustic radiation [38]. The peak level of the TBL spectrum occurs at the wave numbers $k_x = k_c = \omega/u_c$ and $k_y = 0$, where k_c is the convective wavenumber and u_c is the convective flow velocity. The main features of the TBL pressure fluctuations are contained in the Corcos model of the wavevector-frequency spectrum [35] which can be written as [3]:

$$P(k_x, k_y, \omega) = P(\omega) \frac{\alpha_1 \alpha_2 k_c^2}{\left\{\pi^2 \left[(k_x - k_c)^2 + (\alpha_1 k_c)^2\right] \left[k_y^2 + (\alpha_2 k_c)^2\right]\right\}}, \tag{8.22}$$

where α_1 and α_2 are constants, and $P(\omega) \approx \rho^2 v_*^4/\omega$ (Pa^2 s in MKS) where ρ is the density of water and v_* is the friction velocity [3]. $P(k_x, k_y, \omega)$ is a mean squared pressure spectral density with respect to frequency and two wavevector components ($Pa^2 m^2$ s in MKS).

At low frequency the convective flow velocity is approximately equal to the ship speed. Thus, at a speed of 10 m/s ($\sim$ 20 knots) the convective wavenumber is 150 times higher than the acoustic wavenumber. As shown in Fig. 8.14, the majority of the TBL energy is concentrated at high wavenumbers around k_c which makes this part of the TBL spectrum relatively easy to control by absorption in the outer decoupler and by area averaging with large area hydrophones (see Sect. 8.4.3).

For sonar applications the most troublesome part of the TBL spectrum is the low wavenumber region, where k_x approaches the acoustic wavenumber, since outer decouplers must be transparent to the acoustic wavenumbers that the array is designed to receive. Thus a very important feature of the TBL spectrum is the ratio of its value at $k_x = k_c$, $k_y = 0$ to its value at the acoustic wavenumber, $k_x = k = \omega/c$, $k_y = 0$. Using Eq. (8.22) this ratio can be approximated for $k \ll k_c$ by

$$P(k_c, 0, \omega)/P(0, 0, \omega)\ 1 + 1/\alpha_1^2.$$

Measurements have indicated that this ratio is about 40 dB [22, 39] which corresponds to $\alpha_1 = 0.01$. Examples of the wavevector-frequency spectrum are given in Fig. 8.14 for a speed of 20 knots using $\alpha_1 = 0.01$ and $\alpha_2 = 1.0$ [3].

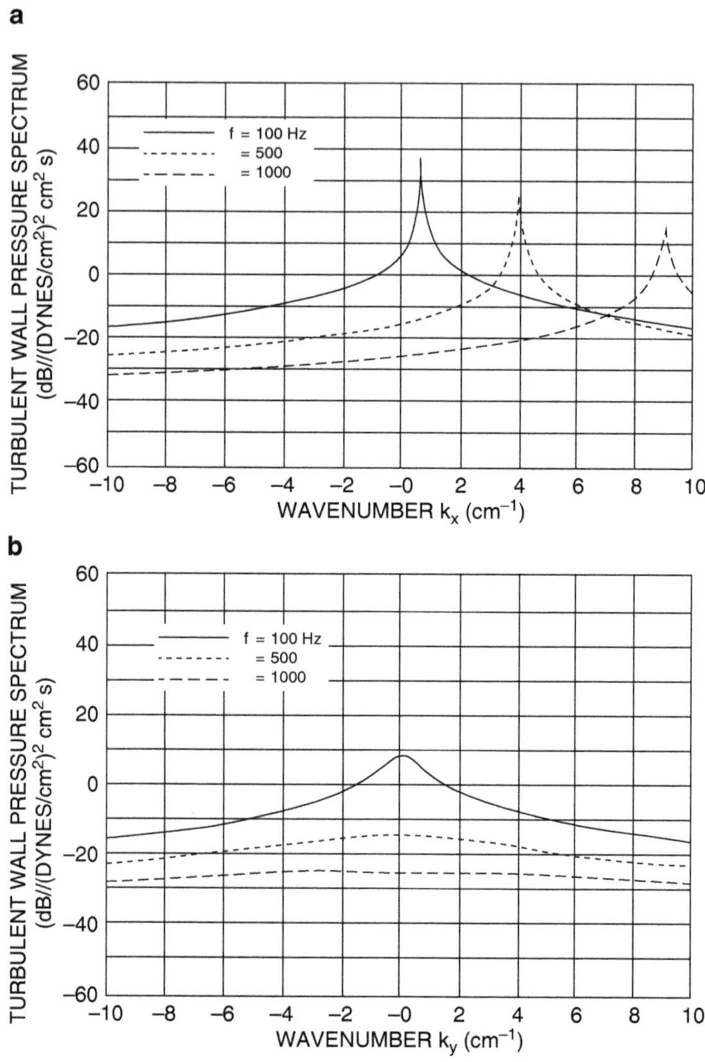

Fig. 8.14 Turbulent wall pressure spectra for flow speed of 20 knots as a function of (**a**) k_x ($k_y = 0$) and (**b**) k_y ($k_x = 0$) for different frequencies [46], with permission from the Acoustical Society of America. Note use of cgs units on this figure

8.4 Reduction of Array Noise

8.4.1 Ambient Noise Reduction

Since the ambient noise is acoustic the main way of controlling it is phasing and shading the array, although it is sometimes possible to use baffles to reflect or absorb noise coming from certain directions, e.g., to keep surface noise from reaching a submarine array. If the ambient acoustic noise is completely incoherent, the array gain is 10 log N, but the noise is often partially coherent which makes the gain less than 10 log N for typical array geometries. This reduction will be illustrated by calculating the array gain for a line array of N hydrophones lying parallel to the ocean surface, simulating a towed array at low speed where ambient isotropic noise or directional surface noise might dominate. Approximate spatial correlation functions for these two types of ambient noise have been given in Eqs. (8.16) and (8.17).

To further simplify the illustration we choose a frequency such that the separation between adjacent hydrophones, d, is one quarter wavelength ($kd = \pi/2$), and include only nearest and next nearest neighbor correlations. The two hydrophones at the ends of the line each have one next neighbor and one next nearest neighbor, the two hydrophones next to the ends have two next neighbors and one next nearest neighbor, and the other ($N - 4$) hydrophones have two next neighbors and two next nearest neighbors. For the isotropic noise field in Eq. (8.16) with $kd = \pi/2$ the next neighbor correlation is $\rho_{ij}^{\mathrm{n}}(\pi/2) = 2/\pi$ and the next nearest neighbor correlation is $\rho_{ij}^{\mathrm{n}}(\pi) = 0$. Thus, including the N values of $\rho_{ii}^{\mathrm{n}} = 1$, the sum of the noise correlations is

$$\sum_i \sum_j \rho_{ij}^{\mathrm{n}} = N + 2(N - 1)(2/\pi).$$

For a plane wave signal from the broadside direction ($\theta = 90°$) with no beam steering ($\phi_{ij} = 0$) each signal correlation is unity, and the sum of the signal correlations is N^2. Thus, from Eq. (8.15), the array gain is

$$\mathrm{AG} = 10\,\log\frac{N^2}{N + 2(N - 1)(2/\pi)} \approx 10\,\log\frac{N}{1 + 4/\pi} = 10\,\log N - 3.6\,\mathrm{dB},$$

where the last form of the result, for large N, can be easily compared with 10 log N for the case of perfectly incoherent noise. We see that the partial coherence associated with the isotropic noise field reduces the array gain by $10\log(1 + 4/\pi) = 3.6\,\mathrm{dB}$. For isotropic noise this result is the same for any other orientation of the array with respect to the surface. Note that if the array spacing was less than $kd = \pi/2$ the noise correlations, and the loss of array gain, would be greater. On the other hand, for half wavelength spacing with $kd = \pi$, the noise correlations would all be zero with no loss of array gain.

We will now calculate the array gain for the same horizontal line array in the directional surface noise field with the spatial correlation given in Eq. (8.17). In this case Fig. 8.13 shows that both nearest neighbor and next nearest neighbor correlations are positive for $kd = \pi/2$ with $\rho(\pi/2) \approx 0.8$ and $\rho(\pi) \approx 0.2$; the more distant correlations are small and will be neglected to simplify this example. For each of the two end hydrophones the sum of the correlations is $(1 + 0.8 + 0.2)$, for each of the two next to end hydrophones the sum is $(1 + 1.6 + 0.2)$, and for each of the other $(N - 4)$ hydrophones the sum is $(1 + 1.6 + 0.4)$. The sum of all the noise correlations being included is then $(3N - 2.4)$, the sum of the signal correlations is again N^2, and the array gain is

$$\mathrm{AG} = 10\log\frac{N^2}{3N - 2.4} \approx 10\log\frac{N}{3} = 10\log N - 4.8\ \mathrm{dB}.$$

The loss of array gain is greater in this case than for isotropic noise because the main beam of the unsteered horizontal array is pointed at the most intense part of the directional surface noise. Barger [32] gives similar results as a function of *kd* for $N = 50$.

The array gain can be increased significantly by steering the beam away from the direction of maximum noise. For example, consider a signal arriving parallel to the array ($\theta = 0°$) with the beam steered to end-fire ($\phi_{ij} = -kd_{ij}$), for which the signal correlations still sum to N^2. But now the noise correlations are changed by the factors $\cos\phi_{ij}$ to

$$\rho(\pi/2) = 0.8\cos(\pi/2) = 0 \quad \text{and} \quad \rho(\pi) = 0.2\cos\pi = -0.2,$$

and they sum to $0.6N + 0.8$. Under these conditions the array gain is increased to

$$\mathrm{AG} = 10\log\frac{N^2}{0.6N + 0.8} \approx 10\log\frac{N}{0.6} = 10\log N + 2.2\,\mathrm{dB}.$$

These simple single frequency examples show how the array gain can be reduced from $10\log N$ by noise coherence, and how, when the noise is directional, the array gain can be increased above $10\log N$ for certain signal reception directions. When the noise is directional, it may be possible to design an array with a null in the direction of the noise or to use signal processing techniques, such as adaptive beamforming, to reduce noise from one or more directions. Faran and Hills [40] developed a method for determining shading coefficients that maximizes the mean square signal-to-noise ratio at the array output in the presence of isotropic noise in the medium or incoherent internal hydrophone noise.

Noise is especially damaging for superdirective arrays of pressure sensitive hydrophones because the differencing of individual outputs reduces signal sensitivity more than it reduces sensitivity to partially coherent noise. The effect can be illustrated by the example of a line array of five hydrophones with quarter wavelength spacing in isotropic noise. First consider uniform shading with all $a_i = 1$; Eq. (8.8a) gives $D_f = 2.5$ and $\mathrm{DI} = 4\,\mathrm{dB}$. With $\rho_{ij}^{n} = \sin kd_{ij}/kd_{ij}$ the sum of the noise correlations is $5 + 8/\pi$, the sum of the signal correlations for the broadside beam ($\theta = 90°$ and $\phi_{ij} = 0$) is 25, and the array gain is $10\log 3.3 = 5.2\,\mathrm{dB}$. If the same array is shaded in a simple superdirective manner with $a_i = 1$ for the center and the two end hydrophones and $a_i = -1$ for the other two hydrophones, the ϕ_{ij}

are either 0 or π. When the $\cos \phi_{ij}$ factors are included in the correlations, the sum of the noise correlations is $5 - 8/\pi$, the sum of the signal correlations is 1, and the array gain is reduced to $10 \log 0.4 = -2.3\,\text{dB}$. The degradation is worse for incoherent noise where the sum of the noise correlations is 5 with both types of shading, and the array gain goes from 7 dB for uniform shading to –7 dB for the simple superdirective shading.

Pressure gradient sensitive and velocity sensitive hydrophones also have high susceptibility to noise because both sensitivities decrease as the frequency decreases. The pressure gradient hydrophone, consisting of two small pressure sensitive hydrophones with a small separation, s, with their outputs subtracted provides an example (see Sect. 8.5.1). For a signal on the MRA of the dipole the sum of the signal correlations is $2(1 - \cos ks)$. For incoherent noise the sum of the noise correlations is 2, and the array gain is given by the ratio as $10 \log(1 - \cos ks) \approx 10 \log\left[(ks)^2/2\right]$, which is negative for small ks (e.g., -23 dB for $ks = 0.1$). Also see Sect. 8.5.2.

8.4.2 Structural Noise Reduction

Usually sonar arrays are mounted in the forward part of a ship as far from the major machinery noise sources as possible. In many cases arrays are also protected from direct exposure to the flow by being mounted inside a faired dome constructed of thin steel or glass reinforced plastic. In that case the array hydrophones must be mounted either on some part of the ship structure inside the dome or on the inside of the dome itself. Thus the hydrophones can receive noise from flexural waves in the dome walls and in other parts of the ship structure inside the dome. Because of the evanescent nature of flexural wave noise, mounting the hydrophones at even a small distance from a vibrating surface is effective in reducing noise if the mounting structure itself can be isolated from the vibration.

Control of structural noise is especially difficult for submarine arrays that are too large to fit in a dome and, therefore, must be mounted directly on the pressure hull. In that case faired protuberances from the pressure hull are necessary, and their thickness and weight are critical, because they increase drag, reduce speed, and influence buoyancy. This type of array consists of several components in addition to the hydrophones as shown in Fig. 8.15.

The low impedance compliant baffle reduces the hull noise reaching the hydrophones, but it also reflects the signal with considerable phase change, which reduces the signal at the hydrophones. Thus a layer of high impedance material (the signal conditioning plate, usually a steel plate) is needed over the baffle to raise the surface impedance and reduce signal degradation [41]. The baffle and signal conditioning plate together comprise the inner decoupler. The array of hydrophones embedded in elastomeric material and the outer decoupler lie between the inner decoupler and the water. All these components must fit in a space of limited thickness and be

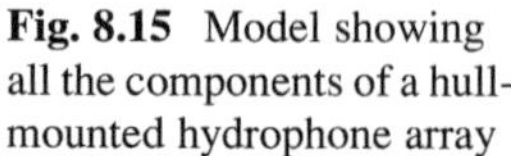

Fig. 8.15 Model showing all the components of a hull-mounted hydrophone array

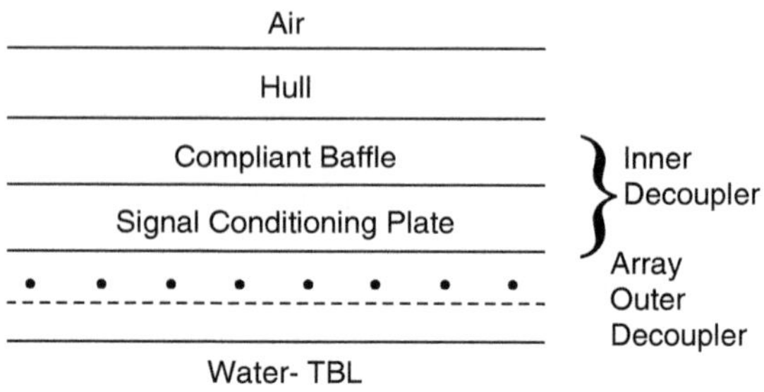

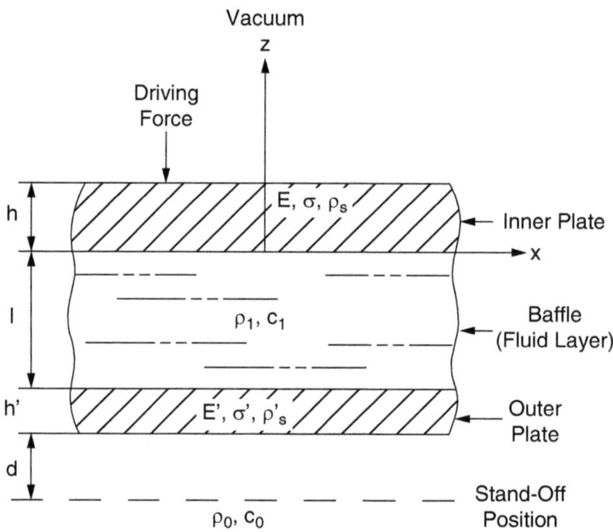

Fig. 8.16 Geometry of the model used for flexural wave baffling [43], with permission from the Acoustical Society of America. (E is Young's modulus, σ is Poisson's ratio, and ρ is the density)

optimized together to achieve the required noise reduction with acceptable weight, thickness, and signal degradation [1]. The effect of the outer decoupler on flow noise will be discussed in the next section.

The wavenumber-frequency spectrum of the structure-borne noise reaching an array mounted on a ship depends on the relative amount of flow and machinery excitation, and that depends on the location of the array within the ship's structure. Here we can only illustrate by discussing analytical modeling of highly simplified cases such as the excitation of an infinite elastic plate by point or line forces or by turbulent flow [3, 5]. The objective of the modeling is to determine the noise reduction provided by layers of compliant, absorbing materials with an outer steel plate (the inner decoupler) placed between a vibrating plate representing the hull and an array as shown in Fig. 8.16.

For a specified vibration of the hull plate the pressure in the water outside the signal conditioning plate, i.e., the noise pressure, is determined as a function of frequency for various inner decoupler materials and thicknesses. The effectiveness of the inner decoupler can be evaluated by its insertion loss, the ratio of noise pressures in the water at the standoff position of a hydrophone, with and without the inner decoupler in place, p_{nid}/p_{no}.

The inner decoupler reduces the structure-borne noise reaching the array hydrophones, but it also usually reduces the signal pressure [41]. For example, with a very soft inner decoupler, which might strongly reduce noise, a plane wave signal would be reflected with a large phase change, and the resulting total signal would be reduced. To minimize signal degradation, the surface impedance of the inner decoupler, which faces the hydrophones, must be increased with a signal conditioning plate, and the ratio of signal pressures with and without the inner decoupler, p_{sid}/p_{so}, the signal gain, must also be considered. The signal-to-noise ratio of an individual hydrophone with and without the inner decoupler can be written in terms of the insertion loss and the signal gain as

$$(s/n)_{id}/(s/n)_0 = [(p_{sid}/p_{nid})/(p_{s0}/p_{n0})] = [(p_{sid}/p_{s0})/(p_{nid}/p_{n0})]. \tag{8.23}$$

This ratio can be called the signal-to-noise gain of the inner decoupler, and the same value applies approximately to each hydrophone in the array.

In the early calculations of flexural wave insertion loss [42, 43], the baffle layer of the inner decoupler in Fig. 8.16 was modeled by a low sound speed fluid with the inner plate representing the vibrating hull and the outer plate a signal conditioning plate. The pressure in the water is calculated at a point, distant $l + h' + d$ (see Fig. 8.16) from the vibrating plate, with and without the baffle/signal conditioning plate combination. Calculated results for the insertion loss, with and without the signal conditioning plate, are given in Fig. 8.17 for the case where the hull plate is driven by a harmonic line force.

It can be seen that 10–15 dB of insertion loss is obtained in this case over frequency bands of about 1 kHz. Many calculations based on models similar to Fig. 8.16, comparing various baffle materials and signal conditioning plates, are included in the book by Ko et al. [3]. The signal gain of the inner decoupler was also evaluated and combined with the insertion loss to determine the signal-to-noise gain of the inner decoupler, as defined in Eq. (8.23) [3]. It was also shown that the flexural wave insertion loss is numerically similar to the acoustic insertion loss for plane waves at normal incidence, a useful result since the acoustic insertion loss is easier to measure [43].

An array may provide some array gain, in addition to the individual hydrophone inner decoupler gain, depending on the coherence and wavenumber-frequency spectrum of the noise passing through the inner decoupler (see Sect. 8.4.4). However, the array gain against structural noise may be degraded if the noise contains significant energy at wavenumbers that coincide with array grating lobes. The wavevector response of an array, which was briefly discussed in Sect. 8.1.4, plays an important part in the overall noise reduction problem for hull-mounted arrays.

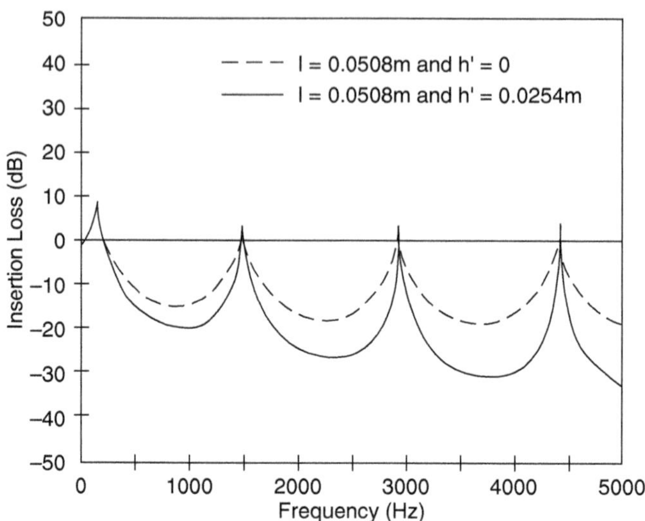

Fig. 8.17 Flexural wave insertion losses for line-force excitation at standoff distance $d = 0.0254$ m for a steel plate with thickness $h = 0.0508$ m. $E = E' = 1.9 \times 10^{10}$, $\sigma = \sigma' = 0.28$, $\rho_s = \rho_s' = 7700$ kg/m^3, $\zeta = \zeta' = 0.1$ (material damping factor), $\rho_1 = 1000$ kg/m^3, $c_1 = 150$ m/s, $\rho_o = 1000$ kg/m^3, $c_o = 1500$ m/s [41], with permission from the Acoustical Society of America

For a plane rectangular array, such as that discussed in Sect. 8.1.1, the wavevector response is illustrated qualitatively in Fig. 8.18 where the central small circle represents the major lobe and the other small circles represent grating lobes.

The minor lobes that lie between the grating lobes are omitted from the diagram for simplicity. For array spacing of D in both dimensions, the grating lobes occur at $k_x = \pm 2\pi n/D$ and $k_y = \pm 2\pi m/D$, where n and m are integers. At the frequency where the acoustic wavelength is $2D$ the acoustic wavevector magnitude is $k_a = \left(k_{ax}^2 + k_{ay}^2\right)^{1/2} = \pi/D$, which is shown in Fig. 8.18 as a circle of radius $17\,\text{m}^{-1}$ representing acoustic grazing waves at 4 kHz coming from all directions in the plane of the array. Acoustic waves from all other directions have values of k_a inside this circle, with $k_a = 0$ being a broadside wave. When the major lobe is steered within the acoustic circle to receive acoustic signals, the entire pattern of grating lobes is also steered by the same wavenumber in the same direction.

Now we will superimpose on this array pattern the noise waves that might result from flexural waves in the hull and in the signal conditioning plate. We will simplify the discussion by using the free plate wavenumbers and ignoring the effects of loading by the compliant part of the inner decoupler, the array, the outer decoupler, and the water. Equation (8.19) was used to calculate the results in Fig. 8.19 which compares acoustic wave numbers with free plate wavenumbers for 1 and 2 in. thick steel plates as models of a signal conditioning plate and a hull, respectively.

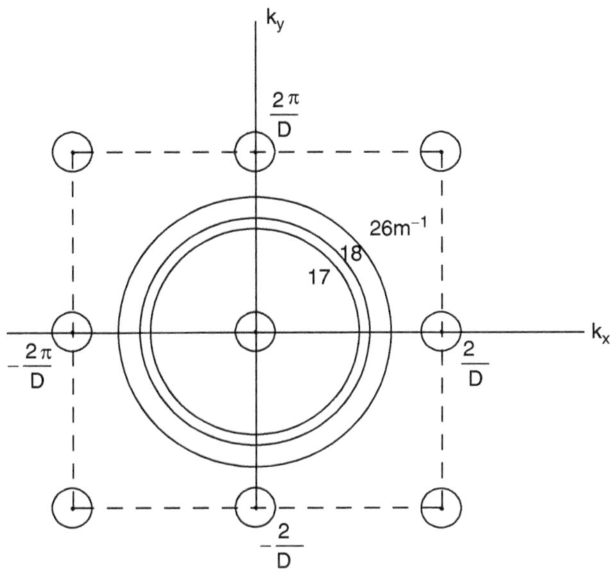

Fig. 8.18 Wavevector response of an unsteered square array with spacing D showing main lobe and first grating lobes. The three circles with radii 17, 18, and 26 m^{-1} represent the acoustic wave numbers and free plate wave numbers for 2 in. and 1 in. thick steel plates at 4 kHz

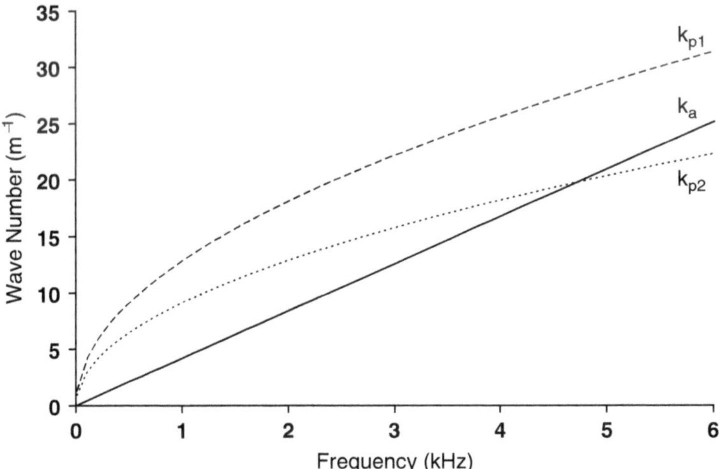

Fig. 8.19 Frequency dependence of wave numbers: k_a for acoustic, k_{p1} for 1 in. thick steel plate, k_{p2} for 2 in. thick steel plate

As a specific example consider that $k_a = \omega/c = \pi/D$ at 4 kHz, then $\pi/D = 17\,\mathrm{m}^{-1}$, $2\pi/D = 34\,\mathrm{m}^{-1}$, etc. in Fig. 8.18. From Fig. 8.19, the plate wavenumbers at 4 kHz for the 1 and 2 in. thick steel plates are $26\,\mathrm{m}^{-1}$ and

$18\,\text{m}^{-1}$, which are shown on the array pattern in Fig. 8.18 as circles representing flexural waves coming from all directions in the plane of the array. When the array is steered the grating lobes move, but the acoustic and flexural wavenumbers are fixed. In this example, the grating lobes of the unsteered array do not intersect either of the flexural wave circles. But if the array was steered to $k_{0x} = +8\,\text{m}^{-1}$, $k_{0y} = 0$ the grating lobe at $k_x = -34\,\text{m}^{-1}$, $k_y = 0$ would coincide with the flexural wave circle for the 1 in. plate. If there were flexural waves in the plate traveling in that direction, they would be received on the grating lobe and the array noise would increase. To minimize such effects the array spacing must be optimized based on detailed analysis of the expected wavevector spectrum of the noise [1, 5]. The wavevector filter and the wavenumber generator are important experimental tools in the analysis of such noise problems (see Sect. 8.1.4).

Inner decouplers must contain compliant material to be effective in reducing the structural noise that reaches the hydrophones. If the compliant material consists of an air-voided elastomer the decoupler insertion loss would degrade with depth as the static pressure compressed the air voids. Thus, in submarine applications, it is necessary to use a tuned decoupler that has high specific acoustic impedance at low frequency but low impedance over the array bandwidth. The compliant tube baffle [1, 44] meets this requirement to some extent. The compliant tube is a long, air-filled, closed-end tube with an oval or a flat-sided cross section. The tube material may be metal or plastic reinforced with glass or carbon fibers depending on depth, frequency, and cost. The width and wall thickness dimensions are determined by two conditions: the fundamental width resonance must be near the middle of the array band, and the maximum stress in the wall must be less than the yield strength of the tube material at the maximum depth. A compliant tube baffle consists of an array of such tubes, closely spaced and embedded in rubber. Insertion losses of 10–20 dB can be obtained over a useful range of depth and over about one octave of bandwidth with a baffle containing one layer of tubes. Multiple layers of tubes and tubes of more than one size can provide greater bandwidth.

8.4.3 Flow Noise Reduction

The control of flow noise can be analyzed somewhat more completely than the control of structural noise, because models of the TBL wavevector-frequency spectrum, such as the Corcos model, can be used for noise reduction calculations that are relevant to ship-mounted sonar arrays. A sample of such results will be discussed from the work of Ko et al. [3] and Ko and Schloemer [45, 46]. The TBL excitation can be considered to act on the outer surface of a layered array structure such as that shown in Fig. 8.15 with the outer decoupler separating the array from the water. Since the wavenumbers of the acoustic signals are much lower than those of the major part of the flow noise, an outer decoupler consisting only of a layer of rubber attenuates the flow noise significantly with very little reduction of the signal.

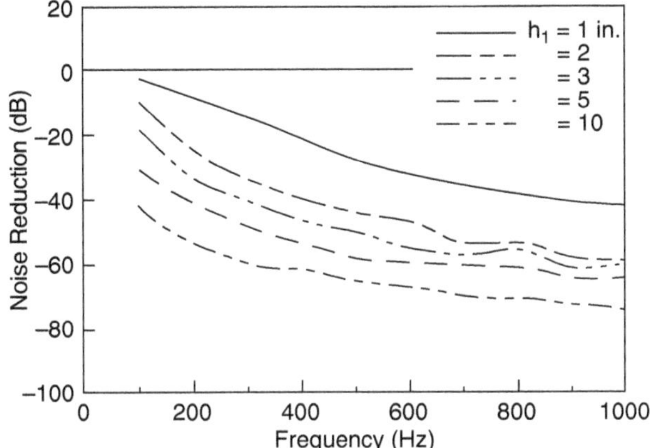

Fig. 8.20 Effect of rubber outer decoupler thickness (h_1) on reduction of TBL noise for point hydrophone [46], with permission from the Acoustical Society of America

Calculated results for the noise reduction (insertion loss) of the outer decoupler are given in Fig. 8.20 for several outer decoupler thicknesses.

These calculations were done without considering the entire structure shown in Fig. 8.15. The outer decouplers were attached to a 2 in. thick signal conditioning plate, and the pressure was calculated at a point within the outer decoupler at a distance of 0.5 in. from the signal conditioning plate to simulate a small embedded hydrophone. The calculation involves transmission of the TBL wavevector-frequency spectrum through the outer decoupler and integration over the two wavevector components to obtain the result at each frequency [3, 47]. It can be seen that large noise reductions are achieved at rather low frequencies. This reduction can be considered to be the insertion loss of the outer decoupler.

Reduction of the flow noise relative to the acoustic signal can also be achieved by the individual hydrophones if they are larger than the flow noise wavelengths because of area averaging. This occurs since any pressure wave traveling parallel to the hydrophone surface causes adjacent regions of increased and decreased pressure, and the electrical responses from these regions partially cancel in the hydrophone output. Thus area averaging causes significant cancellation of those flow noise components with wavelengths much smaller than the width of the hydrophone. Hydrophones made from thin, lightweight, flexible materials with uniform sensitivity, such as the piezoelectric ceramic-polymer composites and polyvinylidene fluoride materials discussed in Sect. 6.3.3, are suitable for this purpose. Figure 8.21 shows examples of the reduction achieved by large area square hydrophones with 1, 5, and 20 in. sides compared to a point hydrophone. Note that this reduction is caused by the size of the hydrophone when no outer decoupler is present.

The hydrophone size noise reduction, in Fig. 8.21, and the outer decoupler noise reduction, in Fig. 8.20, are not independent, because the outer decoupler changes the wavevector-frequency spectrum of the noise reaching the hydrophone. An example of the overall reduction achieved by both an outer decoupler and a large

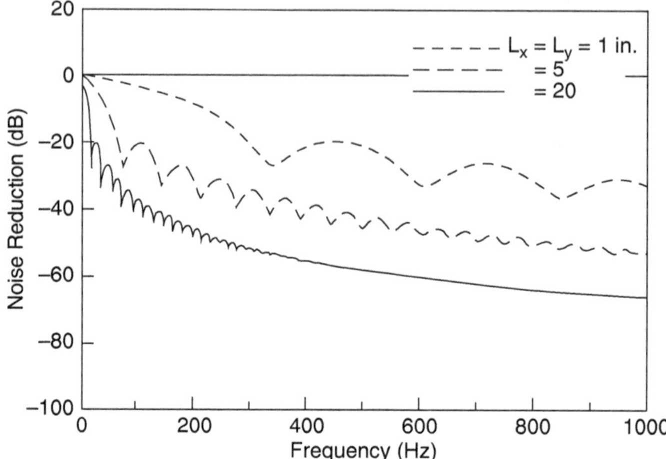

Fig. 8.21 Effect of hydrophone dimensions (flush mounted) on turbulent flow noise reduction [46], with permission from the Acoustical Society of America

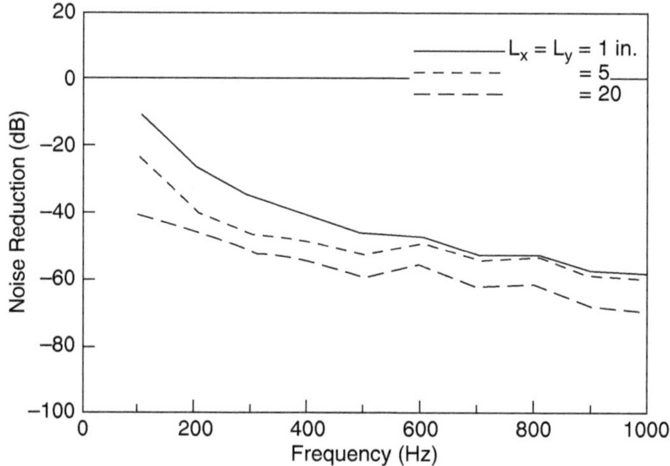

Fig. 8.22 Effect of dimensions of a hydrophone embedded in an elastomer layer 2 in. thick [46], with permission from the Acoustical Society of America

area hydrophone is shown in Fig. 8.22. In this case, hydrophones of the same size as those in Fig. 8.21 are embedded in a 2 in. thick outer decoupler with a standoff distance of 0.5 in. from the signal conditioning plate. It can be seen that the outer decoupler increases the noise reduction significantly for the smallest hydrophone but does not increase it much for the larger hydrophones. This is expected because when most of the high wavenumber noise is removed by one of these methods the other method cannot reduce it much more. To see this in detail, compare Fig. 8.20

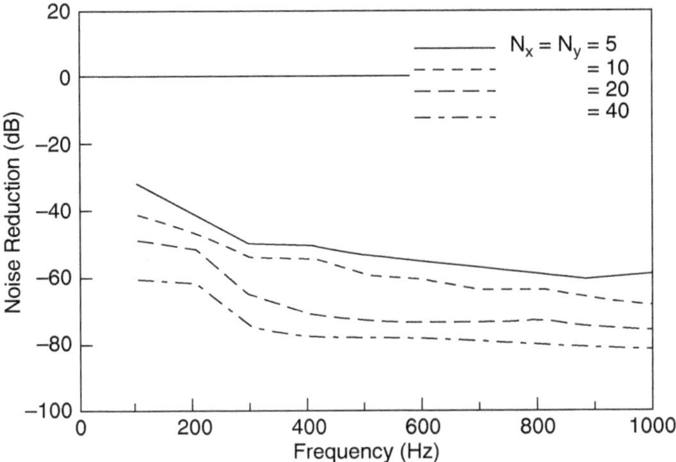

Fig. 8.23 Effect of the number of hydrophones (without inter-hydrophone spacing) embedded within an elastomer layer 2 in. thick, [46], with permission from the Acoustical Society of America

for the outer decoupler alone with point hydrophones, Fig. 8.21 for large area hydrophones alone, and Fig. 8.22 for both outer decoupler and large area hydrophones.

Finally some array gain against the flow noise also occurs with an embedded array behind an outer decoupler. Figure 8.23 gives results for square arrays of 5×5, 10×10, 20×20, and 40×40 close-packed hydrophones, i.e., the calculation assumed no space between the hydrophones. The individual hydrophones are 2 in. squares, embedded in a 2 in. thick outer decoupler with a standoff of 0.5 in. as in Fig. 8.22. An additional noise reduction of about 6 dB can be seen in each case as the number of hydrophones is increased by a factor of 4. It can also be seen, by comparison with the curve for $h_1 = 2$ in. in Fig. 8.20, that the additional reduction by the array, compared to the reduction for one point hydrophone, is most significant at the lower frequencies.

In the calculations of Fig. 8.23 the grating lobes were eliminated by the perfectly close-packed array as discussed in Sect. 7.1.3. These calculations were repeated with 3 in. spacings between hydrophone centers giving 1 in. gaps between adjacent hydrophones, and the results did not differ much. Two opposing effects occurred when the arrays were spread out. First some additional high wavenumber noise was received on grating lobes, but the high wavenumber noise had already been reduced by the outer decoupler; secondly the array area was more than doubled which increased its ability to discriminate against the low wavenumber noise.

The book by Ko et al. [3] includes many other calculations of flow noise reduction by hydrophones, arrays, and outer decouplers, including investigation of the effectiveness of hydrophones with shapes other than rectangular and hydrophones with nonuniform surface sensitivity.

8.4.4 Summary of Noise Reduction

A hull-mounted array of the type considered in the previous sections is subject to the ambient noise, structural noise, and flow noise discussed in this chapter, as well as the internal hydrophone noise discussed in Chap. 6. The relative intensities of each type of noise at the locations of the hydrophones depend on frequency, ship speed, machinery conditions, location on the hull, and the effectiveness of the inner and outer decouplers and the large area hydrophones; furthermore, the inner decoupler may degrade the signal. The overall array gain depends on the relative intensities of the different types of noise at the hydrophones, and therefore it depends on the effectiveness of all the noise reduction measures.

At a particular frequency the acoustic signal reaching each individual hydrophone, s, is related to the free field acoustic signal, s_0, by the signal gain of the inner decoupler, $(p_{sid}/p_{s0}) = G_s$:

$$s = G_s s_0 , \tag{8.24}$$

assuming that the outer decoupler does not degrade the acoustic signal. Consistent with this assumption, the acoustic ambient noise at the hydrophones, n_a, is also not degraded by the outer decoupler, but it is affected by the inner decoupler making $n_a = G_s n_{0a}$, where n_{0a} is the ambient noise in the water. The structural noise reaching each hydrophone, when integrated over all wavenumber components at the frequency in question, n_s, is reduced from its value without the inner decoupler, n_{0s}, by the insertion loss of the inner decoupler, L_{id}:

$$n_s = L_{id} n_{0s}. \tag{8.25}$$

And the flow noise reaching each hydrophone, integrated over all wavenumber components, n_f, is reduced by the combined insertion loss of the outer decoupler and the area averaging of the hydrophone, L_{0dh}:

$$n_f = L_{0dh} n_{0f}. \tag{8.26}$$

The different types of noise are uncorrelated, and the total mean squared noise at each hydrophone, including the internal hydrophone noise, n_h, is

$$\langle n_a^2 \rangle + \langle n_s^2 \rangle + \langle n_f^2 \rangle + \langle n_h^2 \rangle = G_s^2 \langle n_{0a}^2 \rangle + L_{id}^2 \langle n_{0s}^2 \rangle + L_{0dh}^2 \langle n_{0f}^2 \rangle + \langle n_h^2 \rangle, \tag{8.27}$$

where G_s^2, L_{id}^2, and L_{odh}^2 are magnitudes squared.

We will now consider how the individual noise components influence the array gain. Note first from the original definition of array gain in Eq. (8.9) that the signal gain of the inner decoupler cancels out, except for its effect on the ambient noise, as long as we make the reasonable assumption that the inner decoupler signal gain is the same at each hydrophone. But the insertion losses of the inner and outer decouplers and the area averaging do influence the array gain, except in the unlikely case that all types of noise have the same spatial correlation functions. Assuming that the signal, as well as each type of noise, has the same mean squared value at each hydrophone, the array gain in terms of spatial correlation functions, Eq. (8.15), can be written as

$$\mathrm{AG} = 10\log \frac{\sum\sum \rho_{ij}^{\mathrm{s}}}{\sum\sum \frac{\left[G_{\mathrm{s}}^{2}\langle n_{0\mathrm{a}}^{2}\rangle \rho_{ij}^{\mathrm{na}} + L_{\mathrm{id}}^{2}\langle n_{0\mathrm{s}}^{2}\rangle \rho_{ij}^{\mathrm{ns}} + L_{0\mathrm{dh}}^{2}\langle n_{0\mathrm{f}}^{2}\rangle \rho_{ij}^{\mathrm{nf}} + \langle n_{\mathrm{h}}^{2}\rangle \rho_{ij}^{\mathrm{nh}}\right]}{G_{\mathrm{s}}^{2}\langle n_{0\mathrm{a}}^{2}\rangle + L_{\mathrm{id}}^{2}\langle n_{0\mathrm{s}}^{2}\rangle + L_{0\mathrm{dh}}^{2}\langle n_{0\mathrm{f}}^{2}\rangle + \langle n_{\mathrm{h}}^{2}\rangle}}. \tag{8.28}$$

In Eq. (8.28), the summations are over all hydrophones in the array; ρ_{ij}^{s} is the spatial correlation of the signal; and $\rho_{ij}^{\mathrm{na}}, \rho_{ij}^{\mathrm{ns}}, \rho_{ij}^{\mathrm{nf}}$, and ρ_{ij}^{nh} are the spatial correlations of the ambient, structural, flow, and internal noise at the hydrophones where ρ_{ij}^{nh} is incoherent. Note that if any one noise type greatly exceeds the others at the hydrophones, Eq. (8.28) reduces to the simple form in Eq. (8.15). Equation (8.28) can be written in the more compact form

$$\mathrm{AG} = 10\log \frac{\sum\sum \rho_{ij}^{\mathrm{s}}}{\sum\sum \left[f_{\mathrm{a}}\rho_{ij}^{\mathrm{na}} + f_{\mathrm{s}}\rho_{ij}^{\mathrm{ns}} + f_{\mathrm{f}}\rho_{ij}^{\mathrm{nf}} + f_{\mathrm{h}}\rho_{ij}^{\mathrm{nh}}\right]}, \tag{8.29a}$$

where $f_{\mathrm{a}}, f_{\mathrm{s}}, f_{\mathrm{f}}$, and f_{h} are the fractions of total noise intensity at the hydrophones in the form of ambient, structural, flow, and internal hydrophone noise, respectively. These fractions depend on the sources of the noise and also on the effectiveness of the inner and outer decouplers and area averaging.

As a simple example consider the case where the ambient, structural, and flow noise have the same intensity at the hydrophones after the structural noise has been reduced by the inner decoupler and the flow noise has been reduced by the outer decoupler and area averaging. Also consider that the internal hydrophone noise is negligible. Then $f_{\mathrm{a}} = f_{\mathrm{s}} = f_{\mathrm{f}} = 1/3$ and $f_{\mathrm{h}} = 0$. We will also assume, for simplicity in this example, that both structural noise and flow noise are incoherent with

$$\rho_{ij}^{\mathrm{s}} = \rho_{ij}^{\mathrm{f}} = 1 \quad \text{for} \quad i = j = 0 \ \text{for} \ i \neq j,$$

Then the array gain, for an array of N hydrophones, becomes

$$\mathrm{AG} = 10\log \frac{3\sum\sum \rho_{ij}^{\mathrm{s}}}{\sum\sum \rho_{ij}^{\mathrm{na}} + N + N}. \tag{8.29b}$$

We now return to the first example of array gain calculation in Sect. 8.4.1, with a horizontal line array in isotropic ambient noise. But now we consider the array to be mounted on a ship and that the same intensities of incoherent structural noise and incoherent flow noise are also present at the hydrophones in addition to the partially coherent ambient noise. We can use the sum of the ambient noise correlations that was approximated in Sect. 8.4.1 for a line array with quarter wavelength spacing using only nearest and next nearest neighbor correlations. As before, consider that the signal is a plane wave from the broadside direction making the sum of the signal correlations equal to N^2. Under these conditions Eq. (8.29b) gives

$$\mathrm{AG} = 10\log\frac{3N^2}{N + 2(N-1)(2/\pi) + 2N} \approx 10\log\frac{N}{1 + 4/3\pi} = 10\log N - 1.5\ \mathrm{dB}.$$

Comparing with the corresponding result in Sect. 8.4.1 we see that the presence of the additional structural and flow noise has raised the array gain by 2.1 dB from 10 log N – 3.6 dB to 10 log N – 1.5 dB. The increase in array gain results from mixing incoherent structural and flow noise with the partially coherent ambient noise, causing a decrease in the average coherence. The array gain depends on noise coherence and increases when the average coherence decreases; it does not depend on the total intensity of the noise.

The relationship between all the noise control measures described above can be shown in another way by considering the signal-to-noise ratio of the array, which does depend on the total intensity of the noise. Using Eqs. (8.9), (8.24), and (8.27) the signal-to-noise ratio of the array can be written:

$$(S/N)_{\mathrm{array}} = \mathrm{AG} + (s/n)_{\mathrm{hyd}} = \mathrm{AG} + 10\log G_{\mathrm{s}}^2 + 10\log\langle s_0^2\rangle - 10\log I_{\mathrm{N}}, \quad (8.30)$$

where

$$I_{\mathrm{N}} = G_{\mathrm{s}}^2\langle n_{0\mathrm{a}}^2\rangle + L_{\mathrm{id}}^2\langle n_{0\mathrm{s}}^2\rangle + L_{\mathrm{odh}}^2\langle n_{0\mathrm{f}}^2\rangle + \langle n_{\mathrm{h}}^2\rangle$$

is the total intensity of the noise at the hydrophone given in Eq. (8.27). This expression shows, e.g., that an inner decoupler that reduces the structural noise by 10 log L_{id}^2 dB does not raise the signal-to-noise ratio of the array by that number of dB unless structural noise is the only significant noise.

8.5 Arrays of Vector Sensors

Vector sensors are used in large numbers, mainly because of the advantages they have over pressure sensors for applications such as sonobuoys and acoustic intercept receivers on submarines (see Sect. 6.5 for various types of vector sensors). The advantage of vector sensors for these applications is their inherent directionality in a device small compared to the acoustic wavelength. The possibility that this feature may also be useful in other naval applications, and in oceanographic research, has resulted in a strong interest in vector sensors. For example, an American Institute of Physics Conference was held in 1995 on Acoustic Particle Velocity Sensors: Design, Performance and Applications [48], followed in 2001 by an Office of Naval Research Workshop on Directional Acoustic Sensors [49]. In this section, we will discuss some of the features of vector sensor arrays that influence their suitability for sonar applications.

The possible advantage of vector sensors for hull-mounted arrays arises both from their directionality and from the conditions under which they are mounted. When a plane wave signal is reflected from a surface with low acoustic impedance compared to that of water, the pressure at the surface is decreased, but the particle velocity at the surface is increased. For an ideally soft surface the normal component of particle velocity is doubled while the pressure is reduced to zero on the surface. Thus the compliant baffle that is needed for noise reduction in submarine arrays reduces the acoustic pressure on the surface and requires a heavy signal conditioning plate over the baffle to make the surface suitable for pressure sensors (see Sect. 8.4.2). However, if velocity sensors are used, rather than pressure sensors, the compliant baffle may have the advantage of raising the signal. And, since the signal conditioning plate is then not needed, a significant weight reduction would also be achieved for large arrays. In other words, with velocity sensors, the compliant baffle not only provides insertion loss against structural noise, but it may also provide signal gain. However, it was shown in Sect. 6.5.1 that a pressure sensor gives the same, or greater, output as a dipole sensor for the same maximum projection (standoff distance) from an ideal soft baffle. Thus when more realistic mounting conditions and more realistic baffles are considered, the vector sensor advantage in this application appears to be uncertain. This point will be pursued in Sect. 8.5.3 using a model of a compliant baffle that is simple, but more realistic than an ideally soft baffle.

Vector sensors can have an important advantage in towed line arrays by resolving the right-left ambiguity inherent in single line arrays of pressure sensors without requiring a time consuming course change by the towing ship. Their directionality also discriminates against radiated noise from the towing ship, but a heading sensor is also required, since line arrays usually twist and change the orientation of the sensors. It has also been observed that intensity sensors have considerable immunity to ambient noise when it has a significant isotropic component [50, 51].

Whether these vector sensor advantages are realizable in practice depends mainly on noise and the feasibility of controlling noise in each application. It has long been known that vector sensors are more susceptible to some types of noise than scalar sensors, as was briefly discussed in Sects. 6.7.5 and 6.7.6. It was shown there that the equivalent internal thermal noise pressure in dipole sensors is much higher than that in pressure sensors using the same transduction materials because of the decreasing sensitivity of the dipole sensor as the frequency is reduced. The fact that the internal noise is incoherent also means that the signal-to-noise ratio is lower for a dipole than a single pressure sensor.

8.5.1 Directionality

The directional characteristics of vector sensor arrays have been studied in considerable detail by Cray and Nuttall [52]. We will summarize some of their results by considering dipole vector sensors with cosine directivity patterns, although other types of vector sensors with other patterns, such as cardioids, could also be used in arrays. As described in Sect. 6.5.1 the dipole vector sensor consists of two pressure sensitive hydrophones separated by a small distance, s, with their outputs differenced. Equation (6.37a) shows that the voltage output of the dipole is reduced in amplitude by the factor $(2\pi s/\lambda)\cos\gamma$, compared to one of the pressure hydrophones from which it is made, when both are receiving the same plane wave acoustic signal arriving at the angle γ with respect to the dipole axis. The null in the individual dipole pattern at $\gamma = 90°$ makes it necessary to use biaxial or triaxial dipoles, which respond to two or three components of the particle velocity, in order to provide beam steering capability over all directions. Figure 8.24 shows a triaxial sensor formed from three orthogonal dipoles using six hydrophones, with the acoustic center at the same point for all three dipoles. The amplitudes of the voltage outputs of the three dipoles for a plane wave arriving from the direction θ, ϕ are given by Eq. (6.37a) as $(2\pi s/\lambda)M_1 p_i \cos\gamma_n$, where M_1 is the sensitivity of one of the pressure hydrophones, p_i is the amplitude of the plane wave, s is the distance between each hydrophone pair, and γ_n is the angle between the plane wave direction and the nth dipole axis. These angles are given by

$$\cos\gamma_n = \cos\theta\cos\theta_n + \sin\theta\sin\theta_n\cos(\phi - \phi_n),$$

where θ_n and ϕ_n designate the direction of the nth dipole axis. For example, for dipole number 1 pointing in the x-direction $\theta_1 = 90°$, $\phi_1 = 0°$, and $\cos\gamma_1 = \sin\theta \cos\phi$; in the same way, $\cos\gamma_2 = \sin\theta\sin\phi$ and $\cos\gamma_3 = \cos\theta$. Thus the weighted sum of the outputs of a triaxial sensor has a directivity function proportional to

$$P(\theta, \phi) = A\sin\theta\cos\phi + B\sin\theta\sin\phi + C\cos\theta.$$

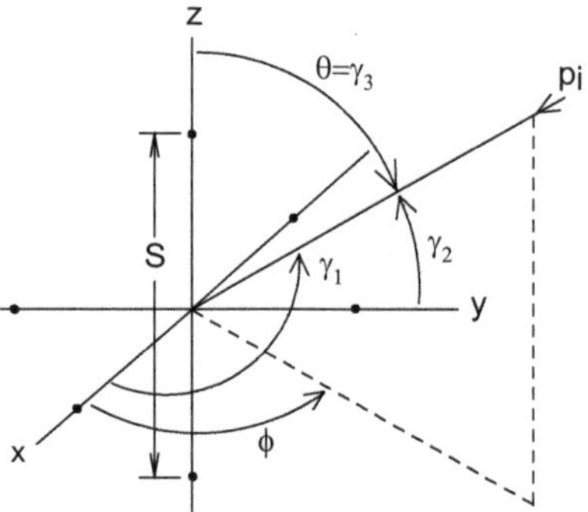

Fig. 8.24 A triaxial dipole vector sensor consisting of six pressure sensors with each pair separated by the distance S

The beam of an individual triaxial dipole can be steered in the direction θ_0, ϕ_0 by adjusting the outputs of the three dipoles such that the values of A, B, and C are:

$$A = \sin\theta_0 \cos\phi_0, \quad B = \sin\theta_0 \sin\phi_0, \quad C = \cos\theta_0.$$

Then the resulting directivity function is

$$P(\theta, \phi, \theta_0\phi_0) = \cos\theta \cos\theta_0 + \sin\theta \sin\theta_0 \cos(\phi - \phi_0) = \cos\alpha_0$$

where α_0 is the angle between the direction of an incident plane wave, θ, ϕ, and the steered direction θ_0, ϕ_0. Thus when the beam is steered to the direction of the incoming wave, i.e., $\theta_0 = \theta, \ \ \phi_0 = \phi$, the response of the triaxial sensor reaches its maximum value.

The directivity pattern of an array of uniaxial dipoles in the xy plane, with their axes oriented parallel to the z-axis, is given by the Product Theorem as $\cos\theta$ times the point sensor array pattern. This array pattern has a null in the xy plane ($\theta = 90°$), the side lobes are reduced, and the main lobe width is decreased slightly compared to the point sensor pattern. These are all desirable characteristics as long as the array is not steered much. For example, 45° steering reduces the main lobe by 3 dB, but steering close to end-fire is impossible because the dipole null destroys the main lobe. However, the same array of triaxial dipoles allows steering in all directions.

Cray and Nuttall considered several approaches to beamforming arrays of vector sensors and presented results showing the increase in directivity index that can be obtained with vector sensors compared to pressure sensors. One of their examples showed that a line array of ten triaxial vector sensors with half wavelength spacing has 5 dB more DI than the same array of pressure sensors for most azimuthal steering angles. To achieve the same DI the array of pressure sensors would have to be about three times longer. This example shows that the inherent directionality of vector sensors is their most important asset, since increasing array aperture is difficult and often impossible. However, such increases in DI do not necessarily imply similar increases in array gain unless the dominant noise is isotropic and incoherent. Cray and Nuttall also considered the effects of ideal baffles and of array curvature, and they pointed out that it would usually be desirable to use sensors that detect pressure in addition to sensors that detect the three components of particle velocity.

8.5.2 Vector Sensor Arrays in Ambient Noise

For individual dipoles partially coherent ambient noise is essentially coherent since the distance between the two pressure hydrophones is very small compared to the acoustic wavelength. For example, see Eqs. (8.16) and (8.17) for two different ambient noise models which both show that the spatial correlation of pressure is close to unity when the separation is a small fraction of a wavelength. Thus the ambient noise power in the dipole would be reduced by nearly the same amount as the signal. Recall from Sect. 8.4.1 that the partial coherence of ambient noise had the undesirable effect of reducing the array gain from the value 10 log$N$ that holds for incoherent noise. Since vector sensors involve differencing within the individual sensor, noise coherence has a desirable effect on the individual sensor output, because noise is then reduced as well as the signal. However, when a dipole sensor is in an inhomogeneous noise field, where the noise pressure amplitude at the two hydrophones is not the same, the differencing does not reduce the noise as much as it would in a homogeneous noise field. (See Sects. 6.7.6 and 8.5.3 for examples of inhomogeneous noise fields.)

Analysis of arrays of vector sensors requires knowledge of the spatial correlation between the outputs of two sensors separated by an arbitrary distance, with specified relative orientation of their axes, for both plane wave signals and noise. General results for the various spatial correlations between the three components of particle velocity at different points and at different times in isotropic noise have been given by Kneipher [53] and Cray and Nuttall [52], and also by Hawkes and Nehorai including the correlations between pressure and the velocity components [54].

Here we will derive the spatial correlations for some of these cases and use them to calculate the array gain for a line array of vector sensors. We will assume the noise is isotropic for simplicity, although ambient noise is usually directional with an isotropic component in some cases. We will consider the vector sensors to be dipoles, but the results apply to some other types of vector sensors. The two pressure hydrophones that comprise each dipole have sensitivity M_1 and are separated by distance s. As the first case, consider two dipoles located on the z-axis of a spherical coordinate system, with centers separated by a distance d (where $d \gg s$), with the center of the line joining them at the origin, and with both sensor axes parallel to the z-axis as shown in Fig. 8.25a. For this orientation, and with a plane wave signal of pressure amplitude p_{i} arriving at the angles θ, ϕ, the angle between the dipole axes and the plane wave direction is equal to θ for both dipoles.

From Eq. (6.37) the voltage outputs of dipoles 1 and 2, both aligned to detect the z component of the plane wave particle velocity, are:

$$V_{z1} = -j(2\pi s/\lambda)M_1 p_i \cos\theta\, \mathrm{e}^{j[\omega t - k(r - d\cos\theta/2)]}, \tag{8.31a}$$

$$V_{z2} = -j(2\pi s/\lambda)M_1 p_i \cos\theta\, \mathrm{e}^{j[\omega t - k(r + d\cos\theta/2) + \phi_{12}]}, \tag{8.31b}$$

where ϕ_{12} is a phase angle introduced between the two dipole outputs. The time average of the product of the real parts of these expressions, as in Sect. 8.2, gives the plane wave signal spatial correlation function between the outputs of the two dipoles:

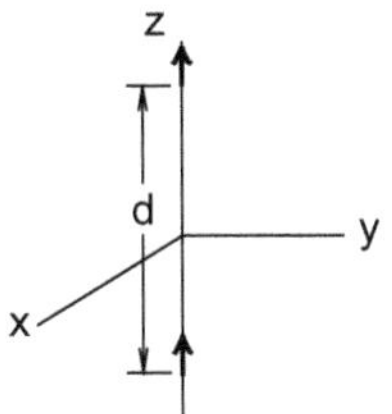

(a) Sensing the z-component of velocity at both points.

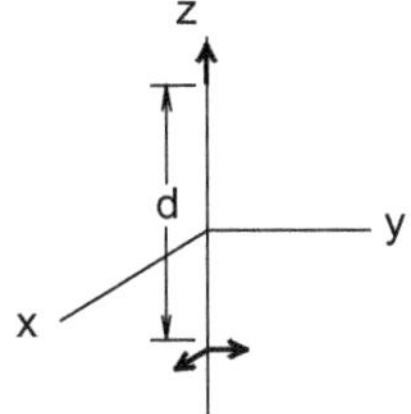

(b) Sensing the z-component at one point, the x or y component at the other point.

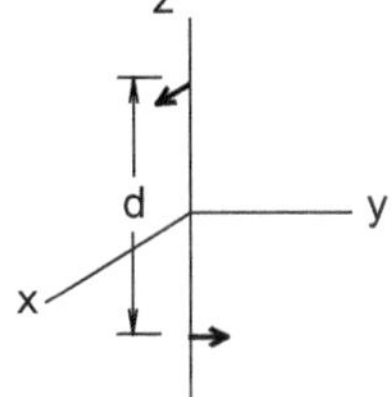

(d) Sensing the x-component at one point, the y-component at the other point.

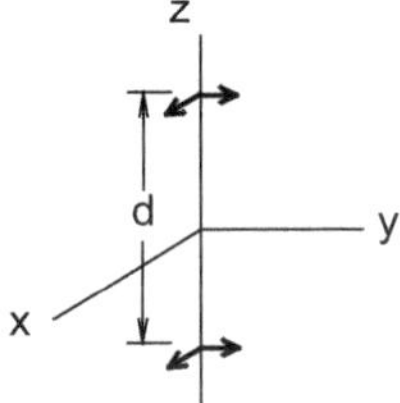

(b) Sensing the x-component or the y component at both points.

Fig. 8.25 Diagrams for calculating the spatial correlation between two uniaxial dipole sensors separated by a distance, *d*, sensing different velocity components (each solid arrow represents a dipole sensor)

$$\langle V_{z1} V_{z2} \rangle^{\mathrm{s}} = \frac{1}{2} \left(\frac{2\pi s}{\lambda} M_1 p_{\mathrm{i}} \right)^2 \cos^2\theta \cos\left(kd\cos\theta + \phi_{12}\right). \tag{8.32}$$

It can be seen that this correlation function is similar to that for two pressure hydrophones given in Eq. (8.12), but that it also depends on the angle between the dipole axes and the plane wave direction, θ, and vanishes when the dipole response is zero at $\theta = 90°$.

The noise correlation between two dipole outputs depends on the type of noise they are receiving, e.g., if the noise is completely incoherent the correlation function is $\rho_{ij}^{\mathrm{n}} = 0$, $i \neq j$ and $\rho_{ii}^{\mathrm{n}} = 1$. But ambient noise is usually partially coherent which will give two dipole outputs a nonzero cross correlation in some cases. A frequently used model for isotropic ambient noise consists of plane waves arriving at any point from all directions, randomly in time and in direction [55]. Since $\langle V_{z1} V_{z2} \rangle^{\mathrm{s}}$ is the spatial correlation between two dipoles receiving one plane wave, the spatial correlation for this isotropic noise model can be found by averaging $\langle V_{z1} V_{z2} \rangle^{\mathrm{s}}$ over all directions, i.e.:

$$\langle V_{z1}V_{z2}\rangle^{\mathrm{n}} = \frac{1}{4\pi}\int_0^{2\pi}\int_0^{\pi}\langle V_{z1}V_{z2}\rangle^{\mathrm{s}} \sin\theta \mathrm{d}\theta \mathrm{d}\phi. \tag{8.33a}$$

Using Eq. (8.32) in Eq. (8.33a), with p_{i} replaced by a noise pressure amplitude, p_{n}, and with the substitution $x = \cos\theta$, gives

$$\begin{aligned}\langle V_{z1}V_{z2}\rangle^{\mathrm{n}} &= \frac{1}{4}\left(\frac{2\pi s}{\lambda}M_1 p_{\mathrm{n}}\right)^2 \int_{-1}^{1} x^2 \cos(xkd + \phi_{12})dx \\ &= \frac{1}{2}\left(\frac{2\pi s}{\lambda}M_1 p_{\mathrm{n}}\right)^2 \cos\phi_{12}\left[\frac{2\cos kd}{(kd)^2} - \frac{2\sin kd}{(kd)^3} + \frac{\sin kd}{kd}\right] \\ &= \frac{1}{2}\left(\frac{2\pi s}{\lambda}M_1 p_{\mathrm{n}}\right)^2 \cos\phi_{12}[j_0(kd) - 2j_1(kd)/kd],\end{aligned} \tag{8.33b}$$

where j_0 and j_1 are spherical Bessel functions of the first kind. Equation (8.33b) is the spatial correlation between the two z-components of velocity at the two locations separated by distance d, as shown in Fig. 8.25a, in isotropic noise.

The spatial correlation between a z-component at one point and an x- or y-component at the other point (see Fig. 8.25b) is found by first replacing one of the $\cos\theta$ factors in Eqs. (8.31a) and (8.31b) by $\sin\theta\cos\phi$ or $\sin\theta\sin\phi$. This leads to a plane wave signal correlation function that differs from Eq. (8.32), and, in either case, the integration over ϕ in the averaging process makes these noise correlations vanish, leaving

$$\langle V_{z1}V_{x2}\rangle^{\mathrm{n}} = \langle V_{z1}V_{y2}\rangle^{\mathrm{n}} = 0. \tag{8.34}$$

The correlation between an x-component at one point and a y-component at the other point (see Fig. 8.25c) involves the product of $\sin\theta\cos\phi$ and $\sin\theta\sin\phi$ which is proportional to $\sin 2\phi$, and also integrates to zero in the averaging, leaving

$$\langle V_{x1}V_{y2}\rangle^{\mathrm{n}} = 0. \tag{8.35}$$

These zero correlations are special properties of vector sensors in isotropic noise, which also hold between the components of a single triaxial vector sensor.

The correlations between x-components at both points, or y-components at both points (Fig. 8.25d), are equal and nonzero. They involve $\cos^2\phi$ or $\sin^2\phi$, which integrate to the value π in the averaging; then with $x = \cos\theta$,

$$\begin{aligned}\langle V_{x1}V_{x2}\rangle^{\mathrm{n}} = \langle V_{y1}V_{y2}\rangle^{\mathrm{n}} &= \frac{1}{8}\left(\frac{2\pi s}{\lambda}M_1p_{\mathrm{n}}\right)^2\int_{-1}^{1}(1-x^2)\cos(xkd+\phi_{12})dx \\ &= \frac{1}{4}\left(\frac{2\pi s}{\lambda}M_1p_{\mathrm{n}}\right)^2\cos\phi_{12}j_0(kd) - \frac{1}{2}\langle V_{z1}V_{z2}\rangle^{\mathrm{n}} \\ &= \frac{1}{2}\left(\frac{2\pi s}{\lambda}M_1p_{\mathrm{n}}\right)^2\cos\phi_{12}[j_1(kd)/kd],\end{aligned} \tag{8.36}$$

where the last two steps follow by use of Eq. (8.33b).

Figure 8.26 shows these spatial correlations of velocity components as a function of *kd*, after normalization by dividing by their value at $kd = 0$, where they are compared with the normalized spatial correlation of pressure for isotropic noise, i.e., $\sin kd/kd = j_0(kd)$ given in Eq. (8.16). Figure 8.26 also contains the correlation between the pressure at one point and the *z*-component of velocity at the other point, $\langle V_{p1}\, V_{z2}\rangle$.

As an example of an unbaffled array of vector sensors consider a horizontal line array of dipoles with separation *d*. Let their axes be parallel to each other and perpendicular to the line as shown in Fig. 8.27, where the *xz* plane is considered to be parallel to the ocean surface.

The dipole axes are oriented in the *x*-direction to simulate a line array of dipole sensors being towed in the *z*-direction with the nulls pointed toward the towing ship. We will compare the array gain and signal-to-noise ratio of this array with the line

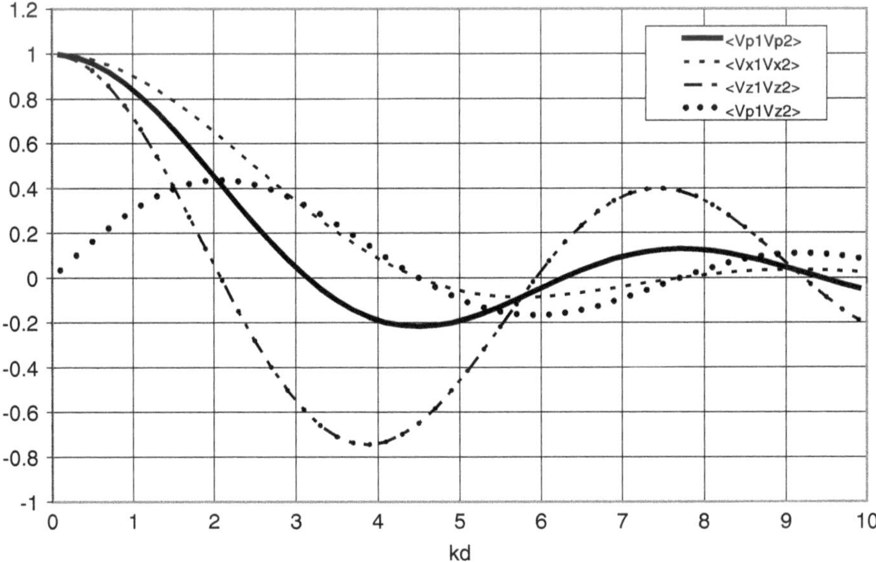

Fig. 8.26 Normalized spatial correlation functions between acoustic pressure and particle velocity components

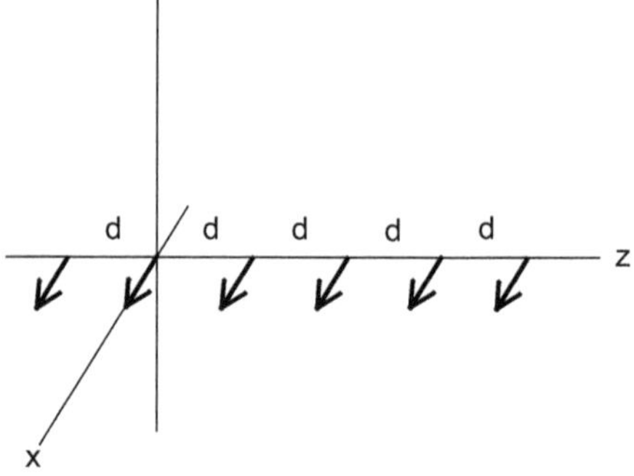

Fig. 8.27 A line array of dipole vector sensors with spacing d

array of pressure sensors given as the first example in Sect. 8.4.1. The plane wave signal spatial correlation for the x-components of velocity is

$$\langle V_{x1} V_{x2} \rangle^{\mathrm{s}} = \frac{1}{2} \left(\frac{2\pi s}{\lambda} M_1 p_{\mathrm{i}} \right)^2 \sin^2\theta \cos^2\phi \cos\left(kd\cos\theta + \phi_{12}\right), \tag{8.37}$$

and the isotropic noise correlation is given in Eq. (8.36). When $d = 0$ and $\phi_{12} = 0$, these expressions reduce to the mean squared signal and noise for one dipole with signal-to-noise ratio given by

$$(\mathrm{SNR})_1 = \frac{\langle V_{x1} V_{x1} \rangle^{\mathrm{s}}}{\langle V_{x1} V_{x1} \rangle^{\mathrm{n}}} = \frac{\sin^2\theta \cos^2\phi \ p_{\mathrm{i}}^2}{(1/3) p_{\mathrm{n}}^2}, \tag{8.38}$$

since $j_1(kd)/kd = 1/3$ when $d = 0$. For a plane wave signal arriving from the direction $\theta = 90°$ and $\phi = 0°$, i.e., along the x-axis, $(\mathrm{SNR})_1 = 3(p_{\mathrm{i}}/p_{\mathrm{n}})^2$, which is three times the SNR for an omnidirectional pressure sensor, showing the effect of the directivity factor of 3 for a dipole in isotropic noise.

Consider the array steered to broadside with $\phi_{12} = 0$ and a plane wave signal arriving along the x-axis ($\theta = 90°$, $\phi = 0°$). These conditions make all the signal correlations in Eq. (8.37) equal, and, for an array of N dipoles, they sum to $N^2[(2\pi s/\lambda)M_1 p_{\mathrm{i}}]^2/2$. As in Sect. 8.4.1 we will consider only noise cross correlations between nearest neighbors and next nearest neighbors; from Eq. (8.36) these cross correlations sum to:

$$\begin{aligned} &\{2(N-1)[j_1(kd)/kd] + 2(N-2)[j_1(2kd)/2kd]\}\frac{1}{2}\left(\frac{2\pi s}{\lambda} M_1 p_{\mathrm{n}}\right)^2 \\ &\approx 2N(0.37)\frac{1}{2}\left(\frac{2\pi s}{\lambda} M_1 p_{\mathrm{n}}\right)^2. \end{aligned}$$

The last approximate form is for N large and for $kd = \pi/2$ using values of $\langle V_{x1} V_{x2} \rangle$ from Fig. 8.26 divided by 3 to remove the normalization factor. In addition the

N noise autocorrelations are each equal to $[(2\pi s/\lambda)M_1 p_n]^2/6$. Combining these results gives the array signal-to-noise ratio:

$$(\text{SNR})_\text{N} = N^2 p_\text{i}^2/[N/3 + 2N(0.37)]p_\text{n}^2 = 0.93N(p_\text{i}/p_\text{n})^2. \tag{8.39}$$

Equation (8.39) shows that the partial coherence of the noise, given by the term $2N$ (0.37) in this specific case, reduces $(\text{SNR})_\text{N}$ significantly. For incoherent noise this term would not exist, and $(\text{SNR})_\text{N}$ would be $3N(p_\text{i}/p_\text{n})^2$.

Using Eqs. (8.38) and (8.39), the array gain is

$$\text{AG} = (\text{SNR})_\text{N}/(\text{SNR})_1 = 10\log(0.93N/3) = 10\log N - 5.1\,\text{dB}, \tag{8.40}$$

which is 1.5 dB lower than the array gain for the same array of pressure sensors under the same conditions, given in Sect. 8.4.1. Note that the array gain in Eq. (8.40) is based on the definition in Sect. 8.2, and in this case, where the single sensor is a dipole, the array gain is not referred to an omnidirectional sensor. If the array gain of the dipole array is referred to an omnidirectional sensor it is 4.8 dB higher, which makes it higher than the array gain for the pressure sensor array. Probably a more meaningful comparison is in terms of the array signal-to-noise ratio; then we find $0.44N(p_\text{i}/p_\text{n})^2$ for the pressure sensor array from Sect. 8.4.1, and $0.93N(p_\text{i}/p_\text{n})^2$ for the dipole sensor array from Eq. (8.39), giving the dipole array a 3 dB advantage. But this result applies only to the broadside beam at the frequency where the array spacing is one quarter wavelength.

8.5.3 Hull-Mounted Arrays in Structural Noise

In hull-mounted arrays, at speeds of interest, structural noise and flow noise exceed ambient sea noise. The hull, when excited by flow and machinery, vibrates with a mixture of flexural waves of different frequencies and wavenumbers, resulting in a partially coherent noise field in the water at the location of an array. Our objective here is comparison of pressure and vector sensors, and since the array SNR depends on the individual sensor SNR, we will develop a simple model for determining the SNR of both types of sensors mounted on a compliant baffle in flexural wave noise. This simplified approach can reveal some of the problems involved in vector sensor arrays and provide a basis for planning the experiments that are necessary to determine their feasibility [1].

Cray clarified one aspect of the pressure versus vector sensor question with a model based on a plane acoustic wave signal in flexural wave noise produced by a vibrating plate [56], and we will start by presenting his argument. The evanescent flexural wave noise travels parallel to the plate (the x-direction) and decays with distance in the water from the plate (the z-direction) as described in Sect. 8.3.2. If it is assumed that the free plate wave number dominates the plate vibration, the pressure in the flexural wave can be described approximately by

$$p_{\mathrm{n}} = P_{\mathrm{n}} \mathrm{e}^{j(\omega t - k_{\mathrm{p}} x)} \mathrm{e}^{-\left(k_{\mathrm{p}}^2 - k_0^2\right)^{1/2} z}, \tag{8.41}$$

where k_0 is the acoustic wave number and k_{p} is the free plate wave number given in Eq. (8.19). The signal is a plane acoustic pressure wave approaching the plate at normal incidence and described by

$$p_{\mathrm{s}} = P_{\mathrm{i}} \mathrm{e}^{j(\omega t - k_0 z)}. \tag{8.42}$$

The z-components of particle velocity associated with the noise and signal are:

$$u_{z\mathrm{n}} = \left[P_{\mathrm{n}} \left(k_{\mathrm{p}}^2 - k_0^2\right)^{1/2} / j\omega\rho_0\right] \mathrm{e}^{-\left(k_{\mathrm{p}}^2 - k_0^2\right)^{1/2} z} \mathrm{e}^{j(\omega t - k_{\mathrm{p}} x)}, \tag{8.43}$$

$$u_{z\mathrm{s}} = (jk_0 P_{\mathrm{i}} / j\omega\rho_0) \mathrm{e}^{j(\omega t - k_0 z)} = (P_{\mathrm{i}} / \rho_0 c_0) \mathrm{e}^{j(\omega t - k_0 z)}. \tag{8.44}$$

At a distance z from the plate, the magnitude of the ratio of signal pressure to noise pressure is

$$p_{\mathrm{s}}(z) / p_{\mathrm{n}}(z) = P_{\mathrm{i}} / P_{\mathrm{n}} \mathrm{e}^{-\left(k_{\mathrm{p}}^2 - k_0^2\right)^{1/2} z}, \tag{8.45}$$

while the magnitude of the ratio of signal velocity to noise velocity is

$$u_{z\mathrm{s}}(z) / u_{z\mathrm{n}}(z) = P_{\mathrm{i}} k_0 / P_{\mathrm{n}} \left(k_{\mathrm{p}}^2 - k_0^2\right)^{1/2} \mathrm{e}^{-\left(k_{\mathrm{p}}^2 - k_0^2\right)^{1/2} z}. \tag{8.46}$$

It can be seen from Eqs. (8.45) and (8.46) that the velocity signal-to-noise ratio is lower than the pressure signal-to-noise ratio by the factor $k_0/(k_{\mathrm{p}}^2 - k_0^2)^{1/2}$ when $k_{\mathrm{p}} > \sqrt{2} k_0$ at any value of z; for k_{p} large compared to k_0 it is lower by the factor k_0/k_{p}. Values of k_0 and k_{p} are given in Fig. 8.19, e.g., for a 2 in. (0.0508 m) steel plate in water $k_0/k_{\mathrm{p}} \approx 0.34$ at 500 Hz and 0.46 at 1000 Hz. Note that this is a comparison of the velocity and pressure fields and does not consider the voltage sensitivity differences that would also be involved in sensing the velocity and pressure. Using the same model, Cray also compared velocity gradient sensors to pressure sensors, and showed that the velocity gradient ratio is lower than the pressure ratio by the factor $(k_0/k_{\mathrm{p}})^2$ when k_{p} is large.

A complete comparison of hull-mounted pressure and velocity sensors must include the effect of the surface impedance and the signal gain of the different inner decouplers that must be used with the two types of sensors (see Fig. 8.15). A relevant comparison can be made by considering two arrays with the same arrangement and number of each type of sensor, but with different inner decouplers designed to improve the performance of each type of sensor:

1. An array of pressure sensors mounted on an inner decoupler consisting of a compliant baffle with a heavy signal conditioning plate that provides a positive pressure signal gain over a certain range of frequency and a reduction of the

structural noise pressure from the hull. The pressure sensors must be embedded in a suitable outer decoupler to reduce flow noise.

2. An array of velocity sensors mounted on an inner decoupler consisting of a compliant baffle with no signal conditioning plate. The baffle provides a positive velocity signal gain over the same range of frequency, and a reduction of the structural noise velocity from the hull. This coating might also be capable of reducing high frequency target strength. The velocity sensors must be embedded in a suitable outer decoupler to reduce flow noise.

Equations (8.28)–(8.30) offer a systematic way of addressing the proposed comparison, with some reinterpretation of some of the parameters in the vector sensor case. However, not much detailed information is available regarding most of these parameters, although some modeling and analysis has been done. Modeling of flexural noise pressure reduction by a compliant layer was discussed in Sect. 8.4.2, while modeling of flow noise pressure reduction by an outer decoupler was discussed in Sect. 8.4.3. Similar modeling of the reduction of noise velocity has been quite limited [57, 58]. An experimental investigation has also been carried out, using accelerometers as velocity sensors, mounted on an air-voided elastomeric compliant baffle covering a portion of a full scale submarine hull fixture [1, 59]. The sensitivity relative to free field sensitivity and the reflection gain (signal gain) were measured as a function of frequency, and both showed approximately the expected 6 dB increase at sufficiently high frequency. However, at lower frequencies in the range of interest for long-range passive sonar, both measures of performance were considerably degraded with the particular 3 in. thick compliant baffle used in these experiments. The velocity reduction (insertion loss) was also measured using vibration generators to activate the hull fixture and simulate structural noise. The measured velocity reduction provided by this particular baffle was also degraded at low frequency. The measurements showed that velocity sensors were feasible for submarine arrays, but baffles with improved low frequency performance were needed.

It is not feasible here to analytically compare velocity and pressure sensor arrays in detail, but we will extend Cray's model, described above [56], by adding a simple model of an inner decoupler consisting of a compliant baffle with no signal conditioning plate. This is the case that would be expected to favor velocity sensors. Consider a large vibrating plate representing the hull in which the free plate wave number is dominant. This hull plate is covered by a layer of material of thickness L that represents a compliant baffle, as shown in Fig. 8.28.

To simplify the calculations we will treat the baffle as a fluid with density and sound speed of ρ_1 and c_1. We will also assume that the evanescent noise wave from the vibrating hull plate extends through the baffle into the water, where the density and sound speed are ρ_0 and c_0. The evanescent waves in the baffle and in the water can be written as

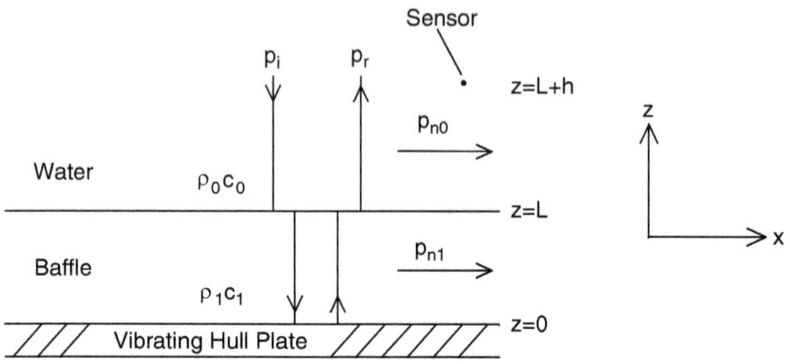

Fig. 8.28 A vibrating plate produces evanescent flexural noise waves traveling parallel to the plate in the baffle and in the water. An acoustic signal arrives at normal incidence and is reflected. An acoustic sensor is located at $z = L + h$

$$p_{n1}(z) = P_{n1}\mathrm{e}^{-\left(k_p^2-k_1^2\right)^{1/2}z}\mathrm{e}^{j\left(\omega t-k_p x\right)}, \quad 0 \le z \le L, \tag{8.47}$$

$$p_{n0}(z) = P_{n0}\mathrm{e}^{-\left(k_p^2-k_0^2\right)^{1/2}z}\mathrm{e}^{j\left(\omega t-k_p x\right)}, \quad z > L. \tag{8.48}$$

We assume that the normal component of noise velocity is continuous at the boundary between the baffle and the water ($z = L$), as it is at the boundary between the plate and the baffle ($z = 0$), which gives the relationship between P_{n1} and P_{n0} :

$$P_{n0} = P_{n1}\frac{\rho_0\left(k_p^2 - k_1^2\right)^{1/2}}{\rho_1\left(k_p^2 - k_0^2\right)^{1/2}}\mathrm{e}^{-\left[\left(k_p^2-k_1^2\right)^{1/2}-\left(k_p^2-k_0^2\right)^{1/2}\right]L}. \tag{8.49}$$

The noise pressure in the water at $z > L$ is given by substituting Eq. (8.49) into Eq. (8.48), and the z-component of noise velocity in the water is then calculated from the pressure with the result

$$u_n(z) = \left(P_{n0}/j\omega\rho_0\right)\left(k_p^2 - k_0^2\right)^{1/2}\mathrm{e}^{-\left(k_p^2-k_0^2\right)^{1/2}z}\mathrm{e}^{j\left(\omega t-k_p x\right)}, \quad z > L. \tag{8.50}$$

Equations (8.48)–(8.50) give the pressure and velocity of the flexural wave noise in the water where sensors would be mounted at some small standoff distance from the surface of the baffle.

Next we consider the signal to be a plane acoustic wave normally incident from the water:

$$p_i = P_i\mathrm{e}^{j(\omega t + k_0 z)},$$

with a reflected wave at the water–baffle boundary

$$p_r = P_r e^{j(\omega t - k_0 z)}.$$

We will assume that the part of the incident wave transmitted into the baffle is perfectly reflected at the plate and that there is no absorption in the baffle. Thus the incident and reflected waves inside the baffle, shown in Fig. 8.28, have the same amplitude. Both pressure and particle velocity must be continuous at the fluid boundary between the baffle and the water, which gives the relationships needed to solve for the amplitude of the reflected acoustic wave in the water:

$$P_r = P_i e^{2jk_0 L} \frac{1 - j\rho_0 c_0 \tan k_1 L / \rho_1 c_1}{1 + j\rho_0 c_0 \tan k_1 L / \rho_1 c_1} = P_i e^{2j(k_0 L - y)}, \tag{8.51}$$

where $\tan y = (\rho_0 c_0 / \rho_1 c_1) \tan k_1 L$.

The reflection coefficient, P_r/P_i, has a magnitude of unity because of the assumptions of perfect reflection at the plate and no absorption in the baffle. In this model, the baffle only changes the phase of the reflected wave in the water, which changes the locations of the maxima and minima in the standing wave formed by the incident and reflected waves. Thus the baffle has a strong effect on the output of both velocity and pressure sensors located near its surface. The resultant signal pressure in the water at $z \geq L$ is given by

$$\begin{aligned} p_s(z) &= p_i + p_r = P_i \left[e^{jk_0 z} + (P_r/P_i) e^{-jk_0 z} \right] e^{j\omega t} \\ &= 2P_i e^{j(k_0 L - y + \omega t)} \cos\left[k_0 (L - z) - y \right]. \end{aligned} \tag{8.52}$$

And the z-component of the resultant signal velocity is

$$\begin{aligned} u_s(z) &= -(P_i/\rho_0 c_0) \left[e^{jk_0 z} - (P_r/P_i) e^{-jk_0 z}) \right] e^{j\omega t} \\ &= -2j(P_i/\rho_0 c_0) e^{j(k_0 L - y + \omega t)} \sin\left[k_0 (L - z) - y) \right]. \end{aligned} \tag{8.53}$$

Equations (8.52) and (8.53) show that the quantity $y = \tan^{-1}[(\rho_0 c_0 / \rho_1 c_1) \tan k_1 L]$ determines the locations of the signal pressure and signal velocity maxima and minima. These locations have a critical effect on the sensor outputs, and they depend only on the baffle parameters $\rho_0 c_0 / \rho_1 c_1$ and $k_1 L$.

We can now compare the signal and noise for both pressure and velocity at a given standoff distance from the surface of the baffle. The following ratio of velocity signal-to-noise ratio to pressure signal-to-noise ratio is one way to make this comparison:

$$R(z) = \left| \frac{u_s(z)/u_n(z)}{p_s(z)/p_{n0}(z)} \right| \tag{8.54}$$

If this ratio is more than unity the conditions may be more favorable for sensing velocity, and, if less than unity, more favorable for sensing pressure, depending on the relative sensitivities of the velocity and pressure sensors.

For zero standoff distance, $z = L$, we have from Eqs. (8.48), (8.50), (8.52), and (8.53), and recalling that $\tan y = (\rho_0 c_0/\rho_1 c_1) \tan k_1 L$:

$$R(L) = \frac{k_0}{\left(k_p^2 - k_0^2\right)^{1/2}} \frac{\rho_0 c_0}{\rho_1 c_1} \tan k_1 L, \tag{8.55a}$$

and for a standoff distance of h, $z = L + h$,

$$R(L+h) = \frac{k_0}{\left(k_p^2 - k_0^2\right)^{1/2}} \tan\left(y + k_0 h\right). \tag{8.55b}$$

The first factor in Eqs. (8.55a) and (8.55b) is given by Cray's model [56] and was discussed following Eq. (8.46); the second factor shows the modification to Cray's model caused by the baffle. It is evident for $h = 0$ that decreasing $\rho_1 c_1$ increases $R(L)$ corresponding to a softer baffle and higher signal velocity, while increasing $\rho_1 c_1$ decreases $R(L)$ corresponding to a harder baffle and higher signal pressure.

As a numerical example to show the effect of standoff distance we will consider a moderately compliant baffle modeled by a fluid with both density and sound speed half that of water, $\rho_1 = 0.5\rho_0$ and $c_1 = 0.5c_0$. Calculated values of $R(L + h)$ are in Table 8.1 for a 2 in. thick steel plate, a baffle thickness of 0.075 m (3 in.), and a range of standoff distances. They show that as the frequency is lowered the pressure SNR exceeds the velocity SNR at a frequency near 800 Hz for $h = 0$ where $R(L + h)$ becomes less than unity. For this baffle, the signal pressure is not zero on the surface; instead it decreases with distance from the surface and goes to zero at $h \approx 0.66$ m at a frequency of 500 Hz. As the pressure decreases, the velocity increases and reaches a maximum at the point where the pressure is zero; thus $R(L + h)$ increases as h increases. At higher frequencies the zero of pressure is nearer the baffle, and for a given h the values of $R(L + h)$ are higher.

The results in Table 8.1 are based on the velocity and pressure signal and noise fields, and do not include the means for sensing those fields. The sensitivities of velocity and pressure sensors are also critical in determining the conditions under which each type of sensor is most effective. For example, it was shown in Sect. 6.5.1 that pressure sensors have as much output as dipole velocity sensors near an ideal soft baffle when the locations and sensitivities of both types of sensor are considered. This occurs as long as the pressure sensor is located at the same

Table 8.1 Signal-to-noise comparison of velocity and pressure fields near a compliant baffle of thickness 0.075 m

Frequency (Hz)	$k_0/(k_p^2 - k_0^2)^{1/2}$	$\rho_0 c_0/\rho_1 c_1$	R (L + h)			
			$h = 0$	$h = 0.025$ m	$h = 0.05$ m	$h = 0.075$ m
500	0.35	4	0.46	0.51	0.58	0.63
750	0.44	4	0.90	1.12	1.43	1.91
1000	0.52	4	1.52	2.17	3.70	–

distance from the baffle surface as the outer part of the dipole, because its greater sensitivity compensates for the pressure being lower. However, when the acoustic centers of the pressure and velocity sensors are colocated and the sensitivities are the same for both signal and noise, the sensitivities cancel out as we will see later. The ratio in Table 8.1 is then a valid signal-to-noise comparison of voltage output.

A comparison that includes the sensitivities, but not the noise, can be made by extracting the ratios of signal velocity to signal pressure from the values of $R(L+h)$ in Table 8.1. It can be seen from Eqs. (8.52), (8.53), and (8.55b) that

$$\frac{\rho_0 c_0 u_s}{p_s} = -j \frac{R(L+h)}{\left[k_0 / \left(k_p^2 - k_0^2\right)^{1/2}\right]}, \tag{8.56}$$

which gives the values of $|\rho_0 c_0 u_s/p_s|$ in Table 8.2. These values are then converted to ratios of velocity sensor voltage output, V_u, to pressure sensor voltage output, V_p, using

$$\frac{V_u}{V_p} = \frac{u_s M_u}{p_s M} = \left(\frac{\rho_0 c_0 u_s}{p_s}\right)\left(\frac{M_u}{\rho_0 c_0 M}\right). \tag{8.57}$$

A pressure hydrophone sensitivity of $M = -193$ dB//V/μPa, as in Fig. 6.11, and a velocity sensor sensitivity of $M_u = (-37 + 20\log\omega)$ dB//V s/m based on the accelerometer discussed in Sect. 6.7.5 (with an acceleration sensitivity of –17 dB//V/g, see [58] of Chap. 6) were used to obtain the voltage ratios in Table 8.2.

Comparison of these two sensors represents a practical case since the hydrophone is a piezoelectric ceramic sphere of about 2 in. diameter, and the velocity sensor is a flexural piezoelectric ceramic disc accelerometer of about 1.5 in. diameter encased in a buoyant body of about 2 in. diameter. Thus both sensors could be mounted at the same distance, h, from the surface of a baffle.

Since the ratio $|\rho_0 c_0 u_s/p_s| = 1$ in a plane wave, and the values in Table 8.2 exceed unity, it is evident that in these cases the baffle has enhanced the velocity relative to the pressure as expected for a compliant baffle. However, the difference in sensor sensitivities makes the voltage output from the pressure sensor exceed that from the velocity sensor at low frequency. The basic cause of this behavior is the low frequency falloff of velocity sensor sensitivity. These results are consistent with the conclusion in Sect. 6.5.1 that in some cases pressure sensors may be superior to velocity sensors, even when mounted on an ideal soft baffle without a signal conditioning plate. The dependence on standoff distance and frequency

Table 8.2 Comparison of voltage outputs of a velocity sensor and a pressure sensor near a compliant baffle

	$h = 0.025$ m		$h = 0.05$ m	
Frequency (Hz)	$\lvert\rho_0 c_0 u_s/p_s\rvert$	V_u/V_p	$\lvert\rho_0 c_0 u_s/p_s\rvert$	V_u/V_p
500	1.46	0.17	1.66	0.19
750	2.54	0.51	3.25	0.65
1000	4.17	1.11	7.12	1.90

shown in Table 8.2 emphasizes the importance of sensor size and the way sensors are mounted on a compliant surface [60, 61].

A more complete comparison of pressure and velocity sensors must consider the voltage sensitivities of the two types of sensors, the values of signal and noise velocity and pressure at the locations of the sensors, including the case where the locations are not the same and also the case where the noise sensitivity of the velocity sensor depends on the type of noise. We will develop a generalization of the ratio $R(L+h)$ in Table 8.1 which can account for all these factors and then apply it to the same compliant baffle model. The signal and noise voltage output magnitudes from a pressure sensor located at h_p and a velocity sensor at h_u can be written:

$$\begin{aligned} V_{us} &= M_u u_s(h_u) = \rho_0 c_0 M_p u_s(h_u), \\ V_{un} &= M_{un} u_n(h_u) = \rho_0 c_0 M_{pn} u_n(h_u), \\ V_{ps} &= M p_s(h_p), \\ V_{pn} &= M p_n(h_p). \end{aligned}$$

The velocity sensor has velocity sensitivity, M_u, and pressure sensitivity, M_p, for signal, but it may have different values for noise (M_{un} and M_{pn}). The pressure sensor has pressure sensitivity, M, for both signal and noise. The general ratio that compares the signal-to-noise performance of velocity and pressure sensors near a baffle is then

$$V_R = \frac{V_{us}/V_{un}}{V_{ps}/V_{pn}} = \left(\frac{M_p}{M_{pn}}\right)\left(\frac{u_s(h_u)p_n(h_p)}{u_n(h_u)p_s(h_p)}\right). \tag{8.58}$$

The sensitivity of the pressure sensor has cancelled out of this ratio, and the first factor depends only on the sensitivities of the velocity sensor to the signal and to the noise and is unity if they are the same. The second factor depends on the velocity and pressure fields at two different distances from the baffle and is equal to $R(L+h)$ in Eq. (8.54) if the acoustic centers of the two sensors are at the same distance from the baffle.

The dipole is an example of a velocity sensor for which the flexural wave noise sensitivity differs from the plane wave signal sensitivity. The flexural wave pressure in Eq. (8.48) shows that the difference between the outputs of the two poles of the dipole is $2Mp_{n0}(k_p^2/k_0^2 - 1)^{1/2}(\pi s/\lambda)$, where s is the dipole separation. Comparison with Eq. (6.37a) shows that this voltage output corresponds to a noise sensitivity that exceeds the plane wave pressure sensitivity by the factor $(k_p^2/k_0^2 - 1)^{1/2}$; for example, this factor has a value of about 2.8 for a 2 in. thick steel plate at 500 Hz (see Fig. 8.19). A similar factor would be expected to apply to other types of velocity sensors, because it arises from the fact that the flexural wave noise field is inhomogeneous, and its amplitude varies with position, i.e., it is similar to the near-field effect discussed in Sect. 6.7.6.

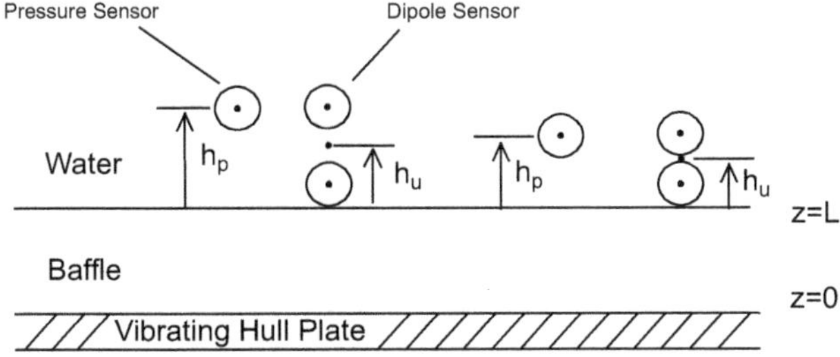

Fig. 8.29 Comparison of pressure sensors and dipole velocity sensors, mounted near a compliant baffle. Both types of sensors have the same standoff distance. Two cases with different standoff distances are shown

Table 8.3 Comparison of velocity and pressure sensors near a compliant baffle

Case I ($h_p = 0.1\,\mathrm{m}$, $h_u = 0.0625\,\mathrm{m}$, $s = 0.075\,\mathrm{m}$):	500 Hz	$V_R = 0.19$
	750 Hz	$V_R = 0.77$
Case II ($h_p = 0.075\,\mathrm{m}$, $h_u = 0.05\,\mathrm{m}$, $s = 0.05\,\mathrm{m}$):	500 Hz	$V_R = 0.18$
	750 Hz	$V_R = 0.63$
	800 Hz	$V_R = 0.96$
	1000 Hz	$V_R = 6.4$

We will now use Eq. (8.58) to compare a pressure sensor consisting of one spherical hydrophone located at distance h_p from a baffle with a dipole velocity sensor consisting of two of the same hydrophones with its acoustic center at h_u as shown in Fig. 8.29.

The case where the two sensors have the same projection from the baffle corresponds to a maximum dipole separation of $s = h_p - a$ where a is the hydrophone radius and to $h_u = a + s/2 = (a + h_p)/2$. The baffle is described by the same parameters that were used for Table 8.1 which, e.g., give tan $y = 1.30$ at 500 Hz. Using Eqs. (8.48), (8.50), (8.52), and (8.53) in Eq. (8.58) gives

$$V_R = \frac{\sin(k_0 h_u + y)}{\left(k_p^2/k_0^2 - 1\right)\cos(k_0 h_p + y)} e^{-\left(k_p^2/k_0^2 - 1\right)^{1/2}\left(k_0 h_p - k_0 h_u\right)},$$

where $M_{pn} = \left(k_p^2/k_0^2 - 1\right)^{1/2} M_p$ was used for the dipole. Numerical results are given in Table 8.3 for two cases, both using spherical hydrophones of 0.05 m diameter. In both cases the single pressure sensor and the outer pole of the dipole are located at the same distance from the baffle, and the inner pole of the dipole is as close to the baffle as possible.

In Table 8.3 values of V_R less than unity mean that the pressure sensor SNR exceeds the velocity sensor SNR. It is evident that for the specific parameters that describe this baffle the velocity sensor is superior above about 800 Hz, while the pressure sensor is superior below that frequency. This behavior is caused by the location of the pressure hull, which moves closer to the baffle, and closer to the pressure sensor, as the frequency increases.

The results in Tables 8.1, 8.2, and 8.3 apply to only one set of baffle parameters, but they illustrate some of the conditions that may arise with velocity sensors mounted on compliant baffles. These results indicate that, for the frequency range of interest, careful evaluation of the baffle properties and of the sensor mounting details is necessary before it can be concluded that velocity sensors are superior to pressure sensors. It should be emphasized that the specific results given here are based on an idealized form of structural noise and a simplified model of a compliant baffle. A better model would include realistic signal reflection from the hull at the inner side of the baffle, absorption in the baffle and signals at other than normal incidence.

Flow noise presents another problem that is likely to be as serious as structural noise in hull-mounted vector sensor arrays, but it has not been considered here. The use of vector sensor arrays on an ocean glider has recently been discussed [62]. A summary of the issues and problems associated with use of vector sensors in towed arrays has been given by Abraham and Berliner [63]. Flow-induced noise resulting from low speed flows in moored and drifting array applications using vector sensors has also been studied [64].

8.6 Steered Planar Circular Arrays

Circular arrays of transducers are useful because with the circular symmetry they can be steered azimuthally in all directions without change in the directivity pattern. They are also useful in that the rings of the array can be separated in a sparse manner without grating lobe problems, typically associated with linear arrays. Analytical results for steered ring arrays of discrete transducer elements are considered first and then approximated by continuous arrays with continuous steering in order to obtain results in a simpler form. The model is based on fixed values for the velocities and does not include array interactions and presumes an equivalent rigid baffle condition. Although array interactions may cause a variation in velocity from ring to ring, the velocities within an unsteered ring alone are identical because of circumferential symmetry. This section has allowed us to also present a quite detailed analysis of another specific acoustical transducer array.

Consider first one transducer in the xy plane with its center at a radial distance a_i from the origin and at the azimuthal angle ϕ_i from the x-axis with an arbitrary far-field point located at r, θ, ϕ where $r \gg a_i$, using spherical coordinates shown in Fig. 8.30.

The distance from the center of the transducer to the field point is

$$r_i = r - a_i \sin\theta \cos(\phi - \phi_i). \tag{8.59}$$

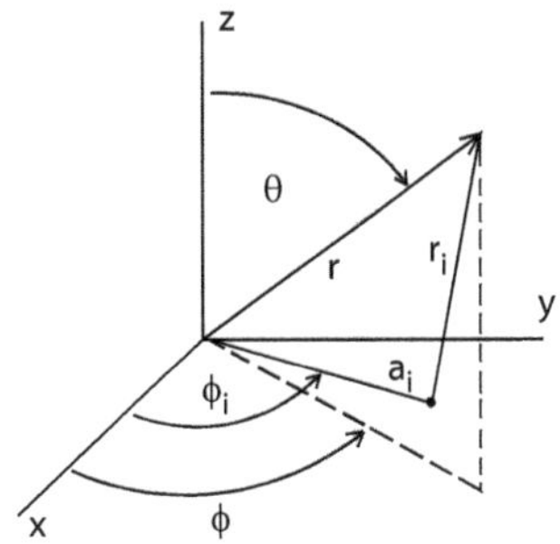

Fig. 8.30 Spherical coordinate system with source in the xy plane at a_i and θ_i

And for the transducer located in the plane of the beam pattern, we get the more familiar far-field condition expression $r_i = r - a_i \sin\theta$, see, also, Fig. 7.5. Here $a_i \sin\theta$ is the difference between far-field parallel rays, $r - r_i$, both at angle θ. The far-field pressure for an array of N identical small omnidirectional transducers with arbitrary velocities u_i arranged in any way on the xy plane is

$$p(\theta,\phi) = (j\rho ckA/2\pi r)\sum_{i=1}^{N} u_i \mathrm{e}^{-jkr_i} \tag{8.60}$$

If the individual transducers have significant directionality, it is necessary to multiply Eq. (8.60) and the following equations for pressure by the appropriate far-field element pattern function.

In a circular array, with the origin at the center of the array, all the $a_i = a$. On using Eq. (8.59), with arbitrary angular spacing of the elements around the circle the field
with array elements of equal velocities, u, evenly spaced around the circle such that $\phi_i = n\phi_0$ with $n = 0, 1, 2, \ldots, N-1$, we have $\phi_0 = 2\pi/N$ leading to

$$p(\theta,\phi) = \left(j\rho ckAu_0\mathrm{e}^{-jkr}/2\pi r\right)\sum_{n=0}^{N-1} \mathrm{e}^{jka[\sin\theta\cos(\phi - n\phi_0) - \sin\theta_s\cos(\phi_s - n\phi_0)]} \tag{8.61}$$

In Eq. (8.61) we have allowed for beam steering θ_s in polar angle and ϕ_s in azimuthal angle by allowing the velocity to be given by $u_n = u_0\mathrm{e}^{-jka\sin\theta_s\cos(\phi_s - n\phi_0)}$. Equation (8.61) gives the pattern as a function of θ in any plane $\phi = \text{constant}$ when the beam is steered by θ_s in one of the symmetry planes given by ϕ_s and reduces to the unsteered case when $\theta_s = 0$. The phase exponent in the sum of Eq. (8.61) simplifies to $ka(\sin\theta - \sin\theta_s)\cos(\varphi_s - n\varphi_o)$ when the steered field is in the plane $\varphi = \varphi_s$ and shows the beam peaking at $\theta = \theta_s$ where the sum equals N. The result in Eq. (8.61) holds in any vertical plane designated by ϕ and reduces to Eq. (7.7) on the symmetry planes for the case of no steering.

The general case, for an even number of elements, for any values of the angle ϕ and the azimuthal steering angle ϕ_s, comes from Eq. (8.61) by noting that the sum of each pair of elements that are separated by 180° have the form $e^{jx} + e^{-jx} = 2\cos x$ because $\cos(\phi - n\phi_o + \pi) \equiv -\cos(\phi - n\phi_o)$. Using $\phi_0 = 2\pi/N$ the result, which holds for an even number of elements, is

$$p(\theta,\phi) = \left(j\rho ckAu_0 e^{-jkr}/2\pi r\right) \times \left\{ 2\sum_{n=0}^{N/2-1} \cos ka\left[\sin\theta\cos(\phi - 2\pi n/N) - \sin\theta_s\cos(\phi_s - 2\pi n/N)\right]\right\} \tag{8.62}$$

The closely spaced discrete ring array results may be approximated by a continuous ring source with continuous steering. Consider a ring source in a rigid plane ($z = 0$) of very narrow width, w, radius, a, and center at the origin. An element of the ring at angular position ϕ_0 is at distance $r - a\sin\theta\cos(\phi - \phi_0)$ from a field point at r, θ, ϕ. If the element has source strength $dq = wad\phi_0 u$, where u is the velocity of the element, its contribution to the far-field pressure is

$$dp(\theta,\phi) = (j\rho ckdq/2\pi)\left(e^{-jkr}/r\right)e^{jka\sin\theta\cos(\phi-\phi_0)}. \tag{8.63}$$

To steer the beam in any direction, θ_s, ϕ_s, in the half space above the plane $z = 0$ the velocity of the ring must have a different phase at each point such that the maximum pressure occurs in the steering direction. This is accomplished, as in the previous discussion, if the velocity is $u = u_0 e^{-jka\sin\theta_s\cos(\phi_s-\phi_0)}$ where u_0 is the velocity magnitude, again considered to be constant. This makes the complete phase factor equal to $ka[\sin\theta\cos(\phi - \phi_0) - \sin\theta_s\cos(\phi_s - \phi_0)]$, and the exponential then has its maximum value of unity when $\theta = \theta_s$ and $\phi = \phi_s$, which is the maximum of the beam. Now the far-field pressure is given by

$$p(\theta,\phi) = (j\rho ck/2\pi)(wau_0)\left(e^{-jkr}/r\right)\int_0^{2\pi} e^{jka[\sin\theta\cos(\phi-\phi_0) - \sin\theta_s\cos(\phi_s-\phi_0)]}d\phi_0. \tag{8.64}$$

Although steering destroys the complete azimuthal symmetry of the unsteered ring, a more limited type of symmetry still exists. Consider the field as a function of θ in the plane at ϕ when the beam is steered to ϕ_s. The field is the same function of θ in the plane at $\phi + \Delta\phi_s$ when the beam is steered to $\phi_s + \Delta\phi_s$, i.e., steering a ring in azimuth rotates the field but otherwise does not change it. This means that the field is a function of $(\phi - \phi_s)$ rather than ϕ and ϕ_s separately.

The exponent in Eq. (8.64) can be rearranged to separate the integration variables, with the result: $ka[\sin\theta\cos(\phi - \phi_0) - \sin\theta_s\cos(\phi_s - \phi_0)] = x\cos\phi_0 + y\sin\phi_0$ where $x = ka(\sin\theta\cos\phi - \sin\theta_s\cos\phi_s)$, $y = ka(\sin\theta\sin\phi - \sin\theta_s\sin\phi_s)$ leading to

$$x^2 + y^2 = (ka)^2 \left[\sin^2\theta + \sin^2\theta_s - 2\sin\theta\sin\theta_s\cos(\phi - \phi_s) \right] \tag{8.65}$$

allowing the value of the integral to be evaluated as

$$\int_0^{2\pi} e^{j(x\cos\phi_0 + y\sin\phi_0)} d\phi_0 = 2\pi J_0\left(\sqrt{x^2 + y^2}\right). \tag{8.66}$$

The complete result for the pressure is then:

$$\begin{aligned} p(\theta, \phi) = {} & (j\rho ck2\pi awu_0)\left(e^{-jkr}/2\pi r\right)J_0 \\ & \times \left(ka\left[\sin^2\theta + \sin^2\theta_s - 2\sin\theta\sin\theta_s\cos(\phi - \phi_s)\right]^{1/2}\right) \end{aligned} \tag{8.67}$$

When $\phi = \phi_s$ in Eq. (8.67), the expression $(x^2 + y^2)^{1/2} = \pm ka(\sin\theta - \sin\theta_s)$ and Eq. (8.67) becomes

$$p(\theta, \theta_s) = (jpck2\pi awu_0)\left(e^{-jbr}/2\pi r\right)J_0(ka[\sin\theta - \sin\theta_s]) \tag{8.68}$$

giving a simple expression for the beam pattern in that plane. Normalized beam patterns for this simpler case are given from Eq. (8.68) in Fig. 8.31 for circumference to wavelength ratio value of $ka = 20$, without and with steering to 60°.

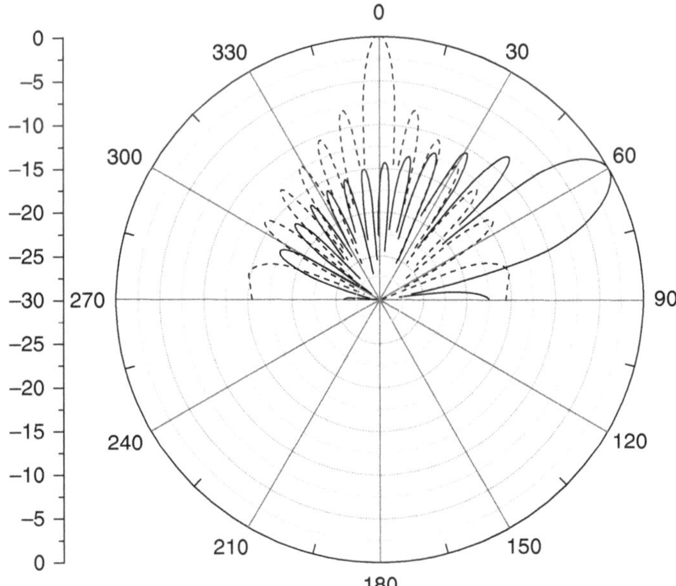

Fig. 8.31 Continuous circular ring $ka = 20$ without (*dashed lines*) and with (*solid lines*) steering to 60°

Table 8.4 Beam widths, BW, as a function of steering angle, θ_s

	$ka = 10$		$ka = 20$	
θ_s	BW(θ)	BW(φ)	BW(θ)	BW(φ)
0°	13°	14°	6.5°	7.0°
10	13	13	6.5	6.5
20	14	13	7.0	6.5
30	15	13	7.5	6.5
40	17	13	8.5	6.5
50	20	13	10	6.5
60	29	13	13	6.5

Figure 8.31 illustrates the distortion in the beam width as it is steered to 60°. Another simple expression occurs in Eq. (8.67) for $\theta = \theta_s$ where $(x^2 + y^2)^{1/2} = \pm \sqrt{2}ka \sin\theta_s[1 - \cos(\phi - \phi_s)]^{1/2}$ which gives the field as a function of $(\phi - \phi_s)$ along an arc of fixed θ (a circle of constant latitude). The −3 dB beam widths, BW, as a function of the steering angle, θ_s, are given in Table 8.4 for circumference to wavelength ratio values $ka = 10$ and 20. And as may be seen, the beam widths as a function of θ, BW(θ), nearly doubles from 0° to 60° for both values of ka while the beam widths as a function of φ, BW(φ), are essentially constant.

The fields of several concentric rings can be added, each with a different radius and source strength. With α being the full bracketed function of the angles in Eq. (8.67), that is, $\alpha = [\sin^2\theta + \sin^2\theta_s - 2\sin\theta\sin\theta_s\cos(\phi - \phi_s)]^{1/2}$, we have

$$p(\theta, \phi) = (j\rho ck)(\mathrm{e}^{-jkr}/2\pi r)\{q_1 J_0(ka_1\alpha) + q_2 J_0(ka_2\alpha) + \cdots\} \tag{8.69}$$

where $q_1 = 2\pi a_1 w_1 u_1$, etc. is the source strength of each ring. Each ring is continuously steered, and usually all rings would be steered in the same direction to get a good beam and all the α's would be the same. But different rings steered in different directions to form multiple simultaneous beams might be considered.

Equation (8.67) holds only for a ring width much less than a wavelength. The field for a wide ring can be obtained from Eq. (8.67) by replacing w by da and integrating over the radius. For a ring of width $(a_2 - a_1)$, we have

$$\begin{aligned} p(\theta, \phi) &= (2\pi u_0 j\rho ck)(\mathrm{e}^{-jkr}/2\pi r)\int_{a_1}^{a_2} J_0(ka\alpha)a\mathrm{d}a \\ &= (2\pi u_0 j\rho ck)(\mathrm{e}^{-jkr}/2\pi r)\left[a_2^2 J_1(ka_2\alpha)/ka_2\alpha - a_1^2 J_1(ka_1\alpha)/ka_1\alpha\right] \end{aligned} \tag{8.70}$$

The case of a circular piston of radius a_2 is given by Eq. (8.70) with $a_1 = 0$ and $\theta_s = 0$, making $\alpha = \sin\theta$. Note that Eq. (8.70) is for a continuously steered annular source for which the velocity phase at a point on the annulus depends on the radial position of that point as well as on the angular position around the circumference of the annulus.

The directivity beam pattern functions of a steered circular discrete and continuous circular ring array have been presented along with the case of an array of continuously steered rings and the case of a ring of finite width. Equation (8.61) gives the beam pattern of a discrete circular array as a function of θ in any plane $\phi =$ constant when the beam is steered by θ_s in one of the symmetry planes given by ϕ_s and reduces to Eq. (8.62) for an even number of elements. Equation (8.67) is for the beam pattern of a steered continuous ring at any azimuth angle and reduces to Eq. (8.68) when steered into the plane of the beam pattern. Equation (8.69) is for the case of a continuous array of concentric rings while Eq. (8.70) is for the case of continuous circular ring of finite width. The continuous ring should be a useful approximation for a close-packed ring of discrete sources.

8.7 Summary

In this chapter on hydrophone arrays we considered directivity functions, beam steering, shading, wavevector response, and array gain as well as ambient, structural, and flow noise sources along with arrays of vector sensors. We also presented sections on steered planar circular arrays and array transparency. The beam pattern response for an N by M array of omnidirectional hydrophone elements with a receiving sensitivity Mo was developed without and with steering. It was then shown how shading, including Dolph–Chebyshev shading, can be used to reduce the side lobes. This was followed by sections on wavevector response using k_x, and k_y and also array gain AG which can account for partial coherency, which cannot be accounted for by the DI.

Sources and properties of noise in arrays include ambient sea noise, structural noise, and flow noise. The spatial correlation for isotropic noise is $\rho(d) = \sin(kd)/kd$, where d is the distance between hydrophones. Ambient noise is important for slow speed vehicles with hydrophones; however, as the speed increases structural and flow noise can dominate. Reduction of array noise was treated in Sect. 8.4 and a summary of means for reducing it was given in Sect. 8.4.4. The directionality of multiple vector sensor arrays and equations for them were developed, including the noise correlation between dipole elements. Noise in hull-mounted arrays was treated in Sect. 8.5.3.

Planar circular arrays were considered in Sect. 8.6 and models for steered discreet elements and steered continuous elements were developed.

Exercises (Degree of Difficulty: *Lowest, **Moderate, ***Highest)

8.1.** Show, starting from the Fourier Transform formulation of the array output in Eq. (8.5b) (see Sect. 13.11), that a linear phase shift in the array sensitivity distribution results in a wavevector displacement that corresponds to steering the beam.

8.2.* What is a non-acoustic plane wave? Why do we need to consider the wavevector response of an array of hydrophones?

8.3.** What are the physical situations in which the array gain and the directivity index could have the same value? What are the basic reasons why they usually are not the same?

8.4.*** Use Eq. (8.8c) to calculate the DI of an unshaded line array of small hydrophones for $N = 6$ and $kD = \pi,\ \pi/2$ and $kD \ll 1$, and for steering angles of 0°, 30, and 90°.

8.5.** Show from Eq. (8.8c), for unshaded line arrays for any values of kD and N, that the D_f is the same when the array is unsteered and when the array is steered to 90° at $2kD$, i.e., at twice the frequency or twice the spacing.

8.6.** Calculate the array gains for the arrays in Exercise 8.4 in isotropic, incoherent noise.

8.7.** Calculate the coincidence frequency of a steel plate with thickness 0.635 cm (0.25 in.). At frequencies below the coincidence frequency flexural waves in a submerged plate produce evanescent pressure waves in the water traveling parallel to the plate that decay with distance from the plate. At 20 kHz how much does the evanescent wave amplitude decay at 1 cm from the surface of the plate?

8.8.*** Develop the spatial correlation function for isotropic noise given by Eq. (8.16).

8.9.** The discussion of the triaxial vector sensor (see Fig. 8.24) makes use of certain relationships between angles. Show that in general $\cos\gamma_n = \cos\theta\cos\theta_n + \sin\theta\sin\theta_n\cos(\varphi - \varphi_n)$ and that specifically for $\varphi = \varphi_n$, $\cos\gamma_n = \cos(\theta - \theta_n)$.

8.10.*** Consider Eq. (8.29a) for an array of N hydrophones under conditions where all four types of noise have the same intensity at each hydrophone, i.e., $f_a = f_s = f_f = f_h = 1/4$. Also consider the structural, flow, and hydrophone internal noise to be incoherent, while ambient noise is isotropic and partially coherent with spatial correlation given by Eq. (8.16), and the signal is a plane wave from the broadside direction. What is the array gain: (1) for any configuration of N hydrophones? (2) for a line array of N with half wavelength spacing? (3) for a line array of $N = 3$ with spacing D? (4) for a line array of $N = 3$ with quarter wavelength spacing?

8.11.*** Note that the development of the generalized diffraction constant in Sect. 11.3.1 suggests that Eq. (8.56) holds for arrays as well as individual transducers, although it does not explicitly show it. Investigate the validity of this suggestion by considering a simple array of two small circular pistons in a plane, rigid baffle and calculating the total radiation resistance of the array, the diffraction constant of the array, and the directivity factor of the array and showing that they satisfy Eq. (8.56).

References

1. H.H. Schloemer, Technology development of submarine sonar hull arrays. Naval Undersea Warfare Center Division Newport, Technical Digest, September 1999 [Distribution authorized to DOD components only]. Also Presentation at Undersea Defense Technology Conference and Exhibition, Sydney, Australia, February 7 (2000)
2. I. Dyer, *"Ocean Ambient Noise" Encyclopedia of Acoustics*, vol. 1 (Wiley, New York, 1997), p. 549
3. S.-H. Ko, S. Pyo, W. Seong, *Structure-Borne and Flow Noise Reductions (Mathematical Modeling)* (Seoul National University Press, Seoul, 2001)
4. D. Ross, *Mechanics of Underwater Noise* (Peninsula, Los Altos Hills, 1987)
5. W.A. Strawderman, *Wavevector-Frequency Analysis with Applications to Acoustics*. U.S. Government Printing Office, undated
6. V.M. Albers, *Underwater Acoustics Handbook* (The Pennsylvania State University Press, University Park, 1960)
7. W.S. Burdic, *Underwater Acoustic System Analysis*, 2nd edn. (Prentice Hall, Upper Saddle River, 1991)
8. J.W. Horton, *Fundamentals of Sonar*, 2nd edn. (U.S. Naval Institute, Annapolis, 1959)
9. A.A. Michelson, A reciprocal relation in diffraction. Philos. Mag. **9**, 506–507 (1905)
10. N. Davids, E.G. Thurston, R.E. Meuser, The design of optimum directional acoustic arrays. J. Acoust. Soc. Am. **24**, 50–56 (1952)
11. R.L. Pritchard, Optimum directivity patterns for linear point arrays. J. Acoust. Soc. Am. **25**, 879–891 (1953)
12. W. Thompson Jr., Higher powers of pattern functions—a beam pattern synthesis technique. J. Acoust. Soc. Am. **49**, 1686–1687 (1971)
13. C.L. Dolph, A current distribution of broadside arrays which optimizes the relationship between beam width and side lobe level. Proc. Inst. Radio Engrs. **34**, 335–348 (1946)
14. R.J. Urick, *Principles of Underwater Sound*, 3rd edn. (Peninsula, Los Altos Hills, 1983)
15. T.T. Taylor, Design of line-source antennas for narrow beam width and low side lobes. IRE Trans. **AP-3**, 316 (1955)
16. O.B. Wilson, *An Introduction to the Theory and Design of Sonar Transducers* (U.S. Government Printing Office, Washington, DC, 1985)
17. R.L. Pritchard, Approximate calculation of the directivity index of linear point arrays. J. Acoust. Soc. Am. **25**, 1010–1011 (1953)
18. R.L. Pritchard, Maximum directivity of a linear point array. J. Acoust. Soc. Am. **26**, 1034–1039 (1954)
19. G. Maidanik, D.W. Jorgensen, Boundary wave-vector filters for the study of the pressure field in a turbulent boundary layer. J. Acoust. Soc. Am. **42**, 494–501 (1967)
20. W.K. Blake, D.M. Chase, Wavenumber-frequency spectra of turbulent-boundary-layer pressure measured by microphone arrays. J. Acoust. Soc. Am. **49**, 862–877 (1971)
21. D.H. Trivett, L.D. Luker, S. Petrie, A.L. VanBuren, J.E. Blue, A planar array for the generation of evanescent waves. J. Acoust. Soc. Am. **87**, 2535–2540 (1990)
22. C.H. Sherman, S.H. Ko, B.G. Buehler, Measurement of the turbulent boundary layer wave-vector spectrum. J. Acoust. Soc. Am. **88**, 386–390 (1990)
23. J.S. Bendat, A.G. Piersol, *Engineering Applications of Correlation and Spectral Analysis* (Wiley, New York, 1993)
24. J.L. Butler, C.H. Sherman, Acoustic radiation from partially coherent line sources. J. Acoust. Soc. Am. **47**, 1290–1296 (1970)
25. D.J. Kewley, D.G. Browning, W.M. Carey, Low-frequency wind-generated ambient noise source levels. J. Acoust. Soc. Am. **88**, 1894–1902 (1990)
26. G.M. Wenz, Acoustic ambient noise in the ocean: spectra and sources. J. Acoust. Soc. Am. **34**, 1936–1956 (1962)

27. V.O. Knudsen, R.S. Alford, J.W. Emling, Underwater ambient noise. J. Mar. Res. **7**, 410 (1948)
28. H.W. Marsh, Origin of the Knudsen spectra. J. Acoust. Soc. Am. **35**, 409 (1963)
29. E.H. Axelrod, B.A. Schoomer, W.A. Von Winkle, Vertical directionality of ambient noise in the deep ocean at a site near Bermuda. J. Acoust. Soc. Am. **37**, 77–83 (1965)
30. B.F. Cron, B.C. Hassel, F.J. Keltonic, Comparison of theoretical and experimental values of spatial correlation. J. Acoust. Soc. Am. **37**, 523–529 (1965). U.S. Navy Underwater Sound Lab. Rept. 596, 1963
31. B.F. Cron, C.H. Sherman, Spatial-correlation functions for various noise models. J. Acoust. Soc. Am. **34**, 1732–1736 (1962). Addendum: J. Acoust. Soc. Am., **38**, 885 (1965)
32. J.E. Barger, "*Sonar Systems*", *Encyclopedia of Acoustics*, vol. 1, Section 3.1 (Wiley, New York, 1997), p. 559
33. R.L. Pritchard, Mutual acoustic impedance between radiators in an infinite rigid plane. J. Acoust. Soc. Am. **32**, 730–737 (1960)
34. M.C. Junger, D. Feit, *Sound, Structures and Their Interaction*, 2nd edn. (MIT Press, Cambridge, MA, 1986)
35. G.M. Corcos, The structure of the turbulent pressure field in boundary layer flows. J. Fluid Mech. **18**(3), 353–378 (1964)
36. D.M. Chase, Modeling the wave-vector frequency spectrum of turbulent boundary wall pressure. J. Sound Vib. **70**, 29–68 (1980)
37. G.C. Lauchle, Calculation of turbulent boundary layer wall pressure spectra. J Acoust. Soc. Am. **98**, 2226–2234 (1995)
38. G.C. Lauchle, Noise generated by axisymmetric turbulent boundary-layer flow. J. Acoust. Soc. Am. **61**, 694–703 (1977)
39. N.C. Martin, P. Leehey, Low wavenumber wall pressure measurements using a rectangular membrane as a spatial filter. J. Sound Vib. **52**(1) (1997)
40. J.J. Faran Jr., R. Hills Jr., Wide-band directivity of receiving arrays. J. Acoust. Soc. Am. **57**, 1300–1308 (1975)
41. S.H. Ko, H.H. Schloemer, Signal pressure received by a hydrophone placed on a plate backed by a compliant baffle. J. Acoust. Soc. Am. **89**, 559–564 (1991)
42. M.A. Gonzalez, Analysis of a composite compliant baffle. J. Acoust. Soc. Am. **64**, 1509–1513 (1978)
43. S.H. Ko, C.H. Sherman, Flexural wave baffling. J. Acoust. Soc. Am. **66**, 566–570 (1979)
44. R.P. Radlinski, R.S. Janus, Scattering from two and three gratings of densely packed compliant tubes. J. Acoust. Soc. Am. **80**, 1803–1809 (1986)
45. S.H. Ko, H.H. Schloemer, Calculations of turbulent boundary layer pressure fluctuations transmitted into a viscoelastic layer. J. Acoust. Soc. Am. **85**(4) (1989)
46. S.H. Ko, H.H. Schloemer, Flow noise reduction techniques for a planar array of hydrophones. J. Acoust. Soc. Am. **92**, 3409–3424 (1992)
47. W. Thompson Jr., R.E. Montgomery, Approximate evaluation of the spectral density integral for a large planar array of rectangular sensors excited by turbulent flow. J. Acoust. Soc. Am. **93**, 3201–3207 (1993)
48. M.J. Berliner, J.F. Lindberg (eds.), *Acoustic Particle Velocity Sensors: Design, Performance and Applications*, AIP Conference Proceedings 368, Mystic CT, September (1995)
49. Proceedings of the Workshop on Directional Acoustic Sensors, Newport, RI, 17–18 April (2001) (Available on CD)
50. E.Y. Lo, M.C. Junger, Signal-to noise enhancement by underwater intensity measurements. J. Acoust. Soc. Am. **82**, 1450–1454 (1987)
51. D. Huang, R.C. Elswick, Acoustic pressure-vector sensor array. J. Acoust. Soc. Am. **115**, 2620 (2004) (Abstract)
52. B.A. Cray, A.H. Nuttall, *A Comparison of Vector-Sensing and Scalar-Sensing Linear Arrays*. Report No. 10632, Naval Undersea Warfare Center, Newport, RI, January 27 (1997)

53. R. Kneipfer, *Spatial Auto and Cross-correlation Functions for Tri-axial Velocity Sensor Outputs in a Narrowband, 3 Dimensional, Isotropic Pressure Field.* Naval Undersea Warfare Center, Newport, RI, Memo. 5214/87, September (1985)
54. M. Hawkes, A. Nehorai, Acoustic vector sensor correlations in ambient noise. IEEE J. Ocean. Eng. **26**, 337–347 (2001)
55. H.W. Marsh, *Correlation in Wave Fields.* U.S. Navy Underwater Sound Laboratory Quart. Rept., 31 March (1950), pp. 63–68
56. B.A. Cray, in *Directional Acoustic Receivers: Signal and Noise Characteristics.* Workshop on Directional Acoustic Sensors, Newport, RI, 17–18 April (2001)
57. S.H. Ko, Performance of velocity sensor for flexural wave reduction, in M.J. Berliner, J.F. Lindberg (eds.), *Acoustic Particle Velocity Sensors: Design, Performance and Applications*, AIP Conference Proceedings 368 (AIP Press, Woodbury, 1996)
58. R.F. Keltie, Signal response of elastically coated plates. J. Acoust. Soc. Am. **103**, 1855–1863 (1998)
59. B.A. Cray, R.A. Christman, Acoustic and vibration performance evaluations of a velocity sensing hull array, in *Acoustic Particle Velocity Sensors: Design, Performance and Applications, AIP Conference Proceedings 368*, ed. by M.J. Berliner, J.F. Lindberg (AIP Press, Woodbury, 1996)
60. N.C. Martin, R.N. Dees, D.A. Sachs, in *Baffle Characteristics: Effects of Sensor Size and Mass*, AIP Conference Proceedings 368, ed. by M.J. Berliner, J.F. Lindberg (AIP Press, Woodbury, 1996)
61. J.J. Caspall, M.D. Gray, G.W. Caille, J. Jarzynski, P.H. Rogers, G.S. McCall II, in *Laser Vibrometer Analysis of Sensor Loading Effects in Underwater Measurements of Compliant Surface Motion*, AIP Conference Proceedings 368, ed. by M.J. Berliner, J.F. Lindberg (AIP Press, Woodbury, 1996)
62. M. Traweek, J. Polcari, D. Trivett, Noise audit model for acoustic vector sensor arrays on an ocean glider. J. Acoust. Soc. Am. **116**(2), 2650 (2004)
63. B.M. Abraham, M.J. Berliner, in *Directional Hydrophones in Towed Systems*, Workshop on Directional Acoustic Sensors, Newport, RI, 17–18 April (2001)
64. G.C. Lauchle, J.F. McEachern, A.R. Jones, J.A. McConnell, in *Flow-Induced Noise on Pressure Gradient Hydrophones*, AIP Conference Proceedings 368, ed. by M.J. Berliner, J.F. Lindberg (AIP Press, Woodbury, 1996)

Chapter 9
Transducer Evaluation and Measurement

Electrical and acoustical measurements are made on transducers to determine performance characteristics and parameters for comparison with goals and theoretical models when the latter are available. In this chapter we discuss the procedures and means for making such measurements [1–11]. Admittance or impedance measurements are normally made first under air loading (which simulates a vacuum) and then under water loading, often at various hydrostatic pressures, temperatures, and drive levels. Measurements of the transmitting response, source level, efficiency, receiving response, beam patterns, and harmonic distortion of transducers and transducer arrays are usually made in the far field, where spherical spreading holds. However, in some cases measurements can be made in the near field and projected to the far field through extrapolation formulas. Measurement is the important final step in evaluating the performance of all transducers. Specific projector and hydrophone designs are given in Chaps. 5 and 6 and the modeling methods used for these designs are discussed in Chaps. 2 and 3.

This chapter discusses means for evaluating the important transducer parameters Q_m, f_r, and k_{eff} and the lumped elements for a simple equivalent circuit representation, including resistances caused by electrical and mechanical dissipation as well as acoustical radiation. The tuning of a transducer is an important last step in its implementation, since the tuning element improves the power factor and reduces volt-ampere requirements in service and during high power testing. The transmitting and receiving responses, as well as the admittance and impedance curves, are discussed for both electric and magnetic field transducers as a guide for evaluating measured results. A development of the reciprocity relation between transmitting and receiving is also presented. Means for near-field measurements along with far-field predictions are presented. A brief review and specifications of some of the commonly used US Navy calibrated transducers is also presented.

J.L. Butler, C.H. Sherman, *Transducers and Arrays for Underwater Sound*,
Modern Acoustics and Signal Processing, DOI 10.1007/978-3-319-39044-4_9

9.1 Electrical Measurement of Transducers in Air

We begin with the measurement of the electrical input impedance, $Z = R + jX$, or admittance, $Y = G + jB$, of a transducer (for complex algebra, see Sect. 13.17). Since $Y = 1/Z$,

$$Y = (R - jX)/\left(R^2 + X^2\right) \quad \text{and} \quad Z = (G - jB)/\left(G^2 + B^2\right).$$

Electrical evaluation allows the determination of a number of important performance parameters as well as the elements of a simplified equivalent circuit. The evaluation is normally made at frequencies near and below the fundamental resonance where a simple lumped equivalent circuit is often an adequate representation. The simplified circuit of Fig. 3.15 includes most of the essential elements of a piezoelectric transducer and will be used to represent all electric field transducers, while the circuit of Fig. 3.20 will be used to represent magnetic field transducers.

9.1.1 *Electric Field Transducers*

We repeat here as Fig. 9.1 the Van Dyke [12] equivalent electric circuit of Fig. 3.16 which is related to the electromechanical equivalent circuit of Fig. 3.15 through the effective electromechanical turns ratio N.

In the Van Dyke representation, R_0 is the electrical dissipation resistance and $1/R_0 = G_0 = \omega C_f \tan\delta$, where $\tan\delta$ is the dissipation factor (often referred to as D). The quantity C_0 is the clamped capacitance, while the motional capacitance $C_e = N^2 C^E$ where C^E is the short circuit mechanical compliance. The free capacity is $C_f \equiv C_0 + C_e$. The inductance $L_e = M/N^2$ where M is the effective mass, including the radiation mass, M_r, and the electrical equivalent resistance $R_e = R_t/N^2$ where R_t is the total mechanical resistance including the internal mechanical resistance, R, and radiation resistance, R_r. Both M_r and R_r are negligible under air loading conditions. The input admittance, $Y = 1/Z$, for the circuit of Fig. 9.1 is

$$Y = G_0 + j\omega C_0 + 1/(R_e + j\omega L_e + 1/j\omega C_e), \tag{9.1}$$

where the third term is the motional part of the admittance.

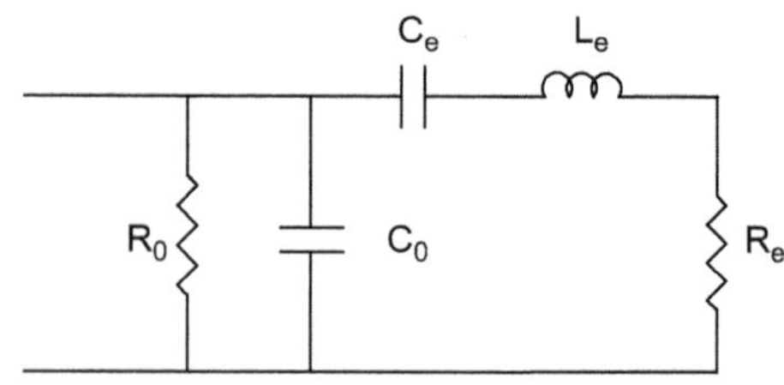

Fig. 9.1 Van Dyke equivalent circuit

Measurement of the admittance magnitude determines the resonance frequency, f_r, antiresonance frequency, f_a, and the maximum admittance $|Y|_{max}$. At resonance (under short circuit conditions) $\omega_r L_e = 1/\omega_r C_e$ where $|Y|$ becomes a maximum and $|Z|$ becomes a minimum, while at antiresonance (under open circuit conditions), $\omega_a L_e = 1/\omega_a C^*$ where $C^* = C_0 C_e/(C_0 + C_e)$, the admittance becomes a minimum and the impedance becomes a maximum. Accordingly, the resonance and antiresonance frequencies are related to the circuit parameters by

$$\omega_r^2 = 1/L_e C_e \quad \text{and} \quad \omega_a^2 = 1/L_e C^* = (C_0 + C_e)/C_e C_0 L_e. \tag{9.2}$$

The impedance and admittance magnitudes, as well as f_r and f_a, are illustrated in Fig. 9.2 for the case of air loading where R_e is typically small.

The measurement of the magnitude of the admittance and the impedance as a function of frequency may be accomplished by means of an "impedometer" or "admittometer." A sketch of the system is shown in Fig. 9.3; it consists of an ac source of known voltage V_0 and a resistor R of known value in series with the transducer of impedance, Z, giving the current $I = V_0/(Z + R)$.

The voltage across the transducer is given by $V_t = IZ = V_0 Z/(R + Z)$. For $R \gg Z$ we then get

$$Z \approx V_t(R/V_0), \tag{9.3}$$

and measurement of the voltage across the transducer gives the magnitude of Z. The admittance magnitude may be obtained by measuring the voltage across the

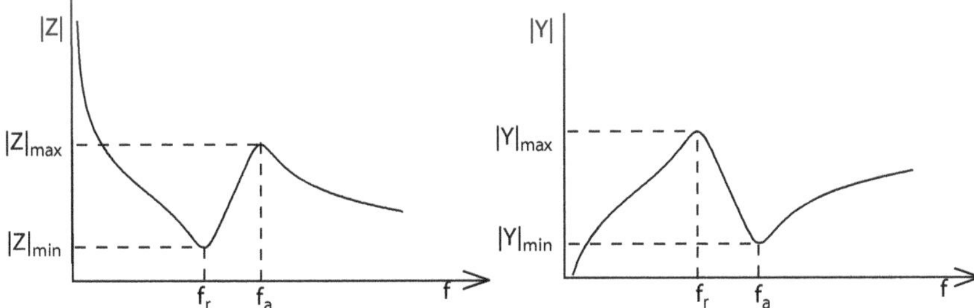

Fig. 9.2 Impedance, Z, and admittance, Y, magnitudes for an electric field transducer

Fig. 9.3 Impedance/admittance circuit for measuring $|Z|$ and $|Y|$

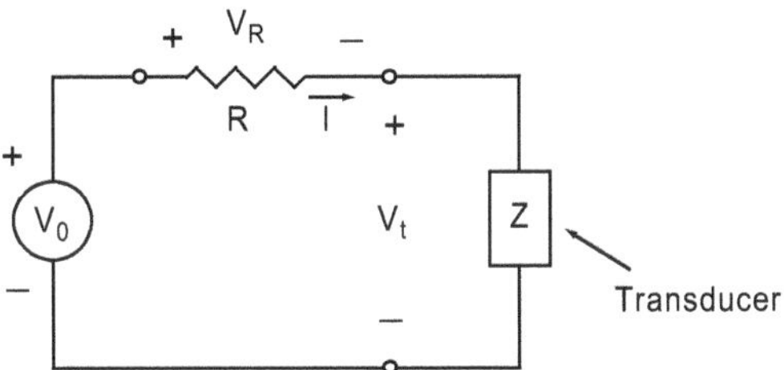

resistor, $V_R = IR = V_0R/(R + Z)$, with a different value of resistance such that $R \ll Z$, which, from V_R, gives the transducer admittance

$$Y = 1/Z \approx V_R/(V_0R). \tag{9.4}$$

A low output impedance, constant voltage source is needed for this method. The impedometer/admittometer may be calibrated with precision resistors.

Measurement of the in-air resonance and antiresonance frequencies gives the effective dynamic coupling coefficient through the often-used formula:

$$k_{\mathrm{eff}}^2 = 1 - (f_r/f_a)^2. \tag{9.5}$$

Equation (9.5) can be obtained from Eq. (9.2) using the ratio $\omega_r^2/\omega_a^2 = C_0/(C_0 + C_e)$ and consequently $1 - \omega_r^2/\omega_a^2 = C_e/(C_0 + C_e) = C_e/C_f = N^2C^E/C_f = k^2$, as discussed in Chaps. 1 and 2. In practice C^E and C_f and possibly, N, are effective values because of added electrical or mechanical components or other modes of vibration, such as bending. The corresponding effective coupling coefficient, k_{eff}, is less than the k of the ideal transducer but it is a measure of the energy converted relative to the energy stored for a specific, complete transducer design (see Sects. 1.4.1 and 4.4).

Equation (9.1) evaluated well below resonance, where $\omega \ll \omega_r$, becomes

$$Y \approx G_0 + j\omega C_0 + j\omega C_e = \omega C_f \tan\delta + j\omega C_f.$$

Piezoelectric ceramics have $\tan\delta$ values typically in the range from 0.004 to 0.02 (see Sect. 13.5). A simple low frequency capacitance and dissipation meter may be used to measure C_f and $\tan\delta$. The circuit of Fig. 9.1 reduces to the simpler circuit of Fig. 9.4 in the low frequency range where C_0 and C_e are in parallel. The free capacitance C_f and k_{eff}^2 give values for $C_0 = C_f(1 - k_{\mathrm{eff}}^2)$ and $C_e = k_{\mathrm{eff}}^2C_f$.

Under air loading, the inductance is proportional to the dynamic mass of the transducer alone. Thus $\omega_r^2 = 1/L_eC_e$ gives $L_e = 1/\omega_r^2C_e$. Also, under air loading R_e is typically small and $|Y|_{\max} \approx 1/R_e$ as may be seen from Eq. (9.1). The value of $|Y|_{\max}$ may then be used to obtain the mechanical Q from

$$Q_m = \omega_rL_e/R_e \approx \omega_rL_e|Y|_{\max}.$$

This is a good approximation for Q_m provided the result is greater than about 30. Typical in-air values of Q_m range from 30 to 300 depending on the mounting conditions and attached components.

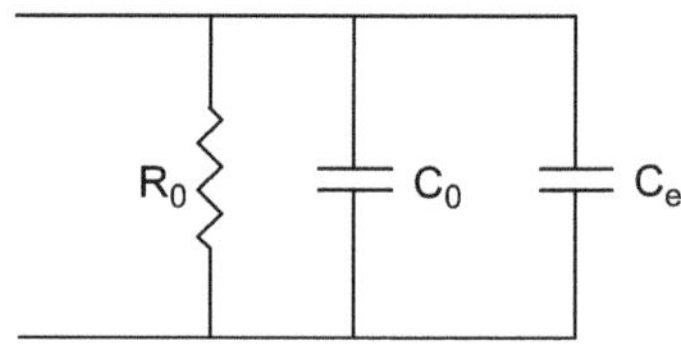

Fig. 9.4 Van Dyke circuit at low frequencies

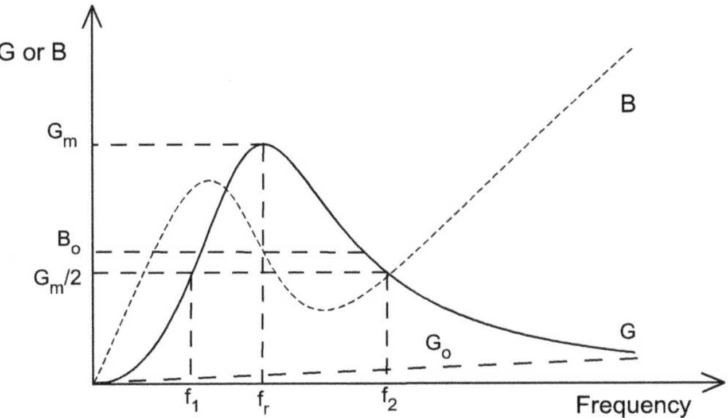

Fig. 9.5 Conductance, G, and susceptance, B, for an electric field transducer

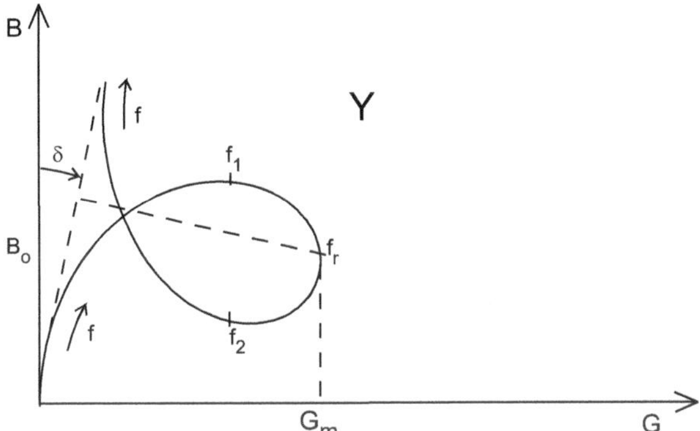

Fig. 9.6 Admittance locus, "loop," for an electric field transducer

To summarize, measurement of the admittance magnitude and the low frequency capacitance and dissipation gives the parameters C_f, $\tan\delta$, f_r, f_a, and $|Y_{max}|$. Then along with k_{eff}, from f_r and f_a, all the parameters of the equivalent circuit of Fig. 9.1 are given by:

$$\begin{aligned} &C_0 = C_f\left(1 - k_{eff}^2\right), \quad C_e = C_f k_{eff}^2, \quad L_e = 1/\omega_r^2 C_e, \\ &R_e = 1/\left|Y_{max}\right|, \quad R_0 = 1/\omega C_f \tan\delta. \end{aligned} \tag{9.6}$$

The parameters may also be obtained if the water-loaded Q_m and Q_e are measured, as will be discussed in connection with Figs. 9.5 and 9.6 and Eq. (9.13). Marshall and Brigham [13] have developed an alternative method for determining the values of the components of Fig. 9.1 by measuring the maximum and minimum values of the transducer capacitance, and the corresponding frequencies, as well as the low frequency dissipation, $\tan\delta$.

The electromechanical turns ratio, N, may be determined from the mass, M, since $N^2 = M/L_e$ (see Fig. 3.16), and M may be calculated or weighed in some well-defined cases. On the other hand, the value of N may be determined if an additional known mass M_a can be firmly attached to the radiating mass M yielding the mechanical resonance frequency $\omega_{ra}^2 = 1/(L_e + M_a/N^2)C_e$ which may be compared with the original resonance $\omega_r^2 = 1/L_eC_e$ to yield

$$N^2 = (M_a/L_e)/\left[(f_r/f_{ra})^2 - 1\right]. \tag{9.7a}$$

In many cases it may be difficult to attach the mass M_a such that M and M_a form a single rigid mass.

A more direct means is to use two lightweight accelerometers to measure the acceleration difference, a_d, between the head mass and tail mass for a given motional current, $N(u - u_1)$, as shown in Fig. 3.18. From 3.18 and Eq. (9.1), we can see that $N(u - u_1) = V(Y - j\omega C_0 - G_0)$ and since the velocity difference $(u - u_1) = a_d/j\omega$, the electromechanical turns ratio N may be determined from

$$N = V(Y - j\omega C_0 - G_0)j\omega/a_d \tag{9.7b}$$

where all the terms on the right side of Eq. (9.7b) are measured quantities. An average value of the acceleration a_d should be used if the motion of the head surface is not uniform.

The electromechanical turns ratio may also be obtained from the Chap. 2, Table 2.1, general formulas written as

$$N = k_{eff}\left(K_m^E C_f\right)^{1/2} \quad \text{and} \quad N = k_{eff}\left(K_m^H L_f\right)^{1/2},$$

for electric and magnetic transducers, respectively. The coupling coefficient is the measured effective coupling coefficient k_{eff}, C_f and L_f are the measured low frequency free capacitance and inductance, and K_m^E and K_m^H are the short circuit and open circuit mechanical stiffness measured at resonance. For example, $K_m^E = M\omega_r^2$ allowing determination of K_m^E if M is known. It can be seen that both mechanical and electrical measurements must be made in order to determine the electromechanical turns ratio, as one might expect.

9.1.2 *Magnetic Field Transducers*

The above discussion has centered on electric field transducers and in particular piezoelectric ceramic transducers. Magnetic field transducers may be represented as the dual of electric field transducers. Figure 9.7 shows a lumped equivalent circuit for a magnetic field transducer based on the circuit of Fig. 3.20.

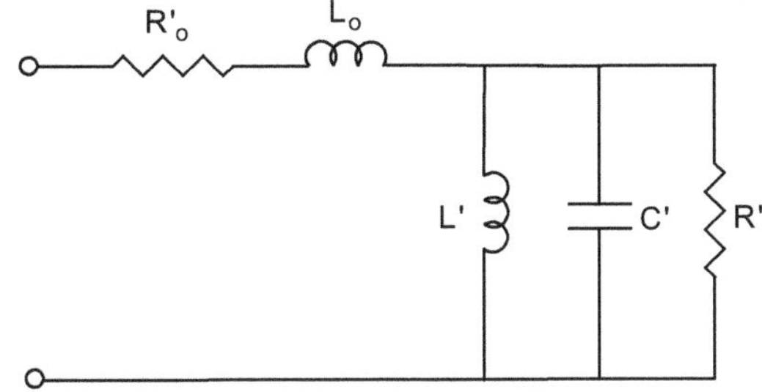

Fig. 9.7 Magnetic field transducer lumped electric circuit

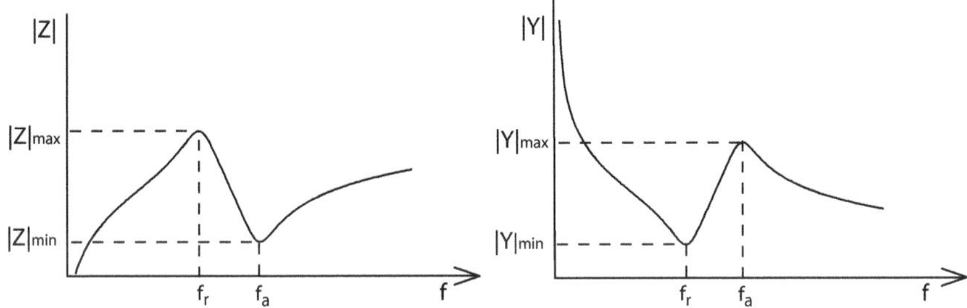

Fig. 9.8 Impedance, *Z*, and admittance, *Y*, magnitudes for a magnetic field transducer

The input impedance is

$$Z = R_0^{'} + j\omega L_0 + 1/\left(1/R^{'} + j\omega C^{'} + 1/j\omega L^{'}\right). \tag{9.8}$$

The motional part of Eq. (9.8), the third term, is in the form of the parallel circuit of Fig. 9.7. In this case the mechanical resonance is a parallel resonance under open circuit conditions where $\omega_r L^{'} = 1/\omega_r C^{'}$, while the antiresonance occurs under short circuit conditions at $\omega_a L^* = 1/\omega_a C^{'}$ where $L^* = L_0 L^{'}/\left(L_0 + L^{'}\right)$. A sketch of the impedance and admittance magnitude is shown in Fig. 9.8 and is seen to be the inverse of Fig. 9.2 as expected since Fig. 9.7 is the dual of Fig. 9.1.

The measurement circuit illustrated in Fig. 9.3 may be used to measure the impedance magnitude for the circuit shown in Fig. 9.7. The coil resistance $R_0^{'}$ and free inductance L_f are measured with an inductance meter at low frequencies where the circuit of Fig. 9.7 reduces to Fig. 9.9 with L_0 and L' in series. The effective dynamic coupling coefficient is

$$k_{\text{eff}} = \left[1 - (f_r/f_a)^2\right]^{1/2},$$

where f_r is the frequency at $|Z|_{\max}$ and f_a is the frequency at $|Z|_{\min}$.

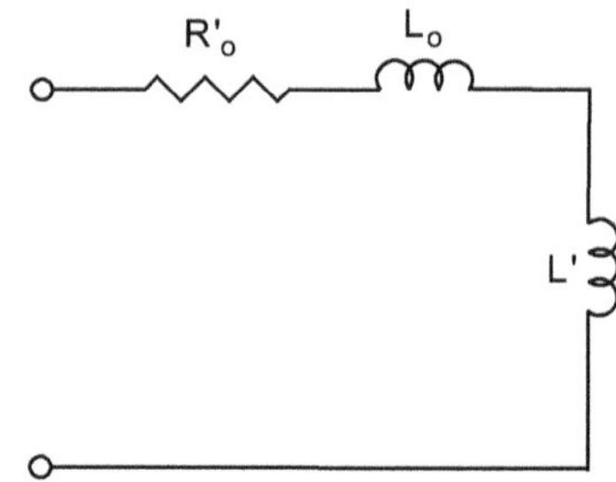

Fig. 9.9 Low frequency electrical representation of a magnetic field transducer

This formula is valid for low loss conditions where $|Z|_{max} \gg |Z|_{min}$ and $|Z_{max}| \approx R'$. In this parallel resonant circuit, $Q_m = R'/\omega_r L' \approx |Z_{max}|/\omega_r L'$ for Q_m greater than 30. The free inductance is $L_f = L_0 + L'$ and the circuit parameters for Fig. 9.7 are

$$L_0 = L_f\left(1 - k_{eff}^2\right), \quad L' = L_f k_{eff}^2, \quad C' = 1/\omega_r^2 L',$$
$$R' \approx |Z_{max}|, \quad R_0' = \text{coil resistance}, \tag{9.9}$$

which may be compared with the electric field results of Eq. (9.6). Although Fig. 9.7 includes a series resistance coil loss, R_0', the equally important eddy current loss has not been included in this simplified impedance magnitude measurement procedure. Measurement of both the real and imaginary parts of the impedance is necessary to obtain the eddy current data.

9.2 Measurement of Transducers in Water

When a transducer is measured in water, some of the quantities discussed above such as f_r, f_a, and Q_m are changed because the radiation impedance in water is much greater than it is in air. Consider the equivalent circuit of Fig. 3.15 which is the basis for Fig. 9.1. The radiation load adds a radiation mass, M_r, and radiation resistance, R_r. In air the medium density is small and the radiation mass loading $M_r \ll M$ for typical underwater sound transducers. In air the mechanical resonance occurs at the frequency at which the mass reactance, ωM, cancels the compliance reactance, $1/\omega C^E$, giving $\omega_r = \left(1/MC^E\right)^{1/2}$. In water $\omega_{rw} = \left[1/C^E(M + M_r)\right]^{1/2}$ leading to

$$\omega_{rw}/\omega_r = [M/(M + M_r)]^{1/2}. \tag{9.10}$$

The reduction of the air resonance frequency when measured in water is typically 10–20 % for piston transducers.

As previously discussed in Chaps. 3 and 4, expressions for Q_m may be written as

$$Q_{mw} = \omega_{rw} L_e / R_e = 1/\omega_{rw} R_e C_e = (L_e/C_e)^{1/2}/R_e. \tag{9.11}$$

In water the added mass, M_r, also increases the stored energy which tends to raise the Q_m, but the radiation resistance is more effective in lowering the Q_m, resulting in $Q_{mw} < Q_{ma}$, where Q_{mw} refers to water and Q_{ma} to air. The relationship between the two, which is found most easily from the last expression in Eq. (9.11), is

$$Q_{mw} = Q_{ma}(1 + M_r/M)^{1/2}/(1 + R_r/R). \tag{9.12}$$

When different methods of determining Q_m do not yield the same results, it may indicate that the transducer is not acting as a single degree of freedom vibrator. The parameter Q_{mw} is normally measured from the power response, W, which for a fixed rms voltage is given by $W = V_{rms}^2 G$. Thus Q_m can be obtained from the frequencies at the peak, f_r, and at half peak, f_1 and f_2, of the conductance curve using Eq. (4.4),

$$Q_m = f_r/(f_2 - f_1). \tag{9.13}$$

The quantity Q_m is also a measure of the number of cycles it takes for a resonator to decay from steady state by the factor $e^{-\pi}$. The transient decay factor for a simple resonator is $R/2M$ and the amplitude decay function may be written as

$$x(t) = x_0 e^{-tR/2M} = x_0 e^{-\pi t/QT},$$

where the period of vibration at resonance is $T = 1/f_r$. At time $t = QT$, $x(t) = e^{-\pi}$. Since the value $e^{-\pi} = 0.043$, one merely counts the number of cycles for the amplitude to decay to a value of about 4 % of the steady state value to get the approximate mechanical Q of the transducer. The number of cycles required to reach steady state conditions may also be used in the same way.

The real and imaginary parts of the impedance or admittance are necessary for an evaluation of the losses and an accurate description of the transducer under water loading conditions where R_e in Fig. 9.1 becomes large and, because of the parallel representation, R' in Fig. 9.7 becomes small. In the electric field model of Figs. 9.1 and 3.15 the resistive load $R_e = (R + R_r)/N^2$ where R is the mechanical loss resistance and R_r is the radiation resistance. With water loading we also have $L = (M + M_r)/N^2$ where M_r is the radiation mass and M is the dynamic mass of the transducer. In the magnetic field cases of Figs. 9.7 and 3.20 the resistive load $R' = N^2/(R + R_r)$ and the mass loading is through $C' = (M + M_r)/N^2$.

A sketch of a typical measurement of G and B as a function of frequency for the electric field model of Fig. 9.1 and Eq. (9.1) is shown in Fig. 9.5. A plot of the corresponding admittance locus is shown in Fig. 9.6.

The electrical loss conductance, $G_0 = \omega C_f \tan\delta$, of Fig. 9.5 is seen to increase the conductance G with frequency and slightly tilt the locus or "loop" in Fig. 9.6.

Since the power $W = V_{\mathrm{rms}}^2 G$, maximum power is achieved, for a given voltage, at resonance, f_r, where G is maximum, G_m. The electrical conductance G_0 does not contribute to the mechanical damping and should be subtracted from the total conductance, G, before evaluating Q_m. However, G_0 is usually quite small and may often be ignored so that under this condition $Q_m \approx f_r/(f_2 - f_1)$, where f_1 and f_2 are the frequencies at $G_m/2$. For a heavily water-loaded transducer the loop is small while for light in-air loading the loop can be large enough to pass through the G axis and look very much like a circle. For the in-air case the conductance $G_m \approx 1/R_e \approx |Y|_{\mathrm{max}}$.

The ratio of the susceptance, B, to conductance, G, may be used to obtain the electrical Q which is defined at resonance as $Q_e = B_0/G_m$. As may be seen from Eq. (9.1), at resonance $B_0 = \omega_r C_0$, and the product $Q_m Q_e = \omega_r C_0 R_e/\omega_r C_e R_e = C_0/C_e = k_{\mathrm{eff}}^2/\left(1 - k_{\mathrm{eff}}^2\right)$. This leads to the important relationship between k_{eff}, Q_m, and Q_e,

$$k_{\mathrm{eff}} = \left(1 + Q_m Q_e\right)^{-1/2}. \tag{9.14a}$$

In cases where the Q_m is low because of high mechanical losses or large radiation resistance, the values of f_r and f_a measured from the maximum and minimum admittance (or impedance) magnitude, and the value of k_{eff} calculated from them, are not accurate. However, in these cases Eq. (9.14a) yields accurate results and is the preferred method of evaluating k_{eff} under water-loaded conditions. Equation (9.14a) yields accurate results for low Q_m since, here, Q_m and Q_e may be accurately determined and the product $Q_m Q_e$ depends on k_{eff} through the ratio of C_0/C_e which is independent of the radiation impedance. Equation (9.14a) is difficult to implement under in-air loading conditions where the sharpness of the resonance makes it difficult to accurately determine Q_m and Q_e from a rapidly changing large diameter B vs. G locus unless very small accurate frequency increments are used. On the other hand, the formula

$$k_{\mathrm{eff}} = \left[1 - \left(f_r/f_a\right)^2\right]^{1/2} \tag{9.14b}$$

is preferred for evaluating k_{eff} for transducers where the Q_m is high and f_r and f_a are readily and accurately determined from the maximum admittance and impedance, respectively.

The admittance locus of Fig. 9.6 has been a key graphical means for the evaluation of electric field transducers. The shape of the locus is based on Eq. (9.1) with the corresponding circuit of Fig. 9.1 and motional impedance,

$$Z_m = R_e + j\omega L_e + 1/j\omega C_e = R + jX = |Z_m| e^{j\varphi}, \tag{9.15}$$

where the phase angle $\varphi = \tan^{-1}(X/R)$. An X vs. R plot of Eq. (9.15), when R is independent of frequency, is illustrated in Fig. 9.10 and seen to be a straight line

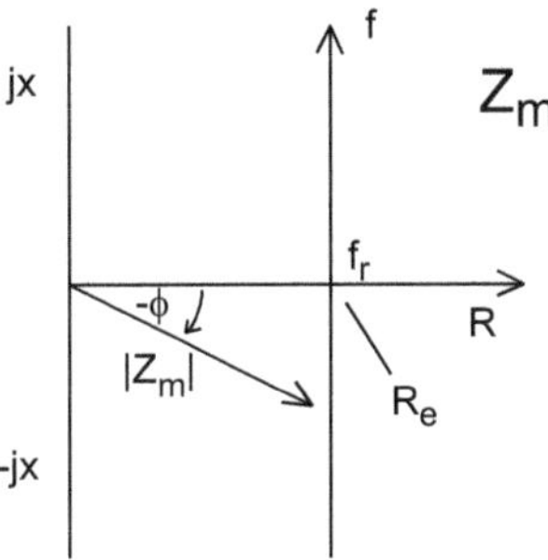

Fig. 9.10 Motional impedance for an electric field transducer

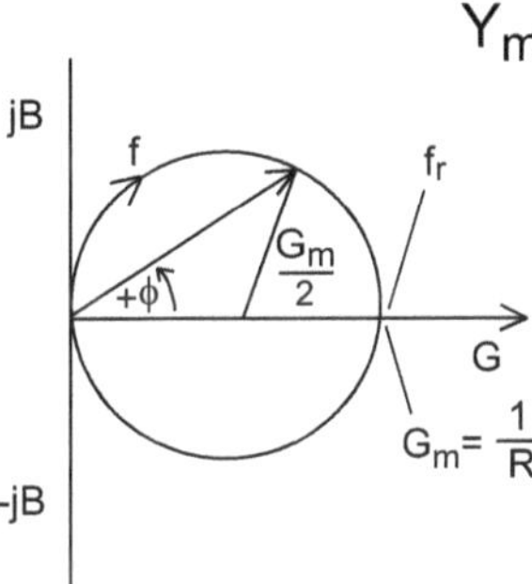

Fig. 9.11 Motional admittance for an electric field transducer

perpendicular to the R axis. As the frequency, f, increases the line crosses the axis at resonance, f_r, where $\omega_r L_e = 1/\omega_r C_e$ and $Z_m = R_e$. The motional admittance $Y_m = 1/Z_m = e^{-j\varphi}/|Z_m|$; thus, a high impedance is translated to a low admittance with an equal but negative phase angle. The result is a circular locus plot in the Y plane since a straight line in the Z plane is mapped into a circle in the Y plane with radius $G_m/2$ as illustrated in Fig. 9.11.

This circle is called the motional admittance circle as it represents the electrical equivalent of the mechanical part of the transducer and obeys the equation of a circle, with radius $G_m/2$ and center at $G = G_m/2$, $B = 0$,

$$(G_m/2)^2 = (G - G_m/2)^2 + B^2, \tag{9.16}$$

where G_m is the value of maximum conductance which occurs at f_r. Since the radiation resistance is a function of frequency, the admittance locus usually does not appear to be a true circle over a wide range of frequencies, especially if the Q_m is low. The diameter of the circle is proportional to the product of the effective coupling coefficient squared and Q_m.

Equation (9.1) and Fig. 9.1 show that the total admittance includes the electrical loss conductance, G_0, and clamped susceptance, $j\omega C_0$, along with the motional circle admittance, Y_m. The addition of these two components leads to the locus of Fig. 9.6 where $B_0 = \omega_r C_0$ and $G_0 = \omega C_f \tan\delta$. The above development assumes an

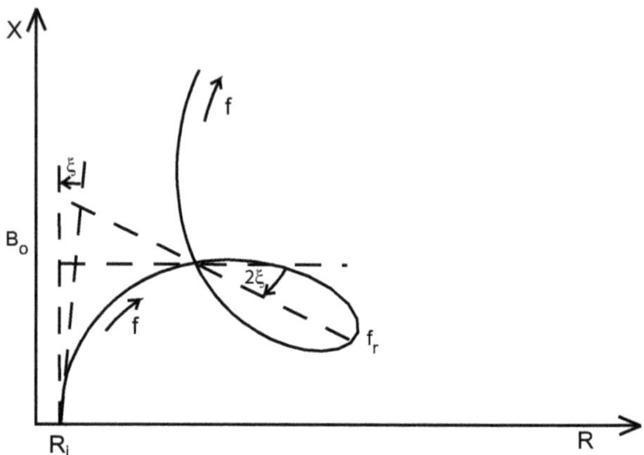

Fig. 9.12 Impedance locus for a magnetic field transducer with eddy current loss angle ξ

ideal single degree of freedom system. In practice there may be a number of modes of vibration above the fundamental mode which appear as additional circles at higher frequencies in the locus plot (and maxima and minima in the $|Y|$ or $|Z|$ plots).

Magnetic field transducers have a similar locus plot but in the impedance plane. That is, the motional part of Fig. 9.7 and of Eq. (9.8), $1/R' + j\omega C' + 1/j\omega L'$, is a straight line in the Y plane that maps into a circle in the Z plane if R' is constant. Inclusion of the clamped inductive reactance and the loss resistance R_e' as well as the eddy current factor $\chi = |\chi|e^{-j\xi}$ then yields the impedance locus shown in Fig. 9.12. As seen, the coil resistance displaces the curve by R_e' and the eddy current loss angle ξ causes the locus to tilt at an angle 2ξ. The factor 2 arises from the eddy current factor appearing in the turns ratio as χN. Since the impedance is transformed by the square of the turns ratio, we get $(\chi N)^2 = |\chi|^2 e^{-2\xi} N^2$ resulting in an angular factor of 2ξ for the motional impedance locus or "loop."

9.3 Measurement of Transducer Efficiency

The acoustical power radiated into the medium is $W_a = |u_{rms}|^2 R_r$ while the power delivered to the mechanical and acoustical sections is $W_m = |u_{rms}|^2 (R + R_r)$ giving a mechano-acoustical efficiency,

$$\eta_{ma} = W_a / W_m = R_r / (R_r + R). \tag{9.17}$$

Thus, a large radiation resistance R_r and low mechanical loss resistance R lead to a high mechanical efficiency (see Sect. 2.8.6). With the in-air mechanical Q written as $Q_m = 1/\omega_r C^E R$, and the in-water mechanical Q written as $Q_{mw} = 1/\omega_{rw} C^E (R + R_r)$, Eq. (9.17) may be rewritten as

$$\eta_{\mathrm{ma}} = [1 - \omega_{\mathrm{rw}} Q_{\mathrm{mw}} / \omega_{\mathrm{r}} Q_{\mathrm{m}}]. \tag{9.18}$$

Thus, the mechano-acoustical efficiency may be obtained from ratios of the mechanical Q's and resonance frequencies measured in air and in water.

Typically the Q_{m} ratio in Eq. (9.18) is more significant than the ω_{r} ratio and is often used alone as a first estimate. For example, the Q_{m} for a transducer in-air might be 30 while the Q_{mw} in water could be as low as 3, but the shift in the water resonance depends on the radiation mass relative to the transducer mass and is usually less than about 20 %. Furthermore, since the diameter of the admittance locus is proportional to the Q_{m},

$$\eta_{\mathrm{ma}} \approx 1 - Q_{\mathrm{mw}}/Q_{\mathrm{m}} = 1 - D_{\mathrm{w}}/D_{\mathrm{a}} = 1 - G_{\mathrm{w}}/G_{\mathrm{a}},$$

where D_{w} is the diameter of the water-loaded circle, D_{a} is the diameter of the air-loaded circle, G_{w} is the water-loaded conductance at resonance, and G_{a} is the air-loaded conductance at resonance. Equation (9.18) often gives a higher value than power response measurements of η_{ma} since it is tacitly assumed that in Eqs. (9.17) and (9.18) the in-water load is only R_{r}. However, the measured Q_{mw} may actually be lowered by other mechanical losses that result from in-water mounting and fluid viscosity. Typically η_{ma} values range from 60 to 90 %. The values for the resonance frequencies and Q's may be readily obtained from the curves of Figs. 9.5 and 9.6.

The power lost to electrical dissipation in electric field transducers is $|V_{\mathrm{rms}}|^2 G_0$. A measure of this is the electromechanical efficiency $\eta_{\mathrm{em}} = W_{\mathrm{m}}/W_{\mathrm{e}}$ where W_{e} is the input electrical power, and the overall electroacoustic efficiency is given by (see Sect. 2.8.6)

$$\eta_{\mathrm{ea}} = \eta_{\mathrm{em}} \eta_{\mathrm{ma}}. \tag{9.19}$$

The overall efficiency may be obtained from measurement of the input electrical power, W_{e}, the transmitting source level, SL(dB/1 μPa@ 1 m), and the directivity index, DI, by use of Eq. (1.25) in which the output power $W = \eta_{\mathrm{ea}} W_{\mathrm{e}}$:

$$10 \log \eta_{\mathrm{ea}} = \mathrm{SL} - \mathrm{DI} - 170.8 \ \mathrm{dB} - 10 \ \log \ W_{\mathrm{e}},$$

where 170.8 dB is the source level at one meter for one acoustic watt from an omnidirectional radiator.

Although the source level and input power can be measured quite accurately (see Sect. 9.4), determination of DI requires measurement of a number of beam patterns in cases where the transducer or array is not axisymmetric. As presented in Chap. 1, the DI = 10logD_{f} where the reciprocal of the directivity factor, D_{f}, is given by

$$D_{\mathrm{f}}^{-1} = (1/4\pi) \int_0^{2\pi} \left\{ \int_0^{\pi} [I(\theta, \phi)/I_0] \sin\theta \ \mathrm{d}\theta \right\} \mathrm{d}\phi,$$

where $I(\theta, \phi)$ is the far-field intensity in spherical coordinates and I_0 is the intensity on the maximum response axis. If the intensity is independent of ϕ, as in the case of a circular piston, the radiator is axisymmetric and the integration over ϕ yields 2π, leaving the inner integral alone for evaluation of the DI (see Sect. 13.13). The case of no symmetry may be handled if the outer integral is approximated by a summation of N terms with $N = 2\pi/\Delta\phi$, yielding

$$D_{\mathrm{f}}^{-1} \approx (1/N)\sum_{n=1}^{N}\left[(1/2I_0)\int_0^{\pi} I(\theta, \phi_n)\sin\ \theta\ \mathrm{d}\theta\right] = (1/N)\sum_{n=1}^{N} D_n^{-1},$$

where D_n^{-1} is given analytically by the quantity in brackets and is the reciprocal of the directivity factor of the pattern at selected angles, ϕ_n. Thus the overall reciprocal directivity factor is approximated by the average of the reciprocals of the directivity factors at the angles, ϕ_n. For example, with a square piston, which has fourfold symmetry, one would choose at least the angles $\phi_n = 0$ and $45°$ with $N = 2$ to get a value of DI with minimal accuracy.

9.4 Acoustic Responses of Transducers

The transmitting response of a projector gives the pressure in the medium per unit of electrical excitation as a function of frequency and is one of the most important means for characterizing the performance of a transducer. Typically, the response is measured in the direction of the maximum response axis (MRA) using a hydrophone at a radial distance in the far field where the pressure variation is proportional to $1/r$ and the beam pattern does not change with distance. The pressure is measured with a calibrated hydrophone and referenced to 1 m and a pressure of $p_0 = 1\ \mu\mathrm{Pa}$ in underwater applications. The response per volt is called the Transmitting Voltage Response or TVR $= 20\log|p/p_0|$ re 1 V @ 1 m, as defined in Chap. 1. Other responses that are often measured are the Transmitting Current Response, TCR, the response per ampere; the Receiving Voltage Response, RVS, (also called FFVS for Free Field Voltage Sensitivity); the Transmitting Volt-Ampere Response, TVAR; and the Transmitting Power Response, TPR. In addition impedance (R and X), admittance (G and B), efficiency, and power factor are often measured as a function of frequency. The measurements above are made at low drive level, but for high power projectors it is also important to measure source level and harmonic distortion as a function of drive level up to the highest level expected in each application. The measurement system must be free from reflections; and normally uses a "pulse gated system" in which the hydrophone is turned on only during the reception of the direct pulse and turned off during the reception of reflections. The pulse length must be long enough for the system to reach steady state (see Sect. 9.7.2). A gated impulse response or maximum length sequence, MLS, measurement technique [14] may also be used to evaluate the TVR. It is important that measurements be performed under bubble free water conditions. Air bubbles can cause scattering and unloading of the radiator surface.

As discussed in Chaps. 2 and 3, the velocity, u, for an electric field transducer is related to the drive voltage, V, by $u = NV/Z^{\mathrm{E}}$ where Z^{E} is the mechanical impedance under short circuit conditions. Similarly, the velocity for a magnetic field transducer is $u = N_{\mathrm{m}}I/Z^{\mathrm{I}}$ where Z^{I} is the mechanical impedance under open circuit conditions and I is the drive current. The far-field pressure radiated from a transducer is inversely proportional to the radial distance, r, and proportional to the density, ρ, of the medium, the radiating area, A, and the acceleration of the radiating surface $j\omega u$ [as can be seen, e.g., from Eq. (10.25a)]. Thus, the pressure is proportional to $j\omega NV/Z^{\mathrm{E}}$ for electric field, and to $j\omega N_{\mathrm{m}}I/Z^{\mathrm{I}}$ for magnetic field, transducers. The impedances may be approximated in the vicinity of resonance, and often well below resonance, as a combination of a mass, M, compliance, C, and resistance, R. Accordingly we may write the far-field pressure for voltage drive of electric field transducers as

$$p = K_1 j\omega\rho ANV/\left[R + j\left(\omega M - 1/\omega C^{\mathrm{E}}\right)\right]r \tag{9.20}$$

and for current drive of magnetic field transducers as

$$p = K_2 j\omega\rho AN_{\mathrm{m}}I/\left[R + j\left(\omega M - 1/\omega C^{\mathrm{I}}\right)\right]r, \tag{9.21}$$

where K_1 and K_2 are different numerical constants which depend on the transducer design.

These direct drive responses are normally the same in form except that one is voltage drive and the other is current drive; thus the TVR for the electric field transducer has the same frequency dependence as the TCR for the magnetic field transducer as shown in Fig. 9.13.

For direct drive conditions, the voltage is held constant for electric field transducers while the current is constant for magnetic field transducers. Well below resonance the pressure varies as ω^2 and increases at 12 dB/octave while well above resonance the pressure does not change with frequency for this simplified lumped model. In practice this plateau, above the fundamental resonance, is often altered by

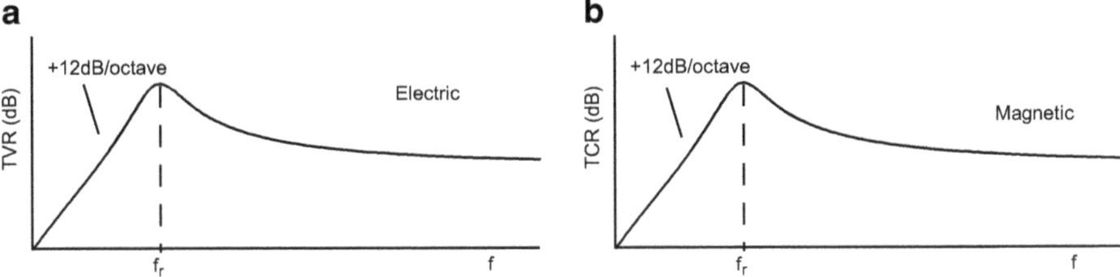

Fig. 9.13 (**a**) Transmitting voltage response for electric field transducer. (**b**) Transmitting current response for magnetic field transducer

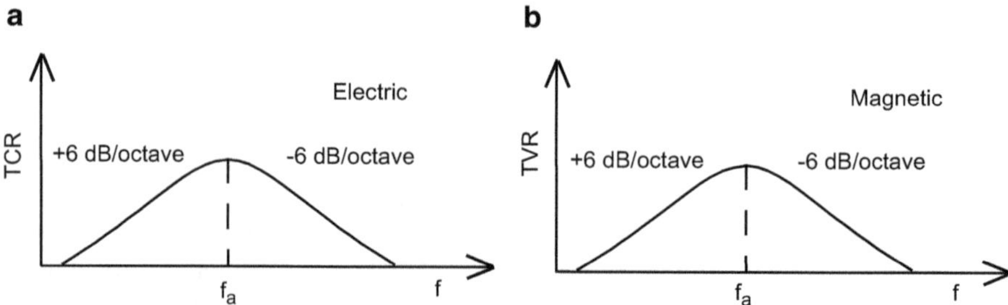

Fig. 9.14 (**a**) Transmitting current response for an electric field transducer. (**b**) Transmitting voltage response for a magnetic field transducer

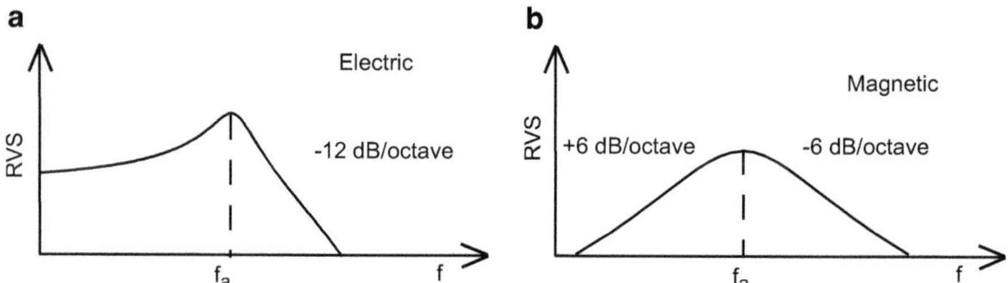

Fig. 9.15 (**a**) Receiving voltage sensitivity for an electric field transducer. (**b**) Receiving voltage sensitivity for a magnetic field transducer

other resonances. The indirect drive cases are related to the direct drive cases through the electrical impedance of the transducer. For electric field transducers the impedance, V/I, is approximately $1/j\omega C_f$ below resonance and $1/j\omega C_0$ above resonance. Ignoring the difference between C_f and C_0, we can approximately replace the voltage V by $I/j\omega C_f$ in Eq. (9.20) and obtain the TCR response shown in Fig. 9.14 with slopes below and above resonance of +6 and −6 dB. Inspection of the frequency dependence of the electrical impedance in the vicinity of resonance shows that the peak in this response is at the antiresonance frequency f_a. For magnetic field transducers, the impedance is approximately $j\omega L_f$ below resonance and $j\omega L_0$ above resonance. Ignoring the difference in the inductances, $V/I = j\omega L_f$ and we may replace I by $V/j\omega L_f$ in Eq. (9.21) resulting in the TVR response of Fig. 9.14. Note that, below resonance, the magnetic field TVR varies at only 6 dB/octave while the electric field TVR of Fig. 9.13 varies at a faster rate of 12 dB/octave.

It will be shown that the open circuit receiving response may be obtained from the TCR through a reciprocity factor, which affects the TCR by −6 dB/octave. The corresponding receiving response curves are illustrated in Fig. 9.15 for electric and magnetic field transducers. It can be seen that both cases resonate at f_a but that the RVS for magnetic field transducers is symmetrical about resonance while for electric field transducers it is frequency independent below resonance, which is particularly desirable for wideband hydrophone performance. Because of the finite input impedance of a preamplifier, the RVS of an electric field transducer actually approaches zero at very low frequencies. The associated -3 dB cutoff frequency is $f_c = 1/(2\pi RC_f)$, where R is the input resistance to the preamplifier and C_f is the free capacity of the hydrophone.

The hydrophone performance of a transducer is usually measured by a substitution method. A broad band projector produces a free field pressure, p_{ff}, in the medium, which is first measured by a calibrated reference hydrophone of precisely known sensitivity M_r, giving a measured voltage of $V_r = M_r p_{ff}$ and $p_{ff} = V_r/M_r$. Then the hydrophone under evaluation is placed in the same location, a voltage V_e is measured, and the sensitivity is given by $M_e = V_e/p_{ff} = M_r V_e/V_r$. The receiving voltage sensitivity, in dB, is

$$\mathrm{RVS} = 20\log M_e = 20\log M_r + 20\log V_e - 20\log V_r. \tag{9.22}$$

9.5 Reciprocity Calibration

In the reciprocity procedure [1, 8] we must assume that the projector is operating over a linear portion of its dynamic range and a negligible amount of power is lost to distortion. Biased electric field and magnetic field transducers are usually quite linear if the drive field is small compared to the bias field. Magnetic transducers are the dual of electric transducers and, while both generally obey reciprocity, the combination of the two may not, as discussed in Sect. 5.3.2 on the magnetostrictive/piezoelectric Hybrid transducer.

The transducer transfer matrix ("ABCD" matrix formulation, see Sect. 3.3.2) may be used to relate the receiving voltage sensitivity, RVS, to the transmitting current response, TCR. Consider the "ABCD" representation illustrated in Fig. 9.16 with a radiation impedance load, Z_R, and an acoustic force, F_b, resulting from an incoming acoustic wave.

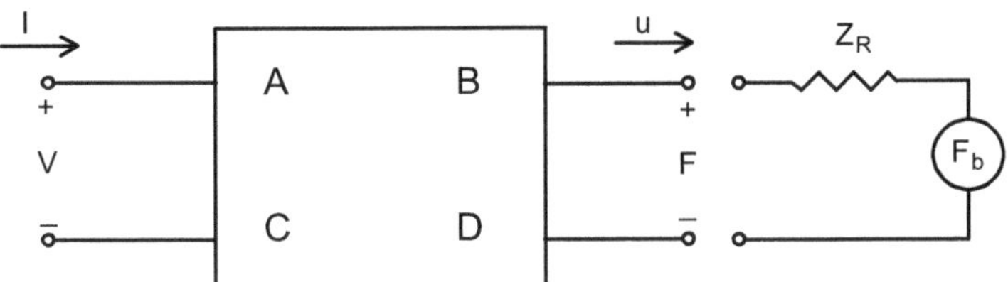

Fig. 9.16 *ABCD* electromechanical representation

As discussed in Chap. 3, the transducer equations may be written as

$$V = AF + Bu, \tag{9.23a}$$

$$I = CF + Du, \tag{9.23b}$$

where V and I are the voltage and current and F and u are the force and velocity which couple to the medium. Note that A, B, C, and D all have different physical units. With the transducer in the water $F = F_\mathrm{b} - uZ_\mathrm{r}$ leading to the set

$$V = AF_\mathrm{b} + (B - AZ_\mathrm{r})u, \tag{9.24a}$$

$$I = CF_\mathrm{b} + (D - CZ_\mathrm{r})u. \tag{9.24b}$$

Under open circuit receiving conditions, where $I = 0$, we get from Eq. (9.24b)

$$u = -CF_\mathrm{b}/(D - CZ_\mathrm{r}), \tag{9.25}$$

and substitution into Eq. (9.24a) yields the open circuit received voltage for an input force F_b.

$$V/F_\mathrm{b}|_{I=0} = (AD - CB)/(D - CZ_\mathrm{r}). \tag{9.26}$$

If the transducer is reciprocal, the determinant of the coefficients of Eqs. (9.23a) and (9.23b) is +1 or –1; then $AD - CB = \pm 1$ so that Eq. (9.26) reduces to

$$V/F_\mathrm{b}|_{I=0} = \pm(D - CZ_\mathrm{r})^{-1}, \tag{9.27}$$

where the sign depends on how the pair of fundamental equations are written. Now, under the conditions of transmitting current response alone where there is no reception from another source the blocked force $F_\mathrm{b} = 0$ and we get from Eq. (9.23b)

$$u/I|_{F_\mathrm{b}=0} = (D - CZ_\mathrm{r})^{-1}. \tag{9.28}$$

Comparing Eqs. (9.28) and (9.29) yields the desired result

$$V/F_\mathrm{b}|_{I=0} = \pm u/I|_{F_\mathrm{b}=0}. \tag{9.29}$$

Equation (9.29) completes the first step in establishing the reciprocity calibration procedure; it is based on the electromechanical reciprocity discussed in Sect. 1.3. We now turn to the second step that requires use of acoustic reciprocity discussed in Sect. 11.2.2.

The open circuit receiving response $M = V/p_\mathrm{ff}$ is the ratio of the open circuit voltage, V, to the free field pressure, p_ff. The free field pressure is the pressure at the hydrophone location with the hydrophone removed. If the hydrophone is small and stiff, which is usually the case below resonance, the force on the surface of the hydrophone is approximately $A_r p_{ff}$. (In this development we have used A_r for the radiating area of the hydrophone to distinguish it from the A of the $ABCD$ parameters.) Otherwise, the force on the surface of the hydrophone $F_b = A_r p_{ff} D_a$ where D_a is the diffraction constant, the ratio of the spatially averaged clamped pressure over the active face of the hydrophone to the free field pressure (see Sect. 6.6). The diffraction constant equals unity for small hydrophones in free space and two for a piston in a rigid baffle. The transmitting current response $S = p_{ff}/I$, and the ratio M/S is given by solving these relationships for V, F_b, and I and substituting into Eq. (9.29):

$$M/S = |A_r D_a u/p_{ff}| = J, \tag{9.30}$$

where J is called the reciprocity constant.

For the case of plane waves, $p_{ff} = \rho c u$ and for an acoustically large piston, $D_a = 2$ yielding $J = 2A_r/\rho c$. The plane wave condition is used in some calibration procedures where a plane wave is generated by a piston in a rigid tube of diameter $D \ll \lambda$. A more often used condition is the case of radiation by a piston of area A_r vibrating with normal velocity u in a rigid baffle where $D_a = 2$. The on-axis pressure at large distance d is, from Eq. (10.25a),

$$p_{ff} = j\omega\rho A_r u e^{-jkd}/2\pi d. \tag{9.31}$$

Substitution of the magnitude of p_{ff} into Eq. (9.30) yields the relation between transmitting current response and the open circuit receiving response.

$$M/S = J = 2d/\rho f. \tag{9.32}$$

(We also note that $M_s/S_s = J = 2d/\rho f$ where M_s is the short circuit receiving response and S_s is the transmitting voltage response.)

The same result is found for any transducer radiating spherical waves into the far field and, therefore, this value of J is known as the spherical wave reciprocity constant. This can be shown from Bobber's expression [1] for the diffraction constant, given in Eq. (6.56), which is based on acoustic reciprocity:

$$D_a^2 = 4\pi c D_f R_r / A_r^2 \omega^2 \rho, \tag{9.33}$$

where R_r is the radiation resistance, D_f is the directivity factor and, again, here A_r is the radiating area. Since $D_f = I_0/I_a$, where the maximum intensity $I_0 = p_{rms}^2/\rho c$, the average intensity $I_a = W/4\pi d^2$ and the power $W = u_{rms}^2 R_r$, $D_a = 2pd/uA_r\rho f$ and substitution into Eq. (9.30) again yields $J = 2d/\rho f$. Equation (9.33) is developed in Sect. 11.3.1.

Equation (9.32) is probably the most important relationship in the measurement and calibration of underwater transducers. It depends on both electromechanical and acoustic reciprocity, a combination that can be called electroacoustic reciprocity. With the response referenced to 1 μPa at $d = 1\,\text{m}$ and the water density $\rho = 1000\,\text{kg/m}^3$, Eq. (9.32) leads to

$$\text{RVS} = \text{TCR} - 20\log f - 294\,\text{dB}. \tag{9.34}$$

Thus, measurement of the TCR yields the RVS and vice versa.

Hydrophone RVS measurements using the substitution method, Eq. (9.22), and reciprocity method, Eq. (9.34), provide an indication of experimental accuracy and reliability of the calibration of the reference hydrophone. If there is an error in the "calibrated" reference hydrophone, the correct RVS value of a hydrophone under test is the average, in dB, of the direct comparative measured value and the value obtained by reciprocity [9]. For example, if the reference hydrophone is actually 1 dB below the published sensitivity value, the comparison method will yield an apparent higher sensitivity for the hydrophone under test by 1 dB. On the other hand, a reciprocity determination will yield a 1 dB lower TCR and corresponding RVS from Eq. (9.34). Consequently, the two errors will cancel out [9] when the two RVS results are averaged, i.e., $[(\text{RVS} + 1) + (\text{RVS} - 1)]/2 = \text{RVS}$.

Equation (9.34) may be rewritten with the TVR instead of the TCR if the magnitude of the electrical impedance is also measured. Since the electrical impedance is $Z = V/I$, the current response $S = p/I$ may be written as $S = Zp/V$, where p/V is the constant voltage transmitting response. In this case Eq. (9.34) becomes

$$\text{RVS} = \text{TVR} + 20\log|Z| - 20\log f - 294\,\text{dB}. \tag{9.35}$$

At very low frequencies the impedance of electric field transducers is $1/j\omega C_f$ and under this restriction

$$\text{TVR} \approx \text{RVS} + 20\log(C_f) + 40\log f + 310\,\text{dB}, \quad f \ll f_r. \tag{9.36}$$

Since the RVS is flat below resonance for piezoelectric ceramic transducers, Eq. (9.36) shows that the TVR rises at a rate of +12 dB/octave in the region below resonance.

Reciprocity principles may be used to calibrate transducers in the free field or, in a limited way, from measurements made in rigid confined enclosures, such as illustrated in Fig. 9.29 and discussed in Sect. 9.7.2. Equation (9.36) may then be used to determine the TVR from RVS measurements made at frequencies below the enclosure resonance frequencies if the frequencies of interest are also well below the fundamental resonance frequency of the transducer. In this range the radiation loading has little effect on the stiffness controlled velocity response as well as the corresponding pressure response. On the other hand, the loading does have a significant effect on power radiated and mechanical efficiency, which cannot be determined from this measurement.

9.6 Tuned Responses

The performance of a transducer can be improved by electrically cancelling out or "tuning out" residual transducer electrical components, which reduces power amplifier volt-ampere requirements by improving the power factor (see Sect. 2.8.4). Electric field transducers are tuned with inductors, while magnetic field transducers are tuned with capacitors. Electrical tuning is usually implemented at the mechanical resonance frequency of the transducer where the mass and stiffness reactance cancel leaving the radiation resistance and internal resistance plus the clamped capacitance or inductance. The electric and magnetic equivalent circuits of Figs. 9.1 and 9.7 show the transducer shunt clamped capacitor, C_0, and the series clamped inductor, L_0, respectively. Corresponding equivalent circuits at mechanical resonance, ω_r, are shown in Fig. 9.17 for electric field transducers and Fig. 9.18 for magnetic field transducers.

9.6.1 *Electric Field Transducers*

A measure of the ratio of the clamped susceptance relative to the conductance at resonance, for the circuits of Figs. 9.1 and 9.17, is the electrical $Q_e = \omega_r C_0 R_e$. Since Eq. (2.81) shows that $Q_e = \left(1 - k_{\mathrm{eff}}^2\right)/k_{\mathrm{eff}}^2 Q_m$, a transducer with a higher effective coupling coefficient has a lower Q_e requiring less tuning reactance for a given Q_m yielding a wider band for effective tuning. The electrical band limits are typically considered to be the frequencies at which the reactance equals the resistance and the impedance or admittance phase angle is 45° (see Sect. 2.8.3).

Consider first the electric field transducer equivalent circuit of Fig. 9.1 along with the admittance response and locus of Figs. 9.5 and 9.6. At mechanical resonance, the circuit of Fig. 9.1 becomes the circuit of Fig. 9.17. The value of the susceptance at mechanical resonance is $B_0 = \omega_r C_0$, where C_0 is the clamped capacitance. For tuning with an inductor, L_p, in parallel with the transducer at mechanical resonance $\omega_r L_p = 1/\omega_r C_0$ or

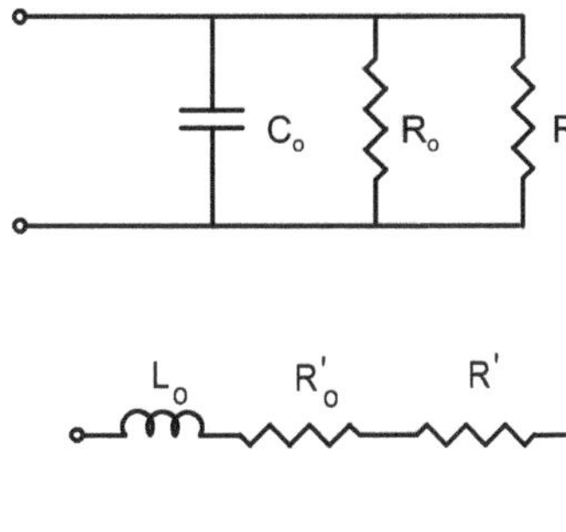

Fig. 9.17 Electric field transducer circuit at resonance

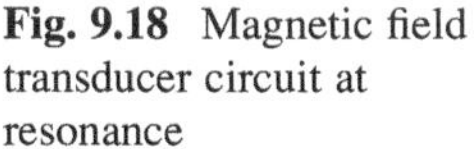

Fig. 9.18 Magnetic field transducer circuit at resonance

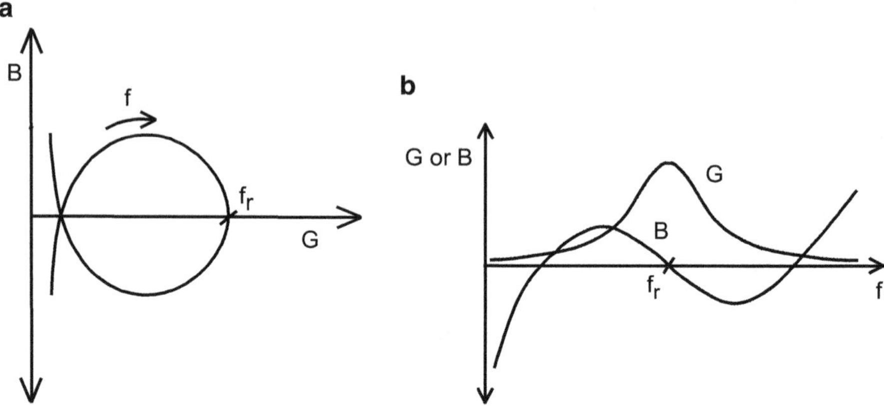

Fig. 9.19 (**a**) Shunt tuned electric field transducer admittance locus. (**b**) Shunt tuned electric field transducer conductance, *G*, and susceptance, *B*

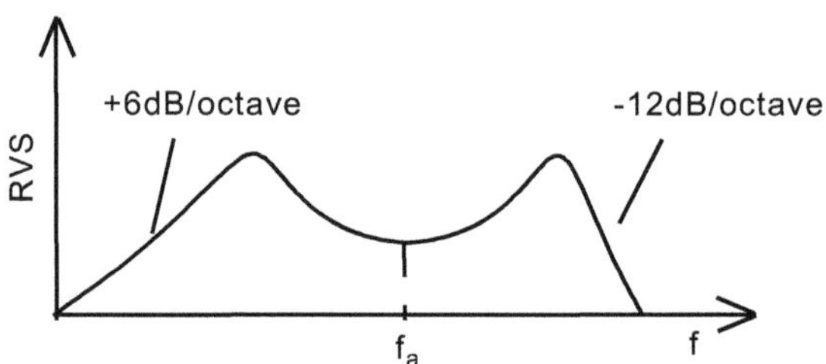

Fig. 9.20 Shunt tuned receiving sensitivity electric field transducer

$$L_p = 1/\omega_r^2 C_0. \tag{9.37}$$

Under this tuned condition, the admittance response and locus of Figs. 9.5 and 9.6 take the form of the curves of Fig. 9.19 with no reactive component at resonance, f_r.

The TVR of Fig. 9.13 does not change since L_p is in parallel with the transducer and does not affect the voltage applied to the transducer. However, the shunt inductor does affect the receiving response output voltage and the RVS of Fig. 9.15 does change and takes the form illustrated in Fig. 9.20. Although the shunt tuning may improve the receiving response in the vicinity of resonance, it reduces the sensitivity well below resonance at a rate of 6 dB/octave, and makes it somewhat like a band-pass filter.

A series inductor, L_s, may also be used to tune out the series reactance term of the impedance at mechanical resonance. At this frequency the reactance of the circuit of Fig. 9.17 is $(1/j\omega_r C_0)Q_e^2/(1+Q_e^2)$ and the required value for the series inductance is

$$L_s = (1/\omega_r^2 C_0)Q_e^2/(1+Q_e^2), \tag{9.38}$$

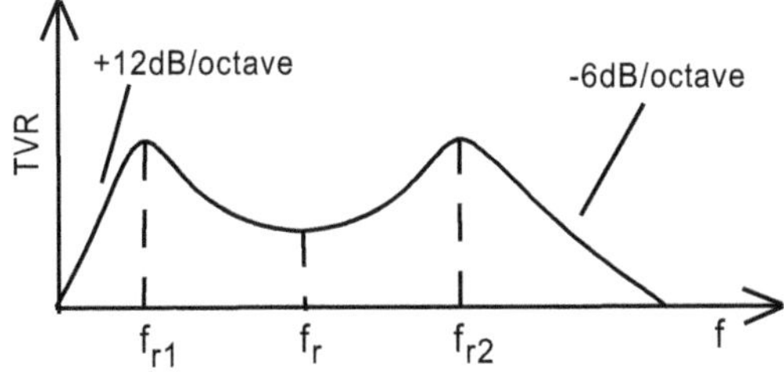

Fig. 9.21 Series tuned electric field transducer transmitting voltage response

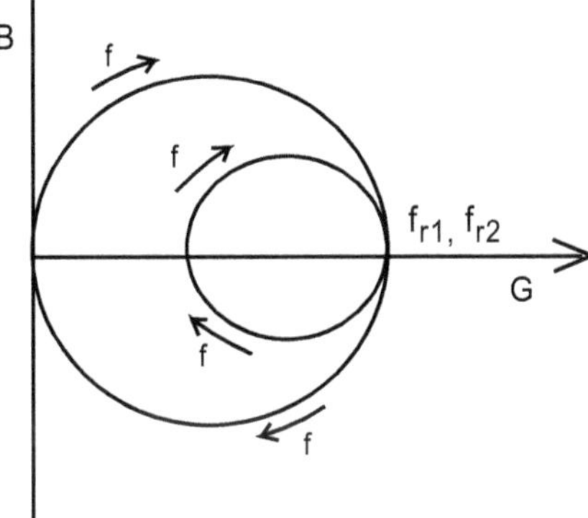

Fig. 9.22 Series tuned admittance locus for an electric field transducer

where $Q_e = \omega_r C_0 R_0 R_e/(R_0 + R_e)$. As seen, L_s approaches L_p for $Q_e \gg 1$; however, for $Q_e \ll 1$, $L_s = Q_e^2 L_p$. The series tuning method lowers transducer input impedance requiring less voltage drive from the amplifier than in the untuned or parallel tuned cases. Parallel tuned cases often need an additional transformer to reduce the electrical input impedance of the transducer. This transformer, however, can also double as a shunt tuning inductor (see Sect. 13.16). The RVS of Fig. 9.15 does not change because L_s is in series with the transducer and the RVS is defined under open circuit conditions which is nearly the case with a high input impedance preamplifier. However, the TVR of Fig. 9.13 does change and takes the form shown in Fig. 9.21.

The depth of the reduction in the response at f_r depends on the values of Q_m and Q_e. The series tuned admittance locus is shown in Fig. 9.22 with the two resonance frequencies f_{r1} and f_{r2}, at which the TVR has maxima, as illustrated in Fig. 9.21. Again, the shape of the response curve corresponds to that of a band-pass filter.

The transducer is often used as both a receiver and projector. This may be accomplished using the transmit/receive, "T/R," circuit illustrated in Fig. 9.23 where V_t is the transmitting voltage, V_r is the received voltage, X is the transducer, T is the transformer (possibly with shunt tuning) with turns ratio, T, and A and B are low voltage reversed diode pairs.

The transmitting power amplifier is usually a constant voltage low impedance source. Under transmitting conditions with voltage drive greater than

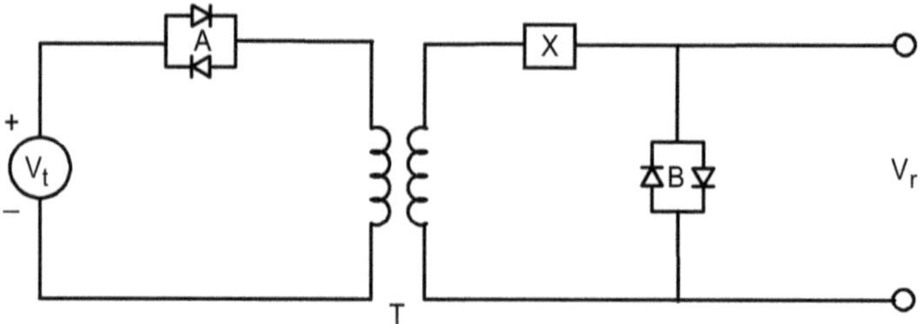

Fig. 9.23 Transmit/receiving, "TR," circuit with back to back diodes A and B

approximately 1 V the diode pairs *A* and *B* conduct in both directions and the voltage TV_t appears across the transducer, *X*. On the other hand, under receive conditions with $V_t = 0$, the transducer acts as a hydrophone with a typical low level output receive voltage, V_r. In this case, the diode pairs *A* and *B* open circuit and the signal passes through the secondary of the transformer and appears as the output received voltage V_r. This voltage is usually amplified through a high input impedance preamplifier, which is coincidentally protected under transmitting by the short circuit conditions of the diode pair, *B*. In some low signal cases, the diode noise could be a problem and should be considered.

9.6.2 Magnetic Field Transducers

The electrical tuning of a magnetic field transducer may be understood with reference to the equivalent circuits of Figs. 9.7 and 9.18. At mechanical resonance, $\omega_r L' = 1/\omega_r C'$ and the series impedance is $(R' + R_0') + j\omega_r L_0$ as shown in Fig. 9.18. The reactive part may be cancelled by a series capacitor with value

$$C_s = 1/\omega_r^2 L_0. \tag{9.39}$$

This series capacitor has no effect on the open circuit RVS or the TCR but it does have an effect on the TVR of Fig. 9.14 and impedance locus of Fig. 9.12. The resulting series tuned impedance locus is shown in Fig. 9.24, and the resulting TVR band-pass filter response is shown in Fig. 9.25.

The magnetic field transducer may also be parallel tuned at mechanical resonance with a capacitor, C_p. With $Q_e = \omega_r L_0/(R' + R_0')$ the equivalent shunt inductive reactance is $j\omega_r L_0(1 + Q_e^2)/Q_e^2$. Thus, for parallel tuning

$$C_p = (1/\omega_r^2 L_0) Q_e^2/(1 + Q_e^2). \tag{9.40}$$

This parallel tuning has no effect on the TVR but does affect the TCR shown in Fig. 9.13 by introducing a 6 dB/octave roll-off above resonance as illustrated in the band-pass filter response of Fig. 9.26.

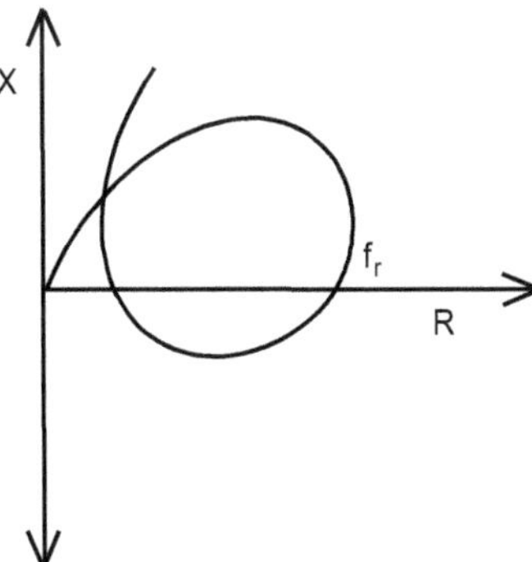

Fig. 9.24 Series tuned impedance locus for magnetic field transducer

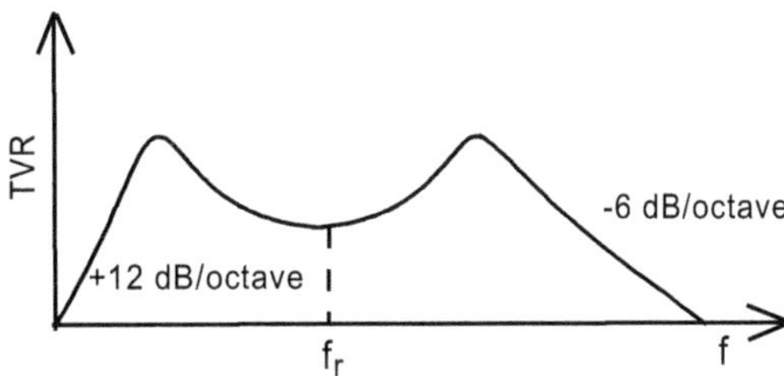

Fig. 9.25 Transmitting voltage response for series tuned magnetic field transducer

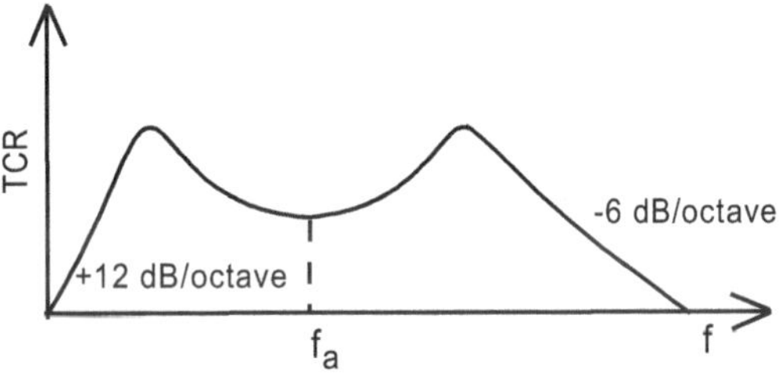

Fig. 9.26 Transmitting current response for shunt tuned magnetic field transducer

Equations (9.37)–(9.40), which give inductance and capacitance values for parallel and series tuning, are based on a lumped equivalent circuit representation of the transducer with effective parameter values and have validity in the vicinity of resonance. A more direct and accurate method is to use measured or accurately modeled reactive values obtained from the impedance and admittance. The reactive part of the impedance would be used to determine the series tuning value while the susceptance part of the admittance would be used to find the parallel tuning value. Thus, if X_r is the reactance of a transducer at mechanical resonance, the series tuning inductance for an electric field transducer would be $L_s = X_r/\omega_r$ while the series capacitance would be $C_s = 1/X_r\omega_r$ for a magnetic field transducer. If B_r is the susceptance of the transducer at resonance, the parallel inductance would be $L_p = 1/B_r\omega_r$ while the parallel capacitance would be $C_p = B_r/\omega_r$ for electric and magnetic field transducers, respectively.

Electrical tuning is useful in reducing the volt-ampere requirements at mechanical resonance particularly under water-loaded conditions. It can also be applied at frequencies above and below resonance; however, if it is not applied in the vicinity of resonance, the resulting tuned bandwidth will be narrow. Since piezoelectric ceramics exponentially age (see Sect. 13.14) with time after poling [15, 16], the tuned frequency will vary with time especially if the piezoelectric ceramic has been recently poled. It is therefore important that the piezoelectric ceramic section of the transducer be adequately aged before being tuned.

9.7 Near-Field Measurements

Acoustic transducer measurements are normally made in the far field where spherical spreading holds, the pressure falls off as $1/r$ with a 6 dB reduction per doubling of distance, and the beam pattern does not change as the distance is increased. Measurements are made in oceans, lakes, quarries, ponds and indoor tanks and pools. Indoor measurements are most convenient but may be impossible at certain frequencies because of limited tank size. However, a variety of near-field measurement techniques have been developed which allow measurements in tanks with dimensions smaller than the far-field distance and permit near-field measurements to be extrapolated to the far field.

9.7.1 Distance to the Far Field

The distance to the far field involves the wavelength and the size of the transducer or array. For example, the distance between a projector being evaluated and a measuring hydrophone must be considerably greater than the projector size to prevent the different distances from the center and ends of the projector from introducing significant phase and amplitude differences at the measuring hydrophone. This is particularly important when beam pattern measurements are made and the transducer is rotated. The far field is established for a projector or hydrophone array of length L at the so-called Rayleigh distance,

$$r \geq L^2/2\lambda. \tag{9.41}$$

This distance may be understood by considering an acoustic wave arriving at a hydrophone array of length L from a small source at a distance r from the ends of the array as illustrated in Fig. 9.27.

If the source is far from the array, the arc of the curved wave front will fall along L and all the hydrophone elements of the array will receive an approximately in-phase wave. The difference between a spherical wave from a nearby small source and a plane wave from a distant small source may be measured by the

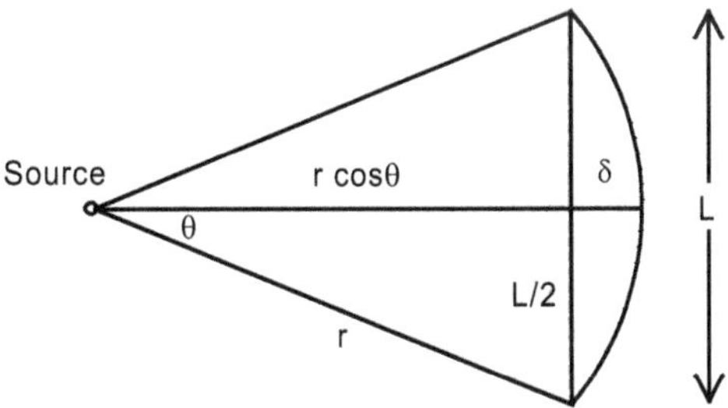

Fig. 9.27 Difference between spherical and plane wave fronts

"sagitta" distance $\delta = r(1 - \cos\theta)$ shown in Fig. 9.27. As the source distance r grows, the angle θ becomes small so that $\delta \approx r\theta^2/2$ and $\theta \approx L/2r$ yielding $\delta \approx L^2/8r$. If the distance δ is less than $\lambda/4$, the phase differences across the array will be small, although not completely negligible. Thus, with $\delta \leq \lambda/4$ we get, from $\delta \approx L^2/8r$, the far-field condition $r \geq L^2/2\lambda$ as in Eq. (9.41). It has been found that the condition of Eq. (9.41) is sufficient in many cases, but that the more accurate condition of $\delta \leq \lambda/8$ or $r \geq L^2/\lambda$ is sometimes necessary.

Note that Eq. (9.41) is based only on a requirement of small phase variation over the surface of the transducer or array of transducers and does not consider the effects of amplitude variation. Amplitude variation can be taken into account by the added condition that $r \gg L$.

An expression similar to Eq. (9.41) may also be obtained from the pressure magnitude along the axis of a radiating piston of radius a set in a rigid baffle (see Sect. 10.3.1). With z the distance from the center of the piston to a point along the axis the pressure magnitude is

$$p(z) = 2\rho cu \sin\left\{k\left[\left(z^2 + a^2\right)^{1/2} - z\right]/2\right\}. \tag{9.42}$$

A plot of this function is shown in Fig. 10.13 illustrating how the interference causes nulls and peaks in the near-field pressure with a smooth $1/z$ pressure variation in the far field. The nulls and peaks only occur when the wavelength is small enough to result in a total phase reversal from different parts of the radiator. In the region where $z \gg a$ we have, from the binomial expansion,

$$\left(z^2 + a^2\right)^{1/2} \approx z + a^2/2z. \tag{9.43}$$

If, in addition to this, the condition $ka^2/4z \ll \pi/4$ is satisfied, Eq. (9.42) reduces approximately to the far-field expression

$$p_{\mathrm{f}} = \rho cuka^2/2z_{\mathrm{f}}, \tag{9.44}$$

where z_{f} indicates that z is a far-field reference distance. With the piston diameter $D = 2a$, the far-field condition above may be written as

$$z_f \geq D^2/2\lambda, \tag{9.45}$$

which is consistent with Eq. (9.41). For a radiating surface of any shape, the far-field distance can be estimated as the square of the maximum dimension divided by 2λ.

9.7.2 Measurements in Tanks

Small tank limitations depend on the size and Q_m of the transducer as well as the frequency of the measurement. If a gated pulse system is used, measurements are made between the time of arrival of the direct pulse and the reflected pulse. This time window should exceed Q_mT, the time for the pulse to reach steady state where T is the period of vibration. In addition a small number, N, of steady state cycles may be necessary to make a reliable measurement, although in some cases $N = 1$ is sufficient. The total time, t, required is then $(Q_mT + NT)$. With Δ equal to the difference in propagation distance between the direct and the first reflected pulses, the time difference Δ/c should be equal to or greater than $(Q_mT + NT)$ to avoid interference with the directly received pulse. Consequently, the difference in the path length between the direct and reflected pulses must be

$$\Delta \geq (Q_m + N)c/f = (Q_m + N)\lambda.$$

Thus, large distances of travel between direct and reflected paths are needed for low frequency, long wavelength, high Q transducers.

Pulse techniques can simulate actual operating conditions allowing high power drive and harmonic distortion evaluation of the transducer. If, however, only the low level response is needed, impulse methods may be preferable since the response curve is displayed almost instantaneously and the pulse length is very short, allowing measurements at lower frequencies.

A large Δ implies a large tank where the reflecting surfaces are far from the receiving hydrophone. A possible arrangement is illustrated in Fig. 9.28 where

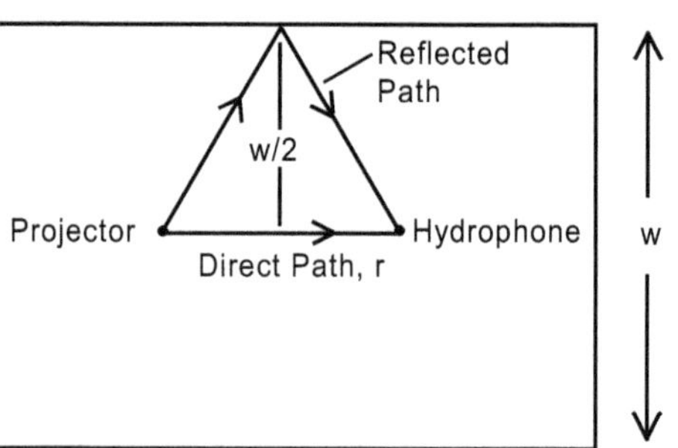

Fig. 9.28 Tank measurements with direct and reflected paths

the direct path distance is r and the reflected path distance is $r_r = (w^2 + r^2)^{1/2}$ giving $\Delta = r_r - r$.

The most favorable situation occurs if $w \gg r$ where $\Delta \approx w - r$ and, in the extreme case, $\Delta \approx w$. On the other hand, for $w \ll r$ the path difference $\Delta \approx w$ $(w/2r)$ and the ability to measure transducers is impaired. For a typical case where $w = r$, the difference $\Delta = 0.414w$. If the transducer $Q = 5$ and we choose $N = 5$ cycles then we need $w \geq 24\lambda$, while with $Q = 1$ and $N = 1$, $w \geq 4.8\lambda$. At 3 kHz where the wavelength is 0.5 m, the former case requires a tank dimension of 12 m while the latter case requires only 2.4 m. Sound absorbing material may be used to line the measuring tanks; however, to be fully effective, the absorbing material usually needs to be at least one-quarter wavelength thick. At higher frequencies, where the material is most effective, pulse gated systems usually work well without the need of absorbers.

Hydrophone measurements in the low frequency band can also be made in the near field. These measurements are usually made within a small rigid container with a maximum dimension, w, less than $\lambda/4$ allowing a nearly uniform pressure field and measurement at frequencies up to $c/4w$. This arrangement is illustrated in Fig. 9.29 with, H, the hydrophone under test, H_r, a calibrated reference hydrophone, and D the projector.

The container may be air filled or fluid filled for testing under uniform pressure. If air filled, a conventional loudspeaker can be used while if water filled, a flexural mode piezoelectric driver could be used to obtain the low frequencies needed. The container must be sealed and the hydrophones must be omnidirectional and operated below their fundamental resonance. In operation the driver is swept over the frequency band creating a pressure, p, in the container while the voltage outputs, V and V_r, from H and H_r are measured. With the sensitivities $M = V/p$ and $M_r = V_r/p$ for the hydrophone under test and reference hydrophone, respectively, the sensitivity of the hydrophone under test is then $M = M_r V/V_r$ or in dB, $20\log M = 20\log M_r + 20\log(V/V_r)$. These results, along with the free capacitance, C_f, may also be used to obtain the low frequency TVR of the transducer through Eq. (9.36).

Directional hydrophones (see Sect. 6.5) may be measured with a system that generates a pressure gradient such as the one illustrated in Fig. 9.30.

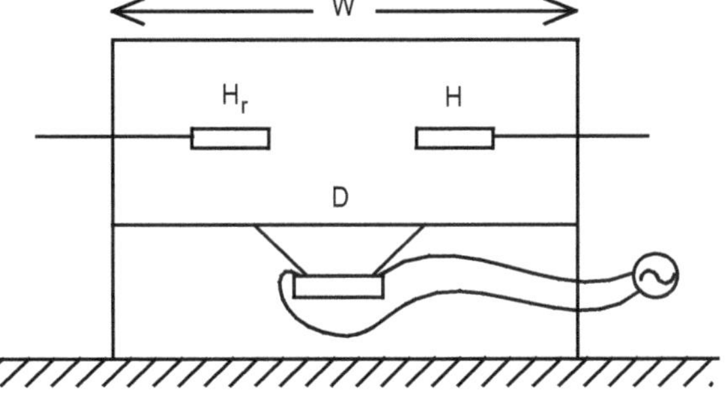

Fig. 9.29 Small closed chamber for testing omnidirectional hydrophone

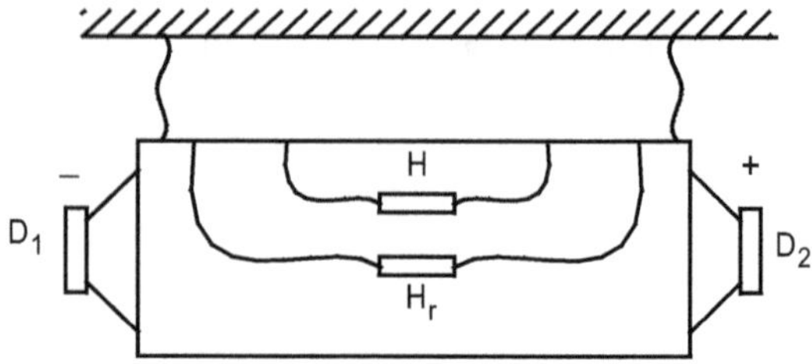

Fig. 9.30 Chamber for testing a directional hydrophone

The gradient is established with the two drivers D_1 and D_2 driven 180° out of phase creating a longitudinal force on the reference hydrophone, H_r, and the hydrophone under test, H. The rigid container as well as the hydrophones H and H_r must be compliantly suspended to allow free horizontal motion. Beam patterns may be measured by rotating the directional hydrophone under test. This system has been extensively developed and analyzed by Bauer et al. [17]. Directional hydrophones are often constructed from accelerometers as described in Sect. 6.5.4. In these cases the accelerometers are initially tested on shake tables with reference accelerometers before in-water tank testing.

9.7.3 Near-to-Far-Field Extrapolation: Small Sources

When a radiating surface is smaller than the wavelength, it can often be approximated by a simpler radiator with a known field. In such cases near-field pressure measurements can be extrapolated to the far field. The most common example is the piston in a rigid baffle where the analytical expression for the pressure on the axis is given in Eq. (9.42). The ratio of Eq. (9.44) to Eq. (9.42),

$$p_f/p(z) = (a/4z_f)ka/\sin\left\{k\left[\left(z^2+a^2\right)^{1/2}-z\right]/2\right\}, \tag{9.46}$$

shows that if the near-field pressure magnitude $p(z)$ is measured at a given location z, then the far-field pressure magnitude on the axis at z_f may be calculated for any frequency. There is, however, the possibility of dividing by a very small number with a serious loss of accuracy when using this equation. Equation (9.46) may be simplified further if the near-field pressure is measured at the center of the piston at $z = 0$ yielding

$$p_f = p(0)(a/2z_f)/\mathrm{Sinc}(ka/2). \tag{9.47}$$

At low frequencies, where $ka \ll 1$, Eq. (9.47) simplifies further to

$$p_f = p(0)(a/2z_f). \tag{9.48}$$

Equation (9.48) was first proposed by D.B. Keel [18] as a means for obtaining the far-field response of low frequency loudspeakers [14] for which far-field conditions are difficult to obtain. Use of Eq. (9.48) requires placement of a small microphone or hydrophone at the center of the radiator, measuring the pressure magnitude and then multiplying by $a/2$ to get the far-field pressure referenced to 1 m ($z_f = 1$). Because the sensor is very close to the vibrator, the direct signal distance is much less than the distance of any reflector and the conditions are favorable for pulse measurement. Moreover, since the acoustic pressure from the direct signal is considerably larger than the pressure from any reflected signal, a pulse signal is not always needed and continuous wave measurements may be sufficient.

Equation (9.48) may be extended to the case where the measurement is made at a small distance, z, from the center of the piston by evaluating Eq. (9.46) for both $ka \ll 1$ and $kz \ll 1$ yielding

$$p_f = p(z)(a/2z_f)/\left\{\left[(z/a)^2 + 1\right]^{1/2} - z/a\right\}. \tag{9.49}$$

Thus, for $z \ll a$, but not necessarily zero,

$$p_f = p(z)(a + z)/2z_f. \tag{9.50}$$

It can also be shown (see Sect. 10.3.1), for the case of an acoustically short thin walled ring of mean radius a, that

$$p_f = p(z)\left(a^2 + z^2\right)^{1/2}/z_f,$$

for the usual case where the radiating surfaces are in-phase [see Eq. (10.36a)]. This is a useful expression for near-field measurement of the common radially vibrating ring transducer.

The above expressions are strictly valid for the case of a piston or ring in a rigid baffle or for a thin piston or ring vibrating on both sides with the same amplitude and phase. These conditions are seldom found in actual practice where the piston is normally housed in a small comparatively rigid container. This practical situation is better approximated by the case of an unbaffled piston vibrating on one side only. Butler and Sherman [19] have shown, through the synthesis of antisymmetric and symmetric vibrators, and using Silberger's oblate spheroidal results [20], that for the unbaffled piston Eq. (9.48) is replaced by

$$p_f = p(0)(a/2z_f)\pi/(\pi + 2) \approx p(0)(a/3.3z_f). \tag{9.51}$$

9.7.4 Near-to-Far-Field Extrapolation: Large Sources

Near-field measurement techniques are readily implemented in the low frequency range where the transducer is smaller than $\lambda/2$. Under this condition there is no cancellation and simple formulas may be established as shown above. In the higher frequency range near-field effects include large variations in pressure as a result of cancellations. In this frequency band, simple formulas are not possible and near-field to far-field extrapolations must be based on series expansion solutions or Helmholtz integral solutions of the wave equation. Both methods are based on enclosing the transducer or array by an imaginary surface and measuring the pressure and/or velocity amplitude and phase on this surface to obtain the far-field pressure. The main difficulty lies in measuring the velocity accurately at a sufficiently large number of points on the surface. Methods have been derived to limit the required number of measuring points and also to eliminate the need for measuring the velocity.

The Helmholtz integral equation (see Sect. 11.2.4) may be written as

$$p(P) = (j\omega\rho/4\pi)\iint u_{\mathrm{s}}\left(\mathrm{e}^{-jkr}/r\right)\mathrm{d}S + (1/4\pi)\iint p_{\mathrm{s}}\partial\left(\mathrm{e}^{-jkr}/r\right)/\partial n\,\mathrm{d}S, \qquad (9.52)$$

where P is a fixed point in space where the pressure is to be determined, ρ is the density of the medium, ω is the angular frequency, $k = \omega/c$ is the wave number, c is the sound speed, u_{s} is the normal surface velocity, $\partial/\partial n$ is the derivative in the direction normal to the surface, p_{s} is the surface pressure, $\mathrm{d}S$ is the element of area on a surface enclosing the source to be measured, and r is the distance from $\mathrm{d}S$ to the point P. In general both the pressure and normal velocity distribution on the surface must be measured to obtain the far-field pressure from this integral equation representation.

A sketch of the coordinate system and transducer is illustrated in Fig. 9.31, where the angle between the surface normal, n, and r is θ.

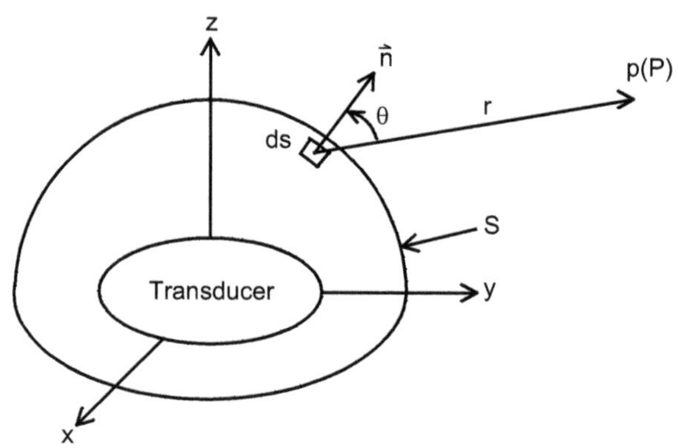

Fig. 9.31 Coordinate system for near-field to far-field evaluation

Equation (9.52) may be written in a more convenient form using [21] (also see Eq. (11.43)

$$\partial\left(\mathrm{e}^{-jkr}/r\right)/\partial n = -\cos\theta\partial\left(\mathrm{e}^{-jkr}/r\right)/\partial r, \tag{9.53}$$

to obtain

$$p(P) = (j\omega\rho/4\pi)\iint[u_{\mathrm{s}} + (p_{\mathrm{s}}/\rho c)(1 + 1/jkr)\cos\theta]\left(\mathrm{e}^{-jkr}/r\right)\mathrm{d}S. \tag{9.54}$$

In the far field where $kr \gg 1$, Eq. (9.54) becomes

$$p(P) = (j\omega\rho/4\pi r)\iint[u_{\mathrm{s}} + (p_{\mathrm{s}}/\rho c)\cos\theta]\mathrm{e}^{-jkr}\,\mathrm{d}S. \tag{9.55}$$

Although this version of the Helmholtz integral equation appears to be in a more convenient form for evaluation, it still requires a measurement of both the normal velocity and the pressure on the surface. A simplification is made by assuming that the curvature of the surface is small and that at each point on the surface $p_{\mathrm{s}} \approx \rho c u_{\mathrm{s}}$, as in plane waves [21]. Equation (9.55) then reduces to

$$p(r) \approx (j/2\lambda r)\iint[1 + \cos\theta]\mathrm{e}^{-jkr}p_{\mathrm{s}}\,\mathrm{d}S. \tag{9.56}$$

With this formulation, under suitable conditions, only measurement of the pressure on the near-field surface enclosing the transducer is needed to obtain the far-field pressure.

The wave function series expansion approach avoids the need for velocity measurement in a fundamental way. In this approach a general solution to the wave equation is written as an expansion in orthogonal wave functions with constant coefficients (see Chaps. 10 and 11). These coefficients are then obtained by appropriate integration of the measured near-field pressure over the chosen coordinate surface as illustrated below. The wave functions are associated with the coordinate systems in which the wave equation is separable (e.g., cylindrical, spherical, spheroidal).

Consider, for example, the simple case of an axisymmetric transducer with an imaginary enclosing spherical surface of radius a as illustrated in Fig. 9.32.

The spherical coordinate wave function solution to the wave equation may be written as

$$p(r,\theta) = \sum_{n=0}^{\infty} b_n h_n^{(2)}(kr)P_n(\cos\theta), \tag{9.57}$$

where $h_n^{(2)}(kr)$ is the spherical Hankel function of the second kind of order n and $P_n(\cos\theta)$ is the Legendre polynomial. If the pressure is measured at a sufficient

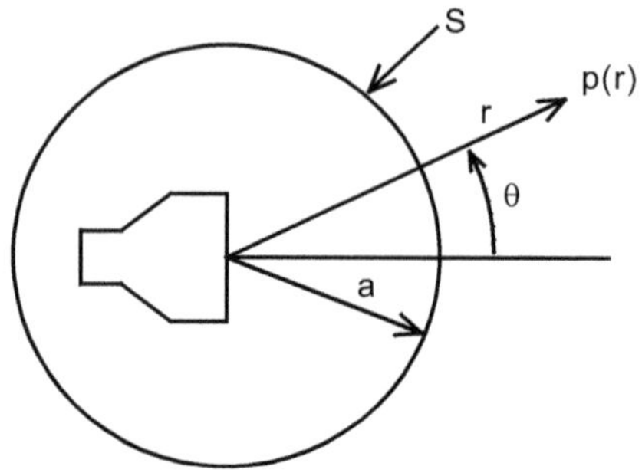

Fig. 9.32 Transducer enclosed by a spherical surface of radius, *a*

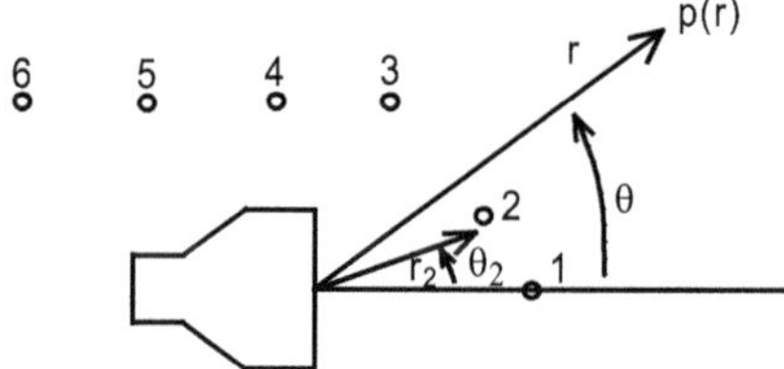

Fig. 9.33 Axisymmetric transducer with pressure measurements taken at near-field points 1–6

number of points on the spherical surface, the coefficients b_n are given by the integral

$$b_n = \left[(2n+1)/h_n^{(2)}(ka)\right]\int_0^{\pi} p(a,\theta)P_n(\cos\theta)\sin\theta\,\mathrm{d}\theta, \tag{9.58}$$

where $p(a, \theta)$ represents the measured near-field pressure. The values of b_n are then substituted back into Eq. (9.57) to calculate the pressure $p(r, \theta)$ at any value of $r \geq a$. Specifically, in the far field Eq. (9.57) becomes

$$p(r,\theta) = \left(\mathrm{e}^{-jkr}/r\right)\sum_{n=0}^{\infty} b_n j^{n+1} P_n(\cos\theta), \tag{9.59}$$

which allows a far-field computation of $p(r, \theta)$ based only on near-field pressure measurements. The integration in Eq. (9.58) assumes a continuous measurement of p at a distance a as θ varies from 0 to π, which may be approximated by a number of measurements at discrete points.

Butler [22] (see also Sect. 11.4.1) has suggested a collocation method which is particularly useful for cases where the pressure is measured at discrete points on a surface in which the wave equation is not separable. The method may be described most simply for an axisymmetric case as illustrated in Fig. 9.33, again using Eq. (9.57), although any wave function expansion which satisfies the wave equation could be used. Six pressure measurement locations and, specifically, location number 2 at r_2, θ_2, are illustrated in Fig. 9.33.

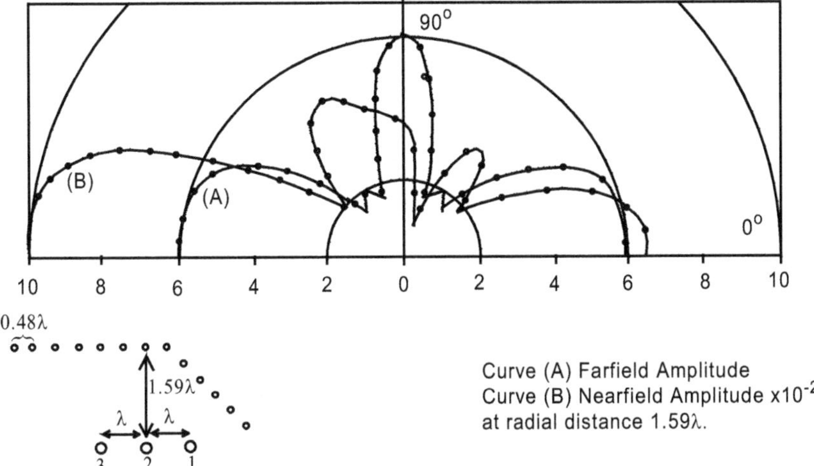

Fig. 9.34 Exact (——) and predicted (···) far-field (*a*) and near-field (*b*) pressure amplitude polar beam patterns for three sources with amplitude ratios 3:2:1 and one wavelength separation with the 13-element hydrophone arrangement [22]

If the pressure is measured at N points Eq. (9.57) becomes, on truncating the series to N terms,

$$p(r_i, \theta_i) = \sum_{n=0}^{N-1} b_n h_n^{(2)}(kr_i) P_n(\cos\theta_i), \quad i = 1, 2, 3, \ldots N. \tag{9.60}$$

The underlying assumption is that N is large enough to include enough terms to describe the field. For example, if it were known that the pattern was omnidirectional, we would need only one measurement, and if the pattern was a cardioid we would need two measurements, with more points required when the unknown pattern is expected to be more complex. Equation (9.60) represents N equations (i from 1 to N) with N unknowns ($n = 0$ to $N - 1$), which can be solved for the unknown coefficients, b_n. In matrix notation

$$\mathbf{p} = \mathbf{A}\mathbf{b}, \tag{9.61}$$

where **p** and **b** are the corresponding pressure and coefficient column matrices and **A** is the square matrix with elements $A_{in} = h_n^{(2)}(kr_i) P_n(\cos\theta_i)$. The rows of A_{in} correspond to measurement positions (r_i,θ_i) while the columns correspond to the orders, n, of the functions. The solution for the coefficients can be written as

$$\mathbf{b} = \mathbf{A}^{-1}\mathbf{p}, \tag{9.62}$$

and obtained through the inverted matrix $\mathbf{A}^{-1}$. The resulting N complex coefficients, b_n, may then be substituted into Eq. (9.60) to obtain the pressure at any distance, r, outside the source.

As an example, 13 near-field, 0.48λ spaced pressure, **p**, values, simulating measuring hydrophones, are shown about a three source axisymmetric array in Fig. 9.34.

A corresponding 13 term spherical wave function expansion, based on the location of the 13 points with origin on source 2, was used to create the matrix **A** from which the coefficients, **b**, were obtained through Eq. (9.62). Calculations of the pressure on a near-field circle at $r = 1.59\lambda$, and on a far-field circle, based on the coefficients **b**, are shown in Fig. 9.34 and seen to compare well with the pressures from an exact three-source calculation.

Other near-field measuring methods such as the Trott array [23] as well as other measurement techniques for underwater sound transducers are discussed in detail by Bobber [1].

9.7.5 *Effect of Transducer Housings*

Tonpilz transducers are typically evaluated individually in watertight housings (see Figs. 1.6, 1.7, and 1.8) before an array (see, for example, Figs. 1.10 and 1.11) is implemented. Normally the housing does not protrude laterally much beyond the face of the piston and has little effect on the free field response. Occasionally, such as during prototyping, the front face of the Tonpilz transducer housing may be large enough to act as a small finite baffle, and in these cases it can have an effect on the transmitting or receiving responses. Such diffraction effects have long been recognized and observed as variations in an otherwise smooth loudspeaker response [10, 24]. The effect of a rigid finite baffle surrounding a small point sensor has been evaluated by Muller et al. [25] by considering a plane wave incident on the end of a rigid cylinder and on a rigid sphere. The results are shown in Fig. 9.35 where the cylinder and sphere have diameter D and the wavelength is λ. The ordinate is the ratio, in dB, of the pressure at the center point of the baffle relative to the free field plane wave pressure, p_0.

Consider the case of a wave arriving at normal incidence where $\phi = 0°$ in Fig. 9.35. In the case of a sphere the pressure approaches a doubling effect (as in a rigid baffle) as D/λ increases beyond about 1.0. However, for the case of the cylinder, pressure doubling occurs at $D/\lambda = 0.4$ and then alternates between 10 and 0 dB without leveling off at 6 dB, as in the case of the sphere. The oscillation is due to diffraction at the edge of the cylinder which creates a scattered wave that arrives at the baffle center delayed by the travel time across the radius. The scattered wave is phase reversed at the edge [26, 27] and travels inward to the central point from all points around the periphery. The first peak is a result of the addition of the scattered wave and the direct wave at the center point when the radius $a = D/2 = \lambda/2$. At this frequency, the 180° phase shift due to travel time plus the additional 180° phase shift at the edge yields the in-phase addition of the scattered and direct waves. At $a = \lambda$, the 180° phase shift at the edge causes a cancellation at the center and yields the same pressure as if there was no baffle.

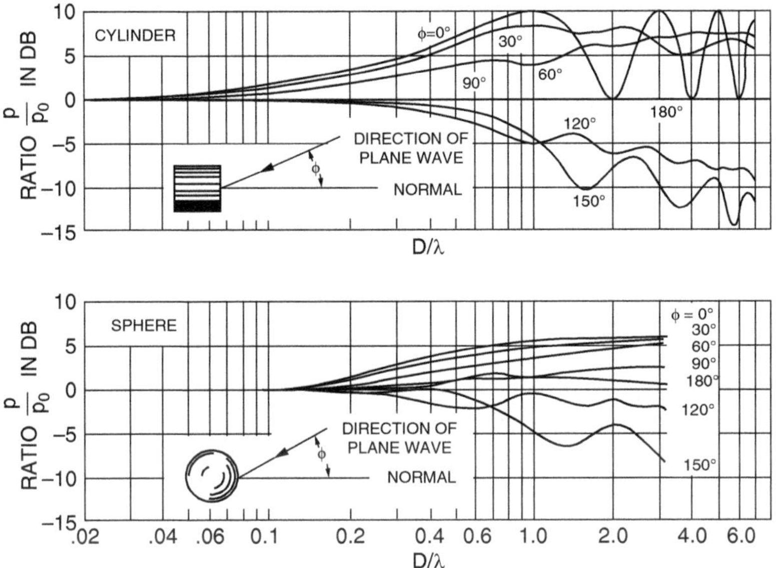

Fig. 9.35 Diffraction of a sound wave by cylinder and sphere

These oscillations continue at half wavelength intervals. However, in practice the finite size of the transducer piston mitigates the effects of diffraction as the piston size approaches one wavelength. At higher frequencies there is pressure doubling due to the size of the piston itself.

The interference effects are most pronounced if the finite baffle is circular, as in Fig. 9.35. In this case, the scattered wave appears to come from a ring source with focused constructive addition occurring at the center of the baffle (or transducer). Accordingly, these effects may be minimized by locating the transducer at an off-center position, use of a baffle with less symmetry or by curving or tapering the edges of the baffle [10, 24]. The best baffle is a large sphere which achieves pressure doubling for $D/\lambda > 1$, or a very small baffle which shows no pressure doubling and no pressure alterations in the frequency band of interest.

9.8 Calibrated Reference Transducers

The US Navy Underwater Sound Reference Division has developed and calibrated a number of transducers [28] for the special purpose of serving as standards in calibrating other transducers. Many of these are available to US Navy contractors and are often used at US Navy facilities. These transducers include:

H52 wideband hydrophone composed of a 5 cm vertical line array of eight lithium sulfate crystals in an oil-filled boot with preamplifier with frequency range 20 Hz to 150 kHz.

H56 high-sensitivity low-noise hydrophone composed of a capped PZT cylinder in an oil-filled boot with preamplifier and with frequency range from 10 Hz to 65 kHz.

F56 high-power omnidirectional transducer composed of a 19.7 cm polyurethane potted PZT spherical shell with frequency range from 1 Hz to 15 kHz as a hydrophone and 1–15 kHz as a projector.

F42 smaller and less powerful PZT spherical shell transducers are also available with diameters 5.0, 3.81, 2.54, and 1.27 cm (with response up to 150 kHz)

F27 directional transducer composed of a 21.4 cm circular array of metaniobate disks with tungsten backing, oil filled with a rubber window, primarily used as a projector from 1 to 40 kHz.

J9 moving coil electrodynamic 28 cm × 11.4 cm projector with passive compensation system for hydrostatic pressure and with a frequency range from 40 Hz to 20 kHz.

9.9 Summary

In this chapter, means for transducer measurement and evaluation were given and techniques for in-air and in-water measurements for piezoelectric and magnetostrictive transducers were described. Use of the in-air impedance allows the evaluation of the resonance, antiresonance yielding the effective coupling coefficient $k_{\mathrm{eff}} = \left[1 - (f_{\mathrm{r}}/f_{\mathrm{a}})^2\right]^{1/2}$ and with a low frequency measurement of the free capacity, C_{f}, yields the clamped (blocked) capacity through $C_0 = C_{\mathrm{f}}(1 - k_{\mathrm{eff}}^2)$ and along with the value of the conductance at resonance it was shown that one can obtain the elements of a Van Dyke equivalent circuit. In-water measurements are usually used to obtain the loaded admittance $Y = G + jB$ and from the conductance, G, the Q_{m} can be obtained and along with B at resonance the Q_{e} may also be obtained yielding $k_{\mathrm{eff}} = (1 + Q_{\mathrm{m}}Q_{\mathrm{e}})^{-1/2}$. It was shown that an approximation to the mechanical efficiency can be obtained from $\eta_{\mathrm{ma}} \approx 1 - Q_{\mathrm{m}}(\mathrm{water})/Q_{\mathrm{m}}(\mathrm{air})$.

It was also shown that the overall acoustic measure of the efficiency, η_{ea}, is from the ratio of the power out over the power in. This ratio may be obtained from the on-axis source level, SL, measured response, and the equation $\mathrm{SL} = 170.8\ \mathrm{dB} + \mathrm{DI} + 10\log\eta_{\mathrm{ea}} + 10\log W$ where W is the power in and DI is the directivity index in dB. Typical voltage, TVR, and current, TCR, transmitting response curves were given in Sect. 9.4 for electric and magnetic field transducer. The receiving voltage sensitivity response, RVS, can be obtained through reciprocity from the TCR and with the impedance from the TVR. It may also be obtained directly with a projector source by a comparison with a calibrated hydrophone.

Means and ways of tuning transducers and their resulting response were discussed in Sect. 9.6. This was followed by far-field and near-field measurement techniques for hydrophones in an enclosure and for free field projectors with measuring hydrophones very near the radiating surface and also by near-field hydrophones surrounding the transducer. For a free transducer small compared to wavelength, and with an effective radius a, the far-field pressure referenced to 1 m was shown to be simply $p_f \approx 0.3 p_n a$, where p_n is the pressure very near the radiating surface. This chapter also includes the effect of a housing on a transducer and a list of US Navy calibrated reference transducers.

Exercises (Degree of difficulty: *lowest, **moderate, ***highest)

9.1.* Show that if $Z = R + jX$, then the conductance $G = R/\left(R^2 + X^2\right)$ and the susceptance $B = -X/\left(R^2 + X^2\right)$. Show that if the electrical Q_e is defined as B/G, then Q_e is also equal to $-X/R$.

9.2.* Show that the Van Dyke equivalent circuit of Fig. 9.1 reduces to the circuit of Fig. 9.4 where the free capacity $C_f = C_0 + C_e$.

9.3.** Convince yourself that the measurements of C_f, $\tan\delta$, f_r, f_a, and $|Y_{max}|$ are sufficient to determine the parameters of the Van Dyke circuit of Fig. 9.1. What measurement or measurements are necessary to evaluate the electromechanical transformer turns ratio N of the equivalent circuit of Fig. 3.15?

9.4.** Explain why the formula $k_{eff} = \left(1 + Q_m Q_e\right)^{-1/2}$ is more suitable for determining k_{eff} under water-loaded conditions (with measurements of B vs. G or response of B and G) while the formula $k_{eff} = \left[1 - \left(f_r/f_a\right)^2\right]$ is more suitable for in-air measurements of $|Y|$ or $|Z|$.

9.5.** Explain the change in the response curve slopes from the TVR curve of Fig. 9.13 to the TCR curve of Fig. 9.14 to the RVS curve of Fig. 9.15.

9.6.** Why does the TCR curve of a magnetostrictive transducer look like the TVR curve of a piezoelectric ceramic transducer?

9.7.* Shunt tuning a transducer does not change the TVR and series tuning does not change the TCR and RVS. Why is this so?

9.8.** The Transmit/Receive, "TR," switch of Fig. 9.23 has at times been criticized for the diode noise introduced in the RVS. With no restrictions on you, how would you avoid this?

9.9.*** What should the smallest dimension of a water-filled tank be for pulsed measurements of transducers which go as low as 2 kHz and have a Q_m of no more than 5, and also for a Q_m of no more than 1. Assume two steady state cycles are needed for measurement.

9.10.* Determine the far-field Rayleigh distance for a 19.35 cm (7.6 in.) diameter piston transducer operating at 10 kHz. Determine the approximate far-field distance for a close-packed seven element circular array of the same transducers.

9.11.* The in-air frequencies of maximum and minimum admittance magnitudes (f_r and f_a, respectively) of a transducer are measured to be 10 and 11.55 kHz. The in-water mechanical Q_m is measured to be 3.0. What is the in-water electrical Q_e?

9.12.* The measured TVR level for a transducer is 140 dB and the measured impedance is 400 – j300 ohms, all at 10 kHz. What is the RVS?

References

1. R.J. Bobber, *Underwater Electroacoustic Measurements* (Naval Research Laboratory, Washington, DC, 1969)
2. L.L. Beranek, *Acoustical Measurements* (American Institute of Physics, New York, 1988)
3. H.B. Miller (ed.), *Acoustical Measurements* (Hutchinson Ross Publishing Company, Stroudsburg, 1982)
4. D. Stansfield, *Underwater Electroacoustic Transducers* (Bath University Press, Bath, 1990)
5. O.B. Wilson, *Introduction to Theory and Design of Sonar Transducers* (Peninsula Publishing Co., Los Altos Hills, 1988)
6. F.V. Hunt, *Electroacoustics* (Harvard University Press, New York, 1954)
7. R. S. Woollett, *Sonar transducer fundamentals* (Naval Underwater Systems Center, Newport), undated
8. L.F. Kinsler, A.R. Frey, A.B. Coppens, J.V. Sanders, *Fundamentals of Acoustics* (Ch. 14 Transduction), 4th edn (Wiley, New York, 2000)
9. Private communication with Frank Massa.
10. H.F. Olson, *Acoustical Engineering* (D. Van Nostrand Company, Inc., Princeton, 1967)
11. L.L. Beranek, *Acoustics* (McGraw-Hill Book Company, Inc., New York, 1954)
12. K.S. Van Dyke, The piezoelectric resonator and its equivalent network. Proc. IRE **16**, 742–764 (1928)
13. W.J. Marshall, G.A. Brigham, Determining equivalent circuit parameters for low figure of merit transducers, J. Acoust. Soc. Am., ARLO **5**(3), 106–110 (2004)
14. D. Rife, J. Vanderkooy, Transfer-function measurements with maximum-length sequences. J. Audio Eng. Soc. **37**, 419–443 (1989). See also, J. D'Appolito, *Testing Loudspeakers* (Audio Amateur Press, Peterborough, 1998)
15. D.A. Berlincourt, D.R. Curran, H. Jaffe, Piezoelectric and piezomagnetic materials and their function in transducers, in *Physical Acoustics,* vol. 1, Part A, ed. by W.P. Mason (Academic Press, New York, 1964)
16. J. deLaunay, P.L. Smith, Aging of barium titanate and lead zirconate-titanate ferroelectric ceramics, Naval Research Laboratory Report 7172, 15 October 1970
17. B.B. Bauer, L.A. Abbagnaro, J. Schumann, Wide-range calibration system for pressure-gradient hydrophones. J. Acoust. Soc. Am. **51**, 1717–1724 (1972)
18. D.B. Keele Jr., Low-frequency loudspeaker assessment by near-field sound pressure measurements. J. Audio. Eng. Soc. **22**, 154–162 (1974)
19. J.L. Butler, C.H. Sherman, Near-field far-field measurements of loudspeaker response. J. Acoust. Soc. Am. **108**, 447–448 (2000)
20. A. Silbiger, Radiation from circular pistons of elliptical profile. J. Acoust. Soc. Am. **33**, 1515–1522 (1961)
21. D.D. Baker, Determination of far-field characteristics of large underwater sound transducers from near-field measurements. J. Acoust. Soc. Am. **34**, 1737–1744 (1962)
22. J.L. Butler, Solution of acoustical-radiation problems by boundary collocation. J. Acoust. Soc. Am. **48**, 325–336 (1970)

23. W.J. Trott, Underwater-sound-transducer calibration from nearfield data. J. Acoust. Soc. Am. **36**, 1557–1568 (1964)
24. H.F. Olson, Direct radiator loudspeaker enclosure. J. Audio Eng. Soc. **17**, 22–29 (1969)
25. G.C. Muller, R. Black, T.E. Davis, The diffraction produced by cylindrical and cubical obstacles and by circular and square plates. J. Acoust. Soc. Am. **10**, 6–13 (1938)
26. J.R. Wright, Fundamentals of diffraction. J. Audio Eng. Soc. **45**, 347–356 (1997)
27. M.R. Urban et al., The distributed edge dipole (DED) model for cabinet diffraction effects. J. Audio Eng. Soc. **52**, 1043–1059 (2004)
28. L.E. Ivey, Underwater electroacoustic transducers, Naval Research Laboratory USRD Report. NRL/PU/5910--94-267, August 31, 1994. For more information contact Underwater Sound Reference Division, NUWC/NPT Newport, RI 02841

Chapter 10
Acoustic Radiation from Transducers

This chapter is concerned with the calculation of acoustic characteristics of transducers, such as directivity function, directivity factor, directivity index, and self-radiation impedance. Convenient formulae and numerical information for frequently used cases will also be given. The classical analytical methods are limited in their ability to calculate the acoustical characteristics of realistic transducers, and it is usually necessary to simplify the details of a transducer and its surrounding structure in order to apply those methods. The dipole coupling to a parasitic monopole radiator is presented as an interesting application of an approximate analytical solution. Finite element numerical modeling of the acoustic field can, however, provide more realistic information to augment the analytical results in many cases. The well-known acoustics books by Kinsler et al. [1], Morse and Ingard [22], Pierce [2], Blackstock [3], Skudrzyk [4], and Beranek and Mello [5] provide excellent background for this chapter as well as additional information.

10.1 The Acoustic Radiation Problem

The acoustic medium is an essential part of electroacoustic transduction. It is usually a fluid, in most cases either water or air, characterized by only two properties, static density, ρ, and bulk modulus, B. But in transducer analysis and design, ρ and sound speed, $c = (B/\rho)^{1/2}$, are the more convenient properties. The product of these two quantities, ρc, the specific acoustic impedance, is much higher for water than for air (~1.5×10^6 vs. 420 kg/m^{2}s) causing water to have a much more significant effect on the operation of a transducer than air. We will consider that the medium surrounding an individual transducer is homogeneous, isotropic, nonviscous, and large enough that its boundaries need not be considered. However, transducers are often used or tested under quite different conditions, e.g., near the water surface, mounted on the hull of a ship, or in a water-filled tank (see Chap. 9).

J.L. Butler, C.H. Sherman, *Transducers and Arrays for Underwater Sound*,
Modern Acoustics and Signal Processing, DOI 10.1007/978-3-319-39044-4_10

In such cases the proximity of other media or structures may strongly affect the operation of a transducer.

When the moveable surface of a transducer vibrates in an acoustic medium it produces a disturbance in that medium called an acoustic field that varies with time and position in the medium. The acoustic radiation problem consists of determining the acoustic field that results from a specified vibration of a particular transducer surface. The acoustic field in a fluid is a scalar field that can be completely described by one quantity, usually the variation of pressure from the static pressure, called the acoustic pressure, p. Other characteristics of the acoustic field that are of interest, such as the components of the particle velocity vector, $\vec{u}$, can be derived from the pressure. The linear acoustic equations can be written [2] as,

$$\rho \frac{\partial \vec{u}}{\partial t} = -\vec{\nabla} p, \tag{10.1a}$$

$$\frac{\partial \rho'}{\partial t} + \rho \vec{\nabla} \cdot \vec{u} = 0, \tag{10.1b}$$

$$p = c^2 \rho', \tag{10.1c}$$

where t is the time, $\vec{\nabla}$ is the gradient operator, and ρ' is the acoustic density, i.e., the variation of density from the static density, ρ. The first equation is the equation of motion for a particle of the medium, the second is the equation of continuity for conservation of mass and the third is the equation of state of the medium.

Substituting the equation of motion into the time derivative of the equation of continuity and using the equation of state to eliminate ρ' gives the scalar wave equation for the acoustic pressure:

$$\nabla^2 p - \frac{1}{c^2} \frac{\partial^2 p}{\partial t^2} = 0, \tag{10.2a}$$

where ∇^2 is the Laplacian operator. Since the specific cases to be considered here all have harmonic time dependence, the symbol p in the following equations will represent the spatial dependence of the acoustic pressure field, e.g., the pressure has the form $p(x, y, z)\, e^{j\omega t}$ in rectangular coordinates. The second term in Eq. (10.2a) then becomes $k^2 p$ where $k = \omega / c = 2\pi / \lambda$ is the acoustic wave number and λ is the acoustic wavelength. The wave equation in this form is known as the Helmholtz differential equation:

$$\nabla^2 p + k^2 p = 0. \tag{10.2b}$$

In the familiar Cartesian or rectangular (x, y, z), cylindrical (r, ϕ, z), and spherical (r, θ, ϕ) coordinate systems (see Fig. 10.1) the Helmholtz equation takes the following forms (note that the symbols, r and ϕ, are used in both the cylindrical

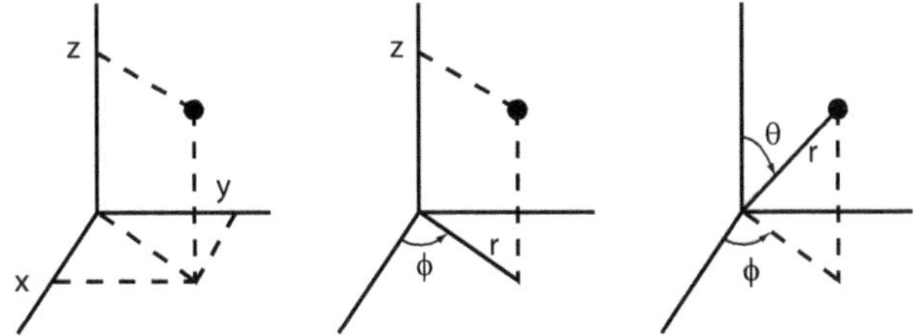

Fig. 10.1 Rectangular (x, y, z), cylindrical (r, φ, z), and spherical (r, θ, φ) coordinate systems

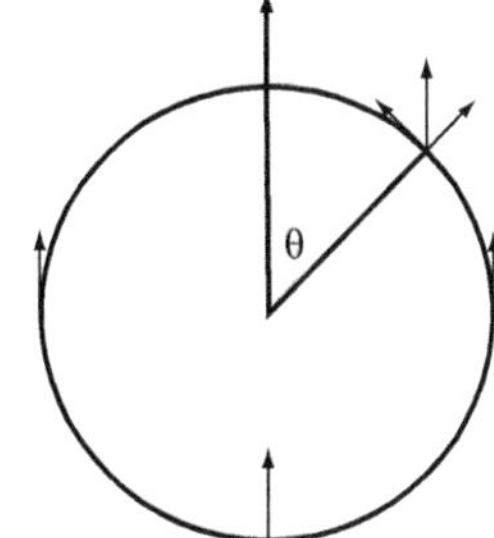

Fig. 10.2 Sphere vibrating as a rigid body showing normal and tangential velocity components

and spherical systems, that ϕ has the same meaning in both systems, but r has different meanings):

$$\frac{\partial^2 p}{\partial x^2} + \frac{\partial^2 p}{\partial y^2} + \frac{\partial^2 p}{\partial z^2} + k^2 p = 0, \tag{10.3}$$

$$\frac{1}{r}\frac{\partial}{\partial r}\left(r\frac{\partial p}{\partial r}\right) + \frac{1}{r^2}\frac{\partial^2 p}{\partial \phi^2} + \frac{\partial^2 p}{\partial z^2} + k^2 p = 0, \tag{10.4}$$

$$\frac{1}{r^2}\frac{\partial}{\partial r}\left(r^2\frac{\partial p}{\partial r}\right) + \frac{1}{r^2 \sin\theta}\frac{\partial}{\partial \theta}\left(\sin\theta\frac{\partial p}{\partial \theta}\right) + \frac{1}{r^2 \sin^2\theta}\frac{\partial^2 p}{\partial \phi^2} + k^2 p = 0, \tag{10.5}$$

In general the velocity at each point of a vibrating surface has a component normal, and a component tangential, to the surface. For example, for a sphere centered on the origin and vibrating as a rigid body parallel to the z-axis the velocity at the two points lying on the z-axis has only a normal component, while at all the points lying in the x-y plane (at $\theta = 90°$) it has only a tangential component. At all other points of the surface the velocity has both normal and tangential components as shown in Fig. 10.2.

Only the normal components of velocity produce an acoustic field in a nonviscous medium, because the tangential components slip without disturbing the medium. If the medium is viscous the tangential components cause a disturbance, but it extends only a short distance from the vibrator and does not contribute to the radiated acoustic field.

The acoustic field can be divided into two spatial regions. In the near field, part of the motion of the medium does not travel far from the vibrator because it corresponds to energy being alternately transferred to the medium and then returned to the vibrator. In Chap. 1 this part of the energy was associated with the radiation mass which increases the effective mass of the transducer. In the far field the energy transferred to the medium never returns, because it is radiated away, and the acoustic power radiated to the far field is proportional to the radiation resistance. Solutions of the wave equation include both parts of the acoustic field. Obviously the radiated far field is the useful part of the acoustic field in most cases, but in other cases the near field is important because it causes problems such as cavitation and acoustic interactions between transducers.

An acoustic radiation problem is given by specifying the normal velocity of vibration of a particular surface. A solution of the wave equation appropriate to the surface is used to calculate the particle velocity in the medium normal to the transducer surface and this velocity is made equal to the specified normal velocity on the transducer surface. This is the boundary condition that determines the specific solution of the specified problem. The general relation between the pressure and the particle velocity vector, $\vec{u}$, in the medium, is given by the equation of motion, Eq. (10.1a), which, for harmonic time dependence, becomes

$$\vec{u} = -\frac{1}{j\omega\rho}\vec{\nabla} p. \tag{10.6}$$

All the characteristics of the acoustic field produced by that particular vibrating surface can then be calculated from that solution of the wave equation.

One of the most useful analytical methods of solving the wave equation is separation of variables. For example, in rectangular coordinates the solution is assumed to be the product of a function of x, a function of y, and a function of z, i.e.,

$$p(x, y, z) = X(x)Y(y)Z(z). \tag{10.7}$$

Substituting this expression into Eq. (10.2b) shows that the functions X, Y, and Z satisfy the equations

$$\frac{\mathrm{d}^2X}{\mathrm{d}x^2} + k_x^2X = 0, \tag{10.8}$$

$$\frac{\mathrm{d}^2Y}{\mathrm{d}y^2} + k_y^2Y = 0, \tag{10.9}$$

$$\frac{\mathrm{d}^2Z}{\mathrm{d}z^2} + k_z^2Z = 0, \tag{10.10}$$

where k_x, k_y, and k_z are constants related to each other by

$$k_x^2 + k_y^2 + k_z^2 = k^2. \tag{10.11}$$

Since Eqs. (10.8)–(10.10) have solutions $e^{\pm j\,k_x x}$, $e^{\pm jk_y y}$, and $e^{\pm jk_z z}$, the complete solution for the acoustic field in rectangular coordinates is given by Eq. (10.7) as

$$p(x, y, z)e^{j\omega t} = P_0 e^{j\left(\omega t \pm k_x x \pm k_y y \pm k_z z\right)}, \tag{10.12}$$

where P_0 is a constant determined by the boundary condition. This expression represents plane waves of amplitude, P_0, traveling in the direction given by a vector with x, y, z components of k_x, k_y, and k_z called the wave vector. For example, for a plane wave traveling in the positive x direction $k_x = k$ and $k_y = k_z = 0$.

Plane waves are a basic concept in acoustics although they exist only as an approximation in limited regions of space. When describing the receiving response of a hydrophone it is usually assumed that the hydrophone is receiving a plane wave, and in calibration of transducers the attempt is usually made to achieve plane wave conditions (see Chap. 9). In a plane wave the particle velocity vector is parallel to the direction of propagation, and an important property of the wave is the ratio of acoustic pressure, p, to particle velocity amplitude, u. The ratio, p/u, is called the specific acoustic impedance and is found from Eq. (10.6) to equal ρc for a plane wave. It follows that the mechanical impedance, the ratio of force to velocity, associated with a section of plane wave front of area A is ρcA, the characteristic mechanical impedance of the medium, which is important for comparison with the radiation impedance of projectors.

Solutions of the wave equation in cylindrical, spherical, and several other coordinate systems can also be found by separation of variables. In these cases the solutions of interest for radiation problems must satisfy the radiation condition, which means that, at large distances from the origin, the solution takes the form of an outgoing wave. The result of separating variables in cylindrical coordinates for an outgoing wave is,

$$p(r, \phi, z)e^{j\omega t} = A_m H_m^{(2)}(k_r r)\, e^{j\,(m\phi \pm k_z z + \omega t)}, \tag{10.13}$$

where A_m is an amplitude constant determined by the boundary conditions, $H_m^{(2)}(k_r r) = J_m(k_r r) - j\,Y_m(k_r r)$ is the cylindrical Hankel function of the second kind, $J_m(k_r r)$ and $Y_m(k_r r)$ are Bessel and Neumann functions, $k_r^2 + k_z^2 = k^2$, and m is a positive or negative integer. The choice of $H_m^{(2)}(k_r r)$, with the time factor $e^{j\omega t}$, gives outgoing waves.

Separation of variables in spherical coordinates gives, for an outgoing wave,

$$p(r, \theta, \phi)e^{j\omega t} = A_{nm}\, h_n^{(2)}(kr) P_n^m(\cos\theta)\, e^{j(\omega t \pm m\phi)}, \tag{10.14}$$

where $h_n^{(2)}(kr) = j_n(kr) - jy_n(kr)$ is the spherical Hankel function of the second kind, $j_n(kr)$ and $y_n(kr)$ are spherical Bessel and Neumann functions, $P_n^m(\cos\theta)$ are the associated Legendre functions, and m and n are positive integers with $n \geq m$. (An example of using the spherical Hankel functions to solve an interesting problem is given in Sect. 10.5.)

These solutions of the wave equation contain two parameters that are independent of the spatial coordinates and time: in rectangular coordinates, where $k_x^2 + k_y^2 + k_z^2 = k^2$, any two of k_x, k_y, or k_z; in cylindrical coordinates, m and k_r or k_z; and in spherical coordinates m and n. Since the wave equation is linear, any combination of these solutions with different values of these parameters is also a solution. Thus more solutions can be constructed by summing over the integer parameters and integrating over the continuous parameters. As a simple example consider a pulsating sphere of radius, a, with every point on its surface vibrating sinusoidally in the radial direction (i.e., normal to the surface) with the same velocity, $u_0 e^{j\omega t}$, as shown in Fig. 10.3.

Thus there are no tangential velocity components in this case. The solution in spherical coordinates, Eq. (10.14), is appropriate since the vibrating surface is a sphere. It is also apparent, since the normal velocity is the same at all points of the surface, that the acoustic field must be the same in all directions, i.e., independent of θ and ϕ. This is only the case if $m = n = 0$ in Eq. (10.14), which then becomes, since $P_0^0(\cos\theta) = 1$,

$$p(r) = A_0 h_0^{(2)}(kr) = jA_0 e^{-jkr}/kr. \tag{10.15a}$$

The amplitude, A_0, can be found from the boundary condition that the vibrating surface of the sphere remains in contact with the medium at all times. Thus the normal velocity of the surface, u_0, must equal the normal velocity in the medium at $r = a$. This condition gives

$$u_0 = -\frac{1}{j\omega\rho}\frac{\partial p}{\partial r}\bigg|_{r=a} = \frac{A_0}{\omega\rho}\left[jk + \frac{1}{a}\right]\frac{1}{ka}e^{-jka}, \tag{10.16}$$

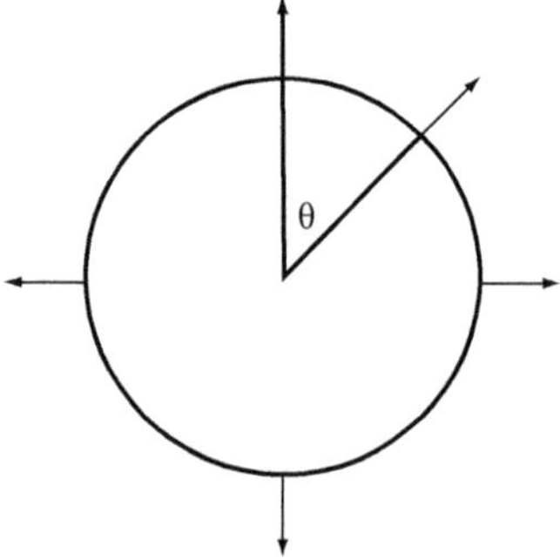

Fig. 10.3 Sphere vibrating with uniform velocity as a pulsating sphere

and it follows that the amplitude is

$$A_0 = \frac{\omega \rho u_0 k a^2}{(1 + jka)} e^{jka}. \tag{10.17}$$

Substitution of A_0 into Eq. (10.15a) gives the complete solution for the acoustic field of the pulsating sphere in terms of the normal velocity on the surface, u_0:

$$p(r) = \frac{j\rho c u_0 k a^2}{1 + jka} \frac{e^{-jk(r-a)}}{r}. \tag{10.15b}$$

The acoustic waves from a pulsating sphere (see Sect. 5.2.3) are called simple spherical waves because they propagate in all directions with the same amplitude. They are a useful approximation for the far field of transducers of any shape that have surface velocities in phase at frequencies where the transducer dimensions are small compared to the wavelength. It is only necessary to replace the source strength of the pulsating sphere by the source strength of the transducer. Source strength is defined as the integral of the normal velocity over the area of the vibrating surface:

$$Q = \iint_A \vec{u}\left(\vec{r}\right) \cdot \hat{n}\, dS, \tag{10.18}$$

where $\hat{n}$ is the unit vector normal to the surface and dS is the area element on the surface of the transducer (see Sect. 10.4.1 regarding this definition of source strength). Note that source strength is equal to volume velocity with units of m^3/s. Thus the source strength of the pulsating sphere is $Q = 4\pi a^2 u_0$. For $ka \ll 1$ the amplitude constant is $A_0 = \omega \rho u_0 k a^2 = \left[\rho c k^2/4\pi\right] Q$, and Eq. (10.15a) becomes

$$p(r) = \frac{j\rho c k}{4\pi r} Q e^{-jkr}. \tag{10.15c}$$

This expression holds approximately for many transducers with shapes other than spherical at sufficiently low frequency, but at higher frequency the pressure amplitude varies with direction.

The pulsating sphere is also the basis for the point source concept. The point source is the idealized case in which the radius of the pulsating sphere approaches zero while the velocity increases to make $Q = 4\pi a^2 u_0$ remain finite. Point sources can be considered non-scattering because of their infinitesimal size and thus can be superimposed to give the fields of other more realistic sources as we will do in the next sections. Since the wave equation is linear, any superposition of solutions that satisfy the same type of boundary conditions on the same surface is also a solution.

The instantaneous acoustic intensity vector, $\vec{I}$, defined as the product of the acoustic pressure and the particle velocity vector, $p\vec{u}$, is a measure of the flow of acoustic energy. It has dimensions of energy per unit area per unit time or power per unit area usually expressed in W/m^2 or W/cm^2. The magnitude of the time average intensity vector, denoted by $\langle I \rangle$, is of most interest. Section 13.3 shows that in general,

$$\langle I \rangle = ½\mathrm{Re}\left(p\vec{u}^*\right), \tag{10.19a}$$

and for plane waves Eqs. (10.6) and (10.12) give

$$\langle I \rangle = \frac{pp^*}{2\rho c} = \frac{|p_{\mathrm{rms}}|^2}{\rho c}. \tag{10.19b}$$

The acoustic energy flows in the direction of propagation of the plane wave. For simple spherical waves the particle velocity vector has only a radial component which is obtained from Eq. (10.15a):

$$u_{\mathrm{r}} = -\frac{1}{j\omega\rho}\frac{\partial p}{\partial r} = \frac{p}{\rho c}\left[1 + \frac{1}{jkr}\right]. \tag{10.20}$$

Then the intensity is $½\mathrm{Re}\left(pu_{\mathrm{r}}^*\right) = pp^*/2\rho c$ as for plane waves. Although this expression for the time average intensity holds exactly only for plane waves and simple spherical waves, it may be used for all projectors in the far field where the radiation is essentially spherical, although it may vary with direction. The energy flow in simple spherical waves is radially outwards. A reactive component of intensity also exists (see Sects. 6.5.7 and 13.3).

10.2 Far-Field Acoustic Radiation

10.2.1 Line Sources

It can be seen from Eq. (10.15c) that the pressure field for the pulsating sphere does not change form as the distance from the sphere increases. However, most acoustic radiators have more complicated pressure distributions in the near field that become approximate spherical waves with directional dependence at sufficient distance from the radiator. A simple example is a uniformly vibrating cylindrical line source of length L and radius a, where a is much smaller than both L and the wavelength. It can be considered to consist of a large number of adjacent infinitesimal point sources each of length $\mathrm{d}z_0$ as shown in Fig. 10.4.

The differential contribution to the pressure field from each point source is given by Eq. (10.15c) as

$$\mathrm{d}p = j\frac{\rho ck}{4\pi R}\mathrm{d}Q\mathrm{e}^{-jkR}, \tag{10.21}$$

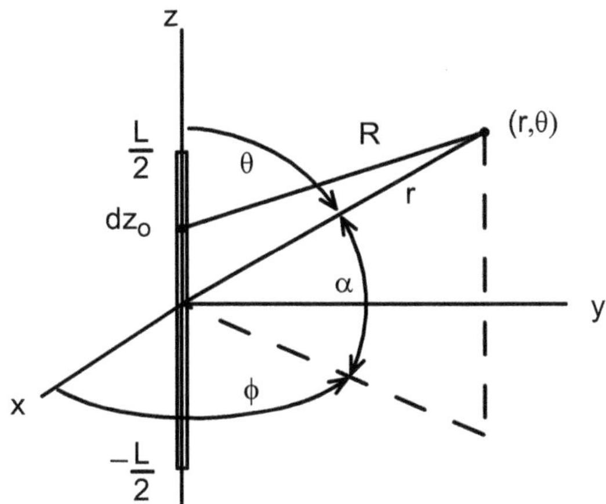

Fig. 10.4 Coordinates for calculating the far field at (r, θ) of a line source of length, L. A source element of length dz_0 is shown. The field is independent of φ

where $\mathrm{d}Q = 2\pi a u_0 \mathrm{d}z_0$ is the differential element of source strength and u_0 is the radial velocity. Although cylindrical coordinates are natural for this line source with cylindrical symmetry, spherical coordinates are more convenient for calculating the far field since it consists of spherical waves. Thus Fig. 10.4 shows that $R = \left[r^2 + z_0^2 - 2rz_0 \cos\theta\right]^{1/2}$ is the distance from $\mathrm{d}Q$ at z_0 to the far-field point (r, θ); the field is independent of ϕ because of the symmetry. The pressure field of the whole line is given by superimposing the fields of all the point sources which is accomplished by integrating Eq. (10.21) over z_0 from $-L/2$ to $+L/2$. The integration can only be done easily in the far field where $r \gg L$ and R in the denominator can be approximated by r. However, R in the exponent must be approximated by $r - z_0 cos\theta$ to preserve the phase relations that are critical in determining the directivity function. The result for the far-field pressure of the continuous line source in terms of the angle $\alpha = (\pi/2) - \theta$ is

$$p(r, \alpha) = \frac{j\rho c k Q_0}{4\pi} \frac{\mathrm{e}^{-jkr}}{r} \frac{\sin\left[(kL/2)\sin\alpha\right]}{(kL/2)\sin\alpha}, \tag{10.22}$$

where $Q_0 = 2\pi a L u_0$ is the source strength of the whole line. The factor e^{-jkr}/r in Eq. (10.22) shows that the far field consists of spherical waves, as it does for all finite size sources. But these are not simple spherical waves, because the pressure amplitude varies with direction from the source as shown by the dependence on the angle α. When spherical waves depend on angle, as they do in most cases, components of particle velocity and intensity exist that are perpendicular to the radial components, but in the far field they are negligible.

The function of α in Eq. (10.22) is the familiar $\sin x/x$, or Sinc(x) function, and its square is the normalized acoustic intensity directivity function or beam pattern with maximum value unity in the plane that bisects the line source where $\alpha = 0$. As α increases the pressure decreases with a null at $(kL/2)\sin\alpha = \pi$, then another lobe

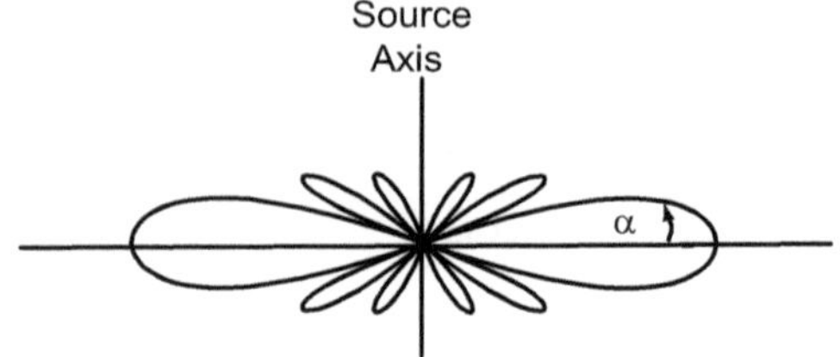

Fig. 10.5 Far-field beam pattern of a line source for $kL = 3\pi$. The symmetrical three-dimensional pattern is given by rotation about the axis of the source

and another null, etc., with the number of nulls depending on the value of kL. A typical pattern in one plane is illustrated in Fig. 10.5.

The most useful part of the beam pattern is the main lobe, and its most important feature is its angular width. Usually the −3 dB points on each side of the main lobe are used as a measure of the beam width (BW). This occurs when $(kL/2)\sin\alpha \approx 1.4$ giving a beam width of $2\sin^{-1}(2.8/kL)$, which, when $L >> \lambda$ and the beams are narrow, simplifies to

$$\text{BW} = 5.6/kL \text{ radians} = 51\lambda/L \text{ degrees.}$$

The side lobes on both sides of the main lobe usually are undesirable, and the first side lobe, being the highest, is the most troublesome. Its peak occurs when $(kL/2)\sin\alpha \approx 3\pi/2$ with a value of $(2/3\pi)^2$ or 13.5 dB below the peak of the main lobe. This result applies to the continuous line source with uniform source strength. Line arrays of individual transducers can be shaded to reduce the side lobes relative to the main lobe (see Chaps. 7 and 8).

The directivity factor, defined in Eq. (1.20), can be calculated from the beam pattern function in Eq. (10.22). The normalized maximum intensity is unity at $\alpha = 0$, and the normalized intensity averaged over all directions is

$$I_a = \frac{1}{4\pi}\int_0^{2\pi} \mathrm{d}\phi \int_0^{2\pi} \left[\frac{\sin\left[(kL/2)\sin\alpha\right]}{(kL/2)\sin\alpha}\right]^2$$

$$\sin\theta \mathrm{d}\theta = \frac{1}{2}\int_{-\pi/2}^{\pi/2} \left[\frac{\sin\left[(kL/2)\sin\alpha\right]}{(kL/2)\sin\alpha}\right]^2 \cos\alpha \mathrm{d}\alpha,$$

where $\alpha = (\pi/2) - \theta$. Using $q = (kL/2)\sin\alpha$ converts this integral to a form that can be approximated, for $kL \gg 1$, by a tabulated definite integral [1, 6] as follows:

$$I_a = \frac{2}{kL}\int_0^{kL/2} \frac{\sin^2 q}{q^2}\mathrm{d}q \approx \frac{2}{kL}\int_0^{\infty} \frac{\sin^2 q}{q^2}\mathrm{d}q = \frac{2}{kL}\frac{\pi}{2}. \tag{10.23a}$$

Thus the directivity factor is

$$D_\mathrm{f} = \frac{I(0)}{I_a} = \frac{kL}{\pi} = \frac{2L}{\lambda}, \tag{10.23b}$$

$$\text{and DI} = 10 \log 2L/\lambda \approx 20 - 10 \ \log \text{BW dB},$$

with the beam width (BW) in degrees. Both Horton [7] and Burdic [8] discuss the line source and other acoustic sources, and give D_f for the line source in terms of the Sine Integral (Si) by exact evaluation of the integral above. Horton also gives an approximation that improves the value in Eq. (10.23b):

$$D_f = \frac{2L}{\lambda} / \left(1 - \frac{\lambda}{\pi^2 L}\right), \tag{10.23c}$$

which shows that $2L/\lambda$ is correct to within about 10 % when $L > \lambda$.

The directivity factor for a line array of N small transducer elements is given by Eqs. (8.84a) and (8.84b) in Sect. 8.1.3 for cases with and without shading, respectively. For half wavelength, $\lambda/2$, spacing $D_f = N$ and for the case of N large $D_f \approx N$.

10.2.2 Flat Sources in a Plane

The far field of flat radiators mounted flush in a large, rigid plane baffle can be formulated from point source fields by first considering two point sources of equal source strength vibrating in phase and close together as shown in Fig. 10.6.

Imagine the infinite plane that bisects the line joining the two sources and is perpendicular to that line. At every point on that plane the pressures from the two equal sources add, but the components of particle velocity normal to the plane are in opposite directions and cancel. Thus the field of the two sources is consistent with an infinite rigid plane baffle lying midway between them. The field of the two sources can be added to the fields of other pairs of equal point sources lying on each side of the same infinite rigid plane to construct the fields of continuous flat sources lying in the plane. The procedure is similar to that used for the line source, but the differential element of source strength is now the sum of two point source fields, each the same as Eq. (10.21):

$$\mathrm{d}p = \frac{j\rho ck}{4\pi} \mathrm{d}Q \left[\frac{1}{R_1} \mathrm{e}^{-jkR_1} + \frac{1}{R_2} \mathrm{e}^{-jkR_2}\right], \tag{10.24}$$

where R_1 and R_2 are the distances from each source to an arbitrary field point (see Fig. 10.6).

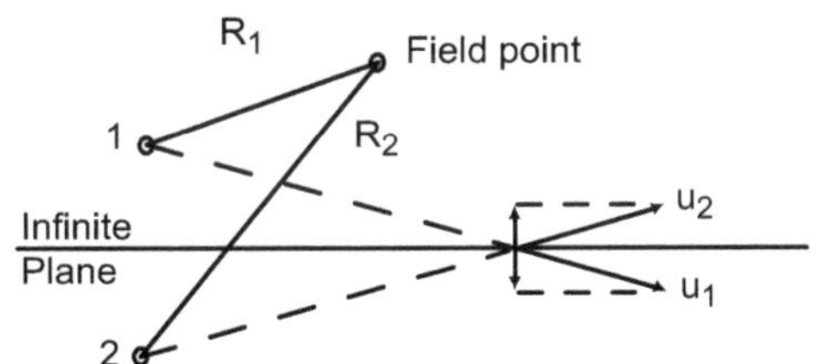

Fig. 10.6 Two point sources of equal strength have an infinite plane between them on which the normal velocity components cancel

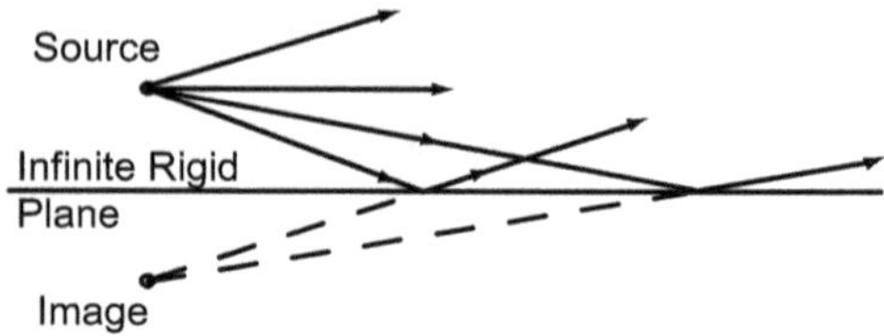

Fig. 10.7 Field of a small source near an infinite rigid plane by method of images

The reasoning that led to Eq. (10.24) is also the basis for the method of images. That method uses the fact that the field of a point source near an infinite rigid plane is the sum of the fields of the point source in free space and its image on the other side of the plane as shown in Fig. 10.7 (also see Sect. 6.5.1).

Now the field of an extended flat radiator mounted in the plane can be found by integrating Eq. (10.24) over the surface of the radiator while letting each pair of sources come together on the plane making $R_1 = R_2 = R$. This can be done in any convenient coordinate system and for a flat radiator of any shape with any fixed velocity distribution, although the integral can seldom be evaluated analytically. Letting $\mathrm{d}Q = u\left(\vec{r}_0\right)\mathrm{d}S_0$ the result for the pressure is

$$p\left(\vec{r}\right) = \frac{j\rho ck}{2\pi}\iint u\left(\vec{r_0}\right)\frac{\mathrm{e}^{-jkR}}{R}\mathrm{d}S_0. \tag{10.25a}$$

Equation (10.25a) was first given by Rayleigh [9] and is often referred to as the Rayleigh integral. It is one of the most frequently used acoustic radiation equations in acoustics.

The circular radiator with uniform normal velocity (called a circular piston because of the uniform velocity) will be discussed first because it is used so often to approximate the sound fields of transducers. For this case cylindrical coordinates are appropriate with the origin at the center of the piston. Letting the normal velocity of the piston be u_0 and writing the differential element of area as $\mathrm{d}S_0 = r_0\mathrm{d}r_0\mathrm{d}\phi_0$, where r_0 and ϕ_0 are source coordinates on the surface of the piston as shown in Fig. 10.8, Eq. (10.25a) becomes

$$p(r,z) = \frac{j\rho cku_0}{2\pi}\int_0^{2\pi}\int_0^{a}\frac{\mathrm{e}^{-jkR}}{R}r_0\mathrm{d}r_0\mathrm{d}\phi_0, \tag{10.25b}$$

where $R^2 = r^2 + z^2 + r_0^2 - 2rr_0 cos\,\phi_0$, a is the radius of the piston, and the field coordinate ϕ does not appear because of the circular symmetry.

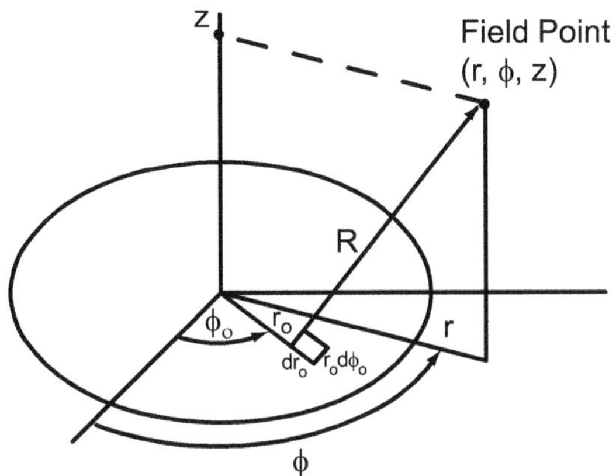

Fig. 10.8 Cylindrical coordinates for calculating the field of a circular piston radiator in an infinite rigid plane

The integration in Eq. (10.25b) can be done analytically only for a limited number of field points: those in the far field, those on the axis of the piston, and those on the edge of the piston. The average pressure over the surface of the piston can also be calculated, which gives the radiation impedance (see Sect. 10.4).

As always, the far field consists of spherical waves and is more conveniently calculated after changing to spherical coordinates where $R^2 = r^2 + r_0^2 - 2rr_0 \sin\theta \cos\phi_0$ and again ϕ does not appear because of symmetry. In the far field, where $r >> r_0 \leq a$, the distance to the field point simplifies to $R = r - r_0 \sin\theta \cos\phi_0$, and Eq. (10.25b) becomes

$$p(r,\theta) = \frac{j\rho cku_0}{2\pi} \frac{e^{-jkr}}{r} \int_0^{2\pi} \int_0^a e^{jkr_0 \sin\theta \cos\phi_0} r_0 dr_0 d\phi_0, \tag{10.26}$$

where R in the denominator was approximated by r. This is a known integral [10] that results in the first-order Bessel function, J_1, and the final result is

$$p(r,\theta) = j\rho cku_0 a^2 \frac{e^{-jkr}}{r} \frac{J_1(ka \sin\theta)}{ka \sin\theta}. \tag{10.27}$$

The function of θ in Eq. (10.27) is the directivity function with a main lobe on the axis of the piston surrounded by a series of side lobes as shown in Fig. 10.9. The -3 dB points on the main lobe occur for kasin $\theta = 1.6$, and the beam width for large ka is given by

$$\mathrm{BW} = 3.2/ka \text{ radians} = 58\lambda/D \text{ degrees},$$

where $D = 2a$ is the piston diameter. Thus the acoustic beam of a piston is 20 % broader than the beam of a line with length equal to the piston diameter, but the first side lobe is 17.8 dB below the main lobe, while the first side lobe for the line is only down 13.5 dB.

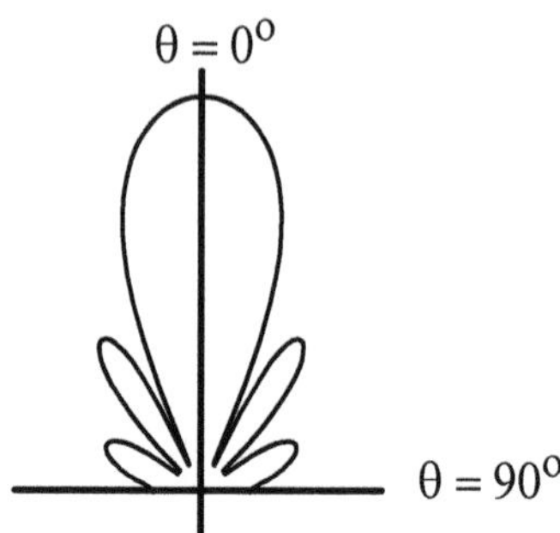

Fig. 10.9 Far-field beam pattern of circular piston in an infinite rigid plane for $ka = 3\pi$. The three-dimensional pattern is given by rotation about $\theta = 0^0$

The maximum intensity occurs at $\theta = 0$, and is obtained from Eq. (10.27) as

$$I_0 = \frac{(pp^*)|_{\theta=0}}{2\rho c} = \frac{k^2 \rho c u_0^2 a^4}{8r^2}. \tag{10.28}$$

The average intensity to be used in calculating the directivity factor, defined in Eq. (1.20), is equal to the total radiated power divided by the area of *a* sphere at a distance *r* in the far field. Since the total radiated power, *W*, can be expressed in terms of the radiation resistance, R_r, the average intensity can be written as

$$I_a = W/4\pi r^2 = \left(R_r u_0^2/2\right)/4\pi r^2, \tag{10.29}$$

and then the directivity factor becomes

$$D_f = I_0/I_a = \left(\pi k^2 \rho c a^4\right)/R_r \tag{10.30}$$

The radiation resistance for the circular piston will be calculated in Sect. 10.4, and when used in Eq. (10.30) it gives

$$D_f = (ka)^2/\left[1 - J_1(2ka)/ka\right]. \tag{10.31a}$$

When ka exceeds π, $J_1(2ka)$ becomes small and

$$D_f \approx (ka)^2 = (2\pi a/\lambda)^2 = 4\pi A/\lambda^2, \tag{10.31b}$$

where A is the area of the piston. The same result is also obtained by letting $R_r = \rho c \pi a^2$ in Eq. (10.30), showing that ρc loading is the basic requirement for its validity. This simple form is a very convenient approximation for the D_f of large piston sources with shapes other than circular (see below for rectangular pistons). When ka is much less than 1, Eq. (10.31a) gives $D_f \sim 2$, because of the rigid baffle and D_f being defined in terms of $4\pi r^2$ rather than $2\pi r^2$. Thus Eq. (10.31a) is a good approximation for small pistons with a large baffle, and it is also good for large pistons with no baffle, because then the piston approximately baffles itself. Figure 10.10 shows DI = 10log D_f as a function of ka for the circular piston in a

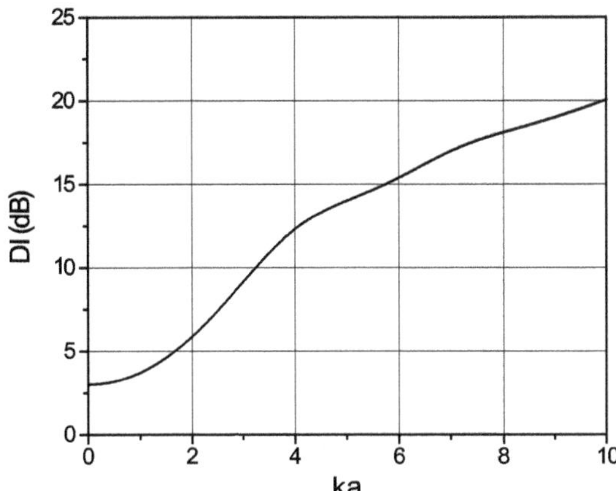

Fig. 10.10 Directivity index vs. *ka* for the circular piston

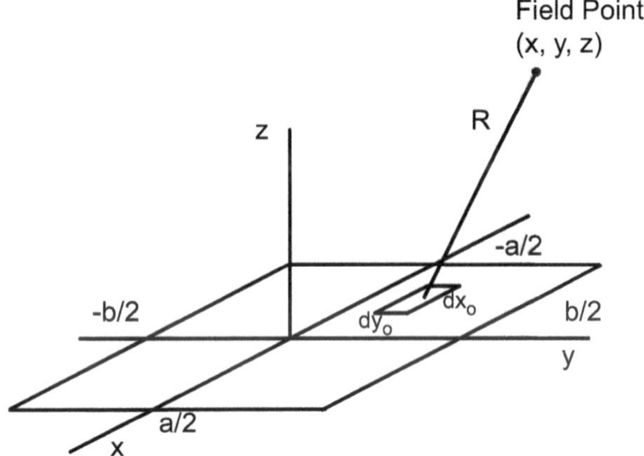

Fig. 10.11 Coordinates for calculating the field of a rectangular piston radiator in an infinite baffle

baffle. Note that DI for a two-dimensional radiating surface, such as the piston, is much greater than that for a one-dimensional line radiator; e.g., a piston with $2ka = 20$ has DI ≈ 20 dB, while a line with $kL = 20$ has DI ≈ 8 dB.

Equation (10.25a) can be applied to other radiators in rigid plane baffles. For a rectangular piston vibrating with uniform normal velocity u_0 in the x, y plane with side lengths a and b (see Fig. 10.11) the far field is given by

$$\begin{aligned} p(r,\theta,\phi) &= \frac{j\rho cku_0}{2\pi}\frac{\mathrm{e}^{-jkr}}{r}\int\limits_{-b/2}^{b/2}\int\limits_{-a/2}^{a/2}\mathrm{e}^{jk\sin\theta(x_0\cos\phi+y_0\sin\phi)}\mathrm{d}x_0\mathrm{d}y_0 \\ &= \frac{j\rho cku_0ab}{2\pi}\frac{\mathrm{e}^{-jkr}}{r}\frac{\sin\left[(ka/2)\sin\theta\cos\phi\right]}{(ka/2)\sin\theta\cos\phi}\frac{\sin\left[(kb/2)\sin\theta\sin\phi\right]}{(kb/2)\sin\theta\sin\phi}. \end{aligned} \tag{10.32a}$$

This expression shows that the far-field beam patterns in the x, z ($\phi = 0$) and y, z ($\phi = \pi/2$) planes are the same as those of a line source. It can also be seen that the patterns in other planes for other values of ϕ are the product of two similar functions. In these cases the far field is the Fourier transform of the source function, since sin x/x is the Fourier transform of a "box car" function. The directivity factor for a large rectangular piston is found to be $4\pi ab/\lambda^2 = 4\pi A/\lambda^2$, by following the procedure used above for the circular piston with $R_r = \rho$cab. Thus Eq. (10.31b) is valid for rectangles as long as both sides are long compared to the wavelength, for only then is the piston approximately ρc loaded with $R_r \approx \rho cab$.

Far-field directivity functions have also been obtained for circular radiators with axially symmetric nonuniform velocity distributions that approximate flexural disk transducers [11]. The results show, as expected, that when the velocity at the center of the source is higher than the average velocity the main lobe is broader and the side lobes are lower. The far field of the wobbling (or rocking) piston has also been calculated [12].

The following convenient approximations for the directivity index of pistons that are large compared to the wavelength can be derived from Eq. (10.31b) and the approximations for beam widths given previously for the line and circular piston:

$$\text{DI} \approx 45\ \text{dB} - 20\log\ \text{BW},\ \ \text{for circular pistons,}$$

$$\text{DI} \approx 45\ \text{dB} - 10\log\ \text{BW}_1 - 10\log\ \text{BW}_2,\ \ \text{for rectangular pistons,}$$

where BW_1 and BW_2 are the beam widths in planes parallel to the sides of the rectangle, and all beam widths are in degrees. (Note: These and other radiation related expressions are listed in Sect. 13.13).

The field of the annular piston in a plane rigid baffle (see Fig. 10.12) provides a good example of the utility of superposition.

The far-field pressure of two circular pistons, one of radius a, the other of radius b with $b < a$, lying in the same plane with centers at the same point is given by Eq. (10.27) as

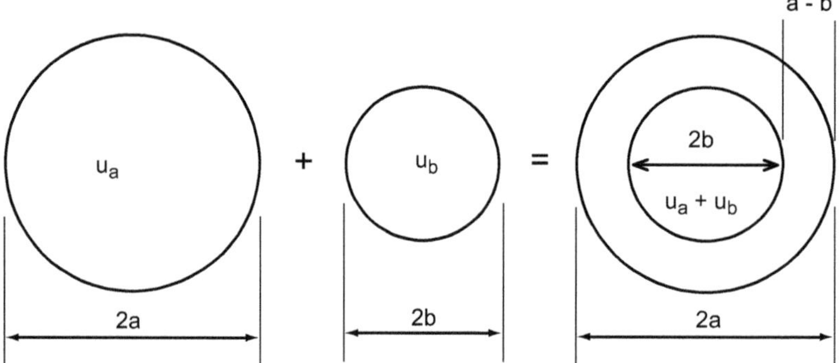

Fig. 10.12 The boundary conditions for two concentric circular pistons combine to give the boundary condition for an annular piston when the velocities are equal and opposite ($u_b = -u_a$). The sound field of the annular piston is the sum of the sound fields of the two pistons

$$p(r,\theta) = j\rho ck \frac{\mathrm{e}^{-jkr}}{r}\left[u_a a^2 \frac{J_1(ka\sin\theta)}{ka\sin\theta} + u_b b^2 \frac{J_1(kb\sin\theta)}{kb\sin\theta}\right], \tag{10.32b}$$

where u_a and u_b are the velocity amplitudes of the two pistons. If $u_b = -u_a$ the combined velocity distribution and pressure field is that of an annular piston of width $(a-b)$ with velocity u_a. When $(a-b) \ll a$, the annular piston becomes a thin ring with the far-field directivity function proportional to $J_0(ka \sin\theta)$. The field of the thin ring can be found directly from Eq. (10.25b) by using a delta function velocity distribution, $u(r_0) = u_0 a\delta\,(r_0 - a)$, or by writing $a = b + \Delta$ in Eq. (10.32b) and letting $\Delta \to 0$ while $u_a b\Delta$ remains equal to the source strength. The result in the far field is

$$p(r,\theta) = j\rho cku_0 2a\Delta J_0\,(ka\sin\theta)\frac{\mathrm{e}^{-jkr}}{r} \tag{10.32c}$$

Superposition holds at all points in the near field and far field, but only the far field can be expressed in a simple way by using Eq. (10.27). As other examples of the use of Eq. (10.32b), $u_b = u_a$ gives the field of a piston with a step function velocity distribution, while more complicated axisymmetric velocity distributions can be stepwise approximated by combining the fields of several concentric pistons. This approach is the basis for an early patent by Massa intended for controlling side lobes [13]. Note that the superposition takes a simple form in these cases because the pistons are concentric, but the solutions for non-concentric pistons in the same plane can also be superimposed.

Although individual transducers often have some surrounding structure that represents a partial baffle it is usually small compared to the wavelength and sometimes far from rigid, making the infinite rigid baffle case of questionable value (see Sect. 9.7.5). The effects of finite size baffles and nonrigid baffles usually must be analyzed by numerical methods (see Sect. 3.4 and 11.4). Piston transducers located in the interior of large, close-packed arrays are baffled by the surrounding transducers, and can be considered to be in a rigid baffle if the interactions between transducers are included (see Chap. 7). See also Chap. 8, Sect. 8.6 for the analysis of a planar circular array of discreet and continuous elements set in a rigid baffle.

10.2.3 Spherical and Cylindrical Sources

Some other cases that are easy to calculate and very useful as approximate models for transducers can be obtained from the spherical coordinate solution of the wave equation in Eq. (10.14). The spherical source that vibrates along the z-axis as a rigid body (see Fig. 10.2) has the same z-component of velocity over its whole surface, but the normal component of velocity varies over the surface as $\cos\theta$. Since this vibration is symmetric about the z-axis, the acoustic field it radiates must have the same symmetry, and the index m in Eq. (10.14) must be zero. Thus the field can be described by the Legendre polynomials, $P_n(\cos\theta)$, and, since $P_1(\cos\theta) = \cos\theta$, it is the only function of θ needed to make the field match the normal velocity on the surface. Thus the solution for the sphere vibrating as a rigid body is

$$p(r,\theta) = A_1 h_1^{(2)}(kr)\cos\theta. \tag{10.33}$$

The $\cos^2\theta$ intensity beam pattern of this transducer is its most important characteristic, because it is the basic pattern of the various vector hydrophones described in Sect. 6.5. It also approximates the far field of any transducer with a rigid oscillating motion or with two similar parts moving out of phase. This pattern is called a dipole because it is the second in a series of multipole radiators corresponding to the index n on the Legendre polynomials, and it follows that the omnidirectional pulsating sphere pattern is called a monopole. Two point sources with small separation compared to the wavelength, and vibrating 180° out of phase, also have the dipole pattern. The dipole pattern is often combined with other patterns to achieve directional radiation or reception (see Sect. 6.5.6). Similar far-field patterns can be obtained by use of the circumferential extensional modes of cylindrical transducers (see Sect. 5.2.6) [14].

The far-field radiation from sources on an infinitely long rigid cylinder has been calculated by Laird and Cohen [15]. A useful simple case is a uniformly vibrating ring of axial length $2L$ that completely encircles the cylinder, for which the far-field pressure in spherical coordinates (r,θ) is

$$p(r,\theta) = \frac{2\rho c u_0 L}{\pi}\,\frac{Sinc(kL\cos\theta)}{\sin\theta H_0'(ka\sin\theta)}\,\frac{e^{-jkr}}{r}. \tag{10.34}$$

Radiation from portions of a cylinder will be discussed more fully in Chap. 11.

10.3 Near-Field Acoustic Radiation

10.3.1 Field on the Axis of a Circular Piston

The part of the acoustic field near a transducer, the near field, is more spatially complicated than the far field and, therefore, more difficult to calculate. One of the few cases that can be easily calculated and expressed in a simple way is the field on the axis of a circular piston in an infinite rigid plane. At a point z on the axis of the piston in Fig. 10.8 the distance R in Eq. (10.25b) is $R^2 = z^2 + r_0^2$, since $r = 0$. Because z remains constant in the integration over the surface of the piston, $RdR = r_0 dr_0$, and Eq. (10.25b) becomes

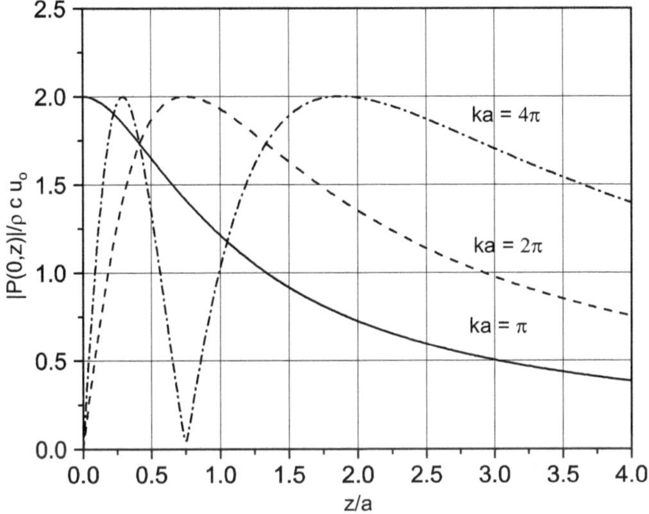

Fig. 10.13 Pressure amplitude on the axis of a circular piston for $ka = \pi$ where the only maximum is on the surface, for $ka = 2\pi$ and $ka = 4\pi$

$$p(0, z) = \frac{j\rho cku_0}{2\pi} \int_0^{2\pi} \int_{R_1}^{R_2} e^{-jkR} dR d\phi_0, \tag{10.35}$$

where $R_1 = z$ and $R_2 = (z^2 + a^2)^{1/2}$. Integrating Eq. (10.35) gives

$$\begin{aligned} p(0, z) &= -\rho cu_0 \left[e^{-jk(z^2+a^2)^{1/2}} - e^{-jkz} \right] \\ &= 2j\rho cu_0 \sin\left[\frac{k}{2}\left(\sqrt{a^2 + z^2} - z\right)\right] e^{-j\frac{k}{2}\left(\sqrt{a^2+z^2}+z\right)}. \end{aligned} \tag{10.36a}$$

Equation (10.36a) shows that the pressure amplitude on the surface of the piston at the center is $2\rho\, cu_0 \sin(ka/2)$, and, if ka is large enough, the pressure varies along the piston axis between maxima of $2\rho\, cu_0$ and zero as shown in Fig. 10.13.

When the diameter of the piston equals λ, the quantity $ka = \pi$ and the only maximum occurs on the surface at the center. When $ka < \pi$ the maximum is at the center with a value less than $2\rho\, cu_0$; when $ka > \pi$ one or more maxima occur along the axis. The maximum on the axis farthest from the piston occurs at $z = a^2/\lambda - \lambda/4$; beyond that the pressure decreases steadily and approaches the far field z dependence of e^{-jkz}/z. The transition from near field to far field is gradual, but this example shows that when z exceeds a^2/λ the far field has been approximately reached. The transitional distance $2a^2/\lambda$ is referred to as the Raleigh distance.

Estimates of where the far field begins are important when making acoustic measurements and can be based on the circular piston case (see Chap. 9).

The axial pressure variations displayed by the circular piston are extreme because of the high degree of symmetry in this case; such large variations occur only on the axis, only for a circular piston and only for sufficiently high frequency. For a rectangular piston the field in a region directly in front of the piston is more uniform and has some resemblance to a plane wave when the piston is large compared to the wavelength. Note that, although the pressure on the axis of the circular piston varies from zero to $2\rho c u_0$, the average value is approximately $\rho c u_0$ as in a plane wave.

The field at any point on the axis of a thin ring can be found from Eq. (10.25a) by using the delta function velocity distribution, $u(r_0) = u_0 a\delta(r_0 - a)$. The result is

$$p(0,z) = j\rho c k u_0 a^2 \left[\mathrm{e}^{-jk\left(z^2+a^2\right)^{1/2}} / \left(z^2+a^2\right)^{1/2} \right]. \tag{10.36b}$$

All the pressure contributions from a thin ring source arrive at a point on the axis in phase; thus there are no pressure amplitude fluctuations along the axis as there are for the piston. The expressions for the pressure at any point on the axis of a piston or ring provide a convenient basis for extrapolating near-field transducer measurements to the far field (see Chap. 9).

10.3.2 The Effect of the Near Field on Cavitation

Cavitation occurs in water when the acoustic pressure amplitude exceeds the hydrostatic pressure. Then, during the negative pressure part of each cycle, tiny bubbles may form around particulate impurities in the water that act as cavitation nuclei. The bubbles in the otherwise homogeneous medium cause scattering and absorption of sound. Thus cavitation in the near field of high power transducers and arrays limits the radiated acoustic power. Urick [16] discusses many practical aspects of cavitation, while here we show how pressure variations in the near field of a transducer may affect the onset of cavitation

For a plane wave near the surface in pure water cavitation may begin when the pressure amplitude exceeds one atmosphere (~10^5 Pa) and the cavitation limited intensity is then

$$I_c = p^2/2\rho\, c = 1/3\ \mathrm{W/cm^2} \tag{10.37}$$

The cavitation limit is increased at a water depth of h feet, which adds h/34 atmospheres to the static pressure at a temperature of 40°F. It is also increased if there is little dissolved air in the water, or other impurities that act as cavitation nuclei; then the water can be considered to have an effective tensile strength of T atmospheres [16]. The limit may also increase with frequency, but this effect is small below about 10 kHz. A factor that decreases the cavitation limit is the spatial variation of pressure in the near field of transducers, such as that illustrated in Fig. 10.13, because cavitation begins at the maxima of pressure amplitude. Thus estimating the cavitation limited output of a transducer requires knowledge of the spatial variations of pressure amplitude in its near field. The radiated power of the transducer can be related to the maximum pressure in the near field, and the cavitation limited intensity at the surface of the transducer resulting from all these effects can be expressed as

$$I_c = (\gamma/3)(1 + h/34 + T)^2 \mathrm{W/cm^2}. \tag{10.38}$$

In Eq. (10.38) γ is the dimensionless near-field cavitation parameter defined as [17]

$$\gamma = (R_r/\rho c A)/(|p_m|/\rho c u_0)^2, \tag{10.39}$$

where R_r is the radiation resistance referred to the velocity u_0, A is the radiating area of the transducer, and p_m is the maximum pressure amplitude in the near field for the velocity u_0. Note that, if a transducer could radiate a plane wave, γ would be unity, since then $R_r = \rho c A$ and $p_m = \rho\, c u_0$. For a pulsating sphere, where the surface pressure is uniform, γ is also unity. However, the near field of most transducers contains a reactive part of the pressure, which does not contribute to radiation but does contribute to forming pressure maxima, which makes γ less than unity.

Numerical results for γ are shown in Fig. 10.14 for flat circular sources in a rigid plane baffle [17].

This figure compares the uniform velocity piston and circular plates vibrating in fundamental flexural modes with different nonuniform velocity distributions. The supported edge and clamped edge plates have the most nonuniform velocity distributions, because the velocity is zero on the edge, and they have the lowest

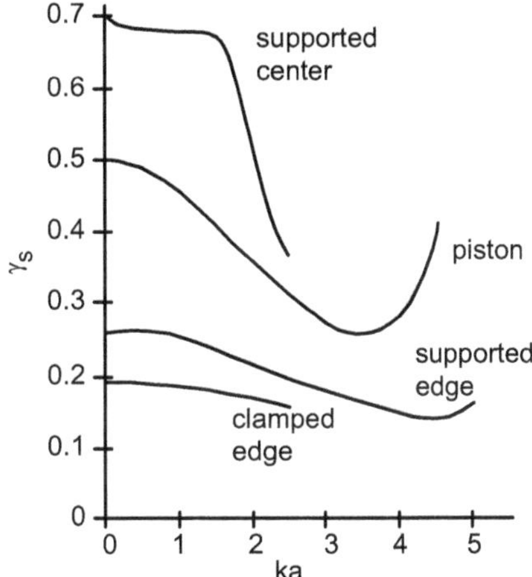

Fig. 10.14 Cavitation parameter for circular radiators with different velocity distributions. All except the piston are flexing plates with different boundary conditions [17]

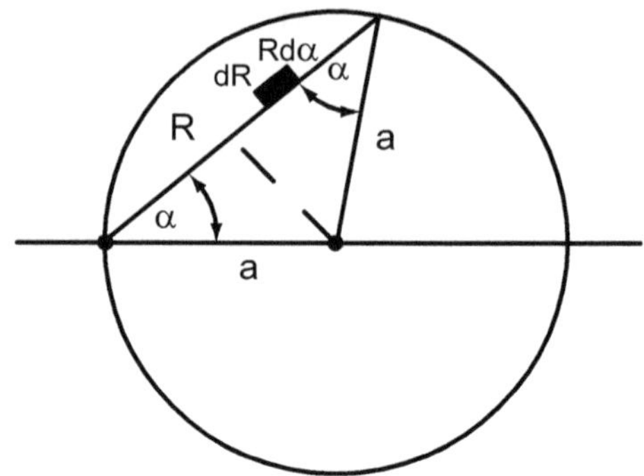

Fig. 10.15 Coordinates for calculating the pressure on the edge of a circular piston. The integration over R goes from 0 to $2a\cos\alpha$, the integration over α goes from $\pi/2$ to $-\pi/2$

values of γ. The supported center case has the highest value of γ but only for small ka. It should be noted that for none of these circular sources does γ approach unity for very small ka. This is related to the fact that the pressure distributions for circular sources do not become uniform, no matter how small the source is compared to the wavelength. This will be shown for the circular piston by comparing the pressure on the edge with the pressure at the center.

The field on the edge of the circular piston can be calculated from Eq. (10.25a) using polar coordinates R and α in the plane of the piston with the origin at an arbitrary point on the piston edge (see Fig. 10.15).

Then the differential area is $RdRd\alpha$, the integration over R goes from 0 to $2a\cos\alpha$ and the integration over α from 0 to $\pi/2$ covers half the area. After integrating over R and multiplying by 2, the pressure on the edge in cylindrical coordinates, (r, z), becomes

$$p(a,0) = \frac{\rho c u_0}{\pi} \int_0^{\pi/2} \left(1 - e^{-2jka\cos\alpha}\right) d\alpha. \tag{10.40}$$

This integral can be evaluated in terms of zero-order Bessel (J_0) and Struve (S_0) functions with the result

$$p(a,0) = \tfrac{1}{2}\rho\, c u_0 [1 - J_0(2ka) + jS_0(2ka)]. \tag{10.41a}$$

At very low frequency, where $ka \ll 1$, the pressure on the edge simplifies to

$$p_1(a) = j\rho\, c u_0(2ka/\pi), \tag{10.41b}$$

while Eq. (10.36a) gives the pressure at the center of the piston for $ka \ll 1$ as

$$p_1(0) = j\rho c u_0(ka). \tag{10.41c}$$

The factor j in Eqs. (10.41b, 10.41c) shows that these are reactive pressures related to radiation mass. The radiation resistance has been neglected in this approximation for $ka \ll 1$. Thus the pressure at the center is higher by the factor $\pi/2$ than the pressure on the edge, even as ka approaches zero. Using $p_1(0)$ for the maximum pressure in Eq. (10.39) with $R_r/\rho cA = ½ (ka)^2$ (as will be shown in Sect. 10.4.2 for small ka) gives a cavitation factor of ½ as shown in Fig. 10.14.

Although these examples of the cavitation factor do not apply exactly to any real transducer, they are useful for estimating cavitation limited power in practical cases.

10.3.3 Near Field of Circular Sources

We pointed out that the near field of the circular piston can only be calculated analytically from Eq. (10.25a) along the axis and on the edge. However, the near field can be calculated from that equation at any point by numerical integration, and that has been done for some specific cases long ago, mainly by Stenzel [18]. These results are still useful and are included in Rschevkin's book [19], where the field on the surface and directly in front of the piston is given for several values of ka.

Comparing the circular piston in an infinite, rigid, plane baffle with two closely related cases gives another example of near fields. It is evident from the way the circular piston problem was solved that it is equivalent to the same piston vibrating equally on both sides without a baffle. Now consider the same piston oscillating as a thin rigid body in which the outward velocities on the two sides are 180° out of phase (see Fig. 10.16).

In this case the pressure on the infinite plane surrounding the piston, including the edge of the piston, is zero because each side of the oscillating piston makes opposite contributions to the pressure at every point on that plane. The field cannot be calculated from Eq. (10.25a) because there is no infinite rigid plane, but this problem has been solved by Silbiger [20] using oblate spheroidal coordinates. His result for the pressure at the center of the oscillating piston for $ka \ll 1$ is

$$p_2(0) = j\rho c u_0(2ka/\pi), \tag{10.41d}$$

$$\text{while } p_2(a) = 0 \tag{10.41e}$$

The third case is the same unbaffled piston vibrating on one side only. This case is given by superimposing the solutions for the previous two cases, since addition of

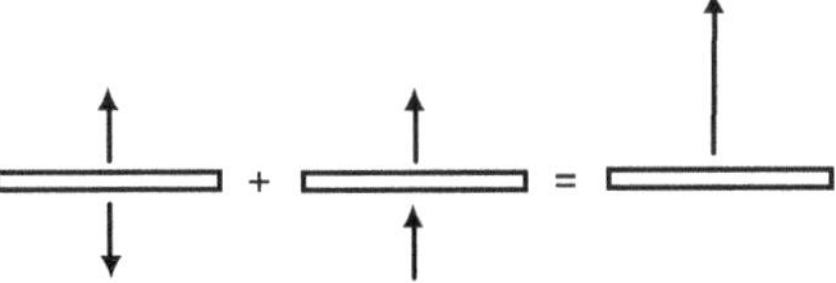

Fig. 10.16 Adding the field of the piston vibrating equally on both sides to the field of the oscillating piston gives the field of the piston vibrating on one side

the surface velocities gives $2u_0$ on one side and zero on the other side as shown in Fig. 10.16. Thus the field of the piston vibrating with velocity amplitude u_0 on one side only is half the sum of the fields in the other two cases. Using Eqs. (10.41b)–(10.41e) the pressures on the edge and at the center of both sides are

$$p_3(a) = j\rho c u_0(ka/\pi),$$
$$p_3(0)_{\text{front}} = j\rho\, c u_0[ka(\pi + 2)/2\pi],$$
$$p_3(0)_{\text{back}} = j\rho\, c u_0[ka(\pi - 2)/2\pi].$$

For the unbaffled piston vibrating on one side with $ka \ll 1$, the pressure at the center of the vibrating side exceeds the pressure on the edge by $(\pi/2 + 1)$, and it exceeds the pressure at the center of the back side by $(\pi + 2)/(\pi - 2)$. Comparing this result with the baffled piston case above shows that the pressure on the surface for $ka \ll 1$ is significantly more nonuniform for the unbaffled piston. These reactive pressure distributions on the piston surface are related to radiation mass distributions. Some of these near-field results will be used in Chap. 9 as an approximate basis for evaluating projectors by near-field measurements.

10.4 Radiation Impedance

The radiation impedance is one of the most important characteristics of the acoustic field of a transducer. It depends directly on the near field since it is the average of the pressure, or of the product of pressure and velocity when the velocity is nonuniform, over the surface of the transducer. The radiation resistance is a measure of the power the transducer is capable of radiating for a given velocity and is the critical factor in determining the efficiency and the effective bandwidth. The radiation reactance is also important because it affects the resonance frequency of the transducer and the bandwidth. In this section the radiation impedance of typical simple radiators will be calculated and numerical results will be given.

10.4.1 Spherical Sources

The general definition of radiation impedance for fixed velocity distribution transducers, given in Eq. (1.4b), is repeated here:

$$Z_r = (1/u_0 u_0^*) \iint_S p(\vec{r}) u^*(\vec{r}) \mathrm{d}S. \tag{1.4b}$$

For the monopole sphere where the velocity is uniform and $u(\vec{r}) = u_0$ the definition reduces to the surface integral of the pressure:

$$Z_{r0} = (1/u_0)\int_0^{2\pi}\int_0^{\pi} p(a,\theta,\phi)a^2 \sin\theta d\theta d\phi, \tag{10.42}$$

where the pressure on the surface is also uniform and given by Eq. (10.15b) as

$$p(a) = \frac{j\rho c u_0 ka}{(1 + jka)}. \tag{10.43}$$

Thus, the integral is evaluated by multiplying by the surface area and, when separated into resistance and reactance, gives

$$Z_{r0} = R_{r0} + jX_{r0} = 4\pi a^2 \rho c \frac{(ka)^2 + jka}{1 + (ka)^2}. \tag{10.44a}$$

Equation (10.44a) illustrates several general characteristics of radiation impedance that hold for any radiators with predominately monopole characteristics. At low frequency R_{r0} is proportional to frequency squared or $(ka)^2$, while at high frequency it approaches $\rho\, c$ times the area, which is the plane wave mechanical impedance, ρcA. The reactance X_{r0} at low frequency is proportional to frequency, and the radiation mass, defined as X_{r0}/ω, is constant and equal to $4\pi a^3 \rho$ for the monopole sphere. This value of the low frequency radiation mass is three times the mass of the water displaced by the sphere or equal to the mass of a layer of water surrounding the sphere $0.59a$ thick. At high frequency jX_{r0} approaches $-4\pi a\rho c^2/j\omega$, and behaves like a negative stiffness reactance. This negative stiffness has been used to cancel the stiffness of a transducer [21]. The radiation resistance and reactance are shown as functions of ka in Fig. 10.17.

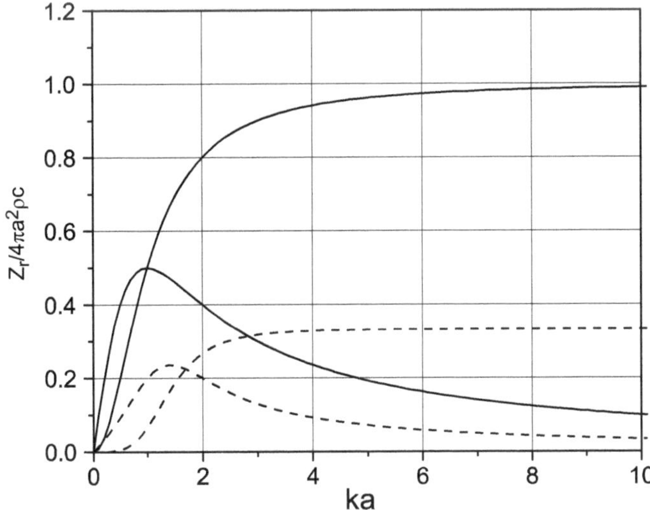

Fig. 10.17 Radiation resistance and reactance for monopole (*solid line*) and dipole (*dashed line*) spherical radiators

The low frequency radiation resistance of the monopole sphere can be written as,

$$R_{r0} = \frac{\rho c k^2 A^2}{4\pi}, \tag{10.44b}$$

where A is the area of the vibrating surface. This is a very useful approximation for the radiation resistance of a small, unbaffled, predominantly monopole radiator of any shape. Equation (10.44b) is a special case of the general relationship between radiation resistance, directivity factor, and the diffraction constant when the latter two quantities are unity (see Sect. 6.6).

For the dipole sphere, where the normal velocity is not uniform, a reference velocity must be chosen. Since $u = u_0 \cos\theta$, a natural choice for reference velocity is u_0, and the radiation impedance referred to u_0 is given by Eq. (1.4b) as

$$Z_{r1} = \frac{1}{u_0} \iint_S p(\vec{r}) \cos\theta \mathrm{d}S, \tag{10.45}$$

where p on the surface is given by Eq. (10.33) with $r = a$. Equation (10.45) then becomes

$$Z_{r1} = \frac{1}{u_0} \int_0^{2\pi} \int_0^{\pi} A_1 h_1^{(2)}(ka) \cos^2\theta \, a^2 \sin\theta \mathrm{d}\theta \mathrm{d}\phi, \tag{10.46}$$

and A_1 remains to be evaluated as a function of ka by using Eqs. (10.33) and (10.6) to satisfy the boundary condition on the velocity. The result is

$$\begin{aligned} Z_{\mathrm{r}1} = R_{\mathrm{r}1} + jX_{\mathrm{r}1} &= 4\pi a^2 \rho c \frac{-jh_1^{(2)}(ka)}{h_0^{(2)}(ka) - 2h_2^{(2)}(ka)} \\ &= \frac{4\pi a^2 \rho c}{3} \left[\frac{(ka)^4 + j\left[2ka + (ka)^3\right]}{4 + (ka)^4} \right]. \end{aligned} \tag{10.47}$$

The last form of Eq. (10.47) is convenient because it is in terms of ka only; it can be obtained by expressing the spherical Hankel functions in terms of trigonometric and algebraic functions [22]. The radiation impedances for the monopole and dipole are shown together in Fig. 10.17.

A similar calculation for the quadrupole sphere, where the sound field is proportional to $P_2(\cos\theta) = (1/4)(3\cos 2\theta + 1)$, gives the radiation impedance [14]

$$Z_{r2} = \frac{4\pi a^2 \rho c}{5} \frac{\left\{(ka)^6 + j\left[27ka + 6(ka)^3 + (ka)^5\right]\right\}}{\left[81 + 9(ka)^2 - 2(ka)^4 + (ka)^6\right]}. \tag{10.48}$$

The monopole sphere is an excellent acoustic radiator, the dipole sphere, with equal portions of the surface vibrating out of phase, is a poor acoustic radiator, and the quadrupole sphere is worse. This is shown by the behavior of the radiation resistances at low and high frequency:

$$\text{For } ka \ll 1: \quad \begin{aligned} R_{r0} &\approx \rho c A\,(ka)^2, \\ R_{r1} &\approx \rho c A (ka)^4/12\,, \\ R_{r2} &\approx \rho c A (ka)^6/405. \end{aligned}$$

$$\text{For } ka \gg 1: \quad \begin{aligned} R_{r0} &\approx \rho c A, \\ R_{r1} &\approx \rho c A/3\,, \\ R_{r2} &\approx \rho c A/5. \end{aligned}$$

These cases also illustrate a problem with the definition of source strength in Eq. (10.18). For the dipole and quadrupole sphere (and higher order multipoles) that definition gives zero for the source strength, although these cases do radiate useful acoustic power (see Sect. 5.2.6). The definition of source strength used in Eq. (10.18) is appropriate only for sources that have predominantly monopole characteristics. A more general definition of source strength could be based on radiated power rather than normal surface velocity.

10.4.2 Circular Sources in a Plane

Although the near-field pressure for the circular piston in a plane cannot be calculated analytically except on the piston axis and on the piston edge it is possible to calculate the average pressure over the surface of the piston which is essentially the radiation impedance. This calculation starts with Eq. (10.25b) using a coordinate system similar to that in Fig. 10.15, but now the origin is placed at an arbitrary point on the surface $r = r$, $z = 0$, between the center and the edge, as shown in Fig. 10.18.

Then, after integrating over R from zero to $R_0 = r\cos\alpha + (a^2 - r^2\sin^2\alpha)^{1/2}$, the pressure at this point on the piston surface becomes

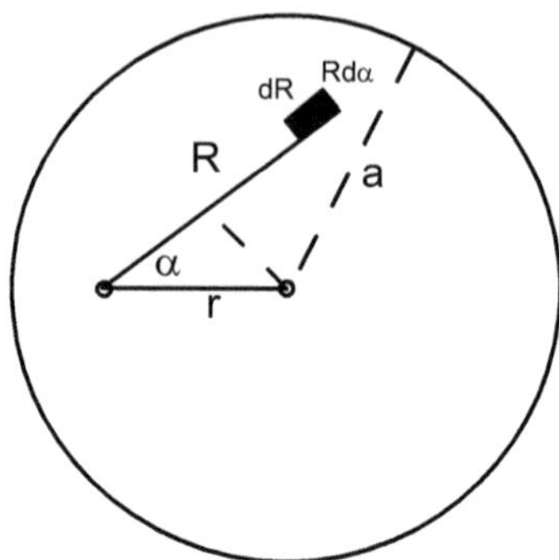

Fig. 10.18 Coordinates for calculating the pressure at an arbitrary point on the surface of a circular piston. The integration over R goes from 0 to $r\ \cos\alpha + (a^2 - r^2\sin^2\alpha)^{1/2}$, the integration over α goes from 0 to 2π

$$p(r,0) = \frac{\rho c u_0}{2\pi}\int_0^{2\pi}\left(1 - \mathrm{e}^{-jkR_0}\right)\mathrm{d}\alpha. \tag{10.49}$$

The radiation impedance for this uniform velocity case is given by integrating Eq. (10.49) over the surface of the piston:

$$Z_r = \frac{1}{u_0}\iint_S p(r,0)\mathrm{d}S = \frac{1}{u_0}\int_0^{2\pi}\int_0^a p(r,0) r\mathrm{d}r\mathrm{d}\phi. \tag{10.50}$$

Since $p(r, 0)$ is not a function of ϕ the integration over ϕ gives the factor 2π, and when Eq. (10.49) is used

$$Z_r = \rho c\int_0^a\int_0^{2\pi}\left(1 - \mathrm{e}^{-jkR_0}\right)\mathrm{d}\alpha\ r\mathrm{d}r = \rho c\pi a^2 - \rho c\int_0^a\int_0^{2\pi}\mathrm{e}^{-jkR_0}\mathrm{d}\alpha\ r\mathrm{d}r. \tag{10.51}$$

The remaining integral can be evaluated in terms of first-order Bessel and Struve functions. Note that a convenient approximation for the first-order Struve function, S_1, is available [23]. The result for the radiation impedance of the circular piston is

$$Z_r = R_r + jX_r = \rho c\ \pi a^2[1 - J_1(2ka)/ka + jS_1(2ka)/ka]. \tag{10.52}$$

At low frequency the resistance is equal to $\rho c\pi a^2(ka)^2/2$, while at high frequency it becomes constant at $\rho c\pi a^2$. The reactance at low frequency is $\rho c\pi a^2(8ka/3\pi)$ corresponding to a radiation mass of $8a^3\rho/3$ or a disk of water with the same radius as the piston and a thickness of $8a/3\pi$. At high frequencies, the reactance goes to zero. The resistance and reactance are shown in Fig. 10.19 as a function of ka; these

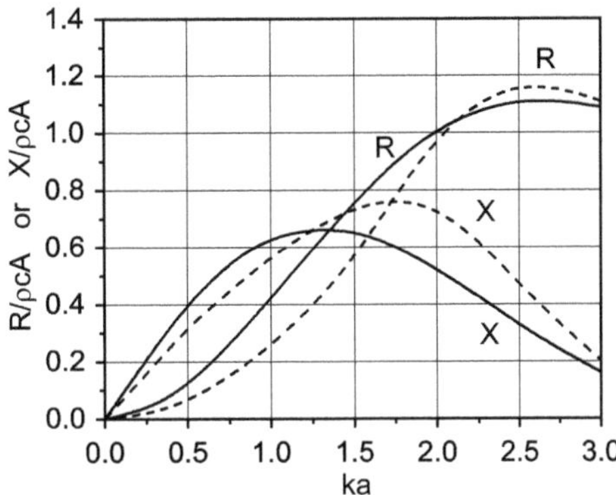

Fig. 10.19 Radiation impedance of a circular piston in an infinite baffle (*solid line*) and with no baffle (*dashed line*). The latter from Nimura and Watanabe [24]

are probably the most frequently used curves for estimating the radiation impedance of transducers.

Figure 10.19 also includes results for a piston vibrating on one side only with no baffle; these curves are similar to those given by Nimura and Watanabe [24] adjusted to reduce small errors. For example, the curves for the baffled piston can be considered exact based on Eq. (10.52), while those for the unbaffled piston come from an oblate spheroidal expansion for the oscillating piston [20, 24] combined with the baffled piston results as explained in Sect. 10.3.3. The curves in reference [24] were adjusted to be consistent with the fact that for small ka the unbaffled resistance is half the baffled resistance. The unbaffled piston curves are very similar to those given by Beranek [25] for a piston mounted in the end of a long tube, which were derived from the Levine–Schwinger calculation for a plane wave radiating from the end of a long tube [26]. The unbaffled case usually gives a better estimate of the radiation impedance of a single transducer, but in a large array the surrounding transducers provide a significant baffle effect.

Radiation impedance for sources in a rigid plane baffle have been calculated for circular sources with nonuniform velocity distributions [11, 12] and for square and rectangular pistons [27, 28]. Results for uniformly vibrating circumferential bands on long rigid cylinders have also been obtained [29, 30].

It is obvious that none of the examples discussed in the preceding sections conform exactly to the geometry of realistic transducers. Indeed, one of the simplest geometrical shapes that does conform fairly well to many transducers, the finite length cylinder, cannot be handled by the methods that have been presented. In spite of this the results obtained by these analytical methods are valuable, because they provide general physical understanding as well as estimates of quantities that are essential for transducer design. Fortunately, finite element modeling, FEM or FEA, of transducer structures, described in Sect. 3.4, can be extended to the acoustic medium [31, 32], and makes possible acoustic calculations that include realistic features of transducers. Some transducer-specific finite element programs, such as Atila [33], avoid the need for large fluid fields by evaluating the pressure and velocity on a closed surface near the transducer, and then using a Helmholtz Integral approach (see Chap. 11) for calculating the far-field pressure and beam patterns. This is an example of the feasibility of combining analytical methods with

FEA to reduce run time or increase the size of the problems that can be handled. A Helmholtz Integral subroutine could be added to other FEA programs after determining the minimum fluid field required. The basics of the finite element method and its application to transducers and transducer radiation are discussed in Sect. 3.4 and other numerical methods are discussed in Sect. 11.4.

10.5 Dipole Coupling to Parasitic Monopole

We conclude this chapter with an interesting example which uses the spherical Hankel function to obtain the radiation from a small dipole transducer in the presence of a small passive monopole radiator. This analysis also illustrates how a simple solution may be obtained when resonators are small compared to the wavelength in the medium.

Transducer arrays may include electrically inactive passive components to modify the performance. The transducer analysis program TRN [34] allows passive resonant or nonresonant components to be used in an array. These components can alter the beam patterns and the impedance of nearby active piezoelectric piston type transducer. The inactive components have been referred to as parasitic elements [35, 36] as they feed off of the field generated by the active elements. In this section we consider the use of a monopole type parasitic radiator as means for enhancing the output of a nearby dipole radiator and develop the equations for the acoustic coupling from a dipole projector to a passive monopole resonator.

Dipole piezoelectric-based bender transducers can be useful for obtaining low frequency resonances at great depths; however, they suffer in output because of the significant cancellation between the oppositely phased front and rear acoustic radiation. We will show that one means for improving their performance is to use a nearby monopole parasitic resonator which does not suffer from out-of-phase cancellation. Since underwater acoustic parasitic elements do not contain any stress-sensitive piezoelectric elements, they can be more easily designed to be an efficient monopole radiator, even under great pressure. A key concept to this system is the unique property of a dipole radiator that provides a near field with strength that is comparable to a monopole source, providing the parasitic element with a strong low frequency signal that can be reradiated into the far field as a more efficient monopole source. A simple approximate spherical Hankel function model will be presented with a small oscillating sphere as the dipole source and a small pulsating sphere as the monopole parasitic resonator. Because the objects are small compared to water wavelengths, the details of their structure and scattering are not as important, and this model may be used to represent other small dipole type transduction sources and monopole type passive parasitic resonators.

Consider the first two terms of an axially symmetric spherical acoustic wave function expansion (see Sect. 10.1) for the dipole pressures $p_a(r,\theta) = c_1 h_1(kr)\cos(\theta)$ and monopole pressure $p_b(r) = c_0 h_0(kr)$ where $h_n(kr)$ is the spherical Hankel function of the second kind of order n and k is the wave number given by $k = \omega/c$ with ω the angular frequency, $2\pi f$, and c the speed of sound in the medium. The constant coefficients c_0 and c_1 can be obtained from the boundary conditions on the velocity. The results for the free field dipole and monopole acoustic pressures are, respectively,

$$p_a(r,\theta) = -j\rho c u_a \left[h_1(kr)/h_1^{'}(ka)\right] \cos(\theta) \tag{10.53}$$

and

$$p_b(r) = -j\rho c u_b \left[h_0(kr)/h_0^{'}(kb)\right], \tag{10.54}$$

where $'$ means derivative with respect to the argument. The dipole and monopole spheres of radii a and b and velocities u_a and u_b are illustrated in Fig. 10.20 with center-to-center distance, d, and distances r and r_0, related by $r_0^2 = r^2 + d^2 - 2rd\cos(\theta)$, yielding a total pressure $p_t(r, \theta)$ at angle θ.

If the monopole were of the same radius and velocity as the dipole, then for $ka \ll 1$, along with $h_0^{'}(ka) = -h_1(ka) \approx -j/(ka)^2$, $h_1^{'}(ka) \approx -2j/(ka)^3$ and $h_1(kr) = h_0(kr)(j + 1/kr)$, we can get

$$p_a(r,\theta) \approx p_b(r)(j + 1/kr)(ka/2)\cos(\theta) \tag{10.55}$$

It can be seen from this important equation that for distances large compared to wavelength, with $kr \gg 1$, the dipole pressure is $p_a(r,\theta) \approx jp_b(r)(ka/2)\cos(\theta)$ and is reduced below that of a monopole by a factor $ka/2$, decreasing by 6 dB/octave as the frequency decreases. That is, at $\theta = 0$, $p_a(r,0)/p_b(r) \approx jka/2$ and the dipole is a progressively weaker radiator into the far field as the frequency decreases or the

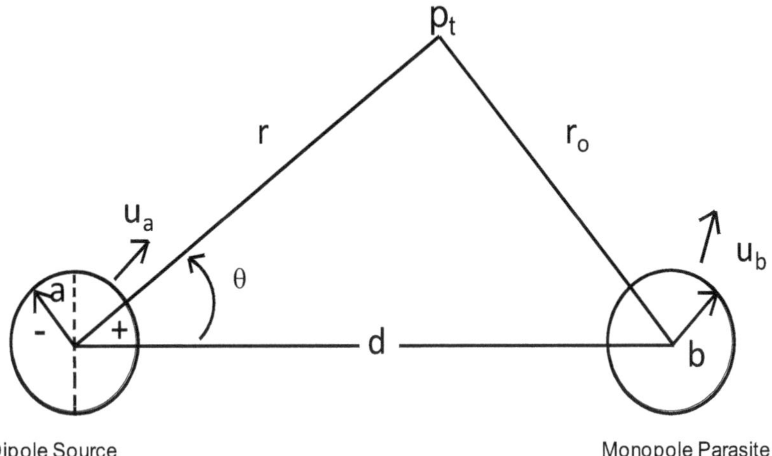

Fig. 10.20 Dipole and monopole spheres of radii a and b and velocities u_a and u_b separated by a center-to-center distance, d, and distances r and r_o arriving together to form the total pressure $p_t(r, \theta)$ at angle θ, measured from the axis of symmetry of the dipole

wavelength increases. However, at near field distances small compared to wavelength, with $kr \ll 1$, the dipole pressure is $p_a(r,\theta) \approx p_b(r)(a/2r)\cos(\theta)$ and does not decrease as the frequency decreases but has the same value as a monopole, reduced by the factor $a/2r$. This near-field effect allows the dipole to significantly excite the parasitic resonator which can reradiate with greater output at $kr \gg 1$.

Consider now the case where b and u_b are not necessarily equal to a and u_a, respectively, as in Fig. 10.20. Here the parasitic velocity, u_b, needs to be determined in terms of the dipole source velocity u_a. This can be done from the impedance of the parasite $Z = F/u_b$ where the force, F, on the parasitic element of radius b, is $F = \int p_a(r,\theta)dA$ and dA is the elemental area on the sphere, of area $A = 4\pi b^2$. This force is due to the dipole source of velocity u_a. If the monopole sphere is small compared to the wavelength of sound $\lambda = 2\pi/k$ in the medium, $F \approx p_a(d,0)A$ and the scattering from the spheres is small and may be ignored. Under these conditions $u_b = p_a(d, 0)A/Z$ and we may write, from Eq. (10.53) with $\theta = 0$ that

$$u_b = -u_a j[\rho cA/Z]\left[h_1(kd)/h_1^{'}(ka)\right] \tag{10.56}$$

Substitution of Eq. (10.56) into Eq. (10.54) and defining the total pressure $p_t(r,\theta) = p_a(r,\theta) + p_b(r)$ yields

$$p_t/p_a(r,0) = \cos(\theta) - [j\rho cA/Z][h_1(kd)/h_0^{'}((kb)][h_0(kr_0)/h_1(kr)] \tag{10.57}$$

If we consider low frequencies and small sizes so that $kb \ll 1$ and $kd \ll 1$ and also wish to evaluate the system in the far field where $kr \gg 1$ and $r_0 \approx r$ we can, on using these conditions and approximations, get the surprisingly simple relation

$$p_t/p_a(r,0) = \cos(\theta) + (\rho cA/Z)(b/d)^2 \tag{10.58}$$

Equations (10.57) and (10.58) are representations of the total pressure, p_t, from both the dipole and associated monopole parasitic source relative to the dipole source at its maximum condition, $p_a(r, 0)$, where $\theta = 0^0$. The first term is from the dipole and shows the dependence on angle, θ, while the second term is due to the parasitic resonator with impedance Z and area $A = 4\pi b^2$. The result from Eq. (10.57) or Eq. (10.58) with $\theta = 90^0$ (where the dipole has a null) would be the radiation of the monopole parasitic element alone while, on the other hand, Eq. (10.58) represents the results for the dipole alone if the distance, d, or the impedance, Z, of the parasitic resonator is very large. Equation (10.58) shows how a low parasitic impedance, Z, can increase the contribution of the monopole parasitic source, overcome the output from the dipole source, and modify the final beam pattern from a dipole angular response to a monopole omnidirectional response.

Equation (10.58) was developed under the conditions: $ka \ll 1$, $kb \ll 1$, and $kd \ll 1$ along with a far-field evaluation of the total pressure at $kr \gg 1$. That is, the monopole and dipole radiators are assumed small compared to wavelength and the center-to-center separation distance d, equal to or greater than $a + b$, is also assumed small compared to wavelength. Although the element size may appear restrictive, it is a common condition for low frequency underwater projectors.

A low frequency lumped-parameter approximation for the impedance, Z, of a small parasitic resonator may be written as

$$Z = R + j\omega M + 1/j\omega C \tag{10.59}$$

where the mass $M = M_m + M_r$, with M_m the mechanical mass and M_r the radiation mass, C is the compliance, the resistance $R = R_m + R_r$ where R_m is the mechanical loss resistance, R_r is the radiation resistance, and the mechanoacoustic efficiency is given by $\eta = R_r/R$. The value of the compliance C is readily calculated when the size is small compare to one-quarter the wavelength of sound in a compliance spring-like structure. In other cases it can be obtained as an effective value from a calculation of the potential energy. Also, at low frequencies, where $kb \ll 1$, and for an equivalent spherical parasitic resonator of area $A = 4\pi b^2$, the radiation resistance is $R_r = \rho c A(kb)^2$ and mass is $M_r = 4\pi b^3 \rho$ (see Sect. 13.13). Resonance occurs as $\omega_o = (MC)^{-1/2}$ and the total mechanical Q may be written, at resonance, as $Q_o = \omega_o M/R_o = 1/\omega_o R_o C$, where R_o is the resistance at resonance. Accordingly, the mass, M, and compliance, C, terms of Eq. (10.59) may be replaced with more general descriptive terms, Q_o, ω_o, η, and the impedance may be rewritten as

$$Z = R[1 + j(\omega/\omega_o - \omega_o/\omega)Q_o R_o/R], \tag{10.60}$$

where the total resistance $R = R_r/\eta$ and the subscript "$_o$" means evaluated at resonance.

Substituting Eq. (10.60) into Eq. (10.58) yields the more generally useful total pressure radiation as

$$p_t/p_a(r,0) = \cos(\theta) + \eta/(k_o d)^2(\omega/\omega_o)^2\left[1 + j(\omega/\omega_o - \omega_o/\omega)Q_o(\eta/\eta_o)(\omega_o/\omega)^2\right]. \tag{10.61}$$

It can be seen from Eq. (10.61) that a desirable parasitic resonator would have a high efficiency at resonance, η_o, and a small value of $k_o d$, along with a resonance in the vicinity of the desired operating band of the active transducer. At resonance Eq. (10.61) yields:

$$p_t/p_a(r,0) = \cos(\theta) + \eta_o/(k_o d)^2, \tag{10.62}$$

The values of Q_o, ω_o, η_o, and $k_o d$ that would be used in Eq. (10.61) depend on the specific design of the parasitic resonator. Figure 10.21 shows the results from Eq. (10.61), plotted as $20\log(p_t/p_a)$ vs. $\omega/\omega_0 = f/f_0$ for three cases with $\eta = \eta_0$. As may be seen, the summed total monopole parasitic and dipole pressure value levels generally exceed the referenced dipole level of 0 dB from $\cos(0) = 1$ in Eq. (10.61). The lowest level is for case (a) with the parasitic resonator at the

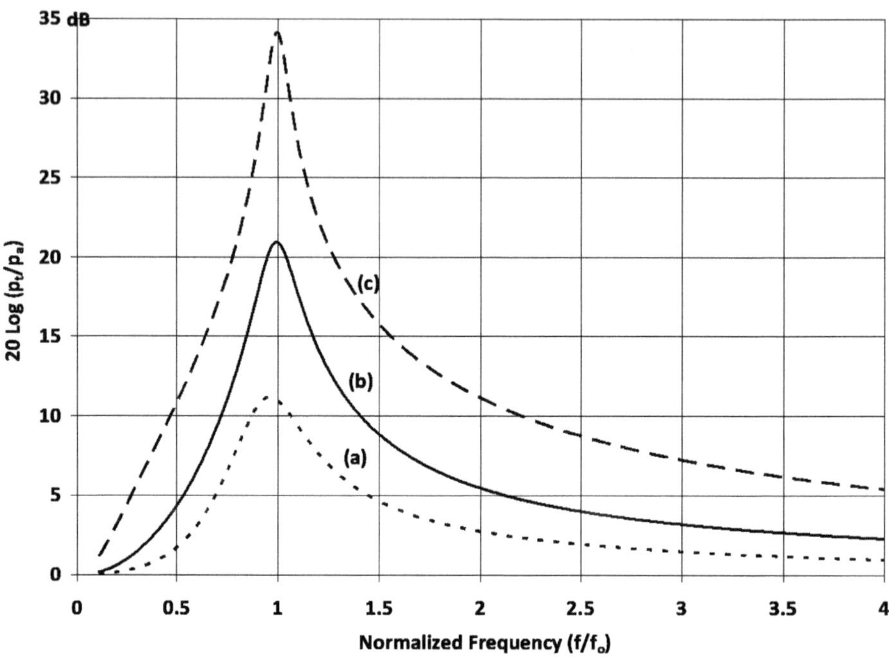

Fig. 10.21 On axis ($\theta = 0$) normalized total pressure level 20log(p_t/p_a) as a function of f/f_o for cases: (*a*) (*dotted line*) with $k_od = 0.2$, $\eta_o = 0.1$, and $Q_o = 2.5$, (*b*) (*solid line*) with $k_od = 0.1$, $\eta_o = 0.1$, and $Q_o = 5.0$, (*c*) (*dashed line*) $k_od = 0.1$, $\eta_o = 0.5$, and $Q_o = 10$

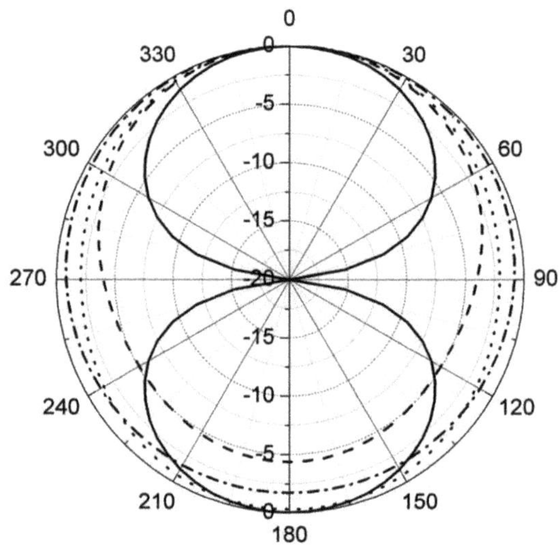

Fig. 10.22 Normalized beam patterns at $f/f_o = 0.5$ (*dotted line*) 1.0 (*dotted dashed line*), and 2.0 (*dashed line*) for case (*b*) of Fig. 10.21 with $k_od = 0.1$, $\eta_o = 0.1$, and $Q_o = 5.0$ which may be compared with the case of no parasitic resonator (*solid line*). The angle θ is measured from the axis of symmetry as illustrated in Fig. 10.20

greatest distance, $k_od = 0.2$, and lowest $Q_o = 2.5$ with an efficiency of $\eta_o = 10\,\%$. The best results are for case (c) with $k_od = 0.1$, $Q_o = 10$, and $\eta_o = 50\,\%$; that is, with a closer separation, higher Q_o and higher efficiency. Figure 10.22 shows the beam pattern results for case (b) of Fig. 10.21, at $f/f_o = 0.5$, 1.0, and

2.0. Various parasitic resonator designs have been considered by Butler et al. [36], with results that show predominately far-field omnidirectional radiation from a dipole-parasitic resonator-pair in the vicinity of the parasitic resonant frequency.

10.6 Summary

In this chapter the fundamentals of radiation from transducers were considered beginning with the Helmholtz differential equation in rectangular, cylindrical, and spherical coordinates. The simplest case of radiation from a uniformly vibration sphere was solved and presented.

Acoustic field solutions for the radiation from line sources, flat sources in a plain along with an extensive analysis of this important case were also developed. It was shown that that the −3 dB beam width from a circular piston of diameter, D, is $\mathrm{BW} \approx 58\lambda/D$ and the $\mathrm{DI} \approx 45\,\mathrm{dB} - 20\log\mathrm{BW}$ for $D > \lambda$. Near-field cavitation was also considered along with the pressure on the edge and at the center of both sides of a disc. The radiation impedance of a sphere operating in monopole, dipole, and quadrupole modes was developed along with the radiation impedance of a uniformly vibrating piston and the directivity index of line array of small sources or sensors.

Finally, spherical Hankel functions were used to develop the acoustic coupling between a dipole transducer and a monopole resonator. It was also shown that a resonate passive-monopole parasitic-element can improve the low frequency response of a dipole radiator and yield an omnidirectional pattern instead of dipole pattern, in the vicinity of the monopole resonance, as a result of mutual radiation coupling.

Exercises (Degree of difficulty: *lowest, **moderate, ***highest)

10.1.* Show that the plane wave representation for the pressure $p = p_0 e^{-j(kx-\omega t)}$ satisfies the one-dimensional Helmholtz wave equation $\partial^2 p/\partial x^2 + k^2 p = 0$. Determine the particle velocity $u = -(1/j\omega\rho)\partial p/\partial x$ and show that, in this case, the characteristic impedance $p/u = \rho c$. What is the characteristic impedance for Type I (PZT-4) piezoelectric ceramic and how does it compare with the value for water?

10.2.** Show the condition under which the first axial null occurs for a continuous line of length L. What is the condition for the first axial null for a line array of two point sources separated by distance s? Why the difference in lengths?

10.3.** Use the approximate formulas to calculate the DI and beam width for a line of length $L = \lambda$. Compare this result with the approximate result for a circular piston in an infinite rigid baffle with diameter $D = \lambda$. Why is the beam width larger and DI higher for a circular piston?

10.4.* Calculate the beam widths and DI for the line and piston in Exercise 10.3 using the exact expressions and compare with the approximate results.

10.5.*** Calculate the beam width for a vibrating ring on an infinite rigid cylinder of length $L = \lambda$ and diameter $D = \lambda$ and compare these results with that of a thin line of length $L = \lambda$. Note in Eq. (10.34) that L is half the length of the vibrating ring, while in Eq. (10.22) L is the whole length of the thin line. Also note that in Eq. (10.34) θ is measured from the axis of the cylinder, while in Eq. (10.22) $\alpha = \pi/2 - \theta$ is measured from the perpendicular to the line.

10.6.* Calculate the values of ka for which there is a pressure null at the center of a circular piston set in an infinite rigid baffle.

10.7.** Consider two spheres vibrating at the same frequency, one as a dipole source with radius a_d and the other as an omni source with radius a_o. Determine the approximate value of ka_o of the omni sphere in order for it to have the same value of radiation resistance as the dipole sphere under the conditions of both ka_o and $ka_d \ll 1$ and also for both ka_o and $ka_d \gg 1$. Why are the radii so different for $ka \ll 1$? Use appropriate approximations.

10.8.* Calculate the diffraction constant for a circular piston in an infinite rigid plane by using Eq. (6.56) with D_f and R_r from Eqs. (10.31a) and (10.52).

10.9.** Calculate the diffraction constant for a sphere vibrating in the dipole mode by using Eq. (6.56). Use R_r from Eq. (10.47) and D_f from the answer to Exercise 1.2 with $A = 0$ and $B = 1$. Find the maximum value of D_a in this case and the value of ka at which the maximum occurs.

10.10.*** An acoustic radiation problem is defined by stating the boundary condition on the normal velocity. In some cases the solution can be completed in terms of known solutions of the wave equation. In other cases a numerical finite element solution is the only practical approach. A clear statement of the boundary condition is the essential starting point for both approaches. Write out the normal velocity boundary condition for a free-flooding ring transducer and consider how it could be solved. Make a sketch showing the various vibrating surfaces in a cylindrical coordinate system.

References

1. L.F. Kinsler, A.R. Frey, A.B. Coppens, J.V. Sanders, *Fundamentals of Acoustics*, 4th edn. (Wiley, New York, 2000)
2. A.D. Pierce, *Acoustics- An Introduction to Its Physical Principles and Applications* (McGraw-Hill Book, New York, 1981)
3. D.T. Blackstock, *Fundamentals of Physical Acoustics* (Wiley, New York, 2000)
4. E. Skudrzyk, *The foundations of Acoustics* (Springer, New York, 1971)
5. L.L. Beranek, T.J. Mellow, *Acoustics: Sound Fields and Transducers* (Academic, Oxford, 2012)

6. R.S. Burington, *Handbook of Mathematical Tables and Formulas* (Handbook Publishers, Sandusky, 1940), p. 88
7. J.W. Horton, *Fundamentals of Sonar*, 2nd edn. (U. S. Naval Institute, Annapolis, 1959)
8. W.S. Burdic, *Underwater Acoustic Systems Analysis*, 2nd edn. (Prentice Hall, Englewood Cliffs, 1991)
9. J.W.S. Rayleigh, *The Theory of Sound*, vol. II (Dover Publications, New York, 1945)
10. P.M. Morse, H. Feshbach, *Methods of Theoretical Physics; Part II* (McGraw-Hill, New York, 1953), pp. 1322–1323
11. D.T. Porter, Self and mutual radiation impedance and beam patterns for flexural disks in a rigid plane. J. Acoust. Soc. Am. **36**, 1154–1161 (1964)
12. V. Mangulis, Acoustic radiation from a wobbling piston. J. Acoust. Soc. Am. **40**, 349–353 (1966)
13. F. Massa, Vibrational energy transmitter or receiver, Patent 2,427,062, 9 Sept 1947
14. J.L. Butler, A.L. Butler, J.A. Rice, A tri-modal directional transducer. J. Acoust. Soc. Am. **115**, 658–667 (2004)
15. D.T. Laird, H. Cohen, Directionality patterns for acoustic radiation from a source on a rigid cylinder. J. Acoust. Soc. Am. **24**, 46–49 (1952)
16. R.J. Urick, *Principles of Underwater Sound*, 3rd edn. (Peninsula, Los Altos Hills, 1983)
17. C.H. Sherman, Effect of the near field on the cavitation limit of transducers. J. Acoust. Soc. Am. **35**, 1409–1412 (1963)
18. H. Stenzel, *Leitfaden zur Berechnung von Schallvorgangen* (Springer, Berlin, 1939)
19. S.N. Rschevkin, *A Course of Lectures on the Theory of Sound* (Pergamon, Oxford, 1963)
20. A. Silbiger, Radiation from circular pistons of elliptical profile. J. Acoust. Soc. Am. **33**, 1515–1522 (1961)
21. J. E. Barger, Underwater acoustic projector, U. S. Patent 5,673,236, 30 Sept 1997
22. P.M. Morse, K.U. Ingard, *Theoretical Acoustics* (McGraw-Hill Book, New York, 1968), pp. 336–337
23. R.M. Aarts, A.J.E.M. Janssen, Approximation of the Struve function $\mathbf{H}_1$ occurring in impedance calculations. J. Acoust. Soc. Am. **113**, 2635–2637 (2003)
24. T. Nimura, Y. Watanabe, Vibrating circular disk with a finite baffle board. J. IEEE Japan. **68**, 263 (1948) (in Japanese). Results available in *Ultrasonic Transducers*, Ed. by Y. Kikuchi, Corona Pub. Co., Tokyo, 1969, p. 348
25. L.L. Beranek, *Acoustics* (McGraw-Hill Book Company, New York, 1954)
26. H. Levine, J. Schwinger, On the radiation of sound from an unflanged circular pipe. Phys. Rev. **73**, 383–406 (1948)
27. E.M. Arase, Mutual radiation impedance of square and rectangular pistons in a rigid infinite baffle. J. Acoust. Soc. Am. **36**, 1521–1525 (1964)
28. J. L. Butler, Self and Mutual Impedance for a Square Piston in a Rigid Baffle, Image Acoustics Report, Contract N66604-92-M-BW19, 20 Mar 1992
29. J.L. Butler, A.L. Butler, A Fourier series solution for the radiation impedance of a finite cylinder. J. Acoust. Soc. Am. **104**, 2773–2778 (1998)
30. D.H. Robey, On the radiation impedance of an array of finite cylinders. J. Acoust. Soc. Am. **27**, 706–710 (1955)
31. ANSYS, Inc., Canonsburg, PA
32. COMSOL, Burlington, MA
33. ATILA, MMech, State College, PA
34. TRN, Transducer Design and Array Analysis Program, NUWC, Newport, RI. Developed by M. Simon and K. Farnham with array analysis module based on the program ARRAY, by J. L. Butler, Image Acoustics, Inc., Cohasset, MA
35. K.F. Lee, *Principles of Antenna Theory* (Wiley, New York, 1984). Ch. 8
36. J.L. Butler, A.L. Butler, V. Curtis, Dipole transducer enhancement from a passive resonator. J. Acoust. Soc. Am. **135**, 2472–2477 (2014)

Chapter 11
Mathematical Models for Acoustic Radiation

This chapter will extend the results in Chap. 10 by using more advanced analytical methods for calculating acoustical quantities such as mutual radiation impedance. Before fast computers were available some of the results obtained by analytical methods had limited usefulness when they were expressed as slowly converging infinite series or integrals that required numerical evaluation. Now such series and integrals can be evaluated more easily. In some cases the analytical methods give more physical insight, or can be reduced to a simpler form, than the strictly numerical methods. Results for several useful cases obtained by analytical methods, and numerically evaluated, will be given in this chapter. However, the most advanced analytical methods cannot handle the geometries presented by practical transducers and arrays; in these cases finite element numerical methods are necessary.

The influence of fast computing increased rapidly from about 1960 when new numerical methods for calculating sound fields began to be developed, methods which have now grown into the large field called boundary element methods (BEM). These numerical methods, when combined with structural finite element analysis, have advanced to the point where it is feasible to include many structural details of a transducer or an array of transducers as well as the acoustics in the surrounding medium. An example of such calculated results for an array of 16 transducers, including the water loading, was given in Sect. 7.6. A brief description of some of these numerical methods will be included in this chapter. It is fortunate that excellent books on acoustical radiation are available [1–6], as well as, books on the boundary element method (BEM) [7, 8] and a document on transducers and arrays that includes Helmholtz integral methods, variational methods, and doubly asymptotic approximations [9].

J.L. Butler, C.H. Sherman, *Transducers and Arrays for Underwater Sound*,
Modern Acoustics and Signal Processing, DOI 10.1007/978-3-319-39044-4_11

11.1 Mutual Radiation Impedance

11.1.1 Piston Transducers on a Sphere

We will now use the general spherical coordinate solution in Eq. (10.14) to solve more complicated problems such as radiation from a piston on a rigid spherical surface. This is the first step in analyzing a spherical array of transducers, such as that often used in sonar (see Fig. 1.11). We will then take the next step and find the mutual radiation impedance between two pistons on a sphere that is needed to analyze a spherical array (see Chap. 7). We use the spherical coordinates (r, θ, ϕ) in Fig. 11.1, and sum the solution in Eq. (10.14) over all values of the integers m and n to obtain a general expression for the pressure:

$$p(r, \theta, \phi)\mathrm{e}^{j\omega t} = \sum_{n=0}^{\infty} \sum_{m=-n}^{n} A_{nm} P_n^m(\cos\theta)\mathrm{e}^{jm\phi} h_n^{(2)}(kr)\mathrm{e}^{j\omega t}. \tag{11.1}$$

This expression can be applied to any normal velocity distribution on a sphere by making the A_{nm} coefficients satisfy the velocity boundary condition. To analyze the case of a single circular piston on a sphere it is convenient to locate the spherical coordinate system such that the $\theta = 0$ direction passes through the center of the piston as shown in Fig. 11.1. The normal velocity on the surface of the rigid sphere can then be written as

$$u(\theta)\mathrm{e}^{j\omega t} = u_0 \mathrm{e}^{j\omega t} \text{ for } 0 < \theta < \theta_{0i}, \text{ and } 0 \text{ elsewhere}, \tag{11.2}$$

where u_0 is the uniform normal velocity amplitude of the piston, and θ_{0i} is the angular radius of the piston. Note that Eq. (11.2) describes a radially pulsating

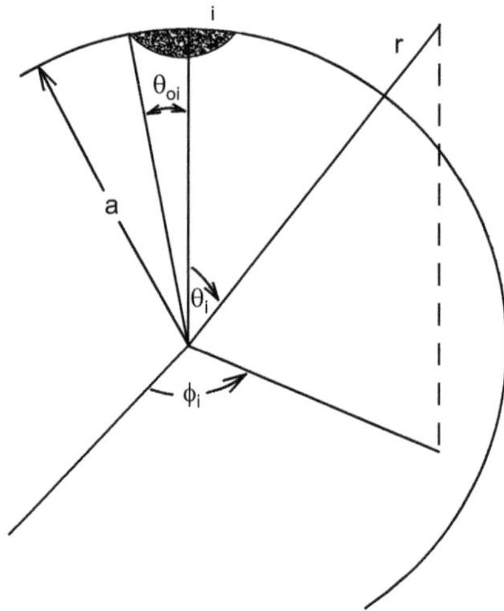

Fig. 11.1 Spherical coordinates for a circular piston on a sphere

piston with a curved surface, rather than the flat surface of most transducers. This is a geometrical approximation that must be made to keep the analysis simple, since a sphere with a flat area does not conform to a constant coordinate surface in any coordinate system. It is a valid approximation acoustically when the piston is small compared to the sphere, i.e., when $ka(1 - \cos\theta_{0i}) \ll 1$ for a sphere of radius a.

The boundary condition requires that the normal velocity in the water on the surface of the sphere be equal to the normal velocity of the sphere, $u(\theta)$, at every point on the spherical surface at $r = a$. Since the velocity distribution in Eq. (11.2) does not depend on ϕ, the radiated pressure field also has no ϕ dependence, and only the case $m = 0$ is needed in Eq. (11.1). Then we have, after omitting the time factor $e^{j\omega t}$,

$$p(r,\theta) = \sum_{n=0}^{\infty} A_n P_n(\cos\theta) h_n^{(2)}(kr), \tag{11.3a}$$

and

$$u(\theta) = -\frac{1}{j\omega\rho}\frac{\partial p}{\partial r}\bigg|_{r=a} = -\frac{k}{j\omega\rho}\sum_{n=0}^{\infty} A_n P_n(\cos\theta) h_n'(ka), \tag{11.3b}$$

where $h_n'(ka) = \partial h_n^{(2)}(x)/\partial x|_{x=ka}$. The A_n coefficients can be determined from Eq. (11.3b) by using the orthogonality of the Legendre polynomials [4] with the result

$$A_n = \frac{\rho c u_0}{2jh_n'(ka)}[P_{n-1}(\cos\theta_{0i}) - P_{n+1}(\cos\theta_{0i})]. \tag{11.4}$$

Substituting Eq. (11.4) into Eq. (11.3a) gives the pressure field at all points outside the sphere:

$$p(r,\theta) = \frac{\rho c u_0}{2j}\sum_{n=0}^{\infty}\left[\frac{P_{n-1}(\cos\theta_{0i}) - P_{n+1}(\cos\theta_{0i})}{h_n'(ka)}\right] P_n(\cos\theta) h_n^{(2)}(kr). \tag{11.5}$$

An interesting discussion of the shadow zone in the space on the backside of the sphere, where $\theta > 90°$, has been given by Junger and Feit [5] for small pistons.

When θ_{0i} is large this model no longer approximates a flat piston on a sphere; instead it represents a sphere with a radially pulsating portion. For $\theta_{0i} = \pi$, where the entire sphere vibrates with uniform normal velocity, these results reduce to the pulsating sphere in Chap. 10, and the coefficients are all zero except $A_0 = \rho c u_0 / jh_0'(ka)$. For $\theta_{0i} = \pi/2$ half the sphere pulsates while the other half is motionless, which could serve as a model for a transducer in a housing. For ka no greater than 2 or 3 only a few terms of the series are needed for a useful approximation.

The far field and the radiation impedance are two practical kinds of information that can be obtained from Eq. (11.5). In the far field where $kr \gg 1$ the spherical Hankel function becomes $(j^{n+1}/kr)\mathrm{e}^{-jkr}$, and Eq. (11.5) becomes

$$p(r,\theta) = \frac{\rho c u_0}{2kr} \mathrm{e}^{-jkr} \sum_{n=0}^{\infty} \left[j^n \frac{P_{n-1}(\cos\theta_{0i}) - P_{n+1}(\cos\theta_{0i})}{h_n'(ka)} \right] P_n(\cos\theta). \quad (11.6)$$

Although this result now shows the usual far-field radial dependence, the far-field directionality can only be found by numerical calculation. For the case of a large spherical array of projectors, where each transducer is small compared to the sphere, the value of ka is typically 20–30, and a large number of terms (equal to more than the value of ka) are required for a good approximation to the series. For very large ka it is possible to use asymptotic methods to approximate infinite series of this type [5]. For example, the Watson transformation converts the series to a residue series that converges in a small number of terms [10].

In an array the mutual impedances between all the pairs of transducers must be calculated for a complete analysis of the array performance as discussed in Chap. 7. For a spherical array we consider another circular piston located on the same sphere at a position given by the angles α_{ij} and β_{ij} as shown in Fig. 11.2.

The mutual radiation impedance between piston transducers was defined in Eq. (7.17) as the ratio of the acoustic force exerted on the jth piston by the ith piston to the velocity amplitude of the ith piston, which in this case can be written as

$$Z_{ij} = \frac{1}{u_i} \int_0^{2\pi} \int_0^{\theta_{0j}} p_i(a,\theta_i) a^2 \sin\theta_j \mathrm{d}\theta_j \mathrm{d}\phi_j, \quad (11.7)$$

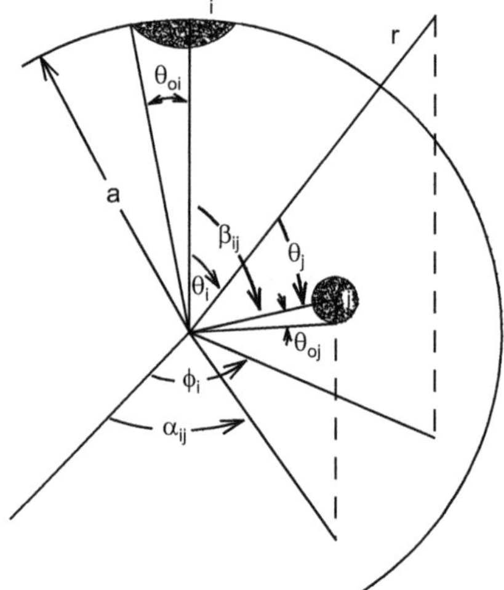

Fig. 11.2 Spherical coordinates for two circular pistons on a sphere [11]

where θ_{0j} is the angular radius of the jth piston. The mutual impedance will be calculated for pistons of different sizes, although in most arrays all transducers have the same size. The integral for Z_{ij} involves two different spherical coordinate systems, the original one used to express p_i in Eq. (11.5) and another rotated system with its polar axis passing through the center of the jth piston. Since the integration is over the surface of the jth piston, the integration variables refer to the jth coordinate system, and $p_i(a, \theta_i)$ must be transformed to a function of θ_j and ϕ_j to evaluate the integral.

The angles in the two coordinate systems are related by

$$\cos\theta_i = \cos\theta_j \cos\beta_{ij} + \sin\theta_j \sin\beta_{ij} \cos\left(\phi_j - \alpha_{ij}\right), \tag{11.8}$$

and the spherical harmonic addition theorem [3] gives the needed transformation:

$$\begin{aligned} P_n(\cos\theta_i) &= P_n(\cos\theta_j)P_n(\cos\beta_{ij}) \\ &\quad + 2\sum_{m=1}^{n} \frac{(n-m)!}{(n+m)!} P_n^m(\cos\theta_j)P_n^m(\cos\beta_{ij}) \cos m(\phi_j - \alpha_{ij}). \end{aligned} \tag{11.9}$$

Note that Eq. (11.8) follows from Eq. (11.9) for $n = 1$. Combining Eqs. (11.9), (11.5), and (11.7) and integrating gives the mutual radiation impedance between two circular pistons of different sizes separated by the angle β_{ij} on the surface of a rigid sphere:

$$\begin{aligned} Z_{ij} &= \rho c \pi a^2 \sum_{n=0}^{\infty} \frac{1}{2n+1} [P_{n-1}(\cos\theta_{0i}) - P_{n+1}(\cos\theta_{0i})] \\ &\quad \times \left[P_{n-1}(\cos\theta_{0j}) - P_{n+1}(\cos\theta_{0j})\right] \frac{h_n^{(2)}(ka)}{jh_n'(ka)} P_n(\cos\beta_{ij}). \end{aligned} \tag{11.10}$$

The self-radiation impedance of a circular piston on a rigid sphere is given by Eq. (11.10) for $\beta_{ij} = 0$ and $\theta_{0j} = \theta_{0i}$:

$$Z_{ii} = \rho c \pi a^2 \sum_{n=0}^{\infty} \frac{1}{2n+1} [P_{n-1}(\cos\theta_{0i}) - P_{n+1}(\cos\theta_{0i})]^2 \frac{h_n^{(2)}(ka)}{jh_n'(ka)}. \tag{11.11}$$

which could also be obtained directly by integrating Eq. (11.5) over the surface of the piston. Equation (11.10) was used to calculate mutual radiation impedance values [11] for comparison with plane and cylindrical cases in Fig. 7.15. As might be expected, for small separations the mutual impedance for small pistons on large baffles is essentially the same for all baffle shapes. Although the transducers in the spherical array in Fig. 1.11 are rectangular, Eqs. (11.10) and (11.11) can be used as approximations after replacing the rectangles by circles of the same area. The case of rectangular pistons on a sphere has been formulated, but numerical results have not been obtained [11].

Rectangular, circular cylinder, and spherical coordinates are the most commonly used coordinate systems for acoustical problems because the special functions involved are familiar, but there are eight other coordinate systems in which the wave equation can be solved by separation of variables [3]. Useful results have been obtained by use of some of them, e.g., Silbiger [12] and Nimura and Watanabe [13] used oblate spheroidal coordinates to calculate the sound field and the radiation impedance of an unbaffled disc vibrating as a rigid body. This is a solvable separation of variables problem because the limiting form of the oblate spheroidal constant coordinate surfaces is an infinitesimally thin disc. This case has special significance because it can be combined with the well-known case of the circular piston in an infinite rigid plane to obtain results for the disc vibrating on one side, as discussed in Sect. 10.3.3. Boisvert and Van Buren [14, 15] calculated the sound fields and radiation impedances of rectangular pistons on prolate spheroids using prolate spheroidal coordinates. The prolate spheroidal models are more closely related to the shape of various underwater vehicles than those in any other coordinate system. McLachlan discussed the application of elliptic cylinder coordinates and Mathieu functions to acoustic radiation problems [16], and Boisvert and Van Buren have calculated the radiation impedance of rectangular pistons on infinite elliptic cylinders [17].

Although the Helmholtz differential equation cannot be solved by separation of variables in the toroidal coordinate system, Weston developed a similar method of solution for this coordinate system [18–20]. It has been applied to calculating the sound field of free-flooding ring transducers, resulting in a toroidal model applicable to short, thin-walled rings that was used to calculate far-field directivity patterns for comparison with measured patterns [21].

11.1.2 Piston Transducers on a Cylinder

In spherical coordinates the constant coordinate surface $r = a$ is a finite surface which makes this coordinate system especially suitable for describing practical problems. Part of the usefulness of spheroidal coordinates also depends on their finite constant coordinate surfaces, and especially the limiting cases, i.e., a thin disc in the oblate system and a thin line of finite length in the prolate system. In rectangular coordinates the constant coordinate surfaces are infinite planes and in circular cylinder coordinates the constant z surfaces are infinite planes while constant r surfaces are infinitely long cylinders. The boundary conditions for practical problems on these infinite surfaces are usually not periodic, and integral transform methods are needed for solving the Helmholtz equation. In this section we will illustrate the integral transform method by using Fourier transforms to calculate the self and mutual radiation impedance for piston transducers on a rigid cylinder.

Consider piston transducers modeled as small, rectangular portions of an infinitely long cylindrical surface of radius a as shown in Fig. 11.3 (also see Fig. 1.10).

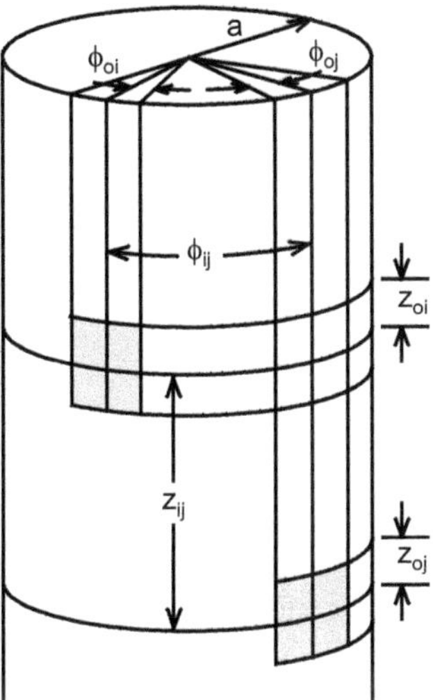

Fig. 11.3 Coordinates for two rectangular pistons on a cylinder [23]

As in the case of pistons on a sphere these pistons are not quite flat, but the departure from flatness is insignificant acoustically if the pistons are small compared to the cylinder radius. If the piston is located in a cylindrical coordinate system (r, ϕ, z) with its center at the point $r = a$, $\phi = 0$, $z = 0$ and is vibrating with velocity $u_0 e^{j\omega t}$, the velocity amplitude distribution on the surface of the cylinder is

$$u(\phi, z) = u_0 \text{ for } (-\phi_0 < \phi < \phi_0 \text{ and } -z_0 < z < z_0) \text{ and } 0, \text{ elsewhere.} \tag{11.12}$$

Since the z dependence of this velocity distribution is not periodic, and the range of z is from $-\infty$ to $+\infty$, Fourier transforms (see Sect. 13.11) provide an appropriate method of solution. The Fourier transform of the pressure, $p(r, \phi, z)$, is defined by

$$\overline{p}(r, \phi, \alpha) = \frac{1}{2\pi} \int_{-\infty}^{\infty} p(r, \phi, z) e^{-j\alpha z} dz, \tag{11.13a}$$

with the inverse transform given by

$$p(r, \phi, z) = \int_{-\infty}^{\infty} \overline{p}(r, \phi, \alpha) e^{j\alpha z} dz. \tag{11.13b}$$

We begin the solution with the Helmholtz differential equation in cylindrical coordinates given in Eq. (10.4), and repeated here:

$$\frac{1}{r}\frac{\partial}{\partial r}\left(\frac{r\partial p}{\partial r}\right)+\frac{1}{r^2}\frac{\partial^2 p}{\partial \phi^2}+\frac{\partial^2 p}{\partial z^2}+k^2 p=0. \tag{11.14a}$$

The Fourier transform of this equation is obtained by multiplying by $e^{-j\alpha z}/2\pi$, integrating each term from ∞ to $-\infty$, evaluating the term involving $\partial^2 p/\partial z^2$ by integrating by parts twice and assuming that p and $\partial p/\partial z$ vanish at infinity. The result is

$$\frac{1}{r}\frac{\partial}{\partial r}\left(\frac{r\partial \overline{p}}{\partial r}\right)+\frac{1}{r^2}\frac{\partial^2 \overline{p}}{\partial \phi^2}+(k^2-\alpha^2)\overline{p}=0, \tag{11.14b}$$

showing that the variable z has been eliminated. Since the ϕ dependence of the velocity distribution in Eq. (11.12) is periodic, it can be expanded in a Fourier series, and the boundary condition on the normal velocity can be written as

$$u(\phi,z)=-\frac{1}{j\omega\rho}\frac{\partial p}{\partial r}\bigg|_{r=a}=u_0\sum_{m=0}^{\infty}\frac{\epsilon_m \sin m\phi_0}{m\pi}\cos m\phi,\ \ |z|\leq z_0, \tag{11.15}$$

and $u(\phi,z) = 0$ for $|z| > z_0$, where $\epsilon_0 = 1$, $\epsilon_m = 2$ for $m > 0$, and $2a\phi_0$ and $2z_0$ are the piston dimensions as shown in Fig. 11.3. Noting that Eq. (11.15) is an even function of z, and taking its transform gives the boundary condition that $\overline{p}$ must satisfy:

$$\overline{u}(\phi,\alpha)=\frac{1}{j\omega\rho}\frac{\partial \overline{p}}{\partial r}\bigg|_{r=a}=\frac{\sin\alpha z_0}{\alpha\pi}u_0\sum_{m=0}^{\infty}\frac{\epsilon_m \sin m\phi_0}{m\pi}\cos m\phi. \tag{11.16}$$

The solution of Eq. (11.14b) that satisfies the radiation condition is obtained from Eq. (10.13):

$$\overline{p}(r,\phi,\alpha)=\sum_{m=0}^{\infty}A_m(\alpha)\cos m\phi\, H_m^{(2)}(\beta r), \tag{11.17}$$

where $\beta=\left(k^2-\alpha^2\right)^{1/2}$. The coefficients $A_m(\alpha)$ are evaluated by making Eq. (11.17) satisfy the boundary condition in Eq. (11.16) with the result

$$A_m(\alpha)=\frac{2\epsilon_m j\omega\rho u_0 \sin\alpha z_0 \sin m\phi_0}{m\pi^2\alpha\beta\left[H_{m-1}^{(2)}(\beta a)-H_{m+1}^{(2)}(\beta a)\right]}.$$

With this expression for $A_m(\alpha)$ the inverse transform of Eq. (11.17) gives the solution for the pressure at any point outside the cylinder:

$$p(r,\phi,z)=\int_{-\infty}^{\infty}\sum_{m=0}^{\infty}A_m(\alpha)\cos m\phi\, H_m^{(2)}(\beta r)e^{j\alpha z}d\alpha. \tag{11.18}$$

A form of Eq. (11.18) was first used by Laird and Cohen to calculate far-field patterns of rectangular piston radiators on an infinitely long rigid cylinder [22] (see Sect. 10.2.3).

The mutual impedance can be calculated by integrating the pressure caused by the ith piston over the surface of the jth piston:

$$Z_{ij} = \frac{1}{u_{0i}} \int_{-z_{0j}}^{z_{0j}} \int_{-\phi_j}^{\phi_{0j}} p_i(a, \phi_i\, z_i) a \, d\phi_j \, dz_j, \tag{11.19}$$

where the jth piston may differ from the ith piston in size, and the integration variables are expressed in a different cylindrical coordinate system in which the center of the jth piston is at the point (a,0,0). If the two pistons are separated in height by z_{ij} and in azimuth by ϕ_{ij}, as shown in Fig. 11.3, the relationships between the variables are

$$\phi_i = \phi_j + \phi_{ij} \quad \text{and} \quad z_i = z_j - z_{ij},$$

and the integrals in Eq. (11.19) can be evaluated with the result:

$$Z_{ij} = \frac{16ja\omega\rho}{\pi^2} \sum_{m=0}^{\infty} \frac{\in_m \sin m\phi_{0i} \sin m\phi_{0j} \cos m\phi_{ij}}{m^2} \times \int_0^{\infty} \frac{H_m^{(2)}(\beta a) \sin \alpha z_{0i} \sin \alpha z_{0j} \cos \alpha z_{ij}}{\alpha^2 \beta \left[H_{m-1}^{(2)}(\beta a) - H_{m+1}^{(2)}(\beta a)\right]} d\alpha. \tag{11.20a}$$

Numerical integration of Eq. (11.20a) has been used to evaluate the mutual impedance for several specific cases [23] (see Figs. 7.15 and 7.16).

The mutual impedance between two rectangular pistons on a cylinder reduces to the self-impedance by setting $\phi_{0i} = \phi_{0j}$, $z_{0i} = z_{0j}$ and $z_{ij} = \phi_{ij} = 0$. For $\phi_{0i} = \phi_{0j} = \pi$ Eq. (11.20a) gives the mutual impedance between two vibrating rings that completely encircle the cylinder. In that case only the term for $m = 0$ is non-zero, and we have

$$Z_{ij}(\text{rings}) = -8ja\omega\rho \int_0^{\infty} \frac{H_0^{(2)}(\beta a) \sin \alpha z_{0i} \sin \alpha z_{0j} \cos \alpha z_{ij}}{\alpha^2 \beta H_1^{(2)}(\beta a)} d\alpha. \tag{11.20b}$$

When the two rings have the same height ($z_{0i} = z_{0j}$) Eq. (11.20b) agrees with the result given by Robey [24]. The self-radiation impedance for a single ring on an infinite cylinder is given for $z_{ij} = 0$ and $z_{0i} = z_{0j}$. Equation (11.20a) also reduces to the mutual impedance per unit length between two infinite strips on a cylinder:

$$Z_{ij}(\text{strips}) = \frac{8ja\omega\rho}{\pi k}\sum_{m=0}^{\infty}\frac{\in_m \sin m\phi_{0i}\sin m\phi_{0j}\cos m\phi_{ij}H_m^{(2)}(ka)}{m^2\left[H_{m-1}^{(2)}(ka) - H_{m+1}^{(2)}(ka)\right]}. \quad (11.20c)$$

This result is obtained by dividing Eq. (11.20a) by the length of a strip, $2z_{0j}$, and then letting $z_{0i} = z_{0j} \rightarrow \infty$ and evaluating the integral by using [23]

$$\lim_{z_{0j}\rightarrow\infty}\frac{\sin^2 \alpha z_{0j}}{\pi\alpha^2 z_{0j}} = \delta(\alpha).$$

The form of the results for transducers mounted on a sphere or cylinder differ mainly in the special functions involved. In both cases extensive numerical calculations are required to achieve adequate convergence when summing infinite series or integrating numerically, and in the cylinder case the results still apply to an infinitely long cylindrical baffle. This situation has been a strong motivation for development of numerical methods that can handle objects with finite dimensions and more realistic shapes. It would be very useful, for example, to do calculations for transducers on a cylinder of finite length, since that would be a good approximation to many sonar arrays. Numerical methods that can handle such cases will be briefly described later in this chapter.

Note that the above solution for the mutual impedance between pistons on a cylinder involves both a Fourier transform and a Fourier series, and that the case of the rings on the cylinder is the $m = 0$ mode of the Fourier series. It is also possible to determine the radiation impedance of one ring on a cylinder in terms of series only [25]. For this purpose imagine vibrating rings periodically placed along the entire length of an infinite cylinder and separated by a distance, d, sufficient to make the acoustic coupling between them insignificant (see Fig. 11.4).

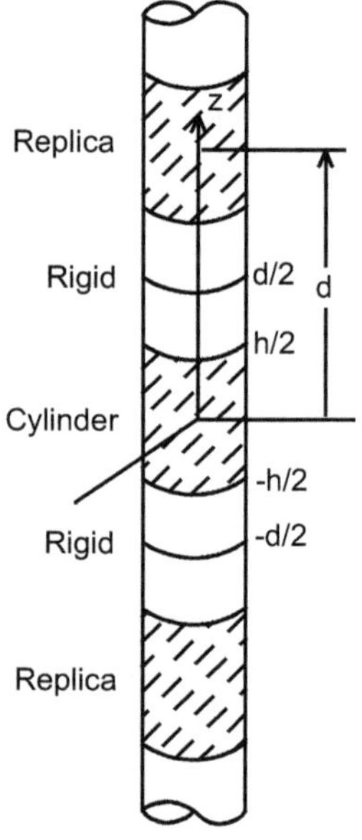

Fig. 11.4 The actual cylinder and the replicated cylinders with period d used in the fourier series solution [25]

The self-radiation impedance of each ring can then be obtained by integrating the pressure over the surface of the ring. It is natural in this approach to determine the radiation impedances for each angular mode, which are the values needed for dealing with transducers such as the multimode cylinder (see Chaps. 5 and 6). Note that results obtained from this model can be considered to apply to an individual vibrating ring only for near-field quantities such as radiation impedance, because the far field consists of contributions from all the rings on the infinitely long cylinder. The general approach used here, which may be worth considering in other cases, replaces a given problem with another, easier to solve, problem that has features that are approximately equal to the features of interest in the original problem.

The final result given by this approach for the modal radiation impedance of the ring is [25]:

$$Z_n = \left(-j\pi a h^2 \rho\omega\delta_n/d\right)\sum_{m=0}^{\infty} \in_m \text{Sinc}^2(\alpha_m h/2)\frac{H_n^{(2)}(\beta_m a)}{\beta_m H_n^{'}(\beta_m a)}, \tag{11.21}$$

where $\beta_m^2 = k^2 - \alpha_m^2$, $\alpha_m = 2\pi m/d$; $\in_0 = 1$, $\in_m = 2$ for $m > 0$, $\delta_0 = 2$, $\delta_n = 1$ for $n > 0$; a is the radius of the cylinder, h is the height of the ring, and d is the replication period. The $n = 0$ mode corresponds to uniform velocity around the cylinder and an omnidirectional sound field in the plane $z = 0$, while $n = 1$ is the dipole mode with both the velocity distribution on the surface, and the sound field in the plane $z = 0$, proportional to $\cos\phi$. Figures 11.5, 11.6, 11.7 and 11.8 show the normalized radiation resistance and reactance calculated from Eq. (11.21) as a function of ka for different values of $h/2a$ for the uniform and dipole modes [25].

The self and mutual radiation impedance between rectangular pistons in an infinite rigid plane [26] has also been calculated by use of Fourier transforms.

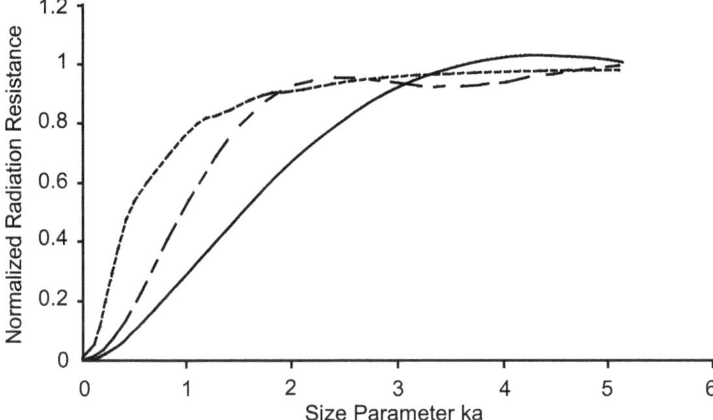

Fig. 11.5 Normalized radiation resistance for uniform motion of the cylinder, mode $n = 0$, for $h/2a = 0.5$ (*solid line*), 1.0 (*dash line*), and 5.0 (*dotted line*) [25]

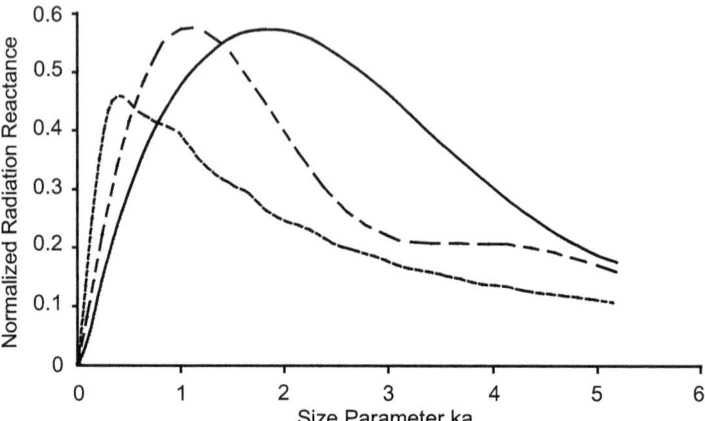

Fig. 11.6 Normalized radiation reactance for uniform motion of the cylinder, mode $n = 0$, for $h/2a = 0.5$ (*solid line*), 1.0 (*dash line*), and 5.0 (*dotted line*) [25]

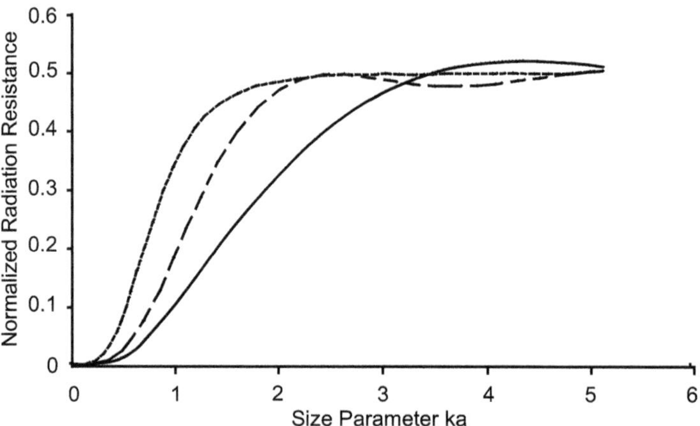

Fig. 11.7 Normalized radiation resistance for dipole motion of the cylinder, mode $n = 1$, for $h/2a = 0.5$ (*solid line*), 1.0 (*dash line*), and 5.0 (*dotted line*) [25]

11.1.3 Hankel Transform

We can illustrate the use of a different integral transform, the Hankel transform (see Sect. 13.11), by the familiar problem of a circular piston in an infinite rigid plane (see Chap. 10). Consider a cylindrical coordinate system with the piston in the plane $z = 0$ and the origin at the center of the piston. The boundary condition on the normal velocity on the plane $z = 0$ is $u(r) = u_0$ for $r \leq a$, the radius of the piston, and $u(r) = 0$ elsewhere. The symmetry of the circular piston gives a sound field

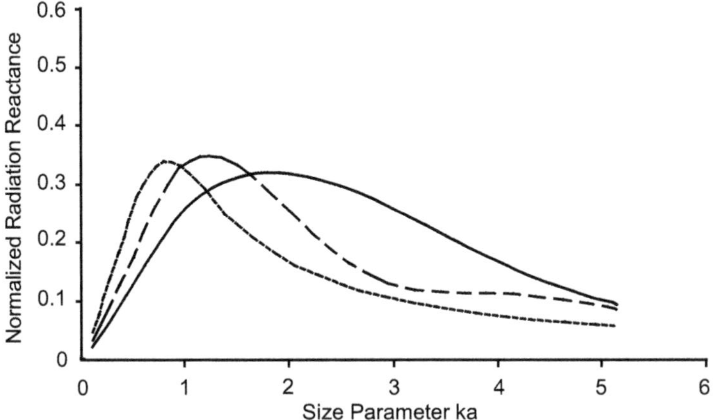

Fig. 11.8 Normalized radiation reactance for dipole motion of the cylinder, mode $n = 1$, for $h/2a = 0.5$ (*solid line*), 1.0 (*dash line*), and 5.0 (*dotted line*) [25]

with no ϕ dependence, and, from Eq. (11.14a), the Helmholtz differential equation reduces to

$$\frac{1}{r}\frac{\partial}{\partial r}\left(r\frac{\partial p}{\partial r}\right) + \frac{\partial^2 p}{\partial z^2} + k^2 p = 0. \tag{11.22}$$

The Hankel transform of $p(r,z)$ is defined by

$$\overline{p}(\gamma, z) = \int_0^\infty p(r, z) J_0(\gamma r) r \mathrm{d}r, \tag{11.23}$$

and, when Eq. (11.22) is multiplied by $rJ_0(\gamma r)$ and integrated over r, it becomes [27]

$$\frac{\partial^2 \overline{p}}{\partial z^2} + \left(k^2 - \gamma^2\right)\overline{p} = 0. \tag{11.24}$$

The solution of Eq. (11.24) is of the form

$$\overline{p}(\gamma, z) = A(\gamma)\mathrm{e}^{-j\beta z}, \tag{11.25}$$

where $\beta = \left(k^2 - \gamma^2\right)^{1/2}$, and $A(\gamma)$ is determined by the boundary condition described above. The Hankel transform of the boundary condition is [28]

$$\overline{u}(\gamma) = u_0 \int_0^a J_0(\gamma r) r \mathrm{d}r = u_0 a J_1(\gamma a)/\gamma = -\frac{1}{j\omega\rho}\frac{\partial \overline{p}(\gamma, z)}{\partial z}\bigg|_{z=0} = \frac{\beta A(\gamma)}{\omega\rho},$$

which can be solved for $A(\gamma)$ and substituted into Eq. (11.25) to give

$$\overline{p}(\gamma, z) = \frac{\omega\rho u_0 a J_1(\gamma a)}{\gamma\beta} \mathrm{e}^{-j\beta z}. \tag{11.26}$$

The solution for the pressure is then given by the inverse transform:

$$p(r, z) = \int_0^\infty \overline{p}(\gamma, z) J_0(\gamma r)\gamma \mathrm{d}\gamma = \omega\rho u_0 a \int_0^\infty \frac{J_1(\gamma a) J_0(\gamma r)}{\beta} \mathrm{e}^{-j\beta z} \mathrm{d}\gamma. \tag{11.27}$$

This integral can be evaluated on the axis of the piston ($r=0$) or used on the surface of the piston ($z=0$) to calculate the radiation impedance [5] as was done in Chap. 10 starting from Rayleigh's Integral.

11.1.4 Hilbert Transform

The radiation impedance is defined in terms of the reaction force exerted on a vibrator by the sound field that it produces; thus calculation of radiation impedance appears to require determination of the near field. We have also often used the fact that the radiation resistance is related to the far field through the total radiated power, and, therefore, that the resistance can be calculated from the far field. We will now discuss the point, noted in Sect. 7.3, that the radiation reactance can be determined from the radiation resistance, which therefore makes it possible to determine the complete impedance from the far field. As shown by Mangulis [29], this approach is based on the Kramers–Kronig relations which depend on the Hilbert transform (see Sect. 13.11). It is also closely related to the result by Bouwkamp [30] that the radiation impedance can be calculated by integrating the far-field directivity function over complex angles.

The general relationships that result from the Hilbert transform when applied to mechanical impedance, where $Z(-\omega) = Z^*(\omega)$ usually holds, are given by Morse and Feshbach [3] as follows:

$$R(\omega) = \frac{2}{\pi} \int_0^\infty \frac{xX(x) - \omega X(\omega)}{(x^2 - \omega^2)} \mathrm{d}x; \quad X(\omega) = \frac{2\omega}{\pi} \int_0^\infty \frac{R(x) - R(\omega)}{(x^2 - \omega^2)} \mathrm{d}x, \tag{11.28}$$

where the mechanical impedance, $Z(\omega) = R(\omega) + jX(\omega)$, is defined as the ratio of force to velocity when both are expressed in complex exponential form. R and X are the Hilbert transforms of each other with frequency as the complex variable.

Integrals of the type in Eq. (11.28) can be evaluated by the calculus of residues, as we will illustrate by calculating the reactance from the resistance for the monopole mode of a sphere. The resistance for this case was given in Eq. (10.44) for a pulsating sphere of radius a, and in normalized form is

$$R(\omega) = \frac{(ka)^2}{1+(ka)^2} = \frac{\omega^2}{\omega^2+\alpha^2} = \frac{\omega^2}{(\omega+j\alpha)(\omega-j\alpha)},$$

where $\alpha = c/a$. From Eq. (11.28) we have

$$X(\omega) = \frac{\omega}{\pi}\int_{-\infty}^{\infty} \frac{[x^2/(x^2+\alpha^2) - \omega^2/(\omega^2+\alpha^2)]}{(x^2-\omega^2)}\mathrm{d}x.$$

This integral can be separated into two integrals, of which the second is zero because it has no singularities in the upper half of the complex plane. The first integral has one singularity in the upper half plane on the imaginary axis at $j\alpha$, and the value of the integral is

$$\begin{aligned} X(\omega) &= \frac{\omega}{\pi}\int_{-\infty}^{\infty} \frac{x^2\mathrm{d}x}{(x+j\alpha)(x-j\alpha)(x^2-\omega^2)} = \left(\frac{\omega}{\pi}\right)(2\pi j)\left[\frac{1}{2j\alpha}\frac{(j\alpha)^2}{\left\{(j\alpha)^2-\omega^2\right\}}\right] \\ &= \frac{\omega}{\pi}\frac{\pi}{\alpha}\frac{\alpha^2}{(\alpha^2+\omega^2)} = \frac{\omega\alpha}{(\alpha^2+\omega^2)} = \frac{ka}{1+(ka)^2}, \end{aligned} \tag{11.29}$$

which is the normalized reactance as given by Eq. (10.44). This example illustrates the fact that the radiation reactance of a sound source can be determined from its radiation resistance and thus from its far field. Since it is almost always easier to calculate the far field than the near field, this approach is sometimes the best way to determine the radiation impedance, and it has often been used [31–34].

11.2 Green's Theorem and Acoustic Reciprocity

11.2.1 Green's Theorem

In 1828 the mostly self-educated George Green of Nottingham, England, published an essay on what we now call Green's theorem and Green's functions, which have found important applications in many fields of physics and engineering including acoustics [35]. We will indicate the mathematical basis for Green's theorem and the acoustic reciprocity theorem, then define Green's functions and show how they can be used to formulate solutions of acoustic radiation problems in a different way from those above.

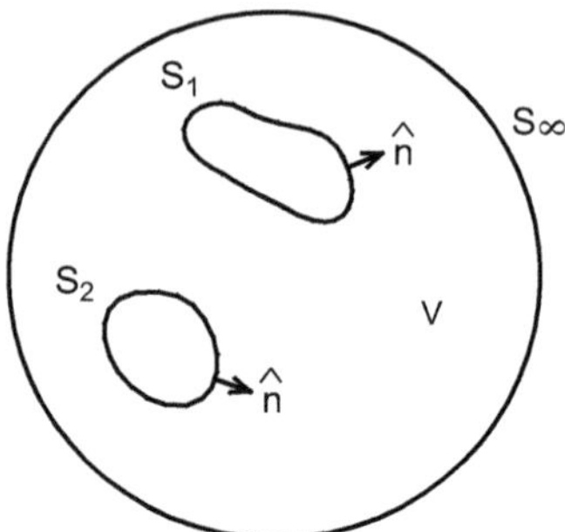

Fig. 11.9 The volume, V, the surface at infinity, $S\,\infty$, and two other surfaces, S_1 and S_2, with unit normal vectors, $\hat{n}$, used in discussing Green's Theorem

Consider two scalar functions of position, p_1 and p_2, defined in a closed volume, V, with the surface, S, where S consists of a sphere at infinity, where p_1 and p_2 vanish, plus two or more closed surfaces, such as S_1 and S_2 in Fig. 11.9.

Green's theorem is concerned with relationships between p_1 and p_2. Applying the divergence theorem of vector analysis to the vector formed by $p_1 \vec{\nabla} p_2$ gives:

$$\iint p_1 \vec{\nabla} p_2 \cdot \hat{n}\, \mathrm{d}S = \iiint \left[p_1 \nabla^2 p_2 + \vec{\nabla} p_1 \cdot \vec{\nabla} p_2\right] \mathrm{d}V. \tag{11.30}$$

Equation (11.30) is the first form of Green's theorem where $\hat{n}$ is the unit vector normal to the surface, S. The presence of $\nabla^2 p_2$ suggests that a useful connection with the Helmholtz differential equation may be possible. Interchanging p_1 and p_2 in Eq. (11.30) gives another equation of the same form; when this equation is subtracted from Eq. (11.30), and $\vec{\nabla} p \cdot \hat{n} = \partial p/\partial n$ is used, we have,

$$\iint (p_1 \partial p_2/\partial n - p_2 \partial p_1/\partial n) \mathrm{d}S = \iiint (p_1 \nabla^2 p_2 - p_2 \nabla^2 p_1) \mathrm{d}V. \tag{11.31}$$

This is the second form of Green's theorem [36] that can be directly applied to the case where p_1 and p_2 represent the pressures of two sound fields radiated by two acoustic sources on surfaces such as S_1 and S_2 in Fig. 11.9 (moving with distributed normal velocities u_1 and u_2). In that case p_1 and p_2 satisfy

$$\nabla^2 p_1 + k_1^2 p_1 = 0 \quad \text{and} \quad \nabla^2 p_2 + k_2^2 p_2 = 0,$$

and the second form of Green's theorem shows that

$$\iint (p_1 \partial p_2/\partial n - p_2 \partial p_1/\partial n) \mathrm{d}S = \iiint p_1 p_2 (k_1^2 - k_2^2) \mathrm{d}V. \tag{11.32}$$

It is evident that the right-hand side of this equation equals zero when $k_1 = k_2$, i.e., when the two sound fields have the same frequency. Since the normal derivative of the pressure on the surface, $\partial p/\partial n$, is equal to $-j\omega\rho u$ where u is the normal component of velocity on the surface, Eq. (11.32) becomes

$$\iint_S [p_1 u_2 - p_2 u_1] \mathrm{d}S = 0. \tag{11.33}$$

The volume, V, may contain more than two closed surfaces with more than two acoustic sources, and Eq. (11.33) applies to any two of the sources on surfaces S_1 and S_2; therefore it can be written as

$$\iint_{S_1} p_2 u_1 \mathrm{d}S = \iint_{S_2} p_1 u_2 \mathrm{d}S, \tag{11.34}$$

since the integral over the sphere at infinity is zero.

11.2.2 Acoustic Reciprocity

Equations (11.33) and (11.34) are two forms of the acoustic reciprocity theorem [1]. Acoustic reciprocity is separate from the electromechanical reciprocity discussed in Sect. 1.3. It is shown in Sect. 9.5 that combination of the two reciprocities is the basis for the reciprocity calibration method that permits an important simplification of transducer measurements. The importance of acoustic reciprocity lies in its generality, since it holds for sources of any shape, with any separation between them, any normal velocity distribution on their surfaces, and with any arrangement of other sources or non-vibrating surfaces in the vicinity. It also holds when the two sources are two modes of vibration of the same transducer. In the latter case, if the two modes are orthogonal, both sides of Eq. (11.34) vanish.

The physical meaning of acoustic reciprocity is most easily seen from the special case of transducers with uniform velocity. Then u_1 and u_2 are constants and Eq. (11.34) can be written as

$$\frac{1}{u_2} \iint_{S_1} p_2 \mathrm{d}S_1 = \frac{1}{u_1} \iint_{S_2} p_1 \mathrm{d}S_2. \tag{11.35}$$

This result means that if the two transducers have the same velocity they exert the same acoustic force on each other no matter what their separation, orientation, or relative size and shape. Each side of Eq. (11.35) is the mutual radiation impedance between two piston transducers, and the equation shows the reciprocity of mutual impedance with $Z_{12} = Z_{21}$. Reciprocity means that for each pair of interacting transducers in an array only one value of mutual impedance is needed to analyze the array performance.

Since Eq. (11.34) holds for nonuniform transducer velocities, it can be seen that acoustic reciprocity is a statement about acoustic power, i.e., the product of pressure and velocity integrated over a surface, rather than acoustic force. The general definition of radiation impedance in Eq. (1.4a) and the mutual radiation impedance derived from it, such as Eq. (7.3.2), are also expressed in terms of acoustic power, which makes them consistent with acoustic reciprocity.

If the two transducers are small enough compared to the wavelength, the pressure produced by each is nearly constant over the surface of the other, and Eq. (11.34) becomes

$$p_{21}\iint_{S_1} u_1 \mathrm{d}S_1 \approx p_{12}\iint_{S_2} u_2 \mathrm{d}S_2, \tag{11.36}$$

where p_{21} is the pressure produced by transducer 2 at transducer 1, and p_{12} is the pressure produced by 1 at 2. Since the integrals are the source strengths of each transducer, this form of the acoustic reciprocity theorem shows that each transducer produces the same pressure per unit source strength at the location of the other.

11.2.3 Green's Function Solutions

In this section we will discuss the use of Green's functions for solving acoustic problems. This method is more general than the separation of variables method given in Chap. 10, since it does not require use of a specific coordinate system and it can be applied when the acoustic sources are either vibrating portions of a boundary or distributed through a volume (see Morse and Ingard, Sect. 7.1 [4] or Baker and Copson [36] for more general discussions). In acoustics the Green's function is usually defined as the spatial factor of the acoustic field at the point $\vec{r}$ produced by a point source of unit source strength at another point, $\vec{r}_0$ (see Fig. 11.10).

It follows from this definition that the Green's function, $G\left(\vec{r}, \vec{r}_0\right)$, is a solution of the inhomogeneous Helmholtz equation,

$$\nabla^2 G(\vec{r}, \vec{r}_0) + k^2 G(\vec{r}, \vec{r}_0) = \delta(\vec{r} - \vec{r}_0). \tag{11.37}$$

The function $\delta(\vec{r} - \vec{r}_0)$ is the three-dimensional Dirac delta function with units of m^{-3}, thus $G\left(\vec{r}, \vec{r}_0\right)$ has units of m^{-1}. The right side of Eq. (11.37) represents a source distribution with harmonic time dependence that has zero strength

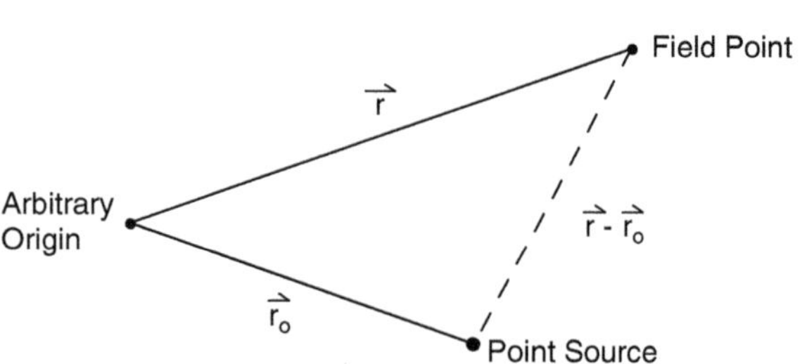

Fig. 11.10 Coordinates for defining Green's functions

everywhere except at the point $\vec{r}_0$, where it has unit integrated source strength. It's evident that the Green's function can also be thought of as a spatial impulse function.

In the case of an unbounded medium the field of a very small spherical source, given in Eq. (10.15a), with the nonspatial factors omitted and the source strength $Q = 1$, also satisfies the definition of a Green's function. In this special case, where no reflecting boundaries are present, we denote the Green's function by $g\left(\vec{r}, \vec{r}_0\right)$ and call it the free space Green's function. From Eq. (10.15a) we then get

$$g(\vec{r}, \vec{r}_0) = \mathrm{e}^{-jkR}/4\pi R, \tag{11.38a}$$

where $R = |\vec{r} - \vec{r}_0|$ is the distance between the field point ($\vec{r}$) and the source point ($\vec{r}_0$). Noting that $|\vec{r} - \vec{r}_0| = |\vec{r}_0 - \vec{r}|$, Eq. (11.38a) shows the reciprocal property of the free space Green's function

$$g(\vec{r}_0, \vec{r}) = g(\vec{r}, \vec{r}_0), \tag{11.38b}$$

which is also satisfied by the general Green's function, $G\left(\vec{r}, \vec{r}_0\right)$ [4].

In the general case, where boundaries are present, acoustic sources may exist in the volume of interest as well as on the boundaries of the volume. Then the equation of continuity, Eq. (10.1b), contains a source term, and, for harmonic time dependence, the wave equation for the pressure, $p\left(\vec{r}\right)$, becomes the inhomogeneous Helmholtz equation,

$$\nabla^2 p(\vec{r}) + k^2 p(\vec{r}) = j\omega\rho S_V\left(\vec{r}\right). \tag{11.38c}$$

To determine specific solutions boundary conditions must be specified on the pressure or its normal derivative depending on the sources on the boundaries. The right side of Eq. (11.38c) describes the sources distributed in the volume. The quantity ρ is the density of the medium and $S_V\left(\vec{r}\right)$ is the source strength density, i.e., the volume velocity per unit volume (units of m^3/s per m^3 or 1/s). The factor $j\omega\rho$ makes the right side consistent with pressure on the left side.

A general solution of Eq. (11.38c) can be expressed in terms of a Green's function that satisfies Eq. (11.37). To show this we multiply Eq. (11.37) by $p(\vec{r})$ and Eq. (11.38c) by $G(\vec{r}, \vec{r}_0)$, subtract the first from the second, interchange $\vec{r}$ and $\vec{r}_0$, utilize the reciprocity relation, and integrate over the volume in the source coordinates, $\vec{r}_0$. The integral over the delta function yields $p(\vec{r})$, and

$$\begin{aligned} p(\vec{r}) = &\iiint \left[- G(\vec{r}, \vec{r}_0)\nabla_0^2 p(\vec{r}_0) + p(\vec{r}_0)\nabla_0^2 G(\vec{r}, \vec{r}_0)\right] \mathrm{d}V_0 \\ &+ j\omega\rho \iiint S_V\left(\vec{r}_0\right) G\left(\vec{r}, \vec{r}_0\right) \mathrm{d}V_0. \end{aligned}$$

The first volume integral can be converted to a surface integral by use of the second form of Green's theorem, Eq. (11.31), resulting in

$$p(\vec{r}) = \iint \left[G(\vec{r}, \vec{r}_0) \frac{\partial p(\vec{r}_0)}{\partial n_0} + p(\vec{r}_0) \frac{\partial G(\vec{r}, \vec{r}_0)}{\partial n_0} \right] \mathrm{d}S_0 + j\omega\rho \iiint S_V(\vec{r}_0) G\left(\vec{r}, \vec{r}_0\right) \mathrm{d}V_0, \tag{11.39}$$

where the differential surface element, $\mathrm{d}S_0$, is expressed in the source coordinates. Equation (11.39) is the general form of the Green's function solution of the inhomogeneous Helmholtz equation. It gives the pressure at any point, $\vec{r}$, in terms of a Green's function plus the values of the pressure and its normal derivative on the boundary and the source strength density function within the volume. The usefulness of this solution depends on finding an appropriate Green's function for each problem.

The remainder of this section will be concerned with explaining and illustrating the use of Eq. (11.39). Since $(\partial p(\vec{r}_0)/\partial n_0) = -j\omega\rho u_n\left(\vec{r}_0\right)$, where $u_n\left(\vec{r}_0\right)$ is the normal velocity, the first term of the surface integral in Eq. (11.39) shows that any points on the boundary that are vibrating contribute a point source value to the pressure at the point $\vec{r}$. But this is not the total pressure at $\vec{r}$, because the second term of the surface integral also contributes from any boundary points where the pressure and the normal derivative of $G(\vec{r}, \vec{r}_0)$ are both non-zero. These are called double source contributions because the derivative is proportional to the difference of two point sources [36]. And finally the third term of Eq. (11.39) gives the contributions to the pressure from other sources distributed throughout the volume. When no boundaries are present only the volume integral contributes to the pressure. When there are sources only on the boundaries, the surface integral alone determines the pressure, which is the case for most transducer and array problems. When boundaries are present, but the only sources are within the volume, both integrals contribute to the pressure.

The physical idea underlying the Green's function solution is construction of the field $p(\vec{r})$ from the fields of point sources, somewhat similar to the approach used in Sect. 10.2, but now formulated in a much more general way. Perhaps the simplest example is a point source of unit source strength ($Q = 1\ \mathrm{m}^3/\mathrm{s}$) located at the origin of spherical coordinates with no boundaries. The surface integral is zero, $S_V\left(\vec{r}_0\right) = \delta\left(\vec{r}_0\right)$ and the pressure, $p\left(\vec{r}\right)$, is known from Eq. (10.15b) with $Q = 1\ \mathrm{m}^3/\mathrm{s}$. Thus we have

$$p\left(\vec{r}\right) = \frac{j\omega\rho}{4\pi r} \mathrm{e}^{-jkr} = j\omega\rho \iiint \delta(\vec{r}_0) g\left(\vec{r}, \vec{r}_0\right) \mathrm{d}V_0 = j\omega\rho g\left(\vec{r}, 0\right),$$

which is consistent with Eq. (11.38a).

As another example of the use of Eq. (11.39) consider the infinitesimally thin line source in a medium with no boundaries discussed in Sect. 10.2.1. Consider each section of the cylinder of radius a and length dz_0 to be a point source with uniform radial velocity u_0 and source strength $2\pi a u_0 \mathrm{d}z_0$. Since the volume element is $\mathrm{d}V_0 = \pi a^2 \mathrm{d}z_0$, the source strength density is $S_V = (2u_0/a)$ along the length of the line and zero elsewhere. Substituting these values and the free space Green's function, Eq. (11.38a), in the volume integral of Eq. (11.39) gives the result obtained somewhat intuitively in Eq. (10.21). Although a cylinder with a vibrating surface is obviously a boundary value problem that requires use of the surface integrals in Eq. (11.39), it appears that when it is infinitesimally thin, it can be treated as a volume distribution of point sources. Another example is the difference frequency parametric array discussed in Sect. 7.7, where the result was also obtained from the volume integral in Eq. (11.39). In that case the sources are distributed throughout the volume of the cylindrical region of finite radius in which the two primary beams interact.

When boundaries are present it is necessary to seek Green's functions that satisfy boundary conditions that will simplify the evaluation of Eq. (11.39). For example, when the pressure on the boundary is not known, it would be useful to have a Green's function that had a zero normal derivative on the boundary. Then the second term in Eq. (11.39) would be eliminated, leaving the solution in terms of known quantities. Such a Green's function for an infinite plane boundary was already used in Sect. 10.2.2, where we noted that the field of two in-phase point sources of unit source strength,

$$G(\vec{r}, \vec{r}_{01}, \vec{r}_{02}) = \frac{1}{4\pi R_1} \mathrm{e}^{-jkR_1} + \frac{1}{4\pi R_2} \mathrm{e}^{-jkR_2}, \tag{11.40}$$

has zero normal velocity at every point on the infinite plane perpendicular to and bisecting the line joining the two sources. Therefore, if this function is used as the Green's function, it will have exactly the property needed to simplify Eq. (11.39) by eliminating the second term of the surface integral for all problems involving acoustic sources on an infinite rigid plane. For source points on the plane, $R_1 = R_2 = R$, the Green's function in Eq. (11.40) becomes equal to $\mathrm{e}^{-jkR}/2\pi R$ and Eq. (11.39) reduces to

$$p(\vec{r}) = \frac{j\omega\rho}{2\pi} \iint u_n(\vec{r}_0) \frac{\mathrm{e}^{-jkR}}{R} \mathrm{d}S_0, \tag{11.41a}$$

which is Eq. (10.25a), Rayleigh's Integral, used in Chap. 10 to calculate radiation from circular and rectangular sources in a plane. In Chap. 10 we obtained Eq. (11.41a) intuitively; here it is proved by the Green's function method for finding solutions of the Helmholtz equation. The Rayleigh Integral has been the starting point for numerous acoustic calculations of great value to transducer development [26, 31, 32, 40].

The simplicity of the Green's function for radiators in an infinite plane appears to be unique. For example, a Green's function with zero normal derivative on any other simple surface, such as a rigid sphere or cylinder, is much more complicated and involves infinite series of the same special functions that occur in separation of variables solutions [4, 5]. An example of such a Green's function will appear in Sect. 11.3.2 where Eq. (11.62), with the factor ($\omega\rho Q/2$) removed, is $G\left(\vec{r}, \vec{r}_0\right)$ for an infinitely long, rigid cylindrical boundary.

Another simple Green's function, similar to Eq. (11.40), can be formed from the difference of two point sources. This function corresponds to zero pressure on an infinite plane midway between the sources, with a doubling of the normal derivative on that plane. This makes the first term of Eq. (11.39) vanish on that plane, and is applicable to acoustic sources in an infinite, soft plane such as an oscillating rigid disc. The resulting expression is

$$p\left(\vec{r}\right) = \frac{1}{2\pi}\iint p\left(\vec{r}_0\right)\frac{\partial}{\partial n_0}\left(\frac{\mathrm{e}^{-jkR}}{R}\right)\mathrm{d}S_0. \tag{11.41b}$$

Equation (11.41b) gives the pressure field of any planar oscillating source that requires only the pressure on the source surface for its evaluation at any point off the surface. The surface pressure can be computed by finite element modeling without requiring an extensive water field. In other cases it can be measured or expressed analytically as a power series and related to the specified velocity [37].

11.2.4 The Helmholtz Integral Formula

Any Green's function can be used in Eq. (11.39), but an important special case is the free space Green's function, Eq. (11.38a), with no distributed sources in the volume. We then have

$$p(\vec{r}) = \frac{1}{4\pi}\iint\left[\frac{\mathrm{e}^{-jkR}}{R}j\omega\rho u(\vec{r}_0) + p(\vec{r}_0)\frac{\partial}{\partial n_0}\left(\frac{\mathrm{e}^{-jkR}}{R}\right)\right]\mathrm{d}S_0, \tag{11.42a}$$

where $u(\vec{r}_0) = -(1/j\omega\rho)(\partial p/\partial n_0)$ is the normal velocity on the boundary. This form of the solution is called the Helmholtz Integral Formula [36, 39]. In most practical cases only one of the boundary functions is known, and that leaves Eq. (11.42a) as an integral equation for the other boundary function. For example, in the usual acoustics problem, where the velocity is known on the boundary but the pressure is not known, the field point can be moved to the boundary leaving an integral equation for the unknown pressure on the boundary. Solving the integral equation then provides the pressure boundary values needed for finding the pressure at any point outside the boundary. This approach is the basis for most of the numerical methods that will be briefly discussed in Sect. 11.4.2.

Equation (11.42a) applies to any boundary shape, but in the case where the boundary is an infinite plane there are some special features that should be noted. When the Rayleigh Integral, Eq. (11.41a), and Eq. (11.42a) are equated for any specific velocity distribution on an infinite rigid plane it can be seen that the Rayleigh Integral is equal to two times the first term of the Helmholtz Integral Formula. It then follows that the first and second terms of the Helmholtz Integral Formula are equal. Thus, for a plane rigid boundary, there are two separate expressions for $p(\vec{r})$, one in terms of the velocity on the plane, the other in terms of the pressure on the plane:

$$p(\vec{r}) = \frac{1}{2\pi}\iint j\omega\rho u_n(\vec{r}_0)\frac{e^{-jkR}}{R}dS_0, \tag{11.42b}$$

$$p(\vec{r}) = \frac{1}{2\pi}\iint p(\vec{r}_0)\frac{\partial}{\partial n_0}\left(\frac{e^{-jkR}}{R}\right)dS_0. \tag{11.42c}$$

Similarly, note that Eq. (11.41b) is twice the second term of the Helmholtz Integral Formula, which leads again to the two expressions in Eqs. (11.42b) and (11.42c). These expressions apply to any velocity distribution on an infinite rigid plane, or to any pressure distribution on an infinite soft (pressure release) plane, or equivalently, to any symmetric or anti-symmetric plane vibrator. The first, Eq. (11.42b), is the Rayleigh Integral and requires knowledge of the surface velocity, while the second, Eq. (11.42c), requires knowledge of the surface pressure. Either expression could be useful for far-field evaluation from near-field FEM computations, or from measurement of the pressure or velocity over a large plane area. Butler and Butler [38] used the near-field measured velocity to determine the far field of a baffled ribbon tweeter.

Some special relationships can be obtained from Eq. (11.42a) since the normal derivative in that equation on certain boundaries is either zero or constant for far-field points in certain directions. The normal derivative can be written as

$$\begin{aligned}\frac{\partial}{\partial n_0}\left(\frac{e^{-jkR}}{R}\right) &= -\vec{\nabla}\left(\frac{e^{-jkR}}{R}\right)\cdot\hat{n}_0 = -\cos\beta\frac{\partial}{\partial R}\left(\frac{e^{-jkR}}{R}\right)\\ &= \cos\beta\frac{e^{-jkR}(jkR+1)}{R^2} \approx jk\cos\beta\frac{e^{-jkR}}{R},\end{aligned} \tag{11.43}$$

where β is the angle between $\hat{n}_0$ and $(\vec{r} - \vec{r}_0)$, and the last form holds when the field point is in the far field. For example, at any source point on the ends of a finite cylinder, for a far-field point on the axis of the cylinder, $\beta = 0$ and R is constant which makes the normal derivative constant. At any source point on the sides of the cylinder, for a far-field point on the axis, $\beta = 90°$, and the normal derivative is zero. Thus, for such field points, the second term in Eq. (11.42a) makes no contribution to the solution from the sides, and it makes a contribution from the ends that is proportional to an integral of the pressure over the ends, i.e., to the radiation impedance. As a specific example consider a cylinder with uniform velocity on one end only. Equation (11.42a) then gives the following relationship between the pressure at a point in the far field on the axis, $p(z)$, and the radiation impedance of the ends of the cylinder [41]:

$$p(z) = \frac{jku_0 e^{-jkz}}{4\pi z}\left[e^{jkb}(\rho cA + Z_{11}) - e^{-jkb}Z_{12}\right],$$

where u_0 is the normal velocity of one end of the cylinder, $2b$ is the length, A is the cross-sectional area, Z_{11} is the self-radiation impedance of one end, and Z_{12} is the mutual radiation impedance between the ends. This relationship holds for right cylinders of any cross-sectional shape. A similar relationship between the far field and the radiation impedance of a piston of arbitrary shape in an infinite, nonrigid baffle with locally reacting normal surface impedance was given by Mangulis [42].

In some situations it may be feasible to use Eq. (11.42a) by approximating the surface pressure in terms of the known velocity. For example, in a large unbaffled source with velocity, $u(r_0)$, the surface pressure in Eq. (11.42a) can sometimes be approximated by a plane wave pressure, i.e., $p(\vec{r}_0) = \rho c u(\vec{r}_0)$. With this approximation, and use of the far-field expression for the normal derivative in Eq. (11.43), the pressure in the far field becomes

$$p(\vec{r}) = \frac{j\omega\rho}{4\pi}\iint u(\vec{r}_0)\frac{e^{-jkR}}{R}(1 + \cos\beta)dS_0. \tag{11.44a}$$

In general, for a fixed far-field point, $\cos\beta$ is a function of position on the surface of a source, but for a plane source it is constant. In that case

$$p\left(\vec{r}\right) = \frac{j\omega\rho}{4\pi}(1 + \cos\beta)\iint u\left(\vec{r}_0\right)\frac{e^{-jkR}}{R}dS_0. \tag{11.44b}$$

Comparison with Eq. (11.41a), which holds for a baffled source, shows that

$$p(\vec{r})_{unbaff} = \frac{1}{2}(1 + \cos\beta)p(\vec{r})_{baff}. \tag{11.44c}$$

For example, the far-field pressure for $\beta = 90°$ of an unbaffled plane source is reduced by 6 dB relative to the same source in a large baffle, because of the effect of the pressure term in Eq. (11.42a).

An estimate of the validity of this approximation can be made by use of results from Mellow and Karkkainen [37] who calculated the fields of piston sources with closed back baffles and finite baffles in the plane of the piston. For the case of no baffle in the plane and a piston diameter of a half wavelength ($ka = \pi/2, a = $ radius) the pressure is down by 7.5 dB at 90° (see their Fig. 17), while for the same case with an infinite rigid baffle the reduction would be 2.8 dB. The additional reduction of 4.7 dB may be compared with the expected 6 dB reduction for the case of no baffle given by Eq. (11.44c). The approximation of Eq. (11.44c) should improve for diameters greater than one-half wavelength. Their Fig. 16 shows beam patterns for the case of a finite rigid baffle of diameter twice the diameter of the radiating piston for *ka* equal to 3 and 5 with levels of −18 and −23 dB at 90°, respectively. The corresponding calculated rigid baffle levels are −12.9 and −17.6 dB revealing additional reductions of −5.1 and −5.4 dB, which are closer to the expected −6 dB from Eq. (11.44c).

In other cases where the pressure can be measured, but the velocity cannot be readily measured, the velocity can be approximated in terms of the pressure. This is the basis for one of the near-field to far-field extrapolation methods to be described in Sect. 9.7.4 [43]. It can be used for calibration of large arrays when the curvature of the measurement surface is small compared to the wavelength, and the velocity in Eq. (11.42a) can be approximated by $u(\vec{r}_0) \approx p(\vec{r}_0)/\rho c$.

11.3 Scattering and the Diffraction Constant

Many different acoustic scattering problems arise in connection with transducers and arrays. Scattering of plane waves from a rigid object such as a sphere is a useful, and relatively simple, model for determining the diffraction constant of hydrophones, but the scattering between transducers in a volume array of projectors is much more difficult. In the latter case the fields being scattered are more like spherical waves than plane waves, because the projectors are close together, and multiple scattering may also be important. The numerical method based on the Helmholtz Integral Equation, to be described in Sect. 11.4.2, handles all aspects of scattering. This type of acoustic interaction problem is especially important for arrays of transducers such as flextensionals, which radiate from more than one side and therefore do not behave as a surface array. In this situation the scattering is likely to excite unintended modes of vibration which usually have undesirable effects.

The volume array problem might also be modeled by scattering of a spherical wave from a sphere, as has been done by Thompson who calculated the change in radiation impedance of a vibrating sphere caused by another nearby sphere [44, 45]. Various cases were considered in which the second sphere was rigid, or had acoustic properties different from the first sphere, or was vibrating in the same mode as the first sphere, or in a different mode. This work required use of the translational addition theorem for spherical wave functions to transform solutions in one spherical coordinate system to another spherical coordinate system with the origin in a different location. The process is similar to the addition theorem used above in Eq. (11.9) to transform solutions from one spherical coordinate system to another rotated spherical coordinate system with the same origin.

11.3.1 The Diffraction Constant

In Sect. 6.6 we discussed the diffraction constant, a measure of how an incident sound wave is disturbed when it is scattered from a hydrophone. It is very important since it strongly affects the effective receiving sensitivity if a hydrophone is not small compared to the wavelength (see Fig. 6.35). Since the diffraction constant depends on the sum of the incident wave pressure and the scattered wave pressure, as shown by Eq. (6.51), the first step in its determination is calculation of the scattered wave. It is necessary to specify a simple shape for the scattering object that represents a hydrophone, one that is compatible with a coordinate system in which the Helmholtz equation is separable. A spherical scatterer is one of the most useful cases because some hydrophones are spherical, and other hydrophone shapes can be approximated by a sphere.

Consider a plane wave traveling in the positive z direction incident on a rigid sphere centered on the origin of a coordinate system as shown in Fig. 11.11.

In this axisymmetric situation the expression for the plane wave can be written in terms of spherical coordinates as follows:

$$p_i = p_0 e^{j(\omega t - kz)} = p_0 e^{\,j(\omega t - kr\cos\theta)}. \tag{11.45}$$

The spatial part of Eq. (11.45) can then be expanded in spherical wave functions:

$$e^{-jkr\cos\theta} = \sum_{n=0}^{\infty} B_n P_n(\cos\theta). \tag{11.46}$$

Use of the orthogonality of the Legendre polynomials gives

$$B_n = (n+1/2)\int_0^{\pi} e^{-jkr\cos\theta} P_n(\cos\theta)\sin\theta d\theta = (n+1/2)[2(-j)^n j_n(kr)], \tag{11.47}$$

where the last step follows from a standard integral representation of the spherical Bessel function, $j_n(kr)$ [3]. Combining these results gives the incident plane wave in terms of spherical wave functions:

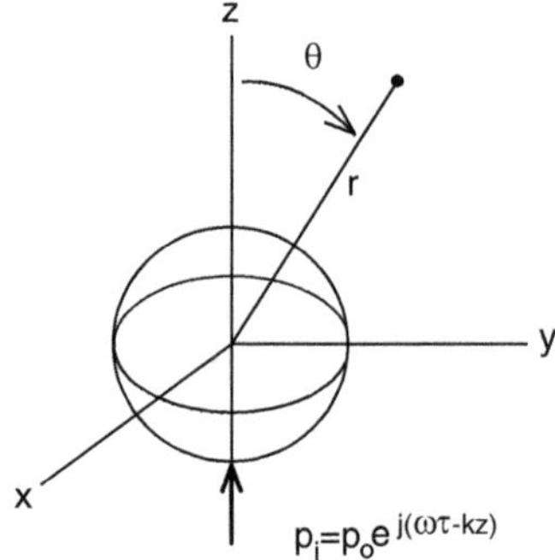

Fig. 11.11 Coordinates for scattering of plane wave from a sphere

$$p_i = p_0 \mathrm{e}^{j\omega t} \sum_{n=0}^{\infty} (2n+1)(-j)^n j_n(kr) P_n(\cos\theta). \tag{11.48}$$

Note that this expression for the plane wave, although expressed in spherical coordinates, is not a series of outgoing spherical waves because of the factor $j_n(kr)$ rather than $h_n^{(2)}(kr)$. However, the scattered wave is assumed to consist of an infinite series of outgoing spherical waves expressed as

$$p_s = \mathrm{e}^{j\omega t} \sum_{n=0}^{\infty} A_n P_n(\cos\theta) h_n^{(2)}(kr). \tag{11.49}$$

The boundary condition on the surface of a rigid sphere requires that the total particle velocity be zero. The total velocity, i.e., the sum of the velocities associated with the incident wave and the scattered wave, can be determined by differentiating the expressions for p_i and p_s and equating the sum to zero to determine the A_n coefficients. Then the scattered wave is completely determined, and the magnitude and directionality of the scattered power can be calculated. However, for calculating the diffraction constant, the force exerted on the surface of the hydrophone by the sum of the incident and scattered pressures is the important quantity, rather than the scattered power, as shown by Eq. (6.51), repeated here:

$$D_a = \frac{F_\mathrm{b}}{Ap_i} = \frac{p_\mathrm{b}}{p_i} = \frac{1}{Ap_i} \iint\limits_A [p_i(\vec{r}) + p_s(\vec{r})] \frac{u^*(\vec{r})}{u_0^*} \mathrm{d}S. \tag{11.50}$$

The area $A = 4\pi a^2$ for the spherical hydrophone of radius a and $\mathrm{d}S = 2\pi a^2 \sin\theta \, \mathrm{d}\theta$ because of the axial symmetry established by the plane wave direction. We will consider a hydrophone in which the sensitive surface is capable only of uniform motion, which makes the normalized velocity distribution $u^*(\vec{r})/u_0^* = 1$. This greatly simplifies the integration, because both p_i and p_s are proportional to infinite series of $P_n(\cos\theta)$, and the orthogonality of the Legendre functions makes all the terms vanish in the integration except for $n = 0$. Thus we only need to consider the zero-order part of the pressure from Eqs. (11.48) and (11.49):

$$[p_i(\vec{r}) + p_s(\vec{r})]_0 = \mathrm{e}^{j\omega t} \left[p_0 j_0(kr) + A_0 h_0^{(2)}(kr) \right]. \tag{11.51}$$

When the radial derivative of this sum is calculated and equated to zero on the surface at $r = a$ the coefficient A_0 is found to be $A_0 = \left[-p_0 j_0^{'}(ka)/h_0^{'}(ka) \right]$ where the primes on the spherical Bessel and Hankel functions indicate the derivative with respect to the argument. This value of A_0 gives the total pressure on the surface of the sphere, and Eq. (11.50) becomes

$$D_a = \frac{2\pi a^2}{p_0 A} \int_0^{\pi} \left[p_0 j_0(ka) - p_0 h_o^{(2)}(ka) j_0'(ka)/h_0'(ka) \right] \sin\theta d\theta$$

$$= j_0(ka) - h_o^{(2)}(ka) j_0'(ka)/h_0'(ka) = \left[1 + (ka)^2\right]^{-1/2}, \tag{11.52}$$

where the final step makes use of relationships between spherical Bessel and Hankel functions [4]. This is the result given originally by Henriques [46], and here in Eq. (6.53). Its simplicity makes it very useful and easy to apply to other shapes, but it only holds exactly for spherical hydrophones. As the frequency increases the large reduction of sensitivity predicted by Eq. (11.52) may not be accurate for other hydrophone shapes.

The diffraction constant for a sphere is an especially simple case because of the symmetry; however, hydrophones of other shapes have diffraction constants that depend on the direction of arrival of the plane wave. A generalized diffraction constant will now be defined that includes this directional dependence, and at the same time the extremely useful relation in Eq. (6.56) will be derived. Consider a transducer of arbitrary shape and area A_1, with normal velocity distribution $u(r_1)$ and reference velocity u_1 located at the origin of spherical coordinates with its MRA in the z direction. The transducer is receiving a plane wave from a small simple spherical source of source strength u_2A_2 located at a large distance, r, in the direction (θ, ϕ). The average clamped pressure exerted on the transducer by the wave from the spherical source, p_b, is defined by Eq. (11.50), and the acoustic reciprocity theorem, Eq. (11.34), shows that p_b is related to the far-field pressure radiated by the transducer at the location of the spherical source, $p_1(r, \theta, \phi)$, by

$$u_1 A_1 p_b = u_2 A_2 p_1(r, \theta, \phi). \tag{11.53}$$

Since $p_1(r,0,0)$ is the radiated pressure amplitude on the MRA of the transducer, it follows from the definitions of the radiation resistance, R_r, (referred to u_1) and the directivity factor, D_f, that

$$\left[p_1^2(r, 0, 0)/2\rho c\right] = D_f\left[R_r u_1^2/8\pi r^2\right]. \tag{11.54}$$

The normalized directivity function of the transducer is defined by

$$P(\theta, \phi) = p_1^2(r, \theta, \phi)/p_1^2(r, 0, 0). \tag{11.55}$$

Combining Eq. (11.55) with Eqs. (11.54) and (11.53) gives the clamped pressure on the transducer as

$$p_b = \left[\rho c D_f R_r P(\theta, \phi)/4\pi\right]^{1/2}\left[u_2 A_2/A_1 r\right], \tag{11.56}$$

Equation (10.15b) gives the free field pressure amplitude produced by the spherical source at the location of the transducer as $p_i = \rho\omega u_2 A_2/4\pi r$. Now, we define the directional plane wave diffraction constant, consistent with the definition in Eq. (11.50), as

$$D(\theta,\phi) = p_b/p_i = \left[\frac{4\pi c D_f R_r P(\theta,\phi)}{\rho\omega^2 A_1^2}\right]^{\frac{1}{2}} = D_a[P(\theta,\phi)]^{\frac{1}{2}}, \tag{11.57a}$$

Equation (11.57a) evaluated on the MRA yields the general relationship between the diffraction constant, D_a, the directivity factor, D_f, and the radiation resistance, R_r, which may be written as

$$D^2(0,0) = D_a^2 = 4\pi c D_f R_r/\rho\omega^2 {A_1}^2 \tag{11.57b}$$

as used in Eqs. (6.56) and (9.33). Using $D(\theta,\phi)$ in place of D_a in any of the previous expressions for receiving sensitivity gives the sensitivity as a function of direction. When $D^2(\theta,\phi))$ is averaged over all directions we get $\overline{D}^2 = D_a^2/D_f$ since the average of $P(\theta,\phi)$ is $1/D_f$ by definition.

11.3.2 *Scattering from Cylinders*

A transducer mounted near a cylinder presents another scattering problem that is often important in transducer and array work [47]. A specific case will now be discussed starting from the general expression for the transform of the pressure in Eq. (11.17), and taking the inverse transform to obtain the pressure in Eq. (11.18), which is then written as

$$p(r,\phi,z) = \sum_{m=0}^{\infty} \cos m\phi \int_{-\infty}^{\infty} A_m(\alpha) H_m^{(2)}(\beta r) e^{j\alpha z} d\alpha. \tag{11.58}$$

Consider the specific problem of a small acoustic source located at a distance r_0 from the axis of a rigid cylinder of radius a as shown in Fig. 11.12.

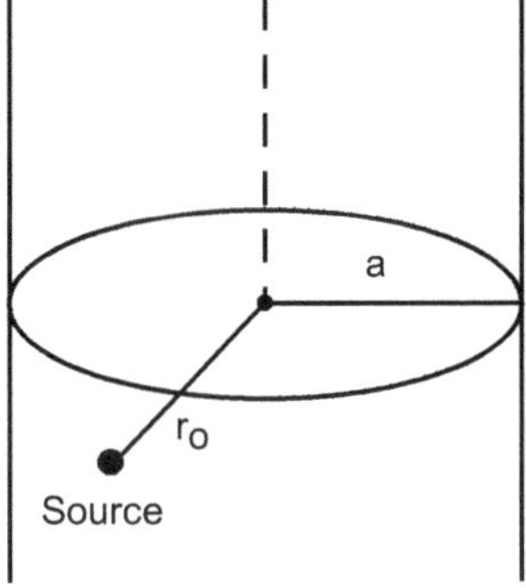

Fig. 11.12 A point source located at a distance $r_o - a$ from the surface of an infinite cylinder

The coordinate system is chosen such that the source is located at $\phi = 0$ and $z = 0$. Then the field of a point source can be expanded in cylindrical wave functions and written as [48]:

$$p_s(r, \phi, z) = -(jq/2)\sum_{m=0}^{\infty} \epsilon_m \cos m\phi \int_{-\infty}^{\infty} \left\{J_m(\beta r_0)H_m^{(2)}(\beta r)\right\} e^{j\alpha z} d\alpha, \quad r_0 \leq r, \tag{11.59}$$

with the bracket replaced by $\{J_m(\beta r)H_m^{(2)}(\beta r_0)\}$ for $r_0 \geq r$, $\epsilon_0 = 1$, $\epsilon_m = 2$ for $m > 0$, and $q = j\omega\rho Q/4\pi$ where Q is the source strength of the point source (see Eq. (10.15b)). In this case the cylinder surface is rigid and motionless, and the sound waves, which originate from the point source, are scattered by the cylinder. The total pressure field is the sum of the point source field and the scattered field which is given by, for $r \leq r_0$,

$$p_t = p + p_s = \sum_{m=0}^{\infty} \cos m\phi \int_{-\infty}^{\infty} \left[A_m(\alpha)H_m^{(2)}(\beta r) - (jq/2)\epsilon_m J_m(\beta r)H_m^{(2)}(\beta r_0)\right] e^{j\alpha z} d\alpha. \tag{11.60}$$

The function, $A_m(\alpha)$, is determined by equating the total normal velocity to zero on the surface of the rigid cylinder, i.e., $u_t = -(1/j\omega\rho)(\partial p_t/\partial r) = 0$ at $r = a$. The result is

$$A_m(\alpha) = (jq\epsilon_m/2)\frac{J_m'(\beta a)H_m^{(2)}(\beta r_0)}{H_m'(\beta a)}, \tag{11.61}$$

where the primes on J_m and H_m mean the derivative with respect to the argument evaluated at $r = a$. Then the final result for the total field at any point where $r \geq r_0$, i.e., outside the location of the point source, is given by

$$p_t(r, \phi, z) = -\frac{\omega\rho Q}{8\pi}\sum_{m=0}^{\infty} \epsilon_m \cos m\phi \int_{-\infty}^{\infty} \left[J_m'(\beta a)H_m^{(2)}(\beta r_0) - J_m(\beta r_0)H_m'(\beta a)\right] \times \frac{H_m^{(2)}(\beta r)}{H_m'(\beta a)} e^{j\alpha z} d\alpha. \tag{11.62}$$

Equation (11.62), for $r_0 = a$, agrees with Eq. (11.18) when specialized to the case of a very small piston, noting that $\left[H_{m-1}^{(2)}(\beta a) - H_{m+1}^{(2)}(\beta a)\right] = 2H_m'(\beta a)$.

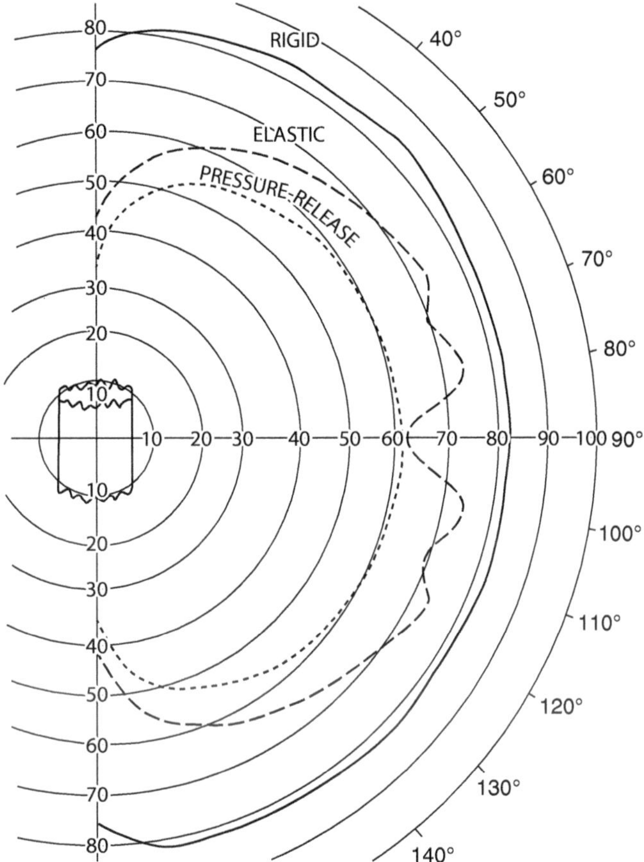

Fig. 11.13 Far-field beam patterns in the vertical plane of the source in Fig. 11.12 at 1800 Hz for rigid (*solid line*), elastic (*dash line*), and pressure release (*dotted line*) cylinders [47]

Such results have also been obtained for cylinders with elastic and pressure release surfaces, and numerical calculations have been made for all three cases [47]. Figures 11.13 and 11.14 show far-field vertical and horizontal beam patterns for rigid, pressure release, and elastic cylinders for the specific case of a cylinder of radius 0.5 m with a small source located 0.01 m from the surface vibrating at a frequency of 1800 Hz. In the elastic case the cylinder is a steel tube with a wall thickness of 0.005 m.

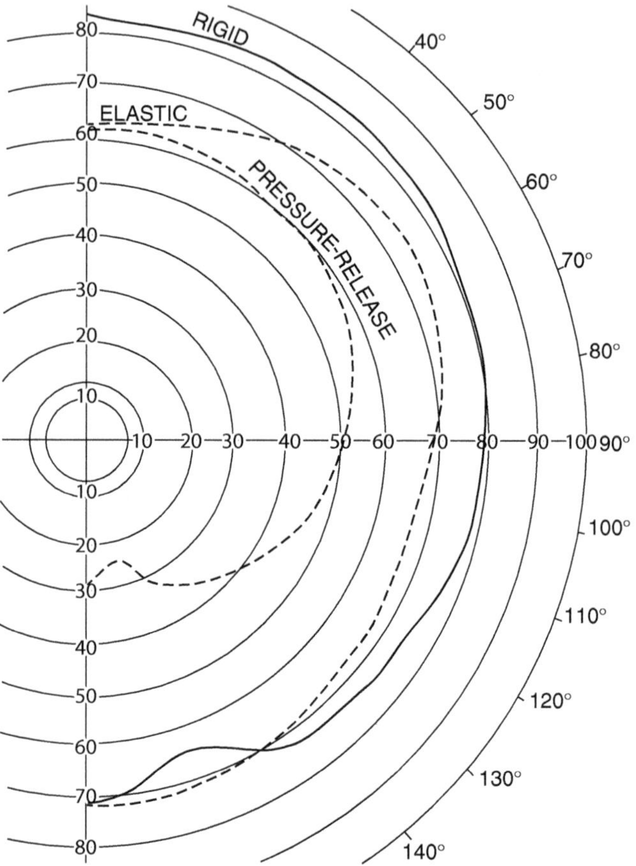

Fig. 11.14 Far-field beam patterns in the horizontal plane of the source in Fig. 11.12 at 1800 Hz for rigid (*solid line*), elastic (*dash line*), and pressure-release (*dotted line*) cylinders [47]

11.4 Numerical Methods for Acoustic Calculations

The analytical methods discussed in this chapter for calculating sound fields are limited to vibrating objects with simple shapes. Problems involving more complex shapes or mixed boundary conditions require numerical methods, and, until about 1960, results were quite limited, although some extensive numerical calculations were made earlier, such as the pressure distribution on the surface of a circular piston [49]. With the availability of high speed computing it became feasible to develop new numerical methods and to extend results obtained by existing methods. The classic problem of radiation from a cylinder of finite length has played an important role in these developments [5, 50–53]. Results were also obtained for other problems, including some cases of mixed boundary conditions, using collocation [54] and least squares analysis [55]. But the most significant progress occurred with the development of finite element analysis and boundary element methods.

11.4.1 Mixed Boundary Conditions: Collocation

The boundary conditions in most transducer radiation problems are not as simple as the idealized boundary conditions considered previously. The vibrating face of an individual piston transducer might be surrounded by a small rigid flange, or by a nonrigid flange, but seldom by a flange large enough to be approximated by an infinite rigid plane. However, it is reasonable to consider the central transducers in a large close-packed plane array to be lying in an infinite rigid plane, but the transducers on or near the edges obviously have quite different surroundings. These are examples of problems in which the velocity can be specified on the vibrating face of a transducer, but it is difficult to specify the velocity on the surfaces immediately surrounding the transducer. Transducers mounted in a housing might be modeled as a sphere with part of the spherical surface vibrating while the other motionless part represents the housing. If the housing could be considered rigid with zero velocity there would be a velocity boundary condition on the entire spherical surface and the problem could be solved as in Sect. 11.1. If the housing was very soft the pressure could be considered zero on the housing, giving a mixed boundary value problem with velocity specified on one part, pressure on the other part, which cannot be solved by the methods of Sect. 11.1. If the housing was flexible it might be described in terms of a locally reacting impedance, which is equivalent to specifying the velocity in terms of the pressure. Problems of this type usually must be handled numerically, although the case of an impedance boundary condition on an infinite plane has been formulated in terms of a Green's function, and approximate results obtained in analytical form by Mangulis [56]. Mellow and Karkkainen [37] obtained extensive results for the mixed boundary value problem of an oscillating disc surrounded by finite rigid baffles using a different numerical method.

We will illustrate a mixed boundary value problem with an approximate numerical solution obtained by the method of boundary collocation. The solution of the Helmholtz equation for any vibrating object that has a closed and finite surface can be approximated by a series of spherical wave functions, even when the object is not a sphere. The series is truncated to N terms, and in the axisymmetric case written as

$$p(r,\theta) = \sum_{n=0}^{N-1} A_n P_n(\cos\theta) h_n^{(2)}(kr). \tag{11.63}$$

If the vibrating object is a sphere with either pressure or velocity boundary conditions the coefficients can be determined as was done in Sect. 11.1. If the boundary values are given on an object that is not a sphere, or if the boundary values are mixed, N boundary values at the points r_j, θ_j are used to form the following N equations, which can be solved for the A_n coefficients:

$$p(r_j, \theta_j) = \sum_{n=0}^{N-1} A_n P_n(\cos\theta_j) h_n^{(2)}(kr_j), \quad j = 1, 2 \cdots N \tag{11.64}$$

If the boundary values are given as values of the pressure Eq. (11.64) can be used directly. If values of the normal velocity are given on the boundary the normal derivative must be found at each point to obtain a different set of equations from Eq. (11.64). Note that the normal derivative is not the radial derivative on portions of the object that are not spherical. If pressure is given at some boundary points and normal velocity at other boundary points a mixed set of N equations can be used. For an impedance boundary condition at some points the ratio of pressure to velocity can be used to form an equation. In any case the set of N equations can be solved by a simultaneous equation routine to obtain the complex A_n coefficients. Using these coefficients in Eq. (11.63) gives an approximate solution that satisfies the N boundary points and has a degree of validity elsewhere that depends on the number of points used and the complexity of the vibrating object.

Butler [54] evaluated the collocation method in detail and gave numerical results for several typical problems. As a specific example we give results that illustrate the effects of nonrigid baffles surrounding a transducer. Consider a spherical surface of radius a, with uniform normal velocity over the portion $0 \leq \theta \leq 60°$, and with three different conditions on the remainder of the surface: normal velocity $= 0$ (rigid), pressure $= 0$ (soft), and pressure $= \rho c$ times normal velocity (impedance condition). The results are shown in Figs. 11.15 and 11.16 for the far-field directivity patterns and the normalized radiation impedance. They show that the main effects of a soft baffle surrounding a transducer are an increase of directivity and a decrease of radiation resistance for values of ka less than about 2. The results in these figures can be used to qualitatively estimate similar effects in other cases.

11.4.2 Boundary Element Methods

In the mid-1960s more powerful numerical methods began to appear based, in most cases, on the Helmholtz Integral Formula. The papers by Chen and Schweikert [57], Chertock [58], Copley [59], and Schenck [60] clearly show the rapid development and improvement of these methods as related to radiation problems of interest in underwater acoustics. The paper by Schenck reviews and evaluates the techniques presented in the preceding papers and shows how to combine them in a practical computational method that reliably gives unique solutions. Schenck's Combined Helmholtz Integral Equation Formulation (CHIEF) has been widely used, often in conjunction with finite element structural analysis, to give comprehensive solutions

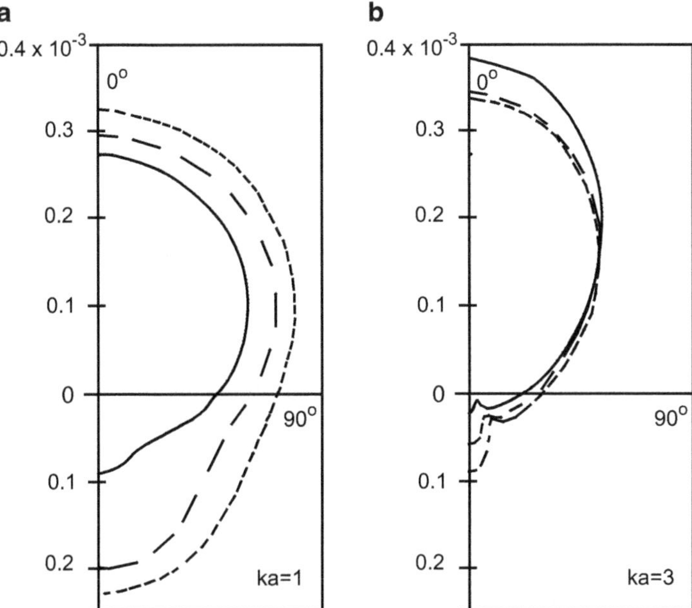

Fig. 11.15 Far-field polar pattern, at $kr = 1000$, of the pressure amplitude relative to ρc for a sphere of radius a vibrating uniformly over the portion from 0° to 60° with a soft, ρc or rigid baffle from 60° to 180° [54]. (**a**) ka is equal to unity and (**b**) $ka = 3$. *solid line* soft: *dash line* ρc; *dotted line* rigid

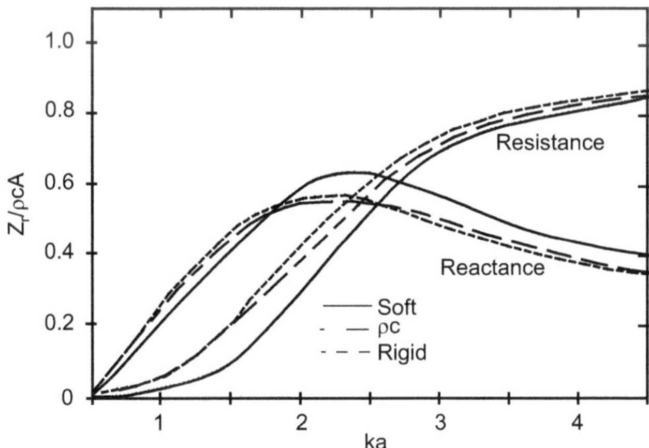

Fig. 11.16 Normalized radiation impedance for the curves of the polar piston on the sphere with the soft, ρc, and rigid baffle in Fig. 11.15 [54]

to complicated transducer vibration-radiation problems. This type of analysis has expanded into a new field, called Boundary Element Methods (BEM) or Boundary Element Acoustics, with a large literature of its own [7, 8].

We will give a brief outline of the CHIEF method based on a version developed especially for transducer and array problems [60]. The method is based on the Helmholtz integral equation, Eq. (11.40), repeated here:

$$p(\vec{r}) = \frac{1}{4\pi}\iint\left[\frac{e^{-jkR}}{R} j\omega\rho u(\vec{r}_0) + p(\vec{r}_0)\frac{\partial}{\partial n_0}\left(\frac{e^{-jkR}}{R}\right)\right] dS_0, \qquad (11.65)$$

where it is understood that the integration is over one or more closed surfaces each representing a transducer. Equation (11.65) gives the field point pressure, $p(\vec{r})$, at all points outside the closed surfaces, but at all points on any surface $p(\vec{r})$ is replaced by ½ $p(\vec{r})$ and at all points inside each surface the right-hand side is null [36, 59, 60]. In the CHIEF algorithm the normal velocity on the surfaces is assumed to be known, the field point is placed on the surface, and Eq. (11.65) becomes an integral equation for the unknown surface pressure:

$$p(\vec{r}_0)/2 - \frac{1}{4\pi}\iint p(\vec{r}_0)\frac{\partial}{\partial n_0}\left(\frac{e^{-jkR}}{R}\right) dS_0 = \frac{j\omega\rho}{4\pi}\iint u(\vec{r}_0)\frac{e^{-jkR}}{R} dS_0, \qquad (11.66)$$

This integral equation can be solved numerically for the pressure on the surface at N positions, r_n, by replacing it by the set of equations:

$$2\pi p(r_n) - \sum_m p(r_m)\frac{\partial}{\partial n_0}\left(\frac{e^{-jkR_{nm}}}{R_{nm}}\right)\Delta S_0 = j\omega\rho\sum_m u(r_m)\frac{e^{-jkR_{nm}}}{R_{nm}}\Delta S_0, \qquad (11.67)$$

where R_{nm} is the distance between the surface point at r_n and the surface point at r_m, and the right-hand side is a known quantity that can be evaluated at each point, r_n. An important part of the CHIEF method requires addition of some interior points, where the right-hand side of Eq. (11.65) is zero, at frequencies corresponding to interior resonances. This gives an overdetermined equation set and a unique solution at all frequencies [60].

When the surface pressure has been found for a specified velocity it can be used in Eq. (11.65) to calculate the pressure at any other point outside the surface such as the far field. The surface pressure for a single transducer can be integrated over the surface to give the acoustic force and the radiation impedance. For more than one transducer the surface pressure on one contains contributions from all the others. Then the integral over the surface gives the force that contains all the mutual radiation impedance contributions for the particular velocity distribution that was originally specified. It is clear that the CHIEF method automatically solves the scattering problem in volume arrays discussed in Sect. 11.4, because the sound field can be calculated for all the transducers operating together, including all scattering events.

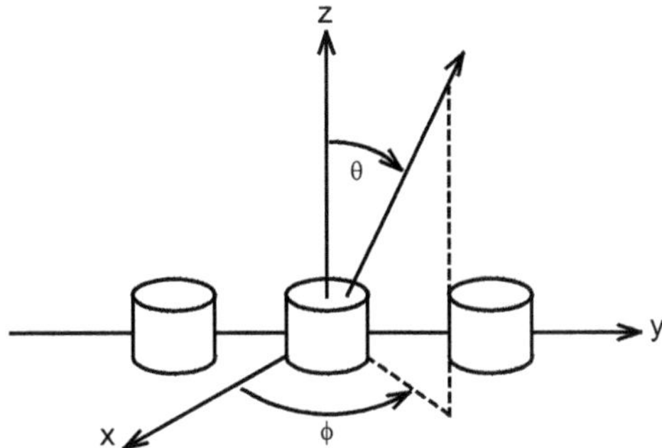

Fig. 11.17 Three-element linear array used for calculations in Table 11.1

Table 11.1 Results for the array in Fig. 11.17

	Voltage		Radiation	Velocity	
Projector	Magnitude	Phase	Resistance	Magnitude	Phase
1	1	0	17,800	0.00010	10
2	1	0	21,200	0.00008	30
3	1	0	17,800	0.00010	10

To solve projector array problems the individual mutual radiation impedances are needed. These can be determined in the CHIEF method by specifying the velocity on only one transducer at a time and calculating the pressure on the others. These impedances can then be used in the array equations in Sect. 7.2.1 that also contain the transducer ABCD parameters. A small example of such results for an array of three projectors with truncated elliptic cylinder shapes, shown in Fig. 11.17, is given in Table 11.1 [61]. The three projectors are driven in phase with unit voltage, and the ABCD parameters are specified at 3300 Hz. Table 11.1 shows the calculated variation of radiation resistance and velocity among the three projectors.

11.5 Summary

In this chapter we discussed and presented solutions to mathematical models for acoustic radiation using series expansions and integral formulations. The mutual radiation impedance for pistons on a sphere and on a cylinder was given from a direction solution to the Helmholtz differential equations through spherical and cylindrical wave function expansions. The case of a ring on a cylinder was also shown to be modeled by a series solution by replicating the ring and expanding the separation along a rigid cylinder. This method was used to obtain results for the radiation impedance operating in monopole and dipole modes. The Hankel transform was presented and shown to offer an alternative solution for a circular piston in an infinite rigid baffle. The Hilbert transform was also introduced and shown to yield the near-field radiation reactance from the radiation resistance, which is easier to obtain from the far field.

Green's theorem was introduced and used to develop the acoustical reciprocity theorem. Green's function and its formulation as solution to the Helmholtz equation with a source was also presented and reduced to the Helmholtz integral equation. Various cases were considered and it was shown that for a free large planar source the far-field pressure $p \approx p_b(1+\cos\theta)/2$, where p_b is the, more easily obtained, far-field pressure for the same source with a rigid baffle and θ is the beam pattern angle to the normal.

The diffraction constant, D_a, was discussed in more detail and the arrival of a wave at different angles is considered. The case of a rigid sphere of radius a was shown to be simply $D_a = [1+(ka)^2]^{-1/2}$. The acoustic scattering from a small source in the vicinity of a rigid, elastic, and pressure release cylinder was presented with results in the axial direction and in a plane normal to the axis, showing a more severe effect for the pressure release cylinder. The chapter concluded with a discussion of mixed pressure and velocity boundary conditions solutions by collocation means and the CHIEF algorithm for the solution of the Helmholtz integral equation.

Exercises (Degree of Difficulty: *Lowest, **Moderate, ***Highest)

11.1.** Of the 11 coordinate systems in which the wave equation is separable (see [3]) only 6 or 7 have been used in modeling practical transducers and arrays. Consider the geometrical shapes that occur in practice and think of specific shapes that can be accommodated by certain coordinate systems as a way of identifying the most useful systems.

11.2.** Show that Eq. (11.8) follows from Eq. (11.9) for $n = 1$. Also see Exercise 8.9 regarding the development of Eq. (11.8).

11.3.** Derive Eq. (11.4) from Eqs. (11.2), (11.3a), and (11.3b) using the orthogonality relation for Legendre polynomials:

$$\int_0^{\pi} P_n(\cos\theta)P_m(\cos\theta)\sin\theta d\theta = 2/(2m+1)$$

$$\text{for} \quad n = m, \quad \text{and} \quad = 0 \quad \text{for} \quad n \neq m.$$

11.4.*** The boundary condition for a free-flooding ring simplifies considerably for an end-capped cylinder where the inside velocity, u_i, is not involved and u_t applies to the entire top. But the problem is still not solvable because of the finite length of the cylindrical surface. In such cases, under low frequency conditions, a useful model can be based on spherical wave functions. In the lowest order approximation the cylinder is replaced by a monopole sphere with the same radiating area as the cylinder. A better approximation is possible by using a sphere with uniform normal velocity in the polar regions, representing the velocity of the end caps, and a different uniform normal velocity in the equatorial region, representing the velocity of the sides of the cylinder. When the end caps are considered motionless this boundary condition is described by:

$$u = 0, \quad r = a, \quad 0 < \theta < \theta_1 \quad \text{and} \quad 180^\circ - \theta_1 < \theta < 180^\circ$$
$$u = u, \quad r = a, \quad \theta_1 < \theta < 180^\circ - \theta_1$$

where a is the radius of the sphere and also the radius of the ring being modeled. Express the solution of this problem as a series of spherical wave functions and determine the coefficients. Consider the solution for ka small enough that only two terms are sufficient. What is the effect of the second term on the beam pattern?

11.5.*** The Hilbert transform was used in Sect. 11.1.4 to obtain the radiation reactance from the radiation resistance of a sphere of radius, a, operating in a uniform omnidirectional mode. Obtain the radiation reactance from the radiation resistance of the same sphere vibrating in a dipole mode given by Eq. (10.47).

11.6.*** Show that for an unbaffled plane array, the pressure in the plane of the array is one-half the pressure for the same array in a large rigid baffle. Use the fact that, for the same array in a large pressure release baffle, the pressure in the plane of the array is zero.

11.7.** Equation (11.52) gives the diffraction constant, D_a, for a sphere of radius, a. Why does D_a decrease from unity as the frequency is increased rather than increase as in the case of a piston. Why does D_a decrease as $1/ka$ for $ka >> 1$?

11.8.** Beam pattern results for a piston set in a pressure release sphere show a significant reduction in the back radiation compared to a rigid sphere as shown in Fig. 11.15. Explain why this happens.

11.9.* Table 11.1 shows the radiation resistance for a linear array of three equally spaced pistons in a rigid baffle. Why do the two outside pistons have the same radiation resistance and the same velocity?

11.10.*** Consider the basis of the important Eq. (6.56) by following its derivation in Sect. 11.3.1 and, specifically, by showing that Eq. (11.53) follows from the acoustic reciprocity relation in Eq. (11.34) and the definition of clamped force in Eq. (11.50).

11.11.** The diffraction constant, D_a, in Eq. (6.56) is for an incoming wave arriving on the MRA of the hydrophone, and the D_f is referred to the same MRA. The directional diffraction constant, $D_a(\theta, \phi)$, in Eq. (11.57a) is D_a referred to the MRA multiplied by the square root of the normalized intensity directivity function. Use of $D_a(\theta, \phi)$ in any of the expressions for hydrophone output [such as Eqs. (6.10) or (6.17)] gives the output as a function of direction. The frequency dependence of D_a and of the hydrophone mechanism together determine the final frequency dependence of the output. In an isotropic noise field the mean squared noise output is proportional to the average of $D_a^2(\theta, \phi)$ which is D_a^2/D_f. What is $D_a(\theta, \phi)$ for a tonpilz hydrophone with a circular head mounted in a rigid baffle, and what is its value at $\theta = 0$?

References

1. P.M. Morse, *Vibration and Sound* (McGraw-Hill, New York, 1948)
2. L.E. Kinsler, A.R. Frey, A.B. Coppens, J.V. Sanders, *Fundamentals of Acoustics*, 4th edn. (Wiley, New York, 2000)
3. P.M. Morse, H. Feshbach, *Methods of Theoretical Physics* (McGraw-Hill, New York, 1953)
4. P.M. Morse, K.U. Ingard, *Theoretical Acoustics* (McGraw-Hill, New York, 1968)
5. M.C. Junger, D. Feit, *Sound, Structures and Their Interaction*, 2nd edn. (The MIT Press, Cambridge, 1986)
6. L.L. Beranek, T.J. Mellow, *Acoustics: Sound Fields and Transducers* (Academic, Oxford, 2012)
7. T.W. Wu (ed.), *Boundary Element Acoustics* (WIT Press, Southampton, 2000)
8. C.A. Brebbia, J. Dominguez, *Boundary Elements: An Introductory Course* (Computational Mechanics Publications, Southampton and McGraw-Hill, New York, 1989)
9. G.W. Benthein, S.L. Hobbs, *Modeling of Sonar Transducers and Arrays*, Technical Document 3181, April 2004 (SPAWAR Systems Center, San Diego, CA) (Available on CD)
10. M.C. Junger, Surface pressures generated by pistons on large spherical and cylindrical baffles. J. Acoust. Soc. Am. **41**, 1336–1346 (1967)
11. C.H. Sherman, Mutual radiation impedance of sources on a sphere. J. Acoust. Soc. Am. **31**, 947–952 (1959)
12. A. Silbiger, Radiation from circular pistons of elliptical profile. J. Acoust. Soc. Am. **33**, 1515–1522 (1961)
13. T. Nimura, Y. Watanabe, Vibrating circular disk with a finite baffle board. Jour. IEEE Japan **68**, 263 (1948) (in Japanese). Results available in *Ultrasonic Transducers*, ed. by Y. Kikuchi (Corona Pub. Co., Tokyo, 1969), p. 348
14. J.E. Boisvert, A.L. Van Buren, Acoustic radiation impedance of rectangular pistons on prolate spheroids. J. Acoust. Soc. Am. **111**, 867–874 (2002)
15. J.E. Boisvert, A.L. Van Buren, Acoustic directivity of rectangular pistons on prolate spheroids. J. Acoust. Soc. Am. **116**, 1932–1937 (2004)
16. N.W. McLachlan, *Theory and Application of Mathieu Functions* (Dover, New York, 1964)
17. J.E. Boisvert, A.L. Van Buren, Acoustic radiation impedance and directional response of rectangular pistons on elliptic cylinders. J. Acoust. Soc. Am. **118**, 104–112 (2005)
18. V.H. Weston, Q. Appl. Math. **15**, 420–425 (1957)
19. V.H. Weston, Q. Appl. Math. **16**, 237–257 (1958)
20. V.H. Weston, J. Math. Phys. **39**, 64–71 (1960)
21. C.H. Sherman, N.G. Parke, Acoustic radiation from a thin torus, with application to the free-flooding ring transducer. J. Acoust. Soc. Am. **38**, 715–722 (1965)
22. D.T. Laird, H. Cohen, Directionality patterns for acoustic radiation from a source on a rigid cylinder. J. Acoust. Soc. Am. **24**, 46–49 (1952)
23. J.E. Greenspon, C.H. Sherman, Mutual radiation impedance and nearfield pressure for pistons on a cylinder. J. Acoust. Soc. Am. **36**, 149–153 (1964)
24. D.H. Robey, On the radiation impedance of an array of finite cylinders. J. Acoust. Soc. Am. **27**, 706–710 (1955)
25. J.L. Butler, A.L. Butler, A Fourier series solution for the radiation impedance of a finite cylinder. J. Acoust. Soc. Am. **104**, 2773–2778 (1998)

26. J.L. Butler, Self and Mutual Impedance for a Square Piston in a Rigid Baffle, Image Acoustics Rept. on Contract N66604-92-M-BW19, March 20, 1992
27. C.J. Tranter, *Integral Transforms in Mathematical Physics*, 3rd edn. (Wiley, New York, 1966), pp. 47–48
28. C.J. Tranter, *Integral Transforms in Mathematical Physics*, 3rd edn. (Wiley, New York, 1966), p.48, Eq. (4.15)
29. V. Mangulis, Kramers-Kronig or dispersion relations in acoustics. J. Acoust. Soc. Am. **36**, 211–212 (1964)
30. C.J. Bouwkamp, A contribution to the theory of acoustic radiation. Phillips Res. Rep. **1**, 251–277 (1946)
31. R.L. Pritchard, Mutual acoustic impedance between radiators in an infinite rigid plane. J. Acoust. Soc. Am. **32**, 730–737 (1960)
32. D.T. Porter, Self and mutual radiation impedance and beam patterns for flexural disks in a rigid plane. J. Acoust. Soc. Am. **36**, 1154–1161 (1964)
33. W. Thompson Jr., The computation of self and mutual radiation impedances for annular and elliptical pistons using Bouwkamp's integral. J. Sound Vib. **17**, 221–233 (1971)
34. G. Brigham, B. McTaggart, Low frequency acoustic modeling of small monopole transducers, in *UDT 1990 Conference Proceedings* (Microwave Exhibitions and Publishers, Tunbridge Wells), pp. 747–755
35. L. Challis, F. Sheard, The green of Green's functions. Phys Today **56**, 41–46 (2003)
36. B.B. Baker, E.T. Copson, *The Mathematical Theory of Huygens' Principle*, 3rd edn. (AMS Chelsea, Providence, 1987)
37. T. Mellow, L. Karkkainen, On the sound field of an oscillating disk in a finite open and closed circular baffle. J. Acoust. Soc. Am. **118**, 1311–1325 (2005)
38. A.L. Butler, J.L. Butler, Near-field and far-field measurements of a ribbon tweeter/midrange. J. Acoust. Soc. Am. **98**(5), 2872 (1995) (A)
39. S.N. Reschevkin, *A Course of Lectures on the Theory of Sound* (Pergamon, Oxford, 1963). Chapter 11
40. E.M. Arase, Mutual radiation impedance of square and rectangular pistons in a rigid infinite baffle. J. Acoust. Soc. Am. **36**, 1521–1525 (1964)
41. C.H. Sherman, Special relationships between the farfield and the radiation impedance of cylinders. J. Acoust. Soc. Am. **43**, 1453–1454 (1968)
42. V. Mangulis, Relation between the radiation impedance, pressure in the far field and baffle impedance. J. Acoust. Soc. Am. **36**, 212–213 (1964)
43. D.D. Baker, Determination of far-field characteristics of large underwater sound transducers from near-field measurements. J. Acoust. Soc. Am. **34**, 1737–1744 (1962)
44. W. Thompson Jr., Radiation from a spherical acoustic source near a scattering sphere. J. Acoust. Soc. Am. **60**, 781–787 (1976)
45. W. Thompson Jr., Acoustic coupling between two finite-sized spherical sources. J. Acoust. Soc. Am. **62**, 8–11 (1977)
46. T.A. Henriquez, Diffraction constants of acoustic transducers. J. Acoust. Soc. Am. **36**, 267–269 (1964)
47. J.L. Butler, D.T. Porter, A Fourier transform solution for the acoustic radiation from a source near an elastic cylinder. J. Acoust. Soc. Am. **89**, 2774–2785 (1991)
48. W. Magnus, F. Oberhettinger, *Formulas and Theorems for the Functions of Mathematical Physics* (Chelsea, New York, 1962), pp. 139–143
49. H. Stenzel, *Leitfaden zur Berechnung von Schallvorgangen* (Springer, Berlin, 1939), pp. 75–79
50. M.C. Junger, Sound radiation from a radially pulsating cylinder of finite length. Harvard Univ. Acoust. Res. Lab. (24 June 1955)
51. M.C. Junger, A Variational Solution of Solid and Free-Flooding Cylindrical Sound Radiators of Finite Length, Cambridge Acoustical Associates Tech. Rept. U-177-48, Contract Nonr-2739(00) (1 March 1964)

52. W. Williams, N.G. Parke, D.A. Moran, C.H. Sherman, Acoustic radiation from a finite cylinder. J. Acoust. Soc. Am. **36**, 2316–2322 (1964)
53. B.L. Sandman, Fluid loading influence coefficients for a finite cylindrical shell. J. Acoust. Soc. Am. **60**, 1256–1264 (1976)
54. J.L. Butler, Solution of acoustical-radiation problems by boundary collocation. J. Acoust. Soc. Am. **48**, 325–336 (1970)
55. C.C. Gerling, W. Thompson Jr., Axisymmetric spherical radiator with mixed boundary conditions. J. Acoust. Soc. Am. **61**, 313–317 (1977)
56. V. Mangulis, On the radiation of sound from a piston in a nonrigid baffle. J. Acoust. Soc. Am. **35**, 115–116 (1963)
57. L.H. Chen, D.G. Schweikert, Sound radiation from an arbitrary body. J. Acoust. Soc. Am. **35**, 1626–1632 (1963)
58. G. Chertock, Sound radiation from vibrating surfaces. J. Acoust. Soc. Am. **36**, 1305–1313 (1964)
59. L.G. Copley, Integral equation method for radiation from vibrating bodies. J. Acoust. Soc. Am. **41**, 807–816 (1967)
60. H.A. Schenck, Improved integral formulation for acoustic radiation problems. J. Acoust. Soc. Am. **44**, 41–58 (1968)
61. J.L. Butler, R.T. Richards, Micro-CHIEF, An interactive desktop computer program for acoustic radiation from transducers and arrays, in *UDT conference Proceedings* (London, 7–9 February 1990)

Chapter 12
Nonlinear Mechanisms and Their Effects

Most natural mechanisms and man-made devices are nonlinear, although linearity is often a good approximation and has been the basis for most engineering developments. In many devices the effects of nonlinearity become apparent only under high drive conditions, while other devices are inherently nonlinear and exhibit nonlinear effects, such as frequency doubling, for the smallest of drives. In the latter cases approximate linearity can only be achieved by imposing a bias. Among the electroacoustic transducers only the piezoelectric and moving coil mechanisms have a linear mechanical response to an applied field, and it remains linear only for small amplitudes.

The electrostrictive, magnetostrictive, electrostatic, and variable reluctance mechanisms differ significantly from the piezoelectric and moving coil mechanisms in that they have no region of linear operation, even for small amplitudes, unless linearization is imposed by applying a bias. In these cases the natural mechanical response, before linearization, is an even function of the applied electric or magnetic fields and therefore follows a square law for small amplitudes. The means of linearization is basically the same in all cases; a large electric or magnetic bias field is applied which establishes a polar axis that gives the material or device a one-way character. Then a superimposed alternating drive field that is smaller than the bias field can only increase and decrease the magnitude of the total field without changing its direction. The bias produces a linear component of the motion, but the nonlinear components are still present and become significant as the drive level is increased.

The first part of this chapter will describe the nonlinear mechanisms in the six major transducer types using the one-dimensional, lumped-parameter transducer models in Chap. 2. Transducers often contain structural features that introduce other nonlinearities in addition to the basic transduction mechanism, and some of these will also be included. Nonlinear mechanisms have observable effects such as distortion of the transducer output waveform, reduction of the output power relative to linear extrapolation from low level or change of resonance frequency and coupling coefficient. The remainder of this chapter will be devoted to analyzing

J.L. Butler, C.H. Sherman, *Transducers and Arrays for Underwater Sound*,
Modern Acoustics and Signal Processing, DOI 10.1007/978-3-319-39044-4_12

such effects to provide quantitative understanding of their causes and to show how they might be controlled.

One of the most important practical effects of nonlinearity is harmonic distortion. It occurs in all the transducer types and is caused by many different nonlinear mechanisms. We will find that the nonlinear mechanical equations for simple lumped-parameter models of all the transducer types can be formulated in a common way that gives solutions for the harmonics that apply to all the types. It will also be shown that these solutions can be used for estimating harmonics in transducers with more complex structures if the active drive part undergoes one-dimensional motion, and if the other moving parts contain no significant nonlinearities; e.g., flextensional transducers may be in this category. Analysis of harmonic distortion in distributed parameter transducers will also be illustrated for the case of nonlinear longitudinal vibrations of a bar. Nonlinear distortion is usually a precursor to the ultimate nonlinear effects of mechanical failure, electrical breakdown, and overheating.

12.1 Nonlinear Mechanisms in Lumped-Parameter Transducers

12.1.1 Piezoelectric Transducers

Materials with coupled elastic and electric or magnetic properties can be described by phenomenological equations of state that relate stress, T, and strain, S, to electric field, E, and electric displacement, D, or magnetic field, H, and magnetic flux density, B. Both electric and magnetic fields could be included in the same set of equations, and temperature and entropy could also be included for a complete description [1]. However, for present purposes it is sufficient to consider materials in which only electric or only magnetic fields are important and to consider thermal effects only in that material properties are understood to be temperature dependent. The equations to be used here are series expansions with the same phenomenological basis as those used in Chap. 2, but here the first nonlinear terms (squares and cross products of the independent variables) will be kept. Although S and D were chosen as dependent variables in the linear equations of Sect. 2.1, it will be more convenient with nonlinear equations to make T and D dependent.

The lumped-parameter longitudinal resonator transducer models used in Chap. 2 will also be used in this section to illustrate nonlinear mechanisms. In the ideal piezoelectric ceramic longitudinal resonator all the nonlinearity resides in the active material and is included in the equations of state. The basic piezoelectric transducer to be discussed first is based on a thin 33 mode bar of piezoelectric ceramic with the electric field parallel to its length. Figure 12.1 shows the basic mechanical components in a schematic form that also applies to the other transducer types for the purposes of this section (Figs. 2.6, 2.7, 2.8, 2.9, and 2.10 show more detail for each type).

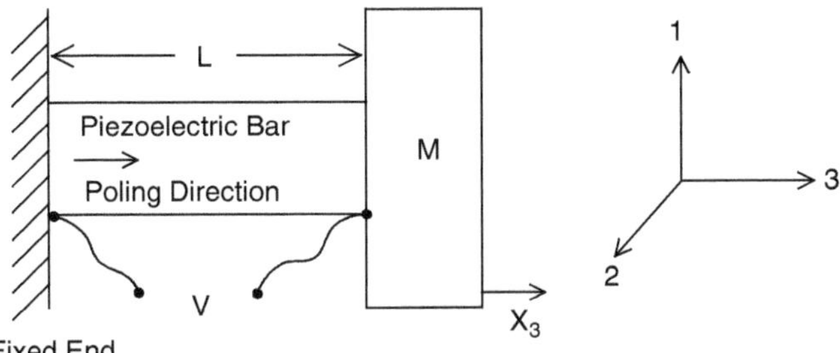

Fig. 12.1 The 33-mode piezoelectric longitudinal vibrator with working strain parallel to poling direction and electrodes on the ends of the bar

The only electric field components are E_3 and D_3, and there is only one stress component, T_3, because of the stress free conditions on the sides of the thin bar. However, there are both transverse, S_1 and S_2, and parallel, S_3, strain components, and all three appear in the equation for T_3 when strain is an independent variable. However, S_1 and S_2 can be eliminated approximately after expressing them as functions of T_3 and E_3 by use of the linear equations [2, 3]. Then the one-dimensional, nonlinear equations of state can be written as a series expansion to second order:

$$T = c_1 S - e_1 E + c_2 S^2 - 2c_a SE - e_2 E^2, \tag{12.1}$$

$$D = e_1 S + \varepsilon_1 E + c_a S^2 + 2e_2 SE + \varepsilon_2 E^2. \tag{12.2}$$

The derivation of equations of state from thermodynamic potentials is described by Berlincourt [1] in the linear case, and by Ljamov [4] and Mason [5] in the nonlinear case. In these phenomenological equations the number of terms to be used depends on the application, and the values of the coefficients are usually determined by measurements (a magnetostrictive example will be mentioned in Sect. 12.1.3).

In these equations *T*, *S*, *E*, and *D* are the components parallel to the bar, but the subscripts are omitted here to simplify the equations, because only the three component of each is involved. The coefficient notation in Eqs. (12.1) and (12.2) is related to the standard linear piezoelectric notation (see Chap. 2), but, since the presence of nonlinear terms requires new coefficients, new notation is necessary. Thus we let the coefficient subscripts refer to the order of the term rather than to directions. For example, the coefficient c_1 is a first-order elastic constant, c_2 is a second-order elastic constant, e_1 is a first-order, and e_2 and c_a are second-order, piezoelectric constants, ε_1 and ε_2 are first- and second-order permittivities, etc. The physical meaning of each coefficient is best seen from its relationship to a partial derivative, e.g., $c_1 = (\partial T/\partial S)_E, c_2 = (1/2)\left(\partial^2 T/\partial S^2\right)_E$, and the others are evident from the equations. Some of the coefficients are the same in both equations because they are derivatives of an energy function that is an exact differential.

Ljamov [4] used essentially the same second-order equations. Note that the coefficients c_1, e_1, and ε_1 are not c_{33}^E, e_{33}, and ε_{33}^S, respectively, because of approximate elimination of the transverse strains as previously mentioned [2]. The relationships in Sect. 13.4 apply to first-order coefficients, and the values in Sect. 13.5 are first-order coefficients measured at low amplitude.

For sufficiently small values of the applied electric field the linear terms in Eqs. (12.1) and (12.2) are dominant, and the operation of a piezoelectric transducer is approximately linear. However, for larger fields some of the other terms become important and nonlinear effects may be significant. Note that Eqs. (12.1) and (12.2) contain seven independent coefficients, while the linear equations that are adequate for most transducer work contain only three. Note also that nonlinearity destroys electromechanical reciprocity, since $|T/E|_{S=0} = e_1 + e_2 E$ while $|D/S|_{E=0} = e_1 + c_a S$, and these two ratios are not equal unless the nonlinear terms are negligible.

The equation of motion for the one-dimensional, lumped-parameter transducer in Fig. 12.1 is basically the same as Eq. (2.7), except that now the nonlinear Eq. (12.1) will be used, where stress is a dependent variable which can be inserted directly into the equation as follows:

$$M\ddot{x} + R\dot{x} = -AT = A\left[c_1 S - e_1 E + c_2 S^2 - 2c_a SE - e_2 E^2\right]. \tag{12.3}$$

In this chapter A, rather than A_0, is the cross-sectional area of the active material, and F_b is omitted because only projector operation will be considered. The strain as a function of displacement, x, and the electric field as a function of x and voltage, V, are the same as in Sect. 2.1:

$$S = x/L_0, \tag{12.4}$$

$$E = V_1 \cos \omega t/(L_0 + x). \tag{12.5}$$

The voltage is written as $V_1 \cos \omega t$, rather than $V_1 e^{j\omega t}$, in Eq. (12.5) because complex exponential notation cannot be used in nonlinear analysis where products of the variables are involved.

Equation (12.5) also shows that voltage drive is being specified (see last paragraph of Sect. 2.10). Nonlinear analysis of electric field transducers, where the drive forces depend on electric field, is easier for voltage drive than for current drive, while the opposite is true for magnetic field transducers, where the drive forces depend on magnetic field. This distinction is significant, and justifies defining voltage drive of electric field transducers and current drive of magnetic field transducers as direct drive, while current drive of electric field transducers and voltage drive of magnetic field transducers is defined as indirect drive. The method of harmonic analysis to be described in Sect. 12.2.1 holds for direct drive of all the transducer types. Nonlinear analysis for indirect drive is more complicated as will be discussed briefly in Sect. 12.2.2. Using Eqs. (12.4) and (12.5) in Eq. (12.3) gives:

$$M\ddot{x} + R\dot{x} = -A\left[c_1\frac{x}{L_0} - e_1\frac{V_1\cos\omega t}{L_0+x} + c_2\left(\frac{x}{L_0}\right)^2 - 2c_a\frac{xV_1\cos\omega t}{L_0(L_0+x)} - e_2\left(\frac{V_1\cos\omega t}{L_0+x}\right)^2\right]. \tag{12.6a}$$

Note that now the denominator of Eq. (12.5) is not being approximated by L_0 as it was in Chap. 2. This means that the variation of the electric field caused by the motion is being included, although the voltage amplitude is maintained constant. Equation (12.6a) is a nonlinear differential equation for x, which can only be solved approximately. An approximate solution can be found more conveniently if we expand the second, fourth, and fifth terms on the right by use of the binomial series. For example,

$$\frac{e_1V_1\cos\omega t}{L_0+x} = e_1V_1\cos\omega t(L_0+x)^{-1} = \frac{e_1V_1\cos\omega t}{L_0}\left(1 - \frac{x}{L_0} + \frac{x^2}{L_0^2} + \cdots\right). \tag{12.6b}$$

It can be seen that the other terms on the right-hand side of Eq. (12.6a) with (L_0+x) in the denominator can be expanded in a similar way, and then the equation can be written in the form

$$\ddot{x} + r\dot{x} = \sum_{n=0}^{n'}\sum_{m=0}^{m'}\gamma_{nm}x^n\cos m\omega t, \tag{12.7}$$

where $r = R/M$, and n' and m' determine the number of terms to be included. The γ_{nm} are the coefficients of each term of the form $x^n\cos m\omega t$ and include the constant factor (A/M). The low order γ_{nm} for Eq. (12.6a), which applies to piezoelectric transducers, are given in Table 12.1. For example, it can be seen that the first parts of the expanded term in Eq. (12.6b) give:

$$\gamma_{01} = Ae_1V_1/ML_0 \quad \text{and} \quad \gamma_{11} = -Ae_1V_1/ML_0^2.$$

The quantity γ_{01} is the amplitude of the linear drive term, while γ_{11} is the coefficient of a nonlinear term that gives the first-order effect caused by the variation of the electric field.

There are several reasons for putting the equation of motion in the form of Eq. (12.7), the most important being that this form will be shown to apply to direct drive of all the other major transducer types with different expressions for the γ_{nm} coefficients for each type. It follows that solutions of Eq. (12.7) in terms of the γ_{nm} coefficients can then be used to describe nonlinear effects for direct drive of all the major transducer types. In addition, the functions $x^n\cos m\omega t$ are convenient for perturbation analysis and facilitate physical interpretation, with the coefficients, γ_{nm}, determining the strength of each term.

Table 12.1 $M\gamma_{nm}$ for piezoelectric ($E_1 = V_1/L_0$) and moving coil transducers (additional results are given in [3])

nm	Piezoelectric	Moving coil
00	$\frac{1}{2}e_2A\,E_1{}^2$	$-\frac{1}{4}L_1I^2$
10	$-c_1A/L_0 - e_2AE_1^2/L_0$	$-K_1 - \frac{1}{2}L_2I^2$
20	$c_2A/L_0^2 + 3e_2A\,E_1^2/2L_0^2$	$-K_2 - \frac{3}{4}L_3I^2$
01	e_1AE_1	B_0l_cI
11	$-AE_1(e_1 - 2c_a)/L_0$	B_1l_cI
21	$AE_1(e_1 - 2c_a)/L_0{}^2$	B_2l_cI
02	$\frac{1}{2}e_2AE_1^2$	$-\frac{1}{4}L_1I^2$
12	$-e_2AE_1^2/L_0$	$-\frac{1}{2}L_2I^2$
22	$3e_2AE_1^2/2L_0^2$	$-\frac{3}{4}L_3I^2$

Of the first nine terms in Eq. (12.7) the three for $n=0$ are a constant force for $m=0$, the fundamental drive force for $m=1$, and a second harmonic drive force for $m=2$. None of these forces for $n=0$ make the equation nonlinear in a mathematical sense, but the constant force and the second harmonic force have nonlinear effects, i.e., they cause a static displacement and a second harmonic displacement. Both of these displacement components are proportional to the square of the drive voltage and are present only because of the second-order terms in the equation of state (see Table 12.1).

The term for $n=1$, $m=0$ is the ordinary linear spring force with amplitude depending mainly on the elasticity of the material but also on the square of the drive voltage as shown in Table 12.1 if the nonlinear constant e_2 is significant. Thus γ_{10} determines the square of the velocity resonance frequency, ω_r, which can be written as

$$-\gamma_{10} = \omega_r^2 = \omega_0^2 + \omega_d^2, \tag{12.8}$$

where ω_0 is the velocity resonance for low amplitude drive and ω_d is the first-order drive-dependent part of the resonance frequency. The quantity ω_d^2 can be identified from γ_{10} in Table 12.1 by its dependence on the drive voltage, showing that $\omega_d^2 = e_2AV_1^2/L_0^3M$, while $\omega_0^2 = c_1A/L_0M$. The term for $n=1$, $m=1$ is linear in x, but the time-dependent coefficient of x makes it as difficult to solve as a term nonlinear in x. Physically this term represents a time-dependent spring force or a displacement-dependent drive force. The remaining terms are all nonlinear in x; some are nonlinear spring terms, some are nonlinear drive terms, others are mixed.

Perturbation solutions of Eq. (12.7) will be deferred to Sect. 12.2.1, following discussion of nonlinear mechanisms in the other transducer types. For direct drive of each transducer type it will be found that the equation of motion can be put in the form of Eq. (12.7), and the perturbation results will then be applicable to all types. The perturbation solutions will give the harmonics in the displacement, from which the harmonics in the strain, velocity, or acceleration can be found. The harmonics in the unspecified electrical variable (current for electric field transducers, voltage for magnetic field transducers) can then be obtained from the time derivative of

Eq. (12.2) by use of the nonlinear solution for the displacement and the specified electrical variable. For example, the current in the piezoelectric case is

$$I = A(\mathrm{d}D/\mathrm{d}t) = A\left[e_1\dot{x}/L_0 + \varepsilon_1\dot{E} + 2c_a x\dot{x}/L_0^2 + (2e_2/L_0)\left(x\dot{E} + \dot{x}E\right) + 2\varepsilon_2 E\dot{E}\right],$$

where x is the nonlinear solution for the displacement and E is given by Eq. (12.5) in terms of x and the applied voltage.

Indirect drive of all the transducer types can also be analyzed by perturbation methods. In that case the analysis is more complicated because the equation of motion contains the unspecified electrical drive variable, while the specified drive variable is in the electrical equation. Thus the two nonlinear transducer equations must be solved simultaneously. Some results for indirect drive have been obtained in order to investigate the influence of drive type on harmonic distortion [6, 7] (see Sect. 12.2.2).

12.1.2 *Electrostrictive Transducers*

In electrostrictive materials, where the stress and strain are even functions of the electric field, appropriate equations of state are obtained by extending the series expansions in Eqs. (12.1) and (12.2) to fourth order [3], and then setting the coefficients of all the terms in E and E^3 equal to zero. Other terms, such as those for third- and fourth-order elasticity, will also be omitted for simplicity. The rationale here is to use equations that have the correct dependence on electric field and have about the same number of terms as the piezoelectric equations. The resulting equations are:

$$T = c_1S + c_2S^2 - e_2E^2 - e_4E^4, \tag{12.9}$$

$$D = 2e_2SE + 4e_4SE^3 + \varepsilon_1E + \varepsilon_2E^2. \tag{12.10}$$

The ferroelectric electrostrictive materials that are of most interest usually exhibit hysteresis and have a remanent polarization which is not described by these equations. As discussed briefly in Sect. 2.2, Piquette and Forsythe [8, 9] have developed a more complete, three-dimensional phenomenological model for electrostrictive ceramics that explicitly includes saturation and remanent polarization. However, some important electrostrictive materials such as ceramic PMN have very little remanent polarization and very narrow hysteresis loops [10]. In these cases T and D as functions of S and E can be approximately described by single average curves corresponding to Eqs. (12.9) and (12.10).

As described in Sect. 2.2 for materials with small remanence, it is necessary to maintain a bias field, E_0, to achieve linearity for small amplitude operation. A compressive prestress, T_0, is also often required to avoid tensile failure when these materials are driven with high fields. These electrical and mechanical bias fields cause a static electric displacement, D_0, and a static strain, S_0. When an alternating electric field, E_3, is applied in addition to the bias it causes alternating values of the other variables, T_3, S_3, and D_3 and the equation for the total stress is given by Eq. (12.9) as

$$T_0 + T_3 = c_1(S_0 + S_3) + c_2(S_0 + S_3)^2 - e_2(E_0 + E_3)^2 - e_4(E_0 + E_3)^4. \quad (12.11)$$

The same longitudinal resonator transducer considered in the previous section will be considered here, but now the active material is a bar of biased electrostrictive ceramic. After expanding the right side of Eq. (12.11) the static terms can be cancelled, because they alone satisfy the equation, since when $E_3 = 0$, S_3 and T_3 also equal zero. This gives the expression for the alternating stress that is needed in the equation of motion:

$$T_3 = (c_1 + 2c_2S_0)S_3 + c_2S_3^2 - \left(2e_2E_0 + 4e_4E_0^3\right)E_3 - \left(e_2 + 6e_4E_0^2\right)E_3^2 \\ - 4e_4E_0E_3^3 - e_4E_3^4. \quad (12.12)$$

T_3 can be expressed in terms of the displacement of the mass, x, the bias voltage, V_0, and the drive voltage, $V_3 = V_{30} \cos \omega t$, by use of the relations: $S_3 = x/L_0$, $E_0 = V_0/L_0$, and $(E_0 + E_3) = (V_0 + V_3)/(L_0 + x)$ which can be solved for $E_3 = (V_3 - xV_0/L_0)/(L_0 + x)$, where L_0 is the length of the bar after application of E_0 and T_0. Then the equation of motion becomes:

$$M_t\ddot{x} + R_t\dot{x} = -AT_3 = -A\Big[(c_1 + 2c_2S_0)\frac{x}{L_0} + c_2\left(\frac{x}{L_0}\right)^2 \\ - e_2\left(2E_0E_3 + E_3^2\right) - e_4\left(4E_0^3E_3 + 6E_0^2E_3^2 + 4E_0E_3^3 + E_3^4\right). \quad (12.13)$$

Equation (12.13) can be written in the form of Eq. (12.7), and some of the low order γ_{nm} that result are given in Table 12.2. The case of γ_{02} will be used in Sect. 12.2.1 to

Table 12.2 $M\gamma_{nm}$ for electrostrictive ($F_0 = E_0 = V_0/L_0$, $F_{30} = E_{30} = V_{30}/L_0$) and magnetostrictive ($F_0 = H_0 = nI_0/2\ L_0$, $F_{30} = H_{30} = nI_{30}/2\,L_0$) transducers with maintained polarization (additional values are given in [3])

nm	$M\gamma_{nm}$
00	$A\left[\tfrac{1}{2}e_2F_{30}^2 + 3e_4\left(F_0^2F_{30}^2 + F_{30}^4/8\right)\right]$
10	$-(A/L_0)\left[c_1 + 2S_0c_2 + 2e_2\left(F_0^2 + \tfrac{1}{2}F_{30}^2\right)\right]$
20	$-(A/L_0^2)\left[c_2 - 3e_2\left(F_0^2 + \tfrac{1}{2}F_{30}^2\right)\right]$
01	$A\left[2e_2F_0F_{30} + e_4\left(4F_0^3F_{30} + 3F_0F_{30}^3\right)\right]$
11	$-(A/L_0)(4e_2F_0F_{30})$
02	$A(\tfrac{1}{2}e_2)F_{30}{}^2$

calculate a second harmonic component for comparison with the electrostrictive theory and measurements of Piquette and Forsythe [9].

The values in Table 12.2 apply to electrostrictive materials with very small remanence in which the bias must be maintained. At the other extreme are the electrostrictive materials with high remanence that, when permanently polarized, become the piezoelectric ceramics. These were discussed in Sect. 2.2 to show how the linear piezoelectric ceramic coefficients are related to the electrostrictive coefficients. They also display nonlinear effects for high voltage drive, and the analysis and results in Sect. 12.1.1 and Table 12.1 for piezoelectric materials apply to them. It should be noted, however, that the piezoelectric ceramic coefficients depend on the remanent polarization, which may be changed by high static stress, high temperature, or high alternating field [3].

12.1.3 Magnetostrictive Transducers

Magnetostriction is, in many ways, the magnetic analog of electrostriction, especially since we are not considering magnetic and electric loss mechanisms. For example, magnetostrictive strain is an even function of the magnetic field, and therefore an approximate phenomenological description of a magnetostrictive material can be obtained simply by changing from electrical to magnetic nomenclature in the electrostrictive equations, Eqs. (12.9) and (12.10). The result is

$$T = c_1 S + c_2 S^2 - e_2 H^2 - e_4 H^4, \tag{12.14}$$

$$B = 2e_2 SH + 4e_4 SH^3 + \mu_1 H + \mu_2 H^2. \tag{12.15}$$

Again the longitudinal resonator transducer, as in Sect. 2.3 and Fig. 2.7, will be used to develop the nonlinear equation of motion. The quantity L_0 is the length of the magnetostrictive bars after application of a prestress, T_0, and a magnetic bias field, H_0, A is the cross-sectional area of both bars, and n is the number of turns in the coil on each bar. With I_0 the bias current, $I_3 = I_{30} \cos \omega t$ the drive current (for direct drive of a magnetic field transducer), and x the displacement of the mass, the relations among the variables are $S = x/L_0$, $H_0 = nI_0/L_0$, and $H_0 + H_3 = n(I_0 + I_3)/(L_0 + x)$ giving

$$H_3 = n(I_3 - xI_0/L_0)/(L_0 + x).$$

Since these relationships and Eqs. (12.14) and (12.15) are analogous to those for electrostriction the equation of motion and the γ_{nm} are also analogous, and Table 12.2 includes the γ_{nm} for magnetostriction as well as electrostriction.

It will be shown in Sect. 12.2.1 that the harmonics generated by nonlinear mechanisms can be expressed in terms of the γ_{nm} coefficients that depend on material parameters such as those in Eq. (12.14). Such material parameters must

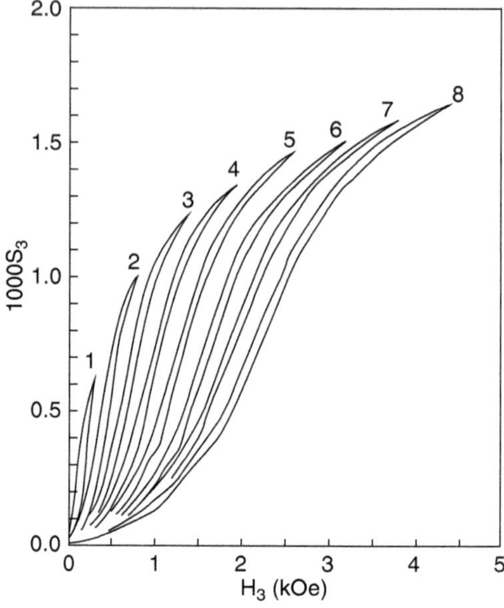

Fig. 12.2 Eight measured hysteresis loops of extensional strain versus applied magnetic field [11] in Terfenol-D for constant stress values of Table 12.3

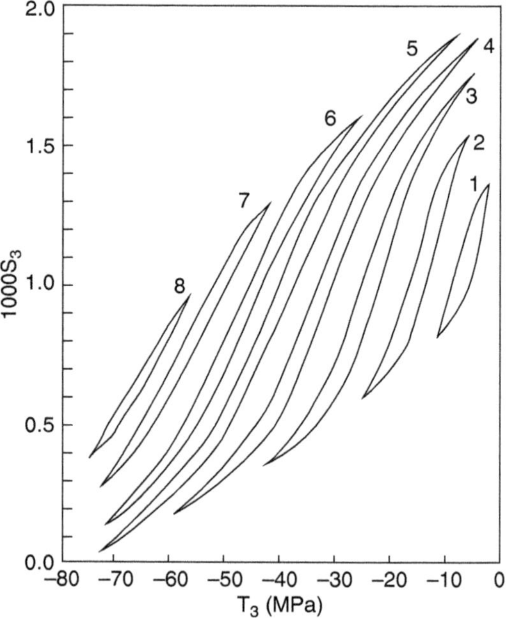

Fig. 12.3 Eight hysteresis loops of extensional strain versus stress [11] in Terfenol-D for constant magnetic fields of Table 12.3

be determined by measurement. Figures 12.2 and 12.3 (Figs. 3 and 4 of [11]) are examples of measurements on Terfenol-D that might be used for this purpose.

They give strain as functions of magnetic field and stress for several bias conditions, each bias condition having different fixed values of stress bias and magnetic field bias as given in Table 12.3.

Table 12.3 Magnetic bias fields as a function of prestress [11]

Bias condition	Compressive prestress		Magnetic bias field	
1	1.01 ksi	6.9 MPa	0.15 kOe	11.9 kA/m
2	2.22	15.3	0.4	31.8
3	3.42	23.6	0.7	55.7
4	4.64	32.0	1.0	712.6
5	5.96	40.4	1.3	103
6	7.07	48.7	1.6	127
7	8.28	57.1	1.9	151
8	12.49	65.4	2.2	175

The measurements in Figs. 12.2 and 12.3 were used to evaluate the coefficients in Eq. (12.14), but the results were incomplete due to a lack of measured values of the static strain for each bias condition [2].

A different approach to using the data in Figs. 12.2 and 12.3 was taken by Moffett et al. [11]. Rather than interpreting the measurements in terms of a nonlinear theory, they used the data to calculate values of the linear constants d_{33}, s^H_{33}, and μ^T_{33} for each bias condition and each level of alternating drive. Because of the nonlinearity these parameters are not constant but depend on the bias and the drive level.

12.1.4 Electrostatic and Variable Reluctance Transducers

The nonlinearity in the simple models of body force transducers discussed in the previous three sections arose only from physical effects included in the equations of state of the active material. In real transducers other sources of nonlinearity often exist associated with structural features, especially those related to mounting and sealing. In the surface force transducers to be discussed now the driving mechanism, the spring, and other components all may include nonlinearities. The specific nonlinear mechanisms to be discussed in the next two sections occur often, but in any particular transducer design there may be other mechanisms that introduce unexpected nonlinear effects [12].

The nonlinearity of the electrostatic transduction mechanism has already been mentioned in Sect. 2.4 because this mechanism is basically nonlinear, and the output motion is second harmonic unless a large bias voltage, V_0, is applied. But the bias leads to another nonlinear effect, the instability that occurs when V_0 exceeds $[8K_mL^3/(27\varepsilon A)]^{1/2}$ as will be shown in Sect. 12.2.3. The equation of motion for voltage drive of the electrostatic transducer, including the nonlinearity associated with the basic mechanism, was given in Eq. (2.47). Now another common nonlinear mechanism in electrostatic transducers will be added to the equation of motion, a nonlinear spring force represented by the power series,

$$F(x) = K_1 x + K_2 x^2 + K_3 x^3 + \cdots. \tag{12.16}$$

Equation (2.47) then becomes, with $x = x_0 + x_1$ and $V = V_0 + V_1 \cos \omega t$,

$$M\ddot{x}_1 + R\dot{x}_1 + K_1(x_1 + x_0) + K_2(x_1 + x_0)^2 + K_3(x_1 + x_0)^3 = -\frac{\varepsilon A(V_0 + V_1 \cos \omega t)^2}{2(L_0 + x_1)^2}, \tag{12.17}$$

where $L_0 = L + x_0$, and x_0 is the static displacement caused by the bias. The number of terms in the spring force is arbitrarily limited here to three; in practice the number required depends on the drive level and specific features of the transducer. When the right side of Eq. (12.17) is expanded it can be put in the form of Eq. (12.7) with the γ_{nm} given in Table 12.4. The physical meaning of some of these terms can be clarified by collecting those that comprise the total nonlinear effective spring constant of the biased transducer, $K_0(x_1)$:

$$K_0(x_1) = \left(K_1 + 2x_0 K_2 + 3x_0^2 K_3 - \frac{\varepsilon A V_0^2}{L_0^3}\right) + \left(K_2 + 3x_0 K_3 + \frac{3\varepsilon A V_0^2}{2L_0^4}\right) x_1 + \left(K_3 - \frac{2\varepsilon A V_0^2}{L_0^5}\right) x_1^2 + \cdots \tag{12.18}$$

It can be seen that the nonlinearity of the spring, expressed by K_2 and K_3, combines with the bias, expressed by V_0 and x_0, to change the effective value of the linear spring constant as well as the coefficients of the nonlinear spring terms. The negative stiffness, that appeared in Sect. 2.4 as part of the linear solution, appears here as part of the linear spring constant, and it can also be seen that, even when the mechanical spring is linear, i.e., when $K_2 = K_3 = 0$, the bias makes the effective spring nonlinear.

Table 12.4 $M\gamma_{nm}$ for electrostatic transducers ($E_0 = V_0/L_0$, $E_1 = V_1/L_0$)

nm	$M\gamma_{nm}$
00	$-\frac{1}{4}\varepsilon A E_1^2$
10	$-(K_1 + 2x_0K_2 + 3x_0^2K_3) + (\varepsilon A/L_0)(E_0^2 + E_1^2/2)$
20	$-(K_2 + 3x_0K_3) - (3\varepsilon A/2L_0^2)(E_0^2 + E_1^2/2)$
01	$-\varepsilon A E_0 E_1$
11	$(2\varepsilon A/L_0)E_0E_1$
02	$-\varepsilon A E_1^2/4$

The results are analogous for Variable Reluctance Transducers (neglecting magnetic saturation) with ε, E_0, and E_1 replaced by μ, $H_0 = nI_0/2L_0$, and $H_1 = nI_1/2L_0$, respectively

In Sect. 2.5 the equation of motion for the variable reluctance transducer was derived assuming that the energy in the gaps in the magnetic circuit was much greater than the energy in the magnetic material. That assumption will be continued here which makes the variable reluctance mechanism analogous to the electrostatic, including the γ_{nm} coefficients in Table 12.4.

Additional results for $M\gamma_{nm}$ are given in [3] including approximate effects of magnetic saturation for the variable reluctance transducer.

12.1.5 Moving Coil Transducers

Three different sources of nonlinearity in the moving coil transducer (see Fig. 2.10) are now added to the linear equation, Eq. (2.65), discussed in Chap. 2 to obtain:

$$M_t\ddot{x} + R_t\dot{x} + K_m(x)x = B(x)l_cI + \frac{1}{2}\frac{dL}{dx}I^2. \tag{12.19}$$

$K_m(x)$ represents a nonlinear spring, as in Eq. (12.16), that is often important in moving coil transducers. The elasticity of the coil suspension and the elasticity of the enclosed air in back-enclosed transducers are two distinct components of $K_m(x)$ which is useful to consider separately because the latter can be readily calculated while the former usually cannot. Thus $K_m(x)$ will be written

$$\begin{aligned} K_m(x) &= K_s(x) + K_a(x) \\ &= (K_{s1} + K_{a1}) + (K_{s2} + K_{a2})x + (K_{s3} + K_{a3})x^2 + \cdots \end{aligned} \tag{12.20}$$

where

$$K_a(x) = \frac{\gamma A^2 p_0}{v_0}\left(1 + \frac{Ax}{v_0}\right)^{-(\gamma+1)} = K_{a1} + K_{a2}x + K_{a3}x^2 + \cdots$$

is the spring constant of the enclosed air for adiabatic variations of pressure and volume. For adiabatic conditions the pressure change, p, and the volume change, Ax, are related to the initial pressure and volume, p_0 and v_0, by $(p_0 + p)(v_0 + Ax)^\gamma = p_0 v_0^\gamma$, where γ is the ratio of specific heats of the air and A is the area of the vibrating surface. Solving for p and calculating $A\partial p/\partial x$ gives $K_a(x)$. The linear and nonlinear spring constants of the air can be found by expanding the expression for $K_a(x)$ with the results

Table 12.5 Linear and nonlinear spring constants (N/m)

i	K_{si}	K_{ai}
1	0.57×10^5	2×10^5
2	6.8×10^6	-0.84×10^6
3	1.3×10^8	0.18×10^7

$$K_{a1} = \gamma A^2 p_0 / v_0,$$

$$K_{a2} = -(\gamma + 1) K_{a1} (A/v_0),$$

$$K_{a3} = \frac{1}{2}(\gamma + 1)(\gamma + 2) K_{a1} (A/v_0)^2.$$

Analysis of measurements on a specific moving coil transducer designed for underwater use showed that the linear elasticity of the enclosed air, K_{a1}, exceeded the linear elasticity of the suspension, K_{s1}, while the nonlinear elasticity of the suspension was much greater than the nonlinear elasticity of the enclosed air. These results are given in Table 12.5 [6]; they are not typical, but they are a good example of how the suspension design can introduce nonlinearity.

In Eq. (12.19) $B(x)$ describes the variation of the magnetic field with position; it can usually be expressed as a power series,

$$B(x) = B_0 + B_1 x + B_2 x^2 + \cdots, \tag{12.21}$$

where B_0 is the radial magnetic field in the central part of the gap, and the other terms are corrections that apply when the coil moves away from the center. For sufficiently small displacement $B(x) \approx B_0$, and this term in Eq. (12.19) is the ordinary linear drive force. However, if the length of the gap is not much greater than the length of the coil, large displacement will bring part of the coil out of the gap into the smaller fringing field and the nonlinearity in this term may become important. In the underwater moving coil transducer mentioned above [6] the magnetic field nonuniformity was measured and found to be an insignificant source of nonlinearity compared to the nonlinearity in the suspension stiffness.

The third source of nonlinearity in Eq. (12.19) is caused by the secondary magnetic field produced by the drive current. This field contains stored energy of ½ LI^2, where L is the inductance of the coil, and, if the inductance varies with position in the gap, it causes an additional force on the coil given by

$$F = -(I^2/2)\mathrm{d}L/\mathrm{d}x,$$

parallel to the primary Lorentz force [13]. This force only exists if the inductance varies as the coil moves, which often occurs, since motion of the coil changes its spatial relationship to the magnet. The inductance can also be expressed as a power series,

$$L(x) = L_c + L_1 x + L_2 x^2 + \cdots \tag{12.22}$$

where L_c is the clamped inductance, and the other coefficients are determined by measurement. When Eqs. (12.20), (12.21), and (12.22) are used in Eq. (12.19), and current drive is specified with $I = I_0 \cos \omega t$, Eq. (12.19) can be put in the form of Eq. (12.7). The γ_{nm} coefficients are given in Table 12.1 where $K_1 = K_{s1} + K_{a1}$, etc. Note the mathematical similarity between the piezoelectric and moving coil mechanisms, the two "linear" transduction mechanisms. Geddes has described a comprehensive, but different, approach to analyzing harmonic distortion in moving coil transducers [14].

12.1.6 Other Nonlinear Mechanisms

The possibility of nonlinearity in resistive mechanisms always exists and has been observed in a significant way in a low frequency underwater moving coil transducer similar to the one mentioned above [6]. In this case the observed harmonic amplitudes were independent of drive level and independent of frequency above resonance, but they did depend on the orientation of the transducer axis with respect to the vertical direction [12]. The orientation dependence suggested that friction was the cause of the harmonics, since the moving parts of this transducer were mounted with bearings on a centering rod. Both even and odd harmonics were observed, whereas only odd harmonics would be expected if the frictional forces obeyed ordinary Coulomb damping where the magnitude of the friction force is constant but changes direction when the velocity changes direction. It was shown that a generalized version of Coulomb damping in which the frictional force changes direction with velocity, but the magnitude of the force changes with time, would generate both even and odd harmonics. The parameters of the generalized model could be adjusted to make the harmonic amplitudes agree approximately with the measurements [12].

When friction is important it is a unique source of harmonic distortion in transducers, because it generates harmonics at low amplitude, while other sources of distortion are usually only important at high amplitude. Thus the friction generated harmonics put a lower bound on the distortion-free dynamic range of projectors which otherwise would not exist.

12.2 Analysis of Nonlinear Effects

Since the nonlinear mechanisms discussed above usually affect transducers in undesirable ways, the purpose of nonlinear analysis is to understand quantitatively how these effects arise and, thus, gain some control over them. Harmonic distortion is an important nonlinear effect that occurs in all transducer types and increases as the drive amplitude increases; it will be discussed in detail in Sects. 12.2.1 and 12.2.2. Instability is another nonlinear effect that occurs in some transducers. As mentioned in Chap. 2, it can interfere with linear operation in some cases, and will be discussed in Sect. 12.2.3. As the drive amplitude is increased, not only does harmonic distortion increase, but other high amplitude effects may also occur. The transducer efficiency decreases as some of the input energy goes into generating harmonics. The resonance frequency, normally considered to be constant, changes with drive level, as can be seen in Eq. (12.8). And other parameters, such as the electromechanical coupling coefficient, may also change as the drive level increases.

12.2.1 Harmonic Distortion: Direct Drive Perturbation Analysis

The harmonic components resulting from any of the nonlinear mechanisms described in the previous sections (except for friction) will now be calculated by finding approximate solutions of Eq. (12.7) using perturbation analysis [15, 16]. These solutions will apply to direct drive of any of the six major transducer types [3]. It will be convenient to rewrite Eq. (12.7) as follows:

$$\ddot{x} + r\dot{x} + \omega_0^2 x = \gamma_{01}\cos\omega t + \gamma_{00} + \omega_{\mathrm{d}}^2 x + \sum_{n,m=1} \gamma_{nm} x^n \cos m\omega t. \tag{12.23}$$

In Eq. (12.23) the terms with coefficients γ_{01} and $\gamma_{10} = \omega_{\mathrm{d}}^2 - \omega_0^2$ have been removed from the summation to clearly separate the usual linear terms from the perturbing nonlinear terms. In addition γ_{10} has been separated into the part independent of drive level, ω_0^2, and the part dependent on drive level, ω_{d}^2. The latter causes a nonlinear change of resonance frequency with drive level; the former is the constant low level resonance frequency. Although γ_{00} is a nonlinear force, it was separated from the sum to emphasize that it is a force independent of both x and t that can only cause a static displacement, i.e., a zeroth-order harmonic. The upper summation limits are not indicated, because the number of terms included is a matter of judgment in each situation. Expressing all the physical perturbations in a series of the functions $\gamma_{nm}x^n \cos m\omega t$ is convenient mathematically, and gives solutions applicable to all transducers, but it should be noted that more than one such term may be associated with an individual physical mechanism. For example, the power series representation of a nonlinear spring in Eqs. (12.16) or (12.20) involves $\gamma_{10}, \gamma_{20}, \gamma_{30}\cdots$, and inclusion of the term with coefficient e_2 in the piezoelectric equations of state involves several different γ_{nm} as can be seen from Table 12.1.

Before starting the analysis a simple description of the growth of harmonics in a nonlinear vibrating system may be helpful. Consider Eq. (12.23) with only one nonlinear term, $\gamma_{20}x^2$:

$$\ddot{x} + r\dot{x} + \omega_0^2 x = \gamma_{01} \cos \omega t + \gamma_{20} x^2.$$

When the applied drive is small, x is small, $\gamma_{20}x^2$ is small enough to be negligible, and the linear solution, $x \approx X_{01} \sin(\omega t - \phi)$, is a good approximation, where X_{01} is the linear approximation of the fundamental amplitude, given in Eq. (12.30). As the drive is increased, x increases, and $\gamma_{20}x^2$ increases relative to x, and may no longer be negligible. The $\gamma_{20}x^2$ term in the equation makes an exact solution impossible, but the physical effect of this term can be approximated by using the linear solution for x to obtain

$$\gamma_{20}x^2 \approx \gamma_{20}X_{01}^2 \sin^2(\omega t - \phi) = \frac{1}{2}\gamma_{20}X_{01}^2[1 - \cos(2\omega t - 2\phi)].$$

This shows that the $\gamma_{20}x^2$ term will have the same effect as a static drive force plus a second harmonic drive force, which will add a static displacement and a second harmonic displacement to the solution. As the drive is increased still more the amplitude of the static and second harmonic components will increase, and they will give rise to another generation of harmonics through the same nonlinear process. Perturbation analysis is a systematic procedure for calculating these generations, or orders, of harmonics.

The perturbation procedure starts by assuming that the solution can be represented as a power series in a dimensionless perturbation parameter δ:

$$x(t) = x_0(t) + \delta x_1(t) + \delta^2 x_2(t) + \cdots \tag{12.24}$$

where x_0 is the linear solution, i.e., the unperturbed or zeroth-order solution, $\delta x_1(t)$ is the first-order solution that contains the first generation of harmonics, $\delta^2 x_2(t)$ is the second-order solution that contains the second generation of harmonics, etc. When the differential equation contains only one perturbing term, such as $\gamma_{20}x^2$ in the case above, δ is usually chosen to be a small quantity proportional to the coefficient of that term. In the present case where the nonlinear differential equation, Eq. (12.23), contains several nonlinear terms with different coefficients it is necessary to choose a perturbation parameter that can be related to all the terms. The parameter,

$$\delta = X_{01}/L_0, \tag{12.25}$$

is a convenient perturbation parameter; X_{01} is the linear approximation to the fundamental amplitude and L_0 is a length that is characteristic of the transducer being analyzed. For example, in the piezoelectric transducer L_0 is the length of the piezoelectric bar. The quantity X_{01}/L_0 is a suitable perturbation parameter in this situation, because it is a small, dimensionless quantity that is independent of any particular nonlinear mechanism. The quantity $\delta L_0/X_{01} = 1$ is then introduced into Eq. (12.23) as follows:

$$\ddot{x} + r\dot{x} + \omega_0^2 x = \gamma_{01} \cos \omega t + \frac{\delta L_0}{X_{01}} \left(\gamma_{00} + \omega_d^2 x \right) + \sum_{n,\, m=1} \gamma_{nm} \left(\frac{\delta L_0}{X_{01}} \right)^{n+m-1} x^n \cos m\omega t. \tag{12.26}$$

Including the perturbation parameter in this way makes the zeroth and second harmonics appear in first-order perturbation, while third harmonics, plus corrections to the linear fundamental, appear in second-order perturbation. The zeroth-order solution is the linear fundamental given by $\delta = 0$.

When Eq. (12.24) is substituted into Eq. (12.26) the terms can be divided into groups where each term in the group is proportional to the same power of δ. Thus each group of terms is the coefficient of a polynomial in δ that remains equal to zero as time changes. Since the coefficients depend on time, each one must separately vanish, which leads to the following three equations:

$$\ddot{x}_0 + r\dot{x}_0 + \omega_0^2 x_0 = \gamma_{01} \cos \omega t, \tag{12.27}$$

$$\ddot{x}_1 + r\dot{x}_1 + \omega_0^2 x_1 = \frac{L_0}{X_{01}} \left[\gamma_{00} + \omega_d^2 x_o + \gamma_{20} x_0^2 + \gamma_{11} x_0 \cos \omega t + \gamma_{02} \cos 2\omega t \right], \tag{12.28}$$

$$\ddot{x}_2 + r\dot{x}_2 + \omega_0^2 x_2 = \left(\tfrac{L_0}{X_{01}} \right)^2 \left[\gamma_{30} x_0^3 + \gamma_{21} x_0^2 \cos \omega t + \gamma_{12} x_0 \cos 2\omega t + \gamma_{03} \cos 3\omega t \right] + \frac{L_0}{X_{01}} \left[\omega_d^2 x_1 + \gamma_{11} x_1 \cos \omega t + 2\gamma_{20} x_0 x_1 \right], \tag{12.29}$$

Equation (12.27) is the zeroth-order equation with the usual linear solution. Equations (12.28) and (12.29) are the first- and second-order equations, which are linear in x_1 and x_2, respectively, with drive terms that depend on the solutions of the lower order equations. The last point is very important, because it means that all the equations can be solved in succession using the solutions of the lower order equations. It is also important to note that perturbation analysis has transformed the original, unsolvable nonlinear equation into a series of easily solvable linear equations. Since the equations are linear the solution for each drive term can be found separately from the others, and they can be added together if necessary, or considered separately to determine which are most important.

Since these equations contain products, such as x_0^2 and $x_0 x_1$, the complex exponential notation used in earlier chapters cannot be used. Therefore, the linear solution obtained by solving Eq. (12.27) must be used in real form, i.e.,

$$x_0 = X_{01} \sin \left(\omega t - \phi_1 \right), \tag{12.30}$$

where

$$X_{01} = \gamma_{01} / \omega_0^2 z_m(\nu), \quad \nu = \omega / \omega_0,$$

$$z_m(N\nu) = \left[\left(1 - N^2\nu^2\right)^2 + \left(N\nu/Q_m\right)^2\right]^{1/2} = \frac{N\omega}{\omega_0^2 M}|Z_m(N\omega)|, \tag{12.31}$$

$$\tan\phi_N = \frac{X_m(N\omega)}{R_m} = \frac{Q_m}{N\nu}\left(N^2\nu^2 - 1\right).$$

$Z_m(N\omega) = R_m + jX_m(N\omega)$ is the mechanical impedance at frequency $N\omega$, $Q_m = \omega_0/r = \omega_0 M/R_m$ is the mechanical quality factor and N is the harmonic order. (Note that [3] used a different definition of $\tan\phi_N$). If ω_d^2 had not been separated from ω_0^2 in Eq. (12.23) these expressions for X_{01}, z_m, Z_m, Q_m, and $\tan\phi_N$ would all have ω_0 replaced by ω_r which depends on drive level. The separation of the level-dependent part of the resonance frequency from the constant part makes it possible to keep using these familiar low level quantities while also obtaining the changes which occur with drive level.

When x_0, the zeroth-order solution, is used in Eq. (12.28), the equation for the first-order solution becomes

$$\ddot{x}_1 + r\dot{x}_1 + \omega_0^2 x_1 = \frac{L_0}{X_{01}}\left[\gamma_{00} + \omega_d^2 X_{01}\sin(\omega t - \phi_1) + \gamma_{20}X_{01}^2\sin^2(\omega t - \phi_1)\right.$$
$$\left. + \gamma_{11}X_{01}\sin(\omega t - \phi_1)\cos\omega t + \gamma_{02}\cos 2\omega t\right]. \tag{12.32}$$

It can be seen that perturbation analysis, not only provides a systematic way of solving for the harmonics, it also illustrates the physical mechanisms of harmonic generation. For example, the first drive term in Eq. (12.32) is a constant force, independent of x or t, and the solution for that term is a static displacement, while the second drive term varies with the fundamental frequency and therefore gives a correction to the linear fundamental. Similarly, the third drive term, when expanded, is a constant force plus a second harmonic force that gives another static displacement and the first approximation to the second harmonic displacement. Thus the mathematical process of perturbation analysis closely follows the physical process of harmonic generation.

Before discussing the complete solution of Eq. (12.32) we should emphasize that the solution for each drive term can be considered separately. For example, if there was reason to think that nonlinearity of the spring was the most important source of harmonics in a particular transducer, the term involving γ_{20} could be considered alone. When expanded this term gives a static force and a second harmonic force. The amplitude of the resulting second harmonic displacement is $\gamma_{20}X_{01}^2/2\omega_0^2 z_m(2\nu)$ which can be compared with the fundamental amplitude to assess its significance.

The complete solution of Eq. (12.32) for the first-order displacement, when multiplied by δ, gives the second term of the perturbation series in Eq. (12.24):

$$\delta x_1 = \frac{\gamma_{00}}{\omega_0^2} - \frac{\omega_d^2 X_{01}}{\omega_0^2 z_m(\nu)} \cos(\omega t - 2\phi_1) + \frac{\gamma_{20}}{2\omega_0^2}\left[X_{01}^2 - \frac{X_{01}^2}{z_m(2\nu)} \sin(2\omega t - 2\phi_1 - \phi_2)\right]$$
$$- \frac{\gamma_{11}}{2\omega_0^2}\left[X_{01}\sin\phi_1 - \frac{X_{01}}{z_m(2\nu)}\cos(2\omega t - \phi_1 - \phi_2)\right] + \frac{\gamma_{02}}{\omega_0^2 z_m(2\nu)}\sin(2\omega t - \phi_2). \tag{12.33}$$

The first, third, and fifth terms are contributions to the static displacement that result from γ_{00}, γ_{20} and γ_{11}. The second term is a correction to the linear fundamental in Eq. (12.30). Note that γ_{20} and γ_{11} both produce zeroth and second harmonics, while γ_{02} produces only second harmonic, because it is the coefficient of a term that does not involve x. The most useful parts of the first-order solution are usually the second harmonic components relative to the amplitude of the linear fundamental. Using the expression for X_{01} following Eq. (12.30) these displacement ratios are:

$$\frac{X_{202}}{X_{01}} = \frac{\gamma_{20}\gamma_{01}}{2\omega_0^4 z_m(2\nu) z_m(\nu)} \sin(2\omega t - 2\phi_1 - \phi_2), \tag{12.34}$$

$$\frac{X_{112}}{X_{01}} = \frac{\gamma_{11}}{2\omega_0^2 z_m(2\nu)} \cos(2\omega t - \phi_1 - \phi_2), \tag{12.35}$$

$$\frac{X_{022}}{X_{01}} = \frac{\gamma_{02} z_m(\nu)}{\gamma_{01} z_m(2\nu)} \sin(2\omega t - \phi_2), \tag{12.36}$$

where the three subscripts on X are n, m, and the harmonic order N. These components are the main contributions to the second harmonic displacement, but small corrections to the second harmonic would appear in higher order solutions.

Recall that these three contributions to the second harmonic arise from three different physical mechanisms: X_{202} from a nonlinear spring, X_{112} from the variation of electric field as displacement varies, and X_{022} from a square law drive. In any particular case one of these contributions may dominate the others. Since X_{01} is proportional to γ_{01} it can be seen from the γ_{nm} (e.g., from Table 12.1) that X_{112} and X_{022} increase with the square of the drive voltage or current, while X_{202} has two parts, one increasing with the square, the other with the fourth power, of the drive. Similar results for the third harmonic components have been obtained by using the zeroth- and first-order solutions, x_0 and x_1, in Eq. (12.29) [3].

It will be more meaningful to show the frequency dependence of the harmonic components after converting the displacement harmonics above to acoustic pressure harmonics. We will assume that the radiation process is linear, and that each harmonic radiates independently of the others. Then any of the results in Chap. 10 for the radiated pressure in terms of the velocity or acceleration of a transducer surface, such as Eqs. (10.18) and (10.15c) for the pulsating sphere, can be used to convert the displacement harmonics to pressure harmonics. Since $u = j\omega x$ the amplitudes of velocity harmonics are enhanced by the harmonic order, N, and the amplitudes of acceleration and pressure harmonics are enhanced by N^2, relative to the displacement harmonics. It follows from Eq. (10.15c) that the pressure amplitude of a particular harmonic component designated by nmN at a distance r is

$$|p_{nmN}(r)| = \frac{\rho A (N\omega)^2 X_{nmN}}{4\pi r}.$$

The ratio of the nmN pressure harmonic to the fundamental pressure is

$$\left|\frac{p_{nmN}(r)}{p_{01}(r)}\right| = \frac{N^2 X_{nmN}}{X_{01}},$$

where the displacement ratios for second harmonic components are given by Eqs. (12.34)–(12.36); e.g., $|p_{202}/p_{01}| = (2\gamma_{20}\gamma_{01})/\left[\omega_0^4 z_{\mathrm{m}}(2\nu) z_{\mathrm{m}}(\nu)\right]$. We can show the frequency dependence of the fundamental and the harmonics separately from the γ_{nm} parameters, which contain the electric or magnetic drive amplitudes, by defining dimensionless pressure amplitudes, P_{nmN}, related to the actual pressure amplitudes, $|p_{nmN}|$, as follows, where $\alpha = 4\pi r/\rho A$:

$$P_{01} = \frac{\alpha |p_{01}|}{\gamma_{01}} = \frac{\nu^2}{z_{\mathrm{m}}(\nu)},$$

$$P_{202} = \frac{2\alpha\omega_0^4 |p_{202}|}{\gamma_{20}\gamma_{01}^2} = \frac{(2\nu)^2}{z_{\mathrm{m}}^2(\nu) z_{\mathrm{m}}(2\nu)},$$

$$P_{112} = \frac{2\alpha\omega_0^2 |p_{112}|}{\gamma_{11}\gamma_{01}} = \frac{(2\nu)^2}{z_{\mathrm{m}}(\nu) z_{\mathrm{m}}(2\nu)},$$

$$P_{022} = \frac{\alpha |p_{022}|}{\gamma_{02}} = \frac{(2\nu)^2}{z_{\mathrm{m}}(2\nu)}.$$

The frequency dependence of the pressure harmonics normalized in this way applies to direct drive of all the transducer types. These quantities are plotted in Fig. 12.4 [3] as functions of the dimensionless frequency, ν, for $Q_{\mathrm{m}} = 10$, 3, and 1. It can be seen that P_{202} and P_{112} depend on the fundamental amplitude, and therefore on both $z_{\mathrm{m}}(\nu)$ and $z_{\mathrm{m}}(2\nu)$. Thus these components peak near $\omega = \omega_0/2$ where the second harmonic of the drive frequency excites the transducer resonance, and also near $\omega = \omega_0$ at the transducer resonance. However, P_{022} peaks only at $\omega_0/2$ because it depends only on $z_{\mathrm{m}}(2\nu)$. It can also be seen that the peaks are greatly diminished for $Q_{\mathrm{m}} = 3$ and completely damped out for $Q_{\mathrm{m}} = 1$.

Similar results for eight different third harmonic components are shown in Fig. 12.5, obtained by solving Eq. (12.29) for the second-order solution [3]. When the solution for x_1 in Eq. (12.33) is used in Eq. (12.29) it can be seen that

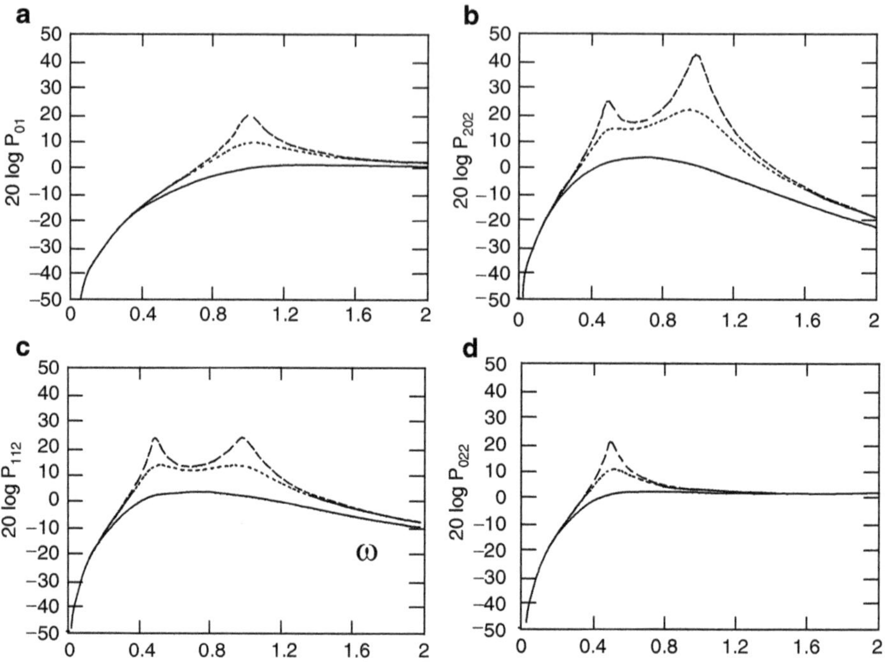

Fig. 12.4 Normalized pressure P (in dB) as a function of the dimensionless frequency, ν, for the (**a**) linear fundamental and second harmonics caused by (**b**) X^2, (**c**) $X \cos \omega t$, and (**d**) $\cos 2\omega t$ for $Q_m = 1$ (*solid line*), 3 (*dotted line*), and 10 (*dashed line*) [3]

terms containing the product of two different γ_{nm} will appear. The third harmonic components that result from these terms sometimes have three peaks because they depend on $z_m(\nu)$, $z_m(2\nu)$, and $z_m(3\nu)$. Measurements of harmonic amplitudes as a function of frequency, when compared with Figs. 12.4 and 12.5, may help determine the type of physical mechanism that causes them.

Results, such as Eqs. (12.34)–(12.36), can be readily incorporated into linear transducer models because the harmonic displacement components can be expressed in terms of the fundamental displacement at the drive frequency and at the harmonic frequencies. For example, Eq. (12.34) can be written as,

$$X_{202} = \frac{\gamma_{20}}{2\gamma_{01}} X_{01}^2(\omega) X_{01}(2\omega) \sin(2\omega t - 2\phi_1 - \phi_2), \qquad (12.37)$$

and the other second and higher harmonic components can also be expressed in terms of the fundamental in similar ways [3]. In Eq. (12.37) X_{202} is expressed entirely in terms of linear parameters except for the nonlinear amplitude parameter γ_{20}. Thus the function X_{01}, which contains all the parameters that determine the linear dynamic mechanical behavior of the transducer, also determines the frequency dependence of the harmonics for direct drive. Of course, the amplitudes of the harmonics are determined by the nonlinear parameters, γ_{nm}.

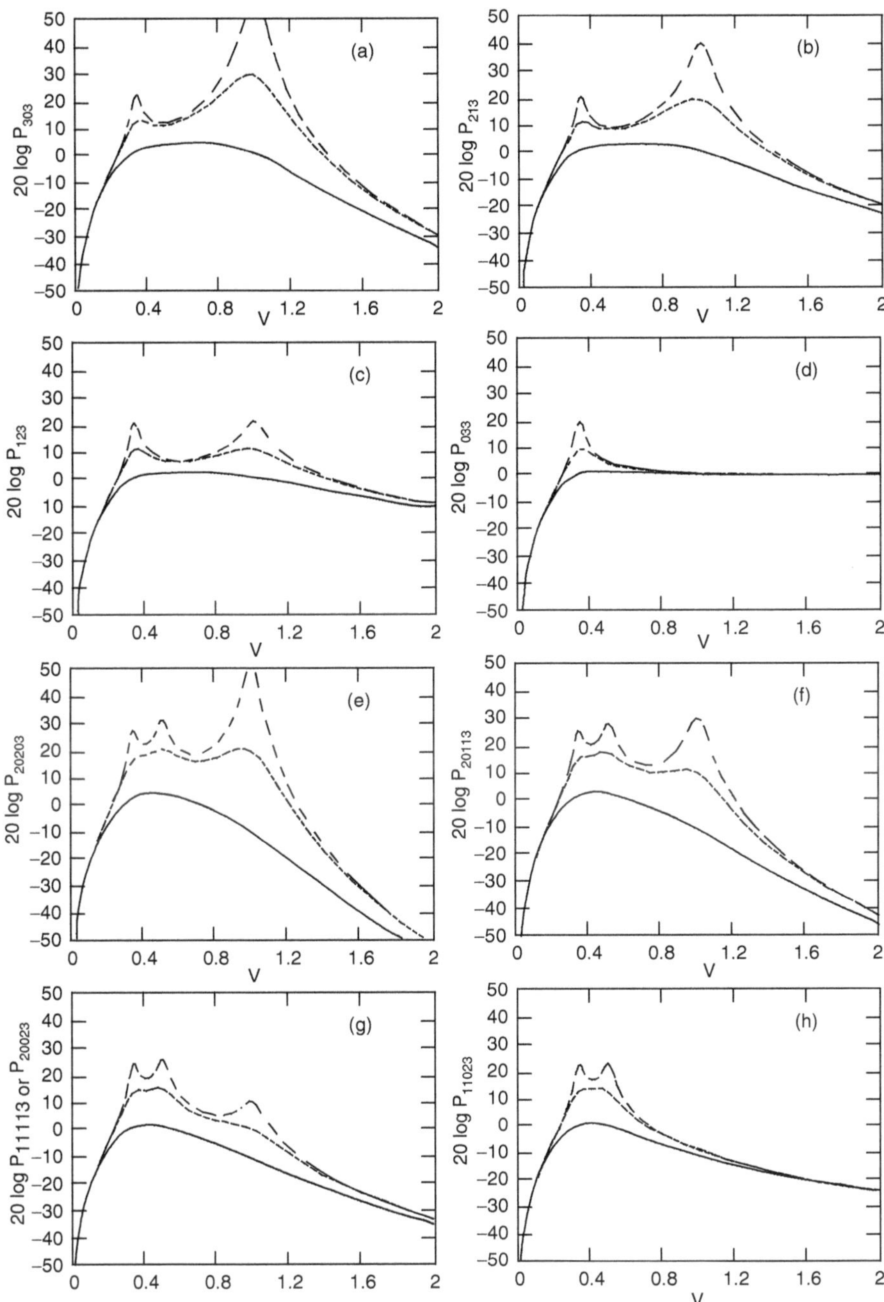

Fig. 12.5 Normalized pressure P (in dB) as a function of the dimensionless frequency v for the third harmonics caused by (**a**) X^3, (**b**) $X^2 \cos \omega t$, (**c**) $X \cos 2\omega t$, (**d**) $\cos 3\omega t$ (**e**) X^2, (**f**) X^2 and $X \cos \omega t$, (**g**) $X \cos \omega t$ or X^2 and $\cos 2\omega t$, (**h**) $X \cos \omega t$ and $\cos 2\omega t$ for $Q_m = 1$ (*solid line*), 3 (*dotted line*), and 10 (*dash line*) [3]

Equations (12.34)–(12.36), although obtained from a one-dimensional transducer model, can also be applied to transducers with more complex structures if the electromechanical drive part of the transducer undergoes one-dimensional motion and if the other moving parts contain no significant nonlinearities. Flextensional transducers driven by a piezoelectric ceramic stack often satisfy these conditions. The harmonics generated in the piezoelectric ceramic drive stack of such a transducer are transmitted through the rest of the transducer structure and radiated into the medium by approximately linear processes. Thus linear models of such transducers can be combined with expressions such as Eq. (12.37) to estimate harmonics [17].

The results for the second harmonic components, which apply to any transduction mechanism, will now be related to the physical properties of electrostrictive ceramic and compared with measurements made by Piquette and Forsythe on PMN [8]. From Table 12.2 it can be seen that the leading terms of the appropriate γ_{nm} are:

$$M\gamma_{01} = 2e_2AE_0E_{30} = 2e_2A\left(V_0V_{30}/L_0^2\right),$$

$$M\gamma_{02} = ½\, e_2AE_{30}^2 = ½\, e_2A(V_{30}/L_0)^2,$$

$$M\gamma_{11} = 4e_2AE_0E_{30}/L_0 = 4e_2A\left(V_0V_{30}/L_0^3\right),$$

$$M\gamma_{20} = c_2A/L_0^2,$$

$$M\omega_0^2 = c_1A/L_0.$$

At low frequency where $z_m(\nu)$ and $z_m(2\nu)$ are approximately unity the magnitudes of the relative second harmonics are given by Eqs. (12.34)–(12.36) as

$$\left|\frac{X_{202}}{X_{01}}\right| = \frac{\gamma_{20}\gamma_{01}}{2\omega_0^4} = \frac{c_2e_2V_0V_{30}}{c_1^2L_0^2}, \tag{12.38}$$

$$\left|\frac{X_{112}}{X_{01}}\right| = \frac{\gamma_{11}}{2\omega_0^2} = \frac{2e_2V_0V_{30}}{c_1L_0^2}, \tag{12.39}$$

$$\left|\frac{X_{022}}{X_{01}}\right| = \frac{\gamma_{02}}{\gamma_{01}} = \frac{V_{30}}{4V_0}. \tag{12.40}$$

Note the different ways in which these relative second harmonic components depend on material properties: X_{202}/X_{01} is zero when the second-order nonlinear elastic constant, c_2, is zero; X_{112}/X_{01} is proportional to the first-order electrostrictive constant, e_2, and X_{022}/X_{01} does not depend on any of the material constants, because X_{022} and X_{01} both depend on e_2. The latter ratio is an example of the fact that any biased square law mechanism produces a fundamental and a second harmonic with relative displacement magnitudes given by Eq. (12.40); other examples will be found in Sect. 12.3. For the γ_{20} and γ_{11} mechanisms decreasing the bias, while keeping the drive level constant, decreases the relative second harmonic, but for the γ_{02} mechanism it increases it. In most cases when the transducer is being driven such that V_{30} is a significant fraction of V_0, the X_{022} component dominates the other second harmonic components.

Piquette and Forsythe used their model for PMN to calculate expressions for the harmonics [9]. They also measured the second harmonic in an experimental PMN transducer with V_{30}/V_0 varying from ~0.2 to almost 1 and with bias fields up to 700 V/cm. Therefore X_{022} was probably the most important component in their measurements. They gave the results in terms of the acceleration ratio which is four times the displacement ratio, which makes the 022 acceleration ratio equal to V_{30}/V_0 from Eq. (12.40). This value of the ratio agrees very closely with the measurements in their Table 1, and also agrees exactly with their Eq. (30) when their saturation parameter is zero and the frequency is well below resonance.

12.2.2 Harmonic Distortion for Indirect Drive

The method of harmonic analysis described in the previous section is restricted to direct drive, i.e., voltage drive of electric field transducers and current drive of magnetic field transducers. Direct drive requires solution only of the mechanical transducer equation, such as Eq. (12.23), where each term represents a force. Indirect drive, i.e., voltage drive of magnetic field transducers and current drive of electric field transducers, requires simultaneous solution of both the mechanical and electrical transducer equations. Thus, harmonic analysis for indirect drive is more complicated than for direct drive. Since many power amplifiers have low internal impedance, approximate voltage drive is common in practice, making direct drive common for electric field transducers, while indirect drive is more common for magnetic field transducers. The two types of drive cause quite different harmonics in some cases, which suggests possibilities for reducing harmonic distortion [6]. For example, current drive has been reported to reduce harmonics in moving coil transducers [18].

A general analysis of indirect drive displacement harmonics has been made [7], but the detailed results are too complicated to present here. The practical question of which drive gives lower harmonic distortion was addressed by comparing the results from [3] and [7]. Indirect drive generates more components of each displacement harmonic than direct drive, because it involves nonlinear mechanisms in both the electrical and mechanical parts of the transducer. Thus indirect drive would be expected to generate higher displacement harmonics than direct drive when electrical, mechanical, and electromechanical nonlinearities are all significant. However, even with no electrical nonlinearities, indirect drive harmonics differ from direct drive harmonics and may be higher [7].

The comparison in [7] did not provide a general answer to the question of which drive gives lower harmonics, but some specific results can be mentioned. For example, it was found that for transducers with low electrical loss, indirect drive usually generates higher second harmonics than direct drive at resonance, except when the coupling coefficient is extremely high ($k > 0.925$) and springs are the dominant nonlinearity. Until recently coupling coefficients this high were found only in moving coil transducers, but they have also been measured in the newest transduction materials such as single crystal PIN-PMN-PT (see Sect. 13.5).

12.2.3 Instability in Electrostatic and Variable Reluctance Transducers

Instability in electrostatic and variable reluctance transducers was mentioned in Chap. 2, because this nonlinear effect could interfere with linear operation of these transducers. Instability in these transducers will be analyzed here in terms of the electrostatic transducer, but the results apply to the variable reluctance transducer if magnetic saturation is neglected. The nonlinearity of the electrostatic transducer may make it unstable if the bias voltage is too high. Letting $y = x_0/L$, Eq. (2.48), which expresses the static balance between the electric force and the spring force, can be written as

$$y^3 + 2y^2 + y + \lambda = 0\,, \qquad (12.41)$$

where $\lambda = \varepsilon A_0 V_0^2/2K_m L^3$ for the electrostatic transducer, and $\lambda = \mu A_0 (nI_0)^2/2K_m L^3$ for the variable reluctance transducer. This cubic equation has been studied extensively, and its implications are well known [19–21]. As the bias voltage is increased the moveable plate moves toward the fixed plate where, for $V_0 = 0$, the two plates were separated by distance L. Thus the physically meaningful range of displacement of the moveable plate is $-1 \leq y \leq 0$. The number of real roots of Eq. (12.41) in this range is as follows: for $\lambda < 4/27$ there are two real roots, for $\lambda = 4/27$ there are two equal real roots given by $y = -1/3$, and for $\lambda > 4/27$ there are no real roots. The situation is illustrated in Fig. 12.6 where the two opposing forces are plotted with the sign of the electric (or magnetic) force reversed to show these roots at the points where the forces are equal and opposite.

As long as the voltage is low enough to make $\lambda < 4/27$ there is a stable equilibrium point between $y = 0$ and $y = -1/3$ where the moveable plate comes to rest. The other root for this voltage, at $y < -1/3$, is an unstable equilibrium point as can be seen from Fig. 12.6 by considering the net force that results by making a small displacement away from this point. If a voltage is applied such that $\lambda > 4/27$ the moveable plate collapses into the fixed plate. Thus the bias voltage must be less than the value that makes $\lambda = 4/27$ and $y_0 = -L/3$, i.e., V_0 must be less than $[8K_m L^3/(27\,\varepsilon A_0)]^{1/2}$ to avoid instability. Note that Eq. (2.55) shows that this value of V_0 corresponds to $k = 1$. For the variable reluctance transducer V_0 and ε are replaced by nI_0 and μ.

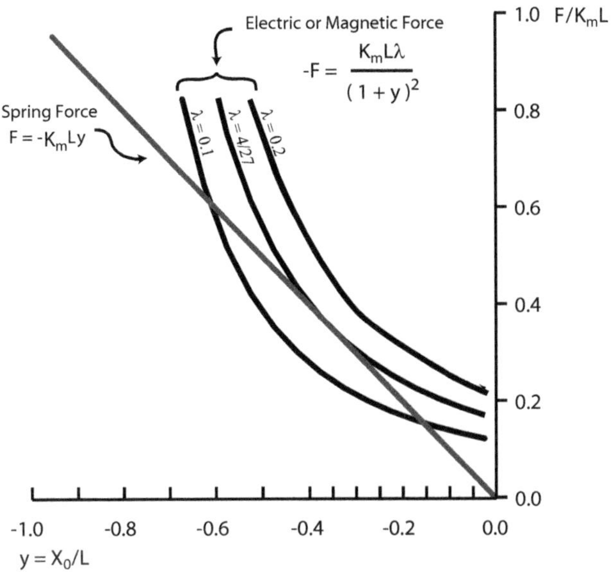

Fig. 12.6 Static equilibrium diagram for electrostatic and variable reluctance transducers. Note the stable equilibrium point at $y=-0.16$ and the unstable point at $y=-0.59$ for the case where $\lambda=0.1$. For $\lambda=4/27$ there is one equilibrium point at $y=-1/3$. For $\lambda>4/27$ there are no equilibrium points [21]

These conclusions hold for a linear spring force of $-K_1x$. Springs are always nonlinear to some extent, usually in a way that makes them stiffer as the displacement increases. Hunt [20] pointed out that for a cubic spring, with force of $-K_3\,x^3$, the gap would close at $x_0=-3\,L/5$ rather than $x_0=-L/3$. He also mentioned experiments on electrostatic transducers in which gap closure occurred before $x_0=-L/3$, probably because of bending of the flexible diaphragm that formed one plate of the transducer. A more realistic spring model is a mixture of linear and cubic terms (a hard spring which stiffens as the displacement increases) with a force of $-K_1x-K_3x^3$. In this case the equilibrium equation, instead of Eq. (12.41), is

$$y+\beta y^3=-\frac{\lambda}{(1+y)^2},\tag{12.42}$$

where $y=x_0/L$ and $\beta=K_3L^2/K_1$. Determination of the equilibrium points for a given value of λ and β now requires solution of a fifth-degree equation, but the displacement at which gap closure occurs can be found without solving that equation. At gap closure not only is Eq. (12.42) satisfied but the slopes of the two forces are also equal as can be seen from Fig. 12.6 in the linear spring case. Since each term of Eq. (12.42) is proportional to force, differentiating with respect to y gives the slope equation:

$$1 + 3\beta y^2 = \frac{2\lambda}{(1+y)^3}. \tag{12.43}$$

Dividing Eq. (12.42) by Eq. (12.43) and rearranging gives

$$\beta y^2 = -\frac{3y+1}{5y+3}, \tag{12.44}$$

which determines the value of y at gap closure for the nonlinear spring. It shows that for $\beta = 0$ (a linear spring) $y = -1/3$ as before, for $\beta \to \infty$ (a cubic spring) $y = -3/5$ in agreement with Hunt, for $\beta = 4$, $y = -½$ and for $\beta = 5/4$, $y = -2/5$. These values of y for a given β can be used in Eq. (12.41) to find the value of λ, and then the voltage, that causes gap closure. Hard nonlinear springs, which stiffen as the displacement increases, raise the bias voltage at which instability occurs in the electrostatic transducer as the results above show quantitatively. In the variable reluctance transducer the same effect raises the bias current at which gap closure occurs.

The results above hold when the bias voltage or current is increased slowly from zero to a value corresponding to a stable equilibrium point. If the increase is rapid the moveable plate will overshoot the intended point, and the overshoot may cause the gap to close and remain closed. For the linear spring this situation has been analyzed [21] for the case where the voltage (or current) is increased instantaneously, with the result that gap closure occurs when λ exceeds 1/8 rather than 4/27. The difference is only about 9 %, since the voltage (or current) is proportional to the square root of λ.

Now we extend the discussion of instability in electrostatic and variable reluctance transducers to include dynamic instability caused by adding the alternating drive voltage (or current) to a stable bias voltage. As the drive voltage is increased, causing the displacement excursions about the equilibrium point to increase, the equilibrium point also changes, and eventually the gap closes when the excursions reach a critical amplitude. It was pointed out above that the first-order solution in Eq. (12.33) contains three static terms that change the static equilibrium displacement, and the most important of these terms is $\gamma_{00}/\omega_0^2 = -\varepsilon A V_1^2/4K_1 L_0^2$ from Table 12.4. This term is an increment of static displacement toward gap closure, which increases with the square of the drive voltage, starting from the static displacement of $-\varepsilon A V_0^2/2K_1 L_0^2$ when the drive voltage is zero. The other two first-order static terms change this increment somewhat and higher order perturbation solutions would change it further. But the perturbation method does not predict whether, or at what drive level, the gap will close.

To investigate dynamic stability in an electrostatic transducer a different approach is necessary [21]. Consider Eq. (12.17) with a linear spring ($K_2 = K_3 = 0$) and with an undetermined phase angle, θ, between the drive voltage and the displacement:

$$M\ddot{x} + R\dot{x} + K_1 x = \frac{-\varepsilon A[V_0 + V_1 \cos(\omega t + \theta)]^2}{2(L+x)^2}. \qquad (12.45)$$

Dividing Eq. (12.45) by L and substituting an approximate solution in the form

$$y = x/L = y_0 + y_1 \cos \omega t, \qquad (12.46)$$

gives a set of three equations for y_0, y_1, and θ, since the static terms, the terms containing $\cos \omega t$, and the terms containing $\sin \omega t$ must separately equal zero, while the terms involving second and third harmonics can be ignored. At resonance these equations simplify considerably, and the equation that determines y_0 becomes

$$y_0^3 + 2y_0^2 + y_0\left(1 + \frac{1}{2}y_1^2\right) + \frac{\varepsilon A}{2K_1L^3}\left[V_0^2 + \frac{1}{2}V_1^2\right] = 0. \qquad (12.47)$$

Equation (12.47) is a cubic equation for y_0 of the same form as the equation that determined static stability, as can be seen by omitting the terms containing y_1 and V_1 and obtaining Eq. (2.48). In the dynamic case the coefficient of the y_0 term depends on the alternating amplitude, y_1, and the term that does not contain y_0 depends on the drive voltage, V_1. This term shows that, because of the nonlinearity, increasing the drive voltage has the same effect as increasing the bias voltage; it drives the transducer closer to the point of collapse where Eq. (12.47) has no real solution. Approximate conditions for dynamic instability to occur can be found by numerical solution of the three equations mentioned above for y_0, y_1, and θ. This has been done at resonance for different values of Q_m [21].

12.3 Nonlinear Analysis of Distributed Parameter Transducers

The analysis in the first part of this chapter was limited to lumped-parameter transducers described by nonlinear ordinary differential equations. As discussed in Chaps. 3 and 4, the lumped-parameter assumption fails when the length of a vibrating section of the transducer exceeds about one quarter wavelength. Distributed parameter analysis is then required based on partial, rather than ordinary, differential equations. Similarly, analysis of nonlinear effects in distributed systems requires solution of nonlinear partial differential equations. Nonlinear effects relevant to transducers are associated with both longitudinal and flexural waves in bars and discs. In this section nonlinear analysis of distributed systems will be illustrated by calculating the second harmonics generated by nonlinear longitudinal waves in the thin segmented piezoelectric ceramic bar described in Sect. 3.2.3.1.

The stress in the bar is related to strain and electric field by Eq. (12.1), but now the stress and strain are not only functions of time but also functions of position along the bar. Using notation consistent with Sect. 3.2.3.1, let ζ be the displacement of an element of the bar parallel to the bar at position z along the bar; then the strain at z is $\partial\zeta/\partial z$. With these variables Eq. (12.1) becomes

$$T(z,t) = c_1 \frac{\partial \zeta}{\partial z} - e_1 E + c_2 \left(\frac{\partial \zeta}{\partial z}\right)^2 - 2c_a E \frac{\partial \zeta}{\partial z} - e_2 E^2. \tag{12.48}$$

Voltage drive will be specified by making the electric field $E = E_3 \cos \omega t$, and it will be assumed, because of the segmented construction, that E is uniform along the length of the bar. The longitudinal forces per unit volume on an infinitesimal element of the bar are given by $\partial T/\partial z$, as in Chap. 3, and the equation of motion of the element is

$$\frac{\partial T}{\partial z} = \rho \ddot{\zeta}, \tag{12.49}$$

where ρ is the density of the bar material. When $\partial T/\partial z$ is calculated from Eq. (12.48) the terms involving only E will drop out, because E is uniform along the bar making these forces equal and opposite on each side of the infinitesimal element. The stresses involving only E will enter the analysis later by way of the boundary conditions. Thus the nonlinear partial differential equation describing wave motion in the bar is

$$\left[c_1 + 2c_2 \frac{\partial \zeta}{\partial z} - 2c_a E\right] \frac{\partial^2 \zeta}{\partial z^2} = \rho \ddot{\zeta}. \tag{12.50}$$

A perturbation parameter will be defined analogous to that used in the lumped-parameter case:

$$\delta = \zeta_{01}/L_0,$$

where ζ_{01} is the amplitude of the linear fundamental at the free end of the bar, and L_0 is the length of the bar. The perturbation parameter could be introduced by multiplying nonlinear terms by powers of $\delta L_0/\zeta_{01}$ as was done in Sect. 12.2.1, but the same result is obtained more easily by multiplying quadratic terms in Eq. (12.50) by δ (cubic terms would be multiplied by δ^2 if they were included) and then letting $\delta = 1$ when the perturbation orders are combined in the end. Thus the wave equation is written as

$$\left[c_1 + 2\delta c_2 \frac{\partial \zeta}{\partial z} - 2\delta c_a E\right] \frac{\partial^2 \zeta}{\partial z^2} = \rho \ddot{\zeta}, \tag{12.51}$$

and the solution is assumed as the perturbation series,

$$\zeta(z,t) = \zeta_0(z,t) + \delta \zeta_1(z,t) + \cdots. \tag{12.52}$$

Substituting this series into the wave equation and separating terms by powers of δ gives

$$c_1 \frac{\partial^2 \zeta_0}{\partial z^2} = \rho \ddot{\zeta}_0, \tag{12.53}$$

$$c_1 \frac{\partial^2 \zeta_1}{\partial z^2} + 2c_2 \frac{\partial \zeta_0}{\partial z} \frac{\partial^2 \zeta_0}{\partial z^2} - 2c_a E \frac{\partial^2 \zeta_0}{\partial z^2} = \rho \ddot{\zeta}_1, \tag{12.54}$$

where only zeroth- and first-order equations are used to be consistent with only linear and quadratic terms in the expression for the stress. The zeroth-order equation, being homogeneous and linear in ξ_0, can be solved as in Chap. 3. The solution for ξ_0 can then be used to evaluate the inhomogeneous terms in the first-order equation, which can also be solved since it is linear in ξ_1.

The general solution of Eq. (12.53) has the form

$$\zeta_0(z,t) = (A \sin kz + B \cos kz)(C \sin \omega t + D \cos \omega t), \tag{12.55}$$

where $k^2 = \omega^2 \rho / c_1$, $(c_1/\rho)^{1/2}$ is the speed of longitudinal waves in the ceramic bar, and A, B, C, and D are constants. As a specific example consider the bar clamped ($\zeta_0 = 0$) at the end $z = 0$ and free ($T = 0$) at the other end, $z = L_0$. Note that the driving electric field enters the solution only by way of the boundary condition on the stress. In real transducers the ends of the bar would be connected to masses, resistances, and the radiation impedance (see Chap. 3), rather than being clamped or free. Applying the boundary conditions above to Eqs. (12.55) and (12.48) gives $B = C = 0$, $AD = e_1 E_3 / c_1 k \cos kL_0$, and

$$\zeta_0(z,t) = \frac{e_1 E_3}{c_1 k \cos kL_0} \sin kz \cos \omega t = \zeta_{01} \sin kz \cos \omega t. \tag{12.56}$$

Only the linear parts of the stress were used in the boundary condition on the zeroth-order solution, but both linear and quadratic terms in the stress will be used in the boundary condition on the first-order solution.

Using this solution for ζ_0, the inhomogeneous drive terms in Eq. (12.54) for ζ_1 can be evaluated giving

$$\begin{aligned} c_1 \frac{\partial^2 \zeta_1}{\partial z^2} &+ c_a E_3 k^2 \zeta_{01} \sin kz (1 + \cos 2\omega t) \\ &- \frac{1}{2} c_2 k^3 \zeta_{01}^2 \sin 2kz (1 + \cos 2\omega t) = \rho \ddot{\zeta}_1. \end{aligned} \tag{12.57}$$

Since Eq. (12.57) is linear in ζ_1 the solutions for the two drive terms can be found separately; e.g., let $\zeta_{11}(z,t)$ be the solution for the drive term containing c_a. Both ζ_{11} and the drive term must be expanded in a series of functions that satisfy the homogeneous part of the equation. Since the homogeneous equation for ζ_1 is the same as that for ζ_0 the appropriate functions are the modal functions associated with the solution for ζ_0 in Eq. (12.56), i.e., sin $k_n z$ where $k_n L_0 = (n + ½)\pi$, $n = 0, 1, 2 \cdots$. The expansion for ζ_{11} must also be consistent with the time dependence of the drive term containing c_a. Thus assume that ζ_{11} has the form

$$\zeta_{11}(z, t) = \sum_{n=0}^{\infty} b_n \sin k_n z + \sum_{n=0}^{\infty} c_n \sin k_n z \cos 2\omega t, \tag{12.58}$$

and expand the z-dependent part of the drive term, sin kz, in the same functions

$$\sin kz = \sum_{n=0}^{\infty} d_n \sin k_n z. \tag{12.59}$$

It can be shown, from the orthogonality of the sin $k_n z$ functions, that

$$d_n = \frac{2k(-1)^n \cos kL_0}{L_0\left(k_n^2 - k^2\right)}. \tag{12.60}$$

Substituting Eqs. (12.58) and (12.59) into Eq. (12.57), with the c_a term only, shows that

$$b_n = \frac{2(-1)^n c_a e_1 E_3^2 (\omega/\omega_n)^4}{c_1^2 k^2 L_0 \left(1 - \omega^2/\omega_n^2\right)}, \quad c_n = \frac{b_n}{\left(1 - 4\omega^2/\omega_n^2\right)}, \tag{12.61}$$

where $\omega_n^2 = c_1 k_n^2/\rho$, and the solution for ζ_{11} is

$$\zeta_{11}(z, t) = \sum_{n=0}^{\infty} b_n \sin k_n z \left[1 + \frac{\cos 2\omega t}{1 - 4\omega^2/\omega_n^2}\right]. \tag{12.62}$$

Let $\zeta_{12}(z, t)$ be the solution for the other drive term in Eq. (12.57), the one containing the coefficient c_2. The same procedure that gave Eq. (12.62) for ζ_{11} gives:

$$\zeta_{12}(z, t) = \sum_{n=0}^{\infty} f_n \sin k_n z \left[1 + \frac{\cos 2\omega t}{1 - 4\omega^2/\omega_n^2}\right], \tag{12.63}$$

where

$$f_n = -\frac{2c_2 e_1^2 E_3^2 (-1)^n (\omega/\omega_n)^4 \cos 2kL_0}{c_1^3 k^2 L_0 \cos^2 kL_0 \left(1 - 4\omega^2/\omega_n^2\right)}. \tag{12.64}$$

Both ζ_{11} and ζ_{12} consist of a static component (or zeroth harmonic) and a second harmonic, all of which have z-dependence expressed as a sum of normal modes of the clamped-free bar. Both also satisfy the same conditions at the ends of the bar, i.e.,

$$\zeta_{11} = \zeta_{12} = 0 \quad \text{at} \quad z = 0, \tag{12.65}$$

$$\partial\zeta_{11}/\partial z = \partial\zeta_{12}/\partial z = 0 \quad \text{at} \quad z = L_0. \tag{12.66}$$

The sum $\zeta_{11} + \zeta_{12}$ is a solution of Eq. (12.57) for ζ_1, but it is not the complete solution because it does not satisfy the boundary condition on the stress at $z = L_0$. The stress contains the term $e_2E^2 = ½e_2E_3^2(1 + \cos 2\omega t)$, making it necessary to add another term to the solution, call it ζ_{13}, that will satisfy the part of the boundary condition that depends on e_2E^2. Since e_2E^2 contains a static term and a second harmonic term, and since ζ_{13} must also satisfy the boundary condition at $z = 0$, a reasonable form for the addition to the solution is

$$\zeta_{13}(z,t) = a_{130} \sin kz + a_{132} \sin 2kz \cos 2\omega t. \tag{12.67}$$

To make the complete first-order solution satisfy the boundary condition to the same order of approximation as it satisfies the original differential equation the perturbation parameter is inserted in the quadratic terms in the expression for the stress, Eq. (12.48):

$$T(z,t) = c_1 \frac{\partial\zeta}{\partial z} - e_1E + \delta c_2 \left(\frac{\partial\zeta}{\partial z}\right)^2 - 2\delta c_a E \frac{\partial\zeta}{\partial z} - \delta e_2 E^2. \tag{12.68}$$

Substituting $\zeta = \zeta_0 + \delta\zeta_1$ and dropping terms in powers of δ higher than the first gives an expression for the stress correct to first order in δ consistent with ζ_1:

$$T(z,t) = c_1 \frac{\partial\zeta_0}{\partial z} - e_1E + \delta c_2 \left(\frac{\partial\zeta_0}{\partial z}\right)^2 - 2\delta c_a E \frac{\partial\zeta_0}{\partial z} - \delta e_2 E^2 + \delta c_1 \frac{\partial\zeta_1}{\partial z}. \tag{12.69}$$

Setting $\delta = 1$, since δ has served its purpose of identifying the first-order terms, and using $\zeta_1 = \zeta_{11} + \zeta_{12} + \zeta_{13}$ and $\partial\zeta_{11}/\partial z = \partial\zeta_{12}/\partial z = 0$ at $z = L_0$ gives the boundary condition on the stress at $z = L_0$:

$$\begin{aligned} T(L_0,t) = 0 = (c_1 - 2c_a E)\frac{\partial\zeta_0}{\partial z}\bigg|_{z=L_0} - e_1E + c_2\left(\frac{\partial\zeta_0}{\partial z}\right)^2\bigg|_{z=L_0} \\ - e_2E^2 + c_1 \frac{\partial\zeta_{13}}{\partial z}\bigg|_{z=L_0}. \end{aligned} \tag{12.70}$$

Using Eq. (12.56) to calculate $\partial\zeta_0/\partial z$ and Eq. (12.67) for ζ_{13} leads to evaluation of the constants a_{130} and a_{132}:

$$a_{130} = \frac{E_3^2}{2c_1 k \cos kL_0}\left(e_2 + \frac{2c_a e_1}{c_1} - \frac{c_2 e_1^2}{c_1^2}\right), \tag{12.71}$$

$$a_{132} = \frac{a_{130}\cos kL_0}{2\cos 2kL_0}. \tag{12.72}$$

The complete solution of the original differential equation, Eq. (12.50), to first order, is

$$\zeta(z,t) = \zeta_0 + \delta\zeta_1 = \zeta_0 + \delta\zeta_{11} + \delta\zeta_{12} + \delta\zeta_{13} = \zeta_{01}\sin kz\cos\omega t + \sum_{n=0}^{\infty}(b_n + f_n)$$
$$\times\ \sin k_n z\left[1 + \frac{\cos 2\omega t}{1 - 4\omega^2/\omega_n^2}\right] + a_{130}\sin kz + a_{132}\sin 2kz\cos 2\omega t. \tag{12.73}$$

Eq. (12.73) also satisfies, to first order, the boundary conditions

$$\zeta(0,t) = 0 \quad \text{and} \quad T(L_0,t) = 0. \tag{12.74}$$

The solution consists of a fundamental component, three separate static components and three separate second harmonic components. Each of the three nonlinear parameters, c_2, c_a, and e_2, gives rise to a static and a second harmonic component; the b_n coefficients depend only on c_a, the f_n coefficients depend only on c_2, but a_{130} and a_{132} depend on all three nonlinear parameters. Carrying the solution to higher order would give corrections to these results plus higher order harmonics.

The fundamental and second harmonic displacements at the free end of the bar should be the same as the lumped-parameter results at low frequency. For $kL_0 \ll 1$ the fundamental component in Eq. (12.56) is

$$\zeta_0(L_0,t) = \frac{e_1 E_3 L_0}{c_1}\cos\omega t = \frac{e_1 E_3 A_0}{K_m}\cos\omega t, \tag{12.75}$$

where $K_m = c_1 A_0/L_0$ is the effective spring constant of the short bar as in Sect. 2.1 with $c_1 = 1/s_{33}^E$ = Young's modulus. The amplitude of $\zeta_0(L_0,t)$ equals X_{01} in Sect. 12.2.1 showing that at very low frequency the linear approximation of the fundamental wave motion in the bar reduces to the lumped-parameter approximation.

Now consider the second harmonic results at very low frequency. For $\omega \ll \omega_0$ the sums in Eq. (12.73) can be approximated by the $n=0$ terms, and the three components of ζ_1 at the end of the bar are:

$$\zeta_{11}(L_0,t) = b_0(1+\cos 2\omega t) = \frac{2c_a e_1 E_3^2 \omega^4}{c_1^2 k^2 L_0 \omega_0^4}(1+\cos 2\omega t), \tag{12.76}$$

$$\zeta_{12}(L_0,t) = f_0(1+\cos 2\omega t) = -\frac{2c_2 e_1^2 E_3^2 \omega^4}{c_1^3 k^2 L_0 \omega_0^4}(1+\cos 2\omega t), \tag{12.77}$$

$$\zeta_{13}(L_0,t) = \frac{E_3^2 L_0}{2c_1}\left(e_2 + \frac{2c_a e_1}{c_1} - \frac{c_2 e_1^2}{c_1^2}\right)(1+\cos 2\omega t). \tag{12.78}$$

It is evident from the frequency dependence that, for $\omega \ll \omega_0$, the components ζ_{11} and ζ_{12} are small compared to ζ_{13}. Denoting the second harmonic part of ζ_{13} by ζ_{132}, the amplitude relative to the fundamental amplitude at the end of the bar is

$$\left|\frac{\zeta_{132}}{\zeta_0}\right|_{z=L_0} = E_3\left(\frac{e_2}{2e_1} + \frac{c_a}{c_1} - \frac{c_2 e_1}{2c_1^2}\right). \tag{12.79}$$

Use of the appropriate terms of the piezoelectric γ_{nm} from Table 12.1 in Eqs. (12.34)–(12.36) shows that the three terms of $|\zeta_{132}/\zeta_0|_{z=L_0}$ are, respectively,

$$|X_{022}|/X_{01}, \quad |X_{112}|/X_{01} \quad \text{and} \quad |X_{202}|/X_{01}.$$

This confirms that the nonlinear distributed parameter results for the displacement at the end of the bar are equal, at low frequency, to the nonlinear lumped-parameter results.

Some of these results can also be applied to biased electrostrictive or magnetostrictive material. In material such as PMN with a maintained bias field, E_0, let $E = E_0 + E_3 \cos \omega t$, $c_a = 0$, and, for simplicity, omit the nonlinear elasticity by making $c_2 = 0$. Under these conditions the first-order solution has only one second harmonic component resulting from the coefficient e_2. Then, following the procedure used in Sect. 2.2, with e_1 replaced by $2e_2E_0$, Eqs. (12.56) and (12.73) give for the fundamental and second harmonic displacement amplitudes at $x = L_0$:

$$\zeta_{01} \sin kL_0 = \frac{2e_2 E_0 E_3 \sin kL_0}{c_1 k \cos kL_0}, \tag{12.80}$$

$$a_{132} \sin 2kL_0 = \frac{e_2 E_3^2 \sin 2kL_0}{4c_1 k \cos 2kL_0}, \tag{12.81}$$

with the ratio

$$\frac{\text{2nd Harmonic}}{\text{Fundamental}} = \frac{E_3 \tan 2kL_0}{8E_0 \tan kL_0}. \tag{12.82}$$

This ratio equals $E_3/4E_0$ when $kL_0 \ll 1$ corresponding to $\omega \ll \omega_0$ as found before in Eq. (12.40) from the lumped-parameter calculation. But Eq. (12.82) is valid for higher frequencies and shows that the ratio increases with frequency and peaks when ω approaches $\omega_0/2$ and $2kL_0$ approaches $\pi/2$. This is consistent with the results in Fig. 12.4 showing that the second harmonic peaks when the drive frequency is near half the fundamental resonance frequency.

A similar analysis can be carried out to find the harmonics for flexural waves in a bar. For the case above of biased electrostrictive or magnetostrictive material the ratio of second harmonic to fundamental displacement amplitudes is similar to Eq. (12.82), and also reduces to $E_3/4E_0$ at low frequency. The value $E_3/4E_0$ for this ratio at low frequency also applies to electrostatic and variable reluctance transducers and to all other biased square law mechanisms.

12.4 Nonlinear Effects on the Electromechanical Coupling Coefficient

The definition of the electromechanical coupling coefficient, k, in Sect. 1.4.1, and the further discussion in Sect. 4.4.1, is limited to linear transducer operation. Under these conditions all energy conversion is assumed to occur at the drive frequency, and the coupling coefficient is considered to be independent of drive level. Since some of the promising modern transduction materials have significant nonlinear characteristics, which at high drive levels would convert some energy to harmonics, it would be useful to consider a more general definition of the coupling coefficient. For example, a definition that included only converted mechanical energy at the drive frequency might be preferable. Then k would be expected to decrease with drive level as increasing amounts of the input energy went into harmonics. Such considerations do not reduce the usefulness of linear definitions of k as a measure of quality in comparing different transducer materials, concepts, or designs, because nonlinear definitions would reduce to the linear values at low drive levels. However, since high drive conditions usually degrade transducer characteristics that are related to k, such as efficiency and bandwidth, a nonlinear generalization that could indicate such degradation would be useful.

The energy-based definitions of k in Sects. 1.4.1 and 4.4.1 are suitable for generalization to nonlinear conditions. Piquette [22] has calculated k under nonlinear conditions by using his nonlinear equations of state for electrostrictive materials, discussed briefly in Sect. 2.2 [8, 9], and by using Eq. (4.25) as the definition of k. The ambiguities associated with this definition and with the concept of mutual energy (see Sect. 4.4.1.2) were avoided by showing that the chosen equations of state gave the expected results for k in the linear case. His nonlinear results for k appear to have reasonable features, such as becoming exactly zero when the maintained bias is zero and diminishing rapidly when the drive amplitude exceeds the fixed bias.

Hom et al. [23] and Robinson [24] have also discussed the coupling coefficient for nonlinear electrostrictive materials. Another approach to estimating nonlinear effects on the coupling coefficient could be based on perturbation calculations such as those described in Sect. 12.2.1. The first approximation to the reduction in the fundamental amplitude associated with increasing harmonic amplitudes is given by second-order perturbation. This reduction in fundamental amplitude corresponds to a reduction in converted fundamental mechanical energy, which could be used to define a coupling coefficient that decreases as a function of drive level.

12.5 Summary

Although linearity is often used as a good approximate representation, most natural and man-made mechanisms are nonlinear. In most transduction mechanisms an internal or external electrical or magnetic bias is needed for linear operation. Even with this, nonlinear effects can appear when the mechanism is driven to the high stress of high displacement conditions. In cases where the mechanism is weak in tension nonlinearity may be mitigated with a compressive bias. Nonlinearity is usually accompanied by second or third harmonic distortion and typically a reduction in the resonance frequency and an asymmetrical response in the vicinity of resonance.

In this chapter nonlinearity in lumped mode transducers was discussed for: piezoelectric, electrostrictive, magnetostrictive, electrostatic, variable reluctance and moving coil transducers. Nonlinear effects were analyzed using a perturbation method and distortion was considered for direct drive and indirect drive condition. Direct drive would be voltage drive for electric field transducers such as piezoelectric, electrostrictive, and electrostatic transducersand current drive for magnetic field transducers such as magnetostrictive, variable reluctance, and moving coil. Indirect drive would be a current drive condition for electric field transducers and voltage drive condition for magnetic field transducers. The indirect drive cases are more difficult to solve since a nonlinear input impedance is needed in addition to the output acoustic response.

The equation set for piezoelectric transducers was taken out to second-order terms S^2, T^2, and SE, where S is the strain, T the stress, and E is the electric field. A double summed harmonic solution was then assumed for the force side of differential equation for the lumped transducer with terms like x^n and cos $(m\omega t)$ and a perturbation solution was sought. A similar procedure was used for the direct drive case of other transducers. Graphical in-water results show a strong dependence on the mechanical Q_m of the transducer. A nonlinear model for a distributed parameter piezoelectric bar was also presented and here a partial differential equation wave equation solution was sought. Finally, the need for further work on the nonlinear effects on the electromechanical coupling coefficient was presented.

Exercises (Degree of Difficulty: *Lowest, **Moderate, ***Highest)

12.1.*** Consider Eq. (2.8) with external force $F_b = 0$ at low frequencies where the acceleration and velocity terms are negligible compared with the displacement term. Show that this leads to the low frequency nonlinear equation for the strain, $S_3^2 + S_3 = d_{33}(V/L)$. Determine the exact solution. Assume $4d_{33}V/L < 1$ and use the binomial series to obtain the three term expansion solution showing terms that could generate second and third harmonics under high sinusoidal drive.

12.2.* Calculate the strain for a Type I piezoelectric material in Exercise 12.1 with a typical maximum electric field of $V/L = 4$ kV/cm (10.2 kV per inch) and compare this result with the linear result given by the first term of the expansion for S_3. Calculate the percentage difference.

12.3.** What are the physical nonlinear conditions for even and for odd harmonic generation?

12.4.*** One general objection to operating below the fundamental resonance of high power Tonpilz transducers is the higher harmonic distortion generated as compared to operating above resonance. What is the physical reason for this? As the most extreme example of this effect explain how operating at one half or one third the fundamental resonance frequency can greatly increase the second or third harmonic distortion.

12.5.** Why does the presence of significant harmonic distortion indicate the possibility of transducer failure under high drive?

12.6.*** The most common nonlinear force in transducers is the square law electric or magnetic drive force. Results that are applicable to many cases can be derived from a simplified form of Eq. (12.9) containing only the linear strain term and the quadratic electric field term, $T = cS - eE^2$. Use this expression in the equation of motion [as in Eq. (12.13)] with a drive voltage of $V = V_0 + V_1 \cos \omega t$, and find the static, fundamental, and second harmonic drive terms and displacements. Show, when $V_0 = 0$, that there is no displacement at the drive frequency, but there is a static displacement and a displacement at twice the drive frequency. Thus a bias voltage, V_0, is necessary to have any linear output with this force law.

12.7.** In Exercise 12.6 calculate the ratio of the second harmonic displacement to the fundamental displacement at the resonance frequency, at one half the resonance frequency, and well below resonance.

12.8.*** When significant harmonic distortion is present a modified definition of the electromechanical coupling coefficient should be considered, because part of the input energy is converted to harmonics which do not contribute to the desired fundamental output. Such a definition might be: ${k_{nl}}^2$ equals the ratio of converted mechanical energy at the fundamental frequency to the total input energy. Express this definition analytically and evaluate it numerically using the second harmonic results from Exercise 12.7.

12.9.** Square law transducers are seldom used without bias, because the frequency content of the output differs so much from that of the input. Using the same equation of motion as Exercise 12.6 show that the output of a square law transducer driven with voltage $V = V_1 \cos \omega t + V_2 \cos 2\omega t$ contains four different frequency components. What are those frequencies and what are the relative displacement amplitudes and velocity amplitudes at low frequency if $V_2 = V_1$?

12.10.* Convert the relative displacement amplitudes calculated in Exercise 12.9 to radiated sound pressure to show the enhancement of the higher frequency pressure components relative to the displacement components (see the discussion in Sect. 12.2.1).

References

1. D.A. Berlincourt, D.R. Curran, H. Jaffe, in *Piezoelectric and Piezomagnetic Materials and Their Function in Transducers*, ed. by W.P. Mason. Physical Acoustics, vol 1, Part A (Academic, New York, 1964)
2. C.H. Sherman, J.L. Butler, Harmonic distortion in magnetostrictive and electrostrictive transducers with application to the flextensional computer program FLEXT, Image Acoustics, Inc. Report on Contract No. N66609-C-0985, 30 Sept 1994
3. C.H. Sherman, J.L. Butler, Analysis of harmonic distortion in electroacoustic transducers. J. Acoust. Soc. Am. **98**, 1596–1611 (1995)
4. V.E. Ljamov, Nonlinear acoustical parameters in piezoelectric crystals. J. Acoust. Soc. Am. **52**, 199–202 (1972)
5. W.P. Mason, *Piezoelectric Crystals and Their Application to Ultrasonics* (Van Nostrand, New York, 1950)
6. C.H. Sherman, J.L. Butler, Perturbation analysis of nonlinear effects in moving coil transducers'. J. Acoust. Soc. Am. **94**, 2485–2496 (1993)
7. C.H. Sherman, J.L. Butler, Analysis of harmonic distortion in electroacoustic transducers under indirect drive conditions. J. Acoust. Soc. Am. **101**, 297–314 (1997)
8. J.C. Piquette, S.E. Forsythe, A nonlinear material model of lead magnesium niobate (PMN). J. Acoust. Soc. Am. **101**, 289–296 (1997)
9. J.C. Piquette, S.E. Forsythe, Generalized material model for lead magnesium niobate (PMN) and an associated electromechanical equivalent circuit. J. Acoust. Soc. Am. **104**, 2763–2772 (1998)
10. W.Y. Pan, W.Y. Gu, D.J. Taylor, L.E. Cross, Large piezoelectric effect induced by direct current bias in PMN-PT relaxor ferroelectric ceramics. Jpn. J. Appl. Phys. **28**, 653–661 (1989)
11. M.B. Moffett, A.E. Clark, M. Wun-Fogle, J.F. Lindberg, J.P. Teter, E.A. McLaughlin, Characterization of Terfenol-D for magnetostrictive transducers. J. Acoust. Soc. Am. **89**, 1448–1455 (1991)
12. C.H. Sherman, J.L. Butler, Harmonic distortion in moving coil transducers caused by generalized Coulomb damping. J. Acoust. Soc. Am. **96**, 937–943 (1994)
13. W.J. Cunningham, Nonlinear distortion in dynamic loudspeakers due to magnetic effects. J. Acoust. Soc. Am. **21**, 202–207 (1949)
14. E. Geddes, *Audio Transducers*, copyright 2002, Chapter 10
15. J.J. Stoker, *Nonlinear Vibrations* (Interscience, New York, 1950)
16. J.A. Murdock, *Perturbations—Theory and Methods* (Wiley, New York, 1991)
17. J.L. Butler, FLEXT, (Flextensional Transducer Program), Contract N66604-87-M-B328 to NUWC, Newport, RI, Image Acoustics, Inc., Cohasset, MA 02025

18. P.G.L. Mills, M.O.J. Hawksford, Distortion reduction in moving coil loudspeaker systems using current-drive technology. J. Audio Eng. Soc. **37**, 129–147 (1989)
19. A.A. Janszen, R.L. Pritchard, F.V. Hunt, Electrostatic Loudspeakers (Harvard University Acoustics Research Laboratory, Cambridge) Tech. Memo. No. 17, 1 Apr 1950
20. F.V. Hunt, *Electroacoustics: The Analysis of Transduction and Its Historical Background* (Wiley, New York, 1954)
21. C.H. Sherman, Dynamic mechanical stability in the variable reluctance and electrostatic transducers. J. Acoust. Soc. Am. **30**, 48–55 (1958). See also C.H. Sherman, Dynamic Mechanical Stability in the Variable Reluctance Transducer, a thesis submitted to the University of Connecticut, 1957
22. J.C. Piquette, Quasistatic coupling coefficients for electrostrictive ceramics. J. Acoust. Soc. Am. **110**, 197–207 (2001)
23. C.L. Hom, S.M. Pilgrim, N. Shankar, K. Bridger, M. Massuda, R. Winzer, Calculation of quasi-static electromechanical coupling coefficients for electrostrictive ceramic materials. IEEE Trans. Ultrason. Ferroelectr. Freq. Control **41**, 542–551 (1994)
24. H.C. Robinson, A comparison of nonlinear models for electrostrictive materials. Presentation to the 1999 I.E. Ultrasonics Symposium, Lake Tahoe, NV, 17–20 Oct 1999

Chapter 13
Appendix

13.1 Conversions and Constants

13.1.1 Conversions

Length (inches)	1 in.	= 0.0254 m
Length (meters)	1 m	= 39.37 in.
Weight (pounds)	1 lb	= 0.4536 kg
Weight (kg)	1 kg	= 2.205 lb
Pressure (psi)	1 psi	= 6.895 kN/m^2
Pressure (Pa)	1 N/m^2	= 0.145 × 10^{-3} psi
Depth and water pressure	1 ft	= 0.444 psi
Depth and water pressure	1 m	= 1.457 psi
Depth and water pressure	1 m	= 10.04 kPa
Magnetic field	1 Oe	= 79.58 A/m
Magnetic field	1 kA/m	= 12.57 Oe

13.1.2 Constants

Free permittivity	ε_0	$10^{-9}/36\pi$	= 8.842 × 10^{-12} C/mV
Free permeability	μ_0	$4\pi \times 10^{-7}$	= 1.2567 × 10^{-6} H/m
Sound speed	c	Sea water	= 1500 m/s @ 13 °C
Sound speed	c	Fresh water	= 1481 m/s @ 20 °C
Sound speed	c	Air	= 343 m/s @ 20 °C
Density	ρ	Sea water	= 1026 kg/m^3 @ 13 °C
Density	ρ	Fresh water	= 998 kg/m^3 @ 20 °C
Density	ρ	Air	= 1.21 kg/m^3 @ 20 °C

J.L. Butler, C.H. Sherman, *Transducers and Arrays for Underwater Sound*,
Modern Acoustics and Signal Processing, DOI 10.1007/978-3-319-39044-4_13

13.2 Materials for Transducers Ordered by Impedance, ρc

Nominal values of Young's modulus, Y, and bulk modulus, B, (10^9 N/m^2,GPa), density, ρ (kg/m^3), bar sound speed, c (m/s), Poisson's ratio, σ, characteristic impedance, ρc, (10^6 kg/m^{2}s, Mrayls). *Note*: $B = Y/3(1 - 2\sigma)$

Material	Y (GPa)	B (GPa)	σ	ρ (kg/m^3)	c (m/s)	ρc (Mrayls)
Tungsten	362	183	0.17	19,350	4320	83.6
Nickel	210	184	0.31	8800	4890	43.0
Carbon steel	207	157	0.28	7860	5130	40.3
Stainless steel	193	146	0.28	7900	4940	39.0
Alumina	300	172	0.21	3690	9020	33.3
Beryllium Cu	125	123	0.33	8200	3900	32.0
Beryllia, BeO	345	338	0.33	2850	11,020	31.4
Brass	104	96	0.32	8500	3500	29.8
Ferrite	140	111	0.29	4800	5400	25.9
PZT-8[1]	74	77	0.34	7600	3120	23.7
PZT-4[1]	65	68	0.34	7550	2930	22.1
Titanium	104	119	0.36	4500	4810	21.6
Galfenol[2]	57	158	0.44	7900	2690	21.2
AlBemet	200	101	0.17	2100	9760	20.5
Terfenol-D[2]	26	62	0.43	9250	1680	15.5
Aluminum	71	70	0.33	2700	5150	13.9
Lead	16.5	46	0.44	11,300	1200	13.6
Macor[3]	66.9	53	0.29	2520	5150	13.0
Glass	62.0	40	0.24	2300	5200	12.0
PIN-PMN-PT[1]	17.5	97	0.47	8000	1480	11.8
PMN-.29PT[1]	16.7	93	0.47	7740	1470	11.4
Magnesium	44.8	45	0.33	1770	5030	8.90
GRP along fiber	16.4	46	0.44	2020	2850	5.76
GRP cross fiber	11.9	15	0.37	2020	2430	4.91
A-2 Epoxy	5.8	6.0	0.34	1770	1810	3.20
PVDF[1]	3.0	3.1	0.34	1600	1370	2.19
Lucite	4.0	6.7	0.40	1200	1800	2.16
Nylon 6,6	3.3	6.1	0.41	1140	1700	1.94
Nylon 6	2.8	4.2	0.39	1130	1570	1.78
Syntactic foam	4.0	4.4	0.35	690	2410	1.66
Hard rubber	2.3	3.8	0.40	1100	1450	1.60
MDF (fiberboard)	3.0	1.7	0.20	800	1940	1.55
Kraft paper	1.03	1.14	0.35	1200	926	1.11
SADM[5]	0.33	0.55	0.40	2000	406	0.81
Corprene[4] (500 psi)	0.47	1.58	0.45	1100	654	0.72
Onionskin paper[6]	0.50	0.56	0.35	1000	707	0.71
Phenolic resin/cotton	0.23	0.55	0.43	1330	416	0.55

(continued)

Material	Y (GPa)	B (GPa)	σ	ρ (kg/m^3)	c (m/s)	ρc (Mrayls)
Corprene[4] (100 psi)	0.15	0.49	0.45	1000	387	0.39
Neoprene, Type A	0.06	0.50	0.48	1400	207	0.29
Polyurethane PR1590	0.036	0.30	0.48	1080	182	0.20
Silicone rubber	0.014	0.12	0.48	1150	110	0.13
RTV 615	0.0015	0.012	0.48	1050	38	0.04
Water (sea)	0.00	2.28	0.50	1026	1500	1.54

Young's modulus, density, and Poisson's ratio are normally used in FEA models and are given here. However, in one-dimensional analysis the bulk value $B = Y/3(1 - 2\sigma)$ is usually needed for thin soft materials ($\rho c < 1$ Mrayls) which are attached to hard surfaces, limiting lateral expansion. Also, the values for soft materials depend on composition, orientation, and values of Poisson's ratio, that may not be exact. *Notes*: [1]short circuit 33 mode, [2]open circuit 33 mode, [3]machinable glass ceramic, [4]DC-100, [5]Syntactic Acoustic Damping Material, [6]paper stack at 1000 psi.

13.3 Time Averages, Power Factor, Complex Intensity

13.3.1 Time Average

The time average of the product of two complex time harmonic variables with the same period is equal to the time average of the product of their real parts. For example, consider any two such variables with a phase angle, ϕ, between them:

$$x(t) = x_0 e^{j\omega t} \tag{13.1}$$

$$y(t) = y_0 e^{j(\omega t + \phi)} \tag{13.2}$$

The time average of xy is

$$\langle xy \rangle = \frac{1}{T}\int_0^T (x_0 \cos \omega t)[y_0 \cos(\omega t + \phi)]dt = \frac{1}{2} x_0 y_0 \cos \phi \tag{13.3}$$

where T is the period. It can be seen that this result is also given by

$$\langle xy \rangle = \frac{1}{2}\mathrm{Re}(xy^*) = \frac{1}{2}\mathrm{Re}(x_0 y_0 e^{-j\phi}) = \frac{1}{2} x_0 y_0 \cos \phi \tag{13.4}$$

13.3.2 Power

In the case of electrical variables, voltage and current,

$$V(t) = V_0 e^{j\omega t} \tag{13.5}$$

$$I(t) = I_0 e^{j(\omega t + \phi)} \tag{13.6}$$

and the time average power is

$$\langle VI \rangle = \frac{1}{2} V_0 I_0 \cos\phi \tag{13.7}$$

where $\cos\phi$ is the electrical power factor.

In the case of radiated acoustic power, the reaction force of the water on the transducer surface is (see Chap. 1, Sect. 1.3)

$$F_r = (R_r + jX_r) u_0 e^{j\omega t} = |Z_r| e^{j\phi} u_0 e^{j\omega t} \tag{13.8}$$

$$u = u_0 e^{j\omega t} \tag{13.9}$$

where Z_r is the radiation impedance, and the time average radiated power is

$$\langle F_r u \rangle = \frac{1}{2} \mathrm{Re}\left[|Z_r| u_0 e^{j(\omega t + \phi)} u_0 e^{-j\omega t} \right] = \frac{1}{2} |Z_r| u_0^2 \cos\phi = \frac{1}{2} R_r u_0^2 \tag{13.10}$$

where $\tan\phi = X_r/R_r$ and $\cos\phi$ is the mechanical power factor defined in the same way as electrical power factor.

13.3.3 Intensity

The acoustic intensity vector is defined as the product of the pressure and the particle velocity (see Sect. 10.1),

$$\vec{I} = p\vec{u}, \tag{13.11}$$

and the time average intensity is, therefore,

$$\left\langle \vec{I} \right\rangle = \frac{1}{2} \mathrm{Re}\left(p\vec{u}^{\,*} \right). \tag{13.12}$$

Each component of $\left\langle \vec{I} \right\rangle$ gives the flow of radiated energy per unit area at a point in the sound field. On the surface of a transducer, the normal component of $\left\langle \vec{I} \right\rangle$ gives the flow of radiated energy per unit area.

13.3.4 Radiation Impedance

For nonuniform velocity of the transducer surface, the radiation impedance referred to a reference velocity, u_0, is defined to make the time average radiated power equal to $\frac{1}{2}R_r u_0^2$, where R_r is the radiation resistance, as for uniform velocity transducers. Equation (1.4b) is consistent with this, as can be seen from the following:

$$R_r = \mathrm{Re}(Z_r) = \frac{1}{uu^*}\iint_S \mathrm{Re}(pu^*)\mathrm{d}S = \frac{1}{u_0^2}\iint_S \mathrm{Re}(pu^*)\mathrm{d}S \tag{13.13}$$

from which

$$\frac{1}{2}R_r u_0^2 = \iint_S \frac{1}{2}\mathrm{Re}(pu^*)\mathrm{d}S = \iint_S \langle I_n\rangle \mathrm{d}S = \text{time average radiated power} \tag{13.14}$$

where $\langle I_n\rangle$ is the time average normal intensity on the transducer surface.

13.3.5 Complex Intensity

In general the quantity (pu*) is complex (see Sect. 13.17), and the imaginary part is called the reactive intensity [see Sect. 6.5.7]. It has a zero time average value and corresponds to oscillatory transport of acoustic energy from one part of the sound field to another part or from the sound field to the transducer. The reactive intensity is zero in an ideal plane wave sound field. In a simple spherical wave, $p = (P/r)\mathrm{e}^{j(\omega t - kr)}$, the velocity has only a radial component, $u_r = (p/\rho c)(1 + 1/jkr)$, and the imaginary part of (pu*) is $P^2/\rho\omega r^3$, which goes to zero as $1/r^3$ in the far field rather than $1/r^2$ as does the time average intensity. Therefore in most practical situations the reactive intensity is considered to be negligible, but it may be measurable and have useful interpretations in some cases. Radiated and scattered fields that have angular dependence also have velocity and intensity components perpendicular to the radial direction. In the far field these intensity components also diminish as $1/r^3$ and become negligible for most purposes.

13.4 Relationships Between Piezoelectric Coefficients

The relationships between the different sets of piezoelectric coefficients are:

$$d_{mi} = \sum_{n=1}^{3} \varepsilon_{nm}^{T} g_{ni} = \sum_{j=1}^{6} e_{mj} s_{ji}^{E} \qquad g_{mi} = \sum_{n=1}^{3} \beta_{nm}^{T} d_{ni} = \sum_{j=1}^{6} h_{mj} s_{ji}^{D}$$

$$e_{mi} = \sum_{n=1}^{3} \varepsilon_{nm}^{S} h_{ni} = \sum_{j=1}^{6} d_{mj} c_{ji}^{E} \qquad h_{mi} = \sum_{n=1}^{3} \beta_{nm}^{S} e_{ni} = \sum_{j=1}^{6} g_{mj} c_{ji}^{D}$$

Since piezoelectric ceramics and piezomagnetic materials have only ten independent coefficients (three are piezoelectric or piezomagnetic, two are permittivities or permeabilities, five are elastic), these relationships simplify for these materials and may be displayed as:

$d_{31} = \varepsilon_{33}^{T} g_{31} = e_{31} s_{11}^{E} + e_{31} s_{12}^{E} + e_{33} s_{13}^{E}$	$g_{31} = \beta_{33}^{T} d_{31} = h_{31} s_{11}^{D} + h_{31} s_{12}^{D} + h_{33} s_{13}^{D}$
$d_{33} = \varepsilon_{33}^{T} g_{33} = e_{31} s_{13}^{E} + e_{31} s_{13}^{E} + e_{33} s_{33}^{E}$	$g_{33} = \beta_{33}^{T} d_{33} = h_{31} s_{13}^{D} + h_{31} s_{13}^{D} + h_{33} s_{33}^{D}$
$d_{15} = \varepsilon_{11}^{T} g_{15} = e_{15} s_{44}^{E}$	$g_{15} = \beta_{11}^{T} d_{15} = h_{15} s_{44}^{D}$
$e_{31} = \varepsilon_{33}^{S} h_{31} = d_{31} c_{11}^{E} + d_{31} c_{12}^{E} + d_{33} c_{13}^{E}$	$h_{31} = \beta_{33}^{S} e_{31} = g_{31} c_{11}^{D} + g_{31} c_{12}^{D} + g_{33} c_{13}^{D}$
$e_{33} = \varepsilon_{33}^{S} h_{33} = d_{31} c_{13}^{E} + d_{31} c_{13}^{E} + d_{33} c_{33}^{E}$	$h_{33} = \beta_{33}^{S} e_{33} = g_{31} c_{13}^{D} + g_{31} c_{13}^{D} + g_{33} c_{33}^{D}$
$e_{15} = \varepsilon_{11}^{S} h_{15} = d_{15} c_{44}^{E}$	$h_{15} = \beta_{11}^{S} e_{15} = g_{15} c_{44}^{D}$
$\beta_{33}^{T} = 1/\varepsilon_{33}^{T}$ and $\beta_{11}^{T} = 1/\varepsilon_{11}^{T}$	$\beta_{33}^{S} = 1/\varepsilon_{33}^{S}$ and $\beta_{11}^{S} = 1/\varepsilon_{11}^{S}$

with similar relations for magnetostrictive parameters, μ_{33}^{T}, μ_{11}^{T}, etc.

In the following relationships between the elastic constants, the superscripts E or D (or H or B) apply to both c and s in each equation:

$$c_{11} = \left(s_{11}s_{33} - s_{13}^{2}\right)/(s_{11} - s_{12})\left[s_{33}(s_{11} + s_{12}) - 2s_{13}^{2}\right] \tag{13.15}$$

$$c_{12} = -c_{11}\left(s_{12}s_{13} - s_{13}^{2}\right)/\left(s_{11}s_{33} - s_{13}^{2}\right) \tag{13.16}$$

$$c_{13} = -c_{33}s_{13}/(s_{11} + s_{12}) \tag{13.17}$$

$$c_{33} = (s_{11} + s_{12})/\left[s_{33}(s_{11} + s_{12}) - 2s_{13}^{2}\right] \tag{13.18}$$

$$c_{44} = 1/s_{44} \tag{13.19}$$

Analogous relationships hold for s_{ij} in terms of c_{ij}. The coefficients c_{66} and s_{66} are sometimes used where $c_{66} = 1/s_{66}$ and $s_{66} = 2(s_{11} - s_{12})$.

With uniform electric field drive in the polarized three direction, the three orthogonal strains are

$$S_1 = s_{11}T_1 + s_{12}T_2 + s_{13}T_3 + d_{13}E_3, \tag{13.20}$$

$$S_2 = s_{21}T_1 + s_{22}T_2 + s_{23}T_3 + d_{23}E_3, \tag{13.21}$$

$$S_3 = s_{31}T_1 + s_{32}T_2 + s_{33}T_3 + d_{33}E_3, \tag{13.22}$$

where $s_{21} = s_{12}$, $s_{22} = s_{11}$, $s_{23} = s_{31} = s_{32} = s_{13}$ and the s_{ij} elastic coefficients are evaluated for constant E field and normally display a superscript E (short circuit condition). For magnetic field drive the E is replaced by H and the s_{ij} elastic coefficients are evaluated for constant H field (open circuit). The relevant Poisson's ratio under 33 mode drive, for $T_1 = T_2 = 0$, is $\sigma = -s_{13}/s_{33}$ while under 31 mode drive, for $T_2 = T_3 = 0$, the ratios are $-s_{12}/s_{11}$ and $-s_{13}/s_{11}$, evaluated under short and open circuit conditions for electric and magnetic field transduction, respectively.

For homogeneous, isotropic materials $s_{13} = s_{12}$, $s_{33} = s_{11}$, and $s_{44} = s_{66}$, leaving only two independent elastic constants, although four different elastic constants are commonly used: Young's modulus, $Y = 1/s_{11}$; shear modulus, $\mu = 1/s_{66}$; Poisson's ratio, $\sigma = -s_{12}/s_{11}$; and bulk modulus $B = [3(s_{11} + 2s_{12})]^{-1}$. The following relationships between these four constants are useful:

$$B = Y/3(1 - 2\sigma), \quad \mu = Y/2(1 + \sigma), \quad Y = 2\mu(1 + \sigma), \quad Y = 9B\mu/(\mu + 3B). \tag{13.23}$$

13.5 Small Signal Properties of Piezoelectric Materials

Small Signal Properties of Piezoelectric Materials[a,b,c]

	PZT-8	PZT-4	PZT-5A	PZT-5H	PMN-.33PT
Quantity	Type III	Type I	Type II	Type VI	Single crystal
k_{33}	0.64	0.70	0.705	0.752	0.9569
k_{31}	0.30	0.334	0.344	0.388	0.5916
k_{15}	0.55	0.513	0.486	0.505	0.3223
k_p	0.51	0.58	0.60	0.65	0.9290
k_t	0.48	0.513	0.486	0.505	0.6326
K_{33}^T	1000	1300	1700	3400	8200
K_{33}^S	600	635	830	1470	679.0
K_{11}^T	1290	1475	1730	3130	1600
K_{11}^S	900	730	916	1700	1434
d_{33} (pC/N)	225	289	374	593	2820
d_{31}	−97	−123	−171	−274	−1335
d_{15}	330	496	584	741	146.1
g_{33} (Vm/N)	25.4×10^{-3}	26.1×10^{-3}	24.8×10^{-3}	19.7×10^{-3}	38.84×10^{-3}
g_{31}	−10.9	−11.1	−11.4	−9.11	−18.39
g_{15}	28.9	39.4	38.2	26.8	10.31
e_{33} (C/m^2)	14.0	15.1	15.8	23.3	20.40
e_{31}	−4.1	−5.2	−5.4	−6.55	−3.390
e_{15}	10.3	12.7	12.3	17.0	10.08
h_{33} (GV/m)	2.64	2.68	2.15	1.80	33.94
h_{31}	−.77	−.92	−.73	−.505	−5.639
h_{15}	1.29	1.97	1.52	1.13	7.938
s_{33}^E (pm^2/N)	13.5	15.5	18.8	20.7	119.6
s_{11}^E	11.5	12.3	16.4	16.5	70.15

(continued)

Quantity	PZT-8 Type III	PZT-4 Type I	PZT-5A Type II	PZT-5H Type VI	PMN-.33PT Single crystal
s_{12}^{E}	−3.7	−4.05	−5.74	−4.78	−13.19
s_{13}^{E}	−4.8	−5.31	−7.22	−8.45	−55.96
s_{44}^{E}	31.9	39.0	47.5	43.5	14.49
s_{33}^{D}	8.5	7.90	9.46	8.99	10.08
s_{11}^{D}	10.1	10.9	14.4	14.05	45.60
s_{12}^{D}	−4.5	−5.42	−7.71	−7.27	−37.74
s_{13}^{D}	−2.5	−2.10	−2.98	−3.05	−4.111
s_{44}^{D}	22.6	19.3	25.2	23.7	12.99
c_{33}^{E} (GPa)	132	115	111	11.7	103.8
c_{11}^{E}	149	139	121	126	115.0
c_{12}^{E}	81.1	77.8	75.4	79.5	103.0
c_{13}^{E}	81.1	74.3	75.2	84.1	102.0
c_{44}^{E}	31.3	25.6	21.1	23.0	69.00
c_{33}^{D}	169	159	147	157	173.1
c_{11}^{D}	152	145	126	130	116.9
c_{12}^{D}	84.1	83.9	80.9	82.8	104.9
c_{13}^{D}	70.3	60.9	65.2	72.2	90.49
c_{44}^{D}	4.46	5.18	3.97	4.22	77.00
ρ (kg/m^3)	7600	7500	7750	7500	8038
Q_m	1000	600	75	65	–
$\tan\delta$	0.004	0.004	0.02	0.02	<0.01
T_c (°C)	300	330	370	195	–

Note: PMN_.33PT listed is laboratory grade and is not as stable as PMN-.28PT and PMN-.29PT.
[a]Important Properties of Morgan Electro Ceramics, Piezoelectric Ceramics Report, TP-226, Bedford, Ohio, 44146. Properties listed are standard original versions of Types I, II, III, VI. See latest Morgan Electro Ceramics brochure for additional information on other versions. Other piezoelectric ceramic manufacturers include Channel Industries, Inc., Santa Barbara, CA 93111, Exelis (EDO Ceramics), Salt Lake City, Utah 84115, Piezo Kinetics, Inc., Bellefonte, PA 16823, and CTS Electronic Components, Inc. Albuquerque, NM 87113. TBS, Belmawr, NJ.
[b]Private communication, Wewu Cao, Penn State University, University Park, PA, laboratory grade PMN-.33PT. For other grades and types see *Piezoelectric Single Crystals and Their Application*, Edited by: S. Trolier-McKinstry, L. Eric Cross, and Y. Yamashita, The Pennsylvania State University and Toshiba Corp., Penn State University Press, PA (2004). See also, TRS Ceramics, Inc., State College, PA 16801 and CTG Advanced Materials, LLC, Bolingbrook, IL 60440, for commercial grade material.
[c]The tabulated single crystal constants are small signal values. The values change with field, prestress, and temperature. Single crystals also go through phase transitions and the locations in temperature-field-stress space depend on these loadings and the particular cut and composition. Different compositions and orientations of single crystals can have dramatically different values. Materials with a lower percentage of PT are not quite as active but have a larger region of linear behavior and may be more suitable for sonar applications. [Private communication, Elizabeth A. McLaughlin, NUWC, Newport, RI.]

13.5.1 Comparison of Small Signal Properties of Textured Ceramic, PZT-8 Ceramic, and Commercial Grade Single Crystal Piezoelectric Materials

Comparison of Small Signal Properties of Textured Ceramic,[a] PZT-8 Ceramic, and Commercial Grade Single Crystal[b,c,d] Piezoelectric Materials

Quantity	PMN-PT textured (undoped)	PMN-PT textured (doped)	PZT-8	PMN-.29PT single crystal (TRS)	PIN-PMN-PT (TRS)	PIN-PMN-PT (CTG)
k_{33}	0.83	0.76	0.64	0.91	0.91	0.89
k_{31}	0.54	0.44	0.30	0.44	0.47	0.46
k_{15}	0.52	0.55	0.55	0.35	0.25	0.26
k_p	0.76	0.63	0.51	–	–	–
k_t	0.58	0.54	0.48	0.60	0.57	0.50
K_{33}^T	3490	2200	1000	5400	4200	4753
K_{11}^T	3300	2830	1290	1560	1335	1728
d_{33} (pC/N)	855	517	225	1540	1320	1285
d_{31}	−393	−207	−97	−699	−634	−646
d_{15}	423	419	330	164	105	122
g_{33} (mV)m/N	27.7	26.5	25.4	32.2	35.6	30.6
g_{31}	−12.7	−10.6	−10.9	−14.6	−17.0	−15.4
g_{15}	14.5	16.7	28.9	11.9	8.8	8.0
s_{33}^E (pm²/N)	34.1	23.5	13.5	59.9	57.3	49.0
s_{11}^E	17.4	11.5	11.5	52.1	49.0	45.8
s_{12}^E	0.1	−0.5	−3.7	−24.6	−20.0	−19.6
s_{13}^E	−14.3	−8.0	−4.8	−26.4	−26.5	−23.2
s_{44}^E	22.5	23.3	31.9	16.0	15.2	14.3
c_{33}^E (GPa)	93	90	132	108	114	124
c_{11}^E	122	133	149	124	119	124
c_{12}^E	64	50	81.1	111	105	109
c_{13}^E	78	63	81.1	104	104	110
c_{44}^E	44	43	31.3	63	66	70
ρ (kg/m³)	8068	8050	7600	7740	8000	8122
Q_m	94	714	1000	164	165	–
$\tan\delta$	0.01	0.005	0.004	0.005	0.005	–
T_c (°C)	129	130	300	129	180	–
T_{rt} (C⁰)	76	–	–	104	125	–

Notes: PMN-PT materials require an electric field bias. Doping element is Mn. Single crystal PMN-PT and PIN-PMN-PT are [001] polled. The quantity $\varepsilon_{ii} = K_{ii}\varepsilon_0$ where $\varepsilon_0 = 8.842 \times 10^{-12}$ C. The transition temperature T_{rt} is the temperature for rhombohedral to tetragonal phase transition, which causes a change in properties. The Curie temperature, T_c, is the temperature at which piezoelectric material is completely depolarized.

(continued)

[a]S. F. Poterala, S. Trolier-McKinstry, R. J. Meyer, Jr., and G. L. Messing, Processing, texture quality and piezoelectric properties of <001>$_c$ textured $(1-x)Pb(Mg_{1/3}Nb_{2/3})TiO_3$-$xPbTiO_3$ ceramics, J. Appl. Phys. **110**, 014105(2011).
[b]S. Zhang, S. Lee, D. Kim, H. Lee, and T. R. Shrout, J. Am. Ceram. Soc. **91**, 683 (2008).
[c]TRS Technologies, Inc. State College, PA 16801.
[d]CTG Advanced Materials, LLC, Bolingbrook, IL 60440.

13.6 Piezoelectric Ceramic Approximate Frequency Constants (See Footnote 1)

Frequency constant	PZT-8 (Type III)	PZT-4 (Type I)	PZT-5A (Type II)	PZT-5H (Type VI)	Thickness, length, diameter
(kHz m)					
Plate, N_t	2.11	2.03	1.98	1.98	$f \times T$
Bar, N_{31}	1.57	1.50	1.40	1.40	$f \times L$
Bar, (31)	1.70	1.65	1.47	1.45	$f \times L$
Bar, (33)	1.57	1.47	1.37	1.32	$f \times L$
Disc, planar	2.34	2.29	1.93	1.96	$f \times D$
Ring, (31)	1.07	1.04	0.914	0.914	$f \times D$
Ring, (33)	0.990	0.927	0.851	0.813	$f \times D$
Sphere, N_{sp}	1.83	1.73	1.55	1.52	$f \times D$
Hemisphere	2.27	2.14	1.92	1.89	$f \times D$
(kHz in.)					
Plate, N_t	83	80	78	78	$f \times T$
Bar, N_{31}	62	59	55	55	$f \times L$
Bar, (31)	67	65	58	57	$f \times L$
Bar, (33)	62	58	54	52	$f \times L$
Disc, planar	92	90	76	77	$f \times D$
Ring, (31)	42	41	36	36	$f \times D$
Ring, (33)	39	36.5	33.5	32	$f \times D$
Sphere, N_{sp}	72	68	61	60	$f \times D$
Hemisphere	89	84	75.5	74.5	$f \times D$

Notes: Frequency constants are under short circuit conditions. Thickness, T, Length, L, Mean Diameter, D. N_t is for thickness mode plate, N_{31} is for an end electroded bar, (31) is for side electroded bar, (33) is for a segmented bar of parallel wired elements, Planar is the radial mode of a disc, N_{sp} is for a hollow sphere.

13.7 Small Signal Properties of Magnetostrictive Materials

13.7.1 Nominal 33 Magnetostrictive Properties

Nominal 33 Magnetostrictive Properties[a]

Property	Terfenol-D	Galfenol	Metglas
ρ (kg/m^3)	9250	7900	7400
k_{33}	0.72	0.61	0.92
d_{33} (nm/A)	15	46	910
Y^H (GN/m^2)	26	57	22
Y^B (GN/m^2)	55	91	140
μ_r^T	9.3	260	17×10^3
μ_r^S	4.5	160	2.6×10^3
ρ_e (μ Ω cm)	60	75	130
c^H (m/s)	1.7×10^3	2.7×10^3	1.7×10^3
c^B (m/s)	2.4×10^3	3.4×10^3	4.4×10^3
ρc^H (kg/m^{2}s)	16×10^6	21×10^6	13×10^6
ρc^B (kg/m^{2}s)	23×10^6	26×10^6	33×10^6

Notes: Permeability $\mu = \mu_r \mu_0$ where $\mu_0 = 4\pi \times 10^{-7}$. Terfenol-D at compressive stress 18 MPa and 500 Oe bias. Galfenol at compressive stress 20 MPa and 23 Oe bias. Metglas 2605 SC annealed with 7 kOe transverse field, at compressive stress 0 MPa, magnetic bias unknown but small.
[a]Information supplied by A. E. Clark and M. Wun-Fogle, Naval Surface Warfare Center, Carderock Division, West Bethesda, MD. See also M. B. Moffett, A. E. Clark, M. Wun-Fogle, J. F. Lindberg, J. P. Teter, and E. A. McLaughlin, "Characterization of Terfenol-D for magnetostrictive transducers," J. Acoust. Soc. Am., **89**, 1448–1455 (1991), J. L. Butler, "Application manual for the design of Etrema Terfenol-D magnetostrictive transducers," Etrema Products, Ames, Iowa (1988), and Etrema Products, Ames Iowa 50010 for additional information on Terfenol-D.

13.7.2 Three-Dimensional Terfenol-D Properties[1]

Measurements made at 30 MPa compressive stress and 1257 Oe (100 kA/m) bias.

[1]Information supplied by F. Claeyssen, Cedrat Technologies S. A., Meylan, France, See also, F. Claeyssen, N. Lhermet, R. Le Letty, "Giant magnetostrictive alloy actuators," J. Applied Electromagnetics in Materials, **5**, 67–73 (1994) and P. Bouchilloux, N. Lhermet, F. Claeyssen, R. Le Letty, "Dynamic shear characterization in a magnetostrictive rare earth-iron alloy," M.R.S. Symp. Proc., **360**, 265–272 (1994) and F. Claeyssen, "Variational formulation of electromechanical coupling," Handbook of Giant Magnetostrictive Materials, Ed. G. Engdahl, Academic Press, 2000 pp 353–376 (ISBN 0-12-238640-X).

$k_{33} = 0.70$	$k_{31} = 0.33$	$k_{15} = 0.33$
$\mu_{33}^{T} = 3.0\mu_0$	$\mu_{11}^{T} = 8.1\mu_0$	$\mu_{33}^{S} = 1.1\mu_0$
$d_{33} = 8.5 \times 10^{-9}$	$d_{31} = -4.3 \times 10^{-9}$	$d_{15} = 16.5 \times 10^{-9}$
$s_{33}^{H} = 3.8 \times 10^{-11}$	$s_{11}^{H} = 4.4 \times 10^{-11}$	$s_{13}^{H} = -1.65 \times 10^{-11}$
$s_{12}^{H} = -1.1 \times 10^{-11}$	$s_{44}^{H} = 24 \times 10^{-11}$	$s_{66}^{H} = 11 \times 10^{-11}$

The values of other properties may be calculated from the relationships given in Sect. 13.4.

13.8 Voltage Divider and Thevenin Equivalent Circuit

Voltage dividers and Thevenin circuits are often used in transducer equivalent circuit representations. Voltage dividers provide means for simple evaluation while Thevenin equivalent circuits provide means for a reduction from a more complex circuit to a simpler two component circuit. The simple example of Fig. 13.1 may be used to develop both concepts.

13.8.1 Voltage Divider

Figure 13.1 shows a voltage source, V, and current I with two series impedances Z_1 and Z_2 developing an output voltage, V_0, across Z_2 at terminals A–B. The source voltage $V = I(Z_1 + Z_2)$ so that $I = V/(Z_1 + Z_2)$ leading to the output voltage as $V_0 = IZ_2$ or

$$V_0 = VZ_2/(Z_1 + Z_2). \tag{13.24}$$

Thus, the output voltage of a voltage divider is simply the input voltage times the ratio of the output impedance to the sum of the impedances. If $Z_1 = Z_2$, $V_0 = V/2$ while if $Z_2 = 100 \times Z_1$, then $V_0 = V[100/(1 + 100)] = V/1.01$ and $V_0 \approx V$. If the impedances are due to capacitances C_1 and C_2 so that $Z_1 = 1/j\omega C_1$ and $Z_2 = 1/j\omega C_2$, we get

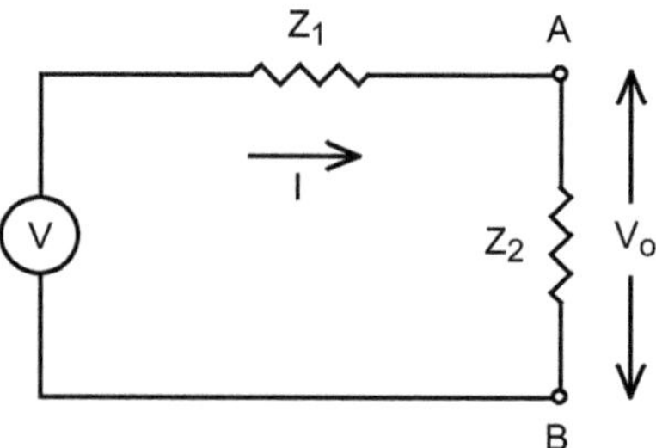

Fig 13.1 A simple voltage divider

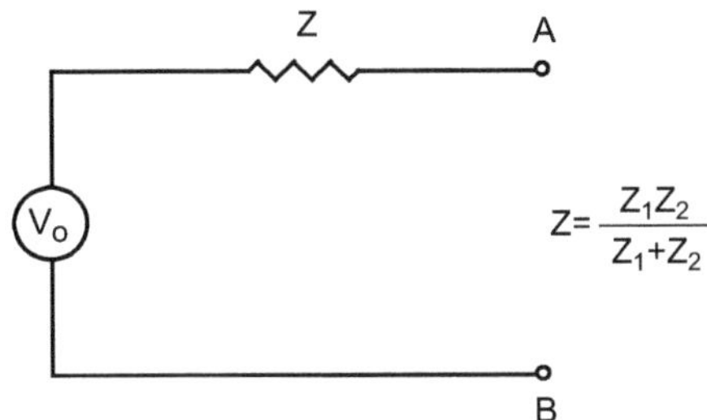

Fig. 13.2 Thevenin equivalent circuit representation

$$V_0 = VC_1/(C_1 + C_2). \tag{13.25}$$

This latter case accounts for the reduction in voltage output from a hydrophone cable with the hydrophone operating below resonance with free capacitance C_1 and a cable of capacitance C_2.

13.8.2 Thevenin Equivalent Circuit

A Thevenin equivalent circuit representation allows a circuit, such as the one shown in Fig. 13.1, to be represented by the simpler circuit shown in Fig. 13.2. Using this example as a basis, Thevenin's Theorem states that the output voltage, V_0, of Fig. 13.1 becomes the source voltage in Fig. 13.2, and the output impedance of Fig. 13.1 at terminals A–B (with V set to zero) becomes the source impedance Z in Fig. 13.2. For the simple example of Fig. 13.1 we get the source impedance $Z = Z_1Z_2/(Z_1 + Z_2)$ from the parallel impedance combination and source voltage $V_0 = V\ Z_2/(Z_1 + Z_2)$ from the output voltage for the Thevenin representation of Fig. 13.2. This procedure may be used no matter how complicated the initial circuit.

13.9 Magnetic Circuit Analysis

13.9.1 Equivalent Circuit

The design of magnetostrictive transducers requires attention to the paths and material of the magnetic circuit. This is particularly true for Terfenol-D for which the relative permeability is only about five times greater than that of free space. Although finite element analysis should be used to accurately evaluate the design of the magnetic circuit, circuit analysis techniques may be used to obtain approximate results during the initial design phase.

A simple circuit analysis model may be based on the electrical equivalent of a magnetic circuit as illustrated in Fig. 13.3. Here the magnetomotive force, F, the

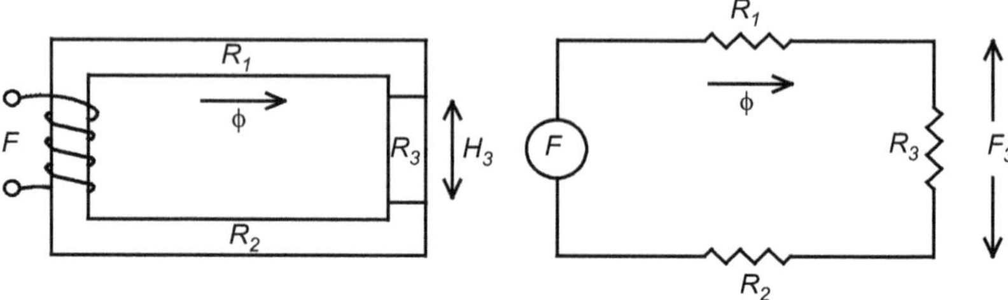

Fig. 13.3 Magnetic equivalent circuit

magnetic flux, φ, and the magnetic reluctance R are analogous to the voltage, current, and resistance, respectively. With A the cross-sectional area, L the length of the section, L_c the coil length of turns N or the length of a permanent magnet, μ the permeability, H the magnetic field intensity, B the flux density, and I the current we have the analogies:

Voltage, V	$\leftrightarrow$	Magnetomotive force	$F = N\,I = H\,L_c$
Current, I	$\leftrightarrow$	Magnetic flux	$\varphi = B\,A$
Resistance, R	$\leftrightarrow$	Reluctance	$R = L/A\mu$

If there are M reluctances around a closed path, the total flux is simply $\varphi = F/\sum R_i$ where the sum goes from 1 to M, and each of the reluctances is calculated according to its respective permeability, length, and cross-sectional area.

13.9.2 Example

A sample model and equivalent circuit is shown in Fig. 13.3 for the case of three reluctances where a magnetomotive force F is generated by a coil of N turns and current I or a magnet of field H and coil length L_c. Here we want to calculate the field intensity, H_3, across the reluctance element R_3 which represents the active part of a transducer, while the reluctances R_1 and R_2 are considered necessary magnetic paths to send the field to element 3 from the source F. First the flux is calculated as $\varphi = F/(R_1 + R_2 + R_3)$ then the magnetomotive force F_3 across the reluctance R_3 is calculated as $F_3 = R_3\varphi$ yielding the magnetic field intensity $H_3 = F_3/L_3$. It can be seen that low reluctance paths 1 and 2 are desirable and these may be obtained by use of a material with high permeability. Other components such as a reluctance, R_4, in parallel with R_3 could also be used to represent magnetic field fringing, reducing the flux in element 3, and thus, reducing the desired magnetic field intensity, H_3. In this case it is desirable that the path 4 should have a high reluctance and low permeability compared to path 3, to reduce the leakage field. Although this type of analysis is useful in the initial design of a magnetostrictive transducer, it should be followed by a magnetic finite element analysis for a more accurate evaluation especially in cases where there is significant magnetic field fringing.

13.10 Norton Circuit Transformations

There are two Norton circuit transformations which have been found helpful in circuit analysis and used by Mason[2] in equivalent circuit representations of transducers. The first transforms an ideal transformer of turns ratio N, shunted at the input by an impedance Z_s to an equivalent "T" network with series impedances Z_a at the input and Z_b at the output with a shunt impedance Z_c at the junction between Z_a and Z_b as shown in Fig. 13.4. The relationships are

$$Z_a = Z_s(1 - N), \quad Z_b = Z_s N(N - 1), \quad Z_c = Z_s N. \tag{13.26}$$

For example consider the common input impedance case of a piezoelectric ceramic transducer where Z_s represents the impedance of the input clamped capacitance C_0, with $Z_s = 1/j\omega C_0$, and where N is the electromechanical turns ratio. In this case we find that the "T" network is given by three impedances represented by three capacitances C_a, C_b, and C_c with values

$$C_a = C_0/(1 - N), \quad C_b = C_0/N(N - 1), \quad C_c = C_0/N, \tag{13.27}$$

with corresponding capacitors replacing the impedances in Fig. 13.4.

Another useful Norton circuit transforms a half T network to a reversed half T network with transformer. This case is illustrated in Fig. 13.5 where T is the turns ratio for the ideal transformer:

$$T = 1 + Z_1/Z_2, \quad Z_A = Z_1 T, \quad Z_B = Z_2 T. \tag{13.28}$$

For example consider the case where the head mass, m, is represented by the mechanical impedance $Z_1 = j\omega m$ and the tail mass, M, by the mechanical impedance $Z_2 = j\omega M$. In this case the impedances Z_A and Z_B are represented by the masses m_A and M_B given by

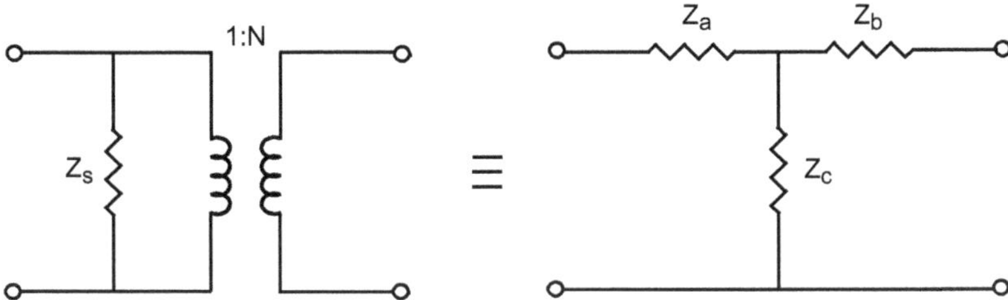

Fig. 13.4 First Norton transformation

[2] W. P. Mason, "An electromechanical representation of a piezoelectric crystal used as a transducer," IRE Proc. **23**, 1252–1263 (1935) and "*Electromechanical Transducers and Wave Filters,*" 2nd Ed. Van Nostrand, NJ (1948).

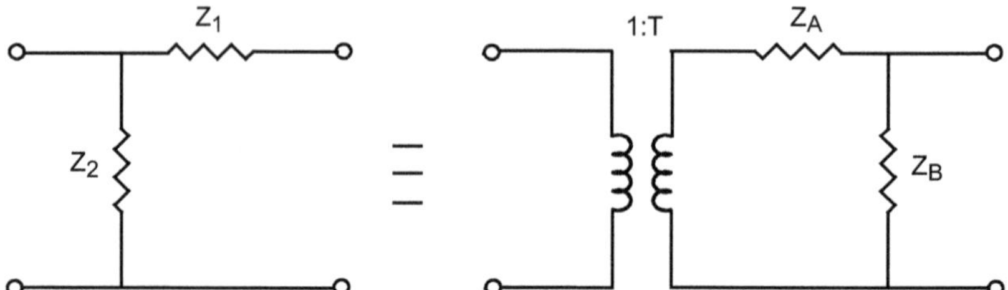

Fig. 13.5 Second Norton transformation

$$T = 1 + m/M, \quad m_{\mathrm{A}} = mT, \quad M_{\mathrm{B}} = MT. \tag{13.29}$$

This transformation may be used to simplify the equivalent circuit of a doubly resonant transducer (with m the center mass) by moving the tail mass to a more favorable position allowing it to be combined, in parallel, with the head mass.

13.11 Integral Transform Pairs

Fourier Transform

$$\overline{p}(\alpha) = \frac{1}{2\pi} \int_{-\infty}^{+\infty} p(z) \mathrm{e}^{-j\alpha z} \mathrm{d}z, \qquad p(z) = \int_{-\infty}^{+\infty} \overline{p}(\alpha) \mathrm{e}^{j\alpha z} \mathrm{d}\alpha \tag{13.30}$$

Hankel Transform

$$\overline{p}(\alpha) = \int_{0}^{\infty} p(r) J_0(r\alpha) r \, \mathrm{d}r, \qquad p(r) = \int_{0}^{\infty} \overline{p}(\alpha) J_0(r\alpha) \alpha \, \mathrm{d}\alpha \tag{13.31}$$

Hilbert Transform (Principal value)

$$\overline{p}(\alpha) = \frac{1}{\pi} \int_{-\infty}^{\infty} \frac{p(x)}{(x-\alpha)} \mathrm{d}x, \qquad p(x) = -\frac{1}{\pi} \int_{-\infty}^{\infty} \frac{\overline{p}(\alpha)}{(x-\alpha)} \mathrm{d}\alpha \tag{13.32}$$

Hilbert Transform for impedances $Z(\omega) = R(\omega) + jX(\omega)$ where $Z(-\omega) = Z^*(\omega)$.

$$X(\omega) = \frac{2\omega}{\pi} \int_{0}^{\infty} \frac{R(\alpha) - R(\omega)}{(\alpha^2 - \omega^2)} \mathrm{d}\alpha, \quad R(\omega) = \frac{2}{\pi} \int_{0}^{\infty} \frac{\alpha X(\alpha) - \omega X(\omega)}{(\alpha^2 - \omega^2)} \mathrm{d}\alpha. \tag{13.33}$$

13.12 Stiffness, Mass, and Resistance

We briefly discuss the fundamentals of stiffness, K, mass, m, and resistance, R, for piezoelectric transducers and their relations and impact on performance and resonance. This section also serves as a very brief description of some of the basic fundamentals of underwater piezoelectric transducers. See Sect. 13.17 for a representation of the complex impedance, Z, and corresponding K, m, and R representation.

13.12.1 *Mechanical Stiffness [K = F/x]*

The stiffness, K, of a material or structure is a measure of the displacement, x, developed for a given force, F. That is $x = F/K$ or equivalently $F = Kx$ and this is known as Hooke's law. The compliance $C = 1/K$ and here $x = CF$ with C another measure of the displacement, x, for a given external force F. The compliance C is often used in equivalent circuit representation of transducers since it is the mechanical analogue of the electrical capacitance. The compliance of a material is given in more general terms as $S = sT$ where s is the elasticity, the stress $T = F/A$ where A is the cross-sectional area, and the strain $S = x/L$ where L is the length of the material being stressed. In simple one-dimensional structures such as a free bar, the Young's modulus $Y = 1/s$. From this and the equation for the compliance we get $C = sL/A$ and $K = YA/L$. That is the compliance, C, of a structure is proportional to the elasticity, s, and length to area ratio and the stiffness, K, is proportional to Y and the area to length ratio. Since the impedance $Z = F/u$ where the velocity $u = j\omega x$ (see Sect. 13.17), at frequency $f = \omega/2\pi$ the associated mechanical impedance is $Z = K/j\omega = 1/j\omega C$ where $1/\omega C$ is the reactance.

13.12.2 *Piezoelectric Compliance [C^E = x/F]*

In the case of piezoelectric transducers where the voltage, V, is the analogue of the force, F, it can be shown, well below resonance, that $x = dV$ where d is the piezoelectric constant (see Sect. 13.5). In this case a piezoelectric material is stretched from an internal force $F = NV$ where N may be thought of as the turns ratio of an electromechanical transformer and given through $N = F/V$ in an equivalent circuit. In one-dimensional piezoelectric models a displacement x can be achieved by both an external stress T and an internal stress generated by an electric field $E = V/x$ resulting in a total strain $S = sT + dE$ or equivalently displacement $x = CF + dV$. Since this includes an elasticity and compliance that would be evaluated for E and V short circuited ($V = E = 0$), s and C are written as s^E and C^E in these equations so as not to be confused with values s^D and C^D which are evaluated under open circuit conditions.

Other inactive mechanical compliances, C_i, which may be necessary to the transducer design, will reduce the electromechanical coupling coefficient, k, based on $k^2 =$ energy converted/total energy stored. Ideally $k^2 = N^2C^E/C_f$, where the free capacity $C_f = C_0 + N^2C^E$ and C_0 is the blocked capacitance which is the electrical capacitance of the transducer while clamped and also is the capacitance of the transducer at resonance.

*13.12.3 Mass [*m = F/a*]*

The above discussion presumed that operation was well below resonance (quasi static condition) where effects of mass are usually negligible. But all materials have mass and this is an important part of a transducer model especially if operation is in the vicinity of resonance or well above resonance where the stiffness may often be ignored. Although piezoelectric materials have contributing mass, it is usually the mass of the piston that is radiating into the water that is of most interest and the effective mass of the piezoelectric material is added to this radiating mass. If the mass, m, moves with acceleration a for a driving force, F, we have $F = ma = mj\omega u$ (see Sect. 13.17) and an impedance $Z = F/u = j\omega m$ for the mass m alone.

The effective mass of the piezoelectric bar or stack which drives the piston of mass m is $m_b/3$, with m_b the mass of the piezoelectric bar and the other end of the bar is fixed (see Sect. 4.2.2). If the radiation mass is m_r, the total mass is $m + m_r + m_b/3$ and not just m alone. At low frequencies where the transducer radiating surface is small compared to the wave of sound in the medium, m_r is a constant but as the frequency increases the value of m_r eventually falls off as $1/f$.

13.12.4 Resonance [$\omega_0 = 1/\sqrt{(mC)}$*]*

Many vibration models are developed by focusing on the mass of interest and discovering the net force on this mass by subtracting out the reaction forces from the driving force. If, for example, we have a mass that is being held in place by stiffness $K = 1/C$ then we can write $ma = F - Kx$ or, in a better form, $ma + Kx = F$. If then F is moving with angular frequency ω, we can write the equation in terms of the velocity, u, as $(j\omega m)u + (1/j\omega C)u = F$ with sinusoidal solution for the velocity as $u = F/j(\omega m - 1/\omega C)$. As seen, the velocity becomes large in the vicinity of resonance $\omega_o = 1/\sqrt{(mC)}$ and infinite at resonance without loss resistance.

13.12.5 *Resistance [R = F/u]*

It is the resistance, R, which stabilizes and reduces the output at resonance. And with this resistance term, the impedance Z is equal to $R + j(\omega m - 1/\omega C)$ resulting in $u = F/R$ at resonance (see Sect. 13.17). The resistance lowers the mechanical $Q = \omega_o m/R$ of the transducer and introduces the smoothness in the frequency response. The reactive terms come from stored potential and kinetic energy and represented by the stiffness and mass respectively while the resistive term represents a loss of energy. This loss is useful if it represents energy leaving the transducer, by means of acoustic radiation, into the intended medium. This transmitted power may be represented by u^2R_r.

The quantity R_r is the radiation resistance and may depend on the size of the transducer radiator compared to the wavelength in the medium. The value of R_r increases as the frequency increases (see Sect. 13.13), as f^2, until the radiator is approximately a wavelength in size. And here the resistance is nearly a constant and the radiator is considered "ρc" loaded and R_r maintains nearly a constant value above that frequency.

The loss resistance, R_l, may contain terms that are constant or terms which depend on frequency. This loss of power, u^2R_l, is not useful and unfortunately heats the transducer. The mechanical efficiency of the transducer is the ratio of the power delivered into the medium to the power delivered to the transducer and may be written as $\eta = R_r/(R_r + R_l)$.

13.13 Frequently Used Formulas

13.13.1 *Transduction*

Relations involving the effective electromechanical coupling coefficient, k_{eff}:

$$k_{\text{eff}}^2 = \left[1 - (f_r/f_a)^2\right], \quad k_{\text{eff}}^2 = (1 + Q_m Q_e)^{-1}, \tag{13.34}$$

where f_r and f_a are resonance and antiresonance frequencies and Q_m and Q_e are mechanical and electrical quality factors:

$$Q_m = f_r/(f_2 - f_1) = \omega_{rw} M/R, \quad Q_e = \omega_{rw} C_0/G_m, \quad \omega_{rw} = (K/M)^{1/2}. \tag{13.35}$$

The frequencies f_2 and f_1 are half power frequencies, M, K, and R are effective mass, stiffness, and resistance, C_0 is the clamped capacitance, G_m is the motional conductance at resonance, and ω_{rw} is the angular resonance frequency in water:

$$C_0 = C_f\left(1 - k_{eff}^2\right), \quad k_{eff}^2 = N_{eff}^2/K^E C_f, \quad \text{or} \quad k_{eff}^2 = N_{eff}^2/\left(N_{eff}^2 + K^E C_0\right) \tag{13.36}$$

where C_f is the free capacitance, N_{eff} is the effective electromechanical turns ratio, and $K^E = 1/C^E$ is the short circuit stiffness and C^E is the short circuit compliance.

If the piezoelectric material of a transducer of turns ratio N has an associated cement joint of compliance total C_i, the effective turns ratio and effective short circuit compliance are

$$N_{eff} = N/\left(1 + C_i/C^E\right), \quad C_{eff}^E = C^E\left(1 + C_i/C^E\right). \tag{13.37}$$

For a small 33 mode block of piezoelectric material with cross-sectional area, A_0, thickness, t, between electrodes, cement of thickness, t_i, and elastic constant s_i

$$N = d_{33}A_0/ts_{33}^E, \quad C^E = s_{33}^E t/A_0, \quad C_i = s_i t_i/A_0. \tag{13.38}$$

Input power per unit volume of piezoelectric ceramic, P, for a given electric field, E, or stress, T, at resonance:

$$P = k_{eff}^2 \omega_{rw} \varepsilon_{33}^T E^2 Q_m/2, \quad P = \omega_{rw} s_{33}^E T^2/2Q_m, \tag{13.39}$$

ε_{33}^T and s_{33}^E are free permittivity and short circuit compliance coefficients which are related to the 33 mode electromechanical coupling coefficient by:

$$\varepsilon_{33}^S = \varepsilon_{33}^T\left(1 - k_{33}^2\right), \quad s_{33}^D = s_{33}^E\left(1 - k_{33}^2\right), \quad k_{33}^2 = d_{33}^2/\varepsilon_{33}^T s_{33}^E. \tag{13.40}$$

Similar relationships exist for biased magnetostrictive transducers with permittivity replaced by permeability and s^E and s^D replaced by s^H and s^B, respectively.

The bar sound speed:

$$c_{bar} = \left(Y/\rho_m\right)^{1/2}, \tag{13.41}$$

where Y is the Young's modulus and ρ_m is the density of the bar material. With f_r the resonance frequency, the frequency constants for a bar of length L and ring of mean diameter D are:

$$f_r L = c_{bar}/2, \quad f_r D = c_{bar}/\pi. \tag{13.42}$$

Internal hydrophone equivalent plane wave thermal noise pressure spectral density:

$$10\log\langle p_n^2\rangle = -198\,\text{dB} - \text{RVS} + 10\log R_h \text{dB}//(\mu\text{Pa})^2/\text{Hz}. \tag{13.43}$$

where RVS is the plane wave receiving sensitivity in dB//V/μPa and R_h is the Thevenin equivalent electrical series resistance of the hydrophone. For piezoelectric ceramic hydrophones at low frequency, the noise becomes

$$10\log\langle p_n^2\rangle = -206\,\text{dB} + 10\log(\tan\delta/C_f) - \text{RVS} - 10\log f\ \text{dB}//(\mu\text{Pa})^2/\text{Hz}, \quad (13.44)$$

where tan δ is the dissipation factor and f is the frequency.

Piezoelectric Matrix Equations of State, with T the stress and E the electric field:

$$\begin{array}{lll} \text{Strain} & S = s^E T + d^t E \\ \text{Electric displacement} & D = dT + \varepsilon^T E \end{array}$$

For $T_1 = T_2 = 0, S_3 = s_{33}^E T_3 + d_{33}E_3$ and for $T_2 = T_3 = 0$, Strain, $S_1 = s_{11}^E T_1 + d_{31}E_3$
For open circuit conditions, $D = 0$ and then $E = -(d^t/\varepsilon^T)T$.

Low Frequency Hydrophone Voltage, V, (well below resonance) may be written as:

$$V = g_{31}tT_1 + g_{32}tT_2 + g_{33}tT_3,$$

where we have used $E_3 = -V/t$ with t the distance between electrodes of a block of piezoelectric material. For $T_1 = T_2 = 0$ and the acoustic pressure $p = T_3$ the receiving sensitivity $M = V/p = g_{33}\,t$. (Note that for a short 31 mode cylinder of mean radius, a, it turns out that $V = g_{31}ap$ and $M = g_{31}\,a$.)

Figure of Merit (FOM) with V_p the volume of the piezoelectric material and V_0 the total volume, W power out, ξ_r rms electric field, η the efficiency, B the buoyancy, and with subscripts 33 or 31 applied:

Projector $\text{FOM}_v = W/(V_0 f_r Q_m) = 2\pi\eta\varepsilon^T k_e^2\xi_r^2 V_p/V_0$, $\quad \text{FOM}_m = B\text{FOM}_v$.

Piezoelectric alone $\text{FOM}_v = 2\pi\xi_r^2 d^2/s^E = 2\pi\xi_r^2 ed$, $\quad$ Hydrophone $\text{FOM}_h = gdV_p$.

13.13.2 Radiation

The far-field normalized pressure beam pattern functions of a line of length L and circular piston of diameter D in a rigid baffle are:

$$P(\theta) = \sin(x)/x, \quad P(\theta) = 2J_1(y)/y, \quad (13.45)$$

where $x = (\pi L/\lambda)\sin\theta$, $y = (\pi D/\lambda)\sin\theta$, and the angle θ is measured from the normal to the line or piston. The function $J_1(y)$ is the Bessel function of order 1 and argument y.

The −3 dB beam width, BW (in degrees):

$$\text{Long line radiator of length } L \gg \text{the wavelength}, \lambda; \quad \text{BW} \approx 51\lambda/L, \tag{13.46}$$

$$\text{Large circular piston radiator of diameter } D \gg \lambda; \quad \text{BW} \approx 58\lambda/D, \tag{13.47}$$

$$\text{The array product theorem for the beam pattern } P(\theta,\varphi) = f(\theta,\varphi)A(\theta,\varphi), \tag{13.48}$$

where $f(\theta, \varphi)$ is the beam pattern of identical elements of a planar array and $A(\theta, \varphi)$ is the beam pattern of the array of point sources located at the centers of the elements.

For a line array of N elements, each of length L, with center-to-center separation s the product theorem gives,

$$P(\theta) = [\sin x_1/x_1][\sin(Nx_2)/N\sin(x_2)], \tag{13.49}$$

where $x_1 = (\pi L/\lambda)\sin\theta$ and $x_2 = (\pi s/\lambda)\sin\theta$. Note that for $L = s$, $P(\theta) = \sin(Nx_2)/Nx_2$.

Directivity Factor, D_f:

For a circular piston of radius a, wave number $k = 2\pi/\lambda$ and in a rigid baffle

$$D_f = (ka)^2/[1 - J_1(2ka)/ka] = (ka)^2/R_n, \tag{13.50a}$$

where the normalized radiation resistance $R_n = R_r/\pi a^2\rho c$ and R_r is the radiation resistance.

For a line array of N small elements separated by a center-to-center distance s,

$$D_f = N/\left[1 + (2/N)\sum(N-q)\text{Sinc}(qks)\right] \tag{13.50b}$$

where the sum, Σ, is from $q = 1$ to $q = N - 1$ and $\text{Sinc}(x) = \sin(x)/x$. For half wavelength, $\lambda/2$, spacing $D_f = N$ and also for N large $D_f \approx N$.

$$\text{For a plane radiator of area } A \text{ and all dimensions} \gg \lambda: \quad D_f \approx 4\pi A/\lambda^2, \tag{13.51}$$

$$\text{For a line radiator of length } L \gg \lambda: \quad D_f \approx 2L/\lambda, \tag{13.52}$$

$$\text{For a line array of } N \text{ small } \lambda/2 \text{ spaced elements:} \quad D_f = N. \tag{13.53}$$

$$\text{Axisymmetric radiator } D_f = 2I(0)/\int_0^\pi I(\theta)\sin\theta\, d\theta \quad \text{where } I(\theta) \text{ is the intensity.} \tag{13.54}$$

Directivity Index, $\mathrm{DI} = 10 \log D_f$

Approximate relations between DI and beam width, BW (in degrees):

$$\text{Line radiator of length } L; \quad \mathrm{DI} \approx 20\,\mathrm{dB} - 10\log \mathrm{BW}, \tag{13.55}$$

$$\text{Circular radiator of diameter } D; \quad \mathrm{DI} \approx 45\,\mathrm{dB} - 20\log \mathrm{BW}, \tag{13.56}$$

$$\text{Rectangular radiator of sides } L_1, L_2 \quad \mathrm{DI} \approx 45\,\mathrm{dB} - 10\log \mathrm{BW}(L_1) - 10\log \mathrm{BW}(L_2).$$

Low frequency approximations, $ka \ll 1$, for the radiation resistance, R_r, reactance, X_r, and radiation mass, M_r, for a circular piston of radius a radiating from one side only, with ρ the density and c the sound speed of the medium, are:

For a piston in a rigid baffle:	$R_r + jX_r = \rho c\pi a^2[(ka)^2/2 + j8ka/3\pi]$,	$M_r = (8/3)\rho a^3$
For a piston with no baffle:	$R_r + jX_r = \rho c\pi a^2[(ka)^2/4 + j2ka/\pi]$,	$M_r = (6/3)\rho a^3$
For a piston in a soft baffle:	$R_r + jX_r = \rho c\pi a^2[8(ka)^4/27\pi^2 + j4ka/3\pi]$,	$M_r = (4/3)\rho a^3$

The quantities R_r and X_r for a pulsating sphere of radius, a, at any frequency are:

$$R_r + jX_r = \rho cA\frac{(ka)^2 + jka}{1 + (ka)^2}, \tag{13.57}$$

and at low frequencies where $ka \ll 1$:

$$R_r + jX_r = \rho cA(ka)^2 + j\omega 3M_w, \tag{13.58}$$

where the surface area $A = 4\pi a^2$, and M_w is the mass of a sphere of water of radius a.

Mutual radiation impedance for small transducers separated by distance d_{12} in a plane:

$$Z_{12} = R_{12} + jX_{12} = R_{11}\left[\frac{\sin kd_{12} + j\cos kd_{12}}{kd_{12}}\right], \tag{13.59}$$

where R_{11} is the self-radiation resistance of one transducer.

General relation between D_a, D_f, and R_r for a transducer of area A at any frequency:

$$D_a^2 = 4\pi R_r D_f/\rho ck^2A^2. \tag{13.60}$$

Diffraction constant for spherical hydrophone of radius a ($k = 2\pi/\lambda$): $D_a = (1 + k^2 a^2)^{-1/2}$.

For a long cylindrical hydrophone of radius a with the wave direction perpendicular to the axis, $D_a = (2/\pi ka)\left[J_1^2(ka) + N_1^2(ka)\right]^{-1/2}$, where J_1 and N_1 are first-order Bessel and Neumann functions. For a thin ring hydrophone of radius a with the wave direction in the plane of the ring, $D_a = J_0\,(ka)$.

Source level for electrical input power, W_i, and transducer electroacoustic efficiency, η_{ea},

$$\mathrm{SL} = 10\log W_i + 10\log \eta_{ea} + \mathrm{DI} + 170.8\,\mathrm{dB}//\ 1\,\mu\mathrm{Pa}\ @\ 1\,\mathrm{m}. \tag{13.61}$$

Reciprocity Factor J:

$$J = M/S = 2d/\rho f, \tag{13.62}$$

where M is the open circuit hydrophone sensitivity, S is the constant current transmitting response, d is the distance between the two transducers, ρ is the medium density, and f the frequency. Reciprocity relations in dB:

$$\mathrm{RVS} = \mathrm{TCR} - 20\log f - 294\,\mathrm{dB} = \mathrm{TVR} + 20\log|Z| - 20\log f - 294,$$

where RVS is the receiving voltage sensitivity in dB//V/μPa, TCR is the transmitting current response in dB//μPa @1 m/A, TVR is the transmitting voltage response in dB//μPa @1 m/V, Z is the electrical impedance of the transducer, and f is the frequency.

Far-field pressure p at distance r for a sphere of radius a vibrating with velocity u and source strength $4\pi a^2 u$

$$p(r) = \left[j\omega\rho 4\pi a^2 u e^{-jk(r-a)}/4\pi r\right][1/(1 + jka)]. \tag{13.63}$$

Note that for $ka \ll 1$ the quantity $1/(1 + jka) = 1$.

Rayleigh distance (far-field distance) for radiator of maximum dimension L is: $R_0 = L^2/2\lambda$. Far-field pressure p at angle θ from the axis at distance r for a piston of radius a vibrating with velocity u and source strength $\pi a^2 u$ while set in a rigid baffle

$$p(r,\theta) = \left[j\omega\rho\pi a^2 u e^{-jkr}/2\pi r\right][2J_1(ka\ \sin\,\theta)/ka\ \sin\,\theta]. \tag{13.64}$$

Note that for $\theta = 0$ or $ka \ll 1$ the quantity $[2J_1(ka\ \sin\,\theta)/ka\ \sin\,\theta] = 1$. For a piston vibrating on one side only (with no baffle) and for $ka \ll 1$

$$p(r) \approx j\omega\rho\pi a^2 u e^{-jkr}/4\pi r \tag{13.65}$$

13.14 Stress, Field Limits, and Aging for Piezoelectric Ceramics

Stress, Field Limits, and Aging for Piezoelectric Ceramics[a,b,c]

Quantity	PZT-8 Type III	PZT-4 Type I	PZT-5A Type II	PZT-5H Type VI
Compressive strength (kpsi)	>75	>75	>75	>75
Max hydrostatic stress (kpsi)	>50	50	20	20
Max* static stress (kpsi) parallel to the three axis	12	12	3	2
Max* static stress (kpsi) perpendicular to the three axis	8	8	2	1.5
Tensile* peak dynamic (kpsi)	5	3.5	4	4
% change in ε_{33}^{T} @ 2 kV/cm	+2	+5	NA	NA
% change in ε_{33}^{T} @ 4 kV/cm	+4	+18	NA	NA
tan δ (low field)	0.003	0.004	0.02	0.02
tan δ (2 kV/cm)	0.005	0.02	NA	NA
tan δ (4 kV/cm)	0.01	0.04	NA	NA
AC depoling field (kV/cm)	15	>10	7	4
% change in k_p/time decade	−1.7	−1.7	−0.0	−0.2
% change in ε_{33}^{T}/time decade	−4.0	−2.5	−0.9	−0.6
% change in d_{33}/time decade	–	−3.9	−6.3	–
Change in k_p (%)				
0–40 °C	–	+4.9	+2.5	+3.2
−60 to +85 °C	–	+9.5	+9.0	+12
Change in K_{33}^{T} (%)				
0–40 °C	–	+2.7	+16	+33
−60 to +85 °C	–	+9.4	+52	+86

Notes: 1 kV/cm = 2.54 kV/in., 2 kV/cm = 5.1 kV/in., 4 kV/cm = 10.2 kV/in. *Maximum allowable one-dimensional stress before a significant effect on performance. The maximum stress is at least 15 kpsi before there is a significant effect on performance with a higher allowable stress value for short transient stress.

[a]Important Properties of Morgan Electro Ceramics, Piezoelectric Ceramics Report, TP-226, Bedford, Ohio, 44146. Properties listed are standard original versions of Types I, II, III, VI. See latest Morgan Electro Ceramics brochure for additional information on other versions. Other piezoelectric ceramic manufacturers include Channel Industries, Inc., Santa Barbara, CA 93111, Exelis (EDO Ceramics), Salt Lake City, Utah 84115, Piezo Kinetics, Inc., Bellefonte, PA 16823, and CTS Electronic Components, Inc. Albuquerque, NM 87113. TBS, Belmawr, NJ.

[b]D. A. Berlincourt and H. Krueger, "Behavior of Piezoelectric Ceramics under Various Environmental and Operation Conditions of Radiating Sound Transducers," Technical Paper TP-228, See also, TP-220, Morgan Electro Ceramics, Bedford, Ohio, 44146. D. Berlincourt, "Piezoelectric Crystals and Ceramics," in Ultrasonic Transducer Materials, O. E. Mattiat, ed., Plenum Press, NY, (1971). H. H. A. Krueger and D. Berlincourt, "Effects of high stress on the piezoelectric properties of transducer materials," J. Acoust. Soc. Am. **33**, 1339–1344 (1961). R. Y. Nishi and R. F. Brown, Behavior of piezoceramic projector materials under hydrostatic pressure," J. Acoust. Soc. Am. **7**, 1292–1296 (1964). H. H. A. Krueger, "Stress sensitivity of piezoelectric ceramic: Part 1. Sensitivity to compressive stress parallel to the polar axis," J. Acoust. Soc. Am. **42**, 636–645 (1967).

(continued)

H. H. A. Krueger, "Stress sensitivity of piezoelectric ceramic: Part 2. Heat treatment," J. Acoust. Soc. Am. **43**, 576–582 (1968). H. H. A. Krueger, "Stress sensitivity of piezoelectric ceramic: Part 3. Sensitivity to compressive stress perpendicular to the polar axis," J. Acoust. Soc. Am. **43**, 583–591 (1968).
[c]R. S. Woollett and C. L. LeBlanc, "Ferroelectric Nonlinearities in Transducer Ceramics" IEEE, Trans. Sonics and Ultrasonics, SU-20, pp. 24–31 (1973). J. deLaunay and P. L. Smith, "Aging of Barium Titanate and Lead Zioconate-Titanate Ferroelectric Ceramics," Naval Research Laboratory Rep. 7172, 15 Oct., 1970.

See Fig. 13.6 for curves of tan δ and per cent change in ε_{33}^{T} as a function of electric field for nonaged and aged materials, showing improvement for aged materials. See Figs. 13.7 and 13.8 of 1 month aged PZT-4 and PZT-8, for curves

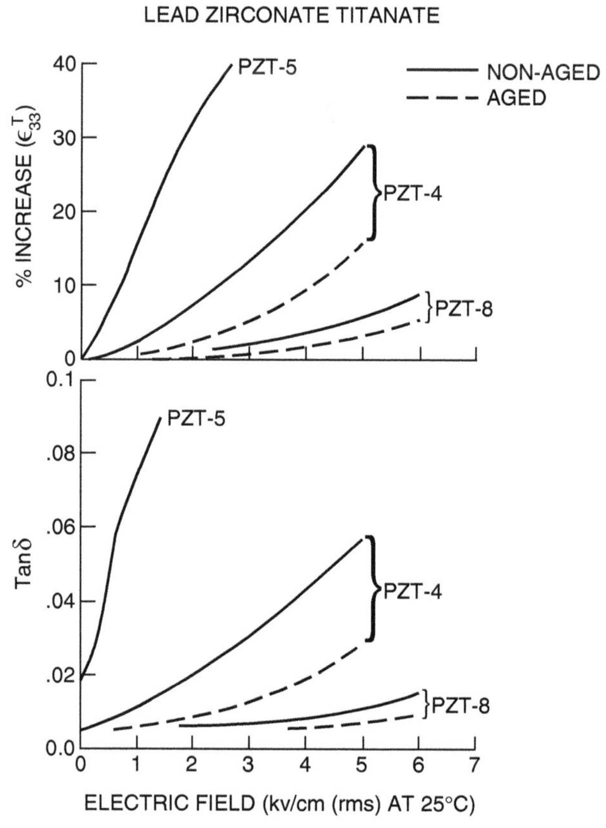

Fig. 13.6 Nonlinear behavior of permittivity and dielectric loss tangent for PZT-4, PZT-5, and PZT-8

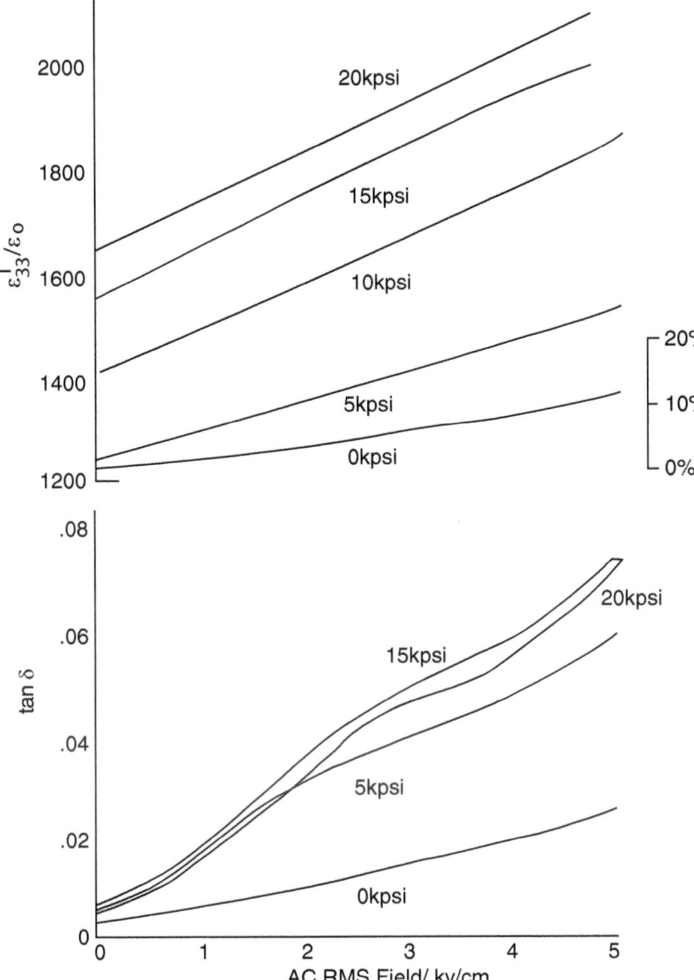

Fig. 13.7 $\varepsilon_{33}^{T}/\varepsilon_{\mathrm{o}}$ and $\tan\delta$ vs. AC field. Parameter; parallel tress (T_3) PZT-4

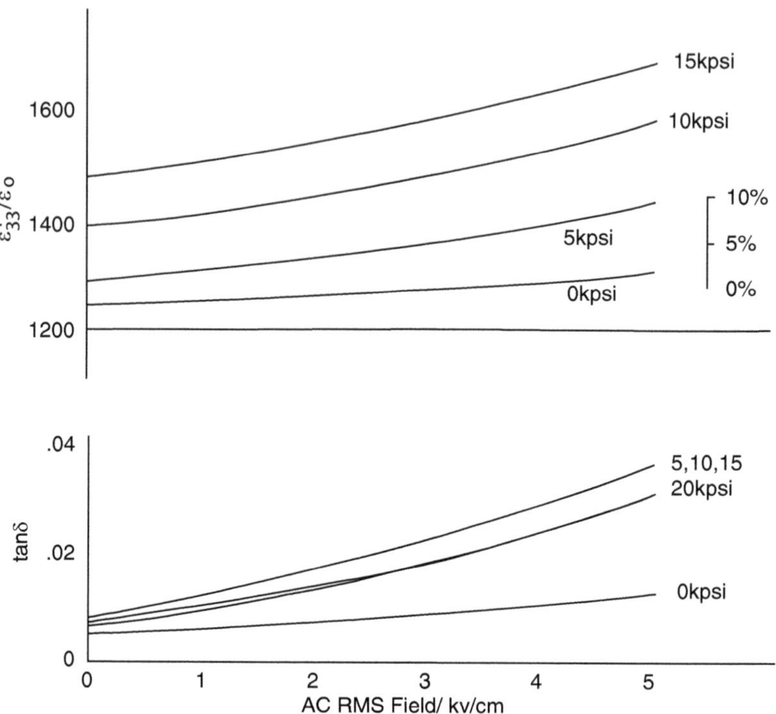

Fig. 13.8 $\varepsilon_{33}^T/\varepsilon_o$ and $\tan\delta$ vs. AC field. At various levels of parallel stress (T_3) PZT-8

of $\tan\delta$ and $\varepsilon_{33}^T/\varepsilon_0$ as a function of electric field at various parallel compressive stresses. See Fig. 13.9 for aging curves. A decrease in the dielectric constant of 5 % per time decade, following polarization, is typical. Stabilization occurs from 10 to 100 days after poling. Aging may be increased by heat treatment, but at the expense of the coupling coefficient value.

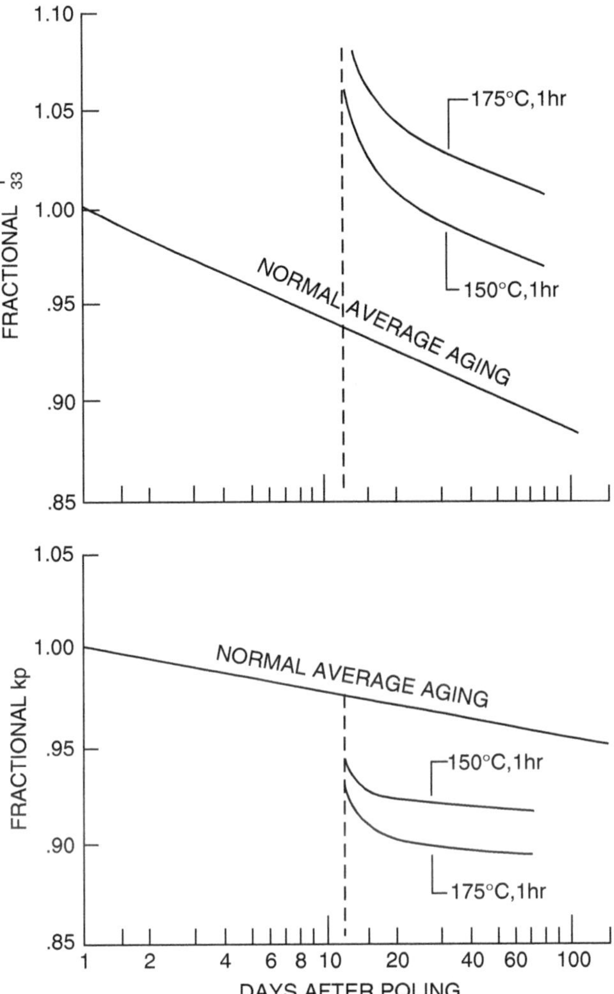

Fig. 13.9 Typical aging of permittivity and planar coupling coefficient of piezoelectric ceramic showing the effects of heat treatments applied 12 days after polarizing

13.15 Development of a Comprehensive Hydrophone Noise Model

A general description of hydrophone noise based on the series electrical resistance, R_h, is given in Sect. 6.7.3. There the electrical noise is determined from R_h through the Johnson thermal noise voltage, given by Eq. (6.57). From this the equivalent noise pressure in the water is evaluated through the hydrophone sensitivity. A different approach is taken in this section by extending the definition of Johnson thermal noise to include mechanical components such as the mechanical resistance

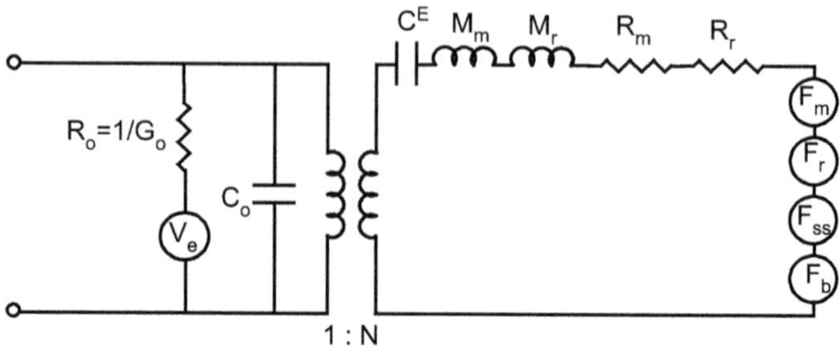

Fig. 13.10 Comprehensive equivalent noise circuit with noise voltage source $\langle V_e^2 \rangle$ and mechanical $\langle F_m^2 \rangle$, radiation $\langle F_r^2 \rangle$ and sea state $\langle F_{ss}^2 \rangle$ noise forces along with acoustic force signal F_b

and radiation resistance, thereby developing a noise force for these components.[3] This noise force is then directly converted to an equivalent noise pressure through the capture area and diffraction constant for the hydrophone. The evaluation of the electrical dissipative noise in the piezoelectric material, due to tan δ, is referred to the mechanical side of the transducer through the hydrophone sensitivity, as before. The total equivalent noise pressure is then the sum of the two mean squared equivalent values. (See also Sect. 6.7.7).

The noise model is based on the lumped equivalent circuit shown in Fig. 13.10 and is consistent with the impedance representation of Fig. 6.38. These circuits may be used to represent commonly used spherical, cylindrical, and flexural hydrophones as well as Tonpilz transducers, all of which may be reduced to an equivalent form.

In the circuit of Fig. 13.10 the electrical dissipative noise alone, in a one Hertz band, is given by

$$\langle V_n^2 \rangle = 4KTR_h, \tag{13.66}$$

with R_h replaced by $R_0 = 1/G_0 = 1/\omega C_f \tan \delta$. The equivalent mechanical noise forces $\langle F_m^2 \rangle$ and $\langle F_r^2 \rangle$ are obtained by replacing R_m and R_r by noiseless resistors in series with the noise forces (in a one Hertz band) resulting in

$$\langle F_m^2 \rangle = 4KTR_m \quad \text{and} \quad \langle F_r^2 \rangle = 4KTR_r. \tag{13.67}$$

With the total mass $M' = M_m + M_r$ and total mechanical resistance $R = R_m + R_r$, the equivalent circuit of Fig. 13.10 may be reduced to Fig. 13.11 where

$$\langle F_{mr}^2 \rangle = \langle F_m^2 \rangle + \langle F_r^2 \rangle = 4KTR = 4KTR_r/\eta_{ma}, \tag{13.68}$$

[3] C. Kittel, *Elementary Statistical Physics*, pp 141–153 John Wiley & Sons, New York, NY, 1958.

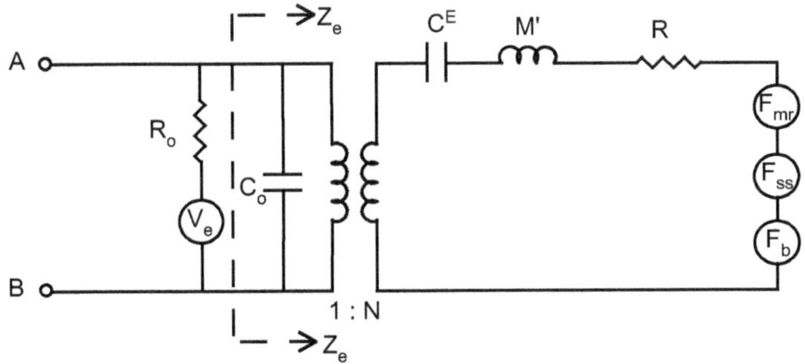

Fig. 13.11 Reduced equivalent circuit where $R = R_m + R_r$ and $M' = M_m + M_r$

and the mechanoacoustic efficiency, η_{ma}, and R_r are generally functions of frequency. Based on the plane wave relation $F = p_{ff}AD_a$, the equivalent plane wave total mechanical noise pressure is given by

$$\langle p_{mr}^2 \rangle = 4KTR_r/\eta_{ma}D_a^2A^2, \tag{13.69}$$

which allows a direct comparison with the free field acoustic signal pressure, p_{ff}. For a signal-to-noise ratio equal to or greater than unity, p_{ff}^2 should be equal to or greater than $\langle p_{mr}^2 \rangle$. Equation (13.69) shows that the thermal noise may be reduced by an increase in the area, the diffraction constant, or the mechanoacoustic efficiency.

Consider first the noise contribution from the radiation resistance alone and temporarily assume $\eta_{ma} = 1$. For a small omnidirectional hydrophone of active area A the radiation resistance $R_r = A^2\omega^2\rho/c4\pi$, $D_a = 1$ and Eq. (13.69) gives an equivalent mean squared noise pressure

$$\langle p_r^2 \rangle = 4KT\pi f^2\rho/c. \tag{13.70}$$

This noise is comparable with SS0 noise at approximately 40 kHz and SS1 noise at approximately 80 kHz increasing at 6 dB/octave as shown in Fig. 6.37 and is consistent with Eq. (6.65) for DI = 0 and $\eta_{ea} = 1$. Equation (13.70) sets a fundamental limit on the noise of a small hydrophone. This minimum limit can be interpreted as due to the radiation resistance, which is fundamental to an electroacoustic transducer.

We note that Eq. (13.70) was developed by Mellen,[4] independently of any transducer considerations, by expressing the thermal motion of the molecules in the medium in terms of modes of vibration and equating the average vibration energy per mode to KT at thermal equilibrium. This thermal noise exists in the sea

[4] R. H. Mellen "Thermal-noise limit in the detection of underwater acoustic signals," J. Acoust. Soc. Am., **24**, 478–480 (1952).

and is received by every hydrophone. It is clear from the numerical examples in Table 4.3 that this internal thermal hydrophone noise is important mainly for underwater sound applications at relatively high frequencies where the ambient noise is dominated by thermal noise, e.g., above 40 kHz as can be seen from Fig. 6.37. For transducers mounted on ships, where flow and structural noise often exceeds ambient noise, the internal noise is seldom significant (see Chap. 8).

For large planar hydrophones, or at high frequencies, the radiation resistance approaches $A\rho c$, while D_a approaches 2 and Eq. (13.69) leads to

$$\langle p_r^2 \rangle = KT\rho c / A\eta_{ma}, \tag{13.71}$$

showing that the thermal noise pressure received by a hydrophone becomes constant at a value that depends on the aperture area, A. Thus in this frequency region, the noise pressure can be lowered for some applications by using a hydrophone of large planar area. An example might be an application that uses a large diameter thickness mode piezoelectric ceramic disc as the hydrophone (see Sect. 5.4.3).

Consider again the more general case where η_{ma} is not necessarily unity. Since $R_r/D_a^2A^2 = \omega^2\rho/D_f 4\pi c$ [see Eq. (6.56)], Eq. (13.69) may also be written in terms of the directivity factor as

$$\langle p_{mr}^2 \rangle = 4\pi KT(\rho/c)f^2/\eta_{ma}D_f. \tag{13.72}$$

Equation (13.72) shows that the equivalent mechanical/radiation thermal noise pressure may be reduced by increasing the directivity factor, as was seen previously. If the pressure is referenced to 1 μPa, Eq. (13.72) may be written as

$$10\log\langle p_{mr}^2 \rangle = 20\log f - 74.8\,\text{dB} - 10\log\eta_{ma} - \text{DI}, \tag{13.73}$$

for a one cycle band. This result is exactly the same as the thermal noise curve of Fig. 6.37 for the case of 100 % mechanoacoustic efficiency and a directivity factor of unity. It is also the same as Eq. (6.65) with η_{ea} replaced by η_{ma}.

In the preceding discussion the mechanical noise force and equivalent noise pressure were calculated directly from the mechanical equivalent of the Johnson thermal noise. The equivalent noise pressure generated from the electrical dissipation, tan δ, $\langle V_e^2 \rangle$ of Fig. 13.11, can be put on the same basis as the mechanical noise and the signal by calculating the noise voltage at the electrical output terminals and transforming it to the mechanical terminals through the sensitivity. As shown in Eq. (6.67a) the electrical noise due to the electrical resistance, R_0, under the typical conditions of a dissipation factor tan $\delta \ll 1$, may be written as

$$\langle V_e^2 \rangle = 4KT \tan\delta/\omega C_f. \tag{13.74}$$

The noise voltage developed across the terminals, due to $\langle V_e^2 \rangle$, is the result of a voltage divider (see Sect. 13.8) reduction, $Z_e/(Z_e + R_0)$, where Z_e of Fig. 13.11 represents the remaining electrical impedance of the transducer in series with the resistor R_0. After considerable algebra the resulting noise at the transducer electrical terminals A, B may be written as

$$\langle V^2 \rangle = \langle V_e^2 \rangle \left[(\omega/\omega_r Q_m)^2 + \left(1 - \omega^2/\omega_r^2\right)^2 \right] / H(\omega), \tag{13.75}$$

where

$$H(\omega) = \left[(\omega/\omega_a Q_a)^2 + \left(1 - \omega^2/\omega_a^2\right)^2 \right] + \tan^2\delta \left[(\omega/\omega_r Q_m)^2 + \left(1 - \omega^2/\omega_r^2\right)^2 \right]. \tag{13.76}$$

In typical piezoelectric ceramic hydrophones $\tan^2\delta \ll 1$, leading to

$$H(\omega) \approx (\omega/\omega_a Q_a)^2 + \left(1 - \omega^2/\omega_a^2\right)^2. \tag{13.77}$$

Under this typical condition Eqs. (13.74), (13.75), and (13.77) yield the open circuit noise voltage

$$\langle V^2 \rangle \approx [4KT \tan \delta/\omega C_f] \left[(\omega/\omega_r Q_m)^2 + \left(1 - \omega^2/\omega_r^2\right)^2 / \left[(\omega/\omega_a Q_a)^2 + \left(1 - \omega^2/\omega_a^2\right)^2 \right]. \tag{13.78}$$

The total electrical noise voltage, $\langle V_n^2 \rangle$, may be obtained from Eq. (13.72) and an expression for the hydrophone voltage sensitivity which may be developed from the equivalent circuit, of Fig. 6.19. This circuit is not restricted to bender hydrophones and the resulting sensitivity, given by Eq. (6.32), is a good approximation for most pressure sensitive hydrophones. Accordingly, we may express the hydrophone sensitivity as

$$|V|^2 = k^2 [pAD_a]^2 C^E / C_f \left[(\omega/\omega_a Q_a)^2 + \left(1 - \omega^2/\omega_a^2\right)^2 \right]. \tag{13.79}$$

Equating Eqs. (13.78) and (13.79) gives the electrically based plane wave equivalent noise pressure

$$\langle p_e^2 \rangle = \left[4KT \tan \delta / A^2 D_a^2 k^2 \omega C^E \right] \left[(\omega/\omega_r Q_m)^2 + \left(1 - \omega^2/\omega_r^2\right)^2 \right]. \tag{13.80}$$

Since $Q_m = \eta_{ma}/\omega_r C^E R_r$ and, from Eq. (6.56), $R_r/D_a^2 A^2 = \omega^2 \rho / c 4\pi D_f$, Eq. (13.80) can also be written as

$$\langle p_{\text{e}}^2 \rangle = \left[4\pi KT \tan \delta f^2 \rho / c\eta_{\text{ma}} k^2 D_{\text{f}}\right] \left[(\omega/\omega_{\text{r}} Q_{\text{m}})^2 + (Q_{\text{m}}\omega_{\text{r}}/\omega)\left(1 - \omega^2/\omega_{\text{r}}^2\right)^2\right]. \tag{13.81}$$

The ratio of this electrically based noise pressure, Eq. (13.81), to the mechanically based noise pressure, Eq. (13.72), is given by

$$\langle p_{\text{e}}^2 \rangle / \langle p_{\text{mr}}^2 \rangle = \left[\tan \delta / k^2\right] \left[\omega/\omega_{\text{r}} Q_{\text{m}} + (Q_{\text{m}}\omega_{\text{r}}/\omega)\left(1 - \omega^2/\omega_{\text{r}}^2\right)^2\right]. \tag{13.82}$$

At low frequencies where $\omega \ll \omega_{\text{r}}$

$$\langle p_{\text{e}}^2 \rangle / \langle p_{\text{mr}}^2 \rangle = \left[\tan \delta / k^2\right] [Q_{\text{m}}\omega_{\text{r}}/\omega], \tag{13.83}$$

which equals the ratio of the two terms in Eq. (6.71) and therefore agrees with Table 6.3 at 1, 10, and 20 kHz. At resonance, $\omega = \omega_{\text{r}}$ and Eq. (13.82) becomes

$$\langle p_{\text{e}}^2 \rangle / \langle p_{\text{mr}}^2 \rangle = \left[\tan \delta / k^2 Q_{\text{m}}\right]. \tag{13.84}$$

For $\tan \delta = 0.01$, $k^2 = 0.5$, and $Q_{\text{m}} = 2.5$ the ratio in Eq. (13.84) is 0.008, and the mechanically based equivalent noise pressure dominates at resonance.

Equations (13.72) and (13.81) give the total equivalent hydrophone noise pressure, $\langle p_{\text{n}}^2 \rangle = \langle p_{\text{mr}}^2 \rangle + \langle p_{\text{e}}^2 \rangle$, as

$$\langle p_{\text{n}}^2 \rangle = \left[4\pi KT(\rho/c) f^2 / D_{\text{f}} \eta_{\text{ma}}\right] \times \left[1 + \left(\tan \delta / k^2\right)\left\{\omega/\omega_{\text{r}} Q_{\text{m}} + (Q_{\text{m}}\omega_{\text{r}}/\omega)\left(1 - \omega^2/\omega_{\text{r}}^2\right)^2\right\}\right], \tag{13.85}$$

where, it should be recalled, η_{ma} is generally a function of frequency. The second bracketed factor in Eq. (13.85) is the reciprocal of the electromechanical efficiency, given in Chap. 2, as can be seen from

$$\eta_{\text{em}} = N^2 R / \left[N^2 R + \left(R^2 + X^2\right)/R_0\right], \tag{13.86}$$

where $X^2 = (\omega M' - 1/\omega C^E)^2$. Then, using $N^2 = k^2 C_{\text{f}}/C^E$, $Q_{\text{m}} = 1/\omega_{\text{r}}\ C^E R$, $\omega_{\text{r}}^2 = 1/M' C^E$, and $1/R_0 = \omega C_{\text{f}} \tan \delta$, Eq. (13.86) becomes

$$\eta_{\text{em}} = 1/\left[1 + \left(\tan \delta / k^2\right)\left\{\omega/\omega_{\text{r}} Q_{\text{m}} + (Q_{\text{m}}\omega_{\text{r}}/\omega)\left(1 - \omega^2/\omega_{\text{r}}^2\right)^2\right\}\right]. \tag{13.87}$$

Since the overall efficiency $\eta_{\text{ea}} = \eta_{\text{em}}\eta_{\text{ma}}$, Eq. (13.85) may now be written as

$$\langle p_{\text{n}}^2 \rangle = 4\pi KT(\rho/c) f^2 / D_{\text{f}} \eta_{\text{ea}}, \tag{13.88}$$

and, in dB//(μPa)^2, as

$$10\log\langle p_n^2\rangle = 20\log f - 74.8 - 10\log\eta_{ea} - \mathrm{DI}, \tag{13.89}$$

which is the same as Eq. (13.73) for the special case of $\eta_{em} = 1$ and identical to Eq. (6.65) developed through reciprocity in Sect. 6.7.1. The incident plane wave signal pressure $|p_i|^2$ must be greater than the value given by Eq. (13.89) to obtain a signal-to-noise ratio greater than unity.

The total electrical noise, V_n, may be obtained from Eq. (13.79) written as

$$\langle V_n^2\rangle = \langle p_n^2\rangle k^2A^2D_a^2C^E/C_f\left[(\omega/\omega_aQ_a)^2 + \left(1-\omega^2/\omega_a^2\right)^2\right]. \tag{13.90}$$

Substitution of Eq. (13.88) and Eq. (6.56) into Eq. (13.90) yields

$$\langle V_n^2\rangle = 4KTR_rk^2C^E/C_f\eta_{ea}\left[(\omega/\omega_aQ_a)^2 + \left(1-\omega^2/\omega_a^2\right)^2\right]. \tag{13.91}$$

The preamplifier electrical noise should be less than this quantity and the hydrophone should be properly matched to the preamplifier for optimum performance. The approximate hydrophone input resistance, R_h, at any frequency, may be obtained from Eq. (13.91) by substitution of Eq. (6.57) (with $\Delta f = 1$) yielding

$$R_h = R_rk^2C^E/C_f\eta_{ea}\left[(\omega/\omega_aQ_a)^2 + \left(1-\omega^2/\omega_a^2\right)^2\right]. \tag{13.92}$$

Equation (13.92) gives the same results as Eq. (6.70) under the condition of the electrical dissipation factor $\tan\delta \ll 1$. This simplifying assumption, used in this section, is normally satisfied in practice and under this condition the results of this section should be quite acceptable. The assumption was initially used in Eq. (6.67a) and is often used in models for an electrical capacitor with typical losses. This acceptable approximation has allowed a comparatively simple but comprehensive self-noise model for hydrophones.

13.16 Cables and Transformers

13.16.1 Cables

The voltage, V_h, from a hydrophone (see Chap. 6) of free capacitance, C_f, operating well below resonance is reduced to V_c by an attached cable capacitance, C_c; the voltage ratio is

$$V_c/V_h = C_f/(C_f + C_c) = 1/(1 + C_c/C_f). \tag{13.93}$$

The cable capacitance can reduce the coupling coefficient, as well as the power factor for a projector (see Chaps. 5 and 6), but has negligible effect for $C_c/C_f \ll 1$.

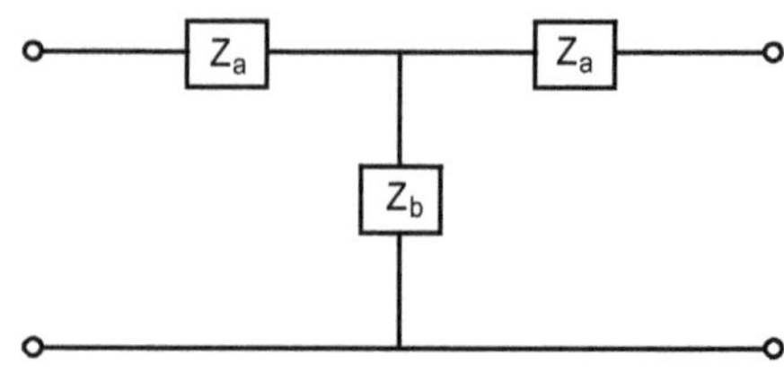

Fig. 13.12 Cable as a transmission line

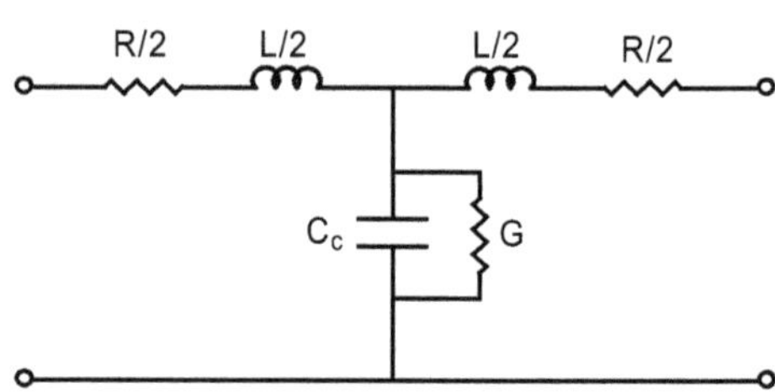

Fig. 13.13 Approximate transmission line model

The inductance and resistance, as well as the capacitance, of the cable can become significant for long cables. Such cables may be treated as transmission lines with impedances Z_a and Z_b as illustrated in Fig. 13.12.

The values of Z_a and Z_b may be approximated, as in Fig. 13.13, by

$$Z_a = R/2 + j\omega L/2 \quad \text{and} \quad Z_b = 1/(G + j\omega C_c), \tag{13.94}$$

where R is the series ohmic resistance, L the inductance, C_c the capacitance, and G the dissipation ωC_c tan δ. The values of C_c and G may be obtained under open circuit conditions while the values of R and L may be obtained under short circuit conditions with both measurements performed at low frequencies where $\omega \ll \omega_0 = (2/LC)^{1/2}$.

In general the values of Z_a and Z_b may be obtained at each frequency by measuring the impedances Z_o and Z_s at one end with the other end under open and short circuit conditions, respectively. The results, in complex form, are:

$$Z_a = Z_o - Z_b \quad \text{and} \quad Z_b = [Z_o(Z_s - Z_o)]^{1/2}. \tag{13.95}$$

13.16.2 *Transformers*

Transformers are typically used to increase the voltage from power amplifiers to a drive voltage suitable for electric field transducers. This is done by means of the turns ratio, N, of the transformer, which is ideally equal to the ratio of the number of secondary turns, n_s, to the number of primary turns, n_p. A conventional transformer which provides isolation between the secondary and primary is illustrated in Fig. 13.14 while an autotransformer, which provides no isolation, is shown in Fig. 13.15.

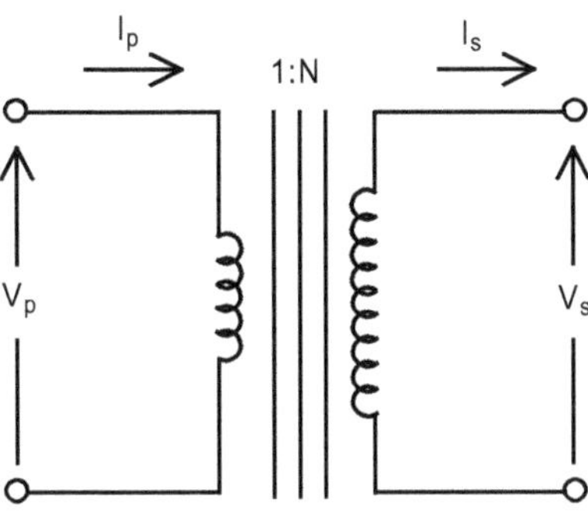

Fig. 13.14 Transformer of turns ratio $N = V_s/V_p$

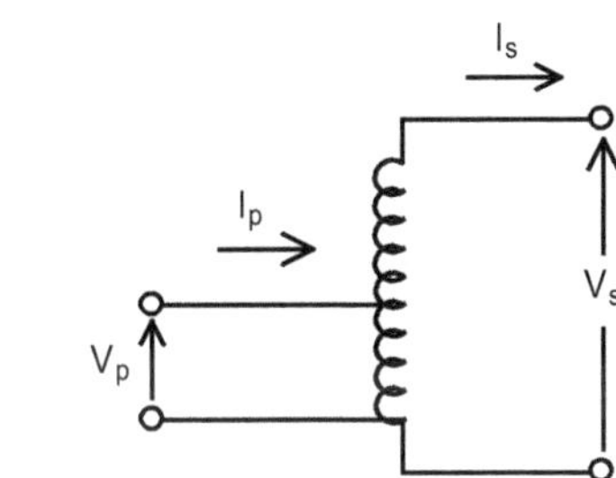

Fig. 13.15 Autotransformer of turns ratio $N = V_s/V_p$

Both provide voltage step-up, current step-down, and transformed load impedance, Z_L, as:

$$V_s = NV_p, \quad I_s = I_p/N, \quad Z_p = Z_L/N^2, \tag{13.96}$$

where the load impedance, Z_L, is across the secondary of the transformer and the primary impedance is $Z_p = V_p/I_p$.

Transformers are often designed such that the additional resistance, inductance, and capacitance associated with the transformer are negligible in the frequency range of operation for a given load impedance. On the other hand, advantage can be taken of the inherent shunt inductance, L, to provide tuning for the clamped capacitance, C_0, of a transducer, in addition to voltage step-up. Autotransformers are more compact than conventional transformers for small values of the turns ratio and, with self-tuning, are often used with electric field transducers.

An equivalent circuit[5] for a typical transformer is shown in Fig. 13.16, with the ideal transformation given by the turns ratio N. The inductors L_p and L_s are due to primary and secondary leakage inductance and the resistors R_p and R_s are due to primary and secondary coil resistance. The shunt capacitance C is due to stray capacitance which is not always well defined in value and circuit placement.

[5] W. M. Flanagan, *Handbook of Transformer Design & Applications*, 2nd ed., McGraw Hill, Boston, MA (1992).

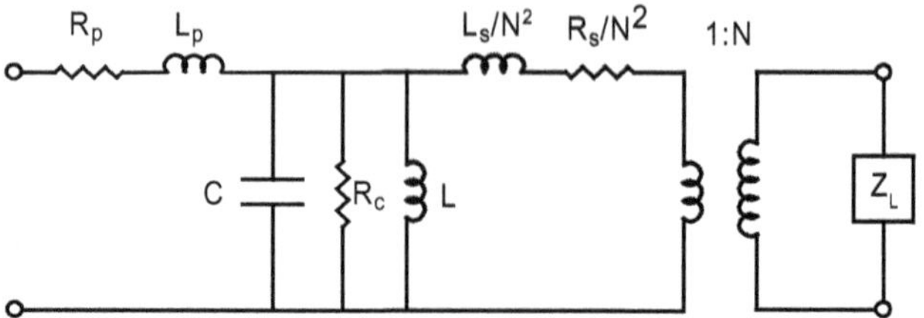

Fig. 13.16 Equivalent circuit for a transformer

The shunt resistance, R_c, is due to both eddy current and hysteresis losses in the core. The core material and its volume must be chosen to operate below saturation with low hysteresis and low eddy current losses in the frequency range of operation.

The shunt inductance, L, is due to the magnetizing current which flows in the primary independent of the load, Z_L, on the secondary. This inductance depends on the number of turns in the primary and is transformed to the secondary through the ideal turns ratio, N, as an inductance N^2L in parallel with the load Z_L. Since L is proportional to n_p^2 and $N = n_s/n_p$, the inductance $L_0 = N^2L$ is proportional to the square of the number of secondary turns, n_s^2. This value of inductance can be adjusted to shunt tune the clamped capacitance of the transducer by setting the value of L_0 equal to $1/\omega_r^2 C_0$.

13.17 Complex Algebra

Complex algebra is used throughout this book because of a need for a representation of acoustical and electrical sinusoidal functions which have both an amplitude and phase (see Sect. 13.3). It is very important since if two of these functions have the same phase, they will add but if 180° out-of-phase, they will subtract. The complex representation occurs for the impedance Z and the admittance $Y = 1/Z$ as well as voltage, V, current, I, force, F, velocity, u, and acoustic pressure p as well as derivatives of these.

Complex numbers are composed of a real part and an imaginary part such as the impedance $Z = R + jX$ where the resistance R is the real part and the imaginary part is X where X is the reactance and $j = \sqrt{(-1)}$. We note here that $j^2 = -1$ and j is commonly used in electrical engineering while $i = -j$ is used in physics and often in wave representations. In the equation for the impedance, X is a reactive result due to energy storage while R is a result of energy dissipated as a loss or acoustic radiation. The impedance Z is illustrated as both a rectangular and polar quantity in Fig. 13.17 and may be represented as $Z = R + jX$ or as $Z = |Z|e^{j\varphi}$. The horizontal axis is real while the vertical axis is considered the imaginary axis and the phase, ϕ, goes from 0 to 360° in a counterclockwise direction.

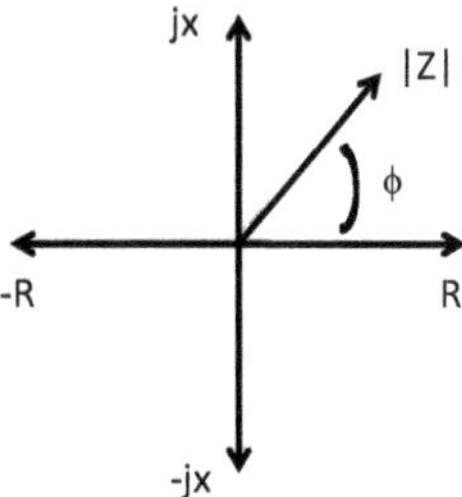

Fig. 13.17 Graph of a complex impedance $Z = R + jx$

The complex conjugate, $Z^* = R - jX$, is simply obtained by replacing j by $-j$ in all occurrences of j. The product

$$ZZ^* = |Z|^2 = (R + jX)(R - jX) = R^2 + X^2 \tag{13.97}$$

is a real quantity and $|Z| = (R^2 + X^2)^{1/2}$. On the other hand, the quantity ZZ is complex and is given by $ZZ = Z^2 = R^2 - X^2 + j2RX$. The ratio of two complex quantities, Z_1/Z_2, may be rationalized by multiplying the numerator and denominator by Z_2^* so that

$$Z_1/Z_2 = Z_1Z_2^*/Z_2Z_2^* = Z_1Z_2^*/|Z_2|^2 \tag{13.98}$$

yielding a complex numerator but real denominator allowing the solution to be written in standard form with a real and imaginary part such as $Z = R + jX$ allowing R and X to be identified. If Z_1 and Z_2 are to be multiplied, we would get

$$Z_1Z_2 = (R_1 + jX_1)(R_2 + jX_2) = R_1R_2 - X_1X_2 + j(R_2X_1 + R_1X_2) \tag{13.99}$$

On the other hand, if Z_1 and the conjugate Z_2^* are to be multiplied, we would get

$$Z_1Z_2^* = (R_1 + jX_1)(R_2 - jX_2) = R_1R_2 + X_1X_2 + j(R_2X_1 - R_1X_2) \tag{13.100}$$

If we were to add, we would simply get

$$Z_1 + Z_2 = (R_1 + R_2) + j(X_1 + X_2) \tag{13.101}$$

As may be seen from Fig. 13.17, the rectangular and polar forms may be related from R and X to ϕ and $|Z|$ as

$$\phi = \tan^{-1}(X/R) \quad \text{and} \quad |Z| = \left(R^2 + X^2\right)^{1/2} \tag{13.102}$$

or from $|Z|$ and ϕ to R and X as

$$R = |Z| \cos \phi \quad \text{and} \quad X = |Z| \sin \phi \tag{13.103}$$

We may also make connections mathematically since $e^{j\varphi} = \cos\,\varphi + j\,\sin\,\varphi$ yielding

$$Z = |Z|e^{j\varphi} = |Z|\cos\varphi + j|Z|\sin\varphi \tag{13.104}$$

The polar form (See Sect. 13.3) is usually used for dynamic variables such as voltage, V, current, I, force, F, velocity, u, and acoustic pressure p. Complex addition in polar variables is usually awkward and best done in rectangular form. However, complex multiplication or division in polar form is easily accomplished, such as $ZZ = |Z|e^{j\phi}|Z|e^{j\phi} = |Z|^2 e^{j2\phi}$ where the magnitudes multiply and the phases add. For division, the magnitudes would divide and the phases would subtract. In evaluating the acoustic pressure in the radiation field from an array, complex pressure values are maintained until the final summing of the pressures, p_i, (allowing the proper addition of the real and imaginary parts). Once this has been accomplished and the total pressure, p, has been obtained, the desired resulting magnitude $|p|$ is the quantity of interest and usually in the form 20 log $(|p|/p_o)$ where p_o is a reference pressure of 1 μPa.

Complex representation of sinusoidal phenomenon allows the reduction of the wave equation to the simpler Helmholtz form. It can also allow the reduction of differential equation of constant coefficients to an algebraic form and solution. For example, if a displacement $x = x_0 e^{j\omega t}$, then the velocity $u = dx/dt = j\omega x$ and the acceleration $a = d^2x/dt^2 = j\omega u = -\omega^2 x$. So that the common second-order differential equation for an alternating force F causing displacement x of a spring of stiffness K, attached mass of M and loss resistance R may be written as

$$M d^2x/dt^2 + R\,dx/dt + Kx = F(\omega).$$

If F is a function of ω of the form, $e^{j\omega t}$, then the motion of x will be of the same form and we may then reduce the differential equation to an algebraic equation in terms of the velocity u, yielding the equation

$$j\omega M u + Ru + (K/j\omega)u = F(\omega).$$

And the algebraic complex solution for u may then be written as

$$u = F/[R + j(\omega M - K/\omega)] = F/Z,$$

where the complex impedance $Z = [R + j(\omega M - K/\omega)] = R + jX$ where $X = \omega M - K/\omega$.

The same equation is also obtained for a simple electrical series circuit with F replaced by the voltage V, u replaced by the current I, M replaced by the inductance L, K replaced by $1/C$ where C is capacitance and finally the mechanical loss, R, replaced by the electrical resistance, R, allowing electrical circuits to be used as equivalent circuits in analyzing mechanical systems.

13.18 Transducer Publications 2000–2015

"Nonuniform piezoelectric circular plate flexural transducers with underwater applications," B. S. Aronov, J. Acoust. Soc. Am., **138**, 1570–1584 (2015).

"Dipole transducer enhancement from a passive resonator," J. L. Butler A. L. Butler, V. Curtis, J. Acoust.Soc. Am. **136**, 2472–2477 (2014).

"Analysis of feedback control of piezoelectric transducers," R. C. Randall, D. A. Brown, J. Acoust. Soc. Am., **135**, 3425–3433 (2014).

"Properties of Corprene, revisited (L)," J. F. Tressler, R. D. Corsaro, J. Acoust. Soc. Am., **135**, 2481–2484 (2014).

"Performance of tonpilz transducers with segmented piezoelectric stacks using materials with high electromechanical coupling coefficient," S. C. Thompson, R. J. Meyer, D. C. Markley, J. Acoust. Soc. Am., **135**, 155–164 (2014).

" Electromechanical properties of stripe-electroded tangentially poled piezoelectric flexural bars," S. Sarangapani, D.A. Brown, J. Acoust. Soc. Am., **133**, 2661–2667 (2013).

"Modified single crystals for high-power underwater projectors," N. P. Sherlock, R. J. Meyer, IEEE-T-UFFC, **59**, 1285–1291 (2012).

"Triple-Resonant Transducers," S. C. Butler, IEEE-T-UFFC, **59**, 1292–1300 (2012).

"Thermal model for piezoelectric transducers," J. Acoust. Soc. Am., J. L. Butler, A. L. Butler, S. C. Butler, J. Acoust. Soc. Am., **132**, 2161–2164 (2012).

"Transducer figure of merit," J. L. Butler, J. Acoust. Soc. Am., **132**, 2158–2160 (2012).

"Improved calculations of the electromechanical properties of tangentially poled stripe-electroded piezoelectric bars and cylinders with nonuniform electric fields," S. Sarangapani, D.A. Brown, J. Acoust. Soc. Am., **132**, 3068–3065 (2012).

"Analysis of unidirectional broadband piezoelectric spherical shell transducers for underwater acoustics," B. Aronov, D. A. Brown, C. L. Bachand , X. Yan, J. Acoust. Soc. Am., **131**, 2079–2090 (2012).

S. F. Poterala, S. Trolier-McKinstry, R. J. Meyer, Jr., and G. L. Messing, Processing, texture quality and piezoelectric properties of $<001>_c$ textured (1-x) $Pb(Mg_{1/3}Nb_{2/3})TiO_3$-$xPbTiO_3$ ceramics, J. Appl. Phys. **110**, 014105 (2011).

"The modal projector," J. L. Butler, A. L. Butler, S. C. Butler, J. Acoust. Soc. Am., **129**, 1881–1889 (2011).

"Effect of coupled vibrations on the parameters of tangentially polarized piezoelectric strip-electroded piezoelectric cylinders," B. S. Aronov, C. L. Bachand, D.A. Brown, J. Acoust. Soc. Am., **129**, 582–584 (2011).

"Analysis of piezoelectric properties of tangentially polarized piezoelectric strip-electroded cylinders," B. S. Aronov, C. L. Bachand, D.A. Brown, J. Acoust. Soc. Am., **129**, 2960–2967 (2011).

"Face shear piezoelectric properties of relaxor-$PbTiO_3$ single crystal," Shujun Zhang, F. Li, W. Jiang, J. Luo, R. J. Meyer, Jr., W. Cao, T. R. Shrout, J. Appl. Phys. **98**, 182903 (2011).

"Power dissipation and temperature distribution in piezoelectric ceramic slabs," D. Thomas, D. D. Ebenezer, S. M. Srinivasan, J. Acoust. Soc. Am., **128**, 1700–1711 (2010).

"Analytical modeling of piezoelectric ceramic transducers based on coupled vibration analysis with application to rectangular thickness poled plates," B. S. Aronov, C. L. Bachand, D.A. Brown, J. Acoust. Soc. Am., **126**, 2983–2990 (2009).

"A brief history of active sonar," A. D'Amico, R. Pittenger, *Aquatic Mammals*, **35**(4), 426–434 (2009).

Single-crystal lead magnesium niobate-lead titanate (PMN-PT) as a broadband high power transduction material. M. Moffett, H. C. Robinson, J. M. Powers, P. D. Baird, J. Acoust. Soc. Am., **121**, 2591–2599 (2007).

"Calibration sphere for low-frequency parametric sonars," K. G. Foote, D. T. I. Francis, P. R. Atkins. J. Acoust. Soc. Am., **121**, 1482–1490 (2007).

"Analysis of axially polarized piezoelectric ceramic rings," R. Ramesh, D. D Ebenezer, Ferroelectrics, **323**, 17–23 (2005).

"Forced response of solid axially polarized piezoelectric ceramic finite cylinders with internal losses," D. D. Ebenezer, K. Ravichandran, R. Ramesh, C. Padmanabhan, J. Acoust. Soc. Am., **117**, 3645–3656 (2005).

"Acoustic radiation impedance and directional response of rectangular pistons on elliptic cylinders," J. E. Boisvert and A. L. Van Buren, J. Acoust. Soc. Am., **118**, 104–112 (2005).

"Forced vibration of solid elastic cylinders," D. D. Ebenezer, K. Ravichandran, C. Padmanabhan, Journal of Sound and Vibration, **282**, 991–1007 (2005).

"The energy method for analyzing the piezoelectric electroacoustic transducers," B. Aronov, J. Acoust. Soc. Am. **117**, 210–220 (2005).

"Protocols for calibrating multibeam sonar," K. G. Foote, D. Chu, T. R. Hammar, K. C. Baldwin, L. A. Mayer, L. C. Hufnagle, Jr., J. M. Jech, J. Acoust. Soc. Am., **117**, 2013–2027 (2005).

"On the sound field of an oscillating disk in a finite open and closed circular baffle," T. Mellow and L. Karkkainen, J. Acoust. Soc. Am., **118**, 1311–1325 (2005).

"A tri-modal directional transducer," J. L. Butler, A. L. Butler, J. A. Rice, J. Acoust. Soc. Am., **115**, 658–665 (2004).

"The distributed edge dipole (DED) model for cabinet diffraction effects," M. R. Urban, et al., J. Audio Eng. Soc., **52**, 1043–1059 (2004).

"Acoustic directivity of rectangular pistons on prolate spheroids," J. E. Boisvert and A. L. Van Buren, J. Acoust. Soc. Am. **116**, 1932–1937 (2004).

"Determining equivalent circuit parameters for low figure of merit transducers," W. J. Marshall and G. A. Brigham, Acoustics Research Letters Online, **5** (3) July 2004.

"On the optimization of the effective electromechanical coupling coefficient of a piezoelectric body," B. S Aronov, J. Acoust. Soc. Am., **114**, 792–800 (2003).

"Approximation of the Struve function H_1 occurring in impedance calculations," R. M. Aarts, A. J. E. M. Janssen, J. Acoust. Soc. Am., **113**, 2635–2637 (2003).

"Energy analysis of a piezoelectric body under nonuniform deformation," B. S. Aronov, J. Acoust. Soc. Am., **113**, 2638–2646 (2003).

"Analysis of axially polarized piezoelectric cylinders with arbitrary boundary conditions on flat surfaces," D. D. Ebenezer, R. Ramesh, J. Acoust. Soc. Am., **113**, 1900–1908 (2003).

"Electromechanical coupling factor of capacitive micromachined ultrasonic transducers," A. Caronti, R. Carotenuto and M. Pappalardo, J. Acoust. Soc. Am., **113**, 279–288 (2003).

"Extraordinary magnetoelasticity and lattice softening in b.c.c. Fe-Ga alloys," A. E. Clark, K. B. Hathaway, M. Wun-Fogle, J. B. Restorff, V. M. Keppens, G. Petculescu and R. A. Taylor, J. Appl. Phys. **93**, 8621 (2003).

"A deep-submergence, very low-frequency, broadband, multiport transducer," A. L. Butler, J. L. Butler, Sea Technology, 31–34, November 2003.

"Analysis of a compliantly suspended acoustic velocity sensor," J. A. McConnell, J. Acoust. Soc. Am. **113**(3), 1395–1405 (2003).

"Analysis of axially polarized ceramic cylindrical shells of finite length with internal losses," D. D. Ebenezer, P. A. Abraham, J. Acoust. Soc. Am., **112**, 1953–1960 (2002).

"Closed-form analysis of thin radially polarized piezoelectric ceramic cylindrical shells with loss," D. D. Ebenezer, P. A. Abraham, *Current Science* **83**, 981–988 (2002).

"An approximated 3-D model of the Langevin transducer and its experimental verification," A. Iula, R. Carotenuto, M. Pappalardo, J. Acoust. Soc. Am., **111**, 2675–2680 (2002).

"Acoustic radiation impedance of rectangular pistons on prolate spheroids," J. E. Boisvert and A. L. Van Buren, J. Acoust. Soc. Am., **111**, 867–874 (2002).

"An approximate 3-D model of the Langevin transducer and its experimental validation," A. Iula, R. Carotenuto and M. Pappalardo, J. Acoust. Soc. Am. **111**, 2675–2680 (2002).

"Phased array element shapes for suppressing grating lobes," F.J. Pompei and S. C. Wooh, J. Acoust. Soc. Am., **111**, 2040–2048 (2002).

"Broadband, multimode, free-flooded, baffled circular ring projectors," B. Aronov, T. Oishi, L. Reinhart, D. A. Brown. J. Acoust. Soc. Am., **109**, 2364 (2001).

"Bender transducer design and operations," J. L. Delany, J. Acoust. Soc. Am. **109**, 544–562 (2001).

"Quasistatic coupling coefficients for electrostrictive ceramics," J. C. Piquette, J. Acoust. Soc. Am., **110**, 197–207 (2001).

"Measurement of electrostrictive coefficients of polymer films," F. M. Guillot, J. Jarzynski and E. Balizer, J. Acoust. Soc. Am., **110**, 2980–2990 (2001).

"Quasistatic coupling coefficients for electrostrictive ceramics," J. C. Piquette, J. Acoust. Soc. Am., **110**, 197–207 (2001).

"Modeling Piezoelectric and Piezomagnetic Devices and Structures via Equivalent Networks," A. Ballato, IEEE-T-UFFC, **48**, 1189–1240, (2001).

"Acoustic Vector Sensor Correlations in Ambient Noise," M. Hawkes and A. Nehorai, IEEE J. Oceanic Eng., **26**, 337–347 (2001).

"Near-field-far-field measurements of loudspeaker response (L)," J. L. Butler, C. H. Sherman, J. Acoust. Soc. Am., **108**, 447–448 (2000).

"Analysis and comparison of four anhysteretic polarization models for lead magnesium niobate," J.C. Piquette and R. C. Smith, J. Acoust. Soc. Am., **108**, 1651–1662 (2000).

"A broadband hybrid magnetostrictive/piezoelectric transducer array," S. C. Butler and F. A. Tito, Oceans 2000 MTS/IEEE Conference Proceedings, Vol. 3 September, 2000.

Answers to Odd-Numbered Exercises

A set of answers to the odd numbered exercises in each chapter is given in this section. In some cases, when the exercises require considerable effort to complete all the numerical and graphical work, it is not practical to present the answers in complete detail. In these cases the answers consist of some discussion of the main issues, some directions for approaching the problem, and enough numerical results to allow the reader to evaluate his understanding. In many cases the answers also contain additional discussion related to the practical relevance of the exercise.

Chapter 1

1.1. Linear differential equations of this type can be solved by assuming a solution in the form $x = \mathrm{e}^{\gamma t}$ and determining γ in terms of the coefficients:

$$\gamma = -R/2M \pm \left[R^2 - 4MK_{\mathrm{m}}\right]^{1/2}/2M = -R/2M \pm j\left[K_{\mathrm{m}}/M - (R/2M)^2\right]^{1/2}$$

$$x = \mathrm{e}^{-\alpha t}\mathrm{e}^{\pm j\omega_0 t}, \quad \alpha = R/2M, \quad \omega_0 = \left[K_{\mathrm{m}}/M - (R/2M)^2\right]^{1/2}$$

where $j = \sqrt{-1}$.

1.3. Note that $\mathrm{Real}(x_1 x_2{}^*) = X_1 X_2 \cos(\varphi_1 - \varphi_2)$.

1.5. $k^2 = 1 - K_{\mathrm{m}}^{\mathrm{H}}/K_{\mathrm{m}}^{\mathrm{B}} = 1 - L_0/L_{\mathrm{f}} = N^2/K_{\mathrm{m}}^{\mathrm{B}}L_0 = N^2/K_{\mathrm{m}}^{\mathrm{H}}L_{\mathrm{f}}$

1.7. The power output $W = \eta_{\mathrm{ea}} W_i$. SL = 30 dB − 3 dB + 0 + 170.8 = 197.8 dB// 1 μPa @ 1 m and SL = 30 dB − 3 dB + 6 + 170.8 = 203.8 dB//1μPa @ 1 m. Because of the 6 dB increase in DI, 6 dB less or ¼ the power would be needed,—truly significant.

1.9. From Sect. 13.13, Eq. (13.64) $p(r, 0) = j\,\omega\,\rho\,\pi a^2\,u\,\mathrm{e}^{-jkr}/2\pi r$ and the intensity, $I_0 = |p|^2/\rho c$, may then be written as $I_0 = (ka)^2 \rho c \pi a^2 u_{\mathrm{r}}^2/4\pi r^2 = (ka)^2 \rho c \pi a^2 N_{\mathrm{em}}^2 V^2/\left|Z_{\mathrm{mr}}^{\mathrm{E}}\right|^2 4\pi r^2$.

J.L. Butler, C.H. Sherman, *Transducers and Arrays for Underwater Sound*, Modern Acoustics and Signal Processing, DOI 10.1007/978-3-319-39044-4

Chapter 2

2.1. The free permittivity $\varepsilon_{33}^{\mathrm{T}} = K^{\mathrm{T}}\varepsilon_0$ with K^{T} the relative dielectric constant found in Sect. 13.5. (The dielectric constant for free space $\varepsilon_0 = 10^{-9}/36\pi$). Capacitances: $C_{\mathrm{f}} = 0.145$ and $C_0 = 0.074$ nF.

2.3. Electromechanical turns ratio $N = 0.235$ (N/V). Velocity $u = 0.0979 \times 10^{-3}$ and 0.4895 m/s for 1 V and 5 kV. Mechanoacoustic efficiency $\eta_{\mathrm{ma}} = 50\,\%$.

2.5. The strain $S_3 = 115.6 \times 10^{-6}$.

2.7. $K_{\mathrm{m}}^{\mathrm{E}}C_{\mathrm{f}} = \left(A_0/s_{33}^{\mathrm{E}}L\right)\left(A_0\varepsilon_{33}^{\mathrm{T}}/L\right) = (N/k)^2$ and $K_{\mathrm{m}}^{\mathrm{I}}L_{\mathrm{f}} = \left(A_0/s_{33}^{\mathrm{H}}L\right)\left(n^2\mu_{33}^{\mathrm{T}}A_0/L\right) = (N/k)^2$.

2.9. Piezoelectric ceramic force = 2.35 N and the magnetostrictive force = 49.1 N.

2.11. At resonance $\omega_{\mathrm{r}}M = 1/\omega_{\mathrm{r}}C_{\mathrm{m}}^{\mathrm{E}}$ and $\omega_{\mathrm{r}} = 1/\left(MC_{\mathrm{m}}^{\mathrm{E}}\right)^{1/2}$. Substitution for $\omega_{\mathrm{r}}M$ leads to the first result and substitution for ω_{r} to the second result with $K_{\mathrm{m}}^{\mathrm{E}} = 1/C_{\mathrm{m}}^{\mathrm{E}}$.

2.13. Use Eqs. (2.83), (2.105), and (2.106). Numerical answers are $Q_{\mathrm{m}} = 11.2$, 4.0, and 3.28, respectively.

2.15. Numerical and analytical evaluation of Eq. (2.91). Limiting expressions of Eq. (2.91) can be obtained from the binomial expansion of $P_{\mathrm{f}}(B/G)$ and $P_{\mathrm{f}}(G/B)$.

2.17. For 100 V rms the input power is 7.85 W. The mechanoacoustic efficiency is 66.7 % and the output power is 5.24 W.

Chapter 3

3.1. For $M_1 = 4$ kg, $f_{\mathrm{r}} = 36$ kHz and $Q_{\mathrm{m}} = 4.7$, while for $M_1 = 8$ kg, $f_{\mathrm{r}} = 34$ kHz and $Q_{\mathrm{m}} = 4.0$. It's probably not worth going from 4 to 8 kg as there is only a 6 % reduction in frequency and a 15 % reduction in Q_{m}. Better to increase the length of the piezoelectric material.

3.3. The sine and binomial expansions yield $1/[\sin\, kL] \approx 1/[kL - (kL)^3/6] \approx [1 + (kL)^2/6]/kL = 1/kL + kL/6$ which when multiplied by $-j\rho cA_0$ yields $-j\omega M/6$ for the second term.

3.5. Eliminate u_2 in Eq. (3.36a) and F_2 in Eq. (3.36b) and then identify coefficients.

3.7. Substitute $Z_{\mathrm{m}} = j(\omega M - 1/\omega C) + R$ and $Y_0 = j\omega C_0 + \omega C_{\mathrm{f}}/\tan\,\delta$ into Eq. (3.70).

Chapter 4

4.1. $f_{\mathrm{r}} = 32$ kHz and $Q_{\mathrm{m}} = 3.3$ when the bar mass is ignored; $f_{\mathrm{r}} = 28$ kHz and $Q_{\mathrm{m}} = 3.8$ with an added bar mass of 0.3 kg.

4.3. For $k = 0.5$, $k_{\mathrm{e}} = 0.482$, for $-3.6\,\%$ change; change in f_{r} is +5 %; original $f_{\mathrm{a}} = f_{\mathrm{r}}/0.87 = 1.15f_{\mathrm{r}}$, new $f_{\mathrm{a}} = 1.05f_{\mathrm{r}}/0.876 = 1.19f_{\mathrm{r}}$ where f_{r} is the original resonance frequency

4.5. For $k = 0.5$, $k_{\mathrm{e}} = 0.458$ for $-8.5\,\%$ change, change in f_{r} is 0.5 %, original $f_{\mathrm{a}} = 1.15\, f_{\mathrm{r}}$ new $f_{\mathrm{a}} = 1.005f_{\mathrm{r}}/0.89 = 1.13f_{\mathrm{r}}$

4.7. For $k = 0.5$ and $n = 1$, $k_{\mathrm{e}} = 0.45$; $n = 2$, $k_{\mathrm{e}} = 0.225$, $n = 3$, $k_{\mathrm{e}} = 0.15$.

4.9. Determine an effective stiffness for each mode from the modal frequency, i.e., $\omega_n^2 = \omega_0^2(1+n^2) = K_n/M = K^E(1+n^2)/M, \quad K_n = K^E(1+n^2)$. Then, assuming C_0 and N are the same for each mode, and following Sect. 4.43 where it is noted that the dynamic increase in stiffness, $(K_n - K^E)$, has the same effect as a stress rod, use of Eq. (4.29) gives

$$k_{edn}^2 = k_{31}^2/\left[1+\left(1-k_{31}^2\right)n^2\right].$$

The values for $n = 0$, 1, 2, 3 are 0.33, 0.24, 0.16, 0.11.

Chapter 5

5.1. Ring: 41/4 = 10.25 kHz. Sphere: 68/4 = 17.0 kHz.

5.3. For the ring: Water resonance ≈ 0.72 × 10.25 = 7.4 kHz and $Q_m \approx 3.9\eta_{ma}$ or ≈ 3 for $\eta_{ma} = 0.8$ using Eqs. (5.14) and (5.16). For the sphere use Eqs. (5.24) and (5.25) and find water resonance ≈ 15kHz and $Q_m \approx 3.05\eta_{ma}$.

5.5. Use equations from Sect. 5.1. Equation (2.112) gives η_{em} at resonance; using k_{31}, $\tan\delta$ and Q_m from Exercises 5.3 and 5.4 show that $\eta_{em} \approx 1$. Assuming $\eta_{ma} = 0.8$ makes $\eta_{ea} \approx 0.8$. The mass and volume of the ring are 0.62 kg and 4×10^{-4} m^3; assume that end caps and waterproofing together increases the total mass and total volume to 1 kg and 6×10^{-4} m^3. Then, assuming a maximum electric field in the ceramic of 4 kV/cm, the equations in Sect. 5.1 give:

$$(\text{FOM})_V = 75\text{W/Hz m}^3$$
$$(\text{FOM})_m = 45\text{W/kHz kg}$$

5.7. Think of the radiation mass as a rigid mass of $M = 8\rho a^3/3$ attached to the end of the spring of length $L/2$ formed by the fluid in the cavity. Alternatively, think of it as an extension of the fluid in the cavity of radius a, length ΔL, and mass $\rho\pi a^2\Delta L$. Equate the two masses and solve for ΔL. These two ways of approximating the radiation mass loading are not equivalent as seen in Exercise 5.6. Because of symmetry, a rigid plane could be inserted through the ring at half the height of the ring, without affecting the radiation loading.

5.9. Calculate the velocity in the mechanical branch for a given input voltage. Then use Eq. (5.17) for the far-field pressure.

5.11. Use the results from Ex. 5.10 with the ring diameter 4 in. For Terfenol-D: $f^H = 5.3$ kHz and $f^B = 7.5$ kHz while for Galfenol: $f^H = 8.6$ kHz and $f^B = 10.6$ kHz. From Ex 3.1 the short circuit PZT-4 resonance, $f^E = 10.25$ kHz, is nearly the same as the Galfenol short circuit resonance but higher than the Terfenol-D short circuit resonance. The free inductance $L_f = \mu_{33}^T n^2 A_c/\pi D$. The cross-sectional area $A_c = 0.258 \times 10^{-3}$ m^2, $\pi D = 0.319$ m, $n = 100$, and $\mu_{33}^T = \mu_0\mu_r^T$. For Terfenol-D the free inductance $L_f = 0.095$ mH, while for Galfenol the considerably higher value $L_f = 2.64$ mH is obtained.

Chapter 6

6.1. This ring has the same dimensions as the ring in Exercise 5.1; therefore, $f_a = 10.25/0.94$ kHz $= 10.9$ kHz; RVS $= -185$ dB/1 V//1 μPa; $C_f = 36.7$ nF; f_r(length mode) $= 65/2 = 32.5$ kHz, f_a (length) $= 32.5/0.94 = 34.6$ kHz. Sphere: $f_a = 17/0.81 = 20.9$ kHz, RVS $= -185$ dB/1 V//1 μPa, $C_f = 73.4$ nF.

6.3. $f_a = 109$ kHz, RVS $= -205$ dB, Yes, the scaling factor is 10 in this case. If summed in parallel RVS $= -205$ dB, while if summed in series -199 dB. If series differenced, RVS $= -199 + 20\log(\pi sf/c_0) = -278.5 + 20\log f$ dB. Deviation expected in the vicinity of quarter wavelength separation at frequency of 7.38 kHz. Axial null summed modes at one-half wavelength at frequency of 14.76 kHz. Axial null differenced modes at one wavelength at frequency of 29.53 kHz.

6.5. From Eq. (6.58), $10\log < V_n^2 >= -198 + 10\log 2R_h$ for both summed and differenced cases since the incoherent noise voltages add. Using $R_h = \tan\delta/\omega C_f$ for low frequency gives $10\log 2R_h = 45 - 10\log f$ and $< V_n{}^2 > = -153 - 10\log f$.

From Eqs. (6.80a) and (6.80b), using RVS $= -199$ dB for the series summed case and $-278.5 + 20\log f$ for the series differenced case from Exercise 6.3 gives

$$10\log <p_{on}^2>= 46 - 10\log f,$$

$$10\log <p_{dn}^2>= 125 - 30\log f.$$

These results illustrate the much higher levels of equivalent noise pressure for the differenced case at low frequency.

6.7. Resonance f_r: 10.25, 14.49, 22.92, 32.41 kHz. From Exercise 6.1 f_r(length mode) $= 32.5$ kHz which is only slightly above the $n = 3$ mode and may seriously distort the directivity pattern of that mode.

6.9. $D_a = [1 + (ka)^2]^{-1/2}$. The direct derivation of this result, which is somewhat lengthy, is given in Sect. 11.3.1.

Chapter 7

7.1. Use $R_1 = R_{11} + R_{12}$ and $R_{12} \approx R_{11}(\sin kd)/kd$ from Eq. (7.24b). A rigid wall creates an image of the piston approaching it, making it equivalent to an array of two elements.

7.3. Packing factor pf $= L^2/(L + L/10)^2 = 0.826$. See Sect. 7.3.1. The average radiation resistance of one piston in the array is ρcL^2pf and the total radiation resistance of an array of N pistons is $\rho cNL^2\text{pf} = \rho cN(L + L/10)^2 (\text{pf})^2 = \rho c\,(\text{total array area})(\text{pf})^2$.

7.5. It allows simplification and further physical interpretation by factoring a common function leaving the product of this function and that for an array of point sources. This interpretation may be helpful for DI calculations, if the integral of the product of the squares of the two functions is easier to evaluate.

It may also be helpful for radiation resistance calculations since radiation resistance is directly related to the far field, and possibly for radiation reactance calculations (see Sect. 11.1.4).

7.7. The function $\text{Sinc}\, x \approx (x - x^3/6 + \cdots)/x = 1 - x^2/6 + \cdots \rightarrow 1$ as $x \rightarrow 0$. Also the function $J_1(x) \approx x/2 - x^3/16 \cdots$ and thus, $2J_1(x)/x \rightarrow 1$ as $x \rightarrow 0$.

7.9. The total radiation resistance of the center transducer for $\lambda = 4d/3$ is

$$R_1 = R_{11}[1 + N \sin kd/kd] = R_{11}[1 - 2N/3\pi]$$

which is negative for $N > 5$, meaning that the center transducer is absorbing power radiated by the other transducers. The total radiation reactance of the center transducer when $d = \lambda/2$ is $X_1 = X_{11} - NR_{11}/\pi$, which could also become negative for large enough N, reducing the total mass on the center transducer and making its resonance frequency higher than that of the others.

Chapter 8

8.1. Substitute $m_1(x_0) = m_0 e^{jkx_0 \sin\gamma}$, where γ is the steering angle and m_0 is a uniform sensitivity, into the one-dimensional version of Eq. (8.5b) where θ is the beam pattern angle. Integrate from $-L/2$ to $L/2$, where L is the line length, and normalize to obtain the beam pattern function Sinc $[(kL/2)\sin\theta - (kL/2)\sin\gamma)]$, which shows that the wave number $k \sin\theta$ is displaced by $k \sin\gamma$ which means the beam is steered from $\theta = 0$ to $\theta = \gamma$.

8.3. The intensity increase indicated by the DI of a projector array is referenced to the average (omnidirectional) radiation. Therefore, in a receiving array only array gains that are determined by isotropic (omnidirectional) noise could be consistent with this definition of DI. Because array gain depends on noise, while DI does not, array gain and DI depend on frequency and array geometry in different ways; therefore, they generally do not have the same value (see Exercises 8.4 and 8.5 and the first paragraph of Sect. 8.2).

8.5. Use $\sin 2x = 2\sin x \cos x$ to show that $D_f(N, kD, 0°) = D_f(N, 2kD, 90°)$.

8.7. Using the properties of carbon steel from Sect. 13.2, the coincidence frequency is approximately 36 kHz. The evanescent pressure wave amplitude decays by a factor of 0.47 at 1 cm from the plate, or 6.6 dB.

8.9. Imagine a coordinate system with two unit vectors, V_1 and V_2, starting from the origin and pointing in different directions given by θ_1, ϕ_1 and θ_2, ϕ_2. The vector dot product of unit vectors is equal to the cosine of the angle between them, γ, and can be evaluated by calculating the sum of the products of the vector components. Thus, using $V_{1x} = \sin\theta_1 \cos\phi_1$, etc., gives

$$\cos\gamma = V_{1x}V_{2x} + V_{1y}V_{2y} + V_{1z}V_{2z} = \cos\theta_1 \cos\theta_2 + \sin\theta_1 \sin\theta_2 \cos(\phi_1 - \phi_2)$$

This general relation also contains the direction cosines of any specified vector. For example, the direction cosine of V_1 with respect to the x-axis is given for $\theta_2 = 90°$ and $\phi_2 = 0°$ as $\cos\gamma_x = \sin\theta_1 \cos\phi_1$.

8.11. Use Eq. (8.23) to determine the total radiation resistance of the array,

$$R_r = 2\left\{\rho c\pi a^2\left[(ka)^2/2\right][1 + \sin kD/kD]\right\}$$

Use Eq. (8.8c) to determine D_f for a two element line array in free space and multiply by 2 for a line array in a plane baffle, $D_f = 2/(1 + \sin kD/kD)$. Determine D_a from Eq. (6.51) for a plane wave arriving on the MRA of the array, $D_a = 2$. Show that these results satisfy Eq. (8.56).

Chapter 9

9.1. The admittance $Y = 1/Z = 1/(R + jX) = (R - jX)/(R^2 + X^2) = G + jB$

9.3. A measurement of both an electrical and mechanical quantity is needed to determine N; such as, voltage and force or voltage and acceleration with a known mass.

9.5. TCR is given by $S = p/I = Zp/V$ where p/V gives the TVR. The electrical impedance is, except at resonance, approximately given by a function like $1/j\omega C$. Thus the TCR response is modified by a function with a slope of -6 dB/ octave. The RVS is obtained from the TVR through reciprocity that includes an additional slope change of -6 dB/octave.

9.7. A parallel inductor does not change the input voltage. A series inductor does not change the input current. RVS is for open circuit conditions which is equivalent to a very high impedance constant current condition.

9.9. At 2 kHz the wavelength $\lambda = 0.75$ m. If we need two cycles to make a measurement, the differential distance, Δ, between the transmitted direct path and the reflected path, would be $\Delta = 0.75(5 + 2) = 5.25$ m. For a mid-tank projector-hydrophone separation, r (see Fig. 9.28), the distance $\Delta = 2H - r$ where $2H$ is the total reflected path and H is the hypotenuse of the right triangles. Thus for a separation of 1 m $H = (5.25 + 1)/2 = 3.125$ m. The distance to the surface, $w/2$, is then $[(3.125)^2 - (0.5)^2]^{1/2} = 3.08$ m, and the water tank depth, w, should be 6.2 m (20 ft) for $Q_m \leq 5$ and frequency ≥ 2 kHz. For $Q_m \leq 1$, $\Delta = 2.25$ m, $H = (2.25 + 1)/2 = 1.625$ m, and the tank depth needs to be only 3.09 m (10.1 ft).

9.11. The effective coupling coefficient is $k = [1 - (f_r/f_a)^2]^{1/2} = 0.5$. Since $Q_m Q_e = (1 - k^2)/k^2$, $Q_e = 1$.

Chapter 10

10.1. Substitution yields $\partial^2 p/\partial x^2 = -k^2 p$. Particle velocity $= u = (jk/j\omega\rho)p = p/\rho c$ showing that characteristic impedance is ρc. Characteristic impedance ratio of PZT to water is $22.2/1.5 = 14.8$.

10.3. Line: BW = 51° and DI = 3 dB. For piston: BW = 58° and DI = 9.9 dB. The higher DI of the circular piston occurs because it radiates mainly in one direction, while the line radiates omnidirectionally in the plane perpendicular to its axis. BW is caused by partial cancellation as the observation point is moved away from the MRA. The fact that the piston has area, while the line does not, causes the piston's greater beam width when the diameter is the same as the line length. This can be seen by considering the piston to be a collection of parallel strips. Note that cancellation for those strips near the diameter is about the same as it is for the line. However, for those shorter strips near the edge of the piston, cancellation is less than for the line, thus giving a broader beam for the piston.

10.5. Normalize Eq. (10.34) by dividing by the value at $\theta = 90°$ giving

$$p(\theta)/p(90°) = \mathrm{Sinc}(kL\cos\theta)H_0'(ka)/\sin\theta H_0'(ka\sin\theta).$$

This ratio can be simplified by use of the approximation, $H_0'(x) = (2/\pi x)^{1/2}$, valid for $x > 1/2$ (see Morse and Ingard, [22], p. 360), and it becomes

$$p(\theta)/p(90°) = \mathrm{Sinc}(kL\cos\theta)(\sin\theta)^{1/2}.$$

Note that the first factor is the line function (for length $2L$). Now with $kL = ka = \pi$, find θ for a pressure amplitude ratio of 0.707 (i.e., -3 dB). The quantity 2θ gives the beam width. This must be done by trial and error, and it is important to start with a good guess. In this case, since the ring is the same length as the line in Exercise 10.4, the beam width will be similar; so 50° is a good initial estimate. Thus $\theta = (\pi - \mathrm{BW})/2 = 65°$, $\mathrm{Sinc}(\pi\cos\theta) = 0.73$, $(\sin\theta)^{1/2} = 0.95$ and the pressure ratio $= 0.69$, close to the desired value. A trial beam width of 46° gives a pressure ratio of 0.74. The correct beam width is about 48°.

10.7. For $ka \ll 1$, $ka_o = (ka_d)^{3/2}/(12)^{1/4}$. For $ka \gg 1$, $ka_o = ka_d/\sqrt{3}$. When the size is small compared to the wavelength the dipole is a much poorer radiator than the monopole because of strong cancellation from its two out-of-phase parts. Thus to radiate the same power the dipole must have much more radiating area or much greater velocity.

10.9. $D_a = (ka)^2/[4 + (ka)^4]$ for a plane wave arriving parallel to the axis of the dipole.
This D_a has a maximum value of 0.25 when $ka = \sqrt{2}$.

Chapter 11

11.1. Spherical coordinates are the most useful because the constant coordinate surfaces are finite and therefore capable of fitting real transducers. Thus exact solutions are available for spherical transducers with any normal velocity distribution and exact solutions for quantities such as radiation impedance have been derived from them. Oblate spheroidal, prolate spheroidal, and ellipsoidal coordinates also have the important advantage of finite constant coordinate surfaces, although the wave functions are not as well developed. Rectangular and cylindrical coordinates are commonly used, but their constant coordinate surfaces are not finite, and radiation from finite flat surfaces such as pistons can only be solved by assuming they are part of an infinite rigid plane. Similarly, cylindrical radiators must be assumed to have infinite rigid extensions. Radiation from common shapes such as a rectangular box or a cylinder of finite length cannot be solved by expansion in wave functions except as an approximation.

11.3. Satisfying the boundary condition in Eq. (11.2) requires the integral:

$$\int_0^{\pi} u(\theta)P_m(\cos\theta)\sin\theta d\theta = u_0\int_0^{\theta_{0i}} P_m(\cos\theta)\sin\theta d\theta$$
$$= \frac{u_0}{(2m+1)}[P_{m+1}(\cos\theta_{0i}) - P_{m-1}(\cos\theta_{0i})].$$

11.5. Follow the procedure in Sect. 11.1.4 but use the radiation resistance of a dipole given by the real part of Eq. (10.47).

11.7. Note that the D_a for the piston holds for a wave at normal incidence. As the wavelength decreases the piston acts as a plane rigid baffle with the reflected and incident waves in phase, which raises the pressure on the surface and increases the D_a. While for the sphere the phase distribution of both the incident and scattered waves becomes more nonuniform as the wavelength decreases, which reduces the average pressure on the surface and decreases the D_a. It is reasonable for D_a to decrease as $1/ka$ as ka increases, because $1/ka$ is proportional to the number of wavelengths contained in the diameter of the sphere; the more wavelengths, the more completely the pressure cancels.

11.9. The radiation resistance and velocity must be equal because of symmetry. The symmetry of the results is a valuable means of checking the validity of the calculation, which can also be used for large arrays. Programs designed for handling large arrays should also be validated by applying them to small arrays for which the results can be estimated by other means.

11.11. $D_a(\theta,\phi) = 4J_1(ka\sin\theta)/ka\sin\theta \rightarrow 2$ for $\theta = 0$.

Chapter 12

12.1. The equation becomes $x_3/L = d_{33}V/(L + x_3/L)$ leading to $(1 + x_3/L)x_3/L = d_{33}V/L$ yielding the quadratic result $S_3^2 + S_3 - d_{33}V/L = 0$ with exact solution $S_3 = [-1 + (1 + 4d_{33}V/L)^{1/2}]/2$. The four term binomial expansion for $(1+y)^n \approx 1 + ny + n\ (n-1)\ y^2/2 + n(n-1)(n-2)\ y^3/6$ leads to $S_3 = d_{33}V/L - (d_{33}V/L)^2 + 2(d_{33}V/L)^3 + \cdots$.

12.3. The strain versus field curve must be symmetric for even harmonic generation and antisymmetric for odd harmonic generation. This can be seen from the nonlinear solution in Exercise 12.1, although that exercise was stated for nearly static conditions. If the applied voltage is $V = V_0\cos\omega t$ the second term of the solution, which is symmetric, gives a strain proportional to

$\cos^2\omega t = (1 + \cos 2\omega t)/2$, i.e., static and second harmonic strain components. Similarly, the third antisymmetric term of the solution, proportional to $\cos^3\omega t = (3\cos\omega t + \cos 3\omega t)/4$, gives a third harmonic and a change in the fundamental.

12.5. Materials generally produce harmonics when the strain approaches and exceeds the elastic limit or when the electric or magnetic field approaches breakdown. In this region the transducer efficiency at the drive frequency decreases since part of the input power is transferred to power in the harmonics. Thus an increase of input power is required to achieve a goal based on linearity, which brings the strain or field closer to the limits of the material and may lead to mechanical or electrical failure of the transducer.

12.7. The ratios can be found from the solutions in Exercise 12.6 or from Eq. (12.36) and Table 12.2:

$$|x_2/x_1| = V_1/4V_0, \quad \text{well below resonance}$$

$$|x_2/x_1| = (V_1/16V_0)\left(4 + 9Q_m^2\right)^{1/2}, \quad \text{at one half the resonance frequency}$$

$$|x_2/x_1| = (V_1/4V_0)/\left(4 + 9Q_m^2\right)^{1/2}, \quad \text{at the resonance frequency}$$

Note that for $Q_m = 10$ the ratio at one half the resonance frequency is 7.5 times greater, and at the resonance frequency 30 times smaller, than it is well below resonance, showing that harmonic distortion and its effects can vary strongly with frequency.

12.9. For the component at

$$\begin{aligned}
\omega:&\quad x_1 \propto V^2, && u_1 \propto \omega V^2\\
2\omega:&\quad x_2 \propto V^2/2, && u_2 \propto \omega V^2\\
3\omega:&\quad x_3 \propto V^2, && u_3 \propto 3\omega V^2\\
4\omega:&\quad x_4 \propto V^2/2, && u_4 \propto 2\omega V^2
\end{aligned}$$

Glossary

Acoustic Beam: Acoustic waves concentrated mainly in a limited direction.

Acoustic Center of a Transducer: The apparent center of the spherical waves in the far field of a transducer.

Acoustic Impedance: The complex ratio of acoustic pressure to amplitude of volume velocity.

Acoustic Medium: Any material through which acoustic waves can travel, usually air or water, but also other fluids and solids.

Acoustic Modem: Underwater device which communicates by means of acoustic waves.

Active Sonar: Detection of a target by projecting acoustic waves and listening for an echo, also called echo ranging.

Admittance, Y: (1) The reciprocal of the impedance, Z. (2) $Y = G + jB$ where G is the conductance and B is the susceptance.

Antiresonance, f_a: (1) For an electric field transducer, the frequency at which the mass and open circuit stiffness cancel or the frequency of maximum impedance or RVS. (2) For a magnetic field transducer, the frequency at which the mass and short circuit stiffness cancel or the frequency of maximum admittance or TVR.

Armature: (1) Part of a magnetic circuit. (2) The moveable part of the magnetic circuit in a variable reluctance transducer.

Array of Transducers: Two or more transducers that are used together usually to achieve a higher intensity and narrower beam.

Array Transparency: A two sided transducer array in which the sound enters one side and reradiates from the opposite second side as if there were no structure.

Astroid: (1) Hypocycloid of four cusps. (2) Described by a point on a small circle of radius, R, rolling around the inside of a larger circle of radius $4R$. Cylindrical-shaped low frequency acoustic transducer with magnified piston motion.

Beam Steering: Changing the principal direction of an acoustic beam by changing the relative phases of the transducers in an array or by rotating a transducer or array.

J.L. Butler, C.H. Sherman, *Transducers and Arrays for Underwater Sound*,
Modern Acoustics and Signal Processing, DOI 10.1007/978-3-319-39044-4

Beam Width, BW: The angular width, in degrees, of an acoustic intensity beam measured between the −3 dB points on each side of the main lobe.

Belleville Spring: (1) A conical disk or washer which provides a compliant member where a washer would otherwise be used. (2) Used to reduce the effective stiffness of a stress bolt. (3) Used to reduce effects of thermal expansion of a stress bolt.

Bias: (1) A fixed (DC) voltage, current or magnetic field applied to a transducer, in addition to an alternating voltage or current, to achieve linearity. (2) A static mechanical compression.

Blocked Condition: (1) A mechanical boundary condition which restricts the motion of a transducer such that the velocity or the displacement is zero. (2) Same as Clamped condition.

Broadside Beam: An acoustic beam directed perpendicular to the surface of an array.

Cardioid Beam: (1) Beam pattern with a null at 180°, −6 dB at ±90°, beam width of 133° and DI = 4.77 dB. (2) Typically formed from the sum of an omni and a dipole source or sensor of equal acoustic far-field pressure. (3) Special case of a limacon beam.

Characteristic Frequency: (1) Any material object, such as a bar, a spherical shell, a room full of air, or an ocean of water, has a series of allowed modes of vibration each with its own characteristic frequency. (2) Frequency at which eddy current effects start to become significant.

Characteristic Impedance: (1) An impedance used to compare media or transducers: for media, usually the specific acoustic impedance, ρc. (2) A value of the mechanical impedance of a transducer in the vicinity of resonance. (3) The square root of the product of the transducer **effective** mass and stiffness.

Clamped Capacitance, C_0: (1) Capacitance of an electric field transducer under clamped conditions. (2) Capacitance of an electric field transducer at mechanical resonance.

Clamped Condition: (1) A mechanical boundary condition which restricts the motion of a transducer such that the velocity or the displacement is zero. (2) Same as Blocked condition.

Coercive Force: (1) The electric field required to reduce the polarization to zero when a ferroelectric material has been polarized to saturation. (2) The magnetic field required to reduce the magnetization to zero when a ferromagnetic material has been magnetized to saturation.

Complex Algebra: Mathematical operations used with complex numbers, with real and imaginary parts.

Compliance, C: (1) The softness of a mechanical element. (2) The reciprocal of the stiffness, K.

Compliance, C_m^E: (1) The mechanical compliance under short circuit conditions for an electric field transducer. (2) The reciprocal of the stiffness, K_m^E.

Compliance, C_m^D: (1) The mechanical compliance under open circuit conditions for an electric field transducer. (2) The reciprocal of the stiffness, K_m^D.

Compliance, C_m^B: (1) The mechanical compliance under short circuit conditions for a magnetic field transducer. (2) The reciprocal of the stiffness, K_m^B.

Compliance, C_m^H: (1) The mechanical compliance under open circuit conditions for a magnetic field transducer. (2) The reciprocal of the stiffness, K_m^H.

Coupling Coefficient, *k*: (1) A measure of the electromechanical activity of a transducer where $0 < k < 1$. (2) $k^2 =$ energy converted/total energy stored where the energy converted is either mechanical or electrical. (3) The quantity $k = 1/(1 + Q_m Q_e)^{1/2}$. (4) The effective dynamic coupling coefficient, for high Q_m, is given by $k = [1 - (f_r/f_a)^2]^{1/2}$.

Curie Temperature, T_c: (1) The temperature at which polarization disappears and piezoelectric ceramics loose their piezoelectric properties. (2) Piezoelectric ceramics are typically operated at temperatures no higher than $Tc/2$.

Dipole: (1) Pair of small sources (or sensors) with equal amplitude but opposite phase and separation distance small compared to the wavelength. (2) Beam pattern with nulls at $\pm 90°$, beam width 90° and DI = 4.77 dB.

Diffraction Constant, D_a: (1) A measure of various effects (scattering, diffraction, reflection, and phase variations) that cause the average clamped acoustic pressure on a hydrophone to differ from the free field pressure at the same location.

Directivity Factor, D_f: (1) Ratio of the far field on axis intensity to the intensity averaged over all directions at the same distance.

Directivity Function: The intensity of acoustic radiation as a function of direction, consisting of a major lobe and minor lobes. For reciprocal transducers the same function holds for receiving acoustic waves from a distant source as the transducer is rotated.

Directivity Index, DI: (1) Increase in the intensity due to the directional characteristics of a transducer or array. (2) DI = 10 log D_f.

Doubly Steered Array: A transducer array in which the transducer elements are steered into the same direction that the array is steered.

Dyadic Sensor: A directional hydrophone with a quadrupole (or higher order) beam pattern.

Eddy Current: Current induced in a conducting material by an alternating magnetic field which causes a reduction in inductance and loss of power in magnetic field transducers.

Electrical Quality Factor, Q_e: (1) The Q at mechanical resonance based on the electrical reactance and resistance components of a transducer. (2) $Q_e = B_0/G_0$ where B_0 is the susceptance ($\omega_r C_0$ for an electric field transducer) and G_0 is the conductance at mechanical resonance.

Electroacoustic Transducer: One in which an electric input gives an acoustic output, or one in which an acoustic input gives an electric output.

Electrostriction: (1) Change of dimension of a material caused by an applied electric field, independent of direction of the field. (2) Approximately linear when electric field bias is also applied.

End-Fire Beam: An acoustic beam directed parallel to the surface of an array.

Equivalent Circuit: (1) An electrical circuit which represents the electrical, mechanical, and acoustical parts of a transducer. (2) Often a lumped-parameter circuit representation using inductors, capacitors, and resistors but can also take the form of distributed elements represented by trigonometric or other functions.

Equivalent Mass: (1) The dynamic mass, M_d, referred to a velocity, u_0, of a distributed system. (2) $M_d = 2(\text{Kinetic Energy})/u_0^2$.

Equivalent Stiffness: (1) The dynamic stiffness, K_d, referred to a displacement, x_0, of a distributed system. (2) $K_d = 2(\text{Potential Energy})/x_0^2$.

Evanescent Wave: A non-propagating wave, usually because the amplitude decays exponentially.

Extensive/Intensive Variables: An extensive variable depends on the internal properties of a system; examples are mechanical strain, electric displacement, and magnetic induction. An intensive variable does not depend on the internal properties; examples are mechanical stress, electric field, and magnetic field. The two types of variables occur in pairs that are related to each other such as $S = sT$, $D = \varepsilon E$, and $B = \mu H$.

Faraday Induction Law: When an electric circuit exists in a region of changing magnetic field a voltage is induced in the circuit equal to the negative of the rate of increase of magnetic flux through the circuit.

Far Field: (1) Radial distance, R, at which the acoustic pressure falls off as $1/R$. (2) For a given transducer or array size L, the far-field distance $R > L^2/2\lambda$ where λ is the wavelength.

FEM: The Finite Element Method, also referred to as FEA for Finite Element Analysis.

Ferroelectricity: Spontaneous polarization of small regions (domains) within crystals of certain materials, analogous to ferromagnetism.

Flexural Vibration: Vibratory motion in which the stiffness component results from bending of a material.

Figure of Merit: (1) Projector power output for a given resonance, mechanical Q and total volume $FOM_v = W/(f_r Q_m Vo)$, (2) Hydrophone sensitivity square free capacity product $FOM_h = M^2 C_f = gdV$ where V is the volume of the material with g and d piezoelectric values.

Free Capacitance, C_f: (1) Electric field transducer capacitance without any mechanical load under a free boundary condition where the stress is zero. (2) Capacitance of an electric field transducer at a frequency well below its fundamental resonance.

Free Condition: (1) A mechanical boundary condition with no restriction on the motion and with zero stress. (2) Opposite of the Blocked or Clamped boundary condition.

Free Field Condition: (1) Acoustic field condition where there are no structures which reflect or scatter the sound. (2) Condition where the measuring hydrophone is small enough that the acoustic field is not disturbed by its presence.

Grating Lobe: (1) Replicate lobe(s) of the main lobe of a beam pattern. (2) Lobes which occur in directions where the output of the array elements add at integer multiples of a wavelength. (3) Lobe where the intensity level is nearly the same as the main lobe. (4) A lobe which cannot be reduced by shading.

Gyrator: (1) A device which acts like a transformer but converts voltage on side 1 to a current on side 2 and current on side 1 to voltage on side 2. (2) An ideal device which "gyrates" a voltage into a current and vice versa. (3) An ideal device which may be used to convert a mobility analogue circuit representation to an impedance analogue circuit representation. (4) Electromechanical gyrators are typically used to transform current to force and voltage to velocity in magnetic field transducer representations.

Harmonic Distortion: A result of nonlinear response where an output contains harmonics of a pure sine wave input.

Helmholtz Resonance Frequency: (1) The fundamental resonance frequency of the fluid in a container that is closed except for a port (opening) or an attached open tube. (2) The frequency at which the compliance of the fluid in the container resonates with the sum of the radiation mass of the opening and the fluid mass in the tube.

Homogeneous Equations: Equations in which all the independent variables are either extensive or intensive and all the dependent variables are either intensive or extensive.

Hybrid Transducer: A transducer containing both electric field and magnetic field transduction sections.

Hydrophone: (1) A transducer used to receive acoustic waves. (2) The underwater analogue of a microphone.

Ideal Electromechanical Transformer: (1) Lossless transformer device without inherent inductive or capacitive values. (2) Often used in equivalent circuits to connect the electrical part of the circuit to the mechanical part of the circuit. (3) For an electric field transducer, the electromechanical transformer turns ratio, N, relates the force, F, to the voltage, V, by $F = NV$ and the velocity, u, to the current, I, by $u = I/N$.

Impedance, Z: (1) Electrical impedance is the complex ratio of the voltage to the current. (2) Mechanical impedance is the complex ratio of the force to the velocity. (3) The impedance $Z = R + jX$ where R is the resistance and X is the reactance.

Impedance Equivalent Circuit: (1) An equivalent circuit representation of a transducer where the force is proportional to the voltage and the velocity is proportional to the current. (2) Typically used for equivalent circuit representation of electric field transducers. (3) May be converted to a mobility representation by use of a gyrator.

Insonify: To project acoustic waves onto an object or into a region.

Intensity, I: (1) Acoustic power per unit area. (2) In the far field $I = p^2/2\rho c$ where p is the amplitude of the pressure, ρ is the density, and c is the sound speed in the medium. (3) Generally a complex quantity in the near field.

Isotropic: A sound field or noise field is said to be isotropic, from the viewpoint of a given observer, when it is seen to be the same in all directions.

Johnson Noise: The mean square voltage fluctuations arising from thermal agitation in an electrical resistor, given quantitatively by the Nyquist Theorem.

Leveraged Transducer: A transducer with radiator that moves with a magnified motion.

Limacon: (1) One of a series of specific mathematical functions. (2) Limacons can describe acoustic pressure as a function of angle and represent useful and achievable directional beam patterns. (3) Consist of omni and dipole contributions. (4) The cardioid is a special case of the limacon with equal omni and dipole contributions.

Lumped Mass, *M*: (1) The mass of an element which is devoid of compliance. (2) The equivalent mass of a distributed element. (3) The mass of an element which is physically small ($\leq\lambda/8$) compared to the wavelength, λ, of sound in the element.

Lumped Mode Equivalent Circuit: An equivalent circuit representation where all the elements are considered physically small ($\leq\lambda/8$) compared to the wavelength, λ, of sound in the element.

Magnetostriction: Change of dimension of a material caused by an applied magnetic field, independent of direction of the field. Approximately linear when magnetic field bias is also applied.

Major Response Axis (MRA): (1) The direction of the acoustic beam, or the major lobe, of a transducer or an array. (2) The direction of maximum response when a transducer or array is used for receiving.

Major Lobe: The main acoustic beam of a transducer or an array.

Mechanical Impedance, Z_m: (1) Complex ratio of the force, F, to velocity, u, of the mechanical part of the transducer, $Z_m = F/u = R_m + jX_m$.

Mechanical Resistance, R_m: (1) Resistance $R_m = F/u$ where F is the force applied to a mechanical loss element and u is the resulting velocity.

Mechanical Resonance, f_r: (1) For an electric field transducer, the frequency at which the mass and the short circuit stiffness cancel or the frequency of maximum admittance or TVR. (2) For a magnetic field transducer, the frequency at which the mass and the open circuit stiffness cancel or the frequency of maximum impedance or TCR. (3) For a distributed system, the frequency at which the equivalent mass and stiffness cancel.

Minor Lobes: (1) Other acoustic beams, lower in amplitude than the major lobe. Also called side lobes, which exist unless the transducer or array is smaller than the wavelength. (2) Minor lobes usually degrade performance, but may be reduced in amplitude by shading.

Mixed Equations: Equations in which the independent variables are both extensive and intensive and the dependent variables are both intensive and extensive.

Mobility Equivalent Circuit: (1) An equivalent circuit representation of a transducer where the force is proportional to the current and the velocity is proportional to the voltage. (2) Typically used for equivalent circuit representation of magnetic field transducers. (3) May be converted to an impedance representation by use of a gyrator.

Modal Transducer: A transducer which operates at modes of vibration simultaneously.

Multimode Transducer: A transducer that uses more than one mode of vibration to achieve specific performance features.

Mutual Radiation Impedance: (1) The complex ratio of the acoustic force exerted on one transducer to the velocity of the second transducer creating the force. (2) The off-diagonal elements of an array impedance matrix.

Near Field: (1) Location near the transducer surface. (2) Usually at a distance less than the Rayleigh distance L^2/λ where λ is the wavelength and L is the size of the transducer or array.

Noise: (1) Any received signal that interferes with the signal being detected. (2) Ambient noise, coherent and partially coherent acoustic waves radiated from ships, from the sea surface and from sea life. (3) Flow noise, turbulent boundary layer pressure fluctuations caused by motion of a ship through the water. (4) Structural noise, machinery and flow excited vibration of a ship's structure. (5) Thermal noise, uncorrelated random electrical, mechanical, or fluid fluctuations arising from ever present thermal agitation.

Octoid: An eight sided Astroid type transducer.

Operating Conditions: (1) Static; No relative motion between parts of a system. (2) Quasistatic; Only small relative displacements or velocities controlled mainly by stiffness components. (3) Dynamic; Larger displacements or velocities controlled by both stiffness and mass components and with a frequency-dependent response.

Open Circuit Resonance: (1) The frequency at which the reactance terms cancel under electrical open circuit conditions. (2) Resonance frequency for a transducer under constant current drive. (3) Maximum open circuit response of an electric field hydrophone.

Parametric Array: An acoustic source that radiates a low frequency beam with no side lobes by means of two collimated high frequency, high amplitude beams of different frequency, in which the nonlinearity of the water generates the difference frequency.

Parasitic Resonator: A passive resonator excited from a nearby active resonator.

Passive Ranging: Use of passive sonar hydrophones to determine range of a target by triangulation or measurement of wave front curvature.

Passive Sonar: Detection of a target by listening for sound radiated by the target.

Passive Transducer: Any transducer that requires no internal energy source for its operation. The carbon button microphone is not considered to be a passive transducer, while a true piezoelectric crystal transducer and a piezoelectric ceramic transducer are considered passive.

Piezoelectricity: Change of dimension of a material caused by an applied electric field, dependent on the direction of the field (inverse effect). Production of electric charge on the surface of a material caused by an applied mechanical stress (direct effect). Piezoelectricity is a linear phenomenon that does not require a bias.

Port (electric/acoustic): (1) The terminals of an electric device. (2) The vibrating surface of an acoustic device.

Power Factor, P_f: (1) Cosine of the phase angle between the voltage and the current. (2) With phase angle $\varphi = \tan^{-1}(X/R)$, the power factor is $\cos\varphi = W/VI$ where W is the power consumed and VI is the product of the input rms voltage and current.

Power Wheel: Cylindrical structure with eight radial tonpilz transducers which shear a central tail mass.

Pressure Release: (1) A pressure release surface consists of material with very low acoustic impedance on which the acoustic pressure must always be approximately zero. (2) An air–water interface satisfies this condition for a wave in water. (3) Soft material such as foam rubber or, in some cases, corprene can often satisfy this condition.

Projector: An electromechanical transducer used to radiate acoustic waves.

Quality Factor, Mechanical, Q_m: (1) A measure of the sharpness of a resonant response curve. (2) $Q_m = f_r/(f_2 - f_1)$ where f_r is the resonance frequency and f_2 and f_1 are the frequencies at half-power relative to the value at f_r. (3) $Q_m = 2\pi$ (Total Energy) /(Energy dissipated per cycle at resonance).

Quartz: Silicon dioxide, silica. Piezoelectricity was discovered in single crystals of quartz in 1880.

Radiation Impedance, Z_r: The complex ratio of the acoustic force on a transducer surface caused by its vibration to the velocity of the surface, $Z_r = R_r + jX_r$.

Radiation Resistance, R_r: (1) The real part, R_r, of the radiation impedance Z_r. (2) At high frequencies, where the size is large compared to the wave length, $R_r \gg X_r$ and R_r becomes a constant.

Radiation Reactance, X_r: (1) The imaginary part, X_r, of the radiation impedance Z_r. (2) At low frequencies where the size is small compared to the wave length, $X_r \gg R_r$ and $X_r \approx j\omega M_r$ where M_r is the radiation mass.

Radiation Mass, M_r: (1) The mass associated with the kinetic energy of the near field during a radiation process. (2) Sometimes referred to as the "accession to inertia." (3) The radiation mass is usually defined as $M_r = X_r/\omega$, which is a constant at low frequencies but decreases in value as the frequency increases.

rayl: MKS unit of specific acoustic impedance in kg/sm^2, named after Lord Rayleigh.

Rayleigh Distance, R: (1) Transitional distance between the near field and the far field. (2) For a given transducer or array size L, the Rayleigh distance $R = L^2/2\lambda$ where λ is the wavelength.

Reactance, X: (1) The imaginary part of the impedance. (2) Reactive impedance associated with the inductive or capacitance components of an electrical circuit. (3) Reactive impedance associated with the mass or stiffness components (or equivalent) of a mechanical harmonic oscillator.

Reciprocity (General): A general relationship that states that for any linear, bidirectional device, the ratio of the output at one port to the input at another port remains the same if the input and output ports are reversed, provided that the boundary conditions at each port remain the same.

Reciprocity (Acoustic): (1) A general relationship between the velocities of two transducers vibrating at the same frequency and the acoustic pressures they exert on each other, from which several important specific results can be derived. For example, (2) the reciprocity of mutual radiation impedance, and (3) a relationship between the diffraction constant, the directivity factor, and the radiation resistance allowing any one of these quantities to be determined from the other two.

Reciprocity (Electromechanical): (1) The magnitude of the ratio of the transducer receive open circuit voltage, V, to the blocked force, F_b, is equal to the magnitude of the ratio of the transmitting velocity, u, to the drive current, I. (2) The relationship $|V/F_b| = |u/I|$ is combined with acoustic reciprocity to formulate electroacoustic reciprocity.

Reciprocity (Electroacoustic): (1) The ratio of the free field open circuit receiving sensitivity, M, to the constant current transmitting response, S, is equal to $J = 2d/\rho f$ where d is a far-field distance, f is the frequency, and ρ is the density of the medium. (2) The known value of J is the basis for reciprocity calibration of transducers.

Reluctance: A property of a portion of a magnetic circuit equal to the length of the portion divided by the product of its cross-sectional area and permeability, analogous to resistance in electric circuits.

Remanence: (1) The polarization that remains when the electric field is reduced to zero after a ferroelectric material has been polarized to saturation. (2) The magnetization that remains when the magnetic field is reduced to zero after a ferromagnetic material has been magnetized to saturation.

Resonance: (1) Frequency at which the kinetic and potential energies are equal. (2) Frequency at which the equivalent mass reactance and equivalent stiffness reactance cancel. (3) Also see, Mechanical Resonance Frequency, f_r, and Antiresonance, f_a.

Resonator: Anything that has both mass and stiffness or, more generally, can store both kinetic and potential energies and therefore is capable of a large response to a small applied force at certain frequencies.

Response: Many different transducer responses are used as measures of transducer performance; each is the ratio of an output to an input as a function of frequency. For frequently used responses see RVS, TVR, and TCR.

RVS: (1) Open circuit receiving voltage sensitivity or open circuit receiving response, ratio of open circuit voltage to free field plane wave pressure. (2) Sometimes referred to as FFVS, for free field voltage sensitivity. (3) Typically expressed in dB below 1 V for a pressure of 1 μPa.

Sagitta: Perpendicular distance from the midpoint of a chord to the arc of a circle.

Sensor: (1) Any device that detects a signal. (2) A hydrophone is a sensor.

Shading: (1) The process of reducing side lobes by means of a variable or tapered velocity amplitude distribution on an array. (2) Does not reduce grating lobes. (3) Generally broadens the width of the major lobe.

Short Circuit Resonance: (1) The frequency at which the reactance terms cancel under electrical short circuit conditions. (2) Resonance for an electric field transducer under constant voltage drive.

SONAR: (1) SOund, NAvigation and Ranging system used to detect and locate an object or target.

Sonobuoy: (1) A small SONAR device dropped into the sea from a helicopter or airplane with the ability to radio the information obtained from the acoustic portion back to the aircraft.

Source Strength: The product of the area of the vibrating surface of a transducer and the average normal velocity over that surface.

Specific Acoustic Impedance: (1) The ratio of acoustic pressure to the amplitude of the acoustic particle velocity. (2) For a plane wave, the product of the density and sound speed, ρc, also referred to as the characteristic impedance of the medium.

Stack: (1) A number of piezoelectric ceramic elements cemented together. (2) A number of magnetostrictive laminations cemented together.

Stiffness, *K*: (1) A measure of the force, F, required to compress an element a distance x. (2) The stiffness $K = F/x$. (3) The reciprocal of the compliance C.

Stress Rod: (1) A high strength rod or bolt used to provide static compression to a piezoelectric ceramic stack. (2) Used to keep a piezoelectric ceramic stack from going into tension under dynamic drive. (3) Also referred to as "Tie Rod" or "Compression Rod."

TBL: (1) Turbulent boundary layer pressure fluctuations, which occur on a surface moving with sufficient speed through a fluid medium. (2) A serious source of noise in hull-mounted arrays.

TCR: (1) Transmitting current response, ratio of far-field pressure to input current. (2) Typically expressed in dB with respect to 1 μPa at a distance of 1 m. (3) TIR is also used.

Tonpilz: A longitudinal resonator transducer with a large head mass driven by a relatively narrow stack with a heavy tail mass.

Transducer: A device that converts energy from one form to another, e.g., electrical to mechanical energy conversion.

Transduction: A process that converts energy from one form to another.

Trioid: A three sided Astroid type transducer.

TVR: (1) Transmitting voltage response, ratio of far-field pressure to input voltage. (2) Typically expressed in dB with respect to 1 μPa at a distance of 1 m.

Vector Sensor: (1) A device that detects a vector signal such as velocity or intensity. (2) A directional hydrophone with a dipole beam pattern.

Volume Velocity: (1) Volume per unit time, m^3/s. (2) Acoustic source strength is measured by volume velocity. (3) The product of the average normal velocity and the radiating area of a transducer.

Wavelength: (1) The distance from any point on a sinusoidal wave to the nearest point at which the amplitude and phase are repeated. (2) The distance from one peak to the next in a sinusoidal wave. (3) Usually denoted by λ.

Wave number: (1) The reciprocal of the wavelength multiplied by 2π. (2) The angular frequency divided by the speed of wave propagation. (3) Usually denoted by $k = 2\pi/\lambda = \omega/c$.

Wave vector: (1) The vector formed from the wave number components of a plane wave traveling in an arbitrary direction with respect to a fixed coordinate system. (2) A means of specifying the direction of a plane wave.

X-Spring Transducer: Transducer where a motion leveraging structure is used between the driving stack and the piston creating greater piston motion.

Index

J.L. Butler, C.H. Sherman, *Transducers and Arrays for Underwater Sound*,
Modern Acoustics and Signal Processing, DOI 10.1007/978-3-319-39044-4

O

P

www.ingramcontent.com/pod-product-compliance
Ingram Content Group UK Ltd.
Pitfield, Milton Keynes, MK11 3LW, UK
UKHW021832270726
14058UKWH00001B/101
* 9 7 8 3 3 1 9 8 1 8 0 2 3 *